Foundations of Quantum Group Theory

Now in paperback, this is a graduate level text for theoretical physicists and mathematicians which systematically lays out the foundations for the subject of quantum groups in a clear and accessible way. The topic is developed in a logical manner with quantum groups (Hopf algebras) treated as mathematical objects in their own right. After formal definitions and basic theory, the book goes on to cover such topics as quantum enveloping algebras, matrix quantum groups, combinatorics, cross products of various kinds, the quantum double, the semi-classical theory of Poisson–Lie groups, the representation theory, braided groups and applications to q-deformed physics. Explicit proofs and many examples will allow the reader quickly to pick up the techniques needed for working in this exciting new field.

Foundations of
Quantum Group Theory

SHAHN MAJID

University of Cambridge

CAMBRIDGE UNIVERSITY PRESS
Cambridge, New York, Melbourne, Madrid, Cape Town, Singapore, São Paulo

Cambridge University Press
The Edinburgh Building, Cambridge CB2 2RU, UK

Published in the United States of America by Cambridge University Press, New York

www.cambridge.org
Information on this title: www.cambridge.org/9780521460323

First published 1995
First paperback edition (with corrections) 2000

Typeset by the author [CRC]

A catalogue record for this publication is available from the British Library

Library of Congress Cataloguing in Publication data
Majid, Shahn.
Foundations of quantum group theory / Shahn Majid.
p.   cm.
ISBN 0–521–46032–8 (hc)
1. Group theory. 2. Quantum theory. I. Title.
QC174. 17.G7M25   1995
512′.55–dc20   95–13641   CIP

ISBN-13 978-0-521-46032-3 hardback
ISBN-10 0-521-46032-8 hardback

ISBN-13 978-0-521-64868-4 paperback
ISBN-10 0-521-64868-8 paperback

Transferred to digital printing 2006

For my friends

Contents

Contents

Introduction

This is an introduction to the foundations of quantum group theory. Quantum groups or *Hopf algebras* are an exciting new generalisation of ordinary groups. They have a rich mathematical structure and numerous roles in situations where ordinary groups are not adequate. The goal in this volume is to set out this mathematical structure by developing the basic properties of quantum groups as objects in their own right; what quantum groups are conceptually and how to work with them. We will also give some idea of the meaning of quantum groups for physics. On the other hand, just as ordinary groups have all sorts of applications in physics, not one specific application but many, in the same way one finds that quantum groups have a wide variety of probably unrelated applications. This diversity is one of the themes in the volume and is a good reason to focus on quantum groups as mathematical objects.

This book is not a survey; many of the most interesting recent results in representation theory, applications in conformal field theory and low-dimensional topology, etc., are not discussed in any detail. In this sense, there is less material here than in my lecture notes [1]. In place of this fashionable material, I have developed the pedagogical side of [1], giving now more details of proofs and solutions to exercises, and in general concentrating more on that part of the theory of quantum groups that can be considered as firmly established. I have also included my more recent work on braided groups.

This text is addressed primarily to theoretical physicists and mathematicians wishing to begin work on quantum groups. For physicists, I have tried to give full details and line-by-line proofs of all the basic results that are needed for research in the field. Also, I have struggled hard to maintain an informal style so that the essential content is not too obscured by inessential formalism. For mathematicians, I have adopted a theorem–proof format so that the main results can be understood clearly, and have endeavoured not to say things that are technically false. This balance between readability and rigour is achieved by taking a completely algebraic line and not discussing in any depth the equally interesting variants of quantum groups based on C^*-algebras and Hopf–von Neumann or Kac algebras. In other words, we limit ourselves primarily to the algebraic theory of quantum groups rather than to the functional-analytic theory.

What is a quantum group? To answer this question, let us first consider what is a group. There are several answers. The most familiar point of view about groups is as collections of transformations. Transformations of a space are assumed invertible, and every closed collection of invertible transformations is, inevitably, a group. This is the role of groups as symmetries. Quantum groups, too, can act on things. However, now the transformations are not all invertible. Instead, quantum groups have a weaker structure, called the antipode S, which provides a nonlocal 'linearised inverse'. It means that now not individual elements but certain linear combinations are invertible. Remarkably, this weaker invertibility is all that is actually used in applications. For example, just as groups can act on themselves in the adjoint representation (which would appear to require an inverse), so quantum groups act on themselves in an adjoint representation. Likewise, just as every representation of a group has a conjugate one (provided by the action of the inverse group element), so every quantum group representation has a conjugate provided by the antipode.

A second point of view about groups is that their representations have a tensor product. This is familiar in particle physics. For example, if J_z is an angular momentum operator, then an element $\Delta J_z = J_z \otimes 1 + 1 \otimes J_z$ provides its action on tensor products (this linear addition is characteristic of ordinary Lie algebras: quantum groups tend to be more complicated). The tensor product is symmetric, the symmetry being implemented by the usual transposition of vector spaces. Quantum group representations, too, have a tensor product. In fact, we will see a theorem that given any collection of objects which can be identified with vector spaces, compatible with the tensor product of vector spaces, we can reconstruct a quantum group and identify the collection as its representations. So, in a certain context, this is a complete characterisation. For strict quantum groups (ones possessing a so-called 'universal R-matrix'), the tensor product of representations is symmetric (just as for representations of groups), but now only up to isomorphism. This isomorphism is not given by the usual transposition but by a weaker structure called a quasisymmetry or 'braiding' Ψ. It is weaker because, in general, Ψ does not obey $\Psi^2 = \text{id}$. Instead, it provides an action of the braid group rather than of the symmetric group. It is this fact that leads to the application of strict quantum groups in low-dimensional topology.

These two points of view cover the most well-known settings in which quantum groups have arisen, namely their connection with quantum inverse scattering, exactly solvable lattice models and low-dimensional topology. In this context, quantum groups arose as symmetries of quantum statistical systems, leading to braid group representations in these systems, as well as in conformal field theories related to their continuum limit. This theory will obviously take up a substantial part of the volume, namely Chapters 2–4 and parts of Chapters 7, 8 and 9. Important, but not the only, examples are the quantum groups $U_q(g)$ introduced by V.G. Drinfeld and M. Jimbo as deformations of the enveloping algebras of complex simple Lie algebras.

However, Hopf algebras in general have a further unusual property or *raison d'être* quite different from their role as generalised symmetry. This gives the third and fourth ideas about what a quantum group or Hopf algebra *is*. These are connected with their duality or self-duality properties and are the author's own reason to be interested in quantum groups. From this point of view, a Hopf algebra is an algebra for which the dual linear space of the algebra is also an algebra. The algebra structure on the dual linear space is expressed in terms of the original algebra A as a coproduct or comultiplication map $\Delta : A \to A \otimes A$. Supplementing an algebra by a comultiplication (forming a coalgebra) restores a kind of input–output symmetry to the system. When A is the algebra of observables of a classical or quantum system, then the ordinary multiplication $A \otimes A \to A$ corresponds to logical deduction (multiplication of projection operators in the quantum case, or simply multiplication of the characteristic functions in the classical case, which is intersection of their underlying sets). By contrast, Δ allows the reverse operation, to 'unmultiply' (comultiply). The comultiplication of an element X in A is the sum of all those things in $A \otimes A$ which could give X when combined according to an underlying group structure. For example, if X is the coordinate function on the real line, $\Delta X = X \otimes 1 + 1 \otimes X$ expresses linear addition on the line. The probabilistic interpretation is that X in A is a random variable, while $X_1 = X \otimes 1$ and $X_2 = 1 \otimes X$ are two independent random variables embedded in $A \otimes A$, which is the system after two steps in a random walk. Embedding X in $A \otimes A$ as $\Delta X = X_1 + X_2$ says precisely that our total position X after two steps is the sum of the two random variables X_1, X_2. This ΔX represents all the ways to obtain X after two steps. Thus, the comultiplication represents 'induction' or possibility rather than deduction. Remarkably, the rules for Δ are just the same as the rules for multiplication, with the arrows reversed. This remarkable way of understanding probability and random walks on groups was one of the classical reasons for interest in Hopf algebras some years ago. This work has naturally had a renaissance with the arrival of quantum groups *en masse.* This is the third idea of what a Hopf algebra is, and is the topic of Chapter 5.

Finally, we must recall that non-Abelian Lie groups are after all the simplest examples of Riemannian geometry with curvature. It is well-known that to do Riemannian geometry on a manifold M it is often convenient to work with the algebra of functions $A = C(M)$. For example, a vector field is a derivation in such algebraic terms. The idea of noncommutative (algebraic) geometry is that even when this algebra A is made into a noncommutative one, one can continue to do geometry, perhaps even Riemannian geometry, provided all our constructions are referred to the algebra A rather than to any manifold M, which need no longer exist. This is an old idea, but one that was developed significantly in recent years by A. Connes and others. If our initial M is phase-space then when A is quantised it becomes noncommutative in just this way (with non-commutativity controlled by $\hbar$). We can still continue to think of it as like '$C(M)$', although, in truth, the points in M no longer exist because the position and momentum coordinates can no longer be measured simultaneously. We can still continue to do geometry in this setting. This is 'quantum Riemannian geometry'. In very general mathematical terms, such a point of view can also be taken in all the above contexts, notably the matrix quantum groups of function algebra type (Chapter 4), where matrix multiplication can be done in the noncommutative setting, or the context of quantum random walks. However, let us ask more specifically about quantum Riemannian geometry. According to our view of Lie groups, noncocommutative Hopf algebras (i.e. with a noncommutative comultiplication) are like non-Abelian groups, i.e. they have curvature. If, at the same time, they are noncommutative as a result of quantisation, then we have a quantum system combined in a consistent way with curvature, i.e. models of quantum-gravity. This was the author's original motivation for Hopf algebras, and is a fourth *raison d'être* for them.

Specifically, it was investigated (by the author) under what conditions the quantum algebra of observables of a particle on a homogeneous space is, in fact, a Hopf algebra. It turns out that it is sufficient for the homogeneous metric to obey a second order 'Einstein' equation. Solving this, one finds a large class of noncommutative and noncocommutative Hopf algebras related to group factorisations – quite different and independent from those of Drinfeld, Jimbo, *et al.* While not connected with braiding, they are characterised instead by a remarkable self-duality property: the dual Hopf (von Neumann) algebra is of just the same type with the roles of position and momentum interchanged. It corresponds to the quantum particle moving on a dual or 'mirror' homogeneous space. We will see this in Chapter 6.

In this setting, Hopf algebra duality takes a very concrete form as a symmetry between quantum observables and quantum states. The dual Hopf algebra is built on A^*, i.e. the algebra of observables of the dual system is the algebra of states (induced by the comultiplication Δ) in terms of the original quantum system A. If a is in A (a quantum observable) and ϕ in A^* (a quantum state) then $\phi(a)$ (the expectation value of a in state ϕ) is interpreted in the dual system as $a(\phi)$ (the expectation value of ϕ in state a from the dual point of view). From this point of view, geometry, in the form of the simplest models of curvature in phase-space, is the dual of quantisation. The noncommutativity of quantisation is mirrored in these models by the noncommutativity of covariant derivatives in Riemannian geometry, expressed Hopf algebraically as noncommutativity of Δ. Thus Hopf algebras provide a unique setting for the unification of quantum mechanics and gravity, in which they appear as the same structure but in dual form; one on the algebra of observables and the other in the algebra of states. It should be appreciated that Hopf algebras provide in this way an example of a general phenomenon: the dual relationship between quantum theory (as an outgrowth of arithmetics and intuitionistic logic) and geometry. In maintaining this duality, our notion of Riemannian geometry must itself be enlarged to the noncommutative geometric setting, i.e. the gravitational field is 'quantised', albeit not in a very conventional sense.

As a concrete demonstration of these ideas, it turns out that a differentiable quantum dynamical system of a particle on a line is a Hopf algebra of self-dual type if and only if the particle moves in something like the background of a black-hole type metric. This approach to quantum-gravity or consistent physics at the Planck scale is one of the themes of Chapter 6 from a physical point of view. Of course, models of this type are quite simple, just as Lie groups are only the simplest Riemannian manifolds. The duality which we describe can be viewed as a generalisation of usual wave–particle duality at least to such spaces, a generalisation made possible by quantum groups. It seems likely that it can be taken further and related also to other duality phenomena running throughout theoretical physics.

Probably the most remarkable thing about Hopf algebras is that any one of the four points of view above would be reason enough to invent Hopf algebras or quantum groups as a generalisation of groups. Yet the same mathematical structure serves all four simultaneously, and therefore provides the framework for some truly remarkable conceptual unifications of these four directions! The comultiplication Δ leads simultaneously to a tensor product of representations (as in particle physics), to a convolution algebra of states expressing random walks, and to quantum-geometric group structure on phase-space expressing curvature. We do not hope to discuss all the myriad applications of Hopf algebras and their generalisa-

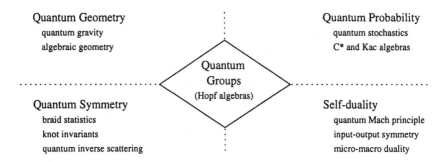

Some physical origins of quantum groups.

tions resulting from this. Let us mention only a few important directions. First, the unification of our third and fourth points of view about quantum groups surely will suggest a better understanding of the deep connection between statistical mechanics (or thermodynamics) and gravity already suggested by Hawking radiation. Indeed, a great deal about entropy, time-reversal symmetries in logic and other ideas can be explored by means of Hopf algebra models, and connected via noncommutative geometry with gravity.

Secondly, the connections between these subjects and quantum groups in the deformation theoretic sense may lead to tools for eliminating or at least regularising the infinities in quantum gravity which occur in the conventional path-integral point of view. To explain this it should be mentioned that a lot of work in quantum groups under the heading of the first and second areas above has been a process of formal q-deformation, i.e. generalising groups of symmetry by putting in a parameter q. There are powerful existence and uniqueness results coming from the work of V.G. Drinfeld, and a general formalism of bialgebra deformations due to M. Gerstenhaber and S.D. Schack. Note that q can have any meaning depending on the application (and is not usually related to physical $\hbar$).

In particular, we can introduce q-deformation as a way to regularise elementary particle quantum field theories, with the usual infinities expressed as poles $1/(q-1)$. Any group symmetries are preserved as quantum group symmetries. Then we renormalise and set q to 1. There may be an anomaly, but the final symmetry should still be a quantum group one. Moreover, as we will see, making a q-deformation is much more systematic and less *ad hoc* than other forms of regularisation because the resulting correlation functions, etc. are simply q-deformed versions of their usual expressions, and have similar algebraic properties. For example, exponentials become q-exponentials. This is related to the fact that differential operators deform to q-difference operators. In this scheme, infinities that arise from the small-scale structure of spacetime are literally controlled by replacing differential operators by finite differences, but in a more systematic way than simply working in a discrete lattice. This is surely an important potential application of quantum group technology under the first and second headings above.

Finally, we can apply such a q-regularisation procedure to the usual approach for quantum gravity itself. The radical suggestion coming from the above is that we need not, after all, renormalise, i.e. that the parameter q can after all be identified with a function of $\hbar$. This means quite simply that the infinities in the path-integral approach to quantum gravity are a product of incorrectly using classical geometry inside the path-integral. In reality the true geometry that we must use is noncommutative geometry, i.e. already partially quantised. If we use this quantum geometry from the start then we may not run into unavoidable infinities in quantum gravity. In some sense, the unifications suggested by Hopf algebras indicate that corrections due to quantum effects to the small-scale structure of spacetime (or phase-space) are of the nature of replacing differential operators by certain kinds of finite differences, with comparably nice properties.

In this respect then, Chapters 5 and 6 are central to one set of applications of Hopf algebras to physics, namely to Planck-scale physics. Chapter 5 develops the probabilistic interpretation as explained above, while Chapter 6 summarises results from the author's Ph.D. thesis about Planck-scale physics. A number of algebraic aspects of general cross product quantisation and extension theory are introduced at the same time, which are surely useful in a wider context as well.

Chapter 7 returns to the theory of quasitriangular Hopf algebras with Drinfeld's quantum double and its generalisations. Chapter 8 gives some of the semiclassical ideas that led to the quantum groups $U_q(g)$. Chapter 9 then proceeds with the formal (category-theoretic) aspects of the representation theory of quantum groups, focusing on the braiding. Here a new idea arises, namely the role of quantum groups as generating categories within which live other algebraic structures that we know and love (and perhaps want to generalise). This is also due to the author and is developed further in Chapter 10. We will see that a certain two-dimensional quantum group $\mathbb{Z}'_{/2}$ has as its category of representations (according to our second point of view above on quantum groups) the category of super-vector spaces. But we can go further: a certain n-dimensional quantum group has as its category of representations the category of anyonic vector spaces, and so on. Moreover, we can start to generalise ideas familiar in supersymmetry to these more general settings. We learn some new things about supersymmetry itself. For example, every Hopf algebra containing a group-like element of order 2 can be turned (transmuted) into a super one. Likewise, if order n, it can be transmuted into an anyonic quantum group. In the reverse direction, every group or quantum group in such categories (such as super or anyonic) can be bosonised to an ordinary quantum group by a process of bosonisation. Thus the category in which an object lives is actually quite fluid: it can be changed like a 'change of coordinates'. In physical terms, it means that quantum and statistical noncommutativity can be interchanged. For example, we need never work with supergroups provided we do not mind working with (their bosonised) quantum groups instead. On the other hand, it might be more natural sometimes to take a quantum system and understand it as the bosonisation of something simpler. For example, the ordinary Weyl algebra of quantum mechanics (which is a certain peculiar Hopf algebra) is nothing other than the bosonisation of a braided version of the real line. Such things as q-difference operators mentioned above are nothing other than the result of bosonising the obvious braided differential operators. Thus, just as quantisation can give braided structures, braidings can give quantum ones.

We can take this point of view to its logical conclusion and view every quantum group in its braided category of representations. All of its quantum aspect now appears in the braiding Ψ in the category. What is left, the resulting 'braided group', is braided commutative and braided cocommutative, i.e. appears more like a (braided) Abelian group. This leads to a number of new results for quantum groups by thinking about them in this way. Braided groups in general also provide a new systematic approach to q-deformation starting from the (braided) addition law on q-deformed $\mathbb{R}^n$. This includes natural q-deformed Euclidean and Minkowski spaces. One of the goals of this volume is to lead up to this theory of braided groups. We will be able to cover in this final Chapter 10 only the basic definitions and ideas of this braided approach to the theory of quantum groups and q-deformation. The theory of braided groups is due to the author and collaborators.

It is hoped that the further theory of braided groups, as well as the more advanced theory of quantum groups and quantum geometry, will be developed in a sequel to the present volume. Important topics which must await this sequel are: the axiomatic theory of differential graded algebras and bicovariant differentials on quantum groups, quantum group gauge theory [2], the general theory of braided groups [3, 4] and braided-Lie algebras [5], and applications to the q-regularisation of quantum field theory. Other major omissions are the general abstract deformation theory and star products, the advanced theory of the quantum Weyl group and the canonical or crystal basis, and the advanced theory of $U_q(g)$ at roots of unity and its connections with Kac–Moody algebras and conformal field theory.

I will try to mention some of these further topics and also give the briefest of historical discussions in the 'Notes' sections at the end of each chapter. My aim in the volume, however, is to give a systematic and pedagogical development rather than a historical one. Moreover, it is inevitable that I have emphasised the points of view developed in my own research work, particularly in the later chapters. Let me apologise therefore in advance for the brevity and, no doubt, incompleteness of these Notes at the end of each chapter. A full discussion of all points of interest would surely be a text in itself.

Finally, it is a pleasure to thank all the friends, colleagues and students who have helped with readings of the manuscript. Prizes go to Arkady Berenstein, Arthur Greenspoon and Konstanze Rietsch for special efforts. The cover design is by the author.

Pembroke College, Cambridge, England

1

Definition of Hopf algebras

Here we collect the definitions and basic constructions that will be needed throughout the book. This material is mostly quite standard for Hopf algebraists, but developed now with simplified or streamlined new proofs. The last section contains newer material. The content of the various definitions will become apparent after numerous examples and exercises. Once the unfamiliar notation is mastered, working with Hopf algebras is no harder than working with algebras alone, and is more fun. More practice will be provided when we turn to the advanced theory in Chapter 2. I would recommend that any reader should work through the present elementary chapter and at least the first part of Chapter 2 in detail since these sections are central to much of the later development. The main exception is Chapter 4 on matrix quantum groups, where one can get quite a long way purely by analogy with ordinary matrices, and perhaps Chapter 3 by analogy with Lie algebras. Thus, a viable alternative is to begin with Chapters 3 or 4 and use the present chapter and its sequel as reference or for clarification.

1.1 Algebras

This section is included to fix our notation and terminology. Any serious attempt to apply the extensive literature on Hopf algebras to physical situations will require familiarity with the basic ideas given here.

Recall that a *group* $(G, \cdot)$ is a set G on which an associative product is defined, with a unit element e such that $e \cdot u = u = u \cdot e$ for all $u \in G$, and such that every element u has an inverse, denoted u^{-1}. If the invertibility condition is relaxed, we have only a (unital) *semigroup*. For an Abelian group we often write the group structure additively as $(G, +)$ and denote the unit element by 0. Otherwise, for a group $(G, \cdot)$ written multiplicatively, we often omit the product, and denote the unit element

1

by 1. A homomorphism $f : G \to H$ between groups means a map such that $f(uv) = f(u)f(v)$ for all $u, v \in G$. It automatically respects the unit element.

A *field* $(k, \cdot, +)$ is a set k for which $(k, +)$ is an Abelian group, and $(k - \{0\}, \cdot)$ is an Abelian group (written multiplicatively), and these two structures are compatible: $\lambda(\mu + \nu) = \lambda\mu + \lambda\nu$ for all $\lambda, \mu, \nu \in k$. For this, we also define $0\lambda = \lambda 0 = 0$. A field has characteristic 0 if $1 + 1 + \cdots + 1$ never vanishes, which is the case for rational, real and complex numbers; otherwise the smallest number p of 1s which give zero is the characteristic, i.e. p and its multiples are zero in the field, but other natural numbers are invertible. We will work with a generic field, assuming the characteristic is not p if we need a particular p to be invertible. Also, if $(k - \{0\}, \cdot)$ is merely a semigroup and not necessarily Abelian, we have only a ring. So a *ring* is a set which is both an Abelian group under $+$ and a compatible multiplicative semigroup.

Now, groups can *act* on other structures. Thus G acts on a set M if for every element $u \in G$ there is a map $M \to M$, say $s \mapsto u \triangleright s$, such that $u \triangleright (v \triangleright s) = (uv) \triangleright s$ for all $s \in M, u, v \in G$. The action shown is a left action. G can also act from the right. An action is thus a *representation* in a set M. This is clearly a very general notion. If M itself has some structure, we can ask that the action respect this. Actions can also be denoted by a period or by omission.

A *vector space* $(V, +; k)$ over a field k is an Abelian group $(V, +)$ and an action of the multiplicative group $k - \{0\}$ on V (scalar multiplication). These should be compatible: $\lambda(v + w) = \lambda v + \lambda w$ and $(\lambda + \mu)v = \lambda v + \mu v$ for all $\lambda, \mu \in k$, $v, w \in V$. For this, we also define $0v = 0$ (the zero in V) so that we can think of all of k as acting.

If $(V, +; k)$ and $(W, +; k)$ are two vector spaces over k, there is a *direct sum* vector space $(V \oplus W, +; k)$ over k, and a *tensor product* vector space $(V \otimes W, +; k)$ over k. The direct sum $(V \oplus W, +; k)$ is defined as the set $V \times W$ consisting of pairs of elements (v, w), where $v \in V, w \in W$. The action of k is defined as $\lambda(v, w) = (\lambda v, \lambda w)$, and the addition structure is defined as $(v, w) + (x, y) = (v + x, w + y)$ for all $(v, w), (x, y) \in V \oplus W$. Elements $(v, 0), (0, w)$ and $(v, w) = (v, 0) + (0, w)$ are written as $v, w, v + w$ when working within $V \oplus W$.

The tensor product $(V \otimes W, +; k)$ is defined as follows. Start with the set of finite strings of pairs in $V \times W$ modulo the relations $(v, w) + (x, y) = (x, y) + (v, w)$, where $+$ denotes concatenation of strings. This is called the free Abelian group on $V \times W$. The tensor product $V \otimes W$ is this modulo the further relations:

$$(\lambda v, w) = (v, \lambda w), \quad (v + x, w) = (v, w) + (x, w), \quad (v, w + y) = (v, w) + (v, y).$$

k acts on this by $\lambda(v, w) = (\lambda v, w) = (v, \lambda w)$, extended by linearity to all of $V \otimes W$. For convenience, we denote pairs (v, w) in $V \otimes W$ by $v \otimes w$, so the defining relations become

$$v \otimes w + x \otimes y = x \otimes y + v \otimes w, \quad \lambda v \otimes w = v \otimes \lambda w,$$

$$(v + x) \otimes w = v \otimes w + x \otimes w, \quad v \otimes (w + y) = v \otimes w + v \otimes y.$$

A more abstract definition of $V \otimes W$ is also possible. Finally, we define the *transposition* map $\tau : V \otimes W \to W \otimes V$ by $\tau(v \otimes w) = w \otimes v$.

Lemma 1.1.1 *Note that k itself can be regarded as a (one-dimensional) vector space over k. There are canonical isomorphisms $k \otimes V \cong V \cong V \otimes k$ for any vector space V.*

Proof: We have to establish isomorphisms, i.e. a linear map $k \otimes V \to V$ which is a one to one correspondence and a linear map $V \otimes k \to V$ which is a one to one correspondence. These are simply given by $\lambda \otimes v \mapsto \lambda v$ etc. The normalisation is fixed by $1 \otimes v \mapsto v$. Such isomorphisms will generally be taken for granted in what follows. ∎

Also, if $(V, +; k)$ is a vector space over k, there is a dual vector space $V^* = \text{Lin}_k(V, k)$. This is the space of k-*linear maps* from V to k, i.e. the maps ϕ such that $\phi(\lambda v + \mu w) = \lambda \phi(v) + \mu \phi(w)$ for all $\lambda, \mu \in k$ and $v, w \in V$. These form a vector space with pointwise addition and the action $(\lambda \phi)(v) = \phi(\lambda v)$. It is clear that $V \subseteq V^{**}$ by defining $v(\phi) = \phi(v)$ for all $\phi \in V^*$ and each $v \in V$. Also, one can see that $V^* \otimes W^* \subseteq (V \otimes W)^*$. Both these inclusions are isomorphisms in the finite-dimensional case.

An *algebra* $(A, \cdot, +; k)$ over a field k is a ring $(A, \cdot, +)$ and an action of k on A (scalar multiplication) which is compatible with both the product and addition. Thus $(A, \cdot, +)$ is a ring, $(A, +; k)$ is a vector space and $\lambda(ab) = (\lambda a)b = a(\lambda b)$ for all $a, b \in A$ and $\lambda \in k$.

Equivalently, an algebra over k is a vector space $(A, +; k)$ for which there is also a compatible product. The compatibility of the product with addition and with the action of k can be concisely expressed as the requirement that the product defines in fact a linear map $\cdot : A \otimes A \to A$. Of course, we want the product to be associative and (usually) for there to be a unit element 1_A. Associativity of the product is depicted on the left in Fig. 1.1. We require that the diagram *commutes*, i.e. it does not matter which route we take. We also require a unit, 1_A. The axioms of a unit can also be expressed as diagrams by the following trick. Any element $a \in A$ defines a linear map $\eta_a : k \to A$ by $\eta_a(\lambda) = \lambda a$. Conversely, from this map we can recover a as $\eta_a(1)$. The axioms that $1_A \in A$ is the unit of A are then expressed in terms of the map $\eta = \eta_{1_A}$ in Fig. 1.1. In

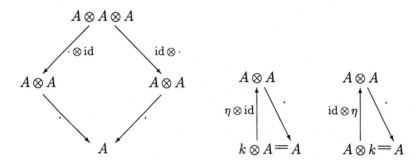

Fig. 1.1.   Associativity and unit element expressed as commutative diagrams.

summary, an algebra is a vector space A with product and unit such that
Fig. 1.1 commutes.

Algebras A, B have a tensor product. The algebra $A \otimes B$ has vector
space given by the tensor product of the vector spaces of A and B, and
product $(a \otimes c)(b \otimes d) = (ab \otimes cd)$ for all $a, b \in A$ and $c, d \in B$. An
algebra map f is a linear map between two algebras that respects the
algebra structure, $f(ab) = f(a)f(b)$ and $f(1) = 1$.

Example 1.1.2 *If V is a vector space, the tensor algebra $(T(V), +, \cdot\, ; k)$
is the noncommutative algebra generated by 1 and the elements of V. It
is defined on the vector space $T(V) = k \oplus V \oplus V \otimes V \oplus V \otimes V \otimes V \cdots$
consisting of linear combinations of finite tensor products of elements of
V.*

Proof: This means simply linear combinations of finite strings of elements
of V, i.e. linear combinations of elements of the form $v_1 \otimes v_2 \cdots \otimes v_n$.
For brevity, we just write $v_1 v_2 \cdots v_n$ for such a string, but we must re-
member that the strings $(\lambda v_1) v_2 \cdots v_n$, $v_1 (\lambda v_2) \cdots v_n, \ldots$, $v_1 v_2 \cdots (\lambda v_n)$
are all equal in the tensor algebra (and equal to $\lambda(v_1 v_2 \cdots v_n)$). The vec-
tor space structure is a direct sum, i.e. just add elements in each space
$V \otimes \cdots \otimes V$ independently. The algebra structure is just concatenation of
strings (which is clearly associative), extended by linearity.   ∎

A left *ideal* I in an algebra A is a subspace which is closed under
multiplication by A from the left. Similarly, for a right ideal under multiplication from the right. A two-sided ideal means from both sides. In
this case A/I is an algebra. Any set of elements generates an ideal, and
making the quotient in this case simply means setting those elements to

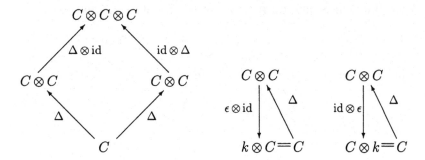

Fig. 1.2. Coassociativity and counit element expressed as commutative diagrams.

zero. For example, many algebras of interest are quotients of the above tensor algebra.

1.2 Coalgebras

The above definition of an algebra in terms of diagrams suggests the definition of a coalgebra. A *coalgebra* $(C, +, \Delta, \epsilon; k)$ over k is a vector space $(C, +; k)$ over k and a linear *coproduct* or *comultiplication* map $\Delta : C \to C \otimes C$, which is *coassociative* and for which there exists a linear *counit* map $\epsilon : C \to k$. The axioms are depicted in Fig. 1.2.

In symbols, coassociativity means $(\Delta \otimes \mathrm{id}) \circ \Delta = (\mathrm{id} \otimes \Delta) \circ \Delta$ and that ϵ is a counit means $(\epsilon \otimes \mathrm{id}) \circ \Delta(c) = c = (\mathrm{id} \otimes \epsilon) \circ \Delta(c)$ for all $c \in C$. There is a nice notation for Δ. We write

$$\Delta(c) = \sum_i c_{i(1)} \otimes c_{i(2)},$$

where the right hand side is a formal sum denoting an element of $C \otimes C$. It denotes how Δ shares out c into linear combinations of a part $_{(1)}$ in the first factor of $C \otimes C$ and a part $_{(2)}$ in the second factor. For brevity, we simply write

$$\Delta(c) = \sum c_{(1)} \otimes c_{(2)}. \tag{1.1}$$

We can leave the summation implicit as well. Coassociativity then says that if we share out again, it does not matter which piece of $\Delta(c)$ we share out. Thus, we write

$$c_{(1)} \otimes c_{(2)(1)} \otimes c_{(2)(2)} = c_{(1)(1)} \otimes c_{(1)(2)} \otimes c_{(2)} = c_{(1)} \otimes c_{(2)} \otimes c_{(3)}, \tag{1.2}$$

say. Physicists should think of c as being like a probability density function: its total probability mass $\epsilon(c)$ is being shared out among different spaces. We will develop such an interpretation in detail in Chapter 5.

Exercise 1.2.1 *Deduce that the six maps $C \to C \otimes C \otimes C \otimes C$ obtained by composing three Δ's give the same result independently of which factors are shared when each Δ is applied. Denote this map by $\Delta^3 : c \mapsto c_{(1)} \otimes c_{(2)} \otimes c_{(3)} \otimes c_{(4)}$. (Similarly for all higher sharings Δ^n.) Show that $\epsilon \otimes \mathrm{id} \otimes \mathrm{id} \otimes \mathrm{id}$ applied to $c_{(1)} \otimes c_{(2)} \otimes c_{(3)} \otimes c_{(4)}$ is $c_{(1)} \otimes c_{(2)} \otimes c_{(3)}$, and similarly for ϵ in any other position.*

Solution: There is a handy way to carry out such computations: on each operation by Δ or ϵ, convert any multiple sharings into a base-ten numbered form, and then into a form suitable for the next operation. Thus, for example, $c_{(1)(1)} \otimes c_{(1)(2)} \otimes c_{(2)} \cdots = c_{(1)} \otimes c_{(2)} \otimes c_{(3)} \cdots$, and hence we have $\epsilon(c_{(1)})c_{(2)} \otimes c_{(3)} \otimes c_{(4)} = \epsilon(c_{(1)(1)})c_{(1)(2)} \otimes c_{(2)} \otimes c_{(3)} = c_{(1)} \otimes c_{(2)} \otimes c_{(3)}$. ∎

If C is a coalgebra, then $\Delta : C \to C \otimes C$ and $\epsilon : C \to k$ define adjoint maps $\cdot : C^* \otimes C^* \to C^*$ and $\eta : k \to C^*$ by

$$(\phi \cdot \psi)(c) = \phi(c_{(1)})\psi(c_{(2)}), \quad \eta(\lambda)(c) = \lambda\epsilon(c) \tag{1.3}$$

for all $\phi, \psi \in C^*$ and $c \in C$. The unit in C^* is $1_{C^*} = \epsilon$, and the coalgebra axioms are just the requirements that make $(C^*, \cdot, \eta; k)$ into an algebra over k. Conversely, a finite-dimensional algebra A defines equivalently a coalgebra A^*. Thus the theory of coalgebras is essentially dual to the theory of algebras.

Coalgebras C, D also have a tensor product. The coalgebra $C \otimes D$ has vector space $C \otimes D$ and $\Delta(c \otimes d) = c_{(1)} \otimes d_{(1)} \otimes c_{(2)} \otimes d_{(2)}$. A *coalgebra map* f is a map between two coalgebras that respects the coalgebra structure, $(f \otimes f) \circ \Delta = \Delta \circ f$ and $\epsilon \circ f = \epsilon$.

1.3   Bialgebras and Hopf algebras

A *bialgebra* $(H, +, \cdot, \eta, \Delta, \epsilon; k)$ over k is a vector space $(H, +; k)$ over k which is both an algebra and a coalgebra, in a compatible way. The compatibility is

$$\Delta(hg) = \Delta(h)\Delta(g), \quad \Delta(1) = 1 \otimes 1, \quad \epsilon(hg) = \epsilon(h)\epsilon(g), \quad \epsilon(1) = 1 \tag{1.4}$$

for all $h, g \in H$. In the first of these, the product on the right takes place in $H \otimes H$; i.e. $\Delta : H \to H \otimes H$ and $\epsilon : H \to k$ are algebra maps. In fact, this is the same as the assertion that $\cdot : H \otimes H \to H$, $\eta : k \to H$ are coalgebra maps, where $H \otimes H$ has the tensor product coalgebra structure. Note

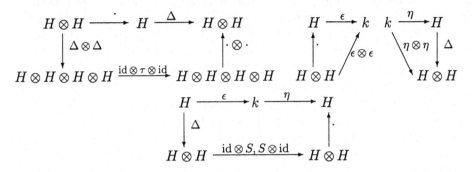

Fig. 1.3. Additional axioms that make the algebra and coalgebra H into a Hopf algebra.

that $\epsilon(1) = 1$ is automatic as k is a field. Bialgebras and Hopf algebras are also studied over k a commutative ring.

A *Hopf algebra* $(H, +, \cdot, \eta, \Delta, \epsilon, S; k)$ over k is a bialgebra over k equipped with a linear *antipode map* $S : H \to H$ obeying

$$\cdot (S \otimes \mathrm{id}) \circ \Delta = \cdot(\mathrm{id} \otimes S) \circ \Delta = \eta \circ \epsilon. \tag{1.5}$$

The role of the antipode is like that of an inverse. However, we do not demand that $S^2 = \mathrm{id}$. We do not even assume that S as a linear map has an inverse S^{-1} (however, this is always so if H is finite dimensional). These axioms are depicted in Fig. 1.3. They have an interesting symmetry, as we shall see in the next section.

Proposition 1.3.1 *The antipode of a Hopf algebra is unique and obeys* $S(hg) = S(g)S(h)$, $S(1) = 1$ *(i.e. S is an* antialgebra *map) and* $(S \otimes S) \circ \Delta h = \tau \circ \Delta \circ Sh$, $\epsilon Sh = \epsilon h$ *(i.e. S is an* anticoalgebra *map).*

Proof: If S, S_1 are two antipodes on a bialgebra H then they are equal because $S_1 h = (S_1 h_{(1)})\epsilon(h_{(2)}) = (S_1 h_{(1)})h_{(2)(1)}Sh_{(2)(2)} = (S_1 h_{(1)})h_{(2)}Sh_{(3)} = (S_1 h_{(1)(1)})h_{(1)(2)}Sh_{(2)} = \epsilon(h_{(1)})Sh_{(2)} = Sh$. Here we wrote $h = h_{(1)}\epsilon(h_{(2)})$ by the counity axiom, and then inserted $h_{(2)(1)}Sh_{(2)(2)}$ knowing that it would collapse to $\epsilon(h_{(2)})$. We then used associativity and (the more novel ingredient) coassociativity to be able to collapse $(S_1 h_{(1)(1)})h_{(1)(2)}$ to $\epsilon(h_{(1)})$. Note that the proof is not any harder than the usual one for uniqueness of group inverses, the only complication being that we are working now with parts of linear combinations and have to take care to keep the order of the coproducts. In practice, the linear ordering as in $(S_1 h_{(1)})h_{(2)}Sh_{(3)}$ is the most convenient for this. We can similarly collapse such expressions as $(S_1 h_{(1)})h_{(2)}$ or $h_{(2)}Sh_{(3)}$ wherever they occur as long as the two

collapsing factors are in linear order. This is just the analogue of cancelling $h^{-1}h$ or hh^{-1} in a group. Armed with such techniques, we return
now to the proof of the proposition. Applying the antipode axiom to hg,
we have $(S(h_{(1)}g_{(1)}))h_{(2)}g_{(2)} = \epsilon(hg) = \epsilon(h)\epsilon(g)$. This is not yet what we
want because we want to learn about $S(hg)$, not about such pieces of
a linear combination as $S(h_{(1)}g_{(1)})$. However, we can apply this identity
not to g but to $g_{(1)}$ and keep $g_{(2)}$ for another purpose. Thus we have
$(S(h_{(1)}g_{(1)(1)}))h_{(2)}g_{(1)(2)} \otimes g_{(2)} = \epsilon(h)\epsilon(g_{(1)}) \otimes g_{(2)} = \epsilon(h) \otimes g$. Applying S
to the second factor and multiplying, we have $(S(h_{(1)}g_{(1)}))h_{(2)}g_{(2)}Sg_{(3)} =$
$\epsilon(h)Sg$. Collapsing $g_{(2)}Sg_{(3)}$ in our expression then gives $(S(h_{(1)}g))h_{(2)} =$
$\epsilon(h)Sg$. We now do the same thing again, namely we apply this result not
to h but to $h_{(1)}$, so $(S(h_{(1)(1)}g))h_{(1)(2)} \otimes h_{(2)} = \epsilon(h_{(1)})Sg \otimes h_{(2)} = Sg \otimes h$.
Applying S to the second factor and multiplying gives $(S(h_{(1)}g))h_{(2)}Sh_{(3)} =$
$(Sg)(Sh)$. Collapsing $h_{(2)}Sh_{(3)}$ then gives the desired result. Comparing
this with the usual proof for group inversion, the reader will see again that
our proof is not fundamentally different in spirit. We obtain $S(1) = 1$
more simply as $S(1) = 1S(1) = 1_{(1)}S1_{(2)} = \epsilon(1) = 1$. For the $\Delta \circ S$ and
$\epsilon \circ S$ identities the proof is similar but 'inside out' to the above (it can be
obtained by a symmetry principle; see below). For a direct proof we have
$\epsilon(Sh) = \epsilon(Sh_{(1)})\epsilon(h_{(2)}) = \epsilon((Sh_{(1)})h_{(2)}) = \epsilon(1)\epsilon(h) = \epsilon(h)$, and finally
we compute $Sh_{(2)} \otimes Sh_{(1)} = (Sh_{(1)})_{(1)}h_{(2)(1)}Sh_{(4)} \otimes (Sh_{(1)})_{(2)}h_{(2)(2)}Sh_{(3)} =$
$(Sh_{(1)})_{(1)}h_{(2)}Sh_{(3)} \otimes (Sh_{(1)})_{(2)} = (Sh)_{(1)} \otimes (Sh)_{(2)}$ using the same techniques
as above.   ∎

Bialgebras and Hopf algebras H, G also have a tensor product. The
Hopf algebra $H \otimes G$ has vector space $H \otimes G$ with the tensor product
algebra structure and the tensor product coalgebra structure. A *Hopf
algebra map* f is a map between two Hopf algebras that respects the
Hopf algebra structure, i.e. it is both an algebra map and a coalgebra
map (a bialgebra map) and $Sf(h) = f(Sh)$. In fact, the last condition
is redundant; a bialgebra map between Hopf algebras is automatically a
Hopf algebra map.

Example 1.3.2 *We consider the algebra generated by 1 and four elements* X, g, g^{-1} *with relations*

$$gg^{-1} = 1 = g^{-1}g, \quad Xg = qgX, \quad Xg^{-1} = q^{-1}g^{-1}X,$$

*where q is a fixed invertible element of the field k. This becomes a Hopf
algebra with*

$$\Delta X = X \otimes 1 + g \otimes X, \quad \Delta g = g \otimes g, \quad \Delta g^{-1} = g^{-1} \otimes g^{-1},$$

$$\epsilon X = 0, \quad \epsilon g = 1 = \epsilon g^{-1}, \quad SX = -g^{-1}X, \quad Sg = g^{-1}, \quad Sg^{-1} = g.$$

The subalgebra generated by X, g (without g^{-1}) is is a sub-bialgebra. Note that $S^2 \neq$ id in this example (because $S^2 X = qX$).

Proof: We have Δ, ϵ on the generators and extend them multiplicatively to products of the generators (so that they are necessarily algebra maps as required). However, we have to check that this is consistent with the relations in the algebra. For example, $\Delta X g = (\Delta X)(\Delta g) = (X \otimes 1 + g \otimes X)(g \otimes g) = Xg \otimes g + g^2 \otimes Xg$, while equal to this must be $\Delta q g X = q(\Delta g)(\Delta X) = q(g \otimes g)(X \otimes 1 + g \otimes X) = qgX \otimes g + qg^2 \otimes gX$. These expressions are equal, using again the relations in the algebra as stated. Similarly for the other relations. For the antipode, we keep in mind the preceding proposition and extend S as an antialgebra map, and check that this is consistent in the same way. Since S obeys the antipode axioms on the generators (an easy computation), it follows that it obeys them also on the products since Δ, ϵ are already extended multiplicatively. ∎

This example, which is well-known to Hopf algebraists, is a key part of the structure of some standard quantum groups (see Chapter 3). It also figures in physics as the *Weyl algebra* and as the bosonisation of the *braided line*. We will see these points of view later. In addition, the subalgebra generated by X, g is sometimes called the *quantum plane* (see Chapters 4.5 and 10.2), but appears here with a bialgebra structure which is not the natural (braided) one for the quantum plane. An even simpler bialgebra is to take X, g with the coproduct shown and no relations at all.

Exercise 1.3.3 *If H is a bialgebra, show that H^{op} defined as H with the opposite multiplication $h \cdot^{op} g = gh$ is also a bialgebra. Likewise for H^{cop} defined as H with the opposite coproduct, $\Delta^{cop} h = h_{(2)} \otimes h_{(1)}$. If H is a Hopf algebra, show that H^{op} is a Hopf algebra if and only if S is invertible. In this case $S^{op} = S^{-1}$. Likewise for H^{cop} with $S^{cop} = S^{-1}$. Finally (if S is invertible), show that the map $S : H \to H^{op/cop}$ (where both are opposite) is a Hopf algebra isomorphism.*

Solution: If S^{-1} exists, then $(S^{-1} h_{(1)}) \cdot^{op} h_{(2)} = h_{(2)} S^{-1} h_{(1)} = \epsilon(h)$, as we see by applying S to both sides and using Proposition 1.3.1. Likewise on the other side; so S^{-1} is the antipode for H^{op}. Conversely, if S^{op} exists, then we have $SS^{op} h = (SS^{op} h_{(1)})(Sh_{(2)}) h_{(3)} = (S(h_{(2)} S^{op} h_{(1)})) h_{(3)} = (S((S^{op} h_{(1)}) \cdot^{op} h_{(2)})) h_{(3)} = h$. Likewise on the other side, so S^{op} is the inverse of S. The arguments for H^{cop} are similar. The last part is just a rewording of Proposition 1.3.1 as far as the bialgebra structure is concerned, while the antipode of $H^{op/cop}$ is just S again. Note that it could happen that H is not a Hopf algebra but that $H^{\mathrm{op}}, H^{\mathrm{cop}}$ are. In this case their antipode S^{op} is called a *skew-antipode* for H.   ∎

1.4   Duality

We come now to one of the distinctive features of Hopf algebras, evident from the structure of the axioms in Figs. 1.1–1.3.

Proposition 1.4.1 *The axioms of a Hopf algebra expressed in Figs. 1.1– 1.3 are self-dual in the sense that reversing arrows and interchanging Δ, ϵ with $\cdot, \eta$ gives the same set of axioms.*

Proof: Recall that the axioms of a coalgebra in Fig. 1.2 are just dual in this way to the axioms of an algebra in Fig. 1.1. Looking now at Fig. 1.3, make the reversal of arrows and interchange Δ, ϵ with $\cdot, \eta$, and then flip the resulting diagrams about a vertical axis on the page. This gives the original diagrams.   ∎

One way to think about this symmetry is in terms of dual linear spaces. Recall that a coalgebra Δ defines an algebra on the dual linear space and, in the finite-dimensional case, an algebra defines a coalgebra on the dual. Thus, from the preceding proposition we conclude that for every finite-dimensional Hopf algebra H there is a dual Hopf algebra H^* built on the vector space H^* dual to H. Note also that since, in the finite-dimensional case, $H^{**} \cong H$ in a canonical way, we should think of H, H^* symmetrically. Thus, instead of writing $\phi(v)$ for the evaluation of a map in H^* on an element of H, we write $\phi(v) = \langle \phi, v \rangle$.

Proposition 1.4.2 *Using this convenient notation, the explicit formulae that determine the Hopf algebra structure on H^* from that on H are as follows. For the bialgebra structure, we have*

$$\langle \phi\psi, h \rangle = \langle \phi \otimes \psi, \Delta h \rangle, \quad \langle 1, h \rangle = \epsilon(h),$$

$$\langle \Delta\phi, h \otimes g \rangle = \langle \phi, hg \rangle, \quad \epsilon(\phi) = \langle \phi, 1 \rangle,$$

for all $\phi, \psi \in H^*$ and $h, g \in H$. In the Hopf algebra case, we have additionally

$$\langle S\phi, h \rangle = \langle \phi, Sh \rangle.$$

Proof: This is an elementary exercise in what we mean by the dual linear space and how maps are 'dualised'. We have already noted that coassociativity of Δ corresponds to associativity of the multiplication in the dual, and similarly associativity to coassociativity in the dual. We verify now that these make H^* into a Hopf algebra. Proceeding from the axioms, we have $\langle \Delta(\phi\psi), h \otimes g \rangle = \langle \phi\psi, hg \rangle = \langle \phi \otimes \psi, \Delta(hg) \rangle = \langle \phi \otimes \psi, (\Delta h)(\Delta g) \rangle = \langle \phi_{(1)} \otimes \psi_{(1)} \otimes \phi_{(2)} \otimes \psi_{(2)}, \Delta h \otimes \Delta g \rangle = \langle (\Delta \phi)(\Delta \psi), h \otimes g \rangle$ as required. We used that H is a bialgebra and the pairing equations. Also, $\langle (S\phi_{(1)})\phi_{(2)}, h \rangle = \langle S\phi_{(1)} \otimes \phi_{(2)}, h_{(1)} \otimes h_{(2)} \rangle = \langle \phi_{(1)} \otimes \phi_{(2)}, Sh_{(1)} \otimes h_{(2)} \rangle = \langle \phi, (Sh_{(1)})h_{(2)} \rangle = \langle \phi, 1 \rangle \epsilon(h) = \epsilon(\phi)\epsilon(h) = \langle 1\epsilon(\phi), h \rangle$. Similarly for S on $\phi_{(2)}$. Since these identities hold for arbitrary test elements h, g, we conclude that H^* is a Hopf algebra, as must be the case from the general remarks above. ∎

In the infinite-dimensional case the correct notion of dual is more intricate. A standard algebraic approach is to restrict to a certain subset $H^\circ \subset H^*$ with the right properties. A different approach is to focus on the pairing.

Definition 1.4.3 *Two bialgebras or Hopf algebras A, H are paired if there is a bilinear map $\langle\ ,\ \rangle : A \otimes H \to k$ obeying the equations displayed in the preceding proposition for all $\phi, \psi \in A$, $h, g \in H$. They are a strictly dual pair if the pairing is nondegenerate in the sense that there are no nonzero null elements in H or in A ($\phi \in A$ is null if $\langle \phi, h \rangle = 0$ for all $h \in H$, and $h \in H$ is null if $\langle \phi, h \rangle = 0$ for all $\phi \in A$).*

In the finite-dimensional case, this is just the same as saying $H = A^*, A = H^*$ with evaluation given by the pairing.

Proposition 1.4.4 *A pairing between bialgebras (or Hopf algebras) can always be made nondegenerate by quotienting out (i.e. setting to zero) those elements that pair as zero with all the elements of the other bialgebra or Hopf algebra. The resulting bialgebras or Hopf algebras are then a strictly dual pair.*

Proof: This needs a little more theory. First, a rewording of Proposition 1.4.2 says that $i_1 : A \to H^*$ and $i_2 : H \to A^*$ given by $i_1(\phi) = \langle \phi, \ \rangle$ and $i_2(h) = \langle\ , h \rangle$ are algebra maps. We are setting to zero, i.e. quotienting by $J_i = \ker i_i$ (the set of elements mapping to zero). Kernels of algebra maps are always ideals, so A/J_1 and H/J_2 are algebras. Secondly, since $\langle \Delta \phi, h \otimes g \rangle = \langle \phi, hg \rangle = 0$ for all h, g if $\phi \in J_1$, we conclude that $\Delta J_1 \subseteq J_1 \otimes A + A \otimes J_1$. Likewise $\Delta J_2 \subseteq J_2 \otimes H + H \otimes J_2$. One says that ideals of bialgebras with this property are *biideals*. We can now quotient by them and have that Δ is still well-defined in the quotient, so the quotients are also bialgebras. Finally, if A, H are Hopf algebras, then $\langle S\phi, h \rangle = \langle \phi, Sh \rangle = 0$ for all h if $\phi \in J_1$ (and similarly on the other side). Hence $SJ_i \subseteq J_i$. This makes our ideals *Hopf ideals* and the quotients Hopf algebras. ∎

1.5   Commutative and cocommutative Hopf algebras

A Hopf algebra is *commutative* if it is commutative as an algebra. It is *cocommutative* if it is cocommutative as a coalgebra, i.e. if

$$\tau \circ \Delta = \Delta. \tag{1.6}$$

Proposition 1.5.1 *The dual of a commutative Hopf algebra is cocommutative, and vice versa. If H is a commutative or cocommutative Hopf algebra, then $S^2 = \mathrm{id}$.*

Proof: The second part follows at once from Exercise 1.3.3, for if H is commutative or cocommutative, then H^{op} or H^{cop}, respectively, coincide with H; hence, $S^{-1} = S$, since the antipode on H is unique. For a direct proof in the cocommutative case, $S^2 h = (S^2 h_{(1)})(S h_{(2)}) h_{(3)} = (S(h_{(1)} S h_{(2)})) h_{(3)} = h$. The commutative case is almost identical. ∎

Example 1.5.2 *Let G be a finite group with identity e. Let $k(G)$ denote the set of functions on G with values in k. This has the structure of a commutative Hopf algebra as follows. The vector space structure is given by pointwise addition and the action of k by $(\lambda.\phi)(u) = \lambda.(\phi(u))$. The algebra product and unit are*

$$(\phi\psi)(u) = \phi(u)\psi(u), \qquad \eta(\lambda)(u) = \lambda, \qquad \phi, \psi \in k(G), \ u \in G.$$

The coproduct, counit and antipode are

$$(\Delta\phi)(u,v) = \phi(uv), \quad \epsilon\phi = \phi(e), \quad (S\phi)(u) = \phi(u^{-1}).$$

Proof: Here we are identifying $k(G) \otimes k(G) = k(G \times G)$ (functions of two group variables) when defining the coproduct. Coassociativity is easily verified as $((\Delta \otimes \mathrm{id})\Delta\phi)(u,v,w) = (\Delta\phi)(uv,w) = \phi((uv)w) = \phi(u(vw)) = (\Delta\phi)(u,vw) = ((\mathrm{id} \otimes \Delta)\Delta\phi)(u,v,w)$. Note that it comes directly from associativity in the group. Likewise $((\epsilon \otimes \mathrm{id})\Delta\phi)(u) = (\Delta\phi)(e,u) = \phi(eu) = \phi(u)$ and $((S\phi_{(1)})\phi_{(2)})(u) = (S\phi_{(1)})(u)\phi_{(2)}(u) = \phi_{(1)}(u^{-1})\phi_{(2)}(u) = \phi(u^{-1}u) = \phi(e) = \epsilon(\phi)$. Similarly on the other side. ∎

Example 1.5.3 *Let G be a finite group. Let kG denote the vector space with basis G, i.e. $\{a = \sum_{u \in G} a(u)e_u\}$, where $\{e_u : u \in G\}$ denotes the basis, and the coefficients have values in k. To simplify notation, we simply write kG as the set of formal k-linear combinations of elements of G, $a = \sum a(u)u$. This has the structure of a cocommutative Hopf algebra. The algebra structure, unit, coproduct, counit and antipode are*

$$\text{product in } G, \quad 1 = e, \quad \Delta u = u \otimes u, \quad \epsilon u = 1, \quad S u = u^{-1}$$

regarded as $u \in G \subset kG$, extended by linearity to all of kG.

Proof: The multiplication is clearly associative because the group multiplication is. The coproduct is coassociative because it is so on each of the basis elements $u \in G$. It is an algebra homomorphism because $\Delta(uv) = uv \otimes uv = (u \otimes u)(v \otimes v) = (\Delta u)(\Delta v)$. The other facts are equally easy.   ∎

   In a general Hopf algebra one can always look for *group-like elements* where the coproduct is of this *group-algebra* type $\Delta u = u \otimes u$. There need not, however, be many of them in general.

Example 1.5.4 *The group Hopf algebra just described can also be thought of as a function space. Namely, the set of coefficients $\{h(u)\}$ of $h \in kG$ can be regarded as a function on G. The Hopf algebra structure is equivalent in terms of these functions to*

$$(hg)(u) = \sum_v h(v)g(v^{-1}u), \quad 1(u) = \delta_e(u),$$

$$(\Delta h)(u,v) = \delta_u(v)h(u), \quad \epsilon(h) = \sum_u h(u), \quad (Sh)(u) = h(u^{-1}),$$

for all $h, g \in kG$. Here $\delta_u(v)$ is 1 if $u = v$ and 0 otherwise. For this reason, the group algebra can also be called the group convolution algebra.

Proof: If $h = \sum h(v)v$, $g = \sum g(u)u$, then $hg = \sum_{u,v} h(v)g(u)vu = \sum_{u,v} h(v)g(v^{-1}u)u$, etc.   ∎

Exercise 1.5.5 *The Hopf algebras $k(G)$ and kG are strictly dual to one another. The pairing is as follows: $\phi \in k(G)$ should be extended by linearity to a function on kG. Thus $\langle \phi, h \rangle = \phi(\sum_u h(u)u) = \sum_u h(u)\phi(u)$.*

Solution: To check the pairing equations it is enough to check on the generators $u \in G \subset kG$, since the structure of kG is just the structure of these, extended by linearity. For example, $\langle \phi \otimes \psi, \Delta u \rangle = \langle \phi \otimes \psi, u \otimes u \rangle = \phi(u)\psi(u) = (\phi\psi)(u) = \langle \phi\psi, u \rangle$. Comparing the proof above that $k(G)$ is a Hopf algebra with the proof of Proposition 1.4.1, we see that we have already done all the work.   ∎

   Here is a simple but important application of these constructions when G is a finite Abelian group. Let $\hat{G}$ denote the character group (Pontryagin dual) of G, i.e. $\hat{G}$ consists of maps from G to $k - \{0\}$ that respect the group structure in the sense $\chi(uv) = \chi(u)\chi(v)$. It is easy to see that $\hat{G}$ becomes a group under pointwise multiplication.

Corollary 1.5.6 *The elements of $\hat{G}$ can be identified with the nonzero algebra maps from kG to k, so that $k\hat{G} = (kG)^*$ as Hopf algebras. Hence, from the previous exercise, we conclude that $k\hat{G} \cong k(G)$ as Hopf algebras. Similarly, we conclude that $kG \cong k(\hat{G})$. These isomorphisms are the* Fourier transform *of the convolution algebra on $\hat{G}$ to functions on G and vice versa. Explicitly, they take the form*

$$\tilde{h}(u) = \sum_{\chi} h(\chi)\chi(u), \quad \tilde{\phi}(\chi) = \frac{1}{|G|} \sum_{u \in G} \chi(u^{-1})\phi(u)$$

for $h, \tilde{\phi} \in k\hat{G}$ and $\phi, \tilde{h} \in k(G)$. Similarly with $G, \hat{G}$ interchanged. Here $|G|$ denotes the number of elements of G or $\hat{G}$, and we assume that it is invertible in our field k.

Proof: We view each element of $\hat{G}$ as extended by linearity to a function on kG. So every element $h = \sum_{\chi} h(\chi)\chi$ of $k\hat{G}$ is viewed by linearity as an element $\tilde{h}$ of $(kG)^* = k(G)$. This is a Hopf algebra mapping and is surjective because $\sum_{\chi} \chi(u) = |G|\delta_e(u)$ gives $\delta_e \in k(G)$ as a sum of characters (the other basis elements δ_u then follow by left-translation on the group). Note that $\Lambda^* \equiv \sum_{\chi'} \chi'$ is characterised up to normalisation by the property $\chi\Lambda^* = \chi\sum_{\chi'}\chi' = \sum_{\chi'}\chi\chi' = \sum_{\chi''}\chi'' = \epsilon(\chi)\Lambda^*$ for all χ in $\hat{G}$, i.e. $h\Lambda^* = \epsilon(h)\Lambda^*$ for all $h \in k\hat{G}$. When this element is viewed in $k(G)$, this becomes the characteristic property (up to normalisation) of δ_e. We remark that one can also write the Fourier transform more explicitly in terms of this element Λ^* and a similar element $\Lambda = \sum_{u \in G} u \in kG$ characterised by $\Lambda\phi = \Lambda\epsilon(\phi)$ for all $\phi \in kG$. Namely, we consider it as a linear isomorphism $k\hat{G} = (kG)^* \to kG$ given by $\tilde{h} = \Lambda_{(1)}\langle h, \Lambda_{(2)}\rangle$ and inverse $\tilde{\phi} = \langle\Lambda^*, \Lambda\rangle^{-1}\Lambda^*_{(1)}\langle\Lambda^*_{(2)}, S\phi\rangle$, where $h, \tilde{\phi} \in (kG)^*$ and $\tilde{h}, \phi \in kG$ as linear spaces. There is also a second product on kG defined by $\phi * \psi = \langle\Lambda^*, \psi S\phi_{(1)}\rangle\phi_{(2)}$, with respect to which the Fourier transform becomes a Hopf algebra isomorphism; in the present case this second product is just the pointwise product of $k(G)$ when we regard the coefficients of ϕ as a function $\phi \in k(G)$. This abstract formulation of Fourier transform works for any finite-dimensional Hopf algebra equipped with such elements Λ, Λ^*, with $\langle\Lambda^*, \Lambda\rangle$ invertible. We will see this later, in Section 1.7. We have covered here just one of the two Fourier isomorphisms: the isomorphism $kG \cong k(\hat{G})$ is proven similarly with the roles of $G, \hat{G}$ reversed. Note also that $(kG)^{**} \cong kG$ reduces, in this setting, to $\hat{\hat{G}} \cong G$ or *Pontryagin's theorem.*

■

If G is not Abelian, then it is natural to extend the definition of $\hat{G}$ as the collection of equivalence classes of irreducible representations. This is not, however, any longer a group. An alternative approach, in view of the second of the above Fourier isomorphisms, is simply to work with the Hopf algebra kG in place of $k(\hat{G})$. This is an example of the philosophy of noncommutative geometry: even when G is non-Abelian and kG is not commutative (and so cannot any longer be isomorphic to the algebra of functions on a space), we can still think of kG as like functions on a 'noncommutative space' generalisation of $\hat{G}$.

Example 1.5.7 *Let g be a Lie algebra over k. We define the* universal *enveloping Hopf algebra of g as follows. As an algebra, $U(g)$ is the noncommutative algebra generated by 1 and elements of g (the tensor algebra over k) modulo the relations $[\xi, \eta] = \xi\eta - \eta\xi$ for all $\xi, \eta \in g$ (i.e. finite strings of elements of g with these relations). The coproduct, counit and antipode are*

$$\Delta\xi = \xi \otimes 1 + 1 \otimes \xi, \quad \epsilon\xi = 0, \quad S\xi = -\xi$$

extended in the case of Δ, ϵ as algebra maps, and in the case of S as an antialgebra map.

Proof: We extend Δ, ϵ as algebra homomorphisms and S as an antialgebra homomorphism, and have to check that this extension is consistent with the relations. For example, $\Delta(\xi\eta) = (\xi \otimes 1 + 1 \otimes \xi)(\eta \otimes 1 + 1 \otimes \eta) = \xi\eta \otimes 1 + 1 \otimes \xi\eta + \xi \otimes \eta + \eta \otimes \xi$. Subtracting from this the corresponding expression for $\Delta\eta\xi$ and using the relations, we obtain $[\xi, \eta] \otimes 1 + 1 \otimes [\xi, \eta] = \Delta[\xi, \eta]$ as required. Similarly for the counit and antipode. ∎

An element in a general Hopf algebra which has this linear form $\Delta\xi = \xi \otimes 1 + 1 \otimes \xi$ is called *primitive*. One can show further that the association of Hopf algebras $k(G)$ and kG to groups G extend as functors from the category of finite groups to the category of Hopf algebras (see Chapter 9). This means that the theory of finite groups precisely embeds in the theory of commutative or cocommutative Hopf algebras. Similarly the association of $U(g)$ to g. In fact, there are theorems which state that essentially all commutative or cocommutative Hopf algebras are of similar form to $k(G)$ or kG (or $U(g)$), respectively.

1.6   Actions and coactions

The definitions of actions and coactions are simply a polarisation of those of algebras and coalgebras, so we include them now among the basic

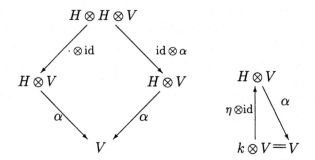

Fig. 1.4. Axioms of an action expressed as commutative diagrams.

definitions. They are, in any case, among the most important of the things that Hopf algebras can do, namely act on other structures. They serve to justify that the definitions above are worthwhile ones. Most importantly of all, they provide some essential exercise in working with abstract Hopf algebras.

A left *action* (or representation) of an algebra H is a pair (α, V), where V is a vector space and α is a linear map $H \otimes V \to V$, say $\alpha(h \otimes v) = \alpha_h(v)$, such that $\alpha_{hg}(v) = \alpha_h(\alpha_g(v))$, $\alpha(1 \otimes v) = v$. Instead of constantly writing α, we often simply denote it by $\triangleright$ (or simply by a period). Thus,

$$h \triangleright v \in V, \qquad (hg) \triangleright v = h \triangleright (g \triangleright v), \qquad 1 \triangleright v = v. \tag{1.7}$$

The axioms are depicted in Fig. 1.4. Compare with Fig. 1.1: the axioms are just the same with some of the multiplications replaced by α. For each h in H, we can also write $\rho(h) = h \triangleright (\ ) = \alpha(h \otimes (\ ))$ viewed as a linear map from $V \to V$. If we write $\mathrm{Lin}(V)$ for the usual algebra of such linear maps (with multiplication given by composition), then clearly the axioms for α just say that the corresponding $\rho : H \to \mathrm{Lin}(V)$ is an algebra map. This is a more conventional way of thinking about actions.

There are two ways of referring to an action or representation. If we want to emphasise the map α itself, then we say that the algebra H acts on the left on the vector space V. The action is α. If, on the other hand, we wish to emphasise the space on which the algebra acts, we say that V is a *left H-module*. Thus, a left H-module is nothing other than a vector space on which the algebra H acts. If two modules are isomorphic, we mean of course not only that they are isomorphic as vector spaces, but that the underlying actions correspond. It is often convenient to place the emphasis on the space that is acted upon in this way. Finally, if there is an algebra map f from another algebra to H, we say that an action or

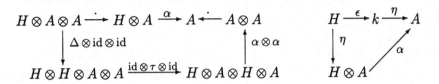

Fig. 1.5. Additional axioms for an H-module algebra expressed as commutative diagrams.

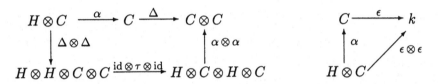

Fig. 1.6. Additional axioms for an H-module coalgebra expressed as commutative diagrams.

module *pulls back* to an action of the other algebra by first mapping it by f and then applying the action of H.

1.6.1 Actions on algebras and coalgebras

More interesting than the action of Hopf algebras on vector spaces is their action on other structures, such as algebras, coalgebras or Hopf algebras. In each case, it is natural to ask that the relevant structure be respected by the action in some way. Here we really need to have a bialgebra or Hopf algebra H and not merely an algebra that is acting. Thus we say that an algebra A is an H-*module algebra* if A is a left H-module (i.e. H acts on it from the left) and

$$h\triangleright(ab) = \sum(h_{(1)}\triangleright a)(h_{(2)}\triangleright b), \qquad h\triangleright 1 = \epsilon(h)1. \tag{1.8}$$

This is depicted in Fig. 1.5. Likewise, we say that a coalgebra C is a left H-*module coalgebra* if

$$\Delta(h\triangleright c) = \sum h_{(1)}\triangleright c_{(1)} \otimes h_{(2)}\triangleright c_{(2)}, \qquad \epsilon(h\triangleright c) = \epsilon(h)\epsilon(c). \tag{1.9}$$

This says that $\triangleright : H \otimes C \to C$ is a coalgebra map, where $H \otimes C$ has the tensor product coalgebra structure, i.e. $\Delta(h\triangleright c) = (\Delta h)\triangleright\Delta c$. The condition is depicted in Fig. 1.6, and is a polarisation of Fig. 1.3. We will see applications of these notions in the next section and in later chapters.

Example 1.6.1 *The left regular action L of a bialgebra or Hopf algebra H on itself is $L_h(g) = hg$, and makes H into an H-module coalgebra.*

Proof: This is an action from associativity and the unit axioms for H. Writing $L_h(g) = h \triangleright g$, we have a module coalgebra because $h \triangleright \Delta g = h_{(1)} g_{(1)} \otimes h_{(2)} g_{(2)} = \Delta(hg)$ and $\epsilon(h \triangleright g) = \epsilon(hg) = \epsilon(h)\epsilon(g)$ as required. ∎

Example 1.6.2 *The left coregular action R^* of a finite-dimensional bialgebra or Hopf algebra H on H^* is $R_h^*(\phi) = \sum \phi_{(1)} \langle h, \phi_{(2)} \rangle$, and makes H^* into an H-module algebra.*

Proof: This is an action from the coassociativity and counit axioms for H^*. It is an H-module algebra because $h \triangleright (\phi\psi) = (\phi\psi)_{(1)} \langle h, (\phi\psi)_{(2)} \rangle = \phi_{(1)}\psi_{(1)} \langle h, \phi_{(2)}\psi_{(2)} \rangle = (h_{(1)} \triangleright \phi)(h_{(2)} \triangleright \psi)$ and $h \triangleright 1 = 1\langle h, 1 \rangle = \epsilon(h)$. Note that when H is not finite dimensional we can still write the action as $\langle R_h^*(\phi), g \rangle = \langle \phi, gh \rangle$ for $\phi \in H^*$; we are in fact making a similar computation to that in the previous example but in a dual language. ∎

Example 1.6.3 *The left adjoint action Ad of a Hopf algebra H on itself is $\mathrm{Ad}_h(g) = \sum h_{(1)} g S h_{(2)}$, and makes H into an H-module algebra.*

Proof: We check $h \triangleright (g \triangleright a) = h \triangleright (g_{(1)} a S g_{(2)}) = h_{(1)} g_{(1)} a (S g_{(2)})(S h_{(2)}) = (hg)_{(1)} a S(hg)_{(2)} = (hg) \triangleright a$ using results about the antipode in Section 1.3. Also $1 \triangleright a = 1 a S(1) = a$. To see that we have a module algebra, we compute $h \triangleright (ab) = h_{(1)} ab (S h_{(2)}) = h_{(1)} a (S h_{(2)}) h_{(3)} b S h_{(4)} = (h_{(1)} \triangleright a)(h_{(2)} \triangleright b)$ and $h \triangleright 1 = h_{(1)} 1 S h_{(2)} = \epsilon(h)$. We insert $(S h_{(2)}) h_{(3)}$, knowing that it collapses using the antipode axioms, and freely renumber (keeping the order) to express coassociativity. ∎

Example 1.6.4 *The left coadjoint action of a finite-dimensional Hopf algebra H on H^* is $\mathrm{Ad}_h^*(\phi) = \sum \phi_{(2)} \langle h, (S\phi_{(1)})\phi_{(3)} \rangle$, and makes H^* into an H-module coalgebra.*

Proof: This is an action since $h \triangleright (g \triangleright \phi) = (g \triangleright \phi)_{(2)} \langle h, (S(g \triangleright \phi)_{(1)})(g \triangleright \phi)_{(3)} \rangle = \phi_{(2)(2)} \langle g, (S\phi_{(1)})\phi_{(3)} \rangle \langle h, (S\phi_{(2)(1)})\phi_{(2)(3)} \rangle$. Expanding the products via the pairing axioms, renumbering and recombining, gives $\phi_{(2)} \langle hg, (S\phi_{(1)})\phi_{(3)} \rangle$, as required. Also $1 \triangleright \phi = \phi_{(2)} \langle 1, (S\phi_{(1)})\phi_{(3)} \rangle = \phi_{(2)} \epsilon((S\phi_{(1)})\phi_{(3)}) = \phi$. That we have a module coalgebra requires similar expanding and recombining via the pairing axioms. Note that when H is not finite dimensional we can write the action as $\langle \mathrm{Ad}_h^*(\phi), g \rangle = \langle \phi, (S h_{(1)}) g h_{(2)} \rangle$; we are making a similar computation to that in Example 1.6.3 but in a dual language. ∎

Exercise 1.6.5 *Show that the adjoint action above for the case of* $H = kG$ *(G a finite group) and* $H = U(\mathfrak{g})$ *(g a Lie algebra) becomes*

$$u \triangleright v = uvu^{-1}, \qquad \xi \triangleright \eta = [\xi, \eta]$$

for all $u, v \in G$ *and* $\xi, \eta \in \mathfrak{g}$. *Show that the axioms of a module algebra reduce to the usual notions of a group acting by automorphisms or a Lie algebra acting by derivations on an algebra* A, *namely*

$$u \triangleright (ab) = (u \triangleright a)(u \triangleright b), \quad u \triangleright 1 = 1,$$

$$\xi \triangleright (ab) = (\xi \triangleright a)b + a(\xi \triangleright b), \quad \xi \triangleright 1 = 0,$$

for all $a, b \in A$. *Verify these directly for the adjoint actions.*

Solution: For the formulae stated, just put $\Delta u = u \otimes u$, $Su = u^{-1}$, $\Delta \xi = \xi \otimes 1 + 1 \otimes \xi$, $S\xi = -\xi$, etc. into the abstract definitions above. Compare your direct proofs for the adjoint actions with the proof of Example 1.6.4 above. ∎

This gives one point of view about the concept of an H-module algebra. It generalises the notion that an algebra A is G-covariant or $\mathfrak{g}$-covariant. When a group acts on an algebra by automorphisms, a standard construction is to make a semidirect or cross product algebra. This is the algebra generated by kG and A, but with commutation relations between them given by $ua = (u \triangleright a)u$ for all $a \in A$ and $u \in G$. Likewise, when a Lie algebra acts, we have the semidirect or cross product algebra with the cross relations $[\xi, a] = \xi \triangleright a$ for all $a \in A$ and $\xi \in \mathfrak{g}$. This principle works just as well for any Hopf algebra.

Proposition 1.6.6 *Let* H *be a bialgebra or Hopf algebra, and let* A *be a left* H-*module algebra. There is a left cross product algebra* $A \rtimes H$ *built on* $A \otimes H$ *with product*

$$(a \otimes h)(b \otimes g) = \sum a(h_{(1)} \triangleright b) \otimes h_{(2)}g, \qquad a, b \in A, \quad h, g \in H.$$

The unit element is $1 \otimes 1$.

Proof: We verify that the product as shown is an associative one. Thus,

$$(a \otimes h)((b \otimes g)(c \otimes f)) = (a \otimes h)b(g_{(1)} \triangleright c) \otimes g_{(2)}f$$
$$= a(h_{(1)} \triangleright (b(g_{(1)} \triangleright c))) \otimes h_{(2)}g_{(2)}f$$
$$= a(h_{(1)} \triangleright b)((h_{(2)}g_{(1)}) \triangleright c) \otimes h_{(3)}g_{(2)}f$$
$$= (a(h_{(1)} \triangleright b) \otimes h_{(2)}g)(c \otimes f)$$
$$= ((a \otimes h)(b \otimes g))(c \otimes f).$$

Here the third equality holds because ▷ respects the algebra structure (so we can compute $h_{(1)}\triangleright(\ )$), and is an action. That $1 \otimes 1$ is the identity is easily verified. Note that $A \otimes 1$ and $1 \otimes H$ appear as subalgebras but with mutual commutation relations

$$ha \equiv (1 \otimes h)(a \otimes 1) = h_{(1)}\triangleright a \otimes h_{(2)}$$
$$= (h_{(1)}\triangleright a \otimes 1)(1 \otimes h_{(2)}) \equiv (h_{(1)}\triangleright a)h_{(2)},$$

where we identify $h \equiv 1 \otimes h$ and $a \equiv a \otimes 1$. This corresponds to the usual way of working with semidirect or cross products, as defined by cross commutation relations.   ∎

Such cross products (also called 'smash products') are motivated by thinking about kG or $U(\mathfrak{g})$ cross products, but can be applied in other situations just as easily. Quite another point of view about module algebras arises from the following example.

Example 1.6.7 *Let G be a finite group and let $H = k(G)$. Then a $k(G)$-module means a vector space V which is G-graded, i.e. $V = \oplus_{u\in G}V_u$. If $v \in V_u$, we say that v is homogeneous of degree $|v| = u$. The action of ϕ is*

$$\phi\triangleright v = \phi(|v|)v.$$

A $k(G)$-module algebra means nothing other than a G-graded algebra, i.e. an algebra for which the multiplication $A \otimes A \to A$ is compatible with the grading in the sense $|ab| = |a|\,|b|$ and $|1| = e$. Likewise, a $k(G)$-module coalgebra means a coalgebra where $\Delta : C \to C \otimes C$ is compatible with the grading.

Proof: Write the action as $\phi\triangleright v = \sum_{u\in G} \phi(u)\beta_u(v)$, say, where $\beta_u(v)$ are vectors in V. To be an action, we need $\beta_u(\beta_w(v)) = \delta_{u,w}\beta_u(v)$ for all $u, w \in G$ and $v \in V$. Hence the $\beta_u : V \to V$ are projection operators and V splits as stated. For A to be a module algebra, we need $\phi\triangleright(ab) = (\phi_{(1)}\triangleright a)(\phi_{(2)}\triangleright b) = (\Delta\phi)(|a|, |b|)ab = \phi(|a|\,|b|)ab$ to equal $\phi(|ab|)ab$ for all ϕ (on homogeneous a, b). Similarly for the other facts. The cross product $A\rtimes k(G)$ is part of a *bosonisation* construction that we will come to later, in Chapter 10; its modules are G-graded vector spaces which are also A-modules in a compatible way.   ∎

Thus, two different ideas, the idea of a G-covariant algebra ($H = kG$) and the idea of a G-graded algebra ($H = k(G)$), are unified in the notion of a module algebra. This is one more of the remarkable unifications in mathematics made possible by working with the notion of Hopf algebras.

There is a simple way of thinking about the conditions of an H-module algebra and H-module coalgebra, namely that the various maps $A \otimes A \to A$, $k \xrightarrow{\eta} A$, $C \xrightarrow{\Delta} C \otimes C$, $C \xrightarrow{\epsilon} k$ are being required to commute with the action of H. Here H acts on k by $h \triangleright \lambda = \epsilon(h)\lambda$ (the trivial action) and on tensor products such as $A \otimes A$ according to the following exercise.

Exercise 1.6.8 *Show that if H acts on V, W then it also acts on $V \otimes W$, by $h \triangleright (v \otimes w) = \sum h_{(1)} \triangleright v \otimes h_{(2)} \triangleright w$ for all $h \in V$, $v \in V$ and $w \in W$.*

Solution: $(hg) \triangleright (v \otimes w) = (hg)_{(1)} \triangleright v \otimes (hg)_{(2)} \triangleright w = (h_{(1)} g_{(1)}) \triangleright v \otimes (h_{(2)} g_{(2)}) \triangleright w$
$= h_{(1)} \triangleright (g_{(1)} \triangleright v) \otimes h_{(2)} \triangleright (g_{(2)} \triangleright w) = h \triangleright (g_{(1)} \triangleright v \otimes g_{(2)} \triangleright w) = h \triangleright (g \triangleright (v \otimes w))$ and
$1 \triangleright (v \otimes w) = (\Delta 1) \triangleright (v \otimes w) = v \otimes w$, as required.   ∎

Thus, two actions have a tensor product, just as for group actions. Also as for group actions, maps $\phi : V \to W$ which commute with the corresponding actions in the sense $\phi(h \triangleright v) = h \triangleright \phi(v)$ are called *intertwiners*. We shall treat them systematically in Chapter 9.

We have concentrated above on the notion of left action. Clearly, there is an analogous notion of right action. This is a map $V \otimes H \to V$, denoted by $v \otimes h \mapsto v \triangleleft h$, such that $(v \triangleleft h) \triangleleft g = v \triangleleft (hg)$ and $v \triangleleft 1 = 1$. For example, if $\triangleright$ is a left action, then $v \triangleleft h = (Sh) \triangleright v$ is a right action because S is an antialgebra map. One can also use S^{-1} here if this exists. When a Hopf algebra H acts on an algebra A from the right, we say that A is a *right H-module algebra* if

$$(ab) \triangleleft h = \sum (a \triangleleft h_{(1)})(b \triangleleft h_{(2)}), \qquad 1 \triangleleft h = 1\epsilon(h). \qquad (1.10)$$

The diagram is given by Fig. 1.5 reflected in a mirror about a vertical axis. Likewise, we say that a coalgebra C is a *right H-module coalgebra* if H acts from the right and

$$\Delta(c \triangleleft h) = \sum c_{(1)} \triangleleft h_{(1)} \otimes c_{(2)} \triangleleft h_{(2)}, \qquad \epsilon(c \triangleleft h) = \epsilon(c)\epsilon(h). \qquad (1.11)$$

The diagrammatic form is Fig. 1.6 reflected in a mirror about a vertical axis.

These axioms say that the algebra or coalgebra lives in the category of right H-modules. Like the left modules above, right modules too are closed under tensor product.

Example 1.6.9 *The right regular representation of a bialgebra or Hopf algebra H on itself is $R_h(g) = gh$, and makes H into a right H-module coalgebra. The right adjoint action of a Hopf algebra on itself is $\mathrm{Ad}_h(g) = \sum (Sh_{(1)}) g h_{(2)}$, and makes H into a right H-module algebra. The right coregular action of finite-dimensional H on H^* is $L_h^*(\phi) = \sum \langle h, \phi_{(1)} \rangle \phi_{(2)}$ and makes H^* also a right H-module algebra.*

Proof: The proofs are strictly analogous to the proofs for the left actions already given above. For example, the proof that $\mathrm{Ad} = \vartriangleleft$ here is a right module is $(a \vartriangleleft h) \vartriangleleft g = ((Sh_{(1)})ah_{(2)}) \vartriangleleft g = (Sg_{(1)})(Sh_{(1)})ah_{(2)}g_{(2)} = (S(hg)_{(1)})a(hg)_{(2)} = a \vartriangleleft (hg)$ and $a \vartriangleleft 1 = (S1)a1 = a$. That it is a right module algebra is $(ab) \vartriangleleft h = (Sh_{(1)})abh_{(2)} = (Sh_{(1)})ah_{(2)}(Sh_{(3)})bh_{(4)} = (a \vartriangleleft h_{(1)})(b \vartriangleleft h_{(2)})$ and $1 \vartriangleleft a = (Sh_{(1)})h_{(2)} = \epsilon(h)$. ∎

The right handed version of Proposition 1.6.6 is

Proposition 1.6.10 *Let H be a bialgebra or Hopf algebra, and let A be a right H-module algebra. There is a right cross product algebra $H \bowtie A$ built on $H \otimes A$ with product*

$$(h \otimes a)(g \otimes b) = \sum hg_{(1)} \otimes (a \vartriangleleft g_{(2)})b, \qquad a, b \in A, \quad h, g \in H.$$

The unit element is $1 \otimes 1$.

Proof: The proof is strictly analogous to the proof of Proposition 1.6.6. If one wrote the proofs there as diagrams, then a left–right reflection about a vertical axis gives the proofs we need now. Note also that $H \otimes 1$ and $1 \otimes A$ appear as subalgebras but with mutual commutation relations

$$ah \equiv (1 \otimes a)(h \otimes 1) = h_{(1)} \otimes a \vartriangleleft h_{(2)}$$
$$= (h_{(1)} \otimes 1)(1 \otimes a \vartriangleleft h_{(2)}) \equiv h_{(1)}(a \vartriangleleft h_{(2)}),$$

where we identify $h \equiv h \otimes 1$ and $a \equiv 1 \otimes a$. This corresponds to our second way of working with cross products as defined by commutation relations. ∎

1.6.2   Coactions

A right *coaction* (or corepresentation) of a coalgebra H is a pair (β, V), where V is a vector space and β is a linear map $V \to V \otimes H$ such that $(\beta \otimes \mathrm{id}) \circ \beta = (\mathrm{id} \otimes \Delta) \circ \beta$ and $\mathrm{id} = (\mathrm{id} \otimes \epsilon) \circ \beta$. This is a polarisation of the definition of a coalgebra, and is depicted in Fig. 1.7. Compare this with Fig. 1.2. Instead of constantly writing β, we sometimes denote it simply by a formal sum notation $\beta(v) = \sum v^{(\bar 1)} \otimes v^{(\bar 2)}$, where the right hand side is an explicit representation of an element of $V \otimes H$. In terms of this notation, the axioms of a right *comodule* are

$$\sum v^{(\bar 1)(\bar 1)} \otimes v^{(\bar 1)(\bar 2)} \otimes v^{(\bar 2)} = \sum v^{(\bar 1)} \otimes v^{(\bar 2)}{}_{(1)} \otimes v^{(\bar 2)}{}_{(2)},$$

$$\sum v^{(\bar 1)} \epsilon(v^{(\bar 2)}) = v. \tag{1.12}$$

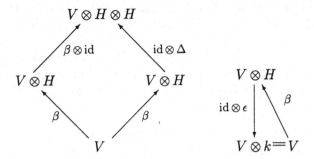

Fig. 1.7.   Axioms of a coaction expressed as commutative diagrams.

If there is a coalgebra map f from H to another coalgebra, we say that a coaction or comodule *pushes out* to a coaction of the other coalgebra by first applying the coaction of H and then f.

The axioms in Fig. 1.7 are obtained from those of Fig. 1.4 by reversing the arrows, interchanging Δ, ϵ, α and $\cdot, \eta, \beta$, and then reflecting about a vertical axis. Hence they are dual to each other, as in Section 1.4. Likewise, one can obtain the diagrams for an H-comodule algebra and H-comodule coalgebra from Figs. 1.5 and 1.6, respectively. In explicit terms, A is a right H-comodule algebra if

$$\beta(ab) = \beta(a)\beta(b), \quad \beta(1) = 1 \otimes 1. \tag{1.13}$$

Here $A \otimes H$ has the tensor product algebra structure and we are requiring β to be an algebra map. A coalgebra C is a right H-comodule coalgebra if

$$\sum c^{(\bar{1})}{}_{(1)} \otimes c^{(\bar{1})}{}_{(2)} \otimes c^{(\bar{2})} = \sum c_{(1)}{}^{(\bar{1})} \otimes c_{(2)}{}^{(\bar{1})} \otimes c_{(1)}{}^{(\bar{2})} c_{(2)}{}^{(\bar{2})},$$
$$\sum \epsilon(c^{(\bar{1})}) c^{(\bar{2})} = \epsilon(c). \tag{1.14}$$

We can summarise the relationship between these axioms and those for modules above as follows.

Proposition 1.6.11 *If H is finite dimensional, then a left action of H corresponds to a right coaction of H^* on the same space. Explicitly, if $\beta(v) = \sum v^{(\bar{1})} \otimes v^{(\bar{2})}$ is the coaction of H^*, then $h \triangleright v = \sum v^{(\bar{1})} \langle h, v^{(\bar{2})} \rangle$ is the corresponding action of H. If A is a left H-module algebra, then it is a right H^*-comodule algebra. If C is a left H-module coalgebra, then it is a right H^*-comodule coalgebra.*

Proof: The proof is similar to that of Proposition 1.4.2, i.e. an elementary exercise in dualisation. We compute $h \triangleright (g \triangleright v) = (g \triangleright v)^{(\bar{1})} \langle h, (g \triangleright v)^{(\bar{2})} \rangle = v^{(\bar{1})(\bar{1})} \langle h, v^{(\bar{1})(\bar{2})} \rangle \langle g, v^{(\bar{2})} \rangle = v^{(\bar{1})} \langle g \otimes h, v^{(\bar{2})}{}_{(1)} \otimes v^{(\bar{2})}{}_{(2)} \rangle = (gh) \triangleright v$ and $1 \triangleright v = v^{(\bar{1})} \langle 1, v^{(\bar{2})} \rangle = v^{(\bar{1})} \epsilon(v^{(\bar{2})}) = v$, so that we have a left action on the same vector space as the right coaction. Similarly for the other facts. ∎

For example, we have seen in Example 1.6.7 that an action of $k(G)$ is the same thing as a G-grading, which is therefore the same thing as a coaction of kG. This second point of view makes sense even when G is not finite.

Example 1.6.12 *The right regular coaction of a bialgebra or Hopf algebra H on itself is given by the coproduct $R = \Delta : H \to H \otimes H$, and makes H into an H-comodule algebra.*

Proof: The axioms of a comodule follow at once from the coassociativity and counity axioms for Δ. That H is a comodule algebra in this case is precisely the axiom that Δ is an algebra homomorphism. Compare with Example 1.6.1. ∎

Example 1.6.13 *The right coregular coaction L^* of a finite-dimensional bialgebra or Hopf algebra H on H^* is $L^*(\phi)(h) = \sum h_{(1)} \langle h_{(2)}, \phi \rangle$, and makes H^* into an H-comodule coalgebra.*

Proof: We write $L^*(\phi)(h) = \langle h, \phi^{(\bar{1})} \rangle \phi^{(\bar{2})}$ in our explicit notation. Then we have $\langle h, \phi^{(\bar{1})(\bar{1})} \rangle \phi^{(\bar{1})(\bar{2})} \otimes \phi^{(\bar{2})} = h_{(1)} \langle h_{(2)}, \phi^{(\bar{1})} \rangle \otimes \phi^{(\bar{2})} = h_{(1)} \otimes h_{(2)} \langle h_{(3)}, \phi \rangle = \langle h, \phi^{(\bar{1})} \rangle \Delta \phi^{(\bar{2})}$ and $\langle h, \phi^{(\bar{1})} \rangle \epsilon(\phi^{(\bar{2})}) = \langle h, \phi \rangle$, as required for a right coaction. Similarly, evaluating against h, g, we have $\langle h, \phi^{(\bar{1})}{}_{(1)} \rangle \langle g, \phi^{(\bar{1})}{}_{(2)} \rangle \phi^{(\bar{2})} = \langle hg, \phi^{(\bar{1})} \rangle \phi^{(\bar{2})} = h_{(1)} g_{(1)} \langle h_{(2)} g_{(2)}, \phi \rangle = \langle h, \phi_{(1)}{}^{(\bar{1})} \rangle \langle g, \phi_{(2)}{}^{(\bar{1})} \rangle \phi_{(1)}{}^{(\bar{2})} \phi_{(2)}{}^{(\bar{2})}$, as required for (1.14). Also $\epsilon(\phi^{(\bar{1})}) \phi^{(\bar{2})} = \langle 1, \phi \rangle = \epsilon(\phi)$, as required. Compare with Example 1.6.2. ∎

Example 1.6.14 *The right adjoint coaction of a Hopf algebra H on itself is $\mathrm{Ad}(h) = \sum h_{(2)} \otimes (Sh_{(1)}) h_{(3)}$, and makes H into an H-comodule coalgebra.*

Proof: We have $(\mathrm{Ad} \otimes \mathrm{id}) \mathrm{Ad}(h) = h_{(2)(2)} \otimes (Sh_{(2)(1)}) h_{(2)(3)} \otimes (Sh_{(1)}) h_{(3)} = h_{(3)} \otimes (Sh_{(2)}) h_{(4)} \otimes (Sh_{(1)}) h_{(5)} = h_{(2)} \otimes (Sh_{(1)})_{(1)} h_{(3)(1)} \otimes (Sh_{(1)})_{(2)} h_{(3)(2)}$ from Proposition 1.3.1. Also $h_{(2)} \otimes \epsilon((Sh_{(1)}) h_{(3)}) = h$, so we have a coaction. This right action then makes H a right H-comodule coalgebra because $h_{(1)(2)} \otimes h_{(2)(2)} \otimes (Sh_{(1)(1)}) h_{(1)(3)} (Sh_{(2)(1)}) h_{(2)(3)} = h_{(2)(1)} \otimes h_{(2)(2)} \otimes (Sh_{(1)}) h_{(3)}$ on cancelling $h_{(1)(3)} Sh_{(2)(1)}$ since they are in the correct order after linear renumbering. Also $\epsilon(h_{(2)})(Sh_{(1)}) h_{(3)} = (Sh_{(1)}) h_{(2)} = \epsilon(h)$. Compare with Example 1.6.3. ∎

Example 1.6.15 *The right coadjoint coaction* Ad^* *of a finite-dimensional Hopf algebra* H *on* H^* *is* $\mathrm{Ad}^*(\phi)(h) = \sum h_{(1)} Sh_{(3)} \langle h_{(2)}, \phi \rangle$, *and makes* H^* *into an* H*-comodule algebra.*

Proof: Write $\mathrm{Ad}^*(\phi)(h) = \langle h, \phi^{(\bar{1})} \rangle \phi^{(\bar{2})}$, say. Then, $\langle h, \phi^{(\bar{1})(\bar{1})} \rangle \phi^{(\bar{1})(\bar{2})} \otimes \phi^{(\bar{2})} = h_{(1)} Sh_{(3)} \langle h_{(2)}, \phi^{(\bar{1})} \rangle \otimes \phi^{(\bar{2})} = h_{(1)} Sh_{(3)} \otimes h_{(2)(1)} Sh_{(2)(3)} \langle h_{(2)(2)}, \phi \rangle$. This equals $\langle h, \phi^{(\bar{1})} \rangle \Delta \phi^{(\bar{2})}$ since S is an anticoalgebra map. We also have $\langle h, \phi^{(\bar{1})} \rangle \epsilon(\phi^{(\bar{2})}) = \langle h, \phi \rangle$. So we have a right coaction. This gives us a right comodule algebra because $\mathrm{Ad}^*(\phi\psi)(h) = h_{(1)} Sh_{(3)} \langle h_{(2)}, \phi\psi \rangle = h_{(1)} Sh_{(4)} \langle h_{(2)}, \phi \rangle \langle h_{(3)}, \psi \rangle = h_{(1)}(Sh_{(3)}) h_{(4)} Sh_{(6)} \langle h_{(2)}, \phi \rangle \langle h_{(5)}, \psi \rangle = \mathrm{Ad}^*(\phi)(h_{(1)}) \mathrm{Ad}^*(\psi)(h_{(2)}) = (\mathrm{Ad}^*(\phi) \mathrm{Ad}^*(\psi))(h)$ and $\mathrm{Ad}^*(1)(h) = h_{(1)} Sh_{(3)} \langle h_{(2)}, 1 \rangle = \langle 1, h \rangle$. Compare with Example 1.6.4. ∎

Proposition 1.6.16 *Let* H *be a bialgebra or Hopf algebra, and let* C *be a right* H*-comodule coalgebra. There is a right cross coproduct coalgebra* $H {\blacktriangleright}{<} C$ *built on* $H \otimes C$ *with the coalgebra structure*

$$\Delta(h \otimes c) = \sum h_{(1)} \otimes c_{(1)}{}^{(\bar{1})} \otimes h_{(2)} c_{(1)}{}^{(\bar{2})} \otimes c_{(2)}, \qquad \epsilon(h \otimes c) = \epsilon(h)\epsilon(c)$$

for $h \in H$, $c \in C$.

Proof: We check coassociativity. Thus,

$(\mathrm{id} \otimes \Delta)\Delta(h \otimes c)$
$$= h_{(1)} \otimes c_{(1)}{}^{(\bar{1})} \otimes (h_{(2)} c_{(1)}{}^{(\bar{2})})_{(1)} \otimes c_{(2)(1)}{}^{(\bar{1})} \otimes (h_{(2)} c_{(1)}{}^{(\bar{2})})_{(2)} c_{(2)(1)}{}^{(\bar{2})} \otimes c_{(2)(2)}$$
$$= h_{(1)} \otimes c_{(1)}{}^{(\bar{1})} \otimes h_{(2)} c_{(1)}{}^{(\bar{2})}{}_{(1)} \otimes c_{(2)}{}^{(\bar{1})} \otimes h_{(3)} c_{(1)}{}^{(\bar{2})}{}_{(2)} c_{(2)}{}^{(\bar{2})} \otimes c_{(3)}$$
$$= h_{(1)} \otimes c_{(1)}{}^{(\bar{1})(\bar{1})} \otimes h_{(2)} c_{(1)}{}^{(\bar{1})(\bar{2})} \otimes c_{(2)}{}^{(\bar{1})} \otimes h_{(3)} c_{(1)}{}^{(\bar{2})} c_{(2)}{}^{(\bar{2})} \otimes c_{(3)}$$
$$= h_{(1)} \otimes c_{(1)(1)}{}^{(\bar{1})(\bar{1})} \otimes h_{(2)} c_{(1)(1)}{}^{(\bar{1})(\bar{2})} \otimes c_{(1)(2)}{}^{(\bar{1})} \otimes h_{(3)} c_{(1)(1)}{}^{(\bar{2})} c_{(1)(2)}{}^{(\bar{2})} \otimes c_{(2)}$$
$$= h_{(1)(1)} \otimes c_{(1)}{}^{(\bar{1})}{}_{(1)}{}^{(\bar{1})} \otimes h_{(1)(2)} c_{(1)}{}^{(\bar{1})}{}_{(1)}{}^{(\bar{2})} \otimes c_{(1)}{}^{(\bar{1})}{}_{(2)} \otimes h_{(2)} c_{(1)}{}^{(\bar{2})} \otimes c_{(2)}$$
$$= (\Delta \otimes \mathrm{id})\Delta(h \otimes c).$$

We used the definition of Δ for the first and last equalities. The second equality uses the homomorphism property of the coproduct of H and some renumbering due to coassociativity in H and C. The third equality uses the comodule property (1.12), while the fourth equality is just a renumbering using coassociativity in C. The fifth equality is the comodule coalgebra property (1.14). The counity axioms are more immediate. Note that our proof looks unfamiliar, but it is nothing other than a dual version of the proof of Proposition 1.6.10, as obtained by a reversal of arrows when the proofs are written diagrammatically. ∎

We have concentrated here on right coactions, but clearly there are analogous notions for left coactions. Thus, H has a left coaction on V

(or V is a left H-comodule) if there is a map $\beta : V \to H \otimes V$ such that $(\mathrm{id} \otimes \beta) \circ \beta = (\Delta \otimes \mathrm{id}) \circ \beta$ and $\mathrm{id} = (\epsilon \otimes \mathrm{id}) \circ \beta$. The relevant diagram is Fig. 1.7 reflected about a vertical axis. If we write the map explicitly as $\beta(v) = \sum v^{(\bar{1})} \otimes v^{(\bar{2})} \in H \otimes V$, then the left comodule property is

$$\sum v^{(\bar{1})} \otimes v^{(\bar{2})(\bar{1})} \otimes v^{(\bar{2})(\bar{2})} = \sum v^{(\bar{1})}{}_{(1)} \otimes v^{(\bar{1})}{}_{(2)} \otimes v^{(\bar{2})},$$

$$\sum \epsilon(v^{(\bar{1})})v^{(\bar{2})} = v. \tag{1.15}$$

For example, since S is an anticoalgebra map, it can be used to convert a right coaction to a left one, and vice versa, by composition with β. Similarly with S^{-1} if it exists. Following the same lines as above, an algebra is a *left H-comodule algebra* if it is a left comodule and β is an algebra map as in (1.13). Finally, a coalgebra C is a *left H-comodule coalgebra* if the left coaction obeys

$$\sum c^{(\bar{1})} \otimes c^{(\bar{2})}{}_{(1)} \otimes c^{(\bar{2})}{}_{(2)} = \sum c_{(1)}{}^{(\bar{1})}c_{(2)}{}^{(\bar{1})} \otimes c_{(1)}{}^{(\bar{2})} \otimes c_{(2)}{}^{(\bar{2})},$$

$$\sum c^{(\bar{1})}\epsilon(c^{(\bar{2})}) = \epsilon(c). \tag{1.16}$$

Left coactions on vector spaces, algebras and coalgebras correspond to right actions of the dual bialgebra, along the same lines as in Proposition 1.6.11.

Example 1.6.17 *The left regular coaction $H \to H \otimes H$ of a bialgebra or Hopf algebra on itself is $L = \Delta$, and gives a comodule algebra. The left adjoint coaction of a Hopf algebra on itself is $\mathrm{Ad}(h) = \sum h_{(1)}Sh_{(3)} \otimes h_{(2)}$, and gives a comodule coalgebra. The left coregular coaction R^* of finite-dimensional H on H^* is defined by $R^*(\phi)(h) = \sum \langle h_{(1)}, \phi \rangle h_{(2)}$, and also gives a comodule coalgebra.*

Proof: The proofs are strictly analogous to those for right coactions already given. For example, that Ad is a left comodule comes out now as $h_{(1)}(Sh_{(3)}) \otimes h_{(2)(1)}Sh_{(2)(3)} \otimes h_{(2)(2)} = h_{(1)(1)}(Sh_{(3)})_{(1)} \otimes h_{(1)(2)}(Sh_{(3)})_{(2)} \otimes h_{(2)}$ and $h_{(1)}Sh_{(3)}\epsilon(h_{(2)}) = h_{(1)}Sh_{(2)} = \epsilon(h)$. ∎

Proposition 1.6.18 *Let H be a bialgebra or Hopf algebra, and let C be a left H-comodule coalgebra. There is a left cross coproduct coalgebra $C \bowtie H$ built on $C \otimes H$ with the coalgebra structure*

$$\Delta(c \otimes h) = \sum c_{(1)} \otimes c_{(2)}{}^{(\bar{1})}h_{(1)} \otimes c_{(2)}{}^{(\bar{2})} \otimes h_{(2)}, \qquad \epsilon(c \otimes h) = \epsilon(c)\epsilon(h)$$

for $h \in H$, $c \in C$.

Proof: The proof is strictly analogous to the right handed case above. If one writes the proofs as diagrams, they are obtained by a left–right reflection in the proof of Proposition 1.6.16, or a reversal of all arrows in the proof of Proposition 1.6.6. To do the proof explicitly, just use the definition of a left comodule coalgebra and verify the coassociativity and counity axioms for Δ, ϵ, as stated. ∎

For general bialgebras and Hopf algebras one frequently needs both the left and right handed versions of modules and comodules. A typical situation is the following, with a similar proposition for comodules.

Proposition 1.6.19 *If V is a left module then V^* is a right module. The correspondence is given by*

$$(f \triangleleft h)(v) = f(h \triangleright v), \qquad \forall v \in V, \ f \in V^*.$$

If A is a finite-dimensional left module algebra, then A^* is a right module coalgebra. If C is a left module coalgebra, then C^* is a right module algebra. Similarly for left–right interchanged and for modules replaced by comodules.

Proof: This is more practice with the definitions. For example, if C is a left module coalgebra, then, acting on elements $\phi, \psi \in C^*$, we have
$$((\phi\psi) \triangleleft h)(c) = (\phi\psi)(h \triangleright c) = \phi((h \triangleright c)_{(1)})\psi((h \triangleright c)_{(2)}) = \phi(h_{(1)} \triangleright c_{(1)} \otimes h_{(2)} \triangleright c_{(2)})$$
$$= (\phi \triangleleft h_{(1)})(c_{(1)})(\psi \triangleleft h_{(2)})(c_{(2)}) = ((\phi \triangleleft h_{(1)})(\psi \triangleleft h_{(2)}))(c) \text{ and } (\epsilon \triangleleft h)(c) = \epsilon(h \triangleright c)$$
$$= \epsilon(h)\epsilon(c), \text{ so that } C^* \text{ is a right module algebra.} \qquad \blacksquare$$

Also, it is sometimes more natural to consider nonunital comodule algebras, where the unit (even if present) is not necessarily respected by the coaction. Similarly for our various other notions.

Example 1.6.20 *The* trigonometric bialgebra *is generated by 1 and indeterminates c, s with the relations and coproduct*

$$c^2 - s^2 = c, \quad cs + sc = s, \quad \Delta c = c \otimes c - s \otimes s, \quad \Delta s = s \otimes c + c \otimes s,$$

and counit $\epsilon c = 1$, $\epsilon s = 0$. The complex number algebra *generated by $1, i$ with the relation $i^2 + 1 = 0$ is a* nonunital comodule algebra *with*

$$\beta(1) = 1 \otimes c - i \otimes s, \quad \beta(i) = 1 \otimes s + i \otimes c$$

Proof: It is easy to check that c, s indeed generate a bialgebra, using similar methods to those in Example 1.3.2. For example, $\Delta(c^2 - s^2) = (c \otimes c - s \otimes s)^2 - (s \otimes c + c \otimes s)^2 = c^2 \otimes c^2 + s^2 \otimes s^2 - cs \otimes cs - sc \otimes sc - s^2 \otimes c^2 - c^2 \otimes s^2 - sc \otimes cs - cs \otimes sc = \Delta c$ after using the relations. That Δ is coassociative and β a coaction on the vector space spanned by $1, i$ is immediate from their general form. That β respects the product in the algebra structure is the assertion that $1' = \beta(1)$, $i' = \beta(i)$ obey the same relations. Thus, $1'1' = (c - is)^2 = c^2 - s^2 - i(cs + sc) = 1'$, $1'i' = (c - is)(s + ic) = cs + sc + i(c^2 - s^2) = i'$, and similarly for $i'1' = i'$. Finally $i'^2 + 1' = (s + ic)^2 + c - is = s^2 + c^2 + i(sc + cs) + c - is = 0$. We work in the tensor product algebra, so that $s \equiv 1 \otimes s$ and $is \equiv i \otimes s$, etc. Note that, over $\mathbb{R}$, the algebra generated by $1, i$ is indeed $\mathbb{C}$ as a 2-dimensional real algebra. Also note that 1 in the bialgebra plays no role, i.e., we could more simply regard the coaction as that of the nonunital bialgebra generated by c, s. ∎

The coproduct here has the same form as the addition rule for sine and cosine. Also note that the coaction on $(1, i)$ has the same form as a rotation in the complex plane. This shows the use of coactions to make 'transformations'. If one wants a unital setting one can work with the algebra generated by $1, i, e$, where the projector e plays the role of 1 in the above example (but this is not particularly natural). We will encounter many more examples of transformation bialgebras in later chapters, notably in Chapter 4.

1.7   Integrals and ∗-structures

In this section we develop two further notions that are familiar in some form for groups. Their versions for Hopf algebras play a key role in generalising our algebraic definitions above to the setting of C^* algebras and Hopf–von Neumann algebras.

Definition 1.7.1 *A* left integral *on a Hopf algebra H is a (not identically zero) linear map $\int : H \to k$ such that*

$$(\mathrm{id} \otimes \int) \circ \Delta = \eta \circ \int .$$

A left integral *in a Hopf algebra H is a nonzero element $\Lambda \in H$ such that $h\Lambda = \epsilon(h)\Lambda$ for all $h \in H$. Similarly for right integrals. Integrals are* normalised *if $\int 1 = 1$ or $\epsilon(\Lambda) = 1$, respectively.*

The motivation behind the definition of $\int$ is in terms of the adjoint of the left action L of H on H. As we saw in Section 1.6.1, its adjoint is a (right) action of H on H^* given by $L_h^*(\phi)(g) = \phi(hg)$. Thus, the left-invariance of the integral on H^* is equivalent to the more familiar

$$\int L_h^*(\phi) = \sum \langle h, \phi_{(1)} \rangle \int \phi_{(2)} = \epsilon(h) \int \phi, \qquad \forall h, \phi. \qquad (1.17)$$

This notion is the Hopf-algebraic version of the notion of Haar measure on a locally compact group. Every finite-dimensional Hopf algebra has an integral (not necessarily normalisable) which is unique up to scale. This applies to both left and right integrals. A Hopf algebra is called *unimodular* if it has left and right integrals which coincide. The notion of an integral Λ in H has some algebraic advantages but is suitable only in the finite-dimensional case. In this case, it is clearly just a left integral on H^*. A normalised integral on a general Hopf algebra with invertible antipode, when it exists, is unique and unimodular.

Example 1.7.2 *A normalised left integral on $k(G)$ is provided by $\int \phi = |G|^{-1} \sum_{u \in G} \phi(u)$. A normalised left integral in kG is provided by $\Lambda = |G|^{-1} \sum_{u \in G} u$. We suppose for the normalisations that $|G|$ is invertible. These examples are evidently unimodular.*

Proof: This is easy and has already appeared in our proof of Corollary 1.5.6 above. ∎

Proposition 1.7.3 *Let H be a finite-dimensional Hopf algebra. If $\int \phi = \operatorname{Tr} L_\phi \circ S^2$ is not identically zero for all $\phi \in H^*$, then it defines a left integral on H^*.*

Proof: We note that $L_{L_h^*\phi} = L_{h_{(1)}}^* \circ L_\phi \circ L_{S^{-1}h_{(2)}}^*$ for any $h \in H$ and $\phi \in H^*$. To see this, we evaluate on elements $\psi \in H^*, g \in H$ and use the duality pairing and property of S^{-1} from Sections 1.3 and 1.4; thus, $\langle L_{h_{(1)}}^* \circ L_\phi \circ L_{S^{-1}h_{(2)}}^* \psi, g \rangle = \langle \phi L_{S^{-1}h_{(2)}}^* \psi, h_{(1)}g \rangle = \langle \phi, h_{(1)}g_{(1)} \rangle \langle L_{S^{-1}h_{(3)}}^* \psi, h_{(2)}g_{(2)} \rangle = \langle \phi, h_{(1)}g_{(1)} \rangle \langle \psi, (S^{-1}h_{(3)})h_{(2)}g_{(2)} \rangle = \langle \phi, hg_{(1)} \rangle \langle \psi, g_{(2)} \rangle = \langle L_h^*\phi, g_{(1)} \rangle \langle \psi, g_{(2)} \rangle = \langle L_{L_h^*\phi} \psi, g \rangle$. Then cyclicity of the trace and that L^* is a right action give $\operatorname{Tr} L_{h_{(1)}}^* \circ L_\phi \circ L_{S^{-1}h_{(2)}}^* \circ S^2 = \operatorname{Tr} L_{h_{(1)}}^* \circ L_\phi \circ S^2 \circ L_{Sh_{(2)}}^* = \operatorname{Tr} L_{Sh_{(2)}}^* \circ L_{h_{(1)}}^* \circ L_\phi \circ S^2 = \operatorname{Tr} L_{h_{(1)}Sh_{(2)}}^* \circ L_\phi \circ S^2 = \epsilon(h) \operatorname{Tr} L_\phi \circ S^2$, as required in (1.17). ∎

We now give some applications of actions and integrals to the theory of Hopf algebras.

Theorem 1.7.4 *Let H be a finite-dimensional Hopf algebra and view $H \subseteq \operatorname{Lin}(H)$ by $h \mapsto h \triangleright = L_h$ (the left regular action in Example 1.6.1). In the algebra $\operatorname{Lin}(H) \otimes \operatorname{Lin}(H)$ there is an invertible element W such that*

$$W_{12}W_{13}W_{23} = W_{23}W_{12},$$

$$\Delta h = W(h \otimes 1)W^{-1}, \quad Sh = (\epsilon \otimes \operatorname{id}) \circ W^{-1}(h \otimes (\ )).$$

Here $W_{12} = W \otimes 1$, $W_{23} = 1 \otimes W$ in $\mathrm{Lin}(H)^{\otimes 3}$ (similarly for W_{13}). Let $H^* \subseteq \mathrm{Lin}(H)$ by $\phi \mapsto \phi \triangleright = R_\phi^*$ (the left coregular action; compare with Example 1.6.2). Then viewed in this algebra we also have, where defined,

$$\Delta \phi = W^{-1}(1 \otimes \phi)W, \quad S\phi = (\mathrm{id} \otimes \phi) \circ W^{-1}((\ ) \otimes 1).$$

The subalgebras $H \subseteq \mathrm{Lin}(H)$ and $H^* \subseteq \mathrm{Lin}(H)$ together generate all of $\mathrm{Lin}(H)$.

Proof: Explicitly, W, W^{-1} are defined as linear maps $H \otimes H \to H \otimes H$ (which we identify with elements of $\mathrm{Lin}(H) \otimes \mathrm{Lin}(H)$ in the standard way) by

$$W(g \otimes h) = g_{(1)} \otimes g_{(2)}h, \quad W^{-1}(g \otimes h) = g_{(1)} \otimes (Sg_{(2)})h. \qquad (1.18)$$

Then we verify the identity

$$
\begin{aligned}
(W(h \otimes 1)W^{-1})(g \otimes f) &= W(h \otimes 1)\triangleright(g_{(1)} \otimes (Sg_{(2)})f) \\
&= W(hg_{(1)} \otimes (Sg_{(2)})f) = h_{(1)}g_{(1)(1)} \otimes h_{(2)}g_{(1)(2)}(Sg_{(2)})f \\
&= h_{(1)}g \otimes h_{(2)}f = (\Delta h)\triangleright(g \otimes f).
\end{aligned}
$$

Likewise we have

$$
\begin{aligned}
(W^{-1}(1 \otimes \phi)W)(g \otimes h) &= W^{-1}(1 \otimes \phi)\triangleright(g_{(1)} \otimes g_{(2)}h) \\
&= W^{-1}(g_{(1)} \otimes g_{(2)(1)}h_{(1)})\langle \phi, g_{(2)(2)}h_{(2)}\rangle \\
&= g_{(1)(1)} \otimes (Sg_{(1)(2)})g_{(2)(1)}h_{(1)}\langle \phi, g_{(2)(2)}h_{(2)}\rangle \\
&= g_{(1)} \otimes h_{(1)}\langle \phi, g_{(2)}h_{(2)}\rangle = \phi_{(1)}\triangleright g \otimes \phi_{(2)}\triangleright h.
\end{aligned}
$$

Explicitly, $\phi \triangleright g = g_{(1)}\langle \phi, g_{(2)}\rangle$. Similarly for the antipodes. As for the equations satisfied by W itself, we evaluate on $H \otimes H \otimes H$ as

$$
\begin{aligned}
W_{12}W_{13}W_{23}(g \otimes h \otimes f) &= W_{12}W_{13}(g \otimes h_{(1)} \otimes h_{(2)}f) \\
&= W_{12}(g_{(1)} \otimes h_{(1)} \otimes g_{(2)}h_{(2)}f) = g_{(1)(1)} \otimes g_{(1)(2)}h_{(1)} \otimes g_{(2)}h_{(2)}f \\
&= g_{(1)} \otimes g_{(2)(1)}h_{(1)} \otimes g_{(2)(2)}h_{(2)}f = W_{23}(g_{(1)} \otimes g_{(2)}h \otimes f) \\
&= W_{23}W_{12}(g \otimes h \otimes f).
\end{aligned}
$$

For the last part, we show that every operator $H \to H$ arises by actions $\triangleright$ of H and H^* on H, at least when H is finite dimensional. In this case, every linear operator can be viewed as an element of $H \otimes H^*$ acting on H in the usual way by evaluation, namely $(h \otimes \phi)(g) = h\langle \phi, g\rangle$. We have to represent this by elements of H, H^* acting via $\triangleright$. Indeed, as operators in $\mathrm{Lin}(H)$ we find $h \otimes \phi = h(S^{-1}e_{a(1)})\triangleright(\langle \phi, e_{a(2)}\rangle f^a \triangleright(\ ))$, where $\{e_a\}$ is a basis of H and $\{f^a\}$ is a dual basis. Thus, $h(S^{-1}e_{a(1)})\triangleright(\langle \phi, e_{a(2)}\rangle \otimes f^a \triangleright g) = h(S^{-1}e_{a(1)})g_{(1)}\langle \phi, e_{a(2)}\rangle\langle f^a, g_{(2)}\rangle = h(S^{-1}g_{(2)(1)})g_{(1)}\langle \phi, g_{(2)(2)}\rangle = h\langle \phi, g\rangle$, as required. We used coassociativity and the properties of S^{-1} in Exercise 1.3.3. ∎

This result indicates a way to construct Hopf algebras. If W is any invertible operator obeying the *pentagonal identity* $W_{12}W_{13}W_{23} = W_{23}W_{12}$, then it is easy to see that $\Delta h = W(h \otimes 1)W^{-1}$, for any operator h, will always be coassociative and an algebra homomorphism. If we can arrange also for a counit and antipode (typically by restricting our operators to some subalgebra), then this gives a Hopf algebra. A Hopf algebra of this type is very concrete, being realised as operators on some vector space, and hence is very suitable for physical applications. Our theorem says that *every* Hopf algebra can be realised concretely in this way by acting on itself. Even more, it says that both the Hopf algebra and its dual can be realised as subalgebras of operators on the same space. W is called the *fundamental operator* of the Hopf algebra, and the concrete realisation in the theorem is called the *operator realisation*. Clearly, we are ready to put this into a Hilbert space setting and discuss Hopf algebras realised as bounded operators.

Our next topic is related to this, and is again motivated from functional analysis. It makes sense whenever our background field is equipped with a 'conjugation'. For simplicity, we fix $k = \mathbb{C}$. In this context the notion of a ∗-algebra is a familiar one. It means an algebra H equipped with an antilinear map $(\;)^* : H \to H$ obeying $*^2 = \mathrm{id}$ and $(hg)^* = g^*h^*$ for all $h, g \in H$.

Definition 1.7.5 *A Hopf ∗-algebra is a ∗-algebra H such that*

$$\Delta h^* = (\Delta h)^{* \otimes *}, \quad \epsilon(h^*) = \overline{\epsilon(h)}, \quad (S \circ *)^2 = \mathrm{id}.$$

If A, H are two ∗-Hopf algebras, they are dually paired if they are dually paired as Hopf algebras and, in addition,

$$\langle \phi^*, h \rangle = \overline{\langle \phi, (Sh)^* \rangle}$$

for all $h \in H$ and $\phi \in A$.

A *real form* of a Hopf algebra over $\mathbb{C}$ is a choice of such a ∗-structure. There are many applications of ∗-structures, particularly in the context of quantum mechanics and operator theory.

Proposition 1.7.6 *In the setting of Theorem 1.7.4, suppose that H is a finite-dimensional Hopf ∗-algebra. Then H^* is also a Hopf ∗-algebra and $\mathrm{Lin}(H)$ (generated by $H \subseteq \mathrm{Lin}(H)$ and $H^* \subseteq \mathrm{Lin}(H)$) becomes a ∗-algebra. With respect to this ∗-structure, the fundamental operator W is unitary,*

$$W^{* \otimes *} = W^{-1}.$$

If $\int$ is a right integral on H, it can be chosen so that it defines a sesquilinear form $(g, h) = \int g^*h = \overline{(h, g)}$, which is compatible with the ∗-structure *on* $\mathrm{Lin}(H)$ *induced by* H, H^* *in the sense*

$$\int (h{\triangleright}g)^* f = \int g^*(h^*{\triangleright}f), \qquad \int (\phi{\triangleright}g)^* f = \int g^*(\phi^*{\triangleright}f)$$

for $g, f, h \in H$, $\phi \in H^*$.

Proof: Since every element of $\mathrm{Lin}(H)$ factorises into the actions of H and H^*, these subalgebras necessarily define a $*$-operation on $\mathrm{Lin}(H)$ by $(h{\triangleright})^* = h^*{\triangleright}$ and $(\phi{\triangleright})^* = \phi^*{\triangleright}$, such that they become $*$-subalgebras. For the first part, we begin by writing W in terms of such elements from H, H^*. Indeed, $W(g \otimes f) = g_{(1)} \otimes g_{(2)}f = g_{(1)}\langle f^a, g_{(2)}\rangle \otimes e_a f = f^a{\triangleright}g \otimes e_a{\triangleright}f$, where $\{e_a\}$ is a basis of H and $\{f^a\}$ is a dual basis. So $W = f^a{\triangleright} \otimes e_a{\triangleright}$ and, hence, by definition, $W^{*\otimes*} = f^{a*}{\triangleright} \otimes e_a{}^*{\triangleright} = f'^a{\triangleright} \otimes (Se'_a){\triangleright}$, where $e'_a = S^{-1}e_a{}^*$, $f'^a = f^{a*}$ form a new basis and dual basis. Dropping the $'$, we have $W^{*\otimes*}(g \otimes f) = f^a{\triangleright}g \otimes (Se_a){\triangleright}f = g_{(1)}\langle f^a, g_{(2)}\rangle \otimes (Se_a)f = g_{(1)} \otimes (Sg_{(2)})f = W^{-1}(g \otimes f)$.

For the second part, we show that this $*$-structure on $\mathrm{Lin}(H)$ really is an adjoint operation with respect to $(\ ,\ )$. Note that we can arrange $(S\int)^* = \int$ without loss of generality: the left hand side is also an integral, so this equality holds up to a scale which (in view of $(* \circ S)^2 = \mathrm{id}$) is a phase; we can then absorb this phase in a rescaling of $\int$. From this, it is easy to see that $(\ ,\ )$ is Hermitian in the form stated. For the operators $h{\triangleright}$ this is easy since $(h{\triangleright}g, f) = \int(h{\triangleright}g)^* f = \int(hg)^* f = \int g^* h^* f = \int g^*(h^*{\triangleright}f) = (g, (h{\triangleright})^* f)$ is automatic. Rather harder is for the operators $\phi{\triangleright}$,

$$
\begin{aligned}
(\phi{\triangleright}g, h) &= \int (\phi{\triangleright}g)^* h = \int (g_{(1)})^* \overline{\langle \phi, g_{(2)}\rangle} h \\
&= \int g^*{}_{(1)} \overline{\langle \phi, (g^*{}_{(2)})^*\rangle} h = \int g^*{}_{(1)} \langle S^{-1}(\phi^*), g^*{}_{(2)}\rangle h \\
&= \int g^*{}_{(1)} h_{(1)} \langle S^{-1}\phi^*{}_{(3)}, g^*{}_{(2)}\rangle \langle (S^{-1}\phi^*{}_{(2)})\phi^*{}_{(1)}, h_{(2)}\rangle \\
&= \int g^*{}_{(1)} h_{(1)} \langle (S^{-1}\phi^*{}_{(2)})_{(1)}, g^*{}_{(2)}\rangle \langle (S^{-1}\phi^*{}_{(2)})_{(2)}, h_{(2)}\rangle \langle \phi^*{}_{(1)}, h_{(3)}\rangle \\
&= \int g^*{}_{(1)} h_{(1)} \langle S^{-1}\phi^*{}_{(2)}, g^*{}_{(2)} h_{(2)}\rangle \langle \phi^*{}_{(1)}, h_{(3)}\rangle \\
&= \int (g^* h_{(1)})_{(1)} \langle S^{-1}\phi^*{}_{(2)}, (g^* h_{(1)})_{(2)}\rangle \langle \phi^*{}_{(1)}, h_{(2)}\rangle \\
&= \int g^* h_{(1)} \langle \phi^*, h_{(2)}\rangle = \int g^*(\phi^*{\triangleright}h) = (g, (\phi{\triangleright})^* h).
\end{aligned}
$$

Here the second equality is from the definition of $\phi{\triangleright}g = g_{(1)}\langle \phi, g_{(2)}\rangle$. The third equality is from the definition of a Hopf $*$-algebra with regard to $\Delta \circ *$. The fourth equality is from the relationship between the $*$-structures in H, H^*, and the axiom $(S \circ *)^2 = \mathrm{id}$. The fifth equality inserts some factors that collapse to $\epsilon(h_{(2)})$ via the fact that S^{-1} is a skew-antipode for H (an antipode for H^{op} as in Exercise 1.3.3). The sixth equality writes the multiplication in H^* in terms of the coproduct in H, and uses coassociativity and that S^{-1} is an anticoalgebra map (compare with Proposition 1.3.1). The seventh equality writes the coproduct in H^* in terms of H to combine some factors, while the eighth equality uses that Δ is an algebra homomorphism. By these manipulations, we are ready, in the ninth equality, to use that $\int$ is a right integral, leading to the required result. ■

In a functional-analytic context, we can take for H a von Neumann algebra, and in place of $\mathrm{Lin}(H)$ we take $B(\mathcal{H}_\phi)$, the bounded operators on the Hilbert space $\mathcal{H}_\phi$ determined by a state or weight ϕ. H is embedded in this by the Gelfand–Naimark–Segal (GNS) construction. We will see more details in Chapter 5.3. With ϕ defined suitably (via the integral on H), the above *-algebra structure on $\mathrm{Lin}(H)$ becomes the usual adjoint operation † on $B(\mathcal{H}_\phi)$, so that $W^\dagger = W^{-1}$. The structure in Theorem 1.7.4 and Proposition 1.7.6 is then characteristic of what is called a *Kac algebra*. This was historically limited to $S^2 = \mathrm{id}$, but such a condition is not needed when formulated along the lines above. Rather a lot of care needs to be taken with the counit (which does not always exist as a finite map) but its role too can be carried to some extent by the integral.

Another, different, way to topologise the situation is by means of C^*-algebras. The problem here is that there is no single good notion of tensor product of C^*-algebras. This can be resolved by fixing a finitely generated dense *-subalgebra and demanding that only this be a Hopf *-algebra. This is the approach of Woronowicz in the context of matrix quantum groups. We will give several examples of matrix quantum groups in Chapter 4.

Without going into details of the functional analysis, we can summarise the situation for the topological version of the group-based examples of Section 1.5, as follows. If G is a Lie group or (more generally), a locally compact group, we can work with the C^*-algebra $C(G)$ of complex continuous functions that vanish at infinity. The *-structure is $\phi^*(u) = \overline{\phi(u)}$. If G is not compact, then the unit (the constant function) is not allowed, but can be approximated. If a unit is required, one can enlarge further to the Hopf–von Neumann or Kac algebra $L^\infty(G)$ defined as pointwise operators on the Hilbert space $L^2(G)$. The fundamental operator acting on $\xi \in L^2(G) \otimes L^2(G)$ is $(W\xi)(u, v) = \xi(u, uv)$. These two approaches provide analogues of $k(G)$.

Also, if G is a locally compact group, we can work with the group C^*-algebra $C^*(G)$ defined as a certain C^* algebra completion of complex continuous functions on G with compact support, and the convolution product and $*$-structure

$$(hg)(u) = \int dv\, h(v)g(v^{-1}u), \quad \epsilon(h) = \int du\, h(u),$$
$$h^*(u) = D(u)^{-1}\overline{h(u^{-1})}.$$
(1.19)

Here summations in G have been replaced by integration with respect to a left-invariant Haar measure (with associated modular function D), and the δ-functions in Example 1.5.4 become left-invariant δ-functions defined with respect to this. In fact, such δ-functions are not technically allowed unless G is discrete, so the unit and coproduct must be approximated. If these are required exactly, one can enlarge to the group Hopf–von Neumann or Kac algebra $\mathcal{M}(G)$ defined as operators on the Hilbert space $L^2(G)$ by $(u \triangleright \xi)(v) = \xi(u^{-1}v)$. The fundamental operator is $(W\xi)(u, v) = \xi(u, u^{-1}v)$. These two approaches provide analogues of kG.

We also have general Hopf algebra results analogous to Fourier theory on groups (see Corollary 1.5.6). As before, we treat the case of finite-dimensional Hopf algebras, but the formulae can also be used in the infinite-dimensional case with suitable topological considerations.

Proposition 1.7.7 *Let $\int, \int'$ be right integrals on a finite-dimensional Hopf algebra H and its dual, respectively, such that $\mu \equiv \int(\int') \neq 0$. The Fourier transform $\mathcal{F} : H \to H^*$ and its inverse are*

$$\mathcal{F}(h) = \sum_a \left(\int e_a h\right) f^a, \quad \mathcal{F}^{-1}(\phi) = \frac{1}{\mu}\sum_a S^{-1}e_a \int' \phi f^a$$

where $\{e_a\}$ is a basis of H and $\{f^a\}$ is a dual basis. One has

$$\mathcal{F}(R^*_{S\phi}(h)) = \mathcal{F}(h)\phi, \quad \forall h \in H, \ \phi \in H^*.$$

If the integrals are normalised then $\mu = 1/\mathrm{Tr}\, S^{-2}$.

Proof: Note firstly the useful identity $(\int g_{(1)}h)g_{(2)} = (\int g_{(1)}h_{(1)})g_{(2)}h_{(2)}Sh_{(3)}$ $= (\int(gh_{(1)})_{(1)})(gh_{(1)})_{(2)}Sh_{(2)} = (\int gh_{(1)})Sh_{(2)}$ for any right integral $\int$. Also, to avoid confusion, we write $\Lambda' = \int'$ when regarded as an element of H and $\Lambda = \int$ as an element of H^*. Then

$$\mathcal{F}^{-1} \circ \mathcal{F}(h) = \frac{S^{-1}e_a}{\mu}\int' f^b f^a \int e_b h = \frac{S^{-1}e_{a(2)}}{\mu}\int' f^a \int e_{a(1)}h$$

$$= \frac{S^{-1}\Lambda'_{(2)}}{\mu}\int \Lambda'_{(1)}h = \mu^{-1}h_{(2)}\int \Lambda' h_{(1)} = \mu^{-1}h\int \Lambda' = h$$

for all $h \in H$. We used the duality pairing, the above identity and, finally, that Λ' is a right integral in H^*. The other side is similar,

$$\mathcal{F} \circ \mathcal{F}^{-1}(\phi) = \frac{f^b}{\mu} \int e_b S^{-1} e_a \int' \phi f^a = \frac{f^b}{\mu} \int e_b e_a \int' \phi S^{-1} f^a$$

$$= \frac{\Lambda_{(1)}}{\mu} \int' \phi S^{-1} \Lambda_{(2)} = \frac{S(S^{-1}\Lambda)_{(2)}}{\mu} \int' \phi(S^{-1}\Lambda)_{(1)}$$

$$= \frac{\phi_{(2)}}{\mu} \int' S^{-1}(\Lambda S\phi_{(1)}) = \mu^{-1} \phi \int' S^{-1}\Lambda = \phi$$

for all $\phi \in H^*$, with the last equality being obtained as follows. Indeed, from our first identity above (applied in H^*), we have

$$\Big(\int' f^a S^{-1}\Lambda_{(2)}\Big)\Lambda_{(1)} \otimes e_a = \Big(\int' f^a (S^{-1}\Lambda)_{(1)}\Big)S(S^{-1}\Lambda)_{(2)} \otimes e_a$$

$$= \Big(\int' f^a{}_{(1)} S^{-1}\Lambda\Big)f^a{}_{(2)} \otimes e_a = \Big(\int' (S^{-1}f^a)_{(1)} S^{-1}\Lambda\Big)(S^{-1}f^a)_{(2)} \otimes Se_a$$

$$= \Big(\int' S^{-1}(\Lambda f^a{}_{(2)})\Big)S^{-1}f^a{}_{(1)} \otimes Se_a = \Big(\int' S^{-1}\Lambda\Big)f^a \otimes e_a.$$

We used that S is an antialgebra and anticoalgebra map, the axioms of a duality pairing and that Λ is a right integral in H^*. Applying $\int' \otimes \int$ to this identity and using that $\int'$ is a right integral now yields

$$\Big(\int' \Lambda\Big)^2 = \int' \Lambda \int' f^a \int e_a = \Big(\int' f^a S^{-1}\Lambda_{(2)}\Big) \int' \Lambda_{(1)} \int e_a = \int' S^{-1}\Lambda \int' \Lambda,$$

as required. Here the second equality is because $\int'$ is a right integral. On the other hand, applying $\langle S^{-2}(\ ), (\ )\rangle$ instead of $\int' \otimes \int$ yields $\int' S^{-1}\Lambda \operatorname{Tr} S^{-2}$ $= \int' f^a S^{-1}\Lambda_{(2)}\langle S^{-2}\Lambda_{(1)}, e_a\rangle = \int'(S^{-2}\Lambda_{(1)})S^{-1}\Lambda_{(2)} = \int 1 \int' 1$ so that $\mu^{-1} = \operatorname{Tr} S^{-2}$ if the integrals are normalised. Also, one can show that one always has $\mu \neq 0$ in the finite-dimensional case, so that this is not really an assumption. Finally, the key property that translation in H Fourier transforms to multiplication in H^* follows easily from our first identity above. Thus $\mathcal{F}(R^*_{S\phi}(h)) = \mathcal{F}(h_{(1)})\langle h_{(2)}, S\phi\rangle = (\int e_a h_{(1)})\langle Sh_{(2)}, \phi\rangle f^a = (\int e_{a(1)}h)\langle e_{a(2)}, \phi\rangle f^a = (\int e_a h)\langle e_b, \phi\rangle f^a f^b = \mathcal{F}(h)\phi$ as required. ∎

There are versions of these results with left-integrals and $\operatorname{Tr} S^2$. Also, reversing the roles of H, H^* in the above gives another and slightly different Fourier transform (the order of the integrands is reversed), the two conventions being related by $*$ in the Hopf $*$-algebra case. Finally, we note that the role of exponential in the usual formulae for Fourier transform is played by the canonical element $\sum_a e_a \otimes f^a$ of $H \otimes H^*$ (otherwise known as the identity matrix). To see this, consider H as the polynomials in one variable x and H' dually paired with it as the polynomials in another variable p (in both cases with the additive coproduct as in

Example 1.5.7). The two are dually paired by $\langle x^n, p^m \rangle = \delta_{n,m} n!$ and the canonical element is $\exp(x \otimes p)$ as a formal powerseries (or living in a suitable topological completion). We will meet these ideas again in a braided context in Chapter 10. More generally, we typically have a duality pairing between enveloping algebras $U(g)$ and the corresponding matrix group coordinate rings $k(G)$ in Chapter 4 (see Exercise 4.2.1). Hence we have a formal 'exponential' and Fourier theory even in this non-Abelian case.

Notes for Chapter 1

Hopf algebras first appeared in the work of H. Hopf in connection with the cohomology of groups. They also appeared in the work of G.I. Kac and co-workers in the study of group duals, e.g. [6]. They have been used extensively by algebraic geometers, who traditionally work with the ring or algebra of functions on a space as a definition of the algebraic space (even when there is no actual space in the usual sense). Thus, commutative Hopf algebras are regarded as the functions on an algebraic group. Hopf algebras have also traditionally been used as a tool in studying Lie algebras over a field k. A general rule of thumb in the theory of commutative or cocommutative Hopf algebras is that a theorem that is true for both group Hopf algebras and universal enveloping algebras can usually be proven for all cocommutative Hopf algebras. In this way, Hopf algebras have become a topic in their own right. Two standard texts are [7] and [8]. R. Heyneman and M.E. Sweedler examined the structure of finite-dimensional Hopf algebras, the complete classification of which is still open and an area of recent research. Several results relate to integrals. For example, a finite-dimensional Hopf algebra is semisimple *iff* the integral in it is normalisable. Another result is a theorem of R.G. Larson and D. Radford which asserts that a finite-dimensional Hopf algebra over a field of characteristic 0 has $S^2 = \mathrm{id}$ *iff* it is semisimple [9]. Central to their proof is a study of the integral as in Proposition 1.7.3. The Fourier theory in Proposition 1.7.7 should also date from this traditional Hopf algebra theory, although the present formulation is new (see also [10, 11] for recent papers). Note also that μ^{-1} in the normalisable case of Proposition 1.7.7 then simplifies slightly in view of the above. A lot of other more recent work centres on actions and coactions of Hopf algebras, and cross products by them. Using a combination of cross products and cross coproducts, one can obtain a large class of noncommutative and nonco-commutative Hopf algebras [12]. We shall come to these in Chapter 6. Other (quite different) examples of noncommutative and noncocommutative Hopf algebras are more often called quantum groups and will be introduced next, in Chapter 2. A third general construction for noncommutative and noncocommutative bialgebras (usually without antipode, however) is the comeasuring or 'transformation' bialgebras of algebras in [13], from which the trigonometric bialgebra in Example 1.6.20 is taken. See also Chapter 4.5. The approach to duality as a pairing in Section 1.4 was emphasised in [14] as the right setting for infinite-dimensional Hopf algebras and quantum groups. Proposition 1.4.4, and a further idea to use the convolution inverse of the pairing in place of an antipode can be found in this work by the author. We will use both constructions extensively in Chapter 4.

A good C^*-algebra setting for Hopf algebras of matrix type was introduced by S.L. Woronowicz [15] under the heading 'compact matrix pseudogroup'. There are also other notions of Hopf C^*-algebra. The Hopf–von Neumann algebra and Kac algebra setting is older, and a seminal work was [16]; see also the text [17]. Essentially the first noncommutative and noncocommutative Hopf–von Neumann and Kac algebras were introduced by the author [18]. These examples and those of Woronowicz were subsequently unified into a modern setting for Kac algebras in [19]. The purely algebraic realisation in Theorem 1.7.4 and Proposition 1.7.6 is taken from [20] by the author. There are other ideas for generalising groups in the context of operator algebras, notably 'paragroups' associated to inclusions of von Neumann algebras in the work of A. Ocneanu and others; see the text [21] and the literature.

2

Quasitriangular Hopf algebras

The theme of this chapter is that when one has an axiom or condition for an algebraic structure, one can consider relaxing it so that it holds only up to some 'cocycle' element, which is then required to obey some consistency conditions which we have to specify. The most important application of this principle will be to the condition of cocommutativity studied in Chapter 1.5. Thus, we now study a class of Hopf algebras that are cocommutative only up to conjugation by an element $\mathcal{R}$, called the 'quasitriangular structure', and obeying some conditions. This simple idea has far-reaching implications which will recur throughout the book. From an algebraic point of view, such Hopf algebras are truly different from the Hopf algebras associated to groups or Lie algebras already encountered in Chapter 1, and yet are so close to being cocommutative that all the familiar results for groups and Lie algebras tend to have analogues here also. Hopf algebras that are almost cocommutative in this way are called *quasitriangular Hopf algebras*. Because their properties are so close to those of groups or Lie algebras, they are also commonly called *quantum groups*. We will often use this term more loosely to apply to Hopf algebras in general, but this is the strict usage. We will give examples that are indeed deformations of familiar groups and Lie algebras later, in Chapters 3 and 4, but it should not be thought that they are the only examples. There are plenty of quasitriangular Hopf algebras that are not based on classical groups or Lie algebras at all.

This chapter is devoted to giving the basic definitions and algebraic properties behind this class of quasitriangular Hopf algebras, both in the form described and in a dual form (the dual of an almost-cocommutative Hopf algebra is almost commutative or 'dual quasitriangular'). The abstract theory is largely due to V.G. Drinfeld and includes a systematic process to obtain examples by means of a process called twisting. This process extends naturally to a variant of the above in which not only the cocommutativity is relaxed up to conjugation, but also the coassociativity of Δ. These are the quasitriangular quasi-Hopf algebras.

In developing the mathematical framework, we must continue in the abstract setting of Chapter 1. A viable alternative is to proceed to the examples in the next chapter and return to the present one for reference. The reader is urged to work through Examples 2.1.6, 2.1.7 and 2.2.6 in the present chapter as a useful warm-up in any case.

2.1 Quasitriangular structures

Recall from Chapter 1.5 that a Hopf algebra H is cocommutative if $\tau \circ \Delta = \Delta$, where τ is the transposition map. We can weaken this by considering a Hopf algebra that is only cocommutative up to conjugation by an element $\mathcal{R} \in H \otimes H$, obeying some further properties. This element $\mathcal{R}$ is called the *quasitriangular structure*. Thus

Definition 2.1.1 *A quasitriangular bialgebra or Hopf algebra is a pair* $(H, \mathcal{R})$*, where H is a bialgebra or Hopf algebra and $\mathcal{R} \in H \otimes H$ is invertible and obeys*

$$(\Delta \otimes \mathrm{id})\mathcal{R} = \mathcal{R}_{13}\mathcal{R}_{23}, \quad (\mathrm{id} \otimes \Delta)\mathcal{R} = \mathcal{R}_{13}\mathcal{R}_{12}, \tag{2.1}$$

$$\tau \circ \Delta h = \mathcal{R}(\Delta h)\mathcal{R}^{-1}, \ \forall h \in H. \tag{2.2}$$

Writing $\mathcal{R} = \sum \mathcal{R}^{(1)} \otimes \mathcal{R}^{(2)}$, the notation used is

$$\mathcal{R}_{ij} = \sum 1 \otimes \cdots \otimes \mathcal{R}^{(1)} \otimes 1 \otimes \cdots \otimes \mathcal{R}^{(2)} \otimes \cdots \otimes 1,$$

the element of $H \otimes H \cdots \otimes H$ which is $\mathcal{R}$ in the ith and jth factors. τ denotes the transposition map.

That this is a good definition will emerge as we consider the various properties of such quasitriangular Hopf algebras or bialgebras. Chapter 9 provides another way of thinking about the meaning of these axioms.

Lemma 2.1.2 *If $(H, \mathcal{R})$ is a quasitriangular bialgebra, then $\mathcal{R}$ as an element of $H \otimes H$ obeys*

$$(\epsilon \otimes \mathrm{id})\mathcal{R} = (\mathrm{id} \otimes \epsilon)\mathcal{R} = 1. \tag{2.3}$$

If H is a Hopf algebra then one also has

$$(S \otimes \mathrm{id})\mathcal{R} = \mathcal{R}^{-1}, \quad (\mathrm{id} \otimes S)\mathcal{R}^{-1} = \mathcal{R}, \tag{2.4}$$

and hence $(S \otimes S)\mathcal{R} = \mathcal{R}$.

Proof: For the first part, apply ϵ to (2.1); thus $(\epsilon \otimes \mathrm{id} \otimes \mathrm{id})(\Delta \otimes \mathrm{id})\mathcal{R} = \mathcal{R}_{23} = (\epsilon \otimes \mathrm{id} \otimes \mathrm{id})\mathcal{R}_{13}\mathcal{R}_{23}$, so that $(\epsilon \otimes \mathrm{id})\mathcal{R} = 1$ because $\mathcal{R}_{23}$ is invertible. Similarly for the other side. For (2.4), $\mathcal{R}^{(1)}{}_{(1)}S\mathcal{R}^{(1)}{}_{(2)} \otimes \mathcal{R}^{(2)} = 1$ by the property of the antipode and (2.3) already proven, but equals $\mathcal{R}(S \otimes \mathrm{id})\mathcal{R}$ by (2.1). Similarly for the other side; hence, $(S \otimes \mathrm{id})\mathcal{R} = \mathcal{R}^{-1}$. Similarly for $(\mathrm{id} \otimes S)\mathcal{R}^{-1} = \mathcal{R}$ once we know that $(\Delta \otimes \mathrm{id})(\mathcal{R}^{-1}) = (\mathcal{R}_{13}\mathcal{R}_{23})^{-1} = \mathcal{R}_{23}^{-1}\mathcal{R}_{13}^{-1}$, etc. since Δ is an algebra homomorphism. See also the next exercise.　∎

Exercise 2.1.3 *Show that if $\mathcal{R}$ is a quasitriangular structure for a bialgebra or Hopf algebra H, then so is $\tau(\mathcal{R}^{-1})$. Also, $\tau(\mathcal{R})$ and $\mathcal{R}^{-1}$ are quasitriangular structures on H^{op} or H^{cop} in Exercise 1.3.3.*

Solution: For the first part, we have $(\Delta \otimes \mathrm{id})(\mathcal{R}^{-1}) = \mathcal{R}_{23}^{-1}\mathcal{R}_{13}^{-1}$ as in the last lemma. After permuting the order in $H \otimes H \otimes H$, we conclude that $(\mathrm{id} \otimes \Delta)\mathcal{R}_{21}^{-1} = \mathcal{R}_{31}^{-1}\mathcal{R}_{21}^{-1}$. Likewise, $(\mathrm{id} \otimes \Delta)(\mathcal{R}^{-1}) = \mathcal{R}_{12}^{-1}\mathcal{R}_{13}^{-1}$ implies, after permuting the order in $H \otimes H \otimes H$, that $(\Delta \otimes \mathrm{id})\mathcal{R}_{21}^{-1} = \mathcal{R}_{31}^{-1}\mathcal{R}_{32}^{-1}$. This is (2.1) for $\mathcal{R}_{21}^{-1} = \tau(\mathcal{R}^{-1})$. Finally, (2.2) implies that $\mathcal{R}^{-1}(\tau \circ \Delta(\ ))\mathcal{R} = \Delta$. After swapping the order in $H \otimes H$, this is $\mathcal{R}_{21}^{-1}(\Delta(\ ))\mathcal{R}_{21} = \tau \circ \Delta$, as required. For the second part, we check only $\tau(\mathcal{R})$ since (by the first part) $\mathcal{R}^{-1}$ is then also a quasitriangular structure for H^{op} or H^{cop}. The proof here comes simply from permuting the order in (2.1) and (2.2), and identifying the opposite multiplication or comultiplication.　∎

Lemma 2.1.4 *Let $(H, \mathcal{R})$ be a quasitriangular bialgebra. Then (2.2) and the second part of (2.1) imply that*

$$\mathcal{R}_{12}\mathcal{R}_{13}\mathcal{R}_{23} = \mathcal{R}_{23}\mathcal{R}_{13}\mathcal{R}_{12},$$

the abstract quantum Yang–Baxter equation (QYBE).

Proof: We compute $(\mathrm{id} \otimes \tau \circ \Delta)\mathcal{R}$ in two ways: using the second of the axioms (2.1) directly, or using the axiom (2.2) and then the second of (2.1). Thus, $(\mathrm{id} \otimes \tau \circ \Delta)\mathcal{R} = (\mathrm{id} \otimes \tau)(\mathrm{id} \otimes \Delta)\mathcal{R} = (\mathrm{id} \otimes \tau)\mathcal{R}_{13}\mathcal{R}_{12} = \mathcal{R}_{12}\mathcal{R}_{13}$ and $(\mathrm{id} \otimes \tau \circ \Delta)\mathcal{R} = \mathcal{R}_{23}((\mathrm{id} \otimes \Delta)\mathcal{R})\mathcal{R}_{23}^{-1} = \mathcal{R}_{23}\mathcal{R}_{13}\mathcal{R}_{12}\mathcal{R}_{23}^{-1}$.　∎

The preceding lemma is one of the original motivations for the definition of quasitriangular Hopf algebras, as we shall see in Chapter 4. For every representation ρ of the algebra of H in matrices, $(\rho \otimes \rho)(\mathcal{R})$ is a matrix solution of the same equations. It is these matrix solutions that are needed in physical applications. Thus, quasitriangular Hopf algebras provide a way of generating many such solutions from a single abstract solution $\mathcal{R}$.

For this reason, the quasitriangular structure $\mathcal{R}$ is sometimes referred to as the 'universal R-matrix'.

Mathematically, the meaning of axiom (2.2) is that, although the quasitriangular Hopf algebra is not usually commutative or cocommutative, the lack of noncocommutativity is under control, being controlled by $\mathcal{R}$. Thus, many of the things that can be done for cocommutative Hopf algebras (e.g. for group algebras or universal enveloping algebras) can be done also for quasitriangular Hopf algebras. The meaning of (2.1) is less clear, but some insight is provided by the following exercise.

Exercise 2.1.5 *Let H be a finite-dimensional bialgebra and let $\mathcal{R}$ be an invertible element of $H \otimes H$ viewed as a linear map $\mathcal{R}_1 : H^* \to H$ by sending $\phi \in H^*$ to $(\mathrm{id} \otimes \phi)(\mathcal{R})$. Show that axioms (2.1) hold iff this linear map is a coalgebra and antialgebra map (i.e. a bialgebra map $H^{*\mathrm{op}} \to H$). Likewise, axioms (2.1) hold iff the map $\mathcal{R}_2 : H^* \to H$ sending ϕ to $(\phi \otimes \mathrm{id})(\mathcal{R})$ is an algebra and anticoalgebra map (i.e. a bialgebra map $H^{*\mathrm{cop}} \to H$).*

Solution: By the pairing relations in Proposition 1.4.2, we have $\mathcal{R}_1(\phi\psi)$ $= \mathcal{R}^{(1)}\langle \phi \otimes \psi, \Delta \mathcal{R}^{(2)}\rangle$, while $\mathcal{R}_1(\psi)\mathcal{R}_1(\phi) = \mathcal{R}'^{(1)}\mathcal{R}^{(1)}\langle \phi \otimes \psi, \mathcal{R}^{(2)} \otimes \mathcal{R}'^{(2)}\rangle$. Here $\mathcal{R}'$ denotes a second copy of $\mathcal{R}$. The equality of these expressions for all ϕ, ψ is just the second of (2.1), so this corresponds to $\mathcal{R}_1$ an antialgebra map. Also, $\Delta(\mathcal{R}_1(\phi)) = \Delta(\mathrm{id} \otimes \phi)\mathcal{R} = \Delta\mathcal{R}^{(1)}\langle \phi, \mathcal{R}^{(2)}\rangle$, while $(\mathcal{R}_1 \otimes \mathcal{R}_1)(\Delta\phi) = \mathcal{R}^{(1)} \otimes \mathcal{R}'^{(1)}\langle \phi, \mathcal{R}^{(2)}\mathcal{R}'^{(2)}\rangle$ by the definition of the pairing. Hence their equality for all ϕ is just the first of (2.1). Similarly for $\mathcal{R}_2$. ∎

Before proceeding with the general theory, we present some of the simplest nontrivial examples. We will return to them in Chapters 9 and 10, respectively – for now they provide familiarity with the definitions above and will be used to further illustrate the theory below.

Example 2.1.6 *Let $\mathbb{Z}_{/n} = \mathbb{Z}/n\mathbb{Z}$ be the finite cyclic group of order n and let $\mathbb{C}\mathbb{Z}_{/n}$ be its group algebra, as in Example 1.5.3, over the complex numbers $\mathbb{C}$. It is generated by $1, g$ and the relation $g^n = 1$, and has coproduct, counit and antipode $\Delta g = g \otimes g$, $\epsilon g = 1$, $Sg = g^{-1}$, respectively. In addition to the trivial quasitriangular structure $1 \otimes 1$, we have also*

$$\mathcal{R} = \frac{1}{n} \sum_{a,b=0}^{n-1} e^{-\frac{2\pi\imath ab}{n}} g^a \otimes g^b.$$

We denote the Hopf algebra $\mathbb{C}\mathbb{Z}_{/n}$ equipped with this nontrivial quasitriangular structure by $\mathbb{Z}'_{/n}$, the anyon-generating quantum group. In this example we have $\tau(\mathcal{R}) = \mathcal{R}$.

Proof: Any cocommutative Hopf algebra, such as this one, can be regarded as trivially quasitriangular with $\mathcal{R} = 1 \otimes 1$. To verify the nontrivial $\mathcal{R}$ shown, we note that $n^{-1} \sum_{b=0}^{n-1} e^{\frac{2\pi i ab}{n}} = \delta_{a,0}$ (i.e. 1 if $a = 0$ and zero otherwise). Then $\mathcal{R}_{13}\mathcal{R}_{23} = n^{-2} \sum e^{-\frac{2\pi i(ab+cd)}{n}} g^a \otimes g^c \otimes g^{b+d} = n^{-2} \sum e^{-\frac{2\pi i b(a-c)}{n}} e^{-\frac{2\pi i cb'}{n}} g^a \otimes g^c \otimes g^{b'} = n^{-1} \sum e^{-\frac{2\pi i ab'}{n}} g^a \otimes g^a \otimes g^{b'}$, where $b' = b + d$ was a change of variables. This equals $(\Delta \otimes \mathrm{id})\mathcal{R}$, as required. Similarly for the second half of (2.1). The remaining axiom (2.2) is automatic because the Hopf algebra is both commutative and cocommutative. Our terminology comes from Chapters 9.2 and 10.1, where this quantum group generates the braided category of anyonic vector spaces. ∎

Example 2.1.7 *Consider the Hopf algebra generated by* $1, g, x$ *with relations*

$$x^2 = 0, \quad g^2 = 1, \quad xg = -gx$$

and Hopf algebra structure

$$\Delta x = x \otimes 1 + g \otimes x, \quad \Delta g = g \otimes g, \quad \epsilon x = 0, \quad \epsilon g = 1, \quad Sx = -gx, \quad Sg = g$$

(this is a quotient of Example 1.3.2 at $q = -1$*). This forms a quasitriangular Hopf algebra* $\mathbb{Z}_{/2,\alpha}$ *with*

$$\mathcal{R} = 1 \otimes 1 - 2p \otimes p + \alpha(x \otimes x + 2xp \otimes xp - 2x \otimes xp); \qquad p = \frac{1-g}{2},$$

where α *is an arbitrary parameter. In this example we have* $\tau(\mathcal{R}^{-1}) = \mathcal{R}$ *(so that the example is in fact triangular in the sense below).*

Proof: We will derive this example in Chapter 10.1 as the *bosonisation* of the superline. For now, we verify it directly. It is easy to verify that we do have a Hopf algebra, along the lines of Example 1.3.2. The new relation $g^2 = 1$ is compatible with the coproduct and other algebra relations because g^2 is group-like and central in the Hopf algebra generated by $xg = -gx$. Likewise, $x^2 = 0$ is compatible with the coproduct because the right hand side of $\Delta x^2 = (x \otimes 1 + g \otimes x)^2 = x^2 \otimes 1 + g^2 \otimes x^2 + xg \otimes x + gx \otimes x = x^2 \otimes 1 + 1 \otimes x^2$, i.e. it is primitive. We verify the axioms of the quasitriangular structure. For this it is convenient to expand $\mathcal{R}$ as

$$\mathcal{R} = \mathcal{R}_g + \frac{\alpha}{2}(x \otimes x + xg \otimes xg + x \otimes xg - xg \otimes x),$$

$$\mathcal{R}_g = 1 \otimes 1 - 2p \otimes p = \frac{1}{2}(1 \otimes 1 + 1 \otimes g + g \otimes 1 - g \otimes g).$$

Note that $\mathcal{R}_g$ is the quasitriangular structure for $\mathbb{Z}'_{/2}$ in Example 2.1.6 and so already obeys (2.1). Note also that, since $x^2 = 0$, the order α^2

terms vanish in $\mathcal{R}_{13}\mathcal{R}_{23}$. Hence, to prove (2.1) for $\mathcal{R}$, we have only to show the part linear in α, i.e.

$$\frac{\alpha}{2}(\Delta \otimes \mathrm{id})(x \otimes x + xg \otimes xg + x \otimes xg - xg \otimes x)$$

$$= \frac{\alpha}{2}(x \otimes 1 \otimes x + xg \otimes 1 \otimes xg + x \otimes 1 \otimes xg - xg \otimes 1 \otimes x)R_{g23}$$

$$+ \frac{\alpha}{2}\mathcal{R}_{g13}(1 \otimes x \otimes x + 1 \otimes xg \otimes xg + 1 \otimes x \otimes xg - 1 \otimes xg \otimes x)$$

on applying Δ and multiplying out. Similarly for the other half. To verify (2.2), we need only verify it on the generators g, x. In the first case, we have already that $\Delta g = g \otimes g$ commutes with $\mathcal{R}_g$. Commutativity with the part of $\mathcal{R}$ linear in α follows at once by multiplying with $g \otimes g$ and using the algebra relations. Finally, we have $\mathcal{R}\Delta x = \mathcal{R}_g(x \otimes 1 + g \otimes x)$ and $(\tau \circ \Delta x)\mathcal{R} = (x \otimes g + 1 \otimes x)\mathcal{R}_g$, again because $x^2 = 0$. The first expression is

$$\frac{1}{2}(x \otimes g + x \otimes 1 + gx \otimes g - gx \otimes 1) + \frac{1}{2}(1 \otimes x + 1 \otimes gx + g \otimes x - g \otimes gx),$$

while the second is

$$\frac{1}{2}(g \otimes x + g \otimes xg + 1 \otimes x - 1 \otimes xg) + \frac{1}{2}(x \otimes 1 + x \otimes g + xg \otimes 1 - xg \otimes g).$$

These are equal in view of the relations $xg = -gx$. Thus we have a quasitriangular structure. The inverse $\mathcal{R}^{-1}$ is most easily obtained from Lemma 2.1.2, equation (2.4). Applying the antipode, it is evidently

$$\mathcal{R}^{-1} = \mathcal{R}_g + \frac{\alpha}{2}(x \otimes x + xg \otimes xg + xg \otimes x - x \otimes xg),$$

where $\mathcal{R}_g^{-1} = \mathcal{R}_g$. Applying τ to this we conclude that the second quasi-triangular structure $\tau(\mathcal{R}^{-1})$ in Exercise 2.1.3 coincides with $\mathcal{R}$ for all α. We assume, for the quasitriangular structure, that 2 is invertible. ∎

There are many interesting objects that can be built from $\mathcal{R}$. Using such elements, one can prove various results which are analogous to those for cocommutative Hopf algebras, but differing by conjugation. Thus Proposition 1.5.1 has the following analogue.

Proposition 2.1.8 *Let $H, \mathcal{R}$ be a quasitriangular Hopf algebra with anti-pode S. Then S is invertible and $S^2(h) = uhu^{-1}$ for all $h \in H$, where u is an invertible element of H obeying*

$$u = \sum(S\mathcal{R}^{(2)})\mathcal{R}^{(1)}, \qquad u^{-1} = \sum \mathcal{R}^{(2)}S^2\mathcal{R}^{(1)}, \qquad \Delta u = Q^{-1}(u \otimes u).$$

Here $Q = \mathcal{R}_{21}\mathcal{R}$. Likewise, $v = Su$ obeys $S^{-2}(h) = vhv^{-1}$, where v is an invertible element of H obeying

$$v = \sum \mathcal{R}^{(1)}S\mathcal{R}^{(2)}, \qquad v^{-1} = \sum(S^2\mathcal{R}^{(1)})\mathcal{R}^{(2)}, \qquad \Delta v = Q^{-1}(v \otimes v).$$

Proof: We do this in several steps. (i) First, we compute $(Sh_{(2)})uh_{(1)} = (Sh_{(2)})(S\mathcal{R}^{(2)})\mathcal{R}^{(1)}h_{(1)} = (S(\mathcal{R}^{(2)}h_{(2)}))\mathcal{R}^{(1)}h_{(1)} = (S(h_{(1)}\mathcal{R}^{(2)}))h_{(2)}\mathcal{R}^{(1)} = (S\mathcal{R}^{(2)})(Sh_{(1)})h_{(2)}\mathcal{R}^{(1)} = \epsilon(h)u$. We used here the antipode properties, axiom (2.2) and the counit property. Then

$$(S^2h)u = (S^2h_{(2)})\epsilon(h_{(1)})u = (S^2h_{(3)})(Sh_{(2)})uh_{(1)} = uh.$$

(ii) Secondly, $u^{-1}u = \mathcal{R}^{(2)}(S^2\mathcal{R}^{(1)})u = \mathcal{R}^{(2)}u\mathcal{R}^{(1)} = \mathcal{R}^{(2)}(S\mathcal{R}'^{(2)})\mathcal{R}'^{(1)}\mathcal{R}^{(1)} = (S\mathcal{R}^{(2)})(S\mathcal{R}'^{(2)})\ \mathcal{R}'^{(1)}S\mathcal{R}^{(1)} = S(\mathcal{R}'^{(2)}\mathcal{R}^{-(2)})\mathcal{R}'^{(1)}\mathcal{R}^{-(1)} = S(1)1 = 1$ using part (i) and the facts already established in Lemma 2.1.2 for the action of S on $\mathcal{R}$. Likewise, $uu^{-1} = u\mathcal{R}^{(2)}S^2\mathcal{R}^{(1)} = (S^2\mathcal{R}^{(2)})uS^2\mathcal{R}^{(1)} = (S\mathcal{R}^{(2)})uS\mathcal{R}^{(1)} = 1$. Hence u, u^{-1} are inverse. (iii) Next, we define $S^{-1}(h) = u^{-1}(Sh)u$ and verify that $(S^{-1}h_{(2)})h_{(1)} = u^{-1}(Sh_{(2)})uh_{(1)} = u^{-1}(Sh_{(2)})(S^2h_{(1)})u = u^{-1}(S((Sh_{(1)})h_{(2)}))u = \epsilon(h)u^{-1}u = \epsilon(h)$, using parts (i) and (ii) and the properties of S. Likewise, we verify that $h_{(2)}S^{-1}h_{(1)} = h_{(2)}u^{-1}(Sh_{(1)})u = u^{-1}(S^2h_{(2)})(Sh_{(1)})u = u^{-1}(S(h_{(1)}Sh_{(2)}))u = \epsilon(h)u^{-1}u = \epsilon(h)$. This means that S^{-1} is an antipode on H^{op} and hence the inverse of the antipode on H according to Exercise 1.3.3. (iv) Finally, we compute

$$\begin{aligned}
\Delta u &= \Delta(S\mathcal{R}^{(2)})\mathcal{R}^{(1)} = (S\mathcal{R}^{(2)}{}_{(2)})\mathcal{R}^{(1)}{}_{(1)} \otimes (S\mathcal{R}^{(2)}{}_{(1)})\mathcal{R}^{(1)}{}_{(2)} \\
&= (S(\mathcal{R}^{(2)}{}_{(2)}\mathcal{R}'^{(2)}{}_{(2)}))\mathcal{R}^{(1)} \otimes (S(\mathcal{R}^{(2)}{}_{(1)}\mathcal{R}'^{(2)}{}_{(1)}))\mathcal{R}'^{(1)} \\
&= (S(\mathcal{R}''^{(2)}\mathcal{R}'''^{(2)}))\mathcal{R}''^{(1)}\mathcal{R}^{(1)} \otimes (S(\mathcal{R}^{(2)}\mathcal{R}'^{(2)}))\mathcal{R}'''^{(1)}\mathcal{R}'^{(1)} \\
&= (S\mathcal{R}'''^{(2)})u\mathcal{R}^{(1)} \otimes (S(\mathcal{R}^{(2)}\mathcal{R}'^{(2)}))\mathcal{R}'''^{(1)}\mathcal{R}'^{(1)} \\
&= (S\mathcal{R}'''^{(2)})\mathcal{R}^{-(1)}u \otimes (S\mathcal{R}'^{(2)})\mathcal{R}^{-(2)}\mathcal{R}'''^{(1)}\mathcal{R}'^{(1)} \\
&= \mathcal{R}^{-(1)}(S\mathcal{R}'''^{(2)})u \otimes \mathcal{R}^{-(2)}(S\mathcal{R}'^{(2)})\mathcal{R}'^{(1)}\mathcal{R}'''^{(1)} \\
&= \mathcal{R}^{-(1)}(S\mathcal{R}'''^{(2)})u \otimes \mathcal{R}^{-(2)}u\mathcal{R}'''^{(1)} \\
&= \mathcal{R}^{-(1)}(S\mathcal{R}'''^{(2)})u \otimes \mathcal{R}^{-(2)}(S^2\mathcal{R}'''^{(1)})u \\
&= \mathcal{R}^{-1}\mathcal{R}_{21}^{-1}(u \otimes u).
\end{aligned}$$

Here the second equality uses the Hopf algebra axioms, the third and fourth equalities use (2.1), and the fifth equality uses the facts about u in part (i) and the action of S on $\mathcal{R}$ in Lemma 2.1.2. The sixth equality is the QYBE established in Lemma 2.1.4, applied in a suitable form. We then recognise another copy of u and use part (i) again. The proof of the corresponding results for v are analogous. They can also be obtained by applying the results for u to the quasitriangular Hopf algebra $(H^{op/cop}, \mathcal{R})$. ∎

Corollary 2.1.9 *The element $uv = vu$ is central and obeys $\Delta(uv) = Q^{-2}(uv \otimes uv)$. The element $uv^{-1} = v^{-1}u$ is group-like and implements S^4 by conjugation.*

Proof: Clearly, $S^2(u) = uuu^{-1} = u$ and $S^2(v) = v$, so that $uv = vS^2(u) = vu$, etc. Moreover, $uvh = S^2(S^{-2}(h))uv = huv$ for all h, so uv is central. From Lemma 2.1.2 we know also that $(S \otimes S)\mathcal{R} = \mathcal{R}$, so that $(S^2 \otimes S^2)Q = Q$. Hence, $\Delta uv = Q^{-1}(u \otimes u)Q^{-1}(v \otimes v) = Q^{-2}(uv \otimes uv)$. Likewise $\Delta uv^{-1} = uv^{-1} \otimes uv^{-1}$ and uv^{-1} implements S^4 because u, v^{-1} each implements S^2. ∎

Let us note that, for a finite-dimensional unimodular Hopf algebra with unimodular dual, we have $S^4 = \mathrm{id}$ and hence that uv^{-1} is central in this case. For a finite-dimensional semisimple Hopf algebra over a field of characteristic zero, we have $S^2 = \mathrm{id}$ and hence that u, v are separately central in this case. We conclude this general theory with some related definitions. They are desirable properties with many applications in the theory of link-invariants and elsewhere, and hold in some form for many familiar quantum groups such as those of the next chapter.

Definition 2.1.10 *A quasitriangular Hopf algebra is called a* ribbon *Hopf algebra if the element uv has a central square root ν, called the* ribbon *element, such that*

$$\nu^2 = vu, \qquad \Delta\nu = Q^{-1}(\nu \otimes \nu), \qquad \epsilon\nu = 1, \qquad S\nu = \nu.$$

These properties are not independent; the latter two can easily be deduced from the first two, while, in the other direction, the first can be deduced from the last three. Also, if a ribbon element does not exist in H, it can always be adjoined by centrally extending H by an element ν with the stated properties. In our elementary examples above, it does exist.

Example 2.1.11 *Examples 2.1.6 and 2.1.7 are both ribbon Hopf algebras according to*

$$u = v = \nu = \frac{1}{n}\sum_{a,b} g^{-b}g^a e^{-\frac{2\pi\imath ab}{n}} = \frac{1}{n}\sum_a g^a \theta_n(a)$$

where $\theta_n(a) = \sum_b e^{-\frac{2\pi\imath(a+b)b}{n}}$ *is the* $\mathbb{Z}_{/n}$ *theta-function. It is the* $\mathbb{Z}_{/n}$*-Fourier transform of a Gaussian. The case for Example 2.1.7 is* $n = 2$, *i.e.* $u = v = \nu = g$.

Proof: Since $\mathbb{Z}'_{/n}$ is commutative, we have at once that $u = v$, hence $\nu = u = v$ is also a ribbon element. Its formula follows at once from the expression for $\mathcal{R}$ and a change of variables. For Example 2.1.7, let us note that, since $x^2 = 0$, only the $1 - 2p \otimes p$ part of $\mathcal{R}$ can contribute to u, v, so these are the same as for $\mathbb{Z}'_{/2}$. ∎

Definition 2.1.12 *Let* $Q = \mathcal{R}_{21}\mathcal{R}$ *as above.* *A quasitriangular Hopf algebra is called* triangular *if* $Q = 1 \otimes 1$ *(i.e.* $\tau(\mathcal{R}^{-1}) = \mathcal{R}$*). It is called* factorisable *if* Q *is nondegenerate in the sense that the linear map* $Q : H^* \to H$ *sending* ϕ *to* $(\phi \otimes \mathrm{id})(Q)$ *is surjective. Equivalently, the linear map sending* ϕ *to* $(\mathrm{id} \otimes \phi)(Q)$ *is surjective.*

Triangular Hopf algebras are the ones most similar to cocommutative Hopf algebras such as $U(g)$ for a Lie algebra g (which can be regarded as quasitriangular in a trivial way, with $\mathcal{R} = 1 \otimes 1$). Their representation theory is not truly braided, and they do not give rise to interesting link- or knot-invariants. We have seen that Example 2.1.7 is of this triangular type for all α.

Example 2.1.13 *The quantum group* $\mathbb{Z}'_{/n}$ *in Example 2.1.6 is factorisable iff* n *is odd and* > 1 *(and triangular for* $n = 2$*).*

Proof: The $n = 2$ case is the $\alpha = 0$ case of Example 2.1.7. For the general case we first compute Q. Let $\{e_a\}$ be a dual basis to the $\{g^a\}$. Then $\sum_b Q^{(1)} e_b(Q^{(2)}) e^{\frac{2\pi i a b}{n}} = g^{2a}$. If n is odd, then $2a$ is a permutation of $\{0, \ldots, n-1\}$ as a runs through this set, and otherwise not. Hence $\mathbb{Z}'_{/n}$ is factorisable *iff* n is odd.   ∎

Factorisable quasitriangular Hopf algebras are at the opposite extreme from triangular Hopf algebras. For them, the second quasitriangular structure $\tau(\mathcal{R}^{-1})$ is 'maximally' distinct from $\mathcal{R}$ in the sense that Q is far from trivial. Note also that Q is not an algebra or antialgebra map, though it will later lead to one in the braided setting in Chapter 9.4. There we will see that it corresponds to a self-duality of the braided groups associated to factorisable quantum groups, which in turn allows such things as Fourier transformation on them. For now, we should view it as a kind of nondegeneracy condition somewhat analogous to the nondegeneracy of the inverse Killing form for a semisimple Lie algebra. We also have a property analogous to the ad-invariance of the Killing form and inverse Killing form. For Hopf algebras, invariance means that the action of an element h is the same as multiplication by $\epsilon(h)$. We have:

Proposition 2.1.14 *For any quasitriangular Hopf algebra* H*, the element* $(S \otimes \mathrm{id})(Q)$ *in* $H \otimes H$ *is invariant under the 'quantum' adjoint action in Example 1.6.3, extended to the tensor square. In view of this, we call* $Q = \mathcal{R}_{21}\mathcal{R}$ *the quantum inverse Killing form. Equivalently, the map* $Q : H^* \to H$ *in Definition 2.1.12 commutes with the action of* H*, where* H *acts on itself by the quantum adjoint action and on* H^* *by the 'quantum' coadjoint action in Example 1.6.4. In addition, we note that*

$$(S \otimes S)(Q) = \tau(Q).$$

Proof: Using the adjoint action in Example 1.6.3 and its extension to an action on $H \otimes H$ via Exercise 1.6.8, we have

$$h \triangleright (S \otimes \mathrm{id})(Q) = h_{(1)}(SQ^{(1)})Sh_{(2)} \otimes h_{(3)}Q^{(2)}Sh_{(4)}$$
$$= h_{(1)}(S\mathcal{R}^{(1)})(S\mathcal{R}'^{(2)})Sh_{(2)} \otimes h_{(3)}\mathcal{R}'^{(1)}\mathcal{R}^{(2)}Sh_{(4)}$$
$$= h_{(1)}(S\mathcal{R}^{(1)})Sh_{(3)}(S\mathcal{R}'^{(2)}) \otimes \mathcal{R}'^{(1)}h_{(2)}\mathcal{R}^{(2)}Sh_{(4)}$$
$$= h_{(1)}Sh_{(2)}(S\mathcal{R}^{(1)})(S\mathcal{R}'^{(2)}) \otimes \mathcal{R}'^{(1)}\mathcal{R}^{(2)}h_{(3)}Sh_{(4)}$$
$$= \epsilon(h)(S \otimes \mathrm{id})(Q).$$

We used that S is an antialgebra map and axiom (2.2) for the third and fourth equalities. The proof of invariance of $Q : H^* \to H$ from the actions in Examples 1.6.3 and 1.6.4 is similar (and equivalent). The final observation follows at once from Lemma 2.1.2 and the fact that S is an antialgebra map. ∎

The last observation, when combined with the result in Proposition 2.1.8 that S is invertible, ensures that the two definitions of factorisability in Definition 2.1.12 are equivalent. We now consider how the quasitriangular structure interacts with the notion of Hopf $*$-algebra in Chapter 1.7. Recall that $*$ is an antialgebra map and coalgebra map, so $* : H \to H^{\mathrm{op}}$ is an (antilinear) bialgebra map. From Exercise 2.1.3, we know that H^{op} has two natural quasitriangular structures induced from $\mathcal{R}$, so we can expect that $\mathcal{R}$ maps under $* \otimes *$ to one or other of them (they coincide in the triangular case).

Definition 2.1.15 *A nontriangular quasitriangular structure in a Hopf $*$-algebra is* real *if $\mathcal{R}^{* \otimes *} = \tau(\mathcal{R})$ (in this case, it is easy to see that $Q^{* \otimes *} = Q$ so that Q is self-adjoint). It is* antireal *if $\mathcal{R}^{* \otimes *} = \mathcal{R}^{-1}$ (in this case, $Q^{* \otimes *} = Q^{-1}$ so that Q is unitary). For triangular Hopf algebras these two notions coincide; we take $\mathcal{R}^{* \otimes *} = \tau(\mathcal{R}) = \mathcal{R}^{-1}$ as an axiom for a triangular Hopf $*$-algebra.*

Example 2.1.16 *The quantum group $\mathbb{Z}'/_n$ in Example 2.1.6 is a real quasitriangular Hopf $*$-algebra with $*$-structure $g^* = g^{-1}$. Also, Example 2.1.7 with α real and $*$-structure $g^* = g, x^* = x$, or with α imaginary and $*$-structure $g^* = g, x^* = \imath x$ are triangular Hopf $*$-algebras.*

Proof: It is easy to see that these do define Hopf $*$-algebras. Thus, for Example 2.1.7 we have to verify that $*$ extended as an antialgebra map is compatible with the relations. For the first case, we have $(xg)^* = g^*x^* = gx$ while $(-gx)^* = -x^*g^* = -xg = gx$, as required. Also, $(S \circ *)^2(x) = (S \circ *)(-gx) = S(gx) = -gxg = x$, etc. That $\mathbb{Z}'/_n$ is antireal follows at once when we remember that $(S \otimes S)\mathcal{R} = \mathcal{R}$ and that $*$ acts on

g in the same way as S, except that it is antilinear on the phase factor. For Example 2.1.7 we note in the first case that $(xp)^* = p^*x^* = px = x(1-p)$ so that $(xp)^* \otimes (xp)^* - x^* \otimes (xp)^* = x(1-p) \otimes x(1-p) - x \otimes x(1-p) = xp \otimes xp - xp \otimes x$, while the other terms of $\mathcal{R}$ are unchanged. For α imaginary, $x^*, (xp)^*$ acquire an extra $\imath$. Hence in both cases we have $\mathcal{R}^{*\otimes *} = \tau(\mathcal{R})$. ∎

We conclude with some other elementary examples. The first two are equivalent in view of the Fourier isomorphism $k(G) \cong k\hat{G}$ for Abelian groups in Corollary 1.5.6. The first is a generalisation of Example 2.1.6.

Example 2.1.17 *Let G be a finite Abelian group and let kG be its group algebra, as in Example 1.5.3. A quasitriangular structure for kG is equivalent to a function $\mathcal{R}$ on $G \times G$ such that*

$$\sum_{cd=v} \mathcal{R}(u,c)\mathcal{R}(w,d) = \delta_{u,w}\mathcal{R}(u,v), \quad \sum_{cd=u} \mathcal{R}(c,v)\mathcal{R}(d,w) = \delta_{v,w}\mathcal{R}(u,v),$$

$$\sum_v \mathcal{R}(u,v) = \delta_{u,e} = \sum_v \mathcal{R}(v,u),$$

for all u, v, w in G, and with e the identity element. This means that a quasitriangular structure for kG corresponds to the Fourier transform of a bicharacter on $\hat{G}$.

Proof: We write an element $\mathcal{R} \in kG \otimes kG$ as $\mathcal{R} = \sum_{u,v} \mathcal{R}(u,v)u \otimes v$. To examine axiom (2.1), we compute $(\Delta \otimes \mathrm{id})\mathcal{R} = \sum \mathcal{R}(u,v)u \otimes u \otimes v$ while $\mathcal{R}_{13}\mathcal{R}_{23} = \sum \mathcal{R}(u,c)\mathcal{R}(w,d)u \otimes w \otimes cd$, giving the stated form for the condition. Similarly for the other half. The condition (2.2) is automatic since kG is assumed to be both cocommutative and commutative. The remaining conditions correspond to (2.3) which, given (2.1), are equivalent to invertibility of $\mathcal{R}$. Note that when $k = \mathbb{C}$ we have $u^* = u^{-1}$ for the natural $*$-structure. Hence $\mathcal{R}^{*\otimes *}(u,v) = \overline{\mathcal{R}(u^{-1}, v^{-1})}$ so that the reality property of $\mathcal{R}$ can also be discussed in terms of these coefficients $\mathcal{R}(u,v)$. ∎

Example 2.1.18 *Let G be a finite Abelian group and let $k(G)$ be its function Hopf algebra, as in Example 1.5.2. Then a quasitriangular structure for $k(G)$ means a function $\mathcal{R} \in H \otimes H$ obeying*

$$\mathcal{R}(uv, w) = \mathcal{R}(u,w)\mathcal{R}(v,w), \quad \mathcal{R}(u, vw) = \mathcal{R}(u,v)\mathcal{R}(u,w),$$

$$\mathcal{R}(u,e) = 1 = \mathcal{R}(e,u),$$

for all u, v, w in G, and with e the identity element. This means that a quasitriangular structure for $k(G)$ means precisely a bicharacter of G.

Proof: As in describing Example 1.5.2, we identify $k(G) \otimes k(G)$ with functions on $G \times G$, with pointwise multiplication. Using the coproduct given in Example 1.5.2, we have at once that (2.1) corresponds to the first two displayed equations. Axiom (2.2) becomes $vu\mathcal{R}(u,v) = \mathcal{R}(u,v)uv$ and so is automatic because the group is Abelian. Given these first two of the stated conditions, the latter two hold *iff* $\mathcal{R}$ is invertible, i.e. nowhere vanishing. Note that when $k = \mathbb{C}$ there is a natural *-structure given by $(f^*)(u) = \overline{f(u)}$. Hence $\mathcal{R}^{* \otimes *}(u,v) = \overline{\mathcal{R}(u,v)}$, so we can have real or antireal quasitriangular structures depending on the reality properties of $\mathcal{R}$ as a function. ∎

Example 2.1.19 *Let $U(1)$ be the universal enveloping algebra of the one-dimensional Lie algebra, as in Example 1.5.7 (with trivial Lie bracket). Its generators are $1, \xi$, with $\Delta\xi = \xi \otimes 1 + 1 \otimes \xi$. We extend this to a Hopf algebra over the ring $\mathbb{C}[[t]]$ of formal power series in a parameter t. In addition to the trivial quasitriangular structure, we have*

$$\mathcal{R} = e^{t\xi \otimes \xi}$$

This is the quantum line *quasitriangular Hopf algebra, denoted $U_t(1)$, or $U_q(1)$ when $q = e^{\frac{t}{2}}$. It forms a real quasitriangular Hopf *-algebra $U_t(\mathbb{R})$ or $U_q(\mathbb{R})$, with $\xi^* = \xi$ and $t^* = t$. It forms an antireal one with $\xi^* = \xi$ and $t^* = -t$.*

Proof: The verification of the axioms is quite straightforward. For example, $(\Delta \otimes \mathrm{id})\mathcal{R} = e^{t(\Delta \otimes \mathrm{id})\xi \otimes \xi} = e^{t(\xi \otimes 1 \otimes \xi + 1 \otimes \xi \otimes \xi)} = \mathcal{R}_{13}\mathcal{R}_{23}$, etc., while (2.2) is automatic because the Hopf algebra is commutative and cocommutative. Note that another *-structure is $\xi^* = -\xi$, which is then precisely an '$n = \infty$' analogue of Example 2.1.6. ∎

This is a variant of the first formulation in Example 2.1.17, but based on the enveloping algebra of $\mathbb{R}$ rather than on a group algebra. Meanwhile, the formulation in Example 2.1.18 in terms of function algebras is convenient for insertion into the functional-analytic setting of Chapter 1.7. Thus, let t be imaginary. Recalling that $L^\infty(\mathbb{R})$ is the usual Hopf–von Neumann or Kac algebra of bounded functions on $\mathbb{R}$ (acting by multiplication on $L^2(\mathbb{R})$), we define $L_t^\infty(\mathbb{R})$ to be this same Hopf–von Neumann algebra but with the nonstandard antireal quasitriangular structure

$$\mathcal{R}(x,y) = e^{txy}.$$

It can be called the *quantum line Kac algebra*. In fact, this formulation is not fundamentally different from $U_t(\mathbb{R})$ when one bears in mind that the functions on $\mathbb{R}$ are in some sense 'generated' by the tautological coordinate function $\xi(x) = x$. The coproduct in $L_t^\infty(\mathbb{R})$ is $(\Delta\xi)(x, y) = \xi(x + y) = (\xi \otimes 1 + 1 \otimes \xi)(x, y)$. Clearly these variants are all based on the same idea, and, moreover, they are built on standard Hopf algebras with only the quasitriangular structure nonstandard. We will give rather more novel examples in Chapter 3.

2.2   Dual quasitriangular structures

In this section we describe the dual results to those above. If a quasitriangular Hopf algebra is almost cocommutative up to conjugation, then its dual Hopf algebra should be almost commutative up to 'conjugation' in a suitable sense. Whereas the former might typically be more suitable in the role of 'quantum symmetry' as a generalisation of a group algebra or enveloping algebra, the latter is more suitable as a generalisation of the function algebra on a group.

In principle, all that we have to say in this section can be obtained by dualising along the lines in Chapter 1.4, by writing out the axioms of a quasitriangular Hopf algebra as diagrams and then reversing all the arrows (and making a left–right reflection) in the usual way. On the other hand, the resulting axioms and the ways of working with these almost-commutative Hopf algebras look somewhat different in practice, so it is surely worthwhile to write them out explicitly in this dual form.

We should also stress that, while the axioms are dual, we are not limiting ourselves to finite-dimensional Hopf algebras or bialgebras. In this case, the two set-ups are not perfectly equivalent, and for some examples the present dual formulation certainly has some technical advantages.

Unlike the Hopf algebra axioms themselves, the axioms of a quasitriangular Hopf algebra are not, of course, self-dual. Thus, an ordinary quasitriangular Hopf algebra H comes equipped with an invertible element $\mathcal{R}$ in $H \otimes H$, which we think of as a map $k \to H \otimes H$, so in the dual formulation we need an 'invertible' map $A \otimes A \to k$, where A will be used to refer to our Hopf algebra in the dual formulation. We can imagine it as dual to H, or H as dual to A, but our goal is to eliminate H itself from the axioms and refer everything to A.

Our first task is to explain in what sense the map is 'invertible'. The sense we need is in terms of the algebra structure on the dual of $A \otimes A$. This algebra structure (corresponding to $H \otimes H$) comes from the coproduct on $A \otimes A$ and is the *convolution algebra* of maps $\mathrm{Hom}(A \otimes A, k)$. More generally, if C is any coalgebra and B is any algebra, then $\mathrm{Hom}(C, B)$ has a convolution algebra structure

$$(\phi\psi)(c) = \sum \phi(c_{(1)}) \cdot_B \psi(c_{(2)}), \qquad 1(c) = \eta_B\epsilon(c), \qquad (2.5)$$

where the product and unit maps on the right are in B. We will be making use of this from time to time. We have already used it in defining the algebra dual to a coalgebra in Chapter 1.2, with $B = k$. Now we use it with $C = A \otimes A$ and $B = k$. Explicitly then, $\mathcal{R} : A \otimes A \to k$ should be invertible in $\mathrm{Hom}(A \otimes A, k)$ in the sense that there exists a map $\mathcal{R}^{-1} : A \otimes A \to k$ such that

$$\sum \mathcal{R}^{-1}(a_{(1)} \otimes b_{(1)}) \mathcal{R}(a_{(2)} \otimes b_{(2)}) = \epsilon(a)\epsilon(b)$$
$$= \sum \mathcal{R}(a_{(1)} \otimes b_{(1)}) \mathcal{R}^{-1}(a_{(2)} \otimes b_{(2)}).$$

Keeping such considerations in mind, it is easy to dualise the remainder of Drinfeld's axioms to obtain the following definition.

Definition 2.2.1 *A dual* quasitriangular *bialgebra or Hopf algebra* $(A, \mathcal{R})$ *is a bialgebra or Hopf algebra A and a convolution-invertible map $\mathcal{R}$: $A \otimes A \to k$ such that*

$$\mathcal{R}(ab \otimes c) = \sum \mathcal{R}(a \otimes c_{(1)}) \mathcal{R}(b \otimes c_{(2)}),$$
$$\mathcal{R}(a \otimes bc) = \sum \mathcal{R}(a_{(1)} \otimes c) \mathcal{R}(a_{(2)} \otimes b), \tag{2.6}$$

$$\sum b_{(1)} a_{(1)} \mathcal{R}(a_{(2)} \otimes b_{(2)}) = \sum \mathcal{R}(a_{(1)} \otimes b_{(1)}) a_{(2)} b_{(2)} \tag{2.7}$$

for all $a, b, c \in A$.

This looks a little unfamiliar, but is in fact obtained by replacing the multiplication in Definition 2.1.1 by the convolution product, and the comultiplication by the product in A. Axiom (2.7) is the dual of (2.2), and says, as promised, that A is almost commutative – up to $\mathcal{R}$. Axioms (2.6) are the dual of (2.1) and say that $\mathcal{R}$ is a 'bialgebra bicharacter'. They should be compared with Example 2.1.18 as we explain below. We also have analogues of the various results in Section 2.1. Again, the new language is perhaps unfamiliar so we give some of the proofs in this dual form in detail.

Lemma 2.2.2 *If* $(A, \mathcal{R})$ *is a dual quasitriangular bialgebra, then*

$$\mathcal{R}(a \otimes 1) = \epsilon(a) = \mathcal{R}(1 \otimes a). \tag{2.8}$$

If A is a Hopf algebra, then, in addition,

$$\mathcal{R}(Sa \otimes b) = \mathcal{R}^{-1}(a \otimes b), \qquad \mathcal{R}^{-1}(a \otimes Sb) = \mathcal{R}(a \otimes b) \tag{2.9}$$

and $\mathcal{R}(Sa \otimes Sb) = \mathcal{R}(a \otimes b)$.

Proof: We have that $\mathcal{R}(a \otimes 1) = (\mathcal{R}^{-1}(a_{(1)} \otimes 1) \mathcal{R}(a_{(2)} \otimes 1)) \mathcal{R}(a_{(3)} \otimes 1) = \mathcal{R}^{-1}(a_{(1)} \otimes 1)(\mathcal{R}(a_{(2)} \otimes 1) \mathcal{R}(a_{(3)} \otimes 1)) = \mathcal{R}^{-1}(a_{(1)} \otimes 1) \mathcal{R}(a_{(2)} \otimes 1.1) = \epsilon(a)$

(using (2.6)). Likewise on the other side. Also, if $\mathcal{R}^{-1}$ exists, it is unique. Hence for A a Hopf algebra it is given by $\mathcal{R}^{-1}(a \otimes b) = \mathcal{R}(Sa \otimes b)$ (use axioms (2.6)). In this case, $a \otimes b \mapsto \mathcal{R}(Sa \otimes Sb)$ is convolution-inverse to $\mathcal{R}^{-1}$ because $\mathcal{R}(Sa_{(1)} \otimes Sb_{(1)}) \, \mathcal{R}(Sa_{(2)} \otimes b_{(2)}) = \mathcal{R}(Sa \otimes (Sb_{(1)})b_{(2)}) = \mathcal{R}(Sa \otimes 1)\epsilon(b) = \epsilon(a)\epsilon(b)$, etc. Hence, $\mathcal{R}(Sa \otimes Sb) = \mathcal{R}(a \otimes b)$, proving the other side and the final remark in the statement. Note from this proof that if $\mathcal{R}$ is a linear map obeying (2.8) and (2.6) and if A is a Hopf algebra, we can use (2.9) as a definition of $\mathcal{R}^{-1}$. ∎

Lemma 2.2.3 *If $(A, \mathcal{R})$ is a dual quasitriangular bialgebra, then*

$$\sum \mathcal{R}(a_{(1)} \otimes b_{(1)})\mathcal{R}(a_{(2)} \otimes c_{(1)})\mathcal{R}(b_{(2)} \otimes c_{(2)})$$
$$= \sum \mathcal{R}(b_{(1)} \otimes c_{(1)})\mathcal{R}(a_{(1)} \otimes c_{(2)})\mathcal{R}(a_{(2)} \otimes b_{(2)}) \tag{2.10}$$

for all a, b, c in A.

Proof: We apply the second of (2.6), (2.7) and the second of (2.6) again, so
$(\mathcal{R}(a_{(1)} \otimes b_{(1)})\mathcal{R}(a_{(2)} \otimes c_{(1)})) \, \mathcal{R}(b_{(2)} \otimes c_{(2)}) = \mathcal{R}(a \otimes c_{(1)}b_{(1)})\mathcal{R}(b_{(2)} \otimes c_{(2)}) = \mathcal{R}(b_{(1)} \otimes c_{(1)})\mathcal{R}(a \otimes b_{(2)}c_{(2)}) = \mathcal{R}(b_{(1)} \otimes c_{(1)}) \, (\mathcal{R}(a_{(1)} \otimes c_{(2)})\mathcal{R}(a_{(2)} \otimes b_{(2)}))$. ∎

Proposition 2.2.4 *Let $(A, \mathcal{R})$ be a dual quasitriangular Hopf algebra. Let $v : A \to k$ and $v^{-1} : A \to k$ be defined by*

$$v(a) = \sum \mathcal{R}(a_{(1)} \otimes Sa_{(2)}), \qquad v^{-1}(a) = \sum \mathcal{R}(S^2 a_{(1)} \otimes a_{(2)}).$$

Then $\sum a_{(1)} v(a_{(2)}) = \sum v(a_{(1)})S^2 a_{(2)}$, and v^{-1} is the inverse of v in the convolution algebra $\mathrm{Hom}(A, k)$. One can likewise define

$$u(a) = \sum \mathcal{R}(a_{(2)} \otimes Sa_{(1)}), \qquad u^{-1}(a) = \sum \mathcal{R}(S^2 a_{(2)} \otimes a_{(1)})$$

obeying $\sum u(a_{(1)})a_{(2)} = \sum S^2 a_{(1)} u(a_{(2)})$. The antipode of A is bijective.

Proof: Again, this is just the dual of Proposition 2.1.8. We will concentrate on v rather than u for a change. (i) To prove the first assertion, we compute

$$a_{(1)} v(a_{(2)}) = a_{(1)} \mathcal{R}(a_{(2)} \otimes Sa_{(3)})$$
$$= (S(Sa_{(3)})_{(1)})(Sa_{(3)})_{(2)} a_{(1)} \mathcal{R}(a_{(2)} \otimes (Sa_{(3)})_{(3)})$$
$$= (S(Sa_{(3)})_{(1)})a_{(2)}(Sa_{(3)})_{(3)} \mathcal{R}(a_{(1)} \otimes (Sa_{(3)})_{(2)})$$
$$= \mathcal{R}(a_{(1)} \otimes Sa_{(4)})(S^2 a_{(5)})a_{(2)} Sa_{(3)} = v(a_{(1)})S^2 a_{(2)}.$$

Here we inserted $\epsilon(b) = (Sb_{(1)})b_{(2)}$ (where $b = Sa_{(3)}$) and then used axiom (2.7) to change the order. (ii) To prove that v^{-1} is the inverse, we first note that $\mathcal{R}(a_{(1)} \otimes a_{(3)}) v(a_{(2)}) = \mathcal{R}(a_{(1)} \otimes a_{(4)}) \mathcal{R}(a_{(2)} \otimes Sa_{(3)}) = \mathcal{R}(a_{(1)} \otimes (Sa_{(2)}) a_{(3)}) = \epsilon(a)$. We used (2.6). Hence using (i) we have

$$\epsilon(a) = \mathcal{R}(a_{(1)} \otimes a_{(3)}) v(a_{(2)}) = v(a_{(1)}) \mathcal{R}(S^2 a_{(2)} \otimes a_{(3)}) = v(a_{(1)}) v^{-1}(a_{(2)})$$

$$= v(a_{(2)}) \mathcal{R}(S^2 a_{(1)} \otimes S^2 a_{(3)}) = \mathcal{R}(S^2 a_{(1)} \otimes a_{(2)}) v(a_{(3)}) = v^{-1}(a_{(1)}) v(a_{(2)}).$$

We used for the latter that $\mathcal{R}(S^2 a \otimes S^2 b) = \mathcal{R}(a \otimes b)$. (iii) To prove that S is invertible, we define $S^{-1}(a) = Sa_{(2)} v(a_{(1)}) v^{-1}(a_{(3)})$ and verify from (i) and (ii) that

$$(S^{-1} a_{(2)}) a_{(1)} = (Sa_{(3)}) a_{(1)} v(a_{(2)}) v^{-1}(a_{(4)})$$
$$= (Sa_{(3)})(S^2 a_{(2)}) v(a_{(1)}) v^{-1}(a_{(4)}) = v(a_{(1)}) v^{-1}(a_{(2)}) = \epsilon(a),$$
$$a_{(2)} S^{-1} a_{(1)} = v^{-1}(a_{(3)}) a_{(4)} Sa_{(2)} v(a_{(1)})$$
$$= (S^2 a_{(3)}) Sa_{(2)} v(a_{(1)}) v^{-1}(a_{(4)}) = v(a_{(1)}) v^{-1}(a_{(2)}) = \epsilon(a).$$

∎

Example 2.2.5 *Let G be an Abelian group and let kG be its group algebra, as in Example 1.5.3. This is dual quasitriangular iff there is a function $\mathcal{R}: G \times G \to k$ obeying the bicharacter conditions in Example 2.1.18.*

Proof: In the group algebra we can work with group-like elements (these form a basis). On such elements the axioms in (2.6) and (2.7) simplify: simply drop the $_{(1)}, _{(2)}$ suffixes! This immediately reduces to the bimultiplicativity, while invertibility corresponds once again (given this) to (2.8). ∎

This example is fitting because kG is dual to $k(G)$ – it is to be expected then that a dual quasitriangular structure on kG is just the same thing as a quasitriangular structure for $k(G)$, namely, as we have seen, a bicharacter. The result is, however, more transparent from the dual quasitriangular point of view. This example is the reason that we can call $\mathcal{R}$ in (2.6) a bialgebra bicharacter. For a concrete example, we can take $G = \mathbb{Z}_{/n}$ and $\mathcal{R}(a, b) = e^{\frac{2\pi\imath ab}{n}}$ to give a nonstandard dual quasitriangular structure on $\mathbb{C}\mathbb{Z}_{/n}$. These Hopf algebras are dual to $\mathbb{Z}'_{/n}$ in Example 2.1.6, as well as isomorphic by $\mathbb{Z}_{/n}$-Fourier transform (so they are both quasitriangular and dual quasitriangular). These Hopf algebras are also quotients of the dual quasitriangular Hopf algebra $\mathbb{Z}_q$ defined by $G = \mathbb{Z}$ and $\mathcal{R}(a, b) = q^{ab}$, where $q \in k$ is invertible. This works over any k, but over $\mathbb{C}[[t]]$ it is just the dual of the quantum line $U_q(1)$ in Example 2.1.19. We will need this example $\mathbb{Z}_q$ in Chapter 10. It can also be obtained as $GL_q(1)$, an example of the matrix quantum groups in Chapter 4.1. For a more complicated example, we consider the dual of Example 2.1.7.

Example 2.2.6 *Let* $A = \text{span}\{1, f, y, fy\}$ *be the four-dimensional dual quasitriangular Hopf algebra with generators* $1, y, f$ *and relations, coproduct and counit*

$$y^2 = 0, \quad f^2 = 1, \quad yf = -fy,$$

$$\Delta y = y \otimes 1 + f \otimes y, \quad \Delta f = f \otimes f, \quad \epsilon(y) = 0, \quad \epsilon(f) = 1.$$

There is also an antipode $Sy = yf$, $Sf = f$. *The dual quasitriangular structure in the basis* $\{1, f, y, yf\}$ *is the bilinear form*

$$\mathcal{R} = \begin{pmatrix} 1 & 1 & 0 & 0 \\ 1 & -1 & 0 & 0 \\ 0 & 0 & \alpha & -\alpha \\ 0 & 0 & \alpha & \alpha \end{pmatrix}.$$

The Hopf algebra is dual to Example 2.1.7 via the Hopf algebra pairing

$$\langle f, g \rangle = -1, \quad \langle f, x \rangle = 0, \quad \langle y, g \rangle = 0, \quad \langle y, x \rangle = 1.$$

It can also be identified with Example 2.1.7 via $y = x, f = g$, *i.e. this Hopf algebra is self-dual.*

Proof: Let us check first that the stated pairing extends as a Hopf algebra one. The pairing with 1 is defined as ϵ, and the pairing between products of the generators is defined via Δ, each according to Proposition 1.4.2. Using this, we show that $\{1, p, x, xp\}$ and $\{1, -2q, y, -2yq\}$ are mutually dual bases, where $p = \frac{1-g}{2}$ and $q = \frac{1-f}{2}$. For example, $\langle yf, g \rangle = \langle y \otimes f, \Delta g \rangle = \langle y, g \rangle \langle f, g \rangle = 0$ so that $\langle -2yq, p \rangle = 0$, etc., while $\langle yf, x \rangle = \langle y \otimes f, \Delta x \rangle = \langle y, x \rangle \langle f, 1 \rangle + \langle y, g \rangle \langle f, x \rangle = 1$ so that $\langle -2yq, x \rangle = \langle yf, x \rangle - \langle y, x \rangle = 1 - 1 = 0$. On the other hand, $\langle -2yq, xp \rangle = \langle \Delta(y(f - 1)), x \otimes p \rangle = \langle (y \otimes 1 + f \otimes y)(f \otimes f - 1 \otimes 1), x \otimes p \rangle = \langle yf \otimes f, x \otimes p \rangle - \langle y \otimes 1, x \otimes p \rangle = \langle yf \otimes f, x \otimes p \rangle = 1$. Thus, we have a linear pairing defined by taking these bases mutually dual, and this is the only possibility for extending the stated pairing of the generators to a Hopf algebra pairing. In the same way, one can then check that all the remaining pairing relations in Proposition 1.4.2 are satisfied. For example, $\langle y, Sx \rangle = \langle y, xg \rangle = \langle y \otimes 1 + f \otimes y, x \otimes g \rangle = 1 = \langle yf, x \rangle = \langle Sy, x \rangle$, etc. Finally, since H in Example 2.1.7 is quasitriangular, A dual to it must be dual quasitriangular. One computes, for example, $\mathcal{R}(y \otimes yf) = \langle y \otimes yf, \mathcal{R} \rangle = \alpha \langle y \otimes yf, x \otimes x - 2x \otimes xp \rangle = -\alpha$ because the other terms in $\mathcal{R}$ fail to contribute as y has zero pairing with $1, p, xp$. Similarly for its other components. $\blacksquare$

Other aspects of Proposition 2.1.8 can clearly be dualised in just the same way. Thus u, v are now almost multiplicative, corresponding to u, v

in Proposition 2.1.8 almost group-like. Meanwhile, uv^{-1} in the convolution algebra is exactly multiplicative, corresponding to uv^{-1} in Corollary 2.1.9 exactly group-like. The reader can easily figure out in the same way what is meant by a dual ribbon structure, etc. For factorisability, central now will be the map $Q : A \otimes A \to k$ defined by

$$Q(a \otimes b) = \sum \mathcal{R}(b_{(1)} \otimes a_{(1)}) \mathcal{R}(a_{(2)} \otimes b_{(2)}). \tag{2.11}$$

Definition 2.2.7 *A dual quasitriangular Hopf algebra* $(A, \mathcal{R})$ *is dual triangular if* $Q(a \otimes b) = \epsilon(a)\epsilon(b)$. *It is* factorisable *if* Q *is nondegenerate in the sense that* $Q(a \otimes b) = 0$ *for all* a *implies that* $b = 0$. *Equivalently, it is factorisable if* $Q(a \otimes b) = 0$ *for all* b *implies* $a = 0$.

Example 2.2.6 is dual triangular, while $k\mathbb{Z}_{/n}$ with the bicharacter corresponding to Example 2.1.6 is factorisable for n odd and > 1. That we have precisely the right definition of factorisability in the infinite-dimensional case is surely only determined by its usefulness, which we shall see later. At least in the finite-dimensional case it is equivalent to $H = A^*$ factorisable in the sense of Definition 2.1.12. One can also prove invariance of $Q(S(\ ) \otimes (\ ))$ under the adjoint coaction of A on itself. Finally, we come to the interaction with $*$-structures. The relationship between the $*$-structures in A and its dual was given in Definition 1.7.5. This involves the antipode S, but note that $\mathcal{R}$ is invariant under $S \otimes S$.

Definition 2.2.8 *A nontriangular dual quasitriangular structure is* real *if* $\mathcal{R}(a^* \otimes b^*) = \overline{\mathcal{R}(b \otimes a)}$ *(in this case* $Q(a^* \otimes b^*) = \overline{Q(a \otimes b)}$*). It is* antireal *if* $\mathcal{R}(a^* \otimes b^*) = \overline{\mathcal{R}^{-1}(a \otimes b)}$ *(in this case,* $Q(a^* \otimes b^*) = \overline{Q^{-1}(a \otimes b)}$*). These two notions coincide in the triangular case.*

These various definitions and properties of dual quasitriangular Hopf algebras will provide the formalism behind the examples in Chapter 4.

2.3 Cocycles and twisting

In Section 2.1, we described the basic theory of quasitriangular Hopf algebras. In this and the next section we will study some more advanced aspects of the theory related to deformation-quantisation and Hopf algebra cohomology. One could proceed directly to the next chapter on a first reading of the book; the main result needed later is Theorem 2.3.4, which provides a natural way of obtaining new examples of quasitriangular Hopf algebras from old ones by introducing a Hopf algebra 2-cocycle. This is the 'twisting' construction and is due ultimately to V.G. Drinfeld. In the next section we will consider the effect of introducing 3-cocycles also in this context. We will then meet cocycles again in the general theory of Hopf algebra extensions in Chapter 6.3.

Cocycles on groups and Lie algebras can easily be formulated at the level of Hopf algebras as follows. Let H be a bialgebra or Hopf algebra. We let

$$\Delta_i : H^{\otimes n} \to H^{\otimes n+1}, \quad \Delta_i = \text{id} \otimes \cdots \otimes \Delta \otimes \cdots \otimes \text{id}, \qquad (2.12)$$

where Δ is in the ith position. Here $i = 1, \ldots, n$, and we add to this the conventions $\Delta_0 = 1 \otimes (\ )$ and $\Delta_{n+1} = (\ ) \otimes 1$ so that Δ_i are defined for $i = 0, \ldots, n+1$. We define an n-cochain χ to be an invertible element of $H^{\otimes n}$ and we define its *coboundary* as the $n+1$-cochain

$$\partial\chi = \left(\prod_{i=0}^{i \text{ even}} \Delta_i \chi \right) \left(\prod_{i=1}^{i \text{ odd}} \Delta_i \chi^{-1} \right), \qquad (2.13)$$

where the even i run $0, 2, \ldots$, and the odd i run $1, 3, \ldots$, and the products are each taken in increasing order. We also write $\partial\chi \equiv (\partial_+\chi)(\partial_-\chi^{-1})$ for the separate even and odd parts. An n-cocycle *for* a Hopf algebra or bialgebra is an invertible element χ in $H^{\otimes n}$ such that $\partial\chi = 1$. Finally, a cochain or cocycle is *counital* if $\epsilon_i \chi = 1$ for all $\epsilon_i = \text{id} \otimes \cdots \otimes \epsilon \otimes \cdots \otimes \text{id}$.

Example 2.3.1 *Let H be a bialgebra or Hopf algebra. Then a 1-cocycle is an invertible element $\chi \in H$ such that*

$$\chi \otimes \chi = \Delta\chi, \quad i.e. \quad \chi \quad \text{group-like.} \qquad (2.14)$$

It is automatically counital. A 2-cocycle is an invertible element $\chi \in H \otimes H$ such that

$$(1 \otimes \chi)(\text{id} \otimes \Delta)\chi = (\chi \otimes 1)(\Delta \otimes \text{id})\chi. \qquad (2.15)$$

It is counital if $(\epsilon \otimes \text{id})\chi = 1$ or equivalently if $(\text{id} \otimes \epsilon)\chi = 1$. A 3-cocycle is an invertible element $\chi \in H \otimes H \otimes H$ such that

$$(1 \otimes \chi)((\text{id} \otimes \Delta \otimes \text{id})\chi)(\chi \otimes 1) = ((\text{id} \otimes \text{id} \otimes \Delta)\chi)((\Delta \otimes \text{id} \otimes \text{id})\chi). \quad (2.16)$$

It is counital if $(\text{id} \otimes \epsilon \otimes \text{id})\chi = 1 \otimes 1$.

Proof: The cocycle formulae are just the formulae for $\partial\chi = 1$ with the inverses moved to the right hand side in each case. The condition $(\epsilon \otimes \text{id})\chi = 1$ implies $(\text{id} \otimes \epsilon)\chi = 1$ if χ is already a 2-cocycle. This is seen by applying ϵ to the middle factor of the 2-cocycle condition, and vice versa. Similarly, one sees from the 3-cocycle condition that the counitarity condition for ϵ in the middle position implies the counitarity condition for ϵ in the other two positions as well. ∎

Example 2.3.2 *Let G be a finite group and let $k(G)$ be its function Hopf algebra, as in Example 1.5.2. Then a counital 1-cocycle means a group homomorphism to $k - \{0\}$. A counital 2-cocycle for $k(G)$ means a nowhere-zero function χ on $G \times G$ such that*

$$\chi(v, w)\chi(u, vw) = \chi(u, v)\chi(uv, w), \quad \chi(e, u) = 1 = \chi(u, e),$$

for all $u, v, w \in G$. Here e is the group identity; i.e. χ is an element of $Z^2(G)$, a normalised 2-cocycle in the usual sense on a group. A 3-cocycle for $k(G)$ means a nowhere-zero function χ on $G \times G \times G$ such that

$$\chi(v, s, t)\chi(u, vs, t)\chi(u, v, s) = \chi(u, v, st)\chi(uv, s, t), \quad \chi(u, e, v) = 1,$$

for all $u, v, s, t \in G$, and with e the group identity, i.e. a normalised 3-cocycle in $Z^3(G)$ in the usual sense.

Proof: This follows at once from the structure of $k(G)$, as in Example 2.1.18, applied in Example 2.3.1. ∎

In general then, the notions above reduce for $H = k(G)$ to the usual theory of group cocycles. A usual group n-cochain is a pointwise invertible function $\chi : G \times G \cdots \times G \to k$ and has coboundary

$$(\partial\chi)(u_1, u_2, \ldots, u_{n+1}) = \prod_{i=0}^{n+1} \chi(u_1, \ldots, u_i u_{i+1}, \ldots, u_{n+1})^{(-1)^i}$$

where, by convention, the first $i = 0$ factor is $\chi(u_2, \ldots, u_{n+1})$ and the last $i = n + 1$ factor is $\chi(u_1, \ldots, u_n)^{\pm 1}$. One has $\partial^2 = 1$, and the group cohomology $\mathcal{H}^n(G, k)$ is then defined as the group of normalised n-cocycles modulo multiplication by coboundaries of the form $\partial(\ )$. These notions are, moreover, the exponentiation of the relevant notions for Lie algebra cohomology. We will see this at the end of the section.

This formulation of the cohomology of groups in terms of Hopf algebras works just as well for any commutative bialgebra or Hopf algebra. One has $\partial^2 = 1$ and cohomology groups $\mathcal{H}^n(k, H)$. One may also try to apply the same definitions to other Hopf algebras. Even when applied to the noncommutative Hopf algebras kG or $U(\mathfrak{g})$, we will see that the above notions become novel and nontrivial constructions. On the other hand, for these and other noncommutative Hopf algebras, we should be careful because we may not have $\partial^2 = 1$ and the attendant cohomological interpretation. The $\mathcal{H}^1$ and $\mathcal{H}^2$ spaces are, however, well-defined for any bialgebra or Hopf algebra. We concentrate on them now. In fact, $\mathcal{H}^1$ is the same as the 1-cocycles because the 0-cocycles are invertible elements of the field k, and their coboundary from (2.13) is always 1. So, $\mathcal{H}^1$ is the group of invertible group-like elements in a bialgebra or Hopf algebra.

Proposition 2.3.3 *Let H be a bialgebra or Hopf algebra. If $\gamma \in H$ is an invertible element with $\epsilon\gamma = 1$ then $\partial\gamma$ is a counital 2-cocycle for H. We say that it is a* coboundary. *More generally, if χ is a counital 2-cocycle, then*

$$\chi^\gamma = (\partial_+\gamma)\chi(\partial_-\gamma^{-1}) = (\gamma \otimes \gamma)\chi\Delta\gamma^{-1}$$

is also a counital 2-cocycle. We say that it is cohomologous to χ. The non-Abelian cohomology space $\mathcal{H}^2(k, H)$ consists of the counital 2-cocycles in H modulo such transformations.

Proof: One can employ several notations to prove this, all of them with some merit. The simplest is just to check $\partial(\partial\gamma) = 1$ from the definition (2.13) using our usual explicit summation notation for coproducts. Similarly to see that $\partial(\chi^\gamma) = 1$ if $\partial\chi = 1$, and to check the counitarity conditions. We proceed here directly in terms of linear maps and compute

$$(1 \otimes \chi^\gamma)(\mathrm{id} \otimes \Delta)\chi^\gamma$$
$$= (1 \otimes (\gamma \otimes \gamma)\chi\Delta\gamma^{-1})(\mathrm{id} \otimes \Delta)((\gamma \otimes \gamma)\chi\Delta\gamma^{-1})$$
$$= (\gamma \otimes \gamma \otimes \gamma)(1 \otimes \chi\Delta\gamma^{-1})(\mathrm{id} \otimes \Delta)(1 \otimes \gamma\chi\Delta\gamma^{-1})$$
$$= (\gamma \otimes \gamma \otimes \gamma)((1 \otimes \chi)(\mathrm{id} \otimes \Delta)\chi)(\mathrm{id} \otimes \Delta)\Delta\gamma^{-1}$$
$$= (\gamma \otimes \gamma \otimes \gamma)((\chi \otimes 1)(\Delta \otimes \mathrm{id})\chi)(\mathrm{id} \otimes \Delta)\Delta\gamma^{-1}$$
$$= (\gamma \otimes \gamma \otimes 1)(\chi\Delta\gamma^{-1} \otimes 1)((\Delta \otimes \mathrm{id})(\gamma \otimes \gamma)\chi)(\Delta \otimes \mathrm{id})\Delta\gamma^{-1}$$
$$= (\chi^\gamma \otimes 1)(\Delta \otimes \mathrm{id})\chi^\gamma$$

as required for the main result. Another method is to write an algebra for the composition of the face maps Δ_i, namely

$$\Delta_i\Delta_j = \Delta_{j+1}\Delta_i, \quad \forall\, i \leq j.$$

This just expresses the elementary axioms for a bialgebra but is convenient for working with the coboundary $\partial\chi = (\Delta_0\chi)(\Delta_2\chi)(\Delta_1\chi^{-1})(\Delta_3\chi^{-1})$. For example, putting in $\chi = (\Delta_0\gamma)(\Delta_2\gamma)(\Delta_1\gamma^{-1})$ and using this algebra gives $\partial\chi = 1$. ∎

This proposition completes our introduction to Hopf algebra non-Abelian cohomology. We now give the main result of the section.

Theorem 2.3.4 *Let $(H, \mathcal{R})$ be a quasitriangular Hopf algebra and let χ be a counital 2-cocycle. Then there is a new quasitriangular Hopf algebra $(H_\chi, \mathcal{R}_\chi)$ defined by the same algebra and counit, and*

$$\Delta_\chi h = \chi(\Delta h)\chi^{-1}, \quad \mathcal{R}_\chi = \chi_{21}\mathcal{R}\chi^{-1}, \quad S_\chi h = U(Sh)U^{-1}$$

for all $h \in H_\chi$. Here $U = \sum \chi^{(1)}(S\chi^{(2)})$ and is invertible.

Proof: (i) We check first that Δ_χ is coassociative. Thus,

$$(\Delta_\chi \otimes \mathrm{id})\Delta_\chi h = \chi_{12}((\Delta \otimes \mathrm{id})(\chi(\Delta h)\chi^{-1}))\chi_{12}^{-1}$$
$$= \chi_{12}((\Delta \otimes \mathrm{id})\chi)((\Delta \otimes \mathrm{id})\Delta h)((\Delta \otimes \mathrm{id})\chi^{-1})\chi_{12}^{-1},$$
$$(\mathrm{id} \otimes \Delta_\chi)\Delta_\chi h = \chi_{23}(\mathrm{id} \otimes \Delta)(\chi(\Delta h)\chi^{-1})\chi_{23}^{-1}$$
$$= \chi_{23}((\mathrm{id} \otimes \Delta)\chi)((\mathrm{id} \otimes \Delta)\Delta h)((\mathrm{id} \otimes \Delta)\chi^{-1})\chi_{23}^{-1}.$$

We see that (2.15) in Example 2.3.1 is just the condition that ensures that these two expressions are equal, given that Δ is already coassociative. Since conjugation by χ is an algebra automorphism, it is clear that Δ_χ is still an algebra map. (ii) We now verify that $\mathcal{R}_\chi$ is a quasitriangular structure for the bialgebra H_χ. It is evidently invertible as $\chi, \mathcal{R}$ are. We compute

$$(\Delta_\chi \otimes \mathrm{id})\mathcal{R}_\chi = \chi_{12}((\Delta \otimes \mathrm{id})(\tau(\chi)\mathcal{R}\chi^{-1}))\chi_{12}^{-1}$$
$$= \chi_{12}((\Delta \otimes \mathrm{id})\tau(\chi))((\Delta \otimes \mathrm{id})\mathcal{R})((\Delta \otimes \mathrm{id})\chi^{-1})\chi_{12}^{-1}$$
$$= \chi_{12}((\Delta \otimes \mathrm{id})\tau(\chi))\mathcal{R}_{13}\mathcal{R}_{23}((\mathrm{id} \otimes \Delta)\chi^{-1})\chi_{23}^{-1}$$
$$= \chi_{12}((\Delta \otimes \mathrm{id})\tau(\chi))\mathcal{R}_{13}((\mathrm{id} \otimes \tau \circ \Delta)\chi^{-1})\mathcal{R}_{23}\chi_{23}^{-1}$$
$$= \chi_{31}(\chi^{(1)}{}_{(2)} \otimes \chi^{(2)} \otimes \chi^{(1)}{}_{(1)})\mathcal{R}_{13}((\mathrm{id} \otimes \tau \circ \Delta)\chi^{-1})\mathcal{R}_{23}\chi_{23}^{-1}$$
$$= \chi_{31}\mathcal{R}_{13}(\chi^{(1)}{}_{(1)} \otimes \chi^{(2)} \otimes \chi^{(1)}{}_{(2)})((\mathrm{id} \otimes \tau \circ \Delta)\chi^{-1})\mathcal{R}_{23}\chi_{23}^{-1}$$
$$= \chi_{31}\mathcal{R}_{13}\chi_{13}^{-1}\chi_{32}((\mathrm{id} \otimes \tau \circ \Delta)\chi)((\mathrm{id} \otimes \tau \circ \Delta)\chi^{-1})\mathcal{R}_{23}\chi_{23}^{-1}$$
$$= \chi_{31}\mathcal{R}_{13}\chi_{13}^{-1}\chi_{32}\mathcal{R}_{23}\chi_{23}^{-1} = (\mathcal{R}_\chi)_{13}(\mathcal{R}_\chi)_{23}.$$

Here the first equality expresses the definitions of $\Delta_\chi, \mathcal{R}_\chi$; the second is that Δ is an algebra map; while the third uses (2.1) and the 2-cocycle condition on the right. The fourth equality is (2.2). For the fifth we use the 2-cocycle condition, to which we make a cyclic rotation of the factors in $H \otimes H \otimes H$ to give $\chi_{12}(\mathrm{id} \otimes \Delta)\tau(\chi) = \chi_{31}(\chi^{(1)}{}_{(2)} \otimes \chi^{(2)} \otimes \chi^{(1)}{}_{(1)})$, which is the form that we use. The sixth equality is (2.2) again. A further permutation in $H \otimes H \otimes H$ also gives $\chi_{13}(\chi^{(1)}{}_{(1)} \otimes \chi^{(2)} \otimes \chi^{(1)}{}_{(2)}) = \chi_{32}(\mathrm{id} \otimes \tau \circ \Delta)\chi$, which we use for the seventh equality. After cancelling inverses, we obtain $(\mathcal{R}_\chi)_{13}(\mathcal{R}_\chi)_{23}$, as required. The proof on the other side is similar, so that we obtain (2.1) for $\mathcal{R}_\chi$. The axiom (2.2) for $\mathcal{R}_\chi, H_\chi$ is automatic as $\tau \circ \Delta_\chi h = \chi_{21}(\tau \circ \Delta h)\chi_{21}^{-1} = \chi_{21}\mathcal{R}(\Delta h)\mathcal{R}^{-1}\chi_{21}^{-1} = \mathcal{R}_\chi\chi(\Delta h)\chi^{-1}\mathcal{R}_\chi^{-1} = \mathcal{R}_\chi(\Delta_\chi h)\mathcal{R}_\chi^{-1}$. (iii) Finally, if H is a Hopf algebra, we check that S_χ is an antipode for H_χ. To do this, we first establish that U is invertible. We define $U^{-1} = (S\chi^{-(1)})\chi^{-(2)}$, where $\chi^{-1} = \chi^{-(1)} \otimes \chi^{-(2)}$ in our notation, and check that

$$UU^{-1} = \chi^{(1)}(S\chi^{(2)})(S\chi^{-(1)})\chi^{-(2)}$$
$$= \chi'^{-(1)}\chi^{(1)}(S(\chi'^{-(2)}{}_{(1)}\chi^{-(1)}\chi^{(2)}))\chi'^{-(2)}{}_{(2)}\chi^{-(2)}$$
$$= \chi^{-(1)}{}_{(1)}(S\chi^{-(1)}{}_{(2)})\chi^{-(2)} = 1,$$
$$U^{-1}U = (S\chi^{-(1)})\chi^{-(2)}\chi^{(1)}S\chi^{(2)}$$

$$= (S(\chi^{-(1)}\chi'^{(1)}))\chi^{-(2)}\chi^{(1)}\chi'^{(2)}{}_{(1)}S(\chi^{(2)}\chi'^{(2)}{}_{(2)})$$
$$= (S\chi^{(1)}{}_{(1)})\chi^{(1)}{}_{(2)}S\chi^{(2)} = 1.$$

We inserted a copy of χ, denoted χ', in a form that collapses via the antipode properties and $(\mathrm{id} \otimes \epsilon)\chi'^{-1} = 1$. We then applied the 2-cocycle condition in the form $(\Delta \otimes \mathrm{id})\chi^{-1} = ((\mathrm{id} \otimes \Delta)\chi^{-1})\chi_{23}^{-1}\chi_{12}$ and $(\epsilon \otimes \mathrm{id})\chi^{-1} = 1$. Similarly for the inverse from the other side using the 2-cocycle identity in the form $\chi_{12}^{-1}\chi_{23}(\mathrm{id} \otimes \Delta)\chi = (\Delta \otimes \mathrm{id})\chi$. We can now proceed to compute

$$
\begin{aligned}
\cdot(S_\chi \otimes \mathrm{id})\Delta_\chi h &= U(S(\chi^{(1)}h_{(1)}\chi^{-(1)}))U^{-1}\chi^{(2)}h_{(2)}\chi^{-(2)} \\
&= U(S\chi^{-(1)})(Sh_{(1)})(S\chi^{(1)})(S\chi'^{-(1)})\chi'^{-(2)}\chi^{(2)}h_{(2)}\chi^{-(2)} \\
&= U(S\chi^{-(1)})(Sh_{(1)})h_{(2)}\chi^{-(2)} = \epsilon(h)UU^{-1} = \epsilon(h), \\
\cdot(\mathrm{id} \otimes S_\chi)\Delta_\chi h &= \chi^{(1)}h_{(1)}\chi^{-(1)}U(S(\chi^{(2)}h_{(2)}\chi^{-(2)}))U^{-1} \\
&= \chi^{(1)}h_{(1)}\chi^{-(1)}(\chi'^{(1)}S\chi'^{(2)})(S\chi^{-(2)})(Sh_{(2)})(S\chi^{(2)})U^{-1} \\
&= \chi^{(1)}h_{(1)}(Sh_{(2)})(S\chi^{(2)})U^{-1} = UU^{-1}\epsilon(h) = \epsilon(h)
\end{aligned}
$$

using the definitions, the elementary antipode properties from Chapter 1 and $UU^{-1} = 1$.   ■

One can study these elements U, U^{-1} along the same lines as for u, u^{-1} in Proposition 2.1.8. For example, they are not group-like but rather one has

$$\Delta U = \chi^{-1}(U \otimes U)(S \otimes S)\chi_{21}^{-1}, \qquad (2.17)$$

which says, in cohomological terms, that $\chi \sim (S \otimes S)\chi_{21}^{-1}$.   ■

Also let us note that the three parts of the proof of Theorem 2.3.4 are independent, i.e. if H is any bialgebra and χ is a 2-cocycle then H_χ is a bialgebra; if H is a Hopf algebra then so is H_χ; and if H is a quasitriangular bialgebra or Hopf algebra then so is H_χ. We also have a cohomological picture. Recall first that a 1-cocycle for H means an invertible group-like element γ. It is easy to see that it defines an inner automorphism $H \to H$ as bialgebras or Hopf algebras by $h \mapsto \gamma h \gamma^{-1}$, i.e. this is a bialgebra map, and in the Hopf case, a Hopf algebra map. If H is quasitriangular, then this automorphism preserves $\mathcal{R}$ as well.

Proposition 2.3.5 *Let χ, ψ be two 2-cocycles. The Hopf algebras given by twisting by them as in the preceding theorem are isomorphic via an inner automorphism if χ, ψ are cohomologous; i.e. there is a map from $\mathcal{H}^2(k, H)$ to the set of twistings of H up to inner automorphism. In particular, if χ is a coboundary then twisting by it can be undone by an inner automorphism.*

Proof: We suppose that χ, ψ are cohomologous in the sense given in Proposition 2.3.3. By definition, this means $\psi = (\gamma \otimes \gamma)\chi \Delta \gamma^{-1}$ for some invertible element $\gamma \in H$. Then we can write the coproduct Δ_ψ in the form $\Delta_\psi(h) = \psi(\Delta h)\psi^{-1} = (\gamma \otimes \gamma)\chi(\Delta \gamma^{-1})(\Delta h)(\Delta \gamma)\chi^{-1}(\gamma^{-1} \otimes \gamma^{-1}) = (\gamma \otimes \gamma)(\Delta_\chi(\gamma^{-1}h\gamma))(\gamma^{-1} \otimes \gamma^{-1})$. As $\gamma(\ )\gamma^{-1}$ is an inner automorphism of the algebra structure, we see that it defines now a bialgebra isomorphism $H_\psi \to H_\chi$. Hence it is also a Hopf algebra isomorphism in the case where H has an antipode. One can also see this directly from the formulae given for the antipode after twisting. Finally, if H is quasitriangular then $\mathcal{R}_\psi = \psi_{21}\mathcal{R}\psi^{-1} = (\gamma \otimes \gamma)\chi_{21}(\Delta^{\mathrm{op}}\gamma^{-1})\mathcal{R}(\Delta \gamma)\chi^{-1}(\gamma^{-1} \otimes \gamma^{-1}) = (\gamma \otimes \gamma)\mathcal{R}_\chi(\gamma^{-1} \otimes \gamma^{-1})$ using (2.2). So the induced isomorphism maps the quasitriangular structures as well, if these are present. ∎

This says that the process of twisting can only give a genuinely new Hopf algebra if the cocycle χ used to twist is cohomologically nontrivial. For example, if $\mathcal{H}^2$ is trivial for a Hopf algebra, then all twists are isomorphic. On the other hand, this gloomy possibility does not seem to occur very often, and the twisting process does generally enable one to obtain new quasitriangular Hopf algebras from old ones.

Example 2.3.6 *Let H be a quasitriangular Hopf algebra and take for a 2-cocycle $\chi = \mathcal{R}$. Then H_χ is the quasitriangular Hopf algebra H^{cop} with the opposite coproduct, as in Exercise 2.1.3.*

Proof: Note that the quasitriangular structure $\mathcal{R}$ for H is manifestly a 2-cocycle in view of (2.1) and the QYBE in Lemma 2.1.4. The result is then obvious from (2.2). On the other hand, we know from Exercise 1.3.3 that H^{cop} has for its antipode S^{-1}, where S is the antipode of H. Hence, as well as recovering the quasitriangular structure $\mathcal{R}_{21}\mathcal{R}\mathcal{R}^{-1} = \mathcal{R}_{21}$ for H^{cop}, we also recover from Theorem 2.3.4 that $S^{-1}h = U(Sh)U^{-1}$ or $S^{-2}h = UhU^{-1}$. Indeed, $U = v$ in Proposition 2.1.8, and we recover the result stated there as an example of the twisting theorem. $\mathcal{R}_{21}^{-1}$ is another 2-cocycle we could twist by, and in this case we recover the results for $U = u$. ∎

Another application of the twisting theorem is to begin with elementary examples of Hopf algebras such as kG, $U(\mathfrak{g})$, which are trivially quasitriangular with $\mathcal{R} = 1 \otimes 1$, and systematically deform their structure by introducing a 2-cocycle χ, possibly depending on one or more parameters. This deformation can then result in a nontrivial quantum group (one says informally that the initial Hopf algebra is deformation-quantised into a quantum group). As a way of obtaining new quantum groups from ordinary groups or enveloping algebras, this twisting procedure by itself will only generate triangular ones. For if our initial $\mathcal{R}$ is trivial, then after twisting we have $\mathcal{R} = \chi_{21}\chi^{-1}$, so that $\tau(\mathcal{R}^{-1}) = \tau(\chi\chi_{21}^{-1}) = \mathcal{R}$. On the other hand, the entire theory of the triangular quantum groups obtained in this way is governed by χ so that many other structures with which our initial cocommutative Hopf algebra interacted can be systematically deformation-quantised at the same time by introducing χ in their definitions also. This gives, at the very least, a systematic approach to deformation-quantisation.

Of course, the initial Hopf algebra for our twisting in Theorem 2.3.4 does not have to be cocommutative. Instead, one could view the theorem as a kind of twisting-equivalence for quantum groups. Two quasitriangular Hopf algebras may look very different, but, if they are related by twisting, their algebras can be identified, and after this their coproducts differ only by conjugation by a 2-cocycle χ. This in turn means that their properties are very similar and differ only by the effects of the 2-cocycle. For example, it means that all the representations of the two Hopf algebras coincide and their tensor products, according to Exercise 1.6.8, also coincide up to isomorphism (with action by the 2-cocycle χ as an intertwiner). Thus, twisting-equivalence is a powerful concept and, at the same time, provides a completely systematic approach to deformation-quantisation.

If we think of twisting as a kind of equivalence or 'gauge transformation', then it is clear that when we twist a Hopf algebra we also have to twist other structures defined on it or with which it interacts, if we want the analogous relationships to be maintained. Likewise from the deformation-quantisation point of view, if we want the quantisation to respect all relationships. Examples are the concept of $*$-structure from Chapter 1.7 and the concept of a module algebra from Chapter 1.6. Recall that the latter means an algebra on which our Hopf algebra acts covariantly.

Proposition 2.3.7 *In the setting of Theorem 2.3.4, if H is a Hopf $*$-algebra over $\mathbb{C}$ in the sense of Definition 1.7.5 and χ is real in the sense $(S \otimes S)(\chi^{*\otimes*}) = \chi_{21}$, then*

$$*_\chi = (S^{-1}U)((\ )^*)S^{-1}U^{-1} \qquad (2.18)$$

makes H_χ in Theorem 2.3.4 into a Hopf $*$-algebra as well. It is (anti)real quasitriangular whenever H is.

Proof: From the stated 'reality' assumption for χ, we see that $U^* = S^{-2}U$, and hence that $S^{-1}U$ is self-adjoint under $*$. This implies that $(*_\chi)^2 = \mathrm{id}$. It is easy to see similarly that $(S_\chi \circ *_\chi)^2 = \mathrm{id}$. For compatibility with the coproduct, we use the result (2.17) analogous to our computations for u, v in Proposition 2.1.8. Thus, from $\Delta S^{-1}U$ we see that $(*_\chi \otimes *_\chi)(\Delta_\chi h) =$

$(S^{-1}U)\chi^{-(1)*}h^*{}_{(1)}\chi^{(1)*}S^{-1}U^{-1}\otimes(S^{-1}U)\chi^{-(2)*}h^*{}_{(2)}\chi^{(2)*}S^{-1}U^{-1}$ is equal to $\chi^{(1)}(S^{-1}U)_{(1)}h^*{}_{(1)}(S^{-1}U^{-1})_{(1)}\chi^{-(1)}\otimes\chi^{(2)}(S^{-1}U)_{(2)}h^*{}_{(2)}(S^{-1}U^{-1})_{(2)}\chi^{-(2)}=$ $\Delta_\chi\circ*_\chi(h)$, as required. If H is real quasitriangular, then $(*_\chi\otimes*_\chi)(\mathcal{R}_\chi)=$ $(S^{-1}U\otimes S^{-1}U)\chi^{-1*\otimes*}\mathcal{R}_{21}\chi_{21}^{*\otimes}{}^*(S^{-1}U^{-1}\otimes S^{-1}U^{-1})=\chi(\Delta S^{-1}U)\mathcal{R}_{21}(\tau$ $\circ\Delta S^{-1}U^{-1})\chi_{21}^{-1}=(\mathcal{R}_\chi)_{21}$ using (2.17) again, and the quasitriangularity assumption. Similarly if $\mathcal{R}$ is antireal. ∎

Proposition 2.3.8 *Let χ be a counital 2-cocycle for a bialgebra or Hopf algebra H. If B is an algebra on which H acts covariantly in the sense of (1.8) then*

$$b\cdot_\chi c=\cdot\left(\chi^{-1}\triangleright(b\otimes c)\right),$$

for all $b,c\in B$, defines a new associative algebra B_χ, which is covariant in the same sense under the Hopf algebra H_χ from Theorem 2.3.4. Likewise, if C is a coalgebra covariant under H in the sense of (1.9) then the new coalgebra C_χ with coproduct $\Delta_\chi=\chi\triangleright\Delta$ is covariant under H_χ.

Proof: Associativity follows from the 2-cocycle condition on χ. We use the same unit. To see the covariance of the new system, one need only to check that $h\triangleright(b\cdot_\chi c)=h\triangleright\cdot(\chi^{-1}\triangleright(b\otimes c))=\cdot((\Delta h)\chi^{-1}\triangleright(b\otimes c))=$ $\cdot(\chi^{-1}(\Delta_\chi h)\triangleright(b\otimes c))=\cdot_\chi((\Delta_\chi h)\triangleright(b\otimes c))$, which is automatic. The proof for the coalgebra case is similar: coassociativity follows from the 2-cocycle condition, the counit is the same, and covariance under H_χ is automatic. ∎

For completeness, we now give the dual version of all these constructions. The methods used to obtain these are the same as those used in Section 2.2 to obtain the theory of dual quasitriangular Hopf algebras. Thus, an n-cochain *on* a bialgebra or Hopf algebra H is a linear functional $\chi:H^{\otimes n}\to k$ which is invertible in the convolution algebra. It is *unital* if $\chi(a\otimes\cdots\otimes1\otimes\cdots\otimes d)=\epsilon(a)\cdots\epsilon(d)$, for 1 in any position. The coboundary $\partial\chi$ is an $n+1$-cochain of the same form as (2.13) but with the product in the convolution algebra,

$$\partial\chi=\left(\prod_{i=0}^{i\text{ even}}\chi\circ\cdot_i\right)\left(\prod_{i=1}^{i\text{ odd}}\chi^{-1}\circ\cdot_i\right),$$

where $\cdot_i$ is the map that multiplies H in the $i,i+1$ positions. The convention is $\chi\circ\cdot_0=\epsilon\otimes\chi$ and $\chi\circ\cdot_{n+1}=\chi\otimes\epsilon$. The first two coboundaries are, explicitly,

$$\partial\chi(a\otimes b)=\sum\chi(b_{(1)})\chi(a_{(1)})\chi^{-1}(a_{(2)}b_{(2)}),$$

$$\partial\chi(a\otimes b\otimes c) = \sum\chi(b_{(1)}\otimes c_{(1)})\chi(a_{(1)}\otimes b_{(2)}c_{(2)})$$
$$\chi^{-1}(a_{(2)}b_{(3)}\otimes c_{(3)})\chi^{-1}(a_{(3)}\otimes b_{(4)}),$$

etc. The corresponding cocycle conditions and the 3-cocycle condition, which we will need later, are:

$$\chi(ab) = \chi(a)\chi(b), \tag{2.19}$$

$$\sum\chi(b_{(1)}\otimes c_{(1)})\chi(a\otimes b_{(2)}c_{(2)}) = \sum\chi(a_{(1)}\otimes b_{(1)})\chi(a_{(2)}b_{(2)}\otimes c), \tag{2.20}$$

$$\sum\chi(b_{(1)}\otimes c_{(1)}\otimes d_{(1)})\chi(a_{(1)}\otimes b_{(2)}c_{(2)}\otimes d_{(2)})\chi(a_{(2)}\otimes b_{(3)}\otimes c_{(3)})$$
$$= \sum\chi(a_{(1)}\otimes b_{(1)}\otimes c_{(1)}d_{(1)})\chi(a_{(2)}b_{(2)}\otimes c_{(2)}\otimes d_{(2)}), \tag{2.21}$$

for all $a, b, c, d \in H$. For cocommutative Hopf algebras, we have $\partial^2 = \epsilon\otimes\cdots\otimes\epsilon$, the trivial cochain, and cohomology groups $\mathcal{H}^n(H, k)$ for cocycles on H modulo convolution with coboundaries. It is easy to see that, in the finite-dimensional case, a cocycle *on* H is nothing other than a cocycle *for* H^* in the previous sense, i.e. the difference is one of notation. On the other hand, the theory of cocycles on Hopf algebras works just as well in the infinite-dimensional case and can have some advantages. For example, the coboundary for cocycles on the enveloping Hopf algebra $U(g)$ of a Lie algebra g reduces on the generators to the usual coboundary operator for Lie algebra cocycles after putting in the linear form $\Delta\xi = \xi\otimes 1 + 1\otimes\xi$ from Example 1.5.7. We recall that a Lie algebra n-cochain is a totally antisymmetric linear map $\chi : g^{\otimes n} \to k$ with coboundary

$$\partial\chi(\xi_1,\ldots,\xi_{n+1}) = \sum_{1\le i<j\le n+1}(-1)^{i+j}\chi([\xi_i,\xi_j],\xi_1,\ldots,\hat{\xi_i},\ldots,\hat{\xi_j},\ldots,\xi_{n+1}) \tag{2.22}$$

where ˆ denotes omission. One has $\partial^2 = 0$ and the Lie algebra cohomology $\mathcal{H}^n(g, k)$ defined as n-cocycles in the sense $\partial\xi = 0$ modulo coboundaries $\partial(\ )$. This completes the task of showing that our Hopf algebra formulation matches what we usually mean by cocycles in the familiar cases. We will return to Lie algebra cohomology in Chapter 8.

Finally, analogous results for twisting etc. apply in this dual formulation. Thus if χ is a unital 2-cocycle on a dual quasitriangular Hopf algebra A, we obtain another one by twisting its product, dual quasitriangular structure and antipode according to

$$a\cdot_\chi b = \sum\chi(a_{(1)}\otimes b_{(1)})a_{(2)}b_{(2)}\chi^{-1}(a_{(3)}\otimes b_{(3)}), \tag{2.23}$$

$$\mathcal{R}_\chi(a\otimes b) = \sum\chi(b_{(1)}\otimes a_{(1)})\mathcal{R}(a_{(2)}\otimes b_{(2)})\chi^{-1}(a_{(3)}\otimes b_{(3)}), \tag{2.24}$$

$$S_\chi(a) = \sum U(a_{(1)})Sa_{(2)}U^{-1}(a_{(3)}), \quad U(a) = \sum\chi(a_{(1)}\otimes Sa_{(2)}). \tag{2.25}$$

We merely dualised Theorem 2.3.4, for the benefit of the reader. The proof of this form is strictly analogous. If the proof is considered diagrammatically by composing maps, one need only reverse all arrows. Likewise, if A is a Hopf $*$-algebra over $\mathbb{C}$ and χ is real in the sense $\overline{\chi(a \otimes b)} = \chi((S^2 b)^* \otimes (S^2 a)^*)$, then

$$a^{*_\chi} = \sum \vartheta^{-1}(a_{(1)}) a_{(2)}{}^* \vartheta(a_{(3)}); \quad \vartheta(a) = \sum U^{-1}(a_{(1)}) U(S^{-1} a_{(2)}) \quad (2.26)$$

makes A_χ a Hopf $*$-algebra as well. It is (anti)real dual quasitriangular if A is. This is the dual of Proposition 2.3.7. Finally, if B is an algebra on which A coacts covariantly from the right in the sense of (1.13) from Chapter 1.6, then

$$c \cdot_\chi d = \sum c^{(\bar{1})} d^{(\bar{1})} \chi^{-1}(c^{(\bar{2})} \otimes d^{(\bar{2})}), \quad (2.27)$$

for all $c, d \in B$, defines a new algebra B_χ coacted upon covariantly by A_χ. If C is a coalgebra on which A coacts covariantly as in (1.14), then

$$\Delta_\chi c = \sum c_{(1)}{}^{(\bar{1})} \otimes c_{(2)}{}^{(\bar{1})} \chi(c_{(1)}{}^{(\bar{2})} \otimes c_{(2)}{}^{(\bar{2})}) \quad (2.28)$$

is a new coalgebra C_χ, coacted upon covariantly by A_χ. This is the dual of Proposition 2.3.8. We will see a generalisation of these ideas in Chapter 6.3, to cocycle cross (co)products.

2.4   Quasi-Hopf algebras

This section is included for completeness because it follows on naturally from the preceding section. It forms, as we shall see, V.G. Drinfeld's motivation for introducing the theory of twisting. The idea is that, as well as relaxing cocommutativity up to conjugation, we can also relax the coassociativity of Δ up to conjugation. This then gives what Drinfeld has called a quasitriangular *quasi-Hopf* algebra.

Thus, a quasitriangular quasi-Hopf algebra means $(H, \Delta, \epsilon, S, \alpha, \beta, \phi, \mathcal{R})$ where first of all H is a unital algebra and $\Delta : H \to H \otimes H$ is an algebra homomorphism such that

$$(\mathrm{id} \otimes \Delta) \circ \Delta = \phi((\Delta \otimes \mathrm{id}) \circ \Delta(\ ))\phi^{-1}. \quad (2.29)$$

The axioms for the counit ϵ are as usual. The element $\phi \in H \otimes H \otimes H$ that controls the nonassociativity is invertible and is required to be a counital 3-cocycle in the sense of Example 2.3.1. Thus, $(H, \Delta, \epsilon, \phi)$ defines a quasibialgebra. Next, $\mathcal{R} \in H \otimes H$ is invertible and still intertwines the coproduct and its opposite as in (2.2), but the other two axioms of a quasitriangular structure are modified by ϕ to

$$(\Delta \otimes \mathrm{id})\mathcal{R} = \phi_{312} \mathcal{R}_{13} \phi_{132}^{-1} \mathcal{R}_{23} \phi, \quad (\mathrm{id} \otimes \Delta)\mathcal{R} = \phi_{231}^{-1} \mathcal{R}_{13} \phi_{213} \mathcal{R}_{12} \phi^{-1} \quad (2.30)$$

in the usual notation. Explicitly, if $\phi = \sum \phi^{(1)} \otimes \phi^{(2)} \otimes \phi^{(3)}$ then $\phi_{213} = \sum \phi^{(2)} \otimes \phi^{(1)} \otimes \phi^{(3)}$, etc. So far, we have defined a quasitriangular quasibialgebra. Finally, the antipode required for a quasi-Hopf algebra consists of elements $\alpha, \beta \in H$ and $S : H \to H$ obeying

$$\sum (Sh_{(1)})\alpha h_{(2)} = \epsilon(h)\alpha, \quad \sum h_{(1)}\beta Sh_{(2)} = \epsilon(h)\beta, \quad \forall h \in H, \quad (2.31)$$

$$\sum \phi^{(1)}\beta(S\phi^{(2)})\alpha\phi^{(3)} = 1, \quad \sum (S\phi^{-(1)})\alpha\phi^{-(2)}\beta S\phi^{-(3)} = 1 \quad (2.32)$$

and is determined uniquely up to a transformation $\alpha \mapsto U\alpha$, $\beta \mapsto \beta U^{-1}$, $Sh \mapsto U(Sh)U^{-1}$, for any invertible $U \in H$. These axioms are due to Drinfeld. They can also be motivated from the representation theory, where the antipode corresponds to the existence of dual representations. We return to this in Chapter 9.3.

Example 2.4.1 *Let H be an ordinary Hopf algebra, and let ϕ be 'ad-invariant' in the sense of (2.29) and obey the counital 3-cocycle condition. Then (H, ϕ) is a quasi-Hopf algebra. The antipode in this case has the same S, and $\beta = 1, \alpha = c^{-1}$, where $c = \sum \phi^{(1)}(S\phi^{(2)})\phi^{(3)}$ is central. We assume that ϕ and c are invertible. Moreover, if $F \in H \otimes H$ is invertible and 'ad-invariant' in the sense $(\Delta h)F = F\Delta h$ for all $h \in H$, then $\phi = \partial F$ is an ad-invariant 3-cocycle of the type required.*

Proof: Note that, in general (when H is noncocommutative), the ad-invariance conditions we use here are not the same as the requirement of invariance under the quantum adjoint action from Example 1.6.3 extended to tensor powers. This is the reason for the quotation marks in the statement. It is clear that if (2.29) does hold, and ϕ is a counital 3-cocycle, then we have a quasibialgebra. For the quasi-Hopf algebra structure, we have to verify the antipode and the 3-cocycle condition. First, the ad-invariance condition (2.29) means that

$$h\phi^{(1)}(S\phi^{(2)})\phi^{(3)} = h_{(1)}\phi^{(1)}(S\phi^{(2)})(Sh_{(2)})h_{(3)}\phi^{(3)}$$
$$= (h_{(1)}\phi^{(1)})(S(h_{(2)}\phi^{(2)}))h_{(3)}\phi^{(3)} = (\phi^{(1)}h_{(1)})(S(\phi^{(2)}h_{(2)}))\phi^{(3)}h_{(3)}$$
$$= \phi^{(1)}(S\phi^{(2)})\phi^{(3)}h,$$

so c as stated is central. Thus we can satisfy (2.31) and the first part of (2.32) with $\alpha = c^{-1}$ and $\beta = 1$. A case of (2.16) then implies that $c = (S\phi^{-(1)})\phi^{-(2)}S\phi^{-(3)}$ also; hence, the second part of (2.32) holds.

Secondly, although the higher cohomology theory for a general Hopf algebra is not so well behaved as in the commutative case, 3-cocycles are not too bad; one can show that if F is ad-invariant in the sense $(\Delta h)F = F\Delta h$, then $\partial^2 F = 1$, i.e. $\phi = \partial F$ is a 3-cocycle as required for this example. The computation is similar to that in Proposition 2.3.5 but more lengthy, and we defer it to the next theorem, of which it is a special case. That ϕ is then ad-invariant is obvious from the definition of ∂F in terms of the $\Delta_i F$, since each of these are ad-invariant. ∎

One can prove the analogues of numerous results familiar for Hopf algebras and quasitriangular Hopf algebras in the quasi-Hopf case, now including conjugation by ϕ and elements built from it. This extension is already evident in the notion of the antipode itself: the more general notion is needed because, in general, α, β cannot be taken to be unity, though they can if ϕ is trivial. In this more general setting we have the following generalisation of Theorem 2.3.4, due to Drinfeld.

Theorem 2.4.2 *Let* $(H, \alpha, \beta, \phi, \mathcal{R})$ *be a quasitriangular quasi-Hopf algebra and let* F *be an arbitrary invertible element of* $H \otimes H$ *such that* $(\epsilon \otimes \mathrm{id})F = 1 = (\mathrm{id} \otimes \epsilon)F$. *Then* H_F, *defined as follows, is also a quasi-Hopf algebra. It has the same algebra and counit as* H, *and*

$$\Delta_F h = F(\Delta h)F^{-1}, \quad \mathcal{R}_F = F_{21}\mathcal{R}F^{-1},$$

$$\phi_F = F_{23}((\mathrm{id} \otimes \Delta)F)\phi((\Delta \otimes \mathrm{id})F^{-1})F_{12}^{-1},$$

$$S_F = S, \quad \alpha_F = \sum (SF^{-(1)})\alpha F^{-(2)}, \quad \beta_F = \sum F^{(1)}\beta S F^{(2)}.$$

Proof: The direct proof is a generalisation of Theorem 2.3.4, involving now ϕ, α, β. We will also see an abstract reason for it Chapter 9.4.1. From a direct point of view, on comparing the first two displayed expressions in the proof of Theorem 2.3.4, we see that, if Δ already fails to be coassociative up to ϕ, then an arbitrary F means that coassociativity of Δ_F also fails, up to a new ϕ_F as stated. The proof of (2.30) for the new $\mathcal{R}_F$ is also similar: where we used the 2-cocycle condition for F in the proof of Theorem 2.3.4, we make now a change from ϕ to ϕ_F. For the antipode, since, in general, we will not be able to keep both α_F, β_F trivial, we may as well keep $S_F = S$ and define α_F and β_F in place of U^{-1}, U. Their direct verification is similar, though more complicated.

Finally, we check that ϕ_F is a 3-cocycle for H_F if ϕ is a 3-cocycle for H. To do this we use the algebraic method mentioned at the end of the proof of Proposition 2.3.3. This time, because H is a quasi-Hopf algebra, the algebra of the face maps Δ_i is

$$\Delta_i\Delta_j = \Delta_{j+1}\Delta_i, \quad i < j, \quad \Delta_i\Delta_i = \phi^{-1}(\Delta_{i+1}\Delta_i)\phi$$

in view of (2.29). The 3-cocycle condition for ϕ is $(\Delta_0\phi)(\Delta_2\phi)(\Delta_4\phi) = (\Delta_3\phi)(\Delta_1\phi)$, and by definition

$$\phi_F = (\partial_+ F)\phi(\partial_- F^{-1}) = (\Delta_0 F)(\Delta_2 F)\phi(\Delta_1 F^{-1})(\Delta_3 F^{-1}).$$

Using this notation, we have

$$((\Delta_F)_0\phi_F)((\Delta_F)_2\phi_F)((\Delta_F)_4\phi_F)$$
$$= (\Delta_0\Delta_0 F)(\Delta_0\Delta_2 F)(\Delta_0\phi)(\Delta_0\Delta_1 F^{-1})(\Delta_0\Delta_3 F^{-1})$$
$$\quad F_{23}(\Delta_2\Delta_0 F)(\Delta_2\Delta_2 F)(\Delta_2\phi)(\Delta_2\Delta_1 F^{-1})(\Delta_2\Delta_3 F^{-1})F_{23}^{-1}$$
$$\quad (\Delta_4\Delta_0 F)(\Delta_4\Delta_2 F)(\Delta_4\phi)(\Delta_4\Delta_1 F^{-1})(\Delta_4\Delta_3 F^{-1})$$
$$= (\Delta_0\Delta_0 F)(\Delta_0\Delta_2 F)(\Delta_0\phi)(\Delta_2\Delta_2 F)(\Delta_2\phi)$$
$$\quad (\Delta_2\Delta_1 F^{-1})(\Delta_4\phi)(\Delta_4\Delta_1 F^{-1})(\Delta_4\Delta_3 F^{-1})$$
$$= (\Delta_0\Delta_0 F)(\Delta_0\Delta_2 F)(\Delta_3\Delta_2 F)(\Delta_0\phi)(\Delta_2\phi)$$
$$\quad (\Delta_4\phi)(\Delta_1\Delta_1 F^{-1})(\Delta_4\Delta_1 F^{-1})(\Delta_4\Delta_3 F^{-1})$$
$$= F_{34}(\Delta_3\Delta_0 F)(\Delta_3\Delta_2 F)(\Delta_3\phi)(\Delta_3\Delta_1 F^{-1})(\Delta_3\Delta_3 F^{-1})F_{34}^{-1}$$
$$\quad F_{12}(\Delta_1\Delta_0 F)(\Delta_1\Delta_2 F)(\Delta_1\phi)(\Delta_1\Delta_1 F^{-1})(\Delta_1\Delta_3 F^{-1})F_{12}^{-1}$$
$$= ((\Delta_F)_3\phi_F)((\Delta_F)_1\phi_F).$$

For the second equality we cancelled various expressions involving F, F^{-1} and their coproducts. For the third we used (2.29) in the notation above. We then used the 3-cocycle condition for ϕ and similar steps in reverse. Finally, it is easy to see that ϕ_F is counital if ϕ, F are.   ∎

As a special case of this theorem, we see that twisting an ordinary Hopf algebra by an arbitrary invertible element F takes us out of the class of ordinary Hopf algebras, by introducing an 'associativity deficit'

$$\phi = \partial F \equiv F_{23}((\mathrm{id}\otimes\Delta)F)((\Delta\otimes\mathrm{id})F^{-1})F_{12}^{-1}$$

as the coboundary of F. It obeys the 3-cocycle condition relative to the twisted coproduct $F(\Delta)F^{-1}$ rather than the original one. This is another way of saying that, for noncommutative H, one does not have a standard 3-cohomology theory but something slightly more complicated. If F is a 2-cocycle, as in the preceding section, then $\phi = 1$ and we remain with a Hopf algebra.

Quasi-Hopf algebras then are a larger class than ordinary Hopf algebras, but one that is closed under arbitrary twisting. This gives a much more general kind of 'twisting-equivalence' than that in the last section. In particular, many important ordinary quasitriangular Hopf algebras can be greatly simplified in their structure by this more general twisting to an equivalent quasi-Hopf algebra. For example, Drinfeld found a certain ad-invariant 3-cocycle ϕ for $U(g)$ in the setting of Example 2.4.1, for all complex simple Lie algebras g. There is also a quasitriangular structure $\mathcal{R}$ obeying (2.30), given by the exponential of the inverse Killing form (which is ad-invariant). This quasitriangular quasi-Hopf algebra $(U(g), \phi, \mathcal{R})$ is then twisting-equivalent (after an isomorphism) to the quasitriangular Hopf algebras $U_q(g)$ in the next chapter. This means, in particular, that these $U_q(g)$ have algebras that are isomorphic to $U(g)$, since twisting changes only the coproduct and quasitriangular structure, etc., but their coproduct is a deformation-quantisation of that $U(g)$. As in Section 2.3, this also means that their representations are just the same as those of $U(g)$, while the tensor products of representations are isomorphic via F to the usual tensor products of $U(g)$ representations.

Finally, the dual of a quasi-Hopf algebra is evidently something that is associative only up to conjugation in a suitable convolution algebra by a 3-cocycle ϕ. Its axioms are found by dualising, as we did for the dual quasitriangular structure in Section 2.2. Explicitly, a dual quasibialgebra is a coalgebra A which also has an almost-associative product $\cdot$ in the sense

$$\sum a_{(1)} \cdot (b_{(1)} \cdot c_{(1)}) \phi(a_{(2)} \otimes b_{(2)} \otimes c_{(2)})$$
$$= \sum \phi(a_{(1)} \otimes b_{(1)} \otimes c_{(1)}) (a_{(2)} \cdot b_{(2)}) \cdot c_{(2)} \tag{2.33}$$

for all $a, b, c \in A$. Here ϕ is a unital 3-cocycle on A, as in (2.21). It is dual quasitriangular if there is a convolution-invertible map $\mathcal{R} : A \otimes A \to k$ such that

$$\mathcal{R}(a \cdot b \otimes c) = \sum \phi(c_{(1)} \otimes a_{(1)} \otimes b_{(1)}) \mathcal{R}(a_{(2)} \otimes c_{(2)}) \phi^{-1}(a_{(3)} \otimes c_{(3)} \otimes b_{(2)})$$
$$\mathcal{R}(b_{(3)} \otimes c_{(4)}) \phi(a_{(4)} \otimes b_{(4)} \otimes c_{(5)}), \tag{2.34}$$

$$\mathcal{R}(a \otimes b \cdot c) = \sum \phi^{-1}(b_{(1)} \otimes c_{(1)} \otimes a_{(1)}) \mathcal{R}(a_{(2)} \otimes c_{(2)}) \phi(b_{(2)} \otimes a_{(3)} \otimes c_{(3)})$$
$$\mathcal{R}(a_{(4)} \otimes b_{(3)}) \phi^{-1}(a_{(5)} \otimes b_{(4)} \otimes c_{(4)}), \tag{2.35}$$

$$\sum b_{(1)} \cdot a_{(1)} \mathcal{R}(a_{(2)} \otimes b_{(2)}) = \sum \mathcal{R}(a_{(1)} \otimes b_{(1)}) a_{(2)} \cdot b_{(2)}. \tag{2.36}$$

Finally, a dual quasibialgebra is a quasi-Hopf algebra if there is a linear map $S : A \to A$ and linear functionals $\alpha, \beta : A \to k$ such that

$$\sum (Sa_{(1)}) a_{(3)} \alpha(a_{(2)}) = 1\alpha(a), \quad \sum a_{(1)} Sa_{(3)} \beta(a_{(2)}) = 1\beta(a), \tag{2.37}$$

$$\sum \phi(a_{(1)} \otimes Sa_{(3)} \otimes a_{(5)}) \beta(a_{(2)}) \alpha(a_{(4)}) = \epsilon(a),$$
$$\sum \phi^{-1}(Sa_{(1)} \otimes a_{(3)} \otimes Sa_{(5)}) \alpha(a_{(2)}) \beta(a_{(4)}) = \epsilon(a). \tag{2.38}$$

These are the dual versions of the axioms appearing at the start of the section. Analogous results for twisting, etc. apply equally well.

Example 2.4.3 *Let G be a group. The group algebra kG can be regarded as a dual quasi-Hopf algebra iff there is an invertible $\phi \in Z^3(G)$ as in Example 2.3.2. If G is Abelian then (kG, ϕ) is a dual quasitriangular dual quasi-Hopf algebra iff there is a quasibicharacter $\mathcal{R} : G \times G \to k$ in* *the sense*

$$R(uv, w) = R(u, w)R(v, w)\frac{\phi(w, u, v)\phi(u, v, w)}{\phi(u, w, v)},$$

$$R(u, vw) = R(u, w)R(u, v)\frac{\phi(v, u, w)}{\phi(v, w, u)\phi(u, v, w)}$$

and $R(u, e) = R(e, u) = 1$ *for all* $u, v, w \in G$.

Proof: This is a generalisation of Example 2.2.5 and proven in just the same way; we omit the $_{(1), (2)}$, etc., suffixes in the axioms and obtain the above. The only subtlety is that given ϕ we have an antipode defined by group inversion and $\alpha = \epsilon$, $\beta(u) = 1/\phi(u, u^{-1}, u)$. ∎

Notes for Chapter 2

The notion of quasitriangular Hopf algebras is due to V.G. Drinfeld [22] as an abstraction of structures implicit in the works of E.K. Sklyanin [23, 24], P.P. Kulish and N.Yu. Reshetikhin [25], M. Jimbo [26] and others working in quantum inverse scattering [27]. We will turn to the standard examples in the next chapters. Some of the computations for Proposition 2.1.8 appeared independently in [1] and [28], but were known to experts at the time; they are part of Drinfeld's theory [29]. The nonstandard quasitriangular Hopf algebras $\mathbb{Z}'_{/n}$ in Example 2.1.6 are from the author's 1991 work [30]. Some applications of the quantum line $U_q(1), L_q^\infty(\mathbb{R})$ in Example 2.1.19 are in [31] and [10]. The use of bicharacters on groups to define quasitriangular structures can be found in the first of these works. Example 2.1.7 is due to D. Radford [32].

The formulation of ribbon Hopf algebras is due to Reshetikhin and V.G. Turaev [28]. The notion of factorisable Hopf algebras is due to Reshetikhin and M.A. Semenov-Tian-Shansky [33]. An application of the ad-invariance of the quantum Killing form in Proposition 2.1.14 can be found in [34]. The classification into real and antireal quasitriangular Hopf *-algebras is taken from [35], by the author.

The notion of dual quasitriangular (or 'coquasitriangular') Hopf algebras is of course just the dual of Drinfeld's notion of quasitriangular ones. Probably the first place that it was formally introduced was in the author's 1989 work [36, 37], in connection with a Tannaka–Krein reconstruction theorem. We will come to this in Chapter 9.4. The basic properties of dual quasitriangular Hopf algebras, such as Proposition 2.2.4, were obtained by the author in the appendix of [38]. Most of Section 2.2 is taken from this work, including the self-dual Example 2.2.6. In between quasitriangular and dual quasitriangular (i.e. making sense for a dual pair of

Hopf algebras H_1, H_2) was the notion of a quasitriangular structure as a map $H_1 \to H_2$, as introduced in [14] on the basis of Exercise 2.1.5.

The theory of twisting, and the related theory of quasi-Hopf algebras in Section 2.4, arose in Drinfeld's work on the deformation of Hopf algebras by means of solutions of the Knizhnik–Zamolodchikov equations [39, 40, 41, 42]. The proof that all the famous quantum groups $U_q(g)$ (see the next chapter) are twistings of $U(g)$ equipped with a nonstandard quasi-Hopf structure ϕ, is in [41]. Also, using this twisting theory in the triangular case, Drinfeld showed that every (triangular) Poisson–Lie group (see Chapter 8) can be deformed to a triangular Hopf algebra [43]. From the dual point of view the coordinate ring is deformation-quantised to a dual triangular Hopf algebra as the twisting is introduced. A recent application of this point of view is in [44] where a theory of twisting of differential calculi on quantum groups is introduced; it provides the corresponding deformation quantisation of the classical differential structure of the Lie group.

The twisting Theorem 2.3.4 for quasitriangular Hopf algebras was formulated in [45] as a special case of Drinfeld's theory; we have provided a new direct proof, and a setting in terms of non-Abelian cohomology as developed by the author [46]. The formula (2.13), and Propositions 2.3.3 and 2.3.5 are from here, while implicit in Drinfeld's earlier works. Proposition 2.3.7 for the twisting of Hopf $*$-algebras is from [47], also by the author. The proof of (2.17) is in [45]. A theory of cocycles and cohomology for commutative Hopf algebras is more standard [48]. We will come to non-Abelian cohomology again in Chapter 6.3, where the 2-cocycles may have values in an algebra and allow the formation of cocycle cross products. The present case with trivial coefficients is also closely related to abstract deformation theory and deformation cohomology in the work of M. Gerstenhaber and S.D. Schack [49]. Recent results in this direction appear in [50] and [51]; see also the useful text [52]. Typically, it is quasi-Hopf algebras rather than Hopf algebras themselves that arise or are convenient in certain contexts, notably (in some form) as symmetries in certain quantum field theories [53]. A quantum double construction to generate a new quasi-Hopf algebra from an old one (cf. Chapter 7 in the case of ordinary Hopf algebras) is in [54]. Finally, the dual theory of dual quasitriangular dual quasi-Hopf algebras was introduced by the author in 1990 [55], in order to extend the Tannaka–Krein reconstruction theory to this case as well (see Chapter 9.4). A recent application of the theory is in [56].

3

Quantum enveloping algebras

After all the abstract algebra in the preceding two chapters, it is clearly high time for some substantial examples. This is the topic of the present chapter and the following one. In this chapter we describe quantum groups (quasitriangular Hopf algebras) that are deformations of the enveloping algebras of Lie algebras. The latter have already been introduced in Example 1.5.7 and are clearly cocommutative. But by deforming the comultiplication (and perhaps also the appearance of the multiplication), we can obtain examples that are not cocommutative and not commutative (if the Lie algebra is non-Abelian). The quasitriangular structure also becomes nontrivial. These examples can therefore be called *quantum groups of enveloping algebra type*. In principle, this deformation tends to be a systematic one related to the twisting construction given in Chapters 2.3–2.4, but this need not concern us as far as describing the resulting structures is concerned.

Lie groups and Lie algebras have their origin as transformations or symmetries of spaces, and this is still how they are most often used. Likewise, the quantum groups of enveloping algebra type are well suited to acting on things. In general, they act on all the things that the initial Lie algebra acts on (the action is deformed along with the quantum group structure). This makes them particularly useful in certain kinds of physical contexts. As quasitriangular Hopf algebras they have many other applications too, notably to the construction of link and three-manifold invariants, as we shall see in Chapter 9.

We begin with the simplest non-Abelian example, based on the Heisenberg algebra. It is nothing other than an ordinary quantum harmonic oscillator with the role of Planck's constant played by a central element. We also add the number operator as an independent operator. These seemingly minor changes allow for a quasitriangular Hopf algebra structure! Its resulting knot-invariants are known to be (the inverse of) the

Alexander–Conway knot polynomial, while in physics it clearly has potential applications along the lines of the usual harmonic oscillator. Then we describe what is perhaps the best-known of all quantum groups of enveloping algebra type, the deformation of the Lie algebra su_2. This arose some years ago in the theory of inverse scattering as a quotient of the Sklyanin algebra, and its simplest knot-invariant recovers the celebrated Jones polynomial. We then proceed to describe the quantum group deformations of general complex simple Lie algebras g. In fact, their structure is determined from the structure of their su_2 subgroups associated to simple roots, and is not much different.

When describing quantum groups of enveloping algebra type we have two natural possibilities. The simplest is to work over $\mathbb{C}$ or another field. This is legitimate so long as we regard the symbols $q^{\frac{H}{2}}, q^N$ below as the actual generators (not H, N). This possibility is fine for the Hopf algebra, but it does, however, run into a problem when it comes to the description of the quasitriangular structure $\mathcal{R}$. We saw, in the one-dimensional case in Example 2.1.19, that to have a quasitriangular structure one needs to work with Gaussians and exponentials. In our case these now appear in the form $q^{\frac{H \otimes H}{2}}, q^{H \otimes N}$ etc., for which we really have to regard H, N as the generators. To do this, we have to write $q = e^{\frac{t}{2}}$, say, and understand all expressions order-by-order in t. It means that we work over the ring $\mathbb{C}[[t]]$ of formal power series in t and demand that all linear spaces are free $\mathbb{C}[[t]]$-modules (along with other nice properties). In practice, this kind of order-by-order work is very familiar to physicists and should not be off-putting. In summary, the Hopf algebras that we describe can be understood either way (we will emphasise the simpler way, over $\mathbb{C}$), but the quasitriangular structure should be understood in the second way, over formal power series in t. Our notation is based on the need to accommodate both of these two points of view. These two approaches are unified in a topological algebra setting where $\mathcal{R}$ lives in a completed tensor product over $\mathbb{C}$.

A third possibility is that we work over $\mathbb{C}$ as above but specialise to q a root of unity. In this situation it turns out that our Hopf algebras have natural quotients which are perfectly finite-dimensional quasitriangular Hopf algebras over $\mathbb{C}$. In this finite-dimensional, case of course, there is no need to take any topological completions, and $\mathcal{R}$ lies perfectly well in the tensor product. This is described briefly in Section 3.4.

3.1   *q*-Heisenberg algebra

One of the most important noncommutative algebras in physics is the quantum harmonic oscillator. We begin by recalling some facts about this which will serve to motivate our discussion. Its algebra $A_\hbar$ is generated by $1, a, a^\dagger$, with relations

$$[a, a^\dagger] = \hbar, \tag{3.1}$$

where $\hbar$ is a real parameter (Planck's constant in physics). This parameter can be absorbed into the normalisation of $a, a^\dagger$, but we prefer not to do so. Moreover, in the physics of n-particle systems (such as quantum optics), one frequently wants to embed the one-particle system into its n-fold tensor product by algebra homomorphisms $\Delta^{(n-1)} : A_\hbar \to A_\hbar^{\otimes n}$. The maps needed here are

$$\Delta^{(n-1)}a = n^{-\frac{1}{2}}(a \otimes \cdots \otimes 1 + 1 \otimes a \otimes \cdots \otimes 1 + \cdots + 1 \otimes \cdots \otimes a),$$
$$\Delta^{(n-1)}a^\dagger = n^{-\frac{1}{2}}(a^\dagger \otimes \cdots \otimes 1 + 1 \otimes a^\dagger \otimes \cdots \otimes 1 + \cdots + 1 \otimes \cdots \otimes a^\dagger).$$

It is easy to see that these are indeed algebra homomorphisms provided we include the $n^{-\frac{1}{2}}$ factor as shown. To this extent then, the harmonic oscillator has a natural bialgebra-like structure. We will see how this kind of map is used in a physical context in Chapter 5. For now, let us note that, if we absorb the $n^{-\frac{1}{2}}$ normalisation into the relevant $a, a^\dagger$, we have

$$\Delta^{(n-1)} : A_\hbar \to A_{\frac{\hbar}{n}} \otimes \cdots \otimes A_{\frac{\hbar}{n}} \tag{3.2}$$

given by $\Delta^{(n-1)}a = a \otimes \cdots \otimes 1 + \cdots + 1 \otimes \cdots a$. This looks like an iteration of the linear coproduct in Example 1.5.7, but note that it does not map a single algebra into tensor powers of itself because of the rescaling of $\hbar$. There are other bialgebra structures for the harmonic oscillator which we will come to later (based on Example 1.3.2), but this bialgebra-like one is our main motivation now.

To proceed, we observe that the generator 1, or its multiple $\hbar$, of course commutes with $a, a^\dagger$ so that these relations can also be regarded as defining a three-dimensional Lie algebra, the Heisenberg Lie algebra. This is the real nilpotent Lie algebra with generators $a, a^\dagger, H$ and relations

$$[a, a^\dagger] = H, \quad [H, a] = 0, \quad [H, a^\dagger] = 0. \tag{3.3}$$

Here the role of $\hbar$ is elevated to an abstract central element H of our Lie algebra. Because it is a Lie algebra, we can also consider its universal enveloping Hopf algebra. Recalling Example 1.5.7, we see that the latter is generated by $a, a^\dagger, H$ and 1 (this last item is the identity in the enveloping algebra, and not to be confused with the identity element above in the harmonic oscillator algebra), with relations (3.3) and the linear coproduct

$$\Delta a = a \otimes 1 + 1 \otimes a, \quad \Delta a^\dagger = a^\dagger \otimes 1 + 1 \otimes a^\dagger, \quad \Delta H = H \otimes 1 + 1 \otimes H.$$

Exercise 3.1.1 *Let $A_\hbar$ be the quantum harmonic oscillator as in (3.1) and let U be the enveloping Hopf algebra of the three-dimensional Heisen-*

berg Lie algebra (3.3). Show that the coproduct of the latter covers the maps $\Delta^{(n-1)}$ in the sense that the algebra homomorphisms

$$
\begin{array}{ccc}
U & \longrightarrow & A_\hbar \\
\Delta^{n-1} \downarrow & & \downarrow \Delta^{(n-1)} \\
U \otimes \cdots \otimes U & \longrightarrow & A_{\frac{\hbar}{n}} \otimes \cdots \otimes A_{\frac{\hbar}{n}} \\
& & \underset{n}{} \qquad\qquad \underset{n}{}
\end{array}
$$

commute. The horizontal maps are given by sending H to the effective value of Planck's constant appropriate to the algebra ($\hbar$ on the top and $\frac{\hbar}{n}$ on the bottom).

Solution: Iterating the linear coproduct in U, we clearly obtain $\Delta^{n-1} a = a \otimes 1 \cdots \otimes 1 + \cdots + 1 \otimes \cdots \otimes a$, etc. $\Delta^{n-1} H$ also has n terms each containing H and 1 elsewhere, and evaluating these according to the lower map gives $\hbar 1 \otimes \cdots \otimes 1 = \hbar \Delta^{(n-1)} 1 = \Delta^{(n-1)} \hbar$ since $\Delta^{(n-1)}$ is an algebra map and so preserves the identity. ∎

This exercise provides one physical motivation for the Heisenberg Lie algebra and its associated universal enveloping algebra (there are numerous others throughout physics and mathematics).

Finally, of particular importance for the harmonic oscillator, is the number operator $N = \hbar^{-1} a^\dagger a$ ($\hbar N$ is the Hamiltonian less the zero-point energy). It obeys $[N, a] = -a$, $[N, a^\dagger] = a^\dagger$ in virtue of (3.1). For the Heisenberg Lie algebra, we can likewise compute (in the enveloping algebra)

$$
[N, a] = -a, \quad [N, a^\dagger] = a^\dagger, \quad [N, H] = 0, \tag{3.4}
$$

at least formally if we suppose that H is invertible and $N = H^{-1} a^\dagger a$. Rather than adjoining such an inverse to the algebra, we can shift our point of view once again and regard $a, a^\dagger, H, N$ as the independent generators now of a four-dimensional Lie algebra (the extended Heisenberg algebra) with the relations (3.3) and (3.4). It is easy to verify that these relations obey the Jacobi identity and hence do define a Lie algebra. As before, we can then view this as a Hopf algebra by considering the universal enveloping algebra of this four-dimensional Lie algebra. It has the coproducts above and, from our new point of view,

$$
\Delta N = N \otimes 1 + 1 \otimes N.
$$

Note that ΔN here is quite different from $\Delta(a^\dagger a) = (\Delta a^\dagger)(\Delta a)$. In other words, there is formally an algebra homomorphism $N \mapsto H^{-1} a^\dagger a$ from our extended Heisenberg algebra onto the unextended one (if we assume H to be invertible), but it is not a Hopf algebra homomorphism.

In summary, we have two Hopf algebras and one bialgebra-like structure related to each other by extending some of the above expressions as independent abstract generators,

$$\text{extended Heisenberg algebra} \to \text{Heisenberg algebra}$$
$$\to \text{harmonic oscillator.}$$

(3.5)

The first map sends N to the number operator if we assume H in the Heisenberg algebra is invertible. The second algebra map takes H to the value of Planck's constant for the harmonic oscillator.

Example 3.1.2 *Let q be a nonzero parameter. The q-Heisenberg algebra is defined with generators $a, a^\dagger, q^{\frac{H}{2}}, q^{-\frac{H}{2}}$ and 1 with the relations $q^{\pm\frac{H}{2}} q^{\mp\frac{H}{2}} = 1$ (the notation is meant to suggest these) and*

$$[q^{\frac{H}{2}}, a] = 0, \quad [q^{\frac{H}{2}}, a^\dagger] = 0, \quad [a, a^\dagger] = \frac{q^H - q^{-H}}{q - q^{-1}}.$$

This forms a Hopf algebra with

$$\Delta a = a \otimes q^{\frac{H}{2}} + q^{-\frac{H}{2}} \otimes a, \quad \Delta a^\dagger = a^\dagger \otimes q^{\frac{H}{2}} + q^{-\frac{H}{2}} \otimes a^\dagger,$$

$$\Delta q^{\pm\frac{H}{2}} = q^{\pm\frac{H}{2}} \otimes q^{\pm\frac{H}{2}}, \quad \epsilon q^{\pm\frac{H}{2}} = 1, \quad \epsilon a = 0 = \epsilon a^\dagger,$$

$$Sa = -a, \quad Sa^\dagger = -a^\dagger, \quad Sq^{\pm\frac{H}{2}} = q^{\mp\frac{H}{2}}.$$

The extended q-Heisenberg algebra is defined with the additional mutually inverse generators q^N, q^{-N} and relations

$$q^N a q^{-N} = q^{-1} a, \quad q^N a^\dagger q^{-N} = q a^\dagger, \quad [q^N, q^{\frac{H}{2}}] = 0.$$

It forms a Hopf algebra with the additional structure

$$\Delta q^N = q^N \otimes q^N, \quad \epsilon q^N = 1, \quad Sq^{\pm N} = q^{\mp N}.$$

If $q = e^{\frac{t}{2}}$ and we work over $\mathbb{C}[[t]]$ rather than $\mathbb{C}$, then we can regard $X_\pm, H, N$ and 1 as the generators. In this case, the extended q-Heisenberg Hopf algebra is quasitriangular with

$$\mathcal{R} = q^{-(N \otimes H + H \otimes N)} e^{(q - q^{-1}) q^{\frac{H}{2}} a \otimes q^{-\frac{H}{2}} a^\dagger}.$$

Proof: First we verify the Hopf algebra structure. We have to show that Δ, ϵ extend as algebra homomorphisms, and that S extends as an anti-algebra homomorphism. Thus,

$$[\Delta a, \Delta a^\dagger] = (a \otimes q^{\frac{H}{2}} + q^{-\frac{H}{2}} \otimes a)(a^\dagger \otimes q^{\frac{H}{2}} + q^{-\frac{H}{2}} \otimes a^\dagger) - (a \leftrightarrow a^\dagger)$$

$$= [a, a^\dagger] \otimes q^H + q^{-H} \otimes [a, a^\dagger]$$

$$= \frac{(q^H - q^{-H}) \otimes q^H}{q - q^{-1}} + \frac{q^{-H} \otimes (q^H - q^{-H})}{q - q^{-1}}$$

$$= \frac{q^H \otimes q^H - q^{-H} \otimes q^{-H}}{q - q^{-1}} = \Delta[a, a^\dagger],$$

as required. Similarly, we have $(\Delta q^N)(\Delta a)(\Delta q^{-N}) = (q^N \otimes q^N)(a \otimes q^{\frac{H}{2}} + q^{-\frac{H}{2}} \otimes a)(q^{-N} \otimes q^{-N}) = q^{-1}(a \otimes q^{\frac{H}{2}} + q^{-\frac{H}{2}} \otimes a) = \Delta(q^{-1}a)$. The other verifications are equally straightforward. Note that the limit as $q \to 1$ gives the enveloping algebra of the (extended) Heisenberg algebra. This is easily seen over $\mathbb{C}[[t]]$ by setting $t = 0$. The coproducts of $q^{\frac{H}{2}}$ and q^N are not deformed at all, in the sense that, over $\mathbb{C}[[t]]$, the generators H, N are primitive as in the undeformed case. Lemma 3.1.3 explains this point in detail. This remark and its proof also helps us to verify the quasitriangular structure according to the axioms in Chapter 2.1. We have

$$(\Delta \otimes \mathrm{id})\mathcal{R} = q^{-(N \otimes 1 \otimes H + 1 \otimes N \otimes H + H \otimes 1 \otimes N + 1 \otimes H \otimes N)}$$

$$\times e^{(q - q^{-1})(q^{\frac{H}{2}} a \otimes q^H \otimes q^{-\frac{H}{2}} a^\dagger + 1 \otimes q^{\frac{H}{2}} a \otimes q^{-\frac{H}{2}} a^\dagger)}$$

$$= q^{-(N \otimes 1 \otimes H + H \otimes 1 \otimes N)} q^{-(1 \otimes H \otimes N)} e^{(q - q^{-1}) q^{\frac{H}{2}} a \otimes q^H \otimes q^{-\frac{H}{2}} a^\dagger}$$

$$\times q^{1 \otimes H \otimes N} q^{-(1 \otimes N \otimes H + 1 \otimes H \otimes N)} e^{(q - q^{-1}) 1 \otimes q^{\frac{H}{2}} a \otimes q^{-\frac{H}{2}} a^\dagger}$$

$$= q^{-(N \otimes 1 \otimes H + H \otimes 1 \otimes N)} e^{(q - q^{-1}) q^{\frac{H}{2}} a \otimes 1 \otimes q^{-\frac{H}{2}} a^\dagger}$$

$$\times q^{-(1 \otimes N \otimes H + 1 \otimes H \otimes N)} e^{(q - q^{-1}) 1 \otimes q^{\frac{H}{2}} a \otimes q^{-\frac{H}{2}} a^\dagger}$$

$$= \mathcal{R}_{13} \mathcal{R}_{23},$$

where we have used that Δ is an algebra homomorphism to compute it in the exponent (as in Lemma 3.1.3 below). The second equality notes that various terms in the exponents commute. The third equality comes from

$$q^{-H \otimes N}((\ ) \otimes f(H) a^\dagger) q^{H \otimes N} = (\ ) q^{-H} \otimes f(H) a^\dagger, \qquad (3.6)$$

for any expression in the empty space, and any function $f(H)$. This follows simply from writing the conjugation as $q^{-[H \otimes N, \ ]}$ and using the relations $[N, a^\dagger] = a^\dagger$ and H central. The proof of the second half of (2.1) is entirely analogous. To prove (2.2), we compute

$$\mathcal{R}(\Delta a)\mathcal{R}^{-1} = q^{-(N \otimes H + H \otimes N)} e^{(q - q^{-1}) q^{\frac{H}{2}} a \otimes q^{-\frac{H}{2}} a^\dagger} (a \otimes q^{\frac{H}{2}} + q^{-\frac{H}{2}} \otimes a)$$

$$\times e^{-(q - q^{-1}) q^{\frac{H}{2}} a \otimes q^{-\frac{H}{2}} a^\dagger} q^{N \otimes H + H \otimes N}$$

$$= q^{-(N \otimes H + H \otimes N)} \left(a \otimes q^{\frac{H}{2}} + q^{-\frac{H}{2}} \otimes a + a \otimes (q^{-\frac{3}{2}H} - q^{\frac{H}{2}}) \right)$$

$$\times q^{N \otimes H + H \otimes N}$$

$$= q^{\frac{H}{2}} \otimes a + a \otimes q^{-\frac{H}{2}} = \tau \circ \Delta a,$$

where the second equality is obtained by writing inner conjugation as $e^{(q-q^{-1})[q^{\frac{H}{2}} a \otimes q^{-\frac{H}{2}} a^\dagger, \ ]}$; its action on $a \otimes q^{\frac{H}{2}}$ is the identity, while its action on $q^{-\frac{H}{2}} \otimes a$ is computed from

$$[q^{\frac{H}{2}} a \otimes q^{-\frac{H}{2}} a^\dagger, q^{-\frac{H}{2}} \otimes a] = a \otimes [q^{-\frac{H}{2}} a^\dagger, a] = a \otimes \left(\frac{q^{-\frac{3}{2}H} - q^{\frac{H}{2}}}{q - q^{-1}} \right).$$

Only the first two terms of the series expansion for the exponentiated commutator contribute. The third equality requires a computation similar to that in (3.6). The proof for $\Delta a^\dagger$ is entirely similar. The proof of (2.2) for $\Delta H = \tau \circ \Delta H$ is automatic since H is central. The proof for ΔN is like the above; the $q^{-(N \otimes H + H \otimes N)}$ commutes automatically, while

$$[q^{\frac{H}{2}} a \otimes q^{-\frac{H}{2}} a^\dagger, N \otimes 1 + 1 \otimes N] = q^{\frac{H}{2}} a \otimes q^{-\frac{H}{2}} a^\dagger - q^{\frac{H}{2}} a \otimes q^{-\frac{H}{2}} a^\dagger = 0,$$

as required. ∎

Lemma 3.1.3 *Let N be an element of a Hopf algebra over $\mathbb{C}[[t]]$. If $\Delta N = N \otimes 1 + 1 \otimes N$, then $\Delta e^{tN} = e^{tN} \otimes e^{tN}$.*

Proof: Since this kind of elementary computation is crucial to the understanding of the preceding proof (and proofs elsewhere in the chapter), we will give it in detail. Thus, $\Delta e^{tN} = \sum_{r=0}^{\infty} \frac{\Delta(tN)^r}{r!} = \sum \frac{(tN \otimes 1 + 1 \otimes tN)^r}{r!} = e^{tN \otimes 1 + 1 \otimes tN} = e^{tN \otimes 1} e^{1 \otimes tN}$ in the tensor product algebra. We used that Δ is an algebra homomorphism to the tensor product algebra, and that in this algebra the operators $tN \otimes 1$ and $1 \otimes tN$ commute. We then note that $e^{tN \otimes 1} e^{1 \otimes tN} = (e^{tN} \otimes 1)(1 \otimes e^{tN}) = e^{tN} \otimes e^{tN}$. Of course, the first step works for any function, $\Delta f(N) = f(N \otimes 1 + 1 \otimes N)$. ∎

The extended q-deformed Heisenberg algebra is factorisable and a ribbon Hopf algebra. Finally, we discuss the $*$-structure. We have

Proposition 3.1.4 *Let q be real. The q-deformed Heisenberg algebra and its extension are Hopf $*$-algebras with*

$$a^* = a^\dagger, \quad H^* = H, \quad N^* = N,$$

and the extended case is real quasitriangular in the sense $\mathcal{R}^{* \otimes *} = \tau(\mathcal{R})$. This $*$-structure justifies the notation $a, a^\dagger$ for the generators.

Proof: We verify first that we have a Hopf $*$-algebra. The stated $*$-structure can also be written $(q^{\frac{H}{2}})^* = q^{\frac{H}{2}}, (q^N)^* = q^N$ if we are working over $\mathbb{C}$ rather than $\mathbb{C}[[t]]$. We have $[a^{\dagger *}, a^*] = [a, a^\dagger] = \frac{q^H - q^{-H}}{q - q^{-1}} =$

$(\frac{q^H - q^{-H}}{q - q^{-1}})^* = [a, a^\dagger]^*$ and $[a^*, N^*] = [a^\dagger, N] = -a^\dagger = [N, a]^*$. Similarly for $[N, a^\dagger]$. Hence $*$ extends as an antialgebra map. It is easy to see that Δ and ϵ are $*$-algebra maps, etc. We show (in the extended case) that $\mathcal{R}$ is real in the sense of Definition 2.1.15,

$$\mathcal{R}^{* \otimes *} = e^{(q - q^{-1})^* a^\dagger q^{\frac{H}{2}} \otimes a q^{-\frac{H}{2}}} q^{-(N \otimes H + H \otimes N)}$$

$$= q^{-(N \otimes H + H \otimes N)} e^{(q - q^{-1}) q^{-\frac{H}{2}} a^\dagger \otimes q^{\frac{H}{2}} a} = \tau(\mathcal{R}),$$

as required. We made here a computation similar to that for (3.6) in the proof of Example 3.1.2. ∎

To justify our study of the q-deformed Heisenberg algebra, we should mention some concrete representations. In fact, this is very easy because we can just proceed in the same way as for an ordinary quantum harmonic oscillator. Recall for this that the Fock representation is created by applying $a^\dagger$ to $|0\rangle$ (the vacuum), which in turn is annihilated by a. For the Heisenberg algebra, we have an additional central element H and we have to specify its value. Thus the vacuum vector has an additional real parameter $\hbar$ and $H|0, \hbar\rangle = \hbar|0, \hbar\rangle$ (or $q^{\frac{H}{2}}|0, \hbar\rangle = q^{\frac{\hbar}{2}}|0, \hbar\rangle$). We can keep this definition also in the q-deformed case. The only difference is a slight change of normalisation due to the modified $[a, a^\dagger]$ relations.

Proposition 3.1.5 *Let q be real. The unitary representations of the q-Heisenberg algebra are labelled by a real positive parameter $\hbar$ and take the form $\mathcal{H}_\hbar = \{|n, \hbar\rangle| \ n = 0, 1, \ldots\}$ where*

$$|n, \hbar\rangle = \frac{a^{\dagger n}}{[\hbar]^{\frac{n}{2}} \sqrt{n!}}|0, \hbar\rangle, \quad H|0, \hbar\rangle = \hbar|0, \hbar\rangle, \quad a|0, \hbar\rangle = 0$$

and $[\hbar] = (q^\hbar - q^{-\hbar})/(q - q^{-1})$. If we take these $\{|n, \hbar\rangle\}$ as an orthonormal basis of a Hilbert space, then $a, a^\dagger$ are adjoint while H is self-adjoint. For the extended q-Heisenberg algebra we define $N|0, \hbar\rangle = 0$ so that N is the number operator.

Proof: Since H has value $\hbar$, the value of $[a, a^\dagger]$ is the expression $[\hbar]$, as stated. Thus, on such representations the generators $a, a^\dagger$ of the q-Heisenberg algebra precisely define an ordinary quantum harmonic oscillator $A_{[\hbar]}$. This means that their Fock representation can be built up in the same way as usual with $\hbar^{-\frac{1}{2}} a^\dagger$ replaced by $[\hbar]^{-\frac{1}{2}} a^\dagger$. This immediately gives the representation shown. The normalisation (as usual) is such that $\langle n, \hbar|(a^\dagger|m, \hbar\rangle) = (\langle n, \hbar|a^\dagger)|m, \hbar\rangle$, etc., for any two states. If we write this in a more mathematical way, we have a sesquilinear inner product $(|n, \hbar\rangle, |m, \hbar\rangle) \equiv \langle n, \hbar|m, \hbar\rangle = \delta_{n,m}$, and the assertion is that $(v, a^\dagger w) = (av, w)$, $(v, aw) = (a^\dagger v, w)$, $(v, Hw) = (Hv, w)$ for any $v, w \in \mathcal{H}_\hbar$. Hence the representation is a unitary one for the $*$-structure in Proposition 3.1.4. This follows at once when we compute on our basis that

$$a^\dagger|n, \hbar\rangle = \sqrt{(n+1)[\hbar]}|n+1, \hbar\rangle, \quad a|n, \hbar\rangle = \sqrt{n[\hbar]}|n-1, \hbar\rangle,$$

$$H|n, \hbar\rangle = \hbar|n, \hbar\rangle$$

from the relations in Example 3.1.2. For N, the simplest possibility is as stated. Then from the relations between $N, a, a^\dagger$ we have $N|n, \hbar\rangle = n|n, \hbar\rangle$ and $(v, Nw) = (Nv, w)$ for all v, w. Note that, since we have declared N to be an independent generator (not necessarily given in terms of $a, a^\dagger$), this choice is not the only possibility; i.e. the extended q-Heisenberg algebra can have other representations where N acts differently. ∎

To complete our study of the q-Heisenberg algebra we should mention that the $*$-structure that we have emphasised above (and reflected in our choice of symbols $a, a^\dagger$ for the generators) is not the only one possible for this algebra.

Example 3.1.6 *If q is real, then another $*$-structure is $a^* = -a^\dagger$, $H^* = H$, $N^* = N$ and makes the extended q-Heisenberg algebra real quasitriangular. A further $*$-structure $a^* = -a$, $a^{\dagger *} = a^\dagger$, $H^* = H$, $N^* = -N$ makes it antireal quasitriangular.*
If q has modulus 1, so that $q^* = q^{-1}$, then

$$a^* = a, \qquad a^{\dagger *} = a^\dagger, \quad H^* = -H, \qquad N^* = -N$$

is antireal quasitriangular.

Proof: Looking at the proof of Proposition 3.1.4, it is evident that an extra minus sign in the action of $*$ on both $a, a^\dagger$ does not change the argument that we have a Hopf $*$-algebra. For $\mathcal{R}$, the expressions involve $a \otimes a^\dagger$ so the minus sign does not enter here at all. We keep $q^{\frac{H}{2}}, q^N$ self-adjoint under $*$ (at least the first of these is needed for $*$ to be compatible with the coproduct). Alternatively, we can take $a^* = -a, a^{\dagger *} = a^\dagger$, which works according to $[a^{\dagger *}, a^*] = [a^\dagger, -a] = [a, a^\dagger] = [a, a^\dagger]^*$ when $H^* = H$. For the $[N, a]$ and $[N, a^\dagger]$ relations, however, we require $N^* = -N$. This time, we have

$$\mathcal{R}^{* \otimes *} = e^{-(q-q^{-1})aq^{\frac{H}{2}} \otimes a^\dagger q^{-\frac{H}{2}}} q^{N \otimes H + H \otimes N} = \mathcal{R}^{-1}$$

so that $\mathcal{R}$ is antireal in the sense of Definition 2.1.15. We could also take

$a^* = a$ and $a^{\dagger *} = -a^\dagger$, but this is more clearly equivalent after a change of notation. We cannot have both with a minus sign (or both without) and still have compatibility with $H^* = H$.

The case $q^* = q^{-1}$ proceeds similarly. Here $q^{\frac{H}{2}}, q^N$ are again self-adjoint when H, N are anti-Hermitian, and $a^* = a, a^{\dagger *} = a^\dagger$ provides a $*$-algebra because $[a^{\dagger *}, a^*] = [a^\dagger, a] = \frac{q^{-H} - q^H}{q - q^{-1}} = (\frac{q^H - q^{-H}}{q - q^{-1}})^* = [a, a^\dagger]^*$ and $[a^*, N^*] = [N, a] = -a = (-a)^* = [N, a]^*$. Similarly for the other verifications. We again have $\mathcal{R}^{* \otimes *} = \mathcal{R}^{-1}$, this time because $(q - q^{-1})^* = -(q - q^{-1})$ in the exponential. The same computation works if we introduce minus signs into the definitions of a^* and $a^{\dagger *}$, but this time it is equivalent after a redefinition of $a, a^\dagger$ as $\imath a, -\imath a^\dagger$. ∎

Note that the notation $a, a^\dagger$ is not at all appropriate when we use any of the new $*$-structures in the preceding example. Another notation should be used for such applications; for example, the antireal quasitriangular version for real q should be denoted by generators D, Z, M, ξ, say, obeying

$$[D, Z] = \frac{q^M - q^{-M}}{q - q^{-1}}, \quad [M, D] = [M, Z] = [M, \xi] = 0,$$

$$[\xi, D] = -D, \quad [\xi, Z] = Z, \tag{3.7}$$

$$D^* = -D, \quad Z^* = Z, \quad M^* = M, \quad \xi^* = -\xi$$

with corresponding coproducts, etc. The q-Heisenberg Hopf $*$-algebra in this case can be interpreted physically as a deformation of the algebra of Hermitian position $M^{-1}Z$, anti-Hermitian momentum D and Hermitian mass M of a quantum particle. For this physical picture, we suppose that M is invertible and consider $\hbar = 1$, since it no longer plays a role. The linear coproduct of Z, D, M in the undeformed case corresponds to the embedding of the one-particle system into the tensor product of two one-particle systems, as the *centre of mass system*. So $\Delta M = M \otimes 1 + 1 \otimes M$ is the total mass, etc. The extended Heisenberg algebra adds to this the anti-Hermitian scaling operator ξ. It can be realised as $M^{-1}ZD$, but we consider it as an independent generator with linear coproduct. The extended q-Heisenberg algebra in this case is a deformation of this system as an antireal quasitriangular Hopf $*$-algebra. This is an alternative motivation for the Heisenberg algebra and its deformation, but not one that we have emphasised above. In terms of our preferred real quasitriangular Hopf $*$-algebra structure and its picture as a q-deformed harmonic oscillator, we can still define position and momentum by making linear combinations of $a, a^\dagger$.

Proposition 3.1.7 *If $a, a^\dagger, H$ are the generators of the q-Heisenberg algebra with the standard $*$-structure for real q from Proposition 3.1.4, then*

$D = 2^{-\frac{1}{2}}(a - a^\dagger), X = 2^{-\frac{1}{2}}(a + a^\dagger)$ *and H are also generators, with the same relations* $[D, X] = \frac{q^H - q^{-H}}{q - q^{-1}}$, *and the $*$-structure* $D^* = -D, X^* = X, H^* = H$. *There is a unitary representation in $L^2(\mathbb{R})$ defined, for each real parameter $\hbar$, by $H\psi = \hbar\psi$, $(D\psi)(x) = [\hbar]\frac{d}{dx}\psi(x)$ and $(X\psi)(x) = x\psi(x)$.*

Proof: The computation is identical to the familiar change of generators for the standard quantum harmonic oscillator, simply with $\hbar$ replaced by the central element $\frac{q^H - q^{-H}}{q - q^{-1}}$. Likewise, once H acts by $\hbar$, the representation is the standard Schrödinger one modified only by replacing $\hbar$ by $[\hbar]$, as in Proposition 3.1.5. By 'unitary' we mean that the $*$-structures stated are compatible with the inner product on $L^2(\mathbb{R})$ in the usual way. ∎

Finally, let us mention a different possibility from the extensions of the harmonic oscillator explored above. We still regard $N = \hbar^{-1}a^\dagger a$ as a new generator, as before, but we do not regard $\hbar$ as an operator. Thus, we work with an algebra with three generators $a, a^\dagger, N$ and 1 (and set $\hbar = 1$ without loss of generality). Its structure may be q-deformed.

Example 3.1.8 *The q-deformed harmonic oscillator is the algebra generated by $a, a^\dagger$, mutually inverse q^N, q^{-N} and 1 with relations*

$$aa^\dagger - q^{-1}a^\dagger a = q^N, \quad aa^\dagger - qa^\dagger a = q^{-N},$$

$$q^N a q^{-N} = q^{-1}a, \quad q^N a^\dagger q^{-N} = qa^\dagger.$$

Over $\mathbb{C}$, this has a $*$-algebra structure $a^* = a^\dagger$, $N^* = N$ for either q real or $|q| = 1$, i.e. $(q^N)^* = q^{\pm N}$, respectively. For real q it is unitarily represented on $\mathcal{H} = \{|n\rangle\}$, with

$$|n\rangle = \frac{a^{\dagger n}}{\sqrt{[n]!}}|0\rangle, \quad a|0\rangle = 0, \quad q^N|0\rangle = |0\rangle); \quad [n] = \frac{q^n - q^{-n}}{q - q^{-1}}$$

and $[n]! = [n][n-1]\cdots[1]$.

Proof: Let us note first that the defining relations are not fully independent. From the first two relations, one deduces at once the useful relations

$$a^\dagger a = [N], \quad aa^\dagger = [N+1],$$

where the notation is similar to that for $[n]$ applied to the relevant operator. These relations imply $[N]a^\dagger = a^\dagger aa^\dagger = a^\dagger[N+1]$ and $a[N] = aa^\dagger a =$

$[N + 1]a$. This is close to the content of the remaining two defining relations, and in fact these are implied in a suitable context. We have chosen to impose them directly. These relations and the definition of $|n\rangle$ (at least for real q) give at once that

$$a^\dagger|n\rangle = \sqrt{[n + 1]}|n + 1\rangle, \qquad a|n\rangle = \sqrt{[n]}|n - 1\rangle, \qquad q^N|n\rangle = q^n|n\rangle.$$

Armed with these, we can compute that $\langle m|(a^\dagger|n\rangle) = \langle m|\sqrt{[n + 1]}|n + 1\rangle = \delta_{m,n+1}\sqrt{[n + 1]} = \delta_{m-1,n}\sqrt{[m]} = (a|m\rangle)^\dagger|n\rangle = (\langle m|a^\dagger)|n\rangle$ etc. This can be said in mathematical notation also, as in the proof of Proposition 3.1.15. We see that the representation is unitary for the stated $*$-structure. That the abstract $*$-structure is well-defined is an easy computation. For real q, we have $a^\dagger{}^*a^* - q^{-1}a^*a^\dagger{}^* = aa^\dagger - q^{-1}a^\dagger a = q^N = (q^N)^*$ etc., while for $|q| = 1$, the relevant defining relations are interchanged. In either case, $*$ extends abstractly as an antialgebra map. Note that the Fock representation constructed here can also be used if $|q| = 1$ with suitable care; for example, if q is a root of unity, we generate a finite-dimensional Hilbert space in the same way. In passing, let us note that the identities above also imply that

$$a^\dagger a + aa^\dagger = \frac{q^{N+\frac{1}{2}} - q^{-N+\frac{1}{2}}}{q^{\frac{1}{2}} - q^{-\frac{1}{2}}}, \tag{3.8}$$

which is useful in some applications. ∎

Although this is not a Hopf algebra, it is relevant to the quantum groups below. It can be used to build representations of them in the same way that the ordinary quantum harmonic oscillator can be used to build representations of ordinary Lie algebras, and has numerous other applications besides.

Proposition 3.1.9 *The q-harmonic oscillator has another $*$-structure. Denoting its generators as D, Z, q^ξ rather than as $a, a^\dagger, q^N$, this is*

$$D^* = -D, \qquad Z^* = Z, \qquad \xi^* = -\xi - 1$$

(i.e. $(q^\xi)^* = q^{\mp 1}q^{\mp\xi}$ for q real or $|q| = 1$, respectively). There is a natural representation on functions in one variable,

$$(Z\psi)(z) = z\psi(z), \qquad D\psi = \partial_q\psi, \qquad q^\xi\psi = L_q\psi,$$

where ∂_q and L_q are the q-differential and degree operators defined by

$$(\partial_q\psi)(z) = \frac{\psi(qz) - \psi(q^{-1}z)}{(q - q^{-1})z}, \qquad (L_q\psi)(z) = \psi(qz),$$

which defines a unitary representation with respect to the L^2 inner product, when q is real.

Proof: We first verify the new $*$-structure. For real q we have $Z^*D^* - q^{-1}D^*Z^* = -ZD + q^{-1}DZ = q^{-1}(DZ - qZD) = q^{-1}q^{-\xi} = (q^\xi)^*$ and $(q^{-\xi})^*D^*(q^\xi)^* = -qq^\xi Dq^{-1}q^{-\xi} = -q^{-1}D = (q^\xi Dq^{-\xi})^*$. Similarly for the other relations. Thus $*$ extends as an antialgebra map. Likewise, for the case of $|q| = 1$. We next verify the stated representation, i.e. we verify

$$\partial_q Z - q^{-1}Z\partial_q = L_q, \quad \partial_q Z - qZ\partial_q = L_q^{-1},$$

$$L_q\partial_q = q^{-1}\partial_q L_q, \quad L_q Z = qZL_q \tag{3.9}$$

as operators. We let $\lambda = q - q^{-1}$. Then,

$$
\begin{aligned}
((\partial_q Z &- q^{-1}Z\partial_q)\psi)(z) \\
&= \lambda^{-1}z^{-1}((Z\psi)(qz) - (Z\psi)(q^{-1}z) - q^{-1}z(\psi(qz) - \psi(q^{-1}z))) \\
&= \lambda^{-1}(q\psi(qz) - q^{-1}\psi(q^{-1}z) - q^{-1}(\psi(qz) - \psi(q^{-1}z))) \\
&= \lambda^{-1}(q\psi(qz) - q^{-1}\psi(qz)) = (L_q\psi)(z),
\end{aligned}
$$

from the definitions. Similarly for the second part of (3.9). We also have

$$
\begin{aligned}
(L_q\partial_q\psi)(z) &= (\partial_q\psi)(qz) = \lambda^{-1}(qz)^{-1}(\psi(q^2z) - \psi(z)) \\
&= q^{-1}\lambda^{-1}z^{-1}((L_q\psi)(qz) - (L_q\psi)(q^{-1}z)) = q^{-1}(\partial_q L_q\psi)(z), \\
(L_q Z\psi)(z) &= (Z\psi)(qz) = qz\psi(qz) = qz(L_q\psi)(z) = q(ZL_q\psi)(z),
\end{aligned}
$$

as required.

This representation may be used in many settings, for example with ψ a polynomial, a holomorphic function or a formal power series; we verify that for functions where it is well-defined, our stated $*$-structure is indeed compatible with the usual L^2 inner product. The (unbounded) operator Z is the multiplication operator, so $Z^\dagger = Z$ with respect to L^2 as usual. Moreover,

$$
\begin{aligned}
\int_{-\infty}^{\infty} \bar{\phi}(z)(L_q\psi)(z)\mathrm{d}z &= \int_{-\infty}^{\infty} \bar{\phi}(z)\psi(qz)\mathrm{d}z \\
&= \int_{-\infty}^{\infty} q^{-1}\bar{\phi}(q^{-1}z')\psi(z')\mathrm{d}z' = \int_{-\infty}^{\infty} q^{-1}\overline{(L_q^{-1}\phi)}(z)\psi(z)\mathrm{d}z,
\end{aligned}
$$

using a change of variables $z' = qz$, and

$$
\begin{aligned}
\int_{-\infty}^{\infty} \bar{\phi}(z)\frac{\psi(qz) - \psi(q^{-1}z)}{(q - q^{-1})z}\mathrm{d}z \\
&= \int_{-\infty}^{\infty} \frac{\bar{\phi}(q^{-1}z')}{(q - q^{-1})z'}\psi(z')\mathrm{d}z' - \int_{-\infty}^{\infty} \frac{\bar{\phi}(qz')}{(q - q^{-1})z'}\psi(z')\mathrm{d}z' \\
&= -\int_{-\infty}^{\infty} \frac{\bar{\phi}(qz) - \bar{\phi}(q^{-1}z)}{(q - q^{-1})z}\psi(z)\mathrm{d}z
\end{aligned}
$$

as well. Hence $L_q^\dagger = q^{-1}L_q^{-1}$ and $\partial_q^\dagger = -\partial_q$ with respect to this inner product, on functions ψ, ϕ where the integrals and changes of variable make sense. Note that this is not the only, and perhaps not even the most natural, inner product to take in this context. Another is based on the so-called Jackson integral, which is a q-deformed or braided version of usual integration. ∎

Let us note that, in addition to satisfying the relations of the q-harmonic oscillator (or q-canonical commutation relations), the above q-differential operator and L_q obey the product rules

$$L_q(\phi\psi) = (L_q\phi)(L_q\psi), \quad \partial_q(\phi\psi) = (\partial_q\phi)L_q\psi + (L_q^{-1}\phi)\partial_q\psi, \quad (3.10)$$

justifying the term 'q-differentiation' for the latter. The first identity is clear, while the second follows from

$$\phi(qz)\psi(qz) - \phi(q^{-1}z)\psi(q^{-1}z)$$
$$= (\phi(qz) - \phi(q^{-1}z))\psi(qz) + \phi(q^{-1}z)(\psi(qz) - \psi(q^{-1}z)).$$

This means that functions in one variable form a module algebra in the sense of Chapter 1.6.1 for the abstract Hopf algebra generated by D, q^ξ with relations $q^\xi D q^{-\xi} = q^{-1}D$ and coproduct

$$\Delta q^\xi = q^\xi \otimes q^\xi, \quad \Delta D = D \otimes q^\xi + q^{-\xi} \otimes D,$$

when D, q^ξ are represented by ∂_q, L_q, as above. This abstract Hopf algebra is a variant of Example 1.3.2, and figures prominently as a sub-Hopf algebra of the quantum group in the next section, as well as being a subalgebra of the q-harmonic oscillator above. It demonstrates how q-derivations typically arise as a consequence of covariance under a quantum group with q-deformed coproduct. Another point of view appears in Chapter 10.4, in terms of braid statistics.

3.2 $U_q(sl_2)$ and its real forms

After the warm-up with the previous section, we come now to what is probably the most best-known quantum group, the deformation of the universal enveloping algebra of the three-dimensional complex Lie algebra, sl_2, and its real form, su_2. Recall that this Lie algebra has generators $X_\pm, H$ say, with relations

$$[H, X_\pm] = \pm 2X_\pm, \quad [X_+, X_-] = H.$$

This is the smallest simple Lie algebra. Moreover (as with much of Lie algebra theory), its properties generalise to all the complex simple Lie algebras. Its universal enveloping Hopf algebra from Example 1.5.7 is $U(sl_2)$. Its standard deformation is as follows.

Example 3.2.1 *Let q be a nonzero parameter. We define $U_q(sl_2)$ as the noncommutative algebra generated by 1 and $X_+, X_-, q^{\frac{H}{2}}, q^{-\frac{H}{2}}$ with the relations $q^{\pm\frac{H}{2}} q^{\mp\frac{H}{2}} = 1$ (the notation suggests these) and*

$$q^{\frac{H}{2}} X_\pm q^{-\frac{H}{2}} = q^{\pm 1} X_\pm, \qquad [X_+, X_-] = \frac{q^H - q^{-H}}{q - q^{-1}}.$$

This forms a Hopf algebra with

$$\Delta q^{\pm\frac{H}{2}} = q^{\pm\frac{H}{2}} \otimes q^{\pm\frac{H}{2}}, \quad \Delta X_\pm = X_\pm \otimes q^{\frac{H}{2}} + q^{-\frac{H}{2}} \otimes X_\pm,$$

$$\epsilon q^{\pm\frac{H}{2}} = 1, \quad \epsilon X_\pm = 0, \quad S X_\pm = -q^{\pm 1} X_\pm, \quad S q^{\pm\frac{H}{2}} = q^{\mp\frac{H}{2}}.$$

Over $\mathbb{C}[[t]]$, we can regard $H, X_\pm$ as the generators, and then the Hopf algebra is quasitriangular with

$$\mathcal{R} = q^{\frac{H \otimes H}{2}} \sum_{n=0}^{\infty} \frac{(1 - q^{-2})^n}{[n]!} (q^{\frac{H}{2}} X_+ \otimes q^{-\frac{H}{2}} X_-)^n q^{\frac{n(n-1)}{2}}, \quad [n] = \frac{q^n - q^{-n}}{q - q^{-1}}$$

and $[n]! = [n][n-1]...[1]$.

Proof: We begin by verifying the Hopf algebra axioms. The two sub-Hopf algebras generated by $q^{\frac{H}{2}}, X_+$ and $q^{\frac{H}{2}}, X_-$ have the same structure as the example at the end of the preceding section and were already verified in an equivalent form in Example 1.3.2, so we concentrate on showing that Δ respects the remaining $[X_+, X_-]$ relations. As in the proof of Example 3.1.2, we have $[\Delta X_+, \Delta X_-] = [X_+ \otimes q^{\frac{H}{2}} + q^{-\frac{H}{2}} \otimes X_+, X_- \otimes q^{\frac{H}{2}} + q^{-\frac{H}{2}} \otimes X_-] = [X_+, X_-] \otimes q^H + q^{-H} \otimes [X_+, X_-] = \Delta[X_+, X_-]$. Although $q^{\frac{H}{2}}$ is not central, the cross terms in the commutator still vanish because

$$X_\pm q^{-\frac{H}{2}} \otimes q^{\frac{H}{2}} X_\pm = q^{-\frac{H}{2}} X_\pm \otimes X_\pm q^{\frac{H}{2}}$$

using the relations in the algebra. Thus, Δ extends as an algebra homomorphism. That the counit and antipode extend is also easily verified. We now verify the quasitriangular structure. An important step for the verification of (2.1) is to write $\mathcal{R}$ in the form

$$\mathcal{R} = q^{\frac{H \otimes H}{2}} e_{q^{-2}}^{(1-q^{-2})q^{\frac{H}{2}} X_+ \otimes q^{-\frac{H}{2}} X_-}, \tag{3.11}$$

where $e_{q^{-2}}$ is the q-exponential. Its basic properties are in Lemma 3.2.2 below. Then, following the same strategy as in the proof of Example 3.1.2, we compute

$$(\Delta \otimes \mathrm{id})\mathcal{R} = q^{\frac{1}{2}(H \otimes 1 \otimes H + 1 \otimes H \otimes H)}$$

$$\times e_{q^{-2}}^{(1-q^{-2})q^{\frac{H}{2}} X_+ \otimes q^H \otimes q^{-\frac{H}{2}} X_- + 1 \otimes q^{\frac{H}{2}} X_+ \otimes q^{-\frac{H}{2}} X_-}$$

$$= q^{\frac{1}{2}(H \otimes 1 \otimes H + 1 \otimes H \otimes H)}$$

$$\times e_{q^{-2}}^{(1-q^{-2})q^{\frac{H}{2}}X_+ \otimes q^H \otimes q^{-\frac{H}{2}}X_-} e_{q^{-2}}^{(1-q^{-2})1 \otimes q^{\frac{H}{2}}X_+ \otimes q^{-\frac{H}{2}}X_-}$$

$$= q^{\frac{1}{2}H \otimes 1 \otimes H} e_{q^{-2}}^{(1-q^{-2})q^{\frac{H}{2}}X_+ \otimes 1 \otimes q^{-\frac{H}{2}}X_-}$$

$$\times q^{\frac{1}{2}1 \otimes H \otimes H} e_{q^{-2}}^{(1-q^{-2})1 \otimes q^{\frac{H}{2}}X_+ \otimes q^{-\frac{H}{2}}X_-}$$

$$= \mathcal{R}_{13}\mathcal{R}_{23}.$$

For the first equality, we use that Δ is an algebra homomorphism. For the second, we note that the elements $A = q^{\frac{H}{2}}X_+ \otimes q^H \otimes q^{-\frac{H}{2}}X_-$ and $B = 1 \otimes q^{\frac{H}{2}}X_+ \otimes q^{-\frac{H}{2}}X_-$ obey $AB = q^2 BA$, and hence we can use the first part of Lemma 3.2.2. For the third equality we use

$$q^{\frac{H \otimes H}{2}}(f(H) \otimes g(H)X_-)q^{-\frac{H \otimes H}{2}} = q^{-H}f(H) \otimes g(H)X_- \qquad (3.12)$$

for any functions f, g to move $q^{\frac{1 \otimes H \otimes H}{2}}$ past the first $e_{q^{-2}}$ factor. Similarly for the second part of (2.1). Next we note that $\mathcal{R}(q^{\frac{H}{2}} \otimes q^{\frac{H}{2}}) = (q^{\frac{H}{2}} \otimes q^{\frac{H}{2}})\mathcal{R}$ because $q^{\frac{H}{2}} \otimes q^{\frac{H}{2}}$ commutes with any function of H and commutes with the exponent of $e_{q^{-2}}$ in (3.11). Hence (2.2) is satisfied on $q^{\frac{H}{2}}$. To show that it is satisfied on X_+, we have to show that $\mathcal{R}(\Delta X_+) = (\tau \circ \Delta X_+)\mathcal{R}$, i.e. we require

$$q^{\frac{H \otimes H}{2}}e_{q^{-2}}^{(1-q^{-2})q^{\frac{H}{2}}X_+ \otimes q^{-\frac{H}{2}}X_-}(X_+ \otimes q^{\frac{H}{2}} + q^{-\frac{H}{2}} \otimes X_+)$$

$$= (q^{\frac{H}{2}} \otimes X_+ + X_+ \otimes q^{-\frac{H}{2}})q^{\frac{H \otimes H}{2}}e_{q^{-2}}^{(1-q^{-2})q^{\frac{H}{2}}X_+ \otimes q^{-\frac{H}{2}}X_-}.$$

We multiply both sides by $q^{-\frac{H \otimes H}{2}}$ and compute $q^{-\frac{H \otimes H}{2}}(\ )q^{\frac{H \otimes H}{2}}$ along the lines of (3.12). Then the required identity becomes

$$e_{q^{-2}}^A(B + C) = (C + D)e_{q^{-2}}^A,$$

where

$$A = (1 - q^{-2})q^{\frac{H}{2}}X_+ \otimes q^{-\frac{H}{2}}X_-, \quad B = X_+ \otimes q^{\frac{H}{2}},$$

$$C = q^{-\frac{H}{2}} \otimes X_+, \quad D = X_+ \otimes q^{-\frac{3}{2}H}.$$

This identity follows from Lemma 3.2.2 below, for which we compute

$$AB = (1 - q^{-2})q^{\frac{H}{2}}X_+^2 \otimes qX_- = (1 - q^{-2})q^2 X_+ q^{\frac{H}{2}}X_+ \otimes X_- = q^2 BA,$$

$$AD = (1 - q^{-2})q^{\frac{H}{2}}X_+^2 \otimes q^{-\frac{H}{2}}X_- q^{-\frac{3}{2}H}$$

$$= (1 - q^{-2})q^{-2}X_+ q^{\frac{H}{2}}X_+ \otimes q^{-2H}X_- = q^{-2}DA,$$

$$AC = (q - q^{-1})X_+ \otimes q^{-\frac{H}{2}}X_- X_+$$

$$= (q - q^{-1})X_+ \otimes q^{-\frac{H}{2}}X_+X_- - X_+ \otimes (q^{\frac{H}{2}} - q^{-\frac{3}{2}H}) = CA - B + D,$$

using the relations in the algebra. The proof of (2.2) for ΔX_- is entirely analogous.   ∎

Lemma 3.2.2 Let $[n; q^{-2}] = \frac{1 - q^{-2n}}{1 - q^{-2}}$ and $[n; q^{-2}]! = [n; q^{-2}] \cdots [1; q^{-2}]$. We define

$$\begin{bmatrix} n \\ m \end{bmatrix}; q^{-2} = \frac{[n; q^{-2}]!}{[m; q^{-2}]![n - m; q^{-2}]!}, \qquad e_{q^{-2}}^x = \sum_{n=0}^{\infty} \frac{x^n}{[n; q^{-2}]!}.$$

(i) If $AB = q^2 BA$, then

$$(A + B)^n = \sum_{m=0}^{n} \begin{bmatrix} n \\ m \end{bmatrix}; q^{-2} A^m B^{n-m}, \qquad e_{q^{-2}}^{A+B} = e_{q^{-2}}^A e_{q^{-2}}^B.$$

(ii) If, in addition, $AD = q^{-2}DA$ and $[C, A] = B - D$, then

$$[C, A^n] = [n; q^{-2}](A^{n-1}B - DA^{n-1}), \qquad [C, e_{q^{-2}}^A] = e_{q^{-2}}^A B - De_{q^{-2}}^A.$$

(iii) Finally, consider the operators

$$(\partial f)(x) = \frac{f(x) - f(q^{-2}x)}{(1 - q^{-2})x}, \qquad (Lf)(x) = f(q^{-2}x)$$

on functions f. Then

$$\partial(fg) = (\partial f)g + (Lf)\partial g, \qquad \partial x^n = [n; q^{-2}]x^{n-1},$$

$$\partial e_{q^{-2}}^{\lambda x} = \lambda e_{q^{-2}}^{\lambda x}, \qquad e_{q^{-2}}^x e_{q^2}^{-x} = 1,$$

for all functions f, g and any constant λ.

Proof: (i) The results about $e_{q^{-2}}$ follow at once from the results on the left for each n. To prove the first of these (the q-binomial theorem), we proceed by induction. Assuming the result for $(A + B)^{n-1}$, we have

$$(A + B)^{n-1}(A + B) = \sum_{m=0}^{n-1} \begin{bmatrix} n-1 \\ m \end{bmatrix}; q^{-2} A^m B^{n-1-m}(A + B)$$

$$= \sum_{m=0}^{n-1} q^{-2(n-1-m)} \begin{bmatrix} n-1 \\ m \end{bmatrix}; q^{-2} A^{m+1} B^{n-1-m}$$

$$+ \sum_{m=0}^{n-1} \begin{bmatrix} n-1 \\ m \end{bmatrix}; q^{-2} A^m B^{n-m}$$

$$= \sum_{m=1}^{n} q^{-2(n-m)} \begin{bmatrix} n-1 \\ m-1 \end{bmatrix}; q^{-2} A^m B^{n-m} + \sum_{m=0}^{n-1} \begin{bmatrix} n-1 \\ m \end{bmatrix}; q^{-2} A^m B^{n-m}$$

$$= A^n + B^n + \sum_{m=1}^{n-1} \left(q^{-2(n-m)} \begin{bmatrix} n-1 \\ m-1 \end{bmatrix} ; q^{-2} \right] + \begin{bmatrix} n-1 \\ m \end{bmatrix} ; q^{-2} \right] \right) A^m B^{n-m}$$

The expression in parentheses combines to $[\begin{smallmatrix} n \\ m \end{smallmatrix}; q^{-2}]$, as required, after an elementary computation using the identity

$$q^{-2(n-m)}[m; q^{-2}] + [n-m; q^{-2}] = [n; q^{-2}]. \tag{3.13}$$

(ii) For the second part, we again proceed by induction. Assuming the result for A^{n-1}, we have

$$
\begin{aligned}
[C, A^n] &= A[C, A^{n-1}] + (B - D)A^{n-1} \\
&= A[n-1; q^{-2}](A^{n-2}B - DA^{n-2}) + (B - D)A^{n-1} \\
&= ([n-1; q^{-2}] + q^{-2(n-1)})A^{n-1}B - (1 + q^{-2}[n-1; q^{-2}])DA^{n-1} \\
&= [n; q^{-2}](A^{n-1}B - DA^{n-1}),
\end{aligned}
$$

where we have used $[n-1; q^{-2}] + q^{-2(n-1)} = [n; q^{-2}] = 1 + q^{-2}[n-1; q^{-2}]$, as in (3.13). (iii) Finally, for the last part (which we will need below), we note that the q-Leibniz rule has already been discussed in an equivalent form in (3.10). This and ∂ on x^n and $e_{q^{-2}}$ are, in any case, immediate. To see that they imply the final statement, we note that $\partial = L \circ \partial_{q^{-1}}$, and compute

$$
\begin{aligned}
\partial(e^x_{q^{-2}} e^{-x}_{q^2}) &= \partial(e^x_{q^{-2}})e^{-x}_{q^2} + L(e^x_{q^{-2}})\partial(e^{-x}_{q^2}) \\
&= e^x_{q^{-2}}e^{-x}_{q^2} + L(e^x_{q^{-2}})L \circ \partial_{q^{-1}}(e^{-x}_{q^2}) \\
&= e^x_{q^{-2}}e^{-x}_{q^2} + L(e^x_{q^{-2}})L(-e^{-x}_{q^2}) = (\mathrm{id} - L)(e^x_{q^{-2}}e^{-x}_{q^2}) \\
&= x(1 - q^{-2})\partial(e^x_{q^{-2}}e^{-x}_{q^2}),
\end{aligned}
$$

from which we conclude that $\partial(e^x_{q^{-2}}e^{-x}_{q^2}) = 0$. But the only power series in x with this property is the constant one (since it means invariance under L), and so we conclude that $e^x_{q^{-2}}e^{-x}_{q^2} = 1$. Our proof here is a little formal, but illustrates some useful techniques. More direct algebraic proofs are also possible. ∎

We now consider ∗-structures for this Hopf algebra. The classical Lie algebra sl_2 has two inequivalent real forms: su_2 and $su(1,1)\cong sl(2,\mathbb{R})$. These are all three-dimensional real Lie algebras contained in sl_2 as sub-Lie algebras and giving sl_2 on complexification, the different cases being specified as the subspaces fixed under an antilinear involution. We denote the latter by $-*$, where ∗ is an antilinear antiinvolution on the complex Lie algebra. It corresponds to a Hopf ∗-algebra structure on the enveloping algebra. This ∗, in turn, determines what it is to be a 'unitary' representation of the real form, namely a representation where ∗ is compatible with the inner product in the usual way (hence, where the elements of the real form are anti-Hermitian operators). In the standard two-dimensional representation of sl_2, the elements of the real form su_2 are indeed anti-Hermitian, and hence exponentiate to unitary matrices. We use the same framework for quantum groups; we consider a real form of a Hopf algebra over $\mathbb{C}$ to be a specification of a Hopf $*$-algebra structure in the sense of Definition 1.7.5. As we have seen already, this determines, among other things, what it is to be a unitary (or $*$-preserving) representation. In the case of $U_q(sl_2)$, we again find two real forms when q is real.

Example 3.2.3 *The real form $U_q(su_2)$ is the Hopf $*$-algebra consisting of the Hopf algebra $U_q(sl_2)$ for q real, and the $*$-structure*

$$H^* = H, \qquad X_\pm^* = X_\mp.$$

This real form is real quasitriangular when we work over $\mathbb{C}[[t]]$, in the sense $\mathcal{R}^{*\otimes*} = \tau(\mathcal{R})$.

Proof: The verification that the $*$-algebra extends as an antialgebra map and that Δ, ϵ are $*$-algebra homomorphisms is analogous to Proposition 3.1.4. For example, $(\Delta X_\pm)^{*\otimes*} = X_\pm^* \otimes (q^{\frac{H}{2}})^* + (q^{-\frac{H}{2}})^* \otimes X_\pm^* = X_\mp \otimes q^{\frac{H}{2}} + q^{-\frac{H}{2}} \otimes X_\mp = \Delta X_\mp = \Delta(X_\pm^*)$. For the antipode we compute $S\circ*(X_\pm) = S(X_\mp) = -q^{\mp 1}X_\mp$, so $(S\circ*)^2(X_\pm) = (S\circ*)(-q^{\mp 1}X_\mp) = X_\pm$, as required for a Hopf $*$-algebra. Similarly for the other facts. For the quasitriangular structure, we note first that

$$q^{-\frac{H\otimes H}{2}}(X_+ q^{-\frac{H}{2}} \otimes X_- q^{\frac{H}{2}})q^{\frac{H\otimes H}{2}}$$
$$= q^{-\frac{H\otimes H}{2}}(X_+ q^{-\frac{H}{2}} \otimes 1)(1 \otimes X_- q^{\frac{H}{2}})q^{\frac{H\otimes H}{2}}$$
$$= (X_+ q^{-\frac{H}{2}} \otimes q^{-H})(q^H \otimes X_- q^{\frac{H}{2}})$$
$$= X_+ q^{\frac{H}{2}} \otimes q^{-H} X_- q^{\frac{H}{2}} = q^{\frac{H}{2}} X_+ \otimes q^{-\frac{H}{2}} X_-,$$

along the lines of (3.12), using the relations in the algebra. This means that an equally good expression for $\mathcal{R}$ is

$$\mathcal{R} = e_{q^{-2}}^{(1-q^{-2})X_+ q^{-\frac{H}{2}} \otimes X_- q^{\frac{H}{2}}} q^{\frac{H\otimes H}{2}}. \qquad (3.14)$$

Starting with our original expression (3.11), we then compute

$$\mathcal{R}^{*\otimes*} = e_{q^{-2*}}^{(1-q^{-2})^*(q^{\frac{H}{2}}X_+)^* \otimes (q^{-\frac{H}{2}}X_-)^*}(q^{\frac{H\otimes H}{2}})^*$$
$$= e_{q^{-2}}^{(1-q^{-2})X_- q^{\frac{H}{2}} \otimes X_+ q^{-\frac{H}{2}}} q^{\frac{H\otimes H}{2}} = \tau(\mathcal{R}).$$

∎

Example 3.2.4 *The real form $U_q(su(1,1))$ is the Hopf $*$-algebra consist-ing of the Hopf algebra $U_q(sl_2)$ for q real and the $*$-structure*

$$H^* = H, \qquad X_\pm^* = -X_\mp.$$

This real form is also real quasitriangular over $\mathbb{C}[[t]]$.

Proof: Changing both X_+^* and X_-^* by a minus sign does not change the computation of $[X_-^*, X_+^*] = [X_+, X_-]^*$ (because the expressions are quadratic) or $(S \circ *)^2 X_\pm = X_\pm$ (because a minus sign enters into the action of each $(S \circ *)$). Other properties are more obviously unaffected by linearity or antilinearity. As a result, we again have a Hopf $*$-algebra. The computation of $\mathcal{R}^{* \otimes *}$ is likewise unaffected because $X_\pm$ enter via $X_+ \otimes X_-$. ∎

Example 3.2.5 *The real form $U_q(sl(2, \mathbb{R}))$ is the Hopf $*$-algebra consist-ing of the Hopf algebra $U_q(sl_2)$ for $|q| = 1$ and the $*$-structure*

$$H^* = -H, \qquad X_\pm^* = -X_\pm.$$

This is antireal quasitriangular over $\mathbb{C}[[t]]$, in the sense $\mathcal{R}^{* \otimes *} = \mathcal{R}^{-1}$, where t is now imaginary.

Proof: With this choice of $*$-structure, we note that $q^{\frac{H}{2}}$ remains self-adjoint. Hence, $\frac{q^H - q^{-H}}{q - q^{-1}}$ now changes sign under $*$, while $[X^*_-, X^*_+]$ also acquires a minus sign. This is as in the proof of Exercise 3.1.6. That Δ is a $*$-algebra map also remains valid because $q^{\frac{H}{2}}$ is self-adjoint. For the quasitriangular structure, we compute

$$\mathcal{R}^{* \otimes *} = e_{q^{-2*}}^{(1-q^{-2})^*(q^{\frac{H}{2}}X_+)^*} \otimes (q^{-\frac{H}{2}}X_-)^* (q^{\frac{H \otimes H}{2}})^*$$

$$= e_{q^2}^{(1-q^2)X_+ q^{\frac{H}{2}}} \otimes X_- q^{-\frac{H}{2}} q^{-\frac{H \otimes H}{2}}$$

$$= e_{q^2}^{-(1-q^{-2})q^{\frac{H}{2}}X_+ \otimes q^{-\frac{H}{2}}X_-} q^{-\frac{H \otimes H}{2}}$$

$$= \left(e_{q^{-2}}^{(1-q^{-2})q^{\frac{H}{2}}X_+ \otimes q^{-\frac{H}{2}}X_-} \right)^{-1} q^{-\frac{H \otimes H}{2}} = \mathcal{R}^{-1}$$

by the relations in the algebra (to organise the exponent of e_{q^2}), followed by the third part of Lemma 3.2.2 with the operator $q^{\frac{H}{2}}X_+ \otimes q^{-\frac{H}{2}}X_-$ formally in the role of x. A less formal proof (not using Lemma 3.2.2) is to compute $\mathcal{R}^{-1} = (S \otimes \mathrm{id})(\mathcal{R})$ directly from the formula for $\mathcal{R}$ in Example 3.2.1 and compare this with the third expression for $\mathcal{R}^{* \otimes *}$ above. Note that the same argument as in the proof of Example 3.2.4 gives a further real form $H^* = -H$ and $X_\pm^* = X_\pm$, but this time it is equivalent after a redefinition of $X_\pm$ by a factor $\pm\imath$.   ∎

We see that the two inequivalent real forms of sl_2 are q-deformed as two real quasitriangular Hopf *-algebras when q is real, while the second of the real forms is q-deformed as an antireal quasitriangular Hopf *-algebra when $|q| = 1$. The su_2 real form also q-deforms when $|q| = 1$, but not exactly as a Hopf *-algebra; one needs an additional transposition in the coproduct axiom in Definition 1.7.5 which is not visible classically (since the undeformed enveloping algebra is cocommutative). We will cover this 'missing' case in Section 3.4. This completes the definition of $U_q(sl_2)$ and its real forms. It acts on objects in much the same way as the Lie algebra sl_2. We focus on the finite-dimensional unitary irreducible su_2 representations, which are classified by a non-negative half-integer j (the spin). They have dimension $2j + 1$.

Proposition 3.2.6 *For real q and each $j = 0, \frac{1}{2}, 1, \ldots$, the quantum group $U_q(su_2)$ has a $2j + 1$-dimensional unitary irreducible representation $V_j = \{|j, m\rangle|\ m = -j, \ldots, j\}$, with*

$$X_\pm|j, m\rangle = \sqrt{[j \mp m][j \pm m + 1]}|j, m \pm 1\rangle, \qquad q^{\frac{H}{2}}|j, m\rangle = q^m|j, m\rangle$$

and $[n] = \frac{q^n - q^{-n}}{q - q^{-1}}$.

Proof: It is easy to verify that this is a representation. For example, along the lines of (3.13), we have that $[X_+, X_-]|j, m\rangle = X_+\sqrt{[j + m][j - m + 1]}$ $|j, m - 1\rangle - X_-\sqrt{[j - m][j + m + 1]}|j, m + 1\rangle = ([j + m][j - m + 1] - [j - m][j + m + 1])|j, m\rangle = [2m]|j, m\rangle$. These normalisations are chosen so that, at least for real q, we have a unitary representation for the *-structure in Example 3.2.3. For example, we have that $\langle j, m|(X_+|j, n\rangle) = \langle j, m|\sqrt{[j - n][j + n + 1]}|j, n + 1\rangle = \delta_{m,n+1}\sqrt{[j - n][j + n + 1]} = \delta_{m-1,n}$ $\sqrt{[j - m + 1][j + m]} = (X_-|j, m\rangle)^\dagger|j, n\rangle$, as required. In mathematical notation, the sesquilinear inner product (,), defined by considering the basis $\{|j, m\rangle\}$ to be orthonormal, obeys $(v, X_+w) = (X_-v, w)$ for all vectors v, w. This is similar to Example 3.1.8 in Section 3.1. We note that this argument also applies when q is a root of unity, provided the allowed range of j is restricted in a suitable way.   ∎

Proposition 3.2.7 *The algebra $U_q(sl_2)$ has a quadratic central element (the quadratic Casimir) given by*

$$C = q^{H-1} + q^{-H+1} + X_+X_-(q - q^{-1})^2.$$

Its value in the spin j representation, the actions of the canonical elements u, v and the value of the central ribbon element are

$$C|_{V_j} = q^{2j+1} + q^{-2j-1}, \quad u|_{V_j} = q^{-2j(j+1)}q^H,$$
$$v|_{V_j} = q^{-2j(j+1)}q^{-H}, \quad \nu|_{V_j} = q^{-2j(j+1)}.$$

Proof: That the quadratic Casimir element is central follows easily from the algebra relations in Example 3.2.1. Its value on V_j is $q^{2m-1}+q^{-2m+1}+ (q^{j+m} - q^{-j-m})(q^{j-m+1} - q^{-j+m-1}) = q^{2j+1}+q^{-2j-1}$, where we have used the actions in the preceding proposition. The actions of the elements $u = \sum(S\mathcal{R}^{(2)})\mathcal{R}^{(1)}$, $v = \sum \mathcal{R}^{(1)}S\mathcal{R}^{(2)}$, are computed in a similar way from the formula for $\mathcal{R}$ in Example 3.2.1 acting on states $|j, m\rangle$. Given this, the square root of their product uv is as stated. ∎

Here we have assumed that $U_q(sl_2)$ is a ribbon Hopf algebra so that uv does have a suitable square root, ν, in the algebra (this is true over $\mathbb{C}[[t]]$). The quasitriangular structure is also factorisable over $\mathbb{C}[[t]]$ as a corollary of a result about quantum doubles in Chapter 7 (see also Section 3.4 below).

To conclude this section, we describe a related algebra which was one of the precursors for the structure of $U_q(sl_2)$ and which has remarkable symmetry and regularity properties of its own. It was discovered by E.K. Sklyanin and has been extensively studied by ring-theorists because of these remarkable properties. This *Sklyanin algebra* depends on three parameters J_{12}, J_{23}, J_{31} subject to the constraint $J_{12}+J_{23}+J_{31}+J_{12}J_{23}J_{31} = 0$ and has four generators S_0, S_α, $\alpha = 1, 2, 3$ subject to the relations

$$[S_0, S_\alpha] = \imath J_{\beta\gamma}\{S_\beta, S_\gamma\}, \quad [S_\alpha, S_\beta] = \imath\{S_0, S_\gamma\}, \tag{3.15}$$

where α, β, γ are from the set 1,2,3 in cyclic order and $\{\ ,\ \}$ denotes anticommutator. There are two quadratic central elements

$$C_1 = S_0^2 + \sum_\alpha S_\alpha^2, \quad C_2 = \sum_\alpha S_\alpha^2 J_\alpha, \tag{3.16}$$

where J_α are such that $J_{\alpha\beta} = -\frac{J_\alpha - J_\beta}{J_\gamma}$. If S_0 is the identity, then the second relation in (3.15) reduces to the familiar vector description of the generators of rotation group in three dimensions; the Sklyanin algebra is an interesting generalisation of this.

Relevant to us is the degenerate case algebra where (say) $J_{12} = 0$. Then $J_{31} = -J_{23}$ and the relations become

$$[S_0, S_1] = \imath J_{23}\{S_2, S_3\}, \quad [S_0, S_2] = -\imath J_{23}\{S_3, S_1\},$$

$$[S_0, S_3] = 0, \quad [S_\alpha, S_\beta] = \imath\{S_0, S_\gamma\}. \tag{3.17}$$

We assume (without loss of generality) that our single remaining parameter, J_{23}, is positive.

Proposition 3.2.8 *The additional relation*

$$S_0^2 - J_{23}S_3^2 = 1$$

reduces the degenerate Sklyanin algebra to the algebra of the quantum group $U_q(sl_2)$. The required identification is

$$q^{\frac{H}{2}} = S_0 + \sqrt{J_{23}}S_3, \quad X_\pm = \frac{1}{2}\sqrt{1 - J_{23}}(S_1 \pm \imath S_2), \quad q = \frac{1 + \sqrt{J_{23}}}{1 - \sqrt{J_{23}}}.$$

Proof: We write $S_\pm = S_1 \pm \imath S_2$ and $K_\pm = S_0 \pm tS_3$, where $t = \sqrt{J_{23}}$ (we fix a square root). Then the relations (3.17) become

$$[K_+, S_\pm] = \pm t\{K_+, S_\pm\}, \quad [K_-, S_\pm] = \mp t\{K_-, S_\pm\},$$

$$[K_+, K_-] = 0, \quad [S_+, S_-] = \frac{1}{t}(K_+^2 - K_-^2).$$

Writing $q = \frac{1+t}{1-t}$ and $Y_\pm = \frac{1}{2}\sqrt{1 - t^2}S_\pm$, we have

$$[K_+, K_-] = 0, \quad [Y_+, Y_-] = \frac{K_+^2 - K_-^2}{q - q^{-1}},$$

$$K_+Y_\pm = q^{\pm 1}Y_\pm K_+, \quad K_-Y_\pm = q^{\mp 1}Y_\pm K_-,$$

while two independent linear combinations of the Casimir elements are (with $J_1 = J_2 = 1, J_3 = 1 + J_{23}$)

$$C_1 - C_2 = K_+K_-, \quad 2\left(\frac{(1 + t^2)}{(1 - t^2)}C_1 - C_2\right) = q^{-1}K_+^2 + qK_-^2 + Y_+Y_-(q - q^{-1})^2.$$

Thus, with these changes of variables, we see that the further quotient $K_+K_- = 1$ gives us the algebra of $U_q(sl_2)$, and the second displayed combination of C_1, C_2 becomes its quadratic Casimir element, as in Proposition 3.2.7. ∎

3.3 $U_q(g)$ for general Lie algebras

In this section, we describe briefly the standard deformations of the enveloping algebras of a general complex simple Lie algebra g. In fact, the main complications here (for generic q or over $\mathbb{C}[[t]]$) are just the usual complexities of working with general simple Lie algebras (such as weights, roots, Weyl chambers, etc.). This means that there is little here from the point of view of abstract quantum group theory beyond what was already seen for $U_q(sl_2)$, and this is one reason that we will keep this section short.

The situation is that the construction of these standard quantum groups $U_q(g)$ depends in a natural way on the Lie algebra g alone, and therefore carries over much of the mathematical beauty and richness of the theory of Lie algebras (as well as its complexities). This means that the theory of $U_q(g)$ can reasonably be viewed as a self-contained outgrowth of Lie algebra theory, and, as such, its results can perhaps be obtained equivalently by other natural constructions based on Lie algebras, such as Kac–Moody and loop-group Lie algebras. This brings us, however, to a small warning. Because the study of $U_q(g)$ is quite beautiful, it is easy to believe that this is what quantum groups are all about, some kind of outgrowth of Lie algebra theory. This is not the point of view of this book, however, where the whole point of Hopf algebra theory is that there are plenty of entirely new examples and phenomena that have no analogy in Lie algebra theory or group theory. With this warning in mind, it is still important for us to describe the standard $U_q(g)$ as a reference point, but we should not stop here.

Let g be a complex simple Lie algebra. Let t be a Cartan subalgebra of g and let t^* be its dual linear space. Recall that the roots α and weights Λ lie in this space. Let $\alpha_i \in t^*$ be a system of positive simple roots. Here i runs from 1 up to $l = \text{rank}(g)$. The positive roots are certain non-negative integral linear combinations of the α_i. The inverse of the Killing form defines a symmetric bilinear form $(\, , \,)$ on t^* in the usual way, and the Cartan matrix $a_{ij} = (\check{\alpha}_i, \alpha_j)$, where $\check{\alpha}_i = 2\alpha_i/(\alpha_i, \alpha_i)$ are the coroots. Let $d_i = (\alpha_i, \alpha_i)/2$. Then

$$d_i \in \mathbb{N}, \quad a_{ii} = 2, \quad d_i a_{ij} = d_j a_{ji}, \quad a_{ij} < 0, \quad \text{for } i \neq j \qquad (3.18)$$

are some general properties of the inner product in terms of the Cartan matrix. Much of the theory can be developed assuming only this data.

With these ingredients, the structure of g and its enveloping algebra $U(g)$ can be described conveniently in terms of generators which are associated to these simple roots. Thus, we define $H_i \in t$ by $K(d_i H_i, h) = \alpha_i(h)$ for all $h \in t$ or, equivalently, $\alpha_i(H_j) = K(d_i H_i, H_j) = a_{ji}$. These are the *Cartan generators*. We choose further *Chevalley generators*, $X_{\pm i}$, such that $K(X_{+i}, X_{-j}) = d_i^{-1}\delta_{ij}$ and such that the Lie algebra g has relations

$$[H_i, H_j] = 0, \quad [H_i, X_{\pm j}] = \pm a_{ij}X_{\pm j}, \quad [X_{+i}, X_{-j}] = \delta_{ij}H_i. \quad (3.19)$$

These 3l generators do not usually span all of g, but they generate it in the sense that the remaining basis elements $X_{\pm\alpha}$, for general positive roots α, are obtained by repeated applications of the Lie bracket of Chevalley generators on themselves. They can be given explicitly by using the Weyl group. This is the sense in which $\{H_i, X_{\pm i}\}$ generate g. The *Serre relations*

$$[X_{\pm i}, \ ]^{1-a_{ij}}(X_{\pm j}) = 0, \qquad i \neq j \quad (3.20)$$

ensure that this process terminates. The general root vectors obey

$$[h, X_{\pm\alpha}] = \pm\alpha(h)X_{\pm\alpha}, \quad (3.21)$$

but for the most part we do not need to work with them directly. One can give a formal definition of g along these lines in terms of the data d_i, a_{ij}.

This description of g is especially useful for constructing the enveloping Hopf algebra $U(g)$. In principle, we should take as generators a basis of g and commutation relations defined by the Lie bracket, as explained in Example 1.5.7. However, since the nonsimple root vectors are generated by repeated applications of the Lie bracket from the simple ones, we can just take $\{H_i, X_{\pm i}\}$ as the generators of $U(g)$ and (3.19)–(3.20) as its commutation relations.

Finally, the Weyl group W associated to g is a discrete group acting on t^* by isometries of the inner product $(,)$, and fixing the set of roots. It is generated by the simple reflections s_i defined as $s_i(\alpha) = \alpha - (\alpha, \check{\alpha}_i)\alpha_i$. Let us also recall that the fundamental weights ω_i are defined by $(\omega_i, \check{\alpha}_j) = \delta_{ij}$. They form a basis of t^*. The weights $\Lambda = \sum_i \Lambda^i\omega_i$ are the elements of t^* with integer coefficients Λ^i. They form an integral lattice in t^*. The dominant weights are those Λ with non-negative coefficients; they label the finite-dimensional representations of g and $U(g)$.

This completes our summary of the situation for the undeformed $U(g)$. For the standard deformations, $U_q(g)$, we use the same data, namely the set of simple roots α_i, the inner product, Cartan matrix, etc. As before, we can either associate to this root system generators $\{q_i^{\pm\frac{H_i}{2}}, X_i, X_{-i}\}$ generating a Hopf algebra over $\mathbb{C}$ or define $U_q(g)$ over formal power series $\mathbb{C}[[t]]$ with the generators $\{H_i, X_{+i}, X_{-i}\}$. We emphasise the latter point of view, with

$$q = e^{\frac{t}{2}}, \quad q_i = q^{d_i}.$$

As elements of a vector space, we identify these generators with the corresponding Cartan–Chevalley generators of g. We do not, however, make use of the ordinary Lie bracket structure on the vector space of g, but rather we define the relations of $U_q(g)$ as

$$[H_i, H_j] = 0, \quad [H_i, X_{\pm j}] = \pm a_{ij} X_{\pm j}, \quad [X_{+i}, X_{-j}] = \delta_{ij} \frac{q_i^{H_i} - q_i^{-H_i}}{q_i - q_i^{-1}}, \quad (3.22)$$

$$\sum_{k=0}^{1-a_{ij}} (-1)^k \begin{bmatrix} 1 - a_{ij} \\ k \end{bmatrix}_{q_i} X_{\pm i}^{1-a_{ij}-k} X_{\pm j} X_{\pm i}^k = 0, \quad \forall i \neq j. \quad (3.23)$$

Here the q-binomial coefficient $\begin{bmatrix} n \\ m \end{bmatrix}_{q_i}$ is defined in terms of $[n]_{q_i} = \frac{q_i^n - q_i^{-n}}{q_i - q_i^{-1}}$ and $[n]_{q_i}! = [n]_{q_i}[n-1]_{q_i} \cdots [1]_{q_i}$. We take for coproduct, counit and antipode the maps

$$\Delta H_i = H_i \otimes 1 + 1 \otimes H_i, \quad \Delta X_{\pm i} = X_{\pm i} \otimes q_i^{\frac{H_i}{2}} + q_i^{-\frac{H_i}{2}} \otimes X_{\pm i},$$

$$\epsilon(H_i) = \epsilon(X_{\pm i}) = 0, \quad SH_i = -H_i, \quad SX_{\pm i} = -q_i^{\pm 1} X_{\pm i} \quad (3.24)$$

extended as algebra and antialgebra maps to products of generators.

Proposition 3.3.1 $U_q(g)$, *as described in (3.22)–(3.24), is a Hopf algebra. The action of the antipode on the generators is −1 times conjugation by $q^{\check{\rho}}$, where $\check{\rho}$ is defined by*

$$K(\check{\rho}, \ ) = \rho, \quad \rho = \frac{1}{2} \sum_{\alpha > 0} \alpha, \quad i.e. \quad \check{\rho} = \frac{1}{2} \sum_{\alpha > 0} d_\alpha H_\alpha.$$

Here $H_\alpha \in t$ is defined by $K(d_\alpha H_\alpha, \ ) = \alpha$ and $d_\alpha = (\alpha, \alpha)/2$. If $\alpha = \sum_i n_i \alpha_i$, then $d_\alpha H_\alpha = \sum_i n_i d_i H_i$.

Proof: We check first that Δ extends consistently to products. That it respects the relations (3.22) follows just the same calculation as for $U_q(sl_2)$ in Example 3.2.1, with a_{ij} in place of the factor 2. For the third of (3.22), we use symmetry of the Cartan matrix in the form $d_i a_{ij} = d_j a_{ji}$. The new relation that we have to check is the q-Serre relation (3.23). From the q-binomial theorem in Lemma 3.2.2, we have

$$\Delta X_{+i}^k = \sum_{n=0}^{k} \begin{bmatrix} k \\ n \end{bmatrix}_{; q_i^{-2}} X_{+i}^n q_i^{-\frac{H_i}{2}(k-n)} \otimes q_i^{\frac{H_i}{2}n} X_{+i}^{k-n},$$

where $A = X_{+i} \otimes q_i^{\frac{H_i}{2}}$, $B = q_i^{-\frac{H_i}{2}} \otimes X_{+i}$ and q is replaced by q_i in the lemma. Similarly for X_{-i}^k. Applying the coproduct to the left hand side of (3.23) and noting the identity $\begin{bmatrix} k \\ n \end{bmatrix}_q = \begin{bmatrix} k \\ n \end{bmatrix}_{; q^{-2}} q^{n(k-n)}$, which we write in terms of q-factorials, we require for the positive roots the vanishing of

$$\sum_{k=0}^{1-a_{ij}}\sum_{m=0}^{1-a_{ij}-k}\sum_{n=0}^{k}\frac{(-1)^k q_i^{k(1-a_{ij}-k)}[1-a_{ij};q_i^{-2}]!}{[m;q_i^{-2}]![1-a_{ij}-k-m;q_i^{-2}]![n;q_i^{-2}]![k-n;q_i^{-2}]!}$$

$$\times X_{+i}^m q_i^{-\frac{H_i}{2}(1-a_{ij}-k-m)}X_{+j}X_{+i}^n q_i^{-\frac{H_i}{2}(k-n)}$$

$$\otimes q_i^{\frac{H_i}{2}m}X_{+i}^{1-a_{ij}-k-m}q_j^{\frac{H_j}{2}}q_i^{\frac{H_i}{2}n}X_{+i}^{k-n}$$

plus a similar expression with the role of $X_{+j}\otimes q_j^{\frac{H_j}{2}}$ played by $q_j^{-\frac{H_j}{2}}\otimes X_{+j}$. The two expressions vanish separately, so we concentrate on this first one. It looks formidable, but multiply top and bottom by $[1-a_{ij}-m-n;q_i^{-2}]!$ and change the order of summation so that k is summed first for each m,n held constant. Also move the $q_i^{\frac{H_i}{2}k}$ in the first tensor factor to the right, picking up $q_i^{\frac{ka_{ij}}{2}}q_i^{kn}$ from the relations (3.22). Likewise, move $q_j^{\frac{H_j}{2}}q_i^{\frac{H_i}{2}n}$ in the second tensor factor to the right, picking up $q_i^{\frac{a_{ij}}{2}(k-n)}q_i^{n(k-n)}$ in view of the symmetry property in (3.18). For each n,m fixed, we then have an expression proportional to

$$\sum_{k=n}^{1-a_{ij}-m}(-1)^k\begin{bmatrix}1-a_{ij}-m-n\\k-n\end{bmatrix};q_i^{-2}\end{bmatrix}q_i^{-k^2+k+2kn-n^2}$$

$$\propto \sum_{r=0}^{1-a_{ij}-m-n}(-1)^r\begin{bmatrix}1-a_{ij}-m-n\\r\end{bmatrix};q_i^{-2}\end{bmatrix}q_i^{-r(r-1)}=0.$$

The vanishing here is itself a q-identity which is most easily proven by Hopf algebra techniques: just consider the antipode axiom for the Weyl Hopf algebra in Example 1.3.2, evaluated on a power of X (or, indeed, consider it on a power of X_{+i} in the Hopf algebra already defined by (3.22) without the q-Serre relation (3.23)). For the second expression mentioned above, we have

$$\sum_{k=0}^{1-a_{ij}}\sum_{m=0}^{1-a_{ij}-k}\sum_{n=0}^{k}\frac{(-1)^k q_i^{k(1-a_{ij}-k)}[1-a_{ij};q_i^{-2}]!}{[m;q_i^{-2}]![1-a_{ij}-k-m;q_i^{-2}]![n;q_i^{-2}]![k-n;q_i^{-2}]!}$$

$$\times X_{+i}^{1-a_{ij}-k-m}q_i^{-\frac{H_i}{2}m}q_j^{-\frac{H_j}{2}}X_{+i}^{k-n}q_i^{-\frac{H_i}{2}n}$$

$$\otimes q_i^{\frac{H_i}{2}(1-a_{ij}-k-m)}X_{+i}^m X_{+j}q_i^{\frac{H_i}{2}(k-n)}X_{+i}^n,$$

where we also interchanged the summation labels $m, 1-a_{ij}-k-m$ and $n, k-n$. We then follow the same steps as above, again getting something proportional to zero for each fixed m,n. There is a strictly analogous computation for the case of negative roots in the q-Serre relations (3.23).

That the counit extends to products is trivial. Finally, the antipode shown clearly works on the generators. That it extends to products of generators holds for (3.22) in just the same way as for $U_q(sl_2)$. It is immediate for (3.23), after interchanging k and $1 - a_{ij} - k$. The characterisation of the antipode in terms of $\check{\rho}$ follows from $\alpha_i(\check{\rho}) = K(\check{\rho}, d_i H_i) = (\rho, \alpha_i) = d_i$ for all α_i. This is deduced, as usual in Lie theory, from $s_i(\rho) = \rho - \alpha_i$ and the fact that Weyl reflections preserve the inner product. We use it to obtain the commutation relation $[\check{\rho}, X_{\pm i}] = \pm d_i X_{\pm i}$.

It is worth noting that, by similar calculations, one can also write the q-Serre relations as

$$\mathrm{Ad}_{X_{\pm i}}^{1-a_{ij}} (X_{\pm j} q_j^{-\frac{H_j}{2}}) = 0. \tag{3.25}$$

These are identities in $U_q(g)$ and nontrivial relations in the Hopf algebra which one has before adding (3.23). This justifies calling the latter an analogue of (3.20). ∎

The definition of $U_q(g)$ is due to V.G. Drinfeld and M. Jimbo. Some other good generators are

$$E_i = X_{+i} q_i^{\frac{H_i}{2}}, \quad F_i = q_i^{-\frac{H_i}{2}} X_{-i}, \quad g_i = q_i^{H_i}, \tag{3.26}$$

so that the relations become

$$[g_i, g_j] = 0, \quad g_i E_j g_i^{-1} = q_i^{a_{ij}} E_j, \quad g_i F_j g_i^{-1} = q_i^{-a_{ij}} F_j,$$

$$[E_i, F_j] = \delta_{ij} \frac{g_i - g_i^{-1}}{q_i - q_i^{-1}},$$

$$\Delta g_i = g_i \otimes g_i, \quad \Delta E_i = E_i \otimes g_i + 1 \otimes E_i, \quad \Delta F_i = F_i \otimes 1 + g_i^{-1} \otimes F_i,$$

$$\epsilon(g_i) = 1, \quad \epsilon(E_i) = \epsilon(F_i) = 0,$$

$$S g_i = g_i^{-1}, \quad S E_i = -E_i g_i^{-1}, \quad S F_i = -g_i F_i,$$

along with q-Serre relations as in (3.23), with $X_{\pm i}$ replaced by E_i, F_i.

There is also a factorisable quasitriangular structure $\mathcal{R}$ which is obtained via Drinfeld's quantum double construction to be covered later, in Chapter 7. It has the form

$$\mathcal{R} = q^{\sum_{i,j} d_i (a^{-1})_{ij} H_i \otimes H_j} \left(\overset{\leftarrow}{\prod_{\alpha>0}} e_{q_\alpha^{-2}}^{(q_\alpha - q_\alpha^{-1}) E_\alpha \otimes F_\alpha} \right), \tag{3.27}$$

where $q_\alpha = q^{d_\alpha}$, and E_α, F_α play the role of $X_+ q^{\frac{H}{2}}, q^{-\frac{H}{2}} X_-$ in (3.11). The product is taken over the set Δ^+ of positive roots in decreasing order. To describe this ordering one has to fix an element w_0 of maximal length in the Weyl group W. This has a reduced form $w_0 = s_{i_1} \cdots s_{i_N}$, where the s_i are the simple reflections (these generate W). Then the sequence

$$\alpha_{i_i}, \; s_{i_1}(\alpha_{i_2}), \; s_{i_1}s_{i_2}(\alpha_{i_3}), \; \ldots, \; s_{i_1}s_{i_2}\cdots s_{i_{N-1}}(\alpha_{i_N}) \tag{3.28}$$

is an enumeration of all the elements of Δ^+; i.e. $\alpha = s_{i_1}s_{i_2}\cdots s_{i_{p-1}}(\alpha_{i_p})$ as $p = 1, 2, \ldots, N$ runs through all the positive roots. This is the ordering that we take. It also provides an explicit description of a general positive root α as obtained from a simple root by a series of simple reflections. Since W acts on the root system and preserves the inner product, it is to be expected that each of its generators s_i has a corresponding automorphism T_i, say, of the algebra $U_q(g)$. These have been introduced by G. Lusztig and take the form

$$T_i(H_j) = H_j - a_{ji}H_i, \quad T_i(E_i) = -q_iF_i, \quad T_i(F_i) = -q_i^{-1}E_i,$$

$$T_i(E_j) = \sum_{k=0}^{-a_{ij}} (-1)^{k-a_{ij}} q_i^{-k} \frac{E_i^{-a_{ij}-k} E_j E_i^k}{[-a_{ij}-k]_{q_i}! [k]_{q_i}!}, \tag{3.29}$$

$$T_i(F_j) = \sum_{k=0}^{-a_{ij}} (-1)^{k-a_{ij}} q_i^{k} \frac{F_i^k F_j F_i^{-a_{ij}-k}}{[k]_{q_i}! [-a_{ij}-k]_{q_i}!},$$

where the last two expressions are for $i \neq j$. One can check directly that the T_i are algebra homomorphisms and invertible. The action on the E_j, F_j has a form which is a variant of the quantum adjoint action. Just as the nonsimple root vectors $X_{\pm \alpha}$ are generated classically by repeated Lie bracket from the simple roots vectors, for our quantum group we likewise define the elements E_α, F_α as obtained by applying the corresponding sequence of the T_i,

$$E_\alpha = T_{i_i} \cdots T_{i_{p-1}}(E_{i_p}), \quad F_\alpha = T_{i_i} \cdots T_{i_{p-1}}(F_{i_p}) \tag{3.30}$$

when $\alpha = s_{i_1}s_{i_2}\cdots s_{i_{p-1}}(\alpha_{i_p})$. For completeness, we can define $H_\alpha = T_{i_i} \cdots T_{i_{p-1}}(H_{i_p})$ in the same way. These are, however, the same elements of t as in Proposition 3.3.1, because the action in (3.29) on the H_j is not deformed. This completes our description of the quasitriangular structure for $U_q(g)$. Note also that, although the T_i are not inner automorphisms, one can extend $U_q(g)$ to a bigger algebra in which they become inner, in the form $T_i = \bar{s}_i( )\bar{s}_i^{-1}$, say. The $\bar{s}_i$ are the called *quantum reflections* and no do not square to unity. The algebra generated by $U_q(g)$ and these quantum reflections is called the *quantum Weyl group*. Its detailed structure requires the extension theory of quantum groups, which we come to in Chapter 6.3. The multiplicative formula (3.27) can be considered as one of its main applications.

Another interesting property of these E_α, F_α is that one may prove identities of the form

$$[H_\alpha, E_\alpha] = 2E_\alpha, \quad [H_\alpha, F_\alpha] = -2F_\alpha, \quad [E_\alpha, F_\alpha] = \frac{q_\alpha^{H_\alpha} - q_\alpha^{-H_\alpha}}{q_\alpha - q_\alpha^{-1}}, \tag{3.31}$$

$$[E_\alpha, E_\beta]_{q^{(\alpha,\beta)}} \equiv E_\alpha E_\beta - q^{(\alpha,\beta)} E_\beta E_\alpha = O(E_{\gamma_p} \cdots E_{\gamma_1}),$$

where $\alpha < \gamma_1 \leq \ldots \leq \gamma_p < \beta$, and similarly for F_α. The left hand side is called a q-*commutator* and the identities are analogous to those (without q) for the commutators in $U(g)$. They give rise to a precise description of a basis for $U(g)$ and $U_q(g)$, the PBW basis. For example, one can take the elements

$$(\overrightarrow{\prod_{\alpha>0}} F_\alpha^{n_\alpha})(\prod_i H_i^{p_i})(\overleftarrow{\prod_{\alpha>0}} E_\alpha^{m_\alpha}); \quad p_i, n_\alpha, m_\alpha \in \mathbb{N}. \tag{3.32}$$

These results can easily be demonstrated for the example of $g = sl_{n+1}$. We identify $t^* = \{x_i | \sum x_i = 0\} \subset \mathbb{R}^{n+1}$, with (,) becoming the Euclidean metric. The simple roots are $\alpha_i = f_i - f_{i+1}$, and they all have length 2. They are also equal to the coroots, $\alpha_i = \check{\alpha}_i$, so

$$d_i = 1, \quad a_{ij} = (\alpha_i, \alpha_j) = \begin{cases} 2 & i = j \\ -1 & i = j+1 \text{ or } i = j-1 \\ 0 & \text{otherwise.} \end{cases} \tag{3.33}$$

The general positive roots take the form $\alpha_{ij} = \alpha_i + \alpha_{i+1} + \cdots + \alpha_j$, where $i \leq j$. The Weyl group is the permutation group S_n. Its generators, the reflections, take the form

$$s_{i-1}(\alpha_{ij}) = \alpha_{i-1 j}, \quad s_{j+1}(\alpha_{ij}) = \alpha_{i j+1}, \quad s_i(\alpha_i) = -\alpha_i,$$

$$s_i(\alpha_{ij}) = \alpha_{i+1 j}, \quad s_j(\alpha_{ij}) = \alpha_{i j-1}.$$

Using this, one arrives at a well-known ordering of Δ^+ as

$$\alpha_1, \alpha_{1 2}, \ldots, \alpha_{1 n}; \alpha_2, \alpha_{2 3}, \ldots, \alpha_{2 n}; \alpha_3, \alpha_{3 4}, \ldots, \alpha_{3 n}; \ldots; \alpha_{n-1}, \alpha_{n-1 n}; \alpha_n.$$

This ordering corresponds to the element

$$w_0 = s_1 s_2 \cdots s_n s_1 s_2 \cdots s_{n-1} \cdots s_1 s_2 s_1.$$

The q-commutators (3.31) take the form

$$[E_{\alpha_1 j-1}, E_j]_{q^{-1}} = -E_{\alpha_1 j}, \quad [F_{\alpha_1 j-1}, F_j]_{q^{-1}} = q^{-1} F_{\alpha_1 j}, \tag{3.34}$$

etc.

Apart from the complication of the higher root vectors and their ordering, we have seen that all of the quantum group structure of $U_q(g)$ is quite analogous to the $U_q(sl_2)$ case. The representation theory is also analogous. Recall that the irreducible representations of g are labelled by the dominant weights $\Lambda \in t^*$. The same is true for $U_q(g)$. As just one sample of a typical result we have, analogous to Proposition 3.2.7,

Proposition 3.3.2 *Let $V(\Lambda)$ be an irreducible $U_q(g)$-module generated by an element $|0\rangle$ such that $X_{+i}|0\rangle = 0$ for all X_{+i} and $h|0\rangle = \Lambda(h)|0\rangle$ for all $h \in t$ and some $\Lambda \in t^*$. Then, acting on $V(\Lambda)$,*

$$u|_{V(\Lambda)} = q^{-(\Lambda, \Lambda + 2\rho)}q^{2\check{\rho}}, \quad v|_{V(\Lambda)} = q^{-(\Lambda, \Lambda + 2\rho)}q^{-2\check{\rho}}, \quad \nu|_{V(\Lambda)} = q^{-(\Lambda, \Lambda + 2\rho)},$$

where $\rho = \frac{1}{2}\sum_{\alpha > 0} \alpha$.

Proof: Recall from Proposition 2.1.8 that u implements the square of the antipode. Hence $q^{-2\check{\rho}}u$ is central. Hence, in an irreducible representation, it is proportional to the identity and can therefore be computed on any vector in $V(\Lambda)$. Computation on the vector $|0\rangle$ is particularly easy because the X_{+i} vanish so that only the lowest part of the series for $\mathcal{R}$ contributes. This lowest part is the Gaussian factor in (3.27), involving only the H_i. Again, on $|0\rangle$ the Cartan subalgebra t acts by Λ, immediately giving the result. This elegant argument is due to Drinfeld. The second part is obtained similarly, while the last part must therefore be the value of the ribbon element (which is known to exist) as the square root of uv.   ∎

3.4   Roots of unity

When we work with the above quantum groups over $\mathbb{C}$, we have to live with the fact that they are infinite dimensional. By working over $\mathbb{C}[[t]]$ (or at 'generic' q), one can avoid the problems arising from this, but this is not really appropriate if we want actually to evaluate the parameter q. When we evaluate q, we soon find that some aspects of the theory depend very critically on its precise value. Of particular interest here is the case when q is a primitive root of unity. We will demonstrate the phenomena that arise on the example $U_q(sl_2)$ in Example 3.2.1. There are (at least) two interesting versions of it.

For the first version, we suppose that $q^{2r} = 1$, or, more precisely, that q^2 is a primitive rth root of unity and $r > 1$. Primitive means that there is no smaller such r. Since we will be tied explicitly to the picture over $\mathbb{C}$ (or another field), we denote the generators of $U_q(sl_2)$ by $K, K^{-1}, X_\pm$ rather than using the ambivalent $q^{\frac{H}{2}}, q^{-\frac{H}{2}}, X_\pm$ notation.

Example 3.4.1 $U_q^{(r)}(sl_2)$ *is the finite-dimensional quasitriangular Hopf algebra generated by 1, mutually inverse K, K^{-1} and $X_\pm$ with relations*

$$KX_\pm K^{-1} = q^{\pm 1}X_\pm, \quad [X_+, X_-] = \frac{K^2 - K^{-2}}{q - q^{-1}}, \quad K^{4r} = 1, \quad X_\pm^r = 0,$$

i.e. we quotient $U_q(sl_2)$ by the last two relations. It has coproduct $\Delta X_\pm = X_\pm \otimes K + K^{-1} \otimes X_\pm$ etc. inherited from $U_q(sl_2)$ and a quasitriangular structure

$$\mathcal{R} = \mathcal{R}_K \sum_{m=0}^{r-1} (KX_+)^m \otimes (K^{-1}X_-)^m \frac{(1-q^{-2})^m}{[m;q^{-2}]!},$$

where

$$\mathcal{R}_K = \frac{1}{4r} \sum_{a,b=0}^{4r-1} q^{-\frac{1}{2}ab} K^a \otimes K^b$$

is the expression already encountered in another context in Example 2.1.6.

Proof: In $U_q(sl_2)$ the element K^{4r} is group-like and central (because $q^{4r} = 1$), so it can certainly be set to unity without spoiling the fact that we have a Hopf algebra (indeed, for this part alone, we could set $K^{2r} = 1$). Next, we set $X_\pm^r = 0$ and check that this too is compatible with Δ (we show that the ideal I generated by the $X_\pm^r$ is a biideal). Indeed, $\Delta X_\pm^r = (X_\pm \otimes K + K^{-1} \otimes X_\pm)^r = \sum_{m=0}^{m=r} \left[{r \atop m};q^{-2}\right](X_\pm \otimes K)^m (K^{-1} \otimes X_\pm)^{m-r} = X_\pm^r \otimes K^r + K^{-r} \otimes X_\pm^r$ using $A = X_\pm \otimes K$ and $B = K^{-1} \otimes X_\pm$ in Lemma 3.2.2. The reason is that, in the q-binomial expansion $(A+B)^r = A^r + [r;q^{-2}]A^{r-1}B + \cdots + [r;q^{-2}]AB^{r-1} + B^r$, only the outer terms contribute because $[r;q^{-2}] = \frac{1-q^{-2r}}{1-q^{-2}} = 0$. We know that $[m;q^{-2}] \neq 0$ for $m = 1, 2, \ldots, r-1$ because the rth root is primitive. We conclude that $\Delta X_\pm^r \subset I \otimes U_q(sl_2) + U_q(sl_2) \otimes I$. Also, $SI \subseteq I$ is clear. Hence we have a Hopf algebra. The quasitriangular structure is more problematic because the Gaussian factor $q^{\frac{H \otimes H}{2}}$ in the formula in Example 3.2.1 has no obvious analogue in terms of K. Its correct analogue turns out to be $\mathcal{R}_K$ as shown (the rest is obtained by setting $X_\pm^r = 0$ in the series for $e_{q^{-2}}$). Looking back over the proof of Example 3.2.1, we see that it is only necessary to ensure that $\mathcal{R}_K$ has the properties used there for $q^{\frac{H \otimes H}{2}}$. ∎

Proposition 3.4.2 *The representations V_j in Proposition 3.2.6 are defined also for $U_q^{(r)}(sl_2)$, for spin in the range $j = 0, \frac{1}{2}, \ldots, \frac{r-1}{2}$. Note that these representations were intended for $U_q(su_2)$ (with real q), but we continue to use the same formulae now for $q^{2r} = 1$.*

Proof: The key point again is that $[r] = 0$ but $[n] \neq 0$ for $n = 1, \ldots, r-1$. Thus, consider the definition $X_+|j, m\rangle = \sqrt{[j-m][j+m+1]}|j, m+1\rangle$, where $m = -j, \ldots, j$. If j is in the allowed range, then $0 \leq j - m \leq r-1$, while $j + m + 1$ can reach r but only when $j = m$, in which case $[j-m] = 0$

anyway (this is when X_+ acts on the highest vector, and this should vanish anyway). Likewise for the action of X_-. This means that the actions of $X_\pm$ (and $K|j,m\rangle = q^m|j,m\rangle$) are well-defined. We still have to check that they represent the additional relations of $U_q^{(r)}(sl_2)$. We have $K^{4r}|j,m\rangle = q^{4rm}|j,m\rangle = |j,m\rangle$ since m is a half-integer and $q^{2r} = 1$. Also $X_+^r|j,m\rangle = 0$ because each action of X_+ raises m by 1, so X_+^r raises m by r, but there are only $2j+1$ states for m, i.e. at most r states for m when j is in the allowed range. Similarly for X_-^r. ∎

Let us recall that (over $\mathbb{C}$) these representations V_j were unitary for the $U_q(su_2)$ *-structure, which now takes the form

$$X_\pm^* = X_\mp, \quad K^* = K^{-1}. \tag{3.35}$$

We can adopt this *-structure, but, as we know from the proofs of Examples 3.2.3–3.2.5, we will *not* any longer have a Hopf *-algebra. Instead, we have

$$(\Delta h)^{* \otimes *} = \tau \circ \Delta(h^*), \qquad \forall h, \tag{3.36}$$

in contrast to Definition 1.7.5. This form works for generic $|q| = 1$ as well.

In any event, the proposition justifies the definition of $U_q^{(r)}(sl_2)$. The standard $U_q(sl_2)$ representations V_j can be used at these roots of unity, but in the allowed range the elements $K^{4r} - 1$ and $X_\pm^r$ are always in the kernel, so we may as well set them equal to zero. The resulting quasitriangular Hopf algebra is ribbon, and was used by Reshetikhin and Turaev (with a different but equivalent formula for $\mathcal{R}$) in their computation of three-manifold invariants. However, it is *not* factorisable.

The nonfactorisability of $U_q^{(r)}(sl_2)$ means that its quantum inverse Killing form (see Definition 2.1.12) is degenerate. This also means that it is not suitable for certain types of purely quantum applications to be developed later (see Chapters 7.4 and 9.4). In fact, this can be fixed by taking a different quotient of $U_q(sl_2)$.

Example 3.4.3 *Let q be a primitive lth root of unity, where $l > 1$ is odd. We define $u_q(sl_2)$ as the finite-dimensional Hopf algebra generated by 1, mutually inverse g, g^{-1} and E, F, with relations, coproduct and quasitriangular structure given by*

$$gE = q^2 Eg, \ gF = q^{-2}Fg, \ [E,F] = \frac{g - g^{-1}}{q - q^{-1}}, \ E^l = 0 = F^l, \ g^l = 1,$$

$$\Delta E = E \otimes g + 1 \otimes E, \quad \Delta F = F \otimes 1 + g^{-1} \otimes F, \quad \Delta g = g \otimes g,$$

$$\mathcal{R} = \mathcal{R}_g \left(\sum_{n=0}^{l-1} \frac{(q - q^{-1})^n}{[n; q^{-2}]!} E^n \otimes F^n \right), \quad \mathcal{R}_g = \frac{1}{l} \sum_{a,b=0}^{l-1} q^{-2ab} g^a \otimes g^b.$$

This quasitriangular Hopf algebra is factorisable.

Proof: We begin with the sub-Hopf algebra of $U_q(sl_2)$ generated by $g = q^H$, $E = X_+ q^{\frac{H}{2}}$ and $F = q^{-\frac{H}{2}} X_-$. Note that, here, g is the square of the K in Example 3.4.1. We now quotient this sub-Hopf algebra by the new relations $g^l = 1, E^l = 0, F^l = 0$. The computation of $\Delta E^l = (E \otimes 1 + g^{-1} \otimes E)^l$ proceeds much as before, with $[l; q^{-2}] = 0$ in the q-binomial theorem (we do not encounter $[\frac{l}{2}; q^{-2}] = 0$ because l is odd). The required $\mathcal{R}$ is also a variant of the previous one, though different. For example, to prove (2.2) on ΔE, we need $\mathcal{R}(E \otimes g + 1 \otimes E) = (E \otimes 1 + g \otimes E)\mathcal{R}$ but $\mathcal{R}_g(E \otimes g^{-1} + 1 \otimes E) = (E \otimes 1 + g \otimes E)\mathcal{R}_g$ so that the required identity is $e^A_{q^{-2}}(B + C) = (C + D)e^A_{q^{-2}}$, as before, but this time with $A = (q - q^{-1})E \otimes F, B = E \otimes g, C = 1 \otimes E, D = E \otimes g^{-1}$. These do have the relations needed in the middle part of Lemma 3.2.2 (the factor $q - q^{-1}$ is crucial for this, in contrast to $1 - q^{-2}$ in a similar role before). A more crucial difference is that $\mathcal{R}$ is now factorisable. To see this we write $\mathcal{R}_g e_{q^{-2}}((q - q^{-1})E \otimes F) = e_{q^{-2}}((q - q^{-1})Eg^{-1} \otimes gF)\mathcal{R}_g$ as usual. Hence, $Q = e_{q^{-2}}((q - q^{-1})gF \otimes Eg^{-1})Q_g e_{q^{-2}}((q - q^{-1})E \otimes F) = \sum_{n,m} a_n a_m (gF)^n Q_g{}^{(1)} E^m \otimes (Eg^{-1})^n Q_g{}^{(2)} F^m$, where $Q_g = \mathcal{R}_g^2$ and the a_n are the coefficients from $e_{q^{-2}}$. A basis of $u_q(sl_2)$ is provided either by expressions of the form $\{(gF)^n g^a E^m\}$ or by expressions of the form $\{(Eg^{-1})^n g^b F^m\}$. Hence, by considering elements of the dual basis to the latter, it is easy to see that $Q = \mathcal{R}_{21}\mathcal{R}$ is nondegenerate *iff* $\mathcal{R}_g^2$ is non-degenerate. We have seen that this is so in Example 2.1.13 when l is odd. ∎

The quantum group $u_q(sl_2)$ is also ribbon. The inverse ribbon element comes out from $\mathcal{R}$ as

$$v = \frac{1}{l} \left(\sum_{n=0}^{l-1} q^{2n^2}\right) \sum_{n,a=0}^{l-1} \frac{(q - q^{-1})^n}{[n; q^{-2}]!} q^{-\frac{1}{2}(l+1)(n-a-1)^2} E^n g^a F^n.$$

Notes for Chapter 3

The q-Heisenberg algebra in Example 3.1.2 (including the quasitriangular structure) was introduced in [57] as a contraction limit of the quantum group $U_q(sl_2)$. It was subsequently studied (in the same context) in [58], where the associated link-invariant was found, as mentioned in the introduction to the present chapter. The more physical motivation for a coproduct in terms of embedding into n-particle harmonic oscillators in

Exercise 3.1.1, the direct proof of the quasitriangular structure and the elucidation of the real and antireal *-structures in Proposition 3.1.4 and Example 3.1.6 are due to the author, presented here for the first time. The q-harmonic oscillator in Example 3.1.8 is due, in the present context, to A.J. Macfarlane [59]. It has also been studied in [60] and elsewhere. The q-differentiation used in Proposition 3.1.9 has a long history in mathematics under the heading of q-analysis; see, for example, [61].

There are still more deformations of the harmonic oscillator or Heisenberg algebra beyond the ones given in Section 3.1. In particular, the latter should not be confused with another q-Heisenberg algebra

$$px - qxp = \imath,$$

which we come to later, in Chapter 10.4, where it corresponds to a natural braided Leibniz rule. For $q \neq 1$ and x invertible, a transformation $y = p + \imath x^{-1}(q - 1)^{-1}$ reduces it to the quantum plane algebra $yx = qxy$. Another point of view on this algebra is to consider the generators as mutually adjoint, so one considers an algebra of the form

$$bb^\dagger - q^2 b^\dagger b = 1.$$

This is a *-algebra with q real and $b^* = b^\dagger$. It is mapped to the q-harmonic oscillator in Example 3.1.8 by $b = q^{\frac{N}{2}} a$ and $b^\dagger = a^\dagger q^{\frac{N}{2}}$, where we assume that q^N has a square root.

The quantum group $U_q(sl_2)$ arose in the theory of quantum inverse scattering in the works of E.K. Sklyanin [23, 24] (in the form of the Sklyanin algebra as above) and [25] (in the Hopf algebra form). We will cover some of this background in the next chapter. The *-structures for $U_q(sl_2)$ were studied in [62] and elsewhere. The direct proof of the quasitriangular structure in Example 3.2.1 is new, whereas its reality properties in Examples 3.2.4 and 3.2.5 are taken from [35]. Other authors have also considered such reality properties [29]. The first and third parts of Lemma 3.2.2 are some well-known elements of q-analysis, while the second part seems to be more novel. For further work on q-analysis, much of it motivated from quantum groups, see [63]. Another well-known element of q-analysis is Jackson or q-integration for which the natural point of view [11] is the braided setting; see the Notes to Chapter 10.

The example of $U_q(sl_2)$ led to V.G. Drinfeld's axioms of a quasitriangular Hopf algebra; see [22]. Drinfeld [22], and M. Jimbo [26] independently introduced the quantum group $U_q(g)$ associated to a general Cartan matrix (including the affine case of loop-group Lie algebras, which are especially important in the construction of exactly solvable lattice models.) Drinfeld's work included a general construction for the quasitriangular structure on $U_q(g)$ as a quotient of the quantum double Hopf algebra, to be covered later, in Chapter 7. The elegant multiplicative form (3.27) was worked out for $U_q(sl_n)$ by N. Burroughs [64] and M. Rosso [65], and for $U_q(g)$ by S.Z. Levendorskii and Ya.S. Soibelman [66] and A.N. Kirillov and N.Yu. Reshetikhin [67]. Some early results on the finite-dimensional representations of $U_q(g)$ appeared in [68] and elsewhere. The element u for $U_q(g)$ is studied further in [69] and [70], in connection with a character formula. The abstract structure of the quantum Weyl group is studied by the author and Ya. S. Soibelman [71]. We will return to some of these results in later chapters. We have not been able to cover the case of affine quantum groups in any detail, nor Yangians; see [72] and [73]. A classification of the *-structures for general $U_q(g)$ can be found in [74] and [75].

The structure of $U_q(g)$ at roots of unity was studied by several authors, for example [76] (but without the quasitriangular structure). The first reduced form $U_q^{(r)}(sl_2)$ of the quantum group $U_q(sl_2)$ at a root of unity in Section 3.4 was studied [77] in the construction of three-manifold invariants. These authors, however, used a different form of $\mathcal{R}$ which was based on Gauss sums; the quasitriangular structure shown in Example 3.4.1 is taken from the 1991 paper [30] by the author, where it was introduced as an application of the anyon-generating quantum group in Example 2.1.6. The second (nonisomorphic) reduced quantum group $u_q(sl_2)$ appears to be more natural. Its factorisability is taken from [78] by V.V. Lyubashenko and the author, where it plays a crucial role for a theory of quantum Fourier transforms on the associated braided groups.

The construction of $U_q(g)$ works more generally than discussed in this chapter, notably it includes q-deformations of Kac-Moody Lie algebras. Also not covered is the canonical or 'crystal' basis of M. Kashiwara [79] and G. Lusztig [80]. This is a remarkable basis of the subalgebra $U_q(n_-) \subset U_q(g)$ (the subalgebra generated by the negative root vectors) with the property that it induces a basis of a highest weight representation when applied to the highest weight vector. The basis also has remarkable positivity and integrality properties arising, in Lusztig's approach, from his construction of $U_q(n_-)$ as certain perverse sheaves and 'shifted' perverse sheaves on the algebraic variety of representations of the quiver associated to g. For example, $U_q(n_-) \subset U_q(sl_3)$ is generated by F_1, F_2 modulo the q-Serre relations of the same form as in (3.23) and has canonical basis given by the three sets

$$\{F_1^{(i)} F_2^{(k)} F_1^{(j)}\}_{i+j<k}, \quad \{F_2^{(i)} F_1^{(k)} F_2^{(j)}\}_{i+j<k}, \quad \{F_1^{(i)} F_2^{(i+j)} F_1^{(j)}\}_{i,j},$$

where $x^{(m)} = x^m/[m]!$ is called a 'divided power'. Note that $U_q(n_-)$ is not itself a quantum group, but rather a braided group of the type introduced in [31], see Chapter 10.

4

Matrix quantum groups

In this chapter, we describe another class of examples of quantum groups, quite different from the deformations of enveloping algebras in the preceding chapter. These are the matrix quantum groups, which generalise the familiar idea of defining groups as groups of matrices. They are dual to the deformed enveloping algebra quantum groups of the Chapter 3 (we will see that they are dual quasitriangular in the sense of Chapter 2.2), and as such they are *dual quantum groups* in the strict sense, or *quantum groups of function algebra type*. In other words, rather than working with the elements of the group or Lie algebra, we work with the functions on the group (see Example 1.5.2) as our classical model. When we deform a Hopf algebra of this type so that it is no longer commutative, we can nevertheless continue to think of its elements as if they were the functions on a group in this way, even though an actual underlying group no longer exists.

This point of view is a well established one in mathematics, and comes under the general heading of noncommutative geometry. We will come to this in later chapters, but, for now, we can explain the general idea at a superficial level. It is only this superficial level that is at all relevant to the present chapter, but it can nevertheless be used as a source of motivation.

This general idea of noncommutative geometry arose in the early half of this century with an important theorem of I. Gelfand and M.A. Naimark about commutative C^* algebras and the related strides in algebraic geometry that were made at the time. Given any space X we can consider the algebra of functions on it, and work directly with this algebra in place of X itself (so the dual quantum groups in this chapter are in this setting). The algebra is necessarily commutative because if f, g are two functions, then $(fg)(x) = f(x)g(x) = g(x)f(x) = (gf)(x)$. Every element has a norm (we will come to this in later chapters) and, over $\mathbb{C}$, there is a $*$-structure. Gelfand and Naimark showed that every commutative

108

*-algebra with a norm (obeying suitable axioms) is necessarily the algebra of continuous functions on some topological space X, which they reconstructed. So there is really an equivalence in this precise setting between a class of commutative algebras and spaces. This in turn means that familiar constructions in the latter context necessarily have equivalent algebraic analogues in terms of the corresponding algebra.

In fact, this kind of principle is also very familiar in differential geometry, where one defines vectors on a manifold as, by definition, derivations on the algebra of functions on the manifold. Even more obviously, perhaps, the notion of manifold itself is defined more precisely in terms of the (locally defined) coordinate functions on it.

This algebraic approach to geometry in terms of coordinate functions motivates two generalisations. The first replaces $\mathbb{C}$ by another field or ring. This leads to many interesting phenomena not possible over $\mathbb{C}$ and is the subject of algebraic geometry. The second, equally obvious, generalisation is to allow the algebra to be noncommutative. As we have seen above, it cannot then be the algebra of functions on any space (because that would be commutative), but many notions defined algebraically can still make sense, so that this need not deter us. Thinking about a noncommutative algebra in this way is the point of view of noncommutative geometry. It is this general point of view that can motivate us now (rather than any serious results of noncommutative geometry). It is not incompatible with the previous idea, so, for the most part, we work over a general field k as well. With a little more care, the same results apply with k a ring. The reader can always keep in mind $k = \mathbb{R}, \mathbb{C}$.

Specifically, we will be interested in the case when the classical model is a matrix group or perhaps some other vector space. Thus, instead of working with the space $M_n(k)$ of $n \times n$ matrices, or the space $V_n(k)$ of vectors in n dimensions, we work with some kind of algebra of functions $k(M_n)$ and $k(V_n)$, respectively. We do not mean to work with all functions here but rather we concentrate on the *linear coordinate functions* $t^i{}_j$ and v^i, respectively, defined by

$$t^i{}_j(M) = M^i{}_j, \qquad v^i(W) = W^i.$$

These assign to a matrix M one of its matrix entries and to a vector W one of its column entries. When $k = \mathbb{R}, \mathbb{C}$, the polynomials in these coordinate functions locally approximate the continuous functions. We can collect the n^2 such matrix coordinate functions $t^i{}_j$ into a single algebra-valued matrix $\mathbf{t} = \{t^i{}_j\}$. Along with 1, its entries, by definition, generate $k(M_n)$ as an algebra. Likewise, we can collect our n vector coordinate functions into a single algebra-valued column vector $\mathbf{v} = \{v^i\}$. Along with 1, its entries, by definition, generate $k(V_n)$ as an algebra. It turns out to be very convenient to collect the generators together in this way.

In these terms, the strategy of Section 4.1 will be to keep these $t^i{}_j$ as generators in our quantum case also, using them now to generate a non-commutative *quantum matrix* algebra, while retaining familiar properties of M_n, even though the $t^i{}_j$ are no longer the actual coordinate functions on any space. In Section 4.5, we extend the picture to include the vector generators v^i. These become generators of a noncommutative *quantum vector* algebra, while we retain some of the usual relationships between vectors and matrices (namely, that matrices act on vectors) by means of Hopf algebra constructions from Chapter 1.

So much for the superficial motivation. Now we come to the business part of the chapter: how to find interesting relations between the $t^i{}_j$, etc. to replace the usual commutativity? The answer here comes from physics and is remarkably simple. It turns out that a sufficient input here is simply a matrix R in $M_n \otimes M_n$ obeying the matrix equation

$$R_{12}R_{13}R_{23} = R_{23}R_{13}R_{12}. \tag{4.1}$$

Here $R_{12} = R \otimes \mathrm{id}$ and $R_{23} = \mathrm{id} \otimes R$ as elements of $M_n \otimes M_n \otimes M_n$ (R_{13} is analogous). This is the matrix *quantum Yang–Baxter equation* (QYBE). Many solutions are known, the earliest ones arising from statistical mechanics. We will see that such a matrix can be used to define a dual quasi-triangular bialgebra, the quantum matrices associated to R. This is the main result of Section 4.1. The situation here is analogous to the way that a Lie algebra can be specified by structure constants obeying a Jacobi identity: the collection of numbers R now define the commutation relations between the generators of the dual quasitriangular bialgebra. There are also quantum vectors of type R and (distinct) quantum covectors of type R. We will meet them in Section 4.5. In between them, in Section 4.2, we explain how the construction can often be modified to obtain actual Hopf algebras with antipodes. Also, Section 4.3 re-examines the quantum enveloping algebras of the preceding chapter in the light of this R-matrix setting. It turns out that, provided $q \neq 1$, these too can be understood as deformations of the coordinate functions on M_n, modulo further relations. This is an unexpected phenomenon which is not possible in the undeformed case. Its proper understanding requires the theory of *braided matrices* in Chapter 10, for which this section provides some preliminary preparation. Section 4.4 provides some historical background from statistical mechanics.

When working with the matrix of generators **t** or the vector of generators **v**, it is very tempting to regard them as if they were actual matrices and vectors. They do have analogous properties, but remember that now the matrix and vector entries are not numbers that we can assign as we like – they are specific fixed elements $t^i{}_j$ and v^i of the relevant noncommutative algebra. In what way, then, are they like arbitrary matrices?

In the commutative case of $k(M_n)$, for example, the answer is that an arbitrary matrix M determines a representation of the algebra $k(M_n)$, namely sending $t^i{}_j$ to $M^i{}_j$. Essentially because the generators $t^i{}_j$ all commute with each other, they can be simultaneously represented as any numbers $M^i{}_j$. Thus, $\mathbf{t} = \{t^i{}_j\}$ should be thought of as the prototypical 'universal matrix' waiting to become any chosen matrix in a representation. In the noncommutative case, the generators $t^i{}_j$ no longer commute amongst themselves. Hence they cannot all be simultaneously represented as numbers. Instead they are represented as operators and $\mathbf{t}$ should be regarded as a prototypical operator-valued matrix which could perhaps become represented as an actual matrix in a commutative limit. This phenomenon is familiar in quantum mechanics, where a collection of operators need not mutually commute and hence cannot be simultaneously diagonalised. Once again, this is no more than a superficial analogy with quantum mechanics. However, it can be useful to keep in mind and perhaps justifies the quantum appellation.

4.1 Quantum matrices

In this section, we introduce the bialgebra $A(R)$ associated to an arbitrary solution R of the quantum Yang–Baxter equation (4.1). The importance of these bialgebras has been stressed by L.D. Faddeev, N.Yu. Reshetikhin and L.A. Takhtajan, so that they are sometimes called FRT bialgebras. We will come to the substantial part of their results in the next section, while, in the present section, we explore the mathematical foundations along lines developed by the author. As outlined in the introduction, we begin with the classical case of the commutative bialgebra $k(M_n)$ and then consider how to modify it to become noncommutative.

Exercise 4.1.1 *Let $k(M_n)$ denote the commutative algebra generated by the coordinate functions $t^i{}_j$ on the space of matrices M_n. It is a bialgebra along the lines of Example 1.5.2. Show that its comultiplication and counit are defined by*

$$\Delta t^i{}_j = \sum_a t^i{}_a \otimes t^a{}_j, \quad \epsilon t^i{}_j = \delta^i{}_j \tag{4.2}$$

extended as algebra maps. There is no antipode since M_n is not exactly a group (not every element is invertible).

Solution: Using the structure in Example 1.5.2 applied to the semi-group M_n, we compute on any matrices M, N,

$$(\Delta t^i{}_j)(M, N) = t^i{}_j(MN) = (MN)^i{}_j = \sum_k M^i{}_k N^k{}_j = \sum_k t^i{}_k(M) \, t^k{}_j(N).$$

Thus, as abstract elements of the algebra of functions, we have the co-multiplication as stated. The counit is $\epsilon(t^i{}_j) = t^i{}_j(\mathrm{id}) = \delta^i{}_j$, where id here is the identity matrix in M_n. ∎

Thus $k(M_n)$ can be characterised abstractly as the commutative bi-algebra generated by n^2 generators $t^i{}_j$ and 1, with the comultiplication and counit shown. We consider now the possibility of more general bi-algebras defined by declaring the $t^i{}_j$ noncommuting, but for which we hope to retain the matrix form in Exercise 4.1.1 for the coproduct. Our guide to how to do this is that we would like our bialgebra to remain dual quasitriangular in the sense of Definition 2.2.1, but now with a nontrivial quasitriangular structure. This means that there should be a linear map $\mathcal{R} : A \otimes A \to k$ controlling the amount of noncommutativity.

Exercise 4.1.2 *Let A be a bialgebra with n^2 generators $\mathbf{t}$ and 1, with matrix form for the coalgebra structure as in (4.2). Show that if A is dual quasitriangular with dual quasitriangular structure $\mathcal{R}$, then*

$$R^i{}_j{}^k{}_l = \mathcal{R}(t^i{}_j \otimes t^k{}_l) \tag{4.3}$$

obeys the matrix quantum Yang–Baxter equation (4.1). Here the R lies in $M_n \otimes M_n$, with the first two indices lying in the first M_n and the second two in the second M_n. Moreover, the relations

$$R^i{}_a{}^k{}_b t^a{}_j t^b{}_l = t^k{}_b t^i{}_a R^a{}_j{}^b{}_l \tag{4.4}$$

hold in the algebra A. Also, the maps $\rho^+(a)^i{}_j = \mathcal{R}(a \otimes t^i{}_j)$ and $\rho^-(a)^i{}_j = \mathcal{R}^{-1}(t^i{}_j \otimes a)$ are matrix representations $\rho^\pm : A \to M_n$.

Solution: We just write the axiom (2.7) shown in Definition 2.2.1 on two elements of the form $t^i{}_j$ to obtain at once the quadratic relations shown in A. Here and below we use the *summation convention* for repeated indices of tensors. Likewise, just write out (2.6) with one of the generators $t^i{}_j$ to obtain at once that ρ^+ obeys $\rho^+(ab) = \rho^+(a)\rho^+(b)$, while (2.8) on $t^i{}_j$ gives $\rho^+(1) = \mathrm{id}$. Writing out the corresponding equations for the convolution-inverse map $\mathcal{R}^{-1}$ gives the corresponding properties for ρ^-. Finally, to show that R obeys the QYBE we merely have to evaluate (2.10) on three generators of the form $t^i{}_j$. Or, arguing directly from the second of (2.6), (2.7) and the second of (2.6) again, we have

$$(R_{12}R_{13}R_{23})^i{}_j{}^k{}_l{}^m{}_n = \Big(\mathcal{R}(t^i{}_a \otimes t^k{}_b)\mathcal{R}(t^a{}_j \otimes t^m{}_c)\Big)\mathcal{R}(t^b{}_l \otimes t^c{}_n)$$

$$= \mathcal{R}(t^i{}_j \otimes t^m{}_c t^k{}_b)\mathcal{R}(t^b{}_l \otimes t^c{}_n) = \mathcal{R}(t^k{}_b \otimes t^m{}_c)\mathcal{R}(t^i{}_j \otimes t^b{}_l t^c{}_n)$$

$$= \mathcal{R}(t^k{}_b \otimes t^m{}_c)\Big(\mathcal{R}(t^i{}_a \otimes t^c{}_n)\mathcal{R}(t^a{}_j \otimes t^b{}_l)\Big) = (R_{23}R_{13}R_{12})^i{}_j{}^k{}_l{}^m{}_n.$$

∎

These elementary exercises show the kind of structure that we are necessarily led to if we want to have a dual quasitriangular bialgebra in matrix form. We still have to know that such objects exist. The universal approach is to impose the relations in the Exercise 4.1.2 without any further relations. Other examples can then be obtained as quotients of this universal one by imposing further relations.

Example 4.1.3 *Let R be an element of $M_n \otimes M_n$. The bialgebra $A(R)$ of* quantum matrices *is defined as generated by 1 and n^2 indeterminates $\mathbf{t} = \{t^i{}_j\}$ with the relations, comultiplication and counit*

$$R\mathbf{t}_1\mathbf{t}_2 = \mathbf{t}_2\mathbf{t}_1 R, \quad \Delta\mathbf{t} = \mathbf{t} \otimes \mathbf{t}, \quad \epsilon\mathbf{t} = \mathrm{id},$$

where $\mathbf{t}_1$ and $\mathbf{t}_2$ denote $\mathbf{t} \otimes \mathrm{id}$ and $\mathrm{id} \otimes \mathbf{t}$ in $M_n \otimes M_n$ with values in $A(R)$. This is a short-hand for the relations (4.4). Likewise, the expressions for Δ, ϵ are short-hand for (4.2).

Proof: The algebra $A(R)$ is by, definition, the associative algebra with generators and relations as stated, for any matrix R determining the relations. We need to check that Δ and ϵ are well-defined when extended multiplicatively. Thus,

$$R^i{}_a{}^k{}_b \Delta(t^a{}_j t^b{}_l) = R^i{}_a{}^k{}_b (\Delta t^a{}_j)(\Delta t^b{}_l) = R^i{}_a{}^k{}_b (t^a{}_c \otimes t^c{}_j)(t^b{}_d \otimes t^d{}_l)$$

$$= R^i{}_a{}^k{}_b t^a{}_c t^b{}_d \otimes t^c{}_j t^d{}_l = t^k{}_b t^i{}_a R^a{}_c{}^b{}_d \otimes t^c{}_j t^d{}_l = t^k{}_b t^i{}_a \otimes t^b{}_d t^a{}_c R^c{}_j{}^d{}_l$$

$$= (t^k{}_b \otimes t^b{}_d)(t^i{}_a \otimes t^a{}_c) R^c{}_j{}^d{}_l = (\Delta t^k{}_d)(\Delta t^i{}_c) R^c{}_j{}^d{}_l = \Delta(t^k{}_d t^i{}_c) R^c{}_j{}^d{}_l$$

as required. We have used the assumption that Δ is an algebra homomorphism, the matrix comultiplication (4.2) and repeated use of the relations (4.4), and then the steps in reverse. The similar computation for ϵ is automatic. Note that, in stating the relations of $A(R)$, we have employed a short-hand notation. To demonstrate the power of the short-hand notation, let us also write out the proof in the compact form. The same proof reads as

$$\Delta R\mathbf{t}_1\mathbf{t}_2 = R(\Delta\mathbf{t}_1)(\Delta\mathbf{t}_2) = R\mathbf{t}_1\mathbf{t}_2 \otimes \mathbf{t}_1\mathbf{t}_2$$
$$= \mathbf{t}_2\mathbf{t}_1 R \otimes \mathbf{t}_1\mathbf{t}_2 = \mathbf{t}_2\mathbf{t}_1 \otimes \mathbf{t}_2\mathbf{t}_1 R = (\Delta\mathbf{t}_2)(\Delta\mathbf{t}_1)R = \Delta\mathbf{t}_2\mathbf{t}_1 R.$$

The power of this matrix notation lies in the fact that $\otimes$ is used only to refer to the abstract tensor product of copies of the algebra (as in defining the axioms of a Hopf algebra, etc.). The matrix tensor product as in $M_n \otimes M_n$ is suppressed, and its role is replaced by the suffixes $_1, _2$ etc. when needed. Thus, $R = R_{12}$ (with the indices suppressed when there are only two M_n in the picture), while $\mathbf{t}_1$ means that the $\mathbf{t}$ is viewed as a matrix in the first M_n (with values in $A(R)$). The rules of the notation are as follows. Matrices are understood as multiplied in the usual order, independently in the $_{1,2}$ etc. copies of M_n. Thus, $\Delta t = t \otimes t$ is an equation taking place in only one copy of M_n (so the distinguishing suffix is not needed), and the two t on the right are understood to have their matrix indices multiplied in the usual way. Likewise, we encounter the same equation $\Delta t_1 = t_1 \otimes t_1$ in the context of a larger computation, where there is more than one copy of M_n in the picture. All of this happens independently of the multiplications or $\otimes$ products going on with respect to the noncommutative algebra in which the entries of t lie. Note that it is not necessary for the construction to require the QYBE for R, though one could impose it to help to ensure that the algebra defined in this way is not too trivial due to the ideal generated by the relations being too big.

■

We also expect this $A(R)$ to be dual quasitriangular when R obeys the QYBE, with $\mathcal{R}$ obtained by extending R. This requires a little more proof, and we shall come to it shortly. First, we check that $\rho^\pm$ work out correctly.

Proposition 4.1.4 *Let R in $M_n \otimes M_n$ be an invertible solution of the QYBE and let $A(R)$ be the associated matrix bialgebra. Define*

$$\rho^+(t^i{}_j)^k{}_l = R^i{}_j{}^k{}_l, \quad \rho^-(t^i{}_j)^k{}_l = R^{-1k}{}_l{}^i{}_j,$$

$$i.e. \quad \rho_2^+(t_1) = R, \quad \rho_2^-(t_1) = R_{21}^{-1}. \tag{4.5}$$

Then $\rho^\pm : A(R) \to M_n$ are matrix representations of $A(R)$. We call ρ^+ the fundamental representation and ρ^- the conjugate fundamental representation.

Proof: To show that ρ^+ is well-defined, we compute

$$\rho^+(R^i{}_a{}^k{}_b t^a{}_j t^b{}_l)^m{}_n = R^i{}_a{}^k{}_b \rho^+(t^a{}_j)^m{}_c \rho^+(t^b{}_l)^c{}_m$$
$$= R^i{}_a{}^k{}_b R^a{}_j{}^m{}_c R^b{}_l{}^c{}_n = R^k{}_b{}^m{}_c R^i{}_a{}^c{}_n R^a{}_j{}^b{}_l$$
$$= \rho^+(t^k{}_b)^m{}_c \rho^+(t^i{}_a)^c{}_n R^a{}_j{}^b{}_l = \rho^+(t^k{}_b t^i{}_a R^a{}_j{}^b{}_l)^m{}_n,$$

where we have used the definition of ρ^+, the relations in $A(R)$ and the QYBE (4.1). The situation for ρ^- is analogous. In compact notation, the same proofs read as

$$\rho_3^+(R_{12} t_1 t_2) = R_{12} \rho_3^+(t_1) \rho_3^+(t_2)$$
$$= R_{12} R_{13} R_{23} = R_{23} R_{13} R_{12}$$
$$= \rho_3^+(t_2) \rho_3^+(t_1) R_{12} = \rho_3^+(t_2 t_1 R_{12}),$$

$$\rho_3^- (R_{12}\mathbf{t}_1\mathbf{t}_2) = R_{12}\rho_3^-(\mathbf{t}_1)\rho_3^-(\mathbf{t}_2)$$
$$= R_{12}R_{31}^{-1}R_{32}^{-1} = R_{32}^{-1}R_{31}^{-1}R_{12}$$
$$= \rho_3^-(\mathbf{t}_2)\rho_3^-(\mathbf{t}_1)R_{12} = \rho_3^-(\mathbf{t}_2\mathbf{t}_1 R_{12}).$$

Note that for ρ^- we need $R_{12}R_{31}^{-1}R_{32}^{-1} = R_{32}^{-1}R_{31}^{-1}R_{12}$, i.e. $R_{31}R_{32}R_{12} = R_{12}R_{32}R_{31}$, which is again the QYBE after a relabelling of the positions in $M_n^{\otimes 3}$. ∎

Theorem 4.1.5 *Let R in $M_n \otimes M_n$ be an invertible solution of the QYBE. Then the associated matrix bialgebra $A(R)$ is dual quasitriangular with $\mathcal{R}$: $A(R) \otimes A(R) \to k$ given by $\mathcal{R}(\mathbf{t} \otimes 1) = \mathrm{id} = \mathcal{R}(1 \otimes \mathbf{t})$ and $\mathcal{R}(\mathbf{t}_1 \otimes \mathbf{t}_2) = R$ extended as a bialgebra bicharacter according to (2.6). Explicitly, the dual quasitriangular structure is*

$$\mathcal{R}(t^{i_1}{}_{j_1} t^{i_2}{}_{j_2} \cdots t^{i_M}{}_{j_M} \otimes t^{k_N}{}_{l_N} t^{k_{N-1}}{}_{l_{N-1}} \cdots t^{k_1}{}_{l_1})$$

$$R^{i_1}{}_{m_{11}}{}^{k_1}{}_{n_{11}} \quad R^{m_{11}}{}_{m_{12}}{}^{k_2}{}_{n_{21}} \quad \cdots \quad R^{m_{1N-1}}{}_{j_1}{}^{k_N}{}_{n_{N1}}$$
$$= R^{i_2}{}_{m_{21}}{}^{n_{11}}{}_{n_{12}} R^{m_{21}}{}_{m_{22}}{}^{n_{21}}{}_{n_{22}}$$

$$\vdots \qquad\qquad\qquad \vdots \qquad\qquad = Z_R \begin{pmatrix} K \\ I \Box J \\ L \end{pmatrix},$$

$$R^{i_M}{}_{m_{M1}}{}^{n_{1M-1}}{}_{l_1} \quad \cdots \qquad \cdots \quad R^{m_{MN-1}}{}_{j_M}{}^{n_{NM-1}}{}_{l_N}$$

where the notation on the right hand side is as a partition function (see Section 4.4). Here $I = (i_1, \ldots, i_M)$ and $K = (k_1, \ldots, k_N)$ etc. There is a similar expression for $\mathcal{R}^{-1}$. If we adopt the notation $\bar{K} = (k_N, \ldots, k_1)$ and $t^{i_1}{}_{j_1} \cdots t^{i_M}{}_{j_M} = t^I{}_J$, then

$$\mathcal{R}(t^I{}_J \otimes t^{\bar{K}}{}_{\bar{L}}) = Z_R \begin{pmatrix} K \\ I \Box J \\ L \end{pmatrix}, \quad \mathcal{R}^{-1}(t^{\bar{I}}{}_{\bar{J}} \otimes t^K{}_L) = Z_{R^{-1}} \begin{pmatrix} K \\ I \Box J \\ L \end{pmatrix}.$$

Proof: $\mathcal{R}(\mathbf{t}_1 \otimes \mathbf{t}_2)$ extends in its first input as a bialgebra bicharacter, as
$\mathcal{R}(R_{12}\mathbf{t}_1\mathbf{t}_2 \otimes \mathbf{t}_3) = R_{12}\mathcal{R}(\mathbf{t}_1 \otimes \mathbf{t}_3)\mathcal{R}(\mathbf{t}_2 \otimes \mathbf{t}_3) = R_{12}R_{13}R_{23} = R_{23}R_{13}R_{12}$
$= \mathcal{R}(\mathbf{t}_2 \otimes \mathbf{t}_3)\mathcal{R}(\mathbf{t}_1 \otimes \mathbf{t}_3)R_{12} = \mathcal{R}(\mathbf{t}_2\mathbf{t}_1 R_{12} \otimes \mathbf{t}_3)$. Note that $\mathcal{R}(\mathbf{t}_1 \otimes \mathbf{t}_2) = \rho_2^+(\mathbf{t}_1)$, and that the proof that $\mathcal{R}$ extends in its first input is just the same computation as our previous proof that ρ^+ extended as a well-defined representation. We therefore have $\mathcal{R}(a \otimes \mathbf{t}) = \rho^+(a)$ for all a. This form makes it clear that $\mathcal{R}(ab \otimes \mathbf{t}) = \mathcal{R}(a \otimes \mathbf{t})\mathcal{R}(b \otimes \mathbf{t})$, as required for the first half of (2.6). In particular,

$$\mathcal{R}(\mathbf{t}_1\mathbf{t}_2 \cdots \mathbf{t}_M \otimes \mathbf{t}_{M+1}) = \rho_{M+1}^+(\mathbf{t}_1\mathbf{t}_2 \cdots \mathbf{t}_M) = R_{1M+1} \cdots R_{MM+1}$$

is well-defined. Next, we saw in Exercise 1.6.8 that the tensor product of representations is also a representation (because $A(R)$ is a bialgebra), and hence there is a well-defined algebra map $\rho^{+\otimes N} : A(R) \to M_n^{\otimes N}$ given by $\rho^{+\otimes N}(a) = (\rho_1^+ \otimes \rho_2^+ \otimes \cdots \otimes \rho_N^+) \circ \Delta^{N-1}(a)$. In particular, we see that the array

$$
\begin{array}{l}
R_{1M+1} R_{1M+2} \cdots R_{1M+N} \\
R_{2M+1} R_{2M+2} \cdots R_{2M+N} = \rho_{M+1}^+(\mathbf{t}_1 \cdots \mathbf{t}_M) \cdots \rho_{M+N}^+(\mathbf{t}_1 \cdots \mathbf{t}_M) \\
\cdots \qquad \cdots \\
R_{MM+1} \cdots \qquad \cdots R_{MM+N}
\end{array}
$$

$$
= (\rho^+)^{\otimes N}(\mathbf{t}_1 \cdots \mathbf{t}_M)
$$

also depends only on $\mathbf{t}_1 \mathbf{t}_2 \cdots \mathbf{t}_M$ as an element of $A(R)$. The array on the left can be read (and multiplied up) column after column (so that the first equality is clear) or row after row (like reading a page of a book). The two are the same when we bear in mind that R living in distinct copies of $M_n \otimes M_n$ commute. The expression is just the array Z_R in our compact notation. If we define $\mathcal{R}(\mathbf{t}_1 \mathbf{t}_2 \cdots \mathbf{t}_M \otimes \mathbf{t}_{M+N} \cdots \mathbf{t}_{M+1})$ as this array, we know that the second half of (2.6) will hold and that $\mathcal{R}$ is well defined in its first input.

Now we repeat the steps above for the second input of $\mathcal{R}$. Thus $\mathcal{R}(\mathbf{t}_1 \otimes \mathbf{t}_2) = R = \bar{\rho}_1^+(\mathbf{t}_2)$ extends in its second input as a bialgebra bicharacter since $\mathcal{R}(\mathbf{t}_1 \otimes R_{23}\mathbf{t}_2\mathbf{t}_3) = R_{23}\mathcal{R}(\mathbf{t}_1 \otimes \mathbf{t}_3)\mathcal{R}(\mathbf{t}_1 \otimes \mathbf{t}_2) = R_{23}R_{13}R_{12} = R_{12}R_{13}R_{23} = \mathcal{R}(\mathbf{t}_1 \otimes \mathbf{t}_2)\mathcal{R}(\mathbf{t}_1 \otimes \mathbf{t}_3)R_{23} = \mathcal{R}(\mathbf{t}_1 \otimes \mathbf{t}_3\mathbf{t}_2 R_{23})$. In terms of $\bar{\rho}^+$, this says that $\bar{\rho}^+$ extends as an antirepresentation $A(R) \to M_n$. Thus we define $\mathcal{R}(\mathbf{t} \otimes a) = \bar{\rho}^+(a)$, and, in particular,

$$
\mathcal{R}(\mathbf{t}_M \otimes \mathbf{t}_{M+N} \cdots \mathbf{t}_{M+2}\mathbf{t}_{M+1}) = \bar{\rho}_M^+(\mathbf{t}_{M+N} \cdots \mathbf{t}_{M+2}\mathbf{t}_{M+1})
$$

$$
= R_{MM+1} R_{MM+2} \cdots R_{MM+N}
$$

is well-defined. Likewise, we can take tensor products of $\bar{\rho}^+$ and will again have well-defined antirepresentations. Hence, we see that the array

$$
\begin{array}{l}
R_{1M+1} R_{1M+2} \cdots R_{1M+N} \\
R_{2M+1} R_{2M+2} \cdots R_{2M+N} = \bar{\rho}_1^+(\mathbf{t}_{M+N} \cdots \mathbf{t}_{M+1}) \cdots \bar{\rho}_M^+(\mathbf{t}_{M+N} \cdots \mathbf{t}_{M+1}) \\
\cdots \qquad \cdots \\
R_{MM+1} \cdots \qquad \cdots R_{MM+N}
\end{array}
$$

$$
= (\bar{\rho}^+)^{\otimes M}(\mathbf{t}_{M+N} \cdots \mathbf{t}_{M+1})
$$

depends only on $\mathbf{t}_{M+N} \cdots \mathbf{t}_{M+2}\mathbf{t}_{M+1}$ as an element of $A(R)$. The first equality comes from writing out the $\bar{\rho}^+$ and rearranging the resulting array (bearing in mind that copies of R in distinct M_n tensor factors commute). The resulting array of matrices then coincides with that above, which we have already defined to be $\mathcal{R}(\mathbf{t}_1 \cdots \mathbf{t}_M \otimes \mathbf{t}_{M+N} \cdots \mathbf{t}_{M+1})$. We see that this array is therefore well-defined as a map $A(R) \otimes A(R) \to k$: in its first input (for fixed $\mathbf{t}_{M+N}, \ldots, \mathbf{t}_{M+1}$) by its realisation as a tensor power of ρ^+ and in its second input (for fixed $\mathbf{t}_1, \ldots, \mathbf{t}_M$) by its realisation as a tensor power of $\bar{\rho}^+$. By its construction, it obeys (2.6).

Next, we note that, when R is invertible, there is a similar construction for $\mathcal{R}^{-1}$. Here $\mathcal{R}^{-1}$ obeys equations similar to (2.6) but with its second input multiplicative and its first input antimultiplicative. We use R^{-1} in the role of R; for example, $\mathcal{R}^{-1}(\mathbf{t} \otimes a) = \rho^-(a)$ extends as a representation, while $\mathcal{R}^{-1}(a \otimes \mathbf{t})$ extends as an antirepresentation. The steps are entirely analogous to those above, and we arrive at the partition function $Z_{R^{-1}}$. We have to show that $\mathcal{R}, \mathcal{R}^{-1}$ are inverse in the convolution algebra of maps $A(R) \otimes A(R) \to k$ (see Chapter 2.2). Explicitly, we need the identity

$$\mathcal{R}(t^I{}_A \otimes t^{\bar{K}}{}_{\bar{B}}) \mathcal{R}^{-1}(t^{\bar{A}}{}_{\bar{J}} \otimes t^B{}_L) = \delta^I{}_J \delta^K{}_L,$$

$$\text{i.e.} \quad Z_R \begin{pmatrix} \bar{K} \\ I \square A \\ \bar{B} \end{pmatrix} Z_{R^{-1}} \begin{pmatrix} B \\ \bar{A} \square \bar{J} \\ L \end{pmatrix} = \delta^I{}_J \delta^K{}_L, \tag{4.6}$$

and similarly on the other side. Writing out the arrays on the left hand side in the compact notation, we have the expression

$R_{1M+N} \cdots R_{1M+1}$

$\cdots \qquad \cdots$

$R_{M-1M+N} \cdots R_{M-1M+1}$
$R_{MM+N} \cdots R_{MM+1} R^{-1}_{MM+1} \cdots R^{-1}_{MM+N}$
$\qquad\qquad R^{-1}_{M-1M+1} \cdots R^{-1}_{M-1M+N}$
$\qquad\qquad\qquad \cdots \qquad \cdots$
$\qquad\qquad\qquad R^{-1}_{1M+1} \cdots R^{-1}_{1M+N}.$

Here, the copies of M_n numbered $1, \ldots, M$ on the left correspond to the index I; those on the right correspond to $\bar{J}$ (they occur reversed). The copies of M_n numbered $M+1, \ldots, M+N$ correspond on the top to $\bar{K}$ (they occur reversed), and on the bottom they correspond to L. In between they are matrix-multiplied as indicated, corresponding to the sum over A, B in (4.6). The overlapping line here collapses after cancellation of inverses, ending in id in the copy of M_n numbered M, and results in a similar picture with one row less. Repeating this, the whole expression collapses to the identity in all the copies of M_n. This is the right hand side of (4.6) when expressed in the multiindex notation.

Finally, we check (2.7), which now takes the form

$$t^{\bar{K}}{}_{\bar{B}} t^I{}_A Z_R \begin{pmatrix} B \\ A \square J \\ L \end{pmatrix} = Z_R \begin{pmatrix} K \\ I \square A \\ B \end{pmatrix} t^A{}_J t^{\bar{B}}{}_{\bar{L}}. \tag{4.7}$$

In the compact notation, we compute:

$$\mathbf{t}_{M+N} \cdots \mathbf{t}_{M+1} \mathbf{t}_1 \cdots \mathbf{t}_M R_{1M+1} \cdots R_{1M+N}$$

$$\cdots \qquad \cdots$$

$$R_{MM+1} \cdots R_{MM+N}$$

$$= R_{1M+1} \cdots R_{1M+N} \mathbf{t}_1 \mathbf{t}_{M+N} \cdots \mathbf{t}_{M+1} \mathbf{t}_2 \cdots \mathbf{t}_M R_{2M+1} \cdots R_{2M+N}$$

$$\cdots \qquad \cdots$$

$$R_{MM+1} \cdots R_{MM+N}$$

$$= \vdots$$

$$= R_{1M+1} R_{1M+2} \cdots R_{1M+N}$$
$$R_{2M+1} R_{2M+2} \cdots R_{2M+N}$$

$$\cdots \qquad \cdots$$

$$R_{MM+1} \cdots \qquad \cdots R_{MM+N} \mathbf{t}_1 \mathbf{t}_2 \cdots \mathbf{t}_M \mathbf{t}_{M+N} \cdots \mathbf{t}_{M+1}.$$

Here the copies of M_n numbered $1, \ldots, M$ on the left correspond to the index I, and the copies of M_n numbered $M+1, \ldots, M+N$ correspond on the top to $\bar{K}$ (they occur reversed), etc. The first equality makes repeated use of the relations (4.4) of $A(R)$ to give

$$\mathbf{t}_{M+N} \cdots \mathbf{t}_{M+1} \mathbf{t}_1 R_{1M+1} \cdots R_{1M+N} = R_{1M+1} \cdots R_{1M+N} \mathbf{t}_1 \mathbf{t}_{M+N} \cdots \mathbf{t}_{M+1}.$$

The $\mathbf{t}_2 \cdots \mathbf{t}_M$ move past the $R_{1M+1} \cdots R_{1M+N}$ freely since they live in different matrix spaces. This argument for the first equality is then applied to move $\mathbf{t}_{M+N} \cdots \mathbf{t}_{M+1} \mathbf{t}_2$, and so on. The method of multiplication of entire rows or columns of matrices is more familiar in the context of exactly solvable statistical mechanics, as we will see in Section 4.4. The above theorem and its proof are due to the author.   ∎

Let us note that, while the relations (4.4) of $A(R)$ do not depend on the normalisation of R, the dual quasitriangular structure does. The elements $t^{i_1}{}_{j_1} \cdots t^{i_M}{}_{j_M}$ of $A(R)$ have a well-defined degree $|t^{i_1}{}_{j_1} \cdots t^{i_M}{}_{j_M}| = M$ (the algebra is graded), and, if $R' = \lambda R$ is a nonzero rescaling of our solution R, then the corresponding dual quasitriangular structure is changed to

$$\mathcal{R}'(a \otimes b) = \lambda^{|a||b|} \mathcal{R}(a \otimes b) \tag{4.8}$$

on homogeneous elements. This is evident from the expression in terms of Z_R that we have obtained in the preceding theorem. We will fix this freedom in the next section when we consider quantum groups with antipodes.

This completes our study of the bialgebra $A(R)$. We have shown that it is dual quasitriangular, and also universal, in the sense that any other dual quasitriangular matrix bialgebra must be the quotient of such an $A(R)$ by further relations. We can also reformulate this result in terms of another bialgebra which is dually paired with $A(R)$ (and is therefore quasitriangular in a certain weak sense). It is also of matrix type.

Example 4.1.6 *Let R be a solution of the QYBE in $M_n \otimes M_n$. There is a bialgebra $\tilde{U}(R)$ generated by 1 and $2n^2$ indeterminates $\mathbf{L}^+, \mathbf{L}^-$, with relations*

$$\mathbf{L}_1^\pm \mathbf{L}_2^\pm R = R\mathbf{L}_2^\pm \mathbf{L}_1^\pm, \quad \mathbf{L}_1^- \mathbf{L}_2^+ R = R\mathbf{L}_2^+ \mathbf{L}_1^-,$$

$$\Delta \mathbf{L}^\pm = \mathbf{L}^\pm \otimes \mathbf{L}^\pm, \qquad \epsilon \mathbf{L}^\pm = \mathrm{id}.$$

Proof: The proof is fully analogous to the one given in detail for $A(R)$ above. The $\mathbf{L}^\pm$ obeys the relations of $A(R)^{\mathrm{op}}$, and, in addition, have the cross relations between $\mathbf{L}^+$ and $\mathbf{L}^-$ (with compatibility with Δ, ϵ verified in the same way). This bialgebra is an example of a general *double cross product* construction given later, in Chapter 7.2. ∎

In terms of this $\tilde{U}(R)$ one may restate (and slightly extend) the various results above. Thus,

Corollary 4.1.7 $\tilde{U}(R)$ *has a canonical representation $\rho : \tilde{U}(R) \to M_n$ given by*

$$\rho(L^{+k}{}_l)^i{}_j = R^i{}_j{}^k{}_l, \qquad \rho(L^{-k}{}_l)^i{}_j = R^{-1k}{}_l{}^i{}_j,$$

$$i.e. \quad \rho_1(\mathbf{L}_2^+) = R, \quad \rho_1(\mathbf{L}_2^-) = R_{21}^{-1}.$$

(4.9)

Proof: The only novel part is to check that ρ represents also the cross relations between $\mathbf{L}^+$ and $\mathbf{L}^-$ (the others are analogous to the relations in Proposition 4.1.4). Using the compact notation, we have

$$\rho_1(\mathbf{L}_2^- \mathbf{L}_3^+ R_{23}) = \rho_1(\mathbf{L}_2^-)\rho_1(\mathbf{L}_3^+)R_{23} = R_{21}^{-1}R_{13}R_{23}$$

$$= R_{23}R_{13}R_{21}^{-1} = R_{23}\rho_1(\mathbf{L}_3^+)\rho_1(\mathbf{L}_2^+).$$

The middle equality is equivalent to (4.1) after a little rearrangement and relabelling of the matrix indices. ∎

Corollary 4.1.8 *Let R be an invertible solution of the QYBE. Then the bialgebra $A(R)$ and the bialgebra $\tilde{U}(R)$ are paired in the sense of Definition 1.4.3. Explicitly, this is*

$$\langle \mathbf{t}_1, \mathbf{L}_2^+ \rangle = R, \qquad \langle \mathbf{t}_1, \mathbf{L}_2^- \rangle = R_{21}^{-1},$$

(4.10)

extended as a bialgebra pairing. The pairing on products of generators is

$$\langle t^{i_1}{}_{j_1} t^{i_2}{}_{j_2} \cdots t^{i_M}{}_{j_M}, L^{+k_1}{}_{l_1} L^{+k_2}{}_{l_2} \cdots L^{+k_N}{}_{l_N}\rangle = Z_R \begin{pmatrix} K \\ I\square J \\ L \end{pmatrix}$$

$$\langle t^{i_1}{}_{j_1} t^{i_2}{}_{j_2} \cdots t^{i_M}{}_{j_M}, L^{-k_1}{}_{l_1} L^{-k_2}{}_{l_2} \cdots L^{-k_N}{}_{l_N}\rangle = Z_{R_{21}^{-1}} \begin{pmatrix} K \\ I\square J \\ L \end{pmatrix},$$

in the notation of Theorem 4.1.5. The pairing for a mixture of $\mathbf{L}^{\pm}$ is analogous, with R or R_{21}^{-1} at the corresponding column of $Z \begin{pmatrix} K \\ I\square J \\ L \end{pmatrix}$.

Proof: This is largely a restatement of Theorem 4.1.5 with $\rho^+ = \langle\,, \mathbf{L}^+\rangle$ and $\rho^- = \langle\,, \mathbf{L}^-\rangle$. The statement that these extend as a bialgebra pairing is analogous to the statement that $\mathcal{R}$ there extends as a dual quasitriangular structure. For example, $\langle R_{12} t_1 t_2, \mathbf{L}_3^+\rangle = R_{12}\langle t_1, \mathbf{L}_3^+\rangle\langle t_2, \mathbf{L}_3^+\rangle = R_{12} R_{13} R_{23} = R_{23} R_{13} R_{12} = \langle t_2, \mathbf{L}_3^+\rangle\langle t_1, \mathbf{L}_3^+\rangle R_{12} = \langle t_2 t_1 R_{12}, \mathbf{L}_3^+\rangle$ is the same computation as the proof that ρ^+ was well-defined. Hence it extends to strings in the generators of $A(R)$. Similarly for $\mathbf{L}^-$, with R_{21}^{-1} in place of R. Likewise for strings in the right hand argument of the pairing, where we use that $\rho = \langle \mathbf{t},\,\rangle$ is a representation as proven in Corollary 4.1.7. Combining these two observations in the same way as in the proof of Theorem 4.1.5 then shows that the pairing stated is well-defined as a map $A(R) \otimes \tilde{U}(R) \to k$. The computation of its explicit form is also entirely analogous. We just repeatedly use the pairing relations as in Proposition 1.4.2. For example,

$$\langle t^{i_1}{}_{j_1} t^{i_2}{}_{j_2} \cdots t^{i_M}{}_{j_M}, L^{+k_1}{}_{l_1} L^{-k_2}{}_{l_2} \cdots L^{+k_N}{}_{l_N}\rangle$$
$$= \langle t^{i_1}{}_{j_1} \otimes \cdots \otimes t^{i_M}{}_{j_M}, (\Delta^{M-1} L^{+k_1}{}_{l_1})(\Delta^{M-1} L^{-k_2}{}_{l_2}) \cdots (\Delta^{M-1} L^{+k_N}{}_{l_N})\rangle$$
$$= \langle (\Delta^{N-1} t^{i_1}{}_{j_1}) \otimes \cdots \otimes (\Delta^{N-1} t^{i_M}{}_{j_M}),$$
$$(\Delta^{M-1} L^{+k_1}{}_{l_1}) \otimes (\Delta^{M-1} L^{-k_2}{}_{l_2}) \cdots \otimes (\Delta^{M-1} L^{+k_N}{}_{l_N})\rangle$$

$$= \left\langle \begin{matrix} t^{i_1}{}_{m_{11}} \otimes \cdots \otimes t^{m_{1N-1}}{}_{j_1} & L^{+k_1}{}_{n_{11}} \otimes L^{-k_2}{}_{n_{21}} \otimes \cdots \otimes L^{+k_N}{}_{n_{N1}} \\ \otimes t^{i_2}{}_{m_{21}} \otimes \cdots \otimes t^{m_{2M-1}}{}_{j_2} & \otimes L^{+n_{11}}{}_{n_{12}} \otimes L^{-n_{21}}{}_{n_{22}} \otimes \cdots \otimes L^{+n_{N1}}{}_{n_{N2}} \\ \vdots & , & \vdots \\ \otimes t^{i_M}{}_{m_{M1}} \otimes \cdots \otimes t^{m_{MN-1}}{}_{j_M} & \otimes L^{+n_{1M-1}}{}_{l_1} \otimes L^{-n_{2M-1}}{}_{l_2} \cdots \otimes L^{+n_{NM-1}}{}_{l_N} \end{matrix} \right\rangle$$

$$= \begin{matrix} R^{i_1}{}_{m_{11}}{}^{k_1}{}_{n_{11}} R^{-1k_2}{}_{n_{21}}{}^{m_{11}}{}_{m_{12}} & \cdots & R^{m_{1N-1}}{}_{j_1}{}^{k_N}{}_{n_{N1}} \\ R^{i_2}{}_{m_{21}}{}^{n_{11}}{}_{n_{12}} R^{-1n_{21}}{}_{n_{22}}{}^{m_{21}}{}_{m_{22}} & & \\ \vdots & & \vdots \\ R^{i_M}{}_{m_{M1}}{}^{n_{1M-1}}{}_{l_1} R^{-1n_{2M-1}}{}_{l_2}{}^{m_{M1}}{}_{m_{M2}} & \cdots & R^{m_{MN-1}}{}_{j_M}{}^{n_{NM-1}}{}_{l_N}. \end{matrix}$$

■

Corollary 4.1.9 $\tilde{U}(R)$ *is universal in the following sense. Let* $(H, \mathcal{R})$ *be a quasitriangular bialgebra or Hopf algebra which is paired with* $A(R)$ *such that* $\langle t_1 \otimes t_2, \mathcal{R} \rangle = R$. *Let*

$$l^+ = (\mathrm{id} \otimes t)(\mathcal{R}), \qquad l^- = (t \otimes \mathrm{id})(\mathcal{R}^{-1}).$$

Here the $l^{\pm i}{}_j$ *are elements of* H. *Then there is a bialgebra map* $\tilde{U}(R) \to H$ *such that* $l^\pm$ *are the images of* $\mathbf{L}^\pm$, *i.e.* H *is a realisation of* $\tilde{U}(R)$.

Proof: We show that $l^\pm$ as defined are indeed a realisation of $\tilde{U}(R)$. We have

$$
\begin{aligned}
l_1^+ l_2^+ R &= \mathcal{R}^{(1)} \mathcal{R}'^{(1)} \langle t_1 \otimes t_2, \mathcal{R}^{(2)} \otimes \mathcal{R}'^{(2)} \rangle R \\
&= \mathcal{R}^{(1)} \langle t_1 \otimes t_2, \mathcal{R}^{(2)}{}_{(2)} \otimes \mathcal{R}^{(2)}{}_{(1)} \rangle R = \mathcal{R}^{(1)} \langle t_2 t_1, \mathcal{R}^{(2)} \rangle R \\
&= R \mathcal{R}^{(1)} \langle t_1 t_2, \mathcal{R}^{(2)} \rangle = R \mathcal{R}^{(1)} \langle t_1 \otimes t_2, \mathcal{R}^{(2)}{}_{(1)} \otimes \mathcal{R}^{(2)}{}_{(2)} \rangle \\
&= R \mathcal{R}^{(1)} \mathcal{R}'^{(1)} \langle t_2 \otimes t_1, \mathcal{R}^{(2)} \otimes \mathcal{R}'^{(2)} \rangle = R l_2^+ l_1^+,
\end{aligned}
$$

where we have used the second part of (2.1) for the quasitriangular structure and the elementary properties of the assumed pairing, as in Chapter 1.4. Likewise for the $l_1^- l_2^-$ relations, using now the first of (2.1). For the mixed $l_1^+ l_2^-$ relations, we have

$$
\begin{aligned}
l_1^- l_2^+ R &= \mathcal{R}^{-(2)} \mathcal{R}^{(1)} \langle t_1 \otimes t_2, \mathcal{R}^{-(1)} \otimes \mathcal{R}^{(2)} \rangle \langle t_1 \otimes t_2, \mathcal{R} \rangle \\
&= \mathcal{R}^{-(2)} \mathcal{R}^{(1)} \langle t_1 \otimes t_2, \mathcal{R}^{-(1)} \mathcal{R}'^{(1)} \otimes \mathcal{R}^{(2)} \mathcal{R}'^{(2)} \rangle \\
&= \mathcal{R}^{(1)} \mathcal{R}^{-(2)} \langle t_1 \otimes t_2, \mathcal{R}'^{(1)} \mathcal{R}^{-(1)} \otimes \mathcal{R}'^{(2)} \mathcal{R}^{(2)} \rangle \\
&= \langle t_1 \otimes t_2, \mathcal{R} \rangle \mathcal{R}^{(1)} \mathcal{R}^{-(2)} \langle t_1 \otimes t_2, \mathcal{R}^{-(1)} \otimes \mathcal{R}^{(2)} \rangle = R l_2^+ l_1^-,
\end{aligned}
$$

using the pairing relations and the QYBE for $\mathcal{R}$ from Lemma 2.1.4, in the form $\mathcal{R}_{21}^{-1} \mathcal{R}_{13} \mathcal{R}_{23} = \mathcal{R}_{23} \mathcal{R}_{13} \mathcal{R}_{21}^{-1}$ (this is the same after some rearrangement and relabelling). The proof is similar to the proof of Corollary 4.1.7, but now holds abstractly as a realisation in H. Moreover, it is easy to see from (2.1) that our realisation also respects the coalgebra structures. Hence we have a bialgebra map $\tilde{U}(R) \to H$, as stated. ∎

Thus, any quasitriangular bialgebra or Hopf algebra dual to $A(R)$ is the image of $\tilde{U}(R)$, and it is in this weak sense that $\tilde{U}(R)$ is weakly quasitriangular. Also, the induced elements $l^\pm$ will often generate all of H. If so, we say that H as generated by $l^\pm$ is in *FRT form*. The reason for this will become clear in the next section when we come to apply this corollary to the standard quantum groups $U_q(g)$. Our present formulation is not tied to these standard quantum groups, however.

On the other hand, $A(R)$ and $\tilde{U}(R)$ themselves are usually not strictly dual, in the sense that the pairing between them may be degenerate.

We recall from Proposition 1.4.4 how to fix this remaining problem: any paired bialgebras can be quotiented in such a way that the pairing becomes nondegenerate. We can do this to the pair $(A(R), \widetilde{U}(R))$ to obtain a nondegenerate pairing of bialgebras $(\check{A}(R), \check{U}(R))$. For some standard R-matrices, they are, in fact, Hopf algebras, and one recovers some of the standard quantum groups in this way. The point is that so far we have constructed only bialgebras with quadratic relations: it is the further quotienting to obtain a nondegenerate pairing that typically introduces the nonquadratic relations familiar for the standard quantum groups.

Lemma 4.1.10 *The quotients making the pairing in Corollary 4.1.8 nondegenerate consist of adding just the relations that hold in all tensor powers of the fundamental and canonical representations $\rho^\pm$ and ρ of $A(R)$ and $\widetilde{U}(R)$, respectively.*

Proof: This is a restatement of Proposition 1.4.2 in our particular context. The relationship between the pairing and tensor powers of $\rho^\pm, \rho$ has already been explained in the proof of Corollary 4.1.8. ∎

We will demonstrate this technique on some standard R-matrices in the next section. For now, we content ourselves with working through an elementary example that is very far from the standard examples. The construction for bialgebras in this section works for any invertible R-matrix, and one should not think that the result need have anything to do with usual enveloping algebras of Lie algebras or any kind of deformation of them. For example, the R-matrix need not have any parameters.

Example 4.1.11 *Let $n > 1$ and $R = P : V_n \otimes V_n \to V_n \otimes V_n$ be the permutation matrix*

$$R^i{}_j{}^k{}_l = \delta^i_l \delta^k_j.$$

Then $A(R)$ is generated by n^2 indeterminates $\mathbf{t}$ and 1, with no relations; i.e. it is the tensor algebra on the n^2-dimensional vector space M_n, as in Example 1.1.2,

$$A(R) = TM_n = k \oplus M_n \oplus M_n^{\otimes 2} \oplus M_n^{\otimes 3} \oplus \cdots.$$

The dual quasitriangular structure is triangular, and the nondegenerate quotients of $A(R), \widetilde{U}(R)$ are also given by TM_n, so TM_n is nondegenerately paired with itself (it is self-dual).

Proof: The relations (4.4) are empty for $R = P$, so $A(R)$ is the algebra generated by 1 and $\mathbf{t}$ with no relations among the $\mathbf{t}$. This is the tensor algebra described (as an algebra) in Example 1.1.2. Its underlying vector space is M_n, and we learn that it becomes a bialgebra with the matrix comultiplication. A general element is $\alpha + \alpha_i{}^j t^i{}_j + \alpha_i{}^j{}_k{}^l t^i{}_j t^k{}_l + \cdots$ for matrices of coefficients $\alpha, \alpha_i{}^j, \alpha_i{}^j{}_k{}^l$ in $k, M_n, M_n^{\otimes 2}$, etc. This is how we identify $A(R)$ with the 'nonsymmetrised Fock space' on M_n.

We compute now the nondegenerate quotients. It is easy to see that in $\tilde{U}(R)$ the elements $\mathbf{L}^+ - \mathbf{L}^-$ have zero pairing with all elements of $A(R)$, where the pairing is from Corollary 4.1.8. So, in forming the non-degenerate quotients, we can begin by setting $\mathbf{L}^+ = \mathbf{L}^-$, denoted $\mathbf{L}$, say. This is clearly always the case when R obeys $R = R_{21}^{-1}$, as it does here. After this quotienting, we obtain again the tensor algebra TM_n for the bialgebra dually paired with $A(R)$. The pairing then takes the form

$$\langle t^{i_1}{}_{j_1} t^{i_2}{}_{j_2} \cdots t^{i_M}{}_{j_M}, L^{k_1}{}_{l_1} L^{k_2}{}_{l_2} \cdots L^{k_N}{}_{l_N} \rangle$$

$$= \begin{cases} \delta^{i_M}_{l_1} \cdots \delta^{i_1}_{l_M} \delta^{k_1}_{l_{M+1}} \cdots \delta^{k_{N-M}}_{l_N} \delta^{k_{N-M+1}}_{j_M} \cdots \delta^{k_N}_{j_1} & M < N \\ \delta^{i_M}_{l_1} \cdots \delta^{i_1}_{l_N} \delta^{k_1}_{j_M} \cdots \delta^{k_N}_{j_1} & M = N \\ \delta^{i_M}_{l_1} \cdots \delta^{i_{M-N+1}}_{l_N} \delta^{i_{M-N}}_{j_M} \cdots \delta^{i_1}_{j_{N+1}} \delta^{k_1}_{j_N} \cdots \delta^{k_N}_{j_1} & M > N, \end{cases}$$

and we show that it is now nondegenerate. The expression is P extended to products of the generators, as in Corollary 4.1.8 (it is also the form of the dual quasitriangular structure in Theorem 4.1.5). Using it, we find that, for any given N, the elements

$$e_{(N)}{}^{j_1' \cdots j_N'}{}_{i_1' \cdots i_N'} = L^{k_1}{}_{i_N'} L^{k_2}{}_{i_{N-1}'} \cdots L^{k_N}{}_{i_1'} L^{j_N'}{}_{k_1} \cdots L^{j_1'}{}_{k_N}, \quad i_1' \notin \{k_1, k_1, \ldots, k_N\}$$

(choose any such k_i for given i', j') have zero pairing with all elements of $A(R)$ of degree $< N$, and

$$\langle t^{i_1}{}_{j_1} \cdots t^{i_N}{}_{j_N}, e_{(N)}{}^{j_1' \cdots j_N'}{}_{i_1' \cdots i_N'} \rangle = \delta^{i_1}_{i_1'} \cdots \delta^{i_N}_{i_N'} \delta^{j_1'}_{j_1} \cdots \delta^{j_N'}_{j_N}$$

on the subspace of $A(R)$ homogeneous of degree N. Now suppose $c \in K_1$ has finite top degree, say $c = c_N + c_{N-1} + \cdots + c_0$ (with c_i homogeneous of degree i). By choosing the vectors $e_{(N)}$, which pair only with c_N, we conclude that $\langle c_N, e_{(N)} \rangle = 0$. But $e_{(N)}$ form a basis for the dual space of the subspace of elements homogeneous of degree N; so $c_N = 0$. Hence, by induction, there are no nonzero elements of finite top degree. Similarly, the kernel in the other input of the pairing is trivial as well. Note that the copy of TM_n generated by the $\mathbf{L}$ is, in some formal sense, (as a formal power series), quasitriangular (because it is dually paired with a dual quasitriangular bialgebra) but if we tried to write it down as an element

$$\mathcal{R} \in (k \oplus M_n \oplus M_n^{\otimes 2} \oplus \cdots) \otimes (k \oplus M_n \oplus M_n^{\otimes 2} \oplus \cdots)$$

we would find that $\mathcal{R}$ has components living in all tensor powers. It consists of $P \in M_n \otimes M_n$ 'amplified' to all tensor powers of M_n. This is a

feature of many quantum groups, namely their quasitriangular structure
is a formal power series, although they have a perfectly well-defined dual
quasitriangular structure.   ∎

   This is our first concrete demonstration of the general theory developed
above. Even the simplest case for $n = 2$ is nontrivial to verify directly.
We have learned that the 'nonsymmetrised Fock space' $TM_2 = k \oplus M_2 \oplus M_2^{\otimes 2} \oplus \cdots$ is a dual quasitriangular bialgebra, nondegenerately paired
with itself.

   What about other solutions R of the QYBE? There are many, as we
shall see next. Let us denote by YB_n the moduli space of all invertible solutions of the QYBE in $M_n \otimes M_n$, with different normalisations
identified. This is the *Yang–Baxter variety*. We have seen that at each
point R of YB_n there is a bialgebra $A(R)$. We will see later that there
are quantum or braided-Lie algebras, quantum vectors, quantum covectors and numerous other natural algebraic constructions at each point R.
Moreover, they fit together into a kind of 'bundle' of quantum algebras
over the Yang–Baxter variety. Most of these constructions are covariant
under conjugation by a matrix of the form $U \otimes U$ in $GL_n \otimes GL_n$, so there
are useful orbits under this GL_n action. In addition, it is clear that if R
obeys the QYBE then so does R_{21}, $R^{t \otimes t}$ (transposition in each matrix
factor) and R with its indices shifted on a circle $1, \ldots, n$. There are other
symmetries of the QYBE as well.

   Altogether, this Yang–Baxter variety or moduli space has a rich structure, and one that it not fully understood. Fig. 4.1 is a schematic im-
pression showing some of the qualitative features. First, it is not at all a
smooth manifold: it is full of singular disconnected points, lines, planes
etc., and jumps in dimension even for fixed n. Secondly, the diagonal
matrices are all solutions and form the central plane shown. This is more
properly a projective plane if we identify solutions up to scale. One very
important point on this plane is the identity matrix $R = I$. This is our reference point since $A(I)$ is the commutative bialgebra of functions on M_n,
i.e. the model for our theory. Radiating out of this solution are various
deformations. Among them there are the standard R-matrices that lead
to the standard quantum groups $SL_q(n)$ etc., as quotients of $A(R)$; we
will come to them in the next section. But there are plenty of others too.
First, any line will tend to be part of a whole manifold of deformations
given by the abovementioned GL_n action. Secondly, many of the lines
are parts of families where the corresponding quantum group is genuinely
changed by twisting in the sense explained at the end of Section 2.3 for
dual quasitriangular bialgebras.

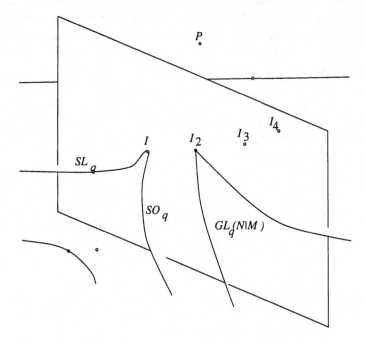

Fig. 4.1. Schematic of the Yang–Baxter variety showing some qualitative features. Many points are deformable by one or many-parameter families while other points are isolated solutions.

Other points on the diagonal plane are the *superidentity* matrices I_2, such as

$$I_2 = \begin{pmatrix} 1 & 0 & 0 & 0 \\ 0 & 1 & 0 & 0 \\ 0 & 0 & 1 & 0 \\ 0 & 0 & 0 & -1 \end{pmatrix},$$

where some of the diagonals are -1. The quantum groups corresponding to their deformations include the nonstandard ones which are also covered in the next section. These look rather strange, and, in fact, such R-matrices are rather more natural when one uses them equivalently to construct superquantum groups such as $GL_q(N|M)$. We will say more about superquantum groups in Chapter 10.1. Among other important deformations of the superidentity, we have in $n = 2$ the 'eight-vertex solutions', which take their name from statistical mechanics, as we will see in Section 4.4.

The solutions which are triangular in the sense $R_{21}R = 1$ form an important subvariety because any triangular point can serve as a base for interesting deformations. We can label other solutions by one or more parameters q, which measure the deviation from the triangular subvariety. Moreover, the semiclassical limit as we approach any point R on the triangular subvariety is associated to the obvious notion of an R-Lie algebra. This is like a super-Lie algebra, but with ± 1 replaced by R. From this point of view, the identity and superidentity are not the only reference points: many quantum groups can be thought of as deformations of R-Lie algebras for some point R on the triangular subvariety. By contrast, the diagonal matrices I_N containing Nth roots of unity are not triangular, and this is the reason that there is not an equally rich semiclassical theory of $\mathbb{Z}_{/N}$-graded or anyonic Lie algebras for $N > 2$.

Also shown in Fig. 4.1 is the isolated point $R = P$ studied in Example 4.1.11 above. We have seen that it is maximally far from $R = I$ in the sense that its $A(R)$ has no relations among its generators. There are plenty of other isolated points, lines and planes, some of them triangular and some not.

The detailed form of the Yang–Baxter variety was computed for $n = 2$ by J. Hietarinta, and demonstrates all of these features. We work up to the symmetries of the QYBE mentioned above and classify according to the number of generically distinct *skew-eigenvalues* of R. These are the eigenvalues of the matrix PR, where P is the permutation matrix. We consider only invertible solutions of the QYBE. There is one type of solution for each of the cases of four and three skew-eigenvalues, namely

$$\begin{pmatrix} 1 & 0 & 0 & 0 \\ 0 & p & 0 & 0 \\ 0 & 0 & s & 0 \\ 0 & 0 & 0 & q \end{pmatrix}, \quad \begin{pmatrix} 0 & 0 & 0 & q \\ 0 & 0 & 1 & 0 \\ 0 & 1 & 0 & 0 \\ q & 0 & 0 & 0 \end{pmatrix},$$

respectively. Here p, q, s are general parameters. All the rest have two skew-eigenvalues when our field is algebraically closed and of characteristic zero. The triangular ones, where the skew-eigenvalues can be chosen to be ± 1, are

$$\begin{pmatrix} 1 & 1 & p & q \\ 0 & 1 & 0 & p \\ 0 & 0 & 1 & 1 \\ 0 & 0 & 0 & 1 \end{pmatrix}, \quad \begin{pmatrix} 1 & 1 & -1 & q \\ 0 & 1 & 0 & q \\ 0 & 0 & 1 & -q \\ 0 & 0 & 0 & 1 \end{pmatrix},$$

$$\begin{pmatrix} 1 & 0 & 0 & 1 \\ 0 & 1 & 0 & 0 \\ 0 & 0 & 1 & 0 \\ 0 & 0 & 0 & 1 \end{pmatrix}, \quad \begin{pmatrix} 1 & 0 & 0 & 1 \\ 0 & -1 & 0 & 0 \\ 0 & 0 & -1 & 0 \\ 0 & 0 & 0 & 1 \end{pmatrix}.$$

These obey $R_{21}R = 1$ for all p, q. The more interesting *Hecke* solutions are those with two skew-eigenvalues $q, -q^{-1}$, say, and have the form

$$\frac{1}{2}\begin{pmatrix} q - q^{-1} + 2 & 0 & 0 & q - q^{-1} \\ 0 & q + q^{-1} & q - q^{-1} & 0 \\ 0 & q - q^{-1} & q + q^{-1} & 0 \\ q - q^{-1} & 0 & 0 & q - q^{-1} - 2 \end{pmatrix},$$

$$\begin{pmatrix} q & 0 & 0 & 0 \\ 0 & p & q - q^{-1} & 0 \\ 0 & 0 & p^{-1} & 0 \\ 0 & 0 & 0 & q \end{pmatrix}, \quad \begin{pmatrix} q & 0 & 0 & 0 \\ 0 & p & q - q^{-1} & 0 \\ 0 & 0 & p^{-1} & 0 \\ 0 & 0 & 0 & -q^{-1} \end{pmatrix},$$

$$\begin{pmatrix} q & 0 & 0 & q \\ 0 & q & q - q^{-1} & 0 \\ 0 & 0 & q^{-1} & 0 \\ 0 & 0 & 0 & -q^{-1} \end{pmatrix}, \quad \frac{-\imath}{\sqrt{2}}\begin{pmatrix} 1 & 0 & 0 & 1 \\ 0 & 1 & 1 & 0 \\ 0 & 1 & -1 & 0 \\ -1 & 0 & 0 & 1 \end{pmatrix},$$

where p, q are free parameters. The last case here is scaled over $\mathbb{C}$ so that it is Hecke with $q = e^{\frac{-\pi\imath}{4}}$. The ones with eight nonzero entries are called eight-vertex solutions. Last, but not least, we have the permutation matrix

$$P = \begin{pmatrix} 1 & 0 & 0 & 0 \\ 0 & 0 & 1 & 0 \\ 0 & 1 & 0 & 0 \\ 0 & 0 & 0 & 1 \end{pmatrix},$$

which was already studied above. In all these matrices, the rows are the upper indices of $R^i{}_j{}^k{}_l$ and the columns are the lower indices, taken in the order 11, 12, 21, 22.

This completes our guided tour of some qualitative features of the Yang–Baxter variety. There is a whole associated algebraic 'universe' at each point R, and our usual theory of Lie algebras is nothing other than the study of some one-parameter families of these near the point $R = I$. From the point of view of quantum group theory, Lie algebras are only a special semiclassical limit of general constructions over the Yang–Baxter variety. We cover this further in Chapter 8. The corresponding notions at or near other points on the triangular subvariety are equally familiar, and include supergroups and super-Lie algebras near I_2. But, as soon as we step off the triangular subvariety, the fact that $R_{21}R \neq 1$ enters nontrivially into our algebra and leads to genuinely new phenomena. We will formalise this remark in Chapter 9, where we see that, in this case, we are working in a braided category. We will also see its effects at various points in the present chapter.

4.2   Quantum determinants and basic examples

So far we have obtained and studied dual quasitriangular bialgebras $A(R)$. They should be thought of as noncommutative versions of the algebra of functions on M_n. They are not generally Hopf algebras, as one should expect, since M_n is not a group. In this section, we consider ways to provide an antipode and thereby make $A(R)$ into an honest dual quasi-triangular Hopf algebra or dual quantum group. We also compute the results explicitly for some of the best-known R-matrices. We begin with the classical situation.

Exercise 4.2.1 *Suppose that $G \subset M_n$ is a group for which the inversion operation is a polynomial in the matrix entries. Let $k(G)$ be the corresponding quotient of the bialgebra $k(M_n)$ from Exercise 4.1.1. Show that it is a Hopf algebra, with antipode operation S obeying*

$$(St^i{}_k)t^k{}_j = \delta^i{}_j = t^i{}_k St^k{}_j. \tag{4.11}$$

Here $t^i{}_j$ are the linear coordinate functions in Exercise 4.1.1, considered in $k(G)$. In a compact notation this says that $St = \mathbf{t}^{-1}$ in the algebra M_n with values in $k(G)$.

Solution: If $G \subset M_n$, then every function on M_n restricts to a function on G; we model the algebra of functions on G by setting to zero the ideal $I \subseteq k(M_n)$ of functions which vanish when restricted to G in this way. This gives the algebra $k(G)$. We show that I is a biideal, and hence that $k(G)$ inherits the (matrix) coproduct of $k(M_n)$. Indeed, consider the group algebra kG from Example 1.5.3 (this makes sense even when G is not finite) as dually paired with $k(M_n)$ in the obvious way (by evaluating the coordinate functions on elements of G, extended linearly). Then I is the kernel of the pairing on the $k(M_n)$ side, hence, as in Proposition 1.4.4, we conclude that the quotient $k(G)$ is a bialgebra. Next, consider the usual matrix inversion as a matrix of functions $S^i{}_j(M) = M^{-1i}{}_j$ defined on G and extended linearly to kG. We then define a map $S : k(M_n) \to (kG)^*$ by $St^i{}_j = S^i{}_j$, extended to products as an (anti)algebra homomorphism. It obeys $((St^i{}_k)t^k{}_j)(M) = M^{-1i}{}_k M^k{}_j = \delta^i{}_j$ for all $M \in G$, and similarly for the inverse on the other side. Also, the map S is compatible with the ideal I since $M^{-1} \in G$ if $M \in G$, so it descends to a map $S : k(G) \to (kG)^*$ and continues to obey the relations (4.11) provided $\mathbf{t}$ and $S\mathbf{t}$ are viewed as functionals on kG. Finally, the additional assumption that the $S^i{}_j$ are polynomial in the coordinate functions $\mathbf{t}$ when restricted to G, just says that $S\mathbf{t}$ lies in the image of $k(G)$ in $(kG)^*$, i.e. is actually given by evaluation against some polynomial in $\mathbf{t}$. Hence, in this case, we have a usual map $S : k(G) \to k(G)$ rather than something weaker. The equations (4.11) tell us that it is an antipode for the matrix coproduct of $k(G)$. ∎

We can still obtain an *algebraic model* $k(G)$ when G does not have polynomial inversion; we just enlarge $k(M_n)$ to a Hopf algebra $k(GL_n)$ by formally adjoining the inverse of the determinant of the matrix generators, as a group-like element. The inversion operation (defined by the matrix minors and the inverse determinant) is a polynomial in this larger set of generators, and defines the antipode of $k(GL_n)$. We can then define $k(G)$ as a quotient of this larger Hopf algebra $k(GL_n)$ by the kernel of its duality pairing with kG, along the same lines as before. Also, given a Lie algebra $g \subseteq M_n$, we can similarly extend its embedding as a duality pairing between $k(M_n)$ or $k(GL_n)$ and the enveloping algebra $U(g)$. This defines $k(G)$, even when G itself is not known. These constructions in the classical case tell us that we should try to impose additional relations among the generators of our matrix bialgebra $A(R)$, or an extension of it, and then, in the quotient by these relations, we might hope to find an antipode S of the matrix form (4.11).

We now make some general remarks about how to obtain the required relations, modelled on the above exercise, but in a general R-matrix setting. First of all, the ideal scenario for us would be that the Hopf algebra that we obtained would still be dual quasitriangular and hence a (dual) quantum group in the strict sense.

Proposition 4.2.2 *Suppose that we can add relations to $A(R)$ such that the dual quasitriangular structure from Theorem 4.1.5 descends to the quotient and gives us a dual quasitriangular Hopf algebra. Then*

$$\mathcal{R}(\mathbf{t} \otimes \mathbf{t}^{-1}) = \tilde{R}$$

obeys

$$\tilde{R}^i{}_a{}^b{}_j R^a{}_l{}^k{}_b = \delta^i{}_l \delta^k{}_j = R^i{}_a{}^b{}_j \tilde{R}^a{}_l{}^k{}_b, \qquad i.e. \quad \tilde{R} = ((R^{t_2})^{-1})^{t_2}, \quad (4.12)$$

where t_2 is transposition in the second matrix factor of R. Moreover,

$$S^2 \mathbf{t} = v^{-1} \mathbf{t} v = u \mathbf{t} u^{-1}; \quad v^i{}_j = \tilde{R}^i{}_a{}^a{}_j, \quad u^i{}_j = \tilde{R}^a{}_j{}^i{}_a, \quad (4.13)$$

$$v_2 = R v_2 \tilde{R}, \quad v_1 = \tilde{R} v_1 R. \quad (4.14)$$

Such a solution R of the QYBE is said to be regular *and in a* quantum group normalisation.

Proof: If we have a Hopf algebra, we know that $R^{-1} = \mathcal{R}^{-1}(\mathbf{t} \otimes \mathbf{t}) = \mathcal{R}(S\mathbf{t} \otimes \mathbf{t})$ from Lemma 2.2.2. From the bicharacter properties (2.6) of a dual quasitriangular structure and the matrix form of the coproduct, we deduce at once that $\mathcal{R}(\mathbf{t} \otimes S\mathbf{t})$ has the properties stated. It is called the *second-inverse* of R. Next, putting this into Proposition 2.2.4 with the matrix form of the coproduct, gives at once the formula (4.13) for S^2. The matrix v obeys some useful identities, also stated here for completeness, in (4.14). The proof of the first is $(R^{-1}v_2)^a{}_b{}^i{}_k = \mathcal{R}(St^a{}_b \otimes t^i{}_j)v^j{}_k = v^i{}_j\mathcal{R}(St^a{}_b \otimes S^2t^j{}_k) = v^i{}_j\mathcal{R}(t^a{}_b \otimes St^j{}_k) = (v_2\tilde{R})^a{}_b{}^i{}_k$ by invariance of $\mathcal{R}$ under $S \otimes S$. The proof of the second is similar: $\mathcal{R}(t^i{}_j \otimes St^a{}_b)v^j{}_k = v^i{}_j\mathcal{R}(S^2t^j{}_k \otimes St^a{}_b) = v^i{}_j R^{-1j}{}_k{}^a{}_b$. ∎

This gives us at least one general way to enlarge $A(R)$ to obtain a Hopf algebra, which could be called $GL(R)$, say: we formally add some generators $\mathbf{t}^{-1}$ with the coproduct and relations one would expect for $S\mathbf{t}$, and define $S\mathbf{t} = \mathbf{t}^{-1}$ and $S\mathbf{t}^{-1} = v^{-1}\mathbf{t}v = u\mathbf{t}u^{-1}$. One can then add further relations compatible with the antipode and obtain other interesting Hopf algebras in this way. On the other hand, not every R-matrix is *biinvertible* in the sense that both R and $\tilde{R}$ exist. For example, the permutation matrix in Example 4.1.11 is not biinvertible. So we cannot always enlarge in this way, and may need more ad-hoc techniques as well. In what follows, we emphasise the quantum matrices $A(R)$, but a suitable extension should be understood in cases where this is required.

We are now ready to outline a general method of obtaining the additional nonquadratic relations which should typically be added to those of the quantum matrices $A(R)$. If we already know the quantum enveloping algebra which we wish to dualise in this matrix form, we consider $A(R)$ as dually paired with it and quotient by the kernel of the pairing. For a general R-matrix, we cannot assume the quantum enveloping algebra either, and therefore begin with the paired bialgebras $A(R), \tilde{U}(R)$, passing to their nondegenerate quotients on both sides. As in the proof of Exercise 4.2.1, we first construct, not exactly an antipode, but something a little weaker called a *weak antipode*. One can think of a usual antipode as, by definition, the inverse to the identity map in the convolution algebra of maps from the Hopf algebra to itself. The weak antipode is likewise defined as a map $S : A(R) \to \tilde{U}(R)^*$ which is inverse in the convolution algebra to the map $A(R) \to \tilde{U}(R)$ defined by the pairing. The notion works for any dually paired bialgebras, and, if the weak antipode exists, it is unique.

Proposition 4.2.3 $A(R)$ *has a weak antipode given on the generators by*

$$(St_1)(\mathbf{L}_2^+) = R^{-1}, \qquad (St_1)(\mathbf{L}_2^-) = R_{21}. \qquad (4.15)$$

Proof: We use just the same method as in the proof of the dual quasitriangular structure in Theorem 4.1.5. Thus, we take this definition on the generators and extend it to products of the $\mathbf{L}^\pm$ as $(S\mathbf{t}_1)(\mathbf{L}_2^\pm \cdots \mathbf{L}_N^\pm) = (S\mathbf{t}_1)(\mathbf{L}_N^\pm) \cdots (S\mathbf{t}_1)(\mathbf{L}_1^\pm)$ and then to products of the $\mathbf{t}$ as an antialgebra map. The general value on arbitrary strings of generators is given by an array of the R and R^{-1} matrices, and consistency of this with the relations is proven in just the same way as in Theorem 4.1.5. For example,

$$(S(R_{12}\mathbf{t}_1\mathbf{t}_2))(\mathbf{L}_3^+) = (R_{12}(S\mathbf{t}_2)(S\mathbf{t}_1))(\mathbf{L}_3^+) = R_{12}(S\mathbf{t}_2)(\mathbf{L}_3^+)(S\mathbf{t}_1)(\mathbf{L}_3^+)$$
$$= R_{12}R_{23}^{-1}R_{13}^{-1} = R_{13}^{-1}R_{23}^{-1}R_{12} = (S\mathbf{t}_1)(\mathbf{L}_3^+)(S\mathbf{t}_2)(\mathbf{L}_3^+)R_{12}$$
$$= ((S\mathbf{t}_1)(S\mathbf{t}_2))(\mathbf{L}_3^+)R_{12} = (S(\mathbf{t}_2\mathbf{t}_1R_{12}))(\mathbf{L}_3^+).$$

This corresponds to checking that $\mathbf{t} \mapsto R^{-1}$ is an antirepresentation. The corresponding computation for $\mathbf{L}^-$ is to check that $\mathbf{t} \mapsto R_{21}$ is an antirepresentation. After this we know that the tensor powers of these two antirepresentations are also well-defined as functions on $A(R)$. Similarly for an antirepresentation of $\tilde{U}(R)$, to show that the expression desired for $S(\mathbf{t}_1 \cdots \mathbf{t}_N)$ is well-defined as a function on $\tilde{U}(R)$. The resulting linear map $A(R) \to \tilde{U}(R)$ then obeys

$$(S\mathbf{t}_1)(\mathbf{L}_2^+ \cdots \mathbf{L}_N^+)\langle \mathbf{t}_1, \mathbf{L}_2^+ \cdots \mathbf{L}_N^+\rangle$$
$$= (S\mathbf{t}_1)(\mathbf{L}_N^+) \cdots (S\mathbf{t}_1)(\mathbf{L}_2^+)\langle \mathbf{t}_1, \mathbf{L}_2^+\rangle \cdots \langle \mathbf{t}_1, \mathbf{L}_N^+\rangle$$
$$= \mathrm{id}^{\otimes N} = \epsilon(\mathbf{t}_1)\epsilon(\mathbf{L}_2^+ \cdots \mathbf{L}_N^+),$$

where $(S\mathbf{t}_1)(\mathbf{L}_2^+)\langle \mathbf{t}_1, \mathbf{L}_2^+\rangle = R^{-1}R = \mathrm{id}$ collapses first, followed by all the others in the same way. When there are some $\mathbf{L}^-$ in the expression, the corresponding step of the collapse is $R_{21}R_{21}^{-1} = \mathrm{id}$. We see that S behaves like an antipode except that, instead of multiplying $S\mathbf{t}$ with $\mathbf{t}$, we evaluate them separately against the coproduct of $\mathbf{L}_1^\pm \cdots \mathbf{L}_N^\pm$. This is the concrete meaning of the requirement that S is a convolution-inverse to the pairing. The proof of this weak antipode property on the other side is analogous. Finally, one has to iterate these arguments to products $\mathbf{t}_1 \cdots \mathbf{t}_M$. Note that an equivalent way to formulate this result is that we have just proven that the duality pairing $\langle \, , \, \rangle$ has a convolution-inverse $\langle \, , \, \rangle^{-1} = \langle S(\,), \, \rangle$ as a linear functional $A(R) \otimes \tilde{U}(R) \to k$. We will emphasise this second point of view later, in Chapter 7. $\blacksquare$

Next, we can pass to the nondegenerate quotients and expect that the weak antipode works at this level also. On the other hand, this generally introduces further relations, and our resulting weak antipodes can become honest antipodes. Sometimes, we should first enlarge to a $GL(R)$ version before this happens. This approach (due to the author) also tends to fix the normalisation of R in the same way as the consideration in Proposition 4.2.2, namely the *quantum group normalisation*.

This general theory can be useful in some examples (we will demonstrate it below for $SL_q(2)$), but, in general, it runs into the difficulty that it is rather hard to know when the pairing is nondegenerate. Alternatively, there is a different approach to obtaining the right relations which is not an algorithm, but is, nevertheless, quite popular. It works when R is close to the identity and consists of identifying the bialgebra $A(R)$ as dually paired with a deformation of a familiar Lie algebra (as in Chapter 3.3). In this case, the classical Lie group G, of which our desired Hopf algebra is a noncommutative analogue, will be known. It will be characterised as some subset $G \subset M_n$. In our dual form, it means that we should seek to quotient $A(R)$ by introducing some corresponding relations among the generators $\mathbf{t}$, based on deforming the classical restrictions on the linear coordinate functions. Whenever our R-matrix yields something like a deformation of a familiar group, we can generally guess the relations that we need to add by means of this analogy.

For example, the subset SL_n is characterised by means of a determinant relation. The general procedure in Exercise 4.2.1 would yield, in this case, the relation $\det(\mathbf{t}) = 1$ in the kernel. On the other hand, we could try to characterise the determinant abstractly as a group-like element of order n in $k(M_n)$; we could therefore look in $A(R)$ and try to identify, by analogy, a suitable group-like element as *quantum determinant*. If it is also central, we could set it equal to unity and obtain a version of $SL_q(n)$. The determinant is only the most obvious kind of restriction; other Lie groups are characterised by means of other restrictions on $k(M_n)$ or $k(GL_n)$ and we would need their analogues as well in the q-deformed case.

Thus, this second *FRT approach* consists in identifying the classical flavour of the underlying quantum group of enveloping algebra type (if it exists) and then looking for the analogues of the determinant and/or other characteristic relations. The first step can be approached by looking for matrices $\mathbf{l}^{\pm}$ of the enveloping algebra generators, typically of some special form (such as upper-triangular or lower-triangular) and studying their properties and semiclassical limits. Some authors have tried to follow similar ansätze for other R-matrices, not related to Lie algebras but of similar form. Of course, there is no guarantee that relations invented by analogy with the undeformed case will be any good in such nonstandard cases.

In the remainder of this section, we demonstrate both the FRT approach and the more general one based on the theory above. We begin with the simplest example (the quantum group $SL_q(2)$) and then give the R-matrices needed for the other quantum deformations of Lie groups (of function algebra type). We also give, for contrast, a nonstandard example. For the dual quantum groups corresponding to the quantum enveloping algebras in Chapter 3, we already know the quasitriangular structure, and, bearing in mind Corollary 4.1.9, it is clear that the required R is simply found by evaluating $\mathcal{R}$ in a suitably canonical representation.

Proposition 4.2.4 *The quasitriangular structure $\mathcal{R}$ for $U_q(sl_2)$ evaluated in the spin $j = \frac{1}{2}$ representation in Proposition 3.2.6 takes the form*

$$R = q^{-\frac{1}{2}} \begin{pmatrix} q & 0 & 0 & 0 \\ 0 & 1 & q - q^{-1} & 0 \\ 0 & 0 & 1 & 0 \\ 0 & 0 & 0 & q \end{pmatrix}.$$

Here, $q \neq 0$ is a complex parameter. The matrix describes $R^i{}_j{}^k{}_l$ at row (i, k) and column (j, l) (numbered $(1,1), (1,2), (2,1), (2,2)$). We call it the sl_2 R-matrix.

Proof: First note that the spin-$\frac{1}{2}$ representation ρ is given as matrices by

$$\rho(H) = \begin{pmatrix} 1 & 0 \\ 0 & -1 \end{pmatrix}, \quad \rho(X_+) = \begin{pmatrix} 0 & 1 \\ 0 & 0 \end{pmatrix}, \quad \rho(X_-) = \begin{pmatrix} 0 & 0 \\ 1 & 0 \end{pmatrix}.$$

Hence, $\rho(X_\pm^2) = 0$, so that only the first two terms of $\mathcal{R}$ in Example 3.2.1 contribute to $(\rho \otimes \rho)(\mathcal{R})$. Putting in the above for $\rho(H), \rho(X_\pm)$ immediately gives R as stated. ∎

Example 4.2.5 *Let R be the sl_2 R-matrix in the preceding proposition. The corresponding bialgebra $A(R)$ is denoted by $M_q(2)$ and takes the form, with $\mathbf{t} = \begin{pmatrix} a & b \\ c & d \end{pmatrix}$,*

$$ac = q^{-1}ca, \quad bd = q^{-1}db, \quad ab = q^{-1}ba, \quad cd = q^{-1}dc,$$

$$bc = cb, \quad ad - da = (q^{-1} - q)bc.$$

Proof: To compute $A(R)$ we use the explicit form (4.4). Since the relations are quadratic, we can ignore the overall factor $q^{-\frac{1}{2}}$. The only nonzero entries in R are then effectively $R^1{}_1{}^1{}_1 = R^2{}_2{}^2{}_2 = q, R^1{}_1{}^2{}_2 = R^2{}_2{}^1{}_1 = 1, R^1{}_2{}^2{}_1 = q - q^{-1}$. There are 16 relations to check for the 16 possible values of $(ijkl)$,

$(1111):\quad R^1{}_1{}^1{}_1 t^1{}_1 t^1{}_1 = t^1{}_1 t^1{}_1 R^1{}_1{}^1{}_1,$

$(1112):\quad R^1{}_1{}^1{}_1 t^1{}_1 t^1{}_2 = t^1{}_2 t^1{}_1 R^1{}_1{}^2{}_2,$

$(1121):\quad R^1{}_1{}^1{}_1 t^1{}_2 t^1{}_1 = t^1{}_1 t^1{}_2 R^2{}_2{}^1{}_1 + t^1{}_2 t^1{}_1 R^1{}_2{}^2{}_1,$

$(1122):\quad R^1{}_1{}^1{}_1 t^1{}_2 t^1{}_2 = t^1{}_2 t^1{}_2 R^2{}_2{}^2{}_2,$

$(1211):\quad R^1{}_1{}^2{}_2 t^1{}_1 t^2{}_1 + R^1{}_2{}^2{}_1 t^2{}_1 t^1{}_1 = t^2{}_1 t^1{}_1 R^1{}_1{}^1{}_1,$

$(1212):\quad R^1{}_1{}^2{}_2 t^1{}_1 t^2{}_2 + R^1{}_2{}^2{}_1 t^2{}_1 t^1{}_2 = t^2{}_2 t^1{}_1 R^1{}_1{}^2{}_2,$

$(1221):\quad R^1{}_1{}^2{}_2 t^1{}_2 t^2{}_1 + R^1{}_2{}^2{}_1 t^2{}_1 t^1{}_2 = t^2{}_2 t^1{}_1 R^1{}_2{}^2{}_1 + t^2{}_1 t^1{}_2 R^2{}_2{}^1{}_1,$

$(1222):\quad R^1{}_1{}^2{}_2 t^1{}_2 t^2{}_2 + R^1{}_2{}^2{}_1 t^2{}_1 t^2{}_2 = t^2{}_2 t^1{}_2 R^2{}_2{}^2{}_2$

and eight similar relations. Most of these relations are redundant, an independent set being as stated.   ∎

Proposition 4.2.6 *In the bialgebra* $M_q(2)$ *the element*

$$\det{}_q(\mathbf{t}) = ad - q^{-1}bc$$

is group-like and central. It is called the quantum determinant *of* $M_q(2)$. *If we make the further quotient* $\det_q(\mathbf{t}) = 1$, *we obtain the Hopf algebra* $SL_q(2)$. *The antipode is*

$$S\mathbf{t} = \begin{pmatrix} d & -qb \\ -q^{-1}c & a \end{pmatrix}.$$

Proof: We first compute in $M_q(2)$ that $\Delta(ad - q^{-1}bc) = (ad - q^{-1}bc) \otimes (ad - q^{-1}bc)$ (so $(ad - q^{-1}bc)$ is group-like). Using this it can be seen that setting $ad - q^{-1}bc = 1$ is consistent with the definition of Δ.   ∎

Thus, by analogy with the classical situation, we can find the correct quotient of $A(R)$ to obtain the Hopf algebra $SL_q(2)$. Clearly, it should be thought of as a deformed version of the algebra of functions on $SL_2 \subset M_2$. The situation for the correct quotient of $\tilde{U}(R)$ (paired with $A(R)$) is not so simple: we would like to use it to obtain a description of the bialgebra $U_q(sl_2)$ (which should be dual to $SL_q(2)$), but we know that $\tilde{U}(R)$ itself has far too many generators, so we will have to quotient it by introducing many further relations. The FRT approach is to make an informed guess for the form of these further relations motivated by the Gauss decomposition of standard Lie groups into upper- and lower-triangular matrices.

A somewhat different method supposes that one knows also the quasitriangular structure of $U_q(sl_2)$ and applies Corollary 4.1.9. It has the advantage of being algorithmic in the sense that one does not need to make any informed guesses.

Example 4.2.7 *For the R-matrix in Proposition 4.2.4 and the quasitriangular Hopf algebra* $(U_q(sl_2), \mathcal{R})$ *in Example 3.2.1 we compute* $\mathbf{l}^\pm$ *in Corollary 4.1.9 as*

$$\mathbf{l}^+ = \begin{pmatrix} q^{\frac{H}{2}} & 0 \\ q^{-\frac{1}{2}}(q - q^{-1})X_+ & q^{-\frac{H}{2}} \end{pmatrix}, \quad \mathbf{l}^- = \begin{pmatrix} q^{-\frac{H}{2}} & q^{\frac{1}{2}}(q^{-1} - q)X_- \\ 0 & q^{\frac{H}{2}} \end{pmatrix}.$$

This identifies $U_q(sl_2)$ *as a quotient of the bialgebra* $\tilde{U}(R)$.

Proof: We already know that $U_q(sl_2)$ is a quasitriangular Hopf algebra. It is paired with $M_q(2)$ (and with $SL_q(2)$) according to $\langle \mathbf{t}, h \rangle = \rho(h)$

for all $h \in U_q(sl_2)$, where ρ is the spin-$\frac{1}{2}$ representation. Since R was defined as $(\rho \otimes \rho)(\mathcal{R})$, we are in the situation of Corollary 4.1.9. As in Proposition 4.2.4, we know that $\rho(X_\pm^2) = 0$, so only the first two terms in the series for $\mathcal{R}$ enter. Putting in the expression at once gives $1^\pm$. These $1^\pm$ obey the relations of $\tilde{U}(R)$ (but also many other relations not belonging to $\tilde{U}(R)$). ∎

Since we know the quasitriangular structure for the standard deformations $U_q(g)$ for complex simple Lie algebras g (see Chapter 3.3), we can follow just the same strategy to find their corresponding dual quantum groups G_q as quotients of the corresponding quantum matrix bialgebra $A(R)$.

Example 4.2.8 *The R-matrix for the dual quantum group* $SL_q(n)$ *(or* $SU_q(n)$ *when we impose a suitable $*$-structure) is*

$$R = q^{-\frac{1}{n}} \left(q \sum_i E_{ii} \otimes E_{ii} + \sum_{i \neq j} E_{ii} \otimes E_{jj} + (q - q^{-1}) \sum_{j > i} E_{ij} \otimes E_{ji} \right),$$

where $\{E_{ij}\}$ is the usual basis of M_n.

Proof: The quasitriangular Hopf algebra $U_q(sl_n)$ was described in Chapter 3.3. It is not hard to see (by analogy with the fundamental representation of the Lie algebra sl_n) that a representation of $U_q(sl_n)$ is given by $\rho(E_i) = q^{-\frac{1}{2}} E_{ii+1}$, $\rho(F_i) = q^{\frac{1}{2}} E_{i+1i}$ and $\rho(H_i) = E_{ii} - E_{i+1i+1}$. Using (3.34), etc., for the nonsimple root vectors E_α, F_α, one has for them the matrices

$$\rho(E_{\alpha_{ij}}) = q^{-\frac{1}{2}}(-1)^{i-j} E_{ij+1}, \qquad \rho(F_{\alpha_{ij}}) = q^{\frac{1}{2}}(-1)^{i-j} E_{j+1i}.$$

Noting that the relevant products of the E_α, F_α, H_i vanish in this representation ρ, we compute from (3.27)

$$(\rho \otimes \rho)(\mathcal{R}) = q^{\sum (a^{-1})_{ij} \rho(H_i) \otimes \rho(H_j)} \left(\prod_{\alpha_{ij}} e_{q^{-2}}^{(q-q^{-1})\rho(E_{\alpha_{ij}}) \otimes \rho(F_{\alpha_{ij}})} \right)$$

$$= q^{-\frac{1}{n}} \left(1 + (q-1) \sum_i E_{ii} \otimes E_{ii} \right) \left(1 + \sum_{i \leq j} (q - q^{-1}) E_{ij+1} \otimes E_{j+1i} \right),$$

giving the form stated, on multiplying out further. ∎

There is a central group-like element of order n in the bialgebra $M_q(n) = A(R)$ for the above R-matrix, called the *quantum determinant*. It is given explicitly by

$$\det {}_q(\mathbf{t}) = \sum_{\sigma \in S_n} (-q)^{-l(\sigma)} t^1_{\sigma(1)} \cdots t^n_{\sigma(n)},$$

where $l(\sigma)$ is the length of the permutation σ in the permutation group S_n. A general R-matrix formula for quantum determinants will be obtained at the end of Chapter 10.4 using braided geometry; for the moment we bring it in by hand by analogy with classical ideas. On quotienting the bialgebra $M_q(n)$ by the relation $\det {}_q(\mathbf{t}) = 1$, we obtain the dual quasitriangular Hopf algebra $SL_q(n)$. Finally, we can apply again the method of Example 4.2.7 via Corollary 4.1.9, to obtain the FRT form of the quantum enveloping algebra $U_q(sl_n)$ as paired with $SL_q(n)$. The formulae are similar to those already given.

Example 4.2.9 *The quantum matrices $M_q(3)$ of su_3 type have the nine generators $t^i{}_j$ and 1, with the 36 relations*

$$0 = [t^1{}_2, t^1{}_1]_q = [t^1{}_3, t^1{}_1]_q = [t^1{}_3, t^1{}_2]_q = [t^2{}_1, t^1{}_1]_q = [t^2{}_2, t^1{}_2]_q = [t^2{}_2, t^2{}_1]_q$$

$$= [t^2{}_3, t^1{}_3]_q = [t^2{}_3, t^2{}_1]_q = [t^2{}_3, t^2{}_2]_q = [t^3{}_1, t^1{}_1]_q = [t^3{}_1, t^2{}_1]_q = [t^3{}_2, t^1{}_2]_q$$

$$= [t^3{}_2, t^2{}_2]_q = [t^3{}_2, t^3{}_1]_q = [t^3{}_3, t^1{}_3]_q = [t^3{}_3, t^2{}_3]_q = [t^3{}_3, t^3{}_1]_q = [t^3{}_3, t^3{}_2]_q,$$

$$0 = [t^2{}_1, t^1{}_2] = [t^2{}_1, t^1{}_3] = [t^2{}_2, t^1{}_3] = [t^3{}_1, t^1{}_2] = [t^3{}_1, t^1{}_3]$$

$$= [t^3{}_1, t^2{}_2] = [t^3{}_1, t^2{}_3] = [t^3{}_2, t^1{}_3] = [t^3{}_2, t^2{}_3],$$

$$[t^2{}_2, t^1{}_1] = t^1{}_2 t^2{}_1 \lambda, \quad [t^2{}_3, t^1{}_1] = t^1{}_3 t^2{}_1 \lambda, \quad [t^2{}_3, t^1{}_2] = t^1{}_3 t^2{}_2 \lambda,$$

$$[t^3{}_2, t^1{}_1] = t^1{}_2 t^3{}_1 \lambda, \quad [t^3{}_2, t^2{}_1] = t^2{}_2 t^3{}_1 \lambda, \quad [t^3{}_3, t^1{}_1] = t^1{}_3 t^3{}_1 \lambda,$$

$$[t^3{}_3, t^1{}_2] = t^1{}_3 t^3{}_2 \lambda, \quad [t^3{}_3, t^2{}_1] = t^2{}_3 t^3{}_1 \lambda, \quad [t^3{}_3, t^2{}_2] = t^2{}_3 t^3{}_2 \lambda,$$

where $[a, b]_q = ab - qba$, $[a, b] = ab - ba$ and $\lambda = q - q^{-1}$. There is a central and group-like q-determinant

$$\det {}_q(\mathbf{t}) = t^1{}_1 t^2{}_2 t^3{}_3 - q^{-1}(t^1{}_2 t^2{}_1 t^3{}_3 + t^1{}_1 t^2{}_3 t^3{}_2)$$
$$+ q^{-2}(t^1{}_3 t^2{}_1 t^3{}_2 + t^1{}_2 t^2{}_3 t^3{}_1) - q^{-3} t^1{}_3 t^2{}_2 t^3{}_1,$$

and setting $\det_q(\mathbf{t}) = 1$ gives the quantum group $SL_q(3)$.

Proof: We compute the relations of $A(R)$ for the sl_3 R-matrix from Example 4.2.8. One can check that $\det_q(\mathbf{t})$ is indeed central and group-like. ∎

Example 4.2.10 *The sl_3 R-matrix gives a description of $U_q(sl_3)$ in FRT form provided by Corollary 4.1.9 as*

$$
l^+ = \begin{pmatrix} q^{\frac{2}{3}H_1+\frac{1}{3}H_2} & 0 & 0 \\ q^{\frac{1}{6}H_1+\frac{1}{3}H_2-\frac{1}{2}}\lambda X_{+1} & q^{-\frac{1}{3}(H_1-H_2)} & 0 \\ q^{\frac{1}{6}(H_1-H_2)-1}\lambda[X_{+1},X_{+2}]_{q^{-1}} & q^{-\frac{1}{6}H_1-\frac{1}{6}H_2-\frac{1}{2}}\lambda X_{+2} & q^{-\frac{1}{3}H_1-\frac{2}{3}H_2} \end{pmatrix},
$$

$$
l^- = \begin{pmatrix} q^{-\frac{2}{3}H_1-\frac{1}{3}H_2} & -\lambda X_{-1}q^{-\frac{1}{6}H_1-\frac{1}{3}H_2+\frac{1}{2}} & -\lambda[X_{-2},X_{-1}]_q q^{\frac{1}{6}(H_2-H_1)+1} \\ 0 & q^{\frac{1}{3}(H_1-H_2)} & -\lambda X_{-2}q^{\frac{1}{3}H_1+\frac{1}{6}H_2+\frac{1}{2}} \\ 0 & 0 & q^{\frac{1}{3}H_1+\frac{2}{3}H_2} \end{pmatrix},
$$

where $\lambda = q - q^{-1}$.

Proof: We use the sl_3 universal R-matrix as in Example 4.2.8 and apply the same ρ to compute $l^\pm$ from Corollary 4.1.9. Explicitly,

$$
\mathcal{R} = q^{\sum_{i,j}(a^{-1})_{ij}H_i \otimes H_j} e_{q^{-2}}^{\lambda E_2 \otimes F_2} e_{q^{-2}}^{\lambda E_{\alpha 12} \otimes F_{\alpha 12}} e_{q^{-2}}^{\lambda E_1 \otimes F_1}
$$

in terms of the preferred generators in Chapter 3.3. The connection with the standard $U_q(sl_3)$ generators comes out as

$$
E_1 = X_{+1}q^{\frac{H_1}{2}}, \quad E_2 = X_{+2}q^{\frac{H_2}{2}}, \quad E_{\alpha 12} = -q^{\frac{1}{2}(H_1+H_2)}q^{-\frac{3}{2}}[X_{+1},X_{+2}]_{q^{-1}},
$$

$$
F_1 = q^{-\frac{H_1}{2}}X_{-1}, \quad F_2 = q^{-\frac{H_2}{2}}X_{-2}, \quad F_{\alpha 12} = -[X_{-2},X_{-1}]_q q^{-\frac{1}{2}(H_1+H_2)}q^{\frac{3}{2}}.
$$

Thus $e_{q^{-2}}^{\lambda E_2 \otimes \rho(F_2)} = \mathrm{id} + \lambda q^{-\frac{1}{2}}q^{\frac{H_2}{2}}X_{+2}E_{32}$, etc. Meanwhile,

$$
q^{\sum_{i,j}(a^{-1})_{ij}H_i\rho(H_j)} = E_{11}q^{\frac{2}{3}H_1+\frac{1}{3}H_2} + E_{22}q^{-\frac{1}{3}(H_1-H_2)} + E_{33}q^{-\frac{1}{3}H_1-\frac{2}{3}H_2}.
$$

One can also apply ρ to the other factor and recover the sl_3 R-matrix as in Example 4.2.8. Putting these calculations together, we obtain l^+ as stated. Similarly for l^-. There are standard formulae for the inverses of the usual Cartan matrix which can be used for this and other examples. ∎

We see that because the standard $U_q(g)$ are known for the classical families of complex simple Lie algebras g, the corresponding R-matrices and matrix quantum groups G_q are not hard to find. The latter are non-commutative analogues of (the algebra of functions on) the usual compact matrix group G with Lie algebra g. Another popular notation is $\mathcal{O}_q(G)$.

But what happens when we are given a nonstandard solution R obeying the QYBE (4.1), and we have no idea what quantum group the corresponding $A(R)$ is paired with, let alone its quasitriangular structure? If there are some analogies with the standard forms above, we can make inspired guesses about how to quotient $A(R)$ and $\tilde{U}(R)$, by means of various ansätze imposed on their generators, as in the FRT approach. Or we can use the general approach explained above, which depends only on R: We begin with the dual pair in Lemma 4.1.10, pass to the nondegenerate quotients, and see if the weak antipodes are actual antipodes. We demonstrate this general method for the standard case of $SL_q(2)$.

Example 4.2.11 *For the sl_2 R-matrix in Proposition 4.2.4, the pairing between $A(R)$ and $\tilde{U}(R)$ in Corollary 4.1.8 is degenerate. The relations*

$$L^{+1}{}_2 = 0 = L^{-2}{}_1, \; L^{+1}{}_1 L^{-1}{}_1 = 1, \; L^{+2}{}_2 L^{-2}{}_2 = 1, \; \det \mathbf{L}^+ = 1$$

lie in the kernel of the pairing on the $\tilde{U}(R)$ side. Similarly, the relation

$$t^1{}_1 t^2{}_2 - q^{-1} t^1{}_2 t^2{}_1 = 1$$

lies in the kernel of the pairing on the $A(R)$ side. The two sets of relations generate biideals in $\tilde{U}(R)$ and $A(R)$, respectively. Setting them to zero, $A(R)$ and $\tilde{U}(R)$ become $SL_q(2)$ and $U_q(sl_2)$, respectively, in the quotients.

Proof: We only have to check the coideal property on the generators. $\langle t^i{}_j, L^{+1}{}_2 \rangle = R^i{}_j{}^1{}_2 = 0$, $\langle t^i{}_j, L^{-2}{}_1 \rangle = R^{-1}{}^2{}_1{}^i{}_j = 0$, so these directions are null. And $\Delta L^{+1}{}_2 = L^{+1}{}_1 \otimes L^{+1}{}_2 + L^{+1}{}_2 \otimes L^{+2}{}_2$, $\Delta L^{-2}{}_1 = L^{-2}{}_1 \otimes L^{-1}{}_1 + L^{-2}{}_2 \otimes L^{-2}{}_1$, so they generate a biideal and thus can consistently be set to zero to obtain a bialgebra. Given this, we also compute that $\langle t^i{}_j, L^{+1}{}_1 L^{-1}{}_1 \rangle = \langle t^i{}_j, 1 \rangle$ and $\Delta(L^{+1}{}_1 L^{-1}{}_1) = L^{+1}{}_1 L^{-1}{}_1 \otimes L^{+1}{}_1 L^{-1}{}_1$, so that $L^{+1}{}_1 L^{-1}{}_1$ can be consistently set to unity. Similarly for the other relations. The explicit identification of the quotient with $U_q(sl_2)$ then follows by incorporating these null relations. For example, if we denote $L^{+1}{}_1$ by the symbol $q^{\frac{H}{2}}$ (say), then we must denote $L^{-1}{}_1$ by the symbol $q^{-\frac{H}{2}}$, etc. Similarly, $L^{+2}{}_1$ is another generator, and we can denote it by $q^{-\frac{1}{2}}(q - q^{-1})X_+$ (say). Thus we are led from the structure of the R-matrix alone to expressions (without loss of generality) of the form obtained by another method in Example 4.2.7. The remaining relations $\mathbf{L}_1^+ \mathbf{L}_2^+ R = R \mathbf{L}_2^+ \mathbf{L}_1^+$, $\mathbf{L}_1^- \mathbf{L}_2^- R = R \mathbf{L}_2^- \mathbf{L}_1^-$ of $\tilde{U}(R)$ give $q^{\frac{H}{2}} X^{\pm} q^{-\frac{H}{2}} = q^{\pm} X^{\pm}$, while $\mathbf{L}_1^- \mathbf{L}_2^+ R = R \mathbf{L}_2^+ \mathbf{L}_1^-$ gives $[X^+, X^-] = \frac{q^H - q^{-H}}{q - q^{-1}}$. To see these, one must in each case write out all 16 relations (cf. Example 4.2.5), most of which are redundant. We see that this approach gives the same answer as the universal R-matrix method based on Corollary 4.1.9. ∎

It is remarkable that this abstract method, depending only on R, gives all the structure of $U_q(sl_2)$ even without building in any knowledge of Lie algebras, roots, quasitriangular structures, etc. In other words, the innocent matrix R in Proposition 4.2.4 encodes the dual quantum group of function algebra type, the quasitriangular Hopf algebra to which it is dual and (in the limit $q \to 1$) the classical Lie algebra of which it is a deformation. All of this is computed from knowledge of R alone. Moreover, the method is quite applicable for general R.

We have seen how the general theory works out in the standard SL_n case. With suitable $*$-structure, it gives the Hopf $*$-algebra $SU_q(n)$, the quantum deformation function algebras of the A series of compact Lie groups. For the record, we mention the other standard families. The B, D series ($SO(n)$ for n odd and even) can be treated together by writing $n = 2s + 1$, $s \in \frac{1}{2}\mathbb{N}$ and $i, j \in -s, -s+1, \ldots, s$ and

$$\bar{i} = \begin{cases} i - \frac{1}{2} & \text{for } i > 0 \\ i & \text{for } i = 0 \\ i + \frac{1}{2} & \text{for } i < 0. \end{cases}$$

Using this notation, the R-matrix for the dual quantum group $SO_q(n)$ is

$$R \propto q \sum_{i \neq 0} E_{ii} \otimes E_{ii} + E_{00} \otimes E_{00} + q^{-1} \sum_{i \neq 0} E_{-i-i} \otimes E_{ii} + \sum_{i \neq j, -j} E_{ii} \otimes E_{jj}$$

$$+ (q - q^{-1}) \sum_{j > i} E_{ij} \otimes E_{ji} + (q^{-1} - q) \sum_{j > i} q^{\bar{i} - \bar{j}} E_{ij} \otimes E_{-i-j},$$

where the $E_{00} \otimes E_{00}$ term only exists if s is integral, i.e. if n is odd.

To obtain the quantum group $SO_q(n)$, one then begins, as before, with the quantum matrices $A(R)$ for this R, and then imposes q-analogues of the defining conditions for the orthogonal group. Thus we impose $\det_q(\mathbf{t}) = 1$, as above, and the additional relation

$$t^i{}_a t^j{}_b \eta^{ba} = \eta^{ji}, \quad \eta_{ab} t^a{}_i t^b{}_j = \eta_{ij}, \quad \eta^{ij} = q^{\bar{i}} \delta_{i,-j}, \quad \eta_{ij} = \delta_{-i,j} q^{\bar{j}}.$$

Here η_{ij} is some kind of *quantum metric* and η^{ij} is its transposed inverse.

Likewise, the R-matrix for the dual quantum group $Sp_q(n)$ corresponding to the C series in the same notation is

$$R \propto q \sum_i E_{ii} \otimes E_{ii} + q^{-1} \sum_i E_{-i-i} \otimes E_{ii} + \sum_{i \neq j, -j} E_{ii} \otimes E_{jj}$$

$$+ (q - q^{-1}) \sum_{j > i} E_{ij} \otimes E_{ji}$$

$$+ (q^{-1} - q) \sum_{j > i} q^{\bar{i} - \bar{j}} q^{\text{sgn}(i) - \text{sgn}(j)} \text{sgn}(i) \text{sgn}(j) E_{ij} \otimes E_{-i-j},$$

where $\text{sgn}(i) = \pm 1$ as $i > 0$ or $i < 0$. The quantum metric is given by $\eta^{ij} = q^{\bar{i}} q^{\text{sgn}(i)} \text{sgn}(i) \delta_{i,-j}$. One can compute the determinant and other relations needed for the quotient of $A(R)$ in these cases also, as well as the FRT form of $U_q(g)$.

Finally, just to show that these standard 'by analogy' quantum groups G_q are not the only interesting ones, we end with some nonstandard families of R-matrices. First, there are multiparameter versions of the sl_n R-matrix given by

$$R \propto q \sum_i E_{ii} \otimes E_{ii} + \sum_{i>j} qq_{ij}^{-1} E_{ii} \otimes E_{jj} + \sum_{i<j} q^{-1}q_{ji} E_{ii} \otimes E_{jj}$$
$$+ (q - q^{-1}) \sum_{j>i} E_{ij} \otimes E_{ji}, \tag{4.16}$$

where the q_{ij} are arbitrary. The constant case $q_{ij} = q$ returns us to Example 4.2.7. The introduction of the further parameters q_{ij} corresponds, at the level of the associated quantum groups, to the process of twisting in Chapter 2.3. Likewise, there are multiparameter versions of the $SO(n)$ and $Sp(n)$ R-matrices.

Secondly, in another direction, we can take the nonstandard R-matrices

$$R \propto \sum_i q_i E_{ii} \otimes E_{ii} + \sum_{i \neq j} E_{ii} \otimes E_{jj} + (q - q^{-1}) \sum_{j>i} E_{ij} \otimes E_{ji}, \tag{4.17}$$

where $q_i \in \{q, -q^{-1}\}$ can be freely chosen. Thus, for any i, a corresponding q_i in the standard sl_n R-matrix can be flipped over from q to $-q^{-1}$. As a result, the limit $q \to 1$ no longer gives $R = 1$, so that these deformations are nonstandard (the corresponding quantum groups are not deformations of familiar Lie groups like those above).

Thus, entire families of nonstandard quantum groups do exist. Moreover, these two directions are, in fact, quite independent: there are multiparameter versions of (4.17) given analogously to (4.16), etc.

When working with nonstandard R, we should be on the lookout for truly unexpected phenomena. Often there is no antipode (only a weak one), even after quotienting to obtain a nondegenerate pairing. Sometimes, more interesting quantum groups are obtained by forgetting about the pairing and making quotients not consistent with the pairing or with the canonical or fundamental representations in Proposition 4.1.4 and Corollary 4.1.7. For example, as we have remarked after Exercise 4.2.1, it may be better to look for a quantum determinant (a group-like element of order n^2 in the generators), but, rather than setting it equal to unity, to adjoin its inverse (or adjoin enough inverses of other elements that the quantum determinant has an inverse). For nonstandard quantum groups, this could be the natural option, e.g if the quantum determinant is not central. Even for the standard R-matrices, it is worth noting that we have potentially distinct $M_q(n)$, $GL_q(n)$ or $SL_q(n)$ of each R-matrix type, before we impose the more characteristic relations of the standard quantum group, i.e. not just with the standard the sl_n R-matrix.

Example 4.2.12 *Let*

$$R = \begin{pmatrix} q & 0 & 0 & 0 \\ 0 & qp^{-1} & q-p^{-1} & 0 \\ 0 & 0 & 1 & 0 \\ 0 & 0 & 0 & q \end{pmatrix}.$$

This gives the simplest multiparameter quantum group. The corresponding $A(R)$ has the form

$$ac = q^{-1}ca, \;\; bd = q^{-1}db, \;\; ab = p^{-1}ba, \;\; cd = p^{-1}dc,$$

$$qbc = pcb, \quad ad - da = (q^{-1} - p)cb.$$

The expression

$$D = ad - p^{-1}bc$$

is group-like but not central. Adjoining its inverse gives the dual quantum group $GL_{p,q}(2)$ with antipode

$$St = \begin{pmatrix} dD^{-1} & -pbD^{-1} \\ -p^{-1}cD^{-1} & aD^{-1} \end{pmatrix} = \begin{pmatrix} D^{-1}d & -qD^{-1}b \\ -q^{-1}D^{-1}c & D^{-1}a \end{pmatrix}.$$

Proof: Here we have introduced a new parameter q_{21} according to (4.16) and then made some redefinitions in terms of a new parameter p, as well as a change of normalisation. The resulting R-matrix then reduces to the standard one when $p = q$. The nonzero entries have exactly the same location as in the standard case, allowing us to read off the relations from the expressions displayed in the proof of Example 4.2.5. A similar computation to that in Proposition 4.2.6 yields that D is group-like, but this time it has the commutations relations indicated by the two expressions for St. The antipode is then easily verified from either side. ∎

Example 4.2.13 *Let*

$$R = \begin{pmatrix} q & 0 & 0 & 0 \\ 0 & 1 & q-q^{-1} & 0 \\ 0 & 0 & 1 & 0 \\ 0 & 0 & 0 & -q^{-1} \end{pmatrix}.$$

This is the simplest nonstandard variant. The corresponding $A(R)$ has the form

$$ac = q^{-1}ca, \;\; bd = -qdb, \;\; ab = q^{-1}ba, \;\; cd = -qdc,$$

$$b^2 = c^2 = 0, \;\; bc = cb, \;\; ad - da = -(q - q^{-1})bc.$$

If we adjoin a^{-1}, d^{-1} then there is an antipode with

$$St = \begin{pmatrix} a^{-1} + a^{-1}bd^{-1}ca^{-1} & -a^{-1}bd^{-1} \\ -d^{-1}ca^{-1} & d^{-1} + d^{-1}ca^{-1}bd^{-1} \end{pmatrix}.$$

Proof: The nonzero elements have the same positions as in the standard case, so we can again read off the relations from the proof of Example 4.2.5. The antipode can be verified directly. Note that this example has nilpotent elements. It is related to the superquantum group $GL_q(1|1)$ by a process of transmutation, which will be covered in Chapters 7.4, 9.4 and 10 ∎

One can apply the above methods to compute the corresponding enveloping algebras in these cases. The enveloping algebra corresponding to the preceding example is

Example 4.2.14 *Inspection of the pairing via R in the preceding example (the method of Example 4.2.11) leads to the quotient of $\tilde{U}(R)$ expressed (without loss of generality) in the form*

$$l^+ \equiv \begin{pmatrix} K_1 & 0 \\ q^{-\frac{1}{2}}(q - q^{-1})X_+ & K_2^{-1} \end{pmatrix}, \quad l^- \equiv \begin{pmatrix} K_1^{-1} & q^{\frac{1}{2}}(q^{-1} - q)X_- \\ 0 & K_2 \end{pmatrix}.$$

The relations of $\tilde{U}(R)$ then give the nonstandard quantum enveloping algebra

$$[K_1, K_2] = 0, \quad K_1 X_\pm = q^{\pm 1} X_\pm K_1, \quad K_2 X_\pm = q^{\mp 1} X_\pm K_2, \quad X_\pm^2 = 0,$$

$$[X_+, X_-] = \frac{K_1 K_2 - K_1^{-1} K_2^{-1}}{q - q^{-1}}, \quad \epsilon K_i = 1, \quad \epsilon X_\pm = 0, \quad \Delta K_i = K_i \otimes K_i,$$

$$\Delta X_+ = X_+ \otimes K_1 + K_2^{-1} \otimes X^+, \quad \Delta X_- = X_- \otimes K_2 + K_1^{-1} \otimes X_-,$$

$$SK_i = K_i^{-1}, \quad SX_+ = -qK_1^{-1} K_2 X_+, \quad SX_- = qK_1 K_2^{-1} X_-.$$

Proof: The computations in the proof of Example 4.2.11 go through in the same way for all except the relation $\det \mathbf{L}^+ = 1$, which we no longer have. This means that their quotients $l^\pm$, after imposing these relations, have the form shown. The rest then follows from the quadratic relations of $\tilde{U}(R)$. ∎

We conclude with some generalities about the ∗-structure. In general, the existence of a ∗-structure corresponds to a real form of the quantum group. This is because we know what we mean by a 'unitary' representation, etc., namely a ∗-representation (this notion is what distinguishes

SU_n from SL_n, as we have seen at the level of representations of the Lie algebra in Chapter 3).

Definition 4.2.15 *We say that a solution R of the QYBE is of real type if R obeys the condition*

$$\overline{R^i{}_j{}^k{}_l} = R^l{}_k{}^j{}_i \quad i.e. \quad \overline{R}^{t\otimes t} = R_{21},$$

where t denotes transposition in the relevant copy of M_n.

If R is of real type, then it is easy to see that the definition

$$t^i{}_j{}^* = St^j{}_i \tag{4.18}$$

is compatible with the relations $Rt_1t_2 = t_2t_1R$ of $A(R)$. In general (and certainly for all the standard examples), it is also compatible with the determinant and other relations forming the matrix quantum group. As a result, all these quantum groups have the $*$-algebra structure (4.18) when their parameters are such that R is of real type. This happens for the standard cases when q is real. The form of (4.18) says that the quantum matrix t is 'unitary' in the sense

$$\sum_i t^i{}_j{}^* t^i{}_k = \delta_{jk}, \quad \sum_j t^i{}_j t^k{}_j{}^* = \delta^{ik}.$$

Such matrix quantum groups can also be called compact.

The standard quantum deformations G_q dual to the $U_q(g)$, where g is semisimple, have as many such generic choices of $*$-structure as the compact real forms of the Lie algebra g. The corresponding quantum groups are, of course, real quasitriangular in the sense of Definition 2.2.8. They are also factorisable in the sense of Definition 2.2.7.

Example 4.2.16 *The quantum group $SU_q(2)$ denotes the Hopf algebra $SL_q(2)$ in Example 4.2.5 with q real and equipped with the $*$-structure*

$$\begin{pmatrix} a^* & b^* \\ c^* & d^* \end{pmatrix} = \begin{pmatrix} d & -q^{-1}c \\ -qb & a \end{pmatrix}.$$

4.3   Matrix quantum Lie algebras

In the preceding section, we have given a variety of standard and nonstandard R-matrices and their corresponding quantum function algebras. These are quotients of the quantum matrices $A(R)$ by determinant and other relations. However, our treatment of the quantum enveloping algebras to which these are dual was a little unnatural, beginning with a bialgebra $\tilde{U}(R)$ with twice as many generators $\mathbf{L}^\pm$ and then cutting them down as explained.

A somewhat more natural way to compute the quantum enveloping algebras, which gives the same answer in the case when the quantum group is factorisable, is the following, based again on a matrix of n^2 generators $\mathbf{u} = (u^i{}_j)$. Just as matrices can be used both for SL_n and for the Lie algebra sl_n, so we can use 'quantum' matrices for the quantum enveloping algebra. Later, in Chapters 9.4 and 10.3, we shall see that these $\mathbf{u}$ are not really quantum matrices like the $\mathbf{t}$; rather, they are braided matrices.

Definition 4.3.1 *Let R be a matrix solution of the QYBE. We define the braided matrices $B(R)$ to be the quadratic algebra with n^2 generators $\{u^i{}_j\}$ and 1, and relations*

$$R^k{}_a{}^i{}_b u^b{}_c R^c{}_j{}^a{}_d u^d{}_l = u^k{}_a R^a{}_b{}^i{}_c u^c{}_d R^d{}_j{}^b{}_l,$$

$$\text{(4.19)}$$

$$i.e. \quad R_{21} \mathbf{u}_1 R \mathbf{u}_2 = \mathbf{u}_2 R_{21} \mathbf{u}_1 R.$$

Proposition 4.3.2 *$B(R)$ is dual to $A(R)$ in the following sense. Let $(H, \mathcal{R})$ be a quasitriangular bialgebra or Hopf algebra which is paired with $A(R)$ such that $\langle \mathbf{t}_1 \otimes \mathbf{t}_2, \mathcal{R} \rangle = R$. Let*

$$\mathbf{l} = (\mathbf{t} \otimes \mathrm{id})(Q), \qquad Q = \mathcal{R}_{21} \mathcal{R}.$$

Here, $l^i{}_j$ are elements of H. Then there is an algebra map $B(R) \to H$ such that $\mathbf{l}$ is the image of $\mathbf{u}$, i.e. H is a realisation of $B(R)$.

Proof: We first show the useful identity

$$\mathbf{l}_1 R \mathbf{l}_2 = R(\mathbf{t}_1 \mathbf{t}_2 \otimes \mathrm{id})(Q), \tag{4.20}$$

This arises out of the transmutation theory of braided groups that relates $A(R)$ to $B(R)$; see Chapter 9.4. For our present purposes, the direct proof is

$$
\begin{aligned}
\mathbf{l}_1 R \mathbf{l}_2 &= \langle \mathbf{t}_1, Q^{(1)} \rangle \langle \mathbf{t}_1 \otimes \mathbf{t}_2, \mathcal{R} \rangle \langle \mathbf{t}_2, Q'^{(1)} \rangle Q^{(2)} Q'^{(2)} \\
&= \langle \mathbf{t}_1, \mathcal{R}^{(2)} \mathcal{R}'^{(1)} \mathcal{R}''''^{(1)} \rangle \langle \mathbf{t}_2, \mathcal{R}''''^{(2)} \mathcal{R}''^{(2)} \mathcal{R}'''^{(1)} \rangle \mathcal{R}^{(1)} \mathcal{R}'^{(2)} \mathcal{R}''^{(1)} \mathcal{R}'''^{(2)} \\
&= \mathcal{R}^{(1)} \mathcal{R}''^{(1)} \mathcal{R}'^{(2)} \mathcal{R}'''^{(2)} \langle \mathbf{t}_1, \mathcal{R}^{(2)} \mathcal{R}''''^{(1)} \mathcal{R}'^{(1)} \rangle \langle \mathbf{t}_2, \mathcal{R}''^{(2)} \mathcal{R}''''^{(2)} \mathcal{R}'''^{(1)} \rangle \\
&= \langle \mathbf{t}_1, \mathcal{R}^{(2)}{}_{(2)} \mathcal{R}''''^{(1)} \mathcal{R}'^{(1)}{}_{(1)} \rangle \langle \mathbf{t}_2, \mathcal{R}^{(2)}{}_{(1)} \mathcal{R}''''^{(2)} \mathcal{R}'^{(1)}{}_{(2)} \rangle \mathcal{R}^{(1)} \mathcal{R}'^{(2)} \\
&= \langle \mathbf{t}_1, \mathcal{R}''''^{(1)} \mathcal{R}^{(2)}{}_{(1)} \mathcal{R}'^{(1)}{}_{(1)} \rangle \langle \mathbf{t}_2, \mathcal{R}''''^{(2)} \mathcal{R}^{(2)}{}_{(2)} \mathcal{R}'^{(1)}{}_{(2)} \rangle \mathcal{R}^{(1)} \mathcal{R}'^{(2)} \\
&= R \langle \mathbf{t}_1 \mathbf{t}_2, \mathcal{R}^{(2)} \rangle \langle \mathbf{t}_1 \mathbf{t}_2, \mathcal{R}'^{(1)} \rangle \mathcal{R}^{(1)} \mathcal{R}'^{(2)} \\
&= R \langle \mathbf{t}_1 \mathbf{t}_2, Q^{(1)} \rangle Q^{(2)}.
\end{aligned}
$$

For the second equality, we recognised the matrix form of the coproduct of the $\mathbf{t}$ as paired to multiplication in H. For the third equality, we used the universal QYBE proven in Lemma 2.1.4. For the fourth, we used the axioms (2.1) to generate coproducts in H, and for the fifth we used axiom

(2.2). We then wrote these as products in $A(R)$ for the sixth equality and recognised the result.

By permuting the matrix position labels, we could equally well have written $l_2 R_{21} l_1 = R_{21}(t_2 t_1 \otimes \mathrm{id})(Q)$. Hence

$$R_{21} l_1 R l_2 = R_{21} R(t_1 t_2 \otimes \mathrm{id})(Q) = R_{21}(t_2 t_1 \otimes \mathrm{id})(Q)R = l_2 R_{21} l_1 R,$$

using the relations $Rt_1 t_2 = t_2 t_1 R$. ■

In this sense then, $B(R)$ is some kind of universal dual algebra to $A(R)$. Just as $A(R)$ has to be cut down by determinant and other relations to obtain an honest Hopf algebra, if H is a Hopf algebra then $B(R)$ is generally a little too big to coincide with H: it too has to be cut down by additional relations. Note that, when H is a Hopf algebra, the elementary identity $\mathcal{R}^{-1} = (S \otimes \mathrm{id})(\mathcal{R})$ from Lemma 2.1.2 means that $l = l^+ S l^-$, thereby relating this description of H to our earlier one in Corollary 4.1.9.

Example 4.3.3 *When R is a standard R-matrix giving matrix quantum group G_q as quotient of $A(R)$, and $H = U_q(g)$, the dual enveloping algebra, then the map $B(R) \to U_q(g)$ has kernel given by 'braided versions' of the determinant and other relations associated to the Lie group G. Hence, formally, $U_q(g)$ can be identified as $B(R)$ modulo such relations.*

Proof: The braided matrices $B(R)$, as well as being dual to the quantum matrices $A(R)$ as above, are also closely related to them as a 'braided variant' by a process of transmutation to be described later in Chapters 7.4 and 9.4. From this general theory one knows that the additional relations for $B(R)$ correspond to the additional relations for $A(R)$. Also, when working over formal power series as in Chapter 3, it is known that $U_q(g)$ is factorisable in the sense that the image of Q, here generated by l, generates $U_q(g)$. ■

We will use this description of $U_q(g)$ in later chapters as well. It should be noted, however, that we do need to work over formal power series in order to assert factorisability; i.e. we have to assume that the appropriate combinations of the l generators have logarithms corresponding to the Cartan generators of $U_q(g)$ in Chapter 3.3. Alternatively, if one wants to work algebraically over a field k, then suitable square roots and inverses have to be adjoined to the matrix generators of $B(R)$ in order to describe $U_q(g)$ in this way. We proceed without making these steps explicitly, considering that in any case $B(R)$ modulo the determinant and other relations provides *a version* of the algebra of $U_q(g)$.

Example 4.3.4 *When R is the sl_2 R-matrix, the braided matrices $B(R)$ consist of the algebra with generators* $\mathbf{u} = \begin{pmatrix} a & b \\ c & d \end{pmatrix}$ *and 1 with relations*

$$ba = q^2 ab, \quad ca = q^{-2} ac, \quad da = ad, \quad bc = cb + (1 - q^{-2})a(d - a),$$

$$db = bd + (1 - q^{-2})ab, \qquad cd = dc + (1 - q^{-2})ca,$$

and denoted $BM_q(2)$. This has a quotient $BSL_q(2)$ defined by $\underline{\det}(\mathbf{u}) = 1$, where $\underline{\det}(\mathbf{u}) = ad - q^2 cb$ is the braided determinant. The quotient provides a version of $U_q(sl_2)$ according to the identification

$$\mathbf{u} \mapsto 1 = \begin{pmatrix} q^H & q^{-\frac{1}{2}}(q - q^{-1})q^{\frac{H}{2}}X_- \\ q^{-\frac{1}{2}}(q - q^{-1})X_+ q^{\frac{H}{2}} & q^{-H} + q^{-1}(q - q^{-1})^2 X_+ X_- \end{pmatrix}.$$

Proof: This follows at once from the form of $\mathbf{l}^\pm$ in Example 4.2.7. Note that we have to adjoin $a^{-\frac{1}{2}}$, or work over $\mathbb{C}[[t]]$ and assume that a has a logarithm, if we want a precise identification with $U_q(sl_2)$ in its standard form in Example 3.2.1. We note also that (with a similar caveat) $BM_q(2)$ coincides with a version of the Sklyanin algebra in Chapter 3.2, while $\underline{\det}(\mathbf{u})$ coincides with $S_0^2 - J_{23}S_3^2$. This is the reason behind Proposition 3.2.8 and is another way of thinking about this example. ∎

Thus, a way of computing $U_q(g)$, or any standard or nonstandard but factorisable quantum enveloping algebra, is to compute the quadratic algebra $B(R)$ and then impose further determinant and other relations. These are such that there exists a braided antipode $\underline{S}$ such that $\mathbf{u}\underline{S}\mathbf{u} = 1 = (\underline{S}\mathbf{u})\mathbf{u}$, giving in fact a braided version BG_q of the corresponding quantum group G_q obtained from $A(R)$. The difference is that the matrix comultiplication $\Delta\mathbf{u} = \mathbf{u} \otimes \mathbf{u}$ which is relevant for this is *not an algebra homomorphism to the usual tensor product*. Instead, the two independent copies of $\mathbf{u}$ in $\mathbf{u} \otimes \mathbf{u}$ must noncommute in a certain way. In physical terms, they must be taken with braid statistics in a similar manner to the way that independent fermionic systems anticommute. Clearly, a discussion of this must await the theory of braided groups in Chapters 9 and 10.

Proposition 4.3.5 *(i) In the factorisable case, 1 and the subspace $\mathcal{L} = \mathrm{span}\langle l^i{}_j \rangle \subset H$ from Proposition 4.3.2 generate all of H.*

(ii) The subspace $\mathcal{L}$ is stable under the quantum adjoint action of H on itself.

(iii) The quantum adjoint action as a map $[\ ,\ ] : \mathcal{L} \otimes \mathcal{L} \to \mathcal{L}$ looks explicitly like

$$[l^I, l^J] = \sum_K c^{IJ}{}_K l^K, \quad c^{IJ}{}_K = \tilde{R}^a{}_{i_1}{}^{j_0}{}_b R^{-1b}{}_{k_0}{}^{i_0}{}_c Q^c{}_a{}^{k_1}{}_{j_1},$$

where $l^I = l^{i_0}{}_{i_1}$ and $I = (i_0, i_1)$ *is a multiindex notation (running from* $(1,1), \ldots, (n,n)$*). Here we suppose that* $\tilde{R} = ((R^{t_2})^{-1})^{t_2}$ *exists (where* t_2 *is transposition in the second matrix factor).*

Proof: (i) The assumption that the image of $A(R)$ under Q in Proposition 4.3.2 gives all of H is just the condition of factorisability from Chapter 2.

(ii) We use the form $1 = 1^+ S 1^-$, valid in the Hopf algebra case, and show first that if $\triangleright = \mathrm{Ad}$ denotes the quantum adjoint action of H on itself as in Example 1.6.3 then

$$1_2^+ \triangleright l_1 = R^{-1} l_1 R, \qquad 1_1^- \triangleright l_2 = R l_2 R^{-1}. \tag{4.21}$$

This follows easily from the theory of braided groups, where $1^\pm$ act on $\mathbf{u}$ in the same way as on $\mathbf{t}$ by the coadjoint action (because Q is an intertwiner, as we know from Proposition 2.1.14). We will see this in a comodule form in Example 4.5.17. For now, however, we give a direct proof using the definition of Ad, properties of the antipode from Chapter 1 and the result of Corollary 4.1.9. Thus,

$$1_2^+ \triangleright (1_1^+ S 1_1^-) = 1_2^+ 1_1^+ S 1_1^- S 1_2^+ = 1_2^+ 1_1^+ S(1_2^+ 1_1^-) = 1_2^+ 1_1^+ R^{-1} S(R 1_2^+ 1_1^-)$$
$$= R^{-1} 1_1^+ 1_2^+ S(1_1^- 1_2^+) R = R^{-1} 1_1^+ (1_2^+ S 1_2^+) S 1_1^- R = R^{-1} 1_1^+ S 1_1^- R,$$
$$1_1^- \triangleright (1_2^+ S 1_2^-) = 1_1^- 1_2^+ S 1_2^- S 1_1^- = 1_1^- 1_2^+ S(1_1^- 1_2^-) = 1_1^- 1_2^+ R S(R^{-1} 1_1^- 1_2^-)$$
$$= R 1_2^+ 1_1^- S(1_2^- 1_1^-) R^{-1} = R 1_2^+ (1_1^- S 1_1^-) S 1_2^- R^{-1} = R 1_2^+ S 1_2^- R^{-1}.$$

(iii) We can also deduce from this the action of $S1^\pm$ from $1^+ S 1^+ = \mathrm{id} = (S1^+)1^+$, etc. (the identity matrix times the action of the identity). In particular,

$$(S l^{-i}{}_j) \triangleright l^k{}_l = \tilde{R}^a{}_j{}^k{}_b l^b{}_c R^i{}_a{}^c{}_l, \tag{4.22}$$

where $\tilde{R}$ obeys $\tilde{R}^i{}_a{}^b{}_l R^a{}_j{}^k{}_b = \delta^i{}_j \delta^k{}_l = R^i{}_a{}^b{}_l \tilde{R}^a{}_j{}^k{}_b$. Combining this with (4.21), we obtain $l^{i_0}{}_{i_1} \triangleright l^{j_0}{}_{j_1} = l^{+i_0}{}_a \triangleright (Sl^{-a}{}_{i_1}) \triangleright l^{j_0}{}_{j_1}$. ∎

Thus, $\mathcal{L}$ is some kind of 'quantum Lie algebra' for H because it is a finite-dimensional subspace that generates H and at the same time is closed under the quantum adjoint action, which provides a kind of 'quantum Lie bracket' $[\xi, \eta] = \xi \triangleright \eta$. From the preceding proposition and the module-algebra property of the quantum adjoint action in Chapter 1.6, we see that it obeys

(L0) $[\xi, \eta] \in \mathcal{L}$ for $\xi, \eta \in \mathcal{L}$,

(L1) $[\xi, [\eta, \zeta]] = \sum [[\xi_{(1)}, \eta], [\xi_{(2)}, \zeta]]$,

(L2) $[[\xi, \eta], \zeta] = \sum [\xi_{(1)}, [\eta, [S\xi_{(2)}, \zeta]]]$.

As explained above, the standard quantum groups $U_q(g)$ are essentially factorisable, so we obtain a description of these as 'generated' (in a suitable sense) by $\mathcal{L}$ as a kind of quantum Lie algebra. As seen from even the simplest example, this finite-dimensional subspace is a mixture of 'group-like' elements with coproduct $\Delta\xi = \xi \otimes \xi$ and more usual Lie-algebra-like elements where $\Delta\xi \sim \xi \otimes 1 + 1 \otimes \xi$ (with a suitable deformation). The latter are how off-diagonal elements of $l^i{}_j$ tend to behave, while the former are how diagonal elements tend to behave. Because of this, one could also consider as the 'quantum Lie algebra' the subspace

$$\mathrm{Lie}(H) = \mathrm{span}\langle \chi^i{}_j, \ \chi^i{}_j = l^i{}_j - \delta^i{}_j\rangle \subset H. \tag{4.23}$$

This is a matter of taste. The subset is also closed under $[\ ,\ ] = \mathrm{Ad}$, which now has structure constants

$$[\chi^I, \chi^J] = [l^I, l^J] + \delta^I \delta^J - \delta^I \delta^J - \delta^I l^J = \sum_K (c^{IJ}{}_K - \delta^I \delta^J{}_K)\chi^K \tag{4.24}$$

using the elementary properties of the quantum adjoint action (notably $l^I \triangleright 1 = \epsilon(l^I) = \delta^I$). Here $\delta^I = \delta^{i_0}{}_{i_1}$ and $\delta^J{}_K$ are Kronecker δ-functions. The last equality uses the identity

$$\sum_K c^{IJ}{}_K \delta^K = \delta^I \delta^J, \tag{4.25}$$

which follows at once from the expression for c in the proposition. These χ^I equally generate H along with 1, and have a better-behaved semiclassical limit in the standard cases.

We note also that some combinations of the basis elements of $\mathcal{L}$ or $\mathrm{Lie}(H)$ can have trivial quantum Lie bracket and can be decoupled if we want to have the minimum number of generators. For the case of $U_q(sl_2)$, for example, the redundant element is $qd + q^{-1}a$, which computes as the Casimir of $U_q(sl_2)$.

We conclude with some generalities.

Proposition 4.3.6 *The generators of $A(R)$ define matrix elements of the representation ρ of $B(R)$ given by*

$$\rho_2(\mathbf{u}_1) = \langle l_1, \mathbf{t}_2 \rangle = Q_{12}, \quad Q = R_{21}R.$$

Proof: We have to show that this extends consistently to all of $B(R)$ as an algebra representation,

$$\begin{aligned}
\rho_3(R_{21}\mathbf{u}_1 R_{12}\mathbf{u}_2) &= R_{21}\rho_3(\mathbf{u}_1)R_{12}\rho_3(\mathbf{u}_2) \\
&= R_{21}Q_{13}R_{12}Q_{23} = Q_{23}R_{21}Q_{13}R_{12} \\
&= \rho_3(\mathbf{u}_2)R_{21}\rho_3(\mathbf{u}_1)R_{12} = \rho_3(\mathbf{u}_2 R_{21}\mathbf{u}_1 R_{12}).
\end{aligned}$$

The middle equality follows from repeated use of the QYBE. Thus the extension is consistent with the algebra relations of $B(R)$. ∎

Proposition 4.3.7 *If R is of real type, then $B(R)$ has a natural $*$-algebra structure given by the Hermitian type*

$$u^i{}_j{}^* = u^j{}_i.$$

Proof: We take this as a definition of $*$ and check that it extends anti-multiplicatively to $B(R)$. Thus, applying it to the component form of the relations of $B(R)$ in Definition 4.3.1, we have

$$u^l{}_d R^d{}_a{}^j{}_c u^c{}_b R^b{}_i{}^a{}_k = R^l{}_b{}^j{}_d u^d{}_c R^c{}_i{}^b{}_a u^a{}_k$$

if R is of real type. These are just the relations of $B(R)$ again. ∎

For the standard R-matrices, both these results project to the level of $U_q(g)$ in the form obtained as a quotient of $B(R)$. Thus we learn, for example, that the standard $U_q(g)$ with q real, have real quasitriangular Hopf $*$-algebra structures, with $*$ given by

$$l^i{}_j{}^* = l^j{}_i. \tag{4.26}$$

For example, putting this into the form of $l^i{}_j$ in Example 4.3.4 recovers precisely the real quasitriangular $*$-structure $U_q(su_2)$ in Example 3.2.3. This is how quantum enveloping algebras can be dealt with from the matrix approach. One can also easily obtain the Casimirs of $U_q(g)$ in this approach. There is a lot more to this theory that we must postpone for the moment: the $\mathcal{L}$ above are more properly understood as *braided-Lie* algebras and $B(R)$ as their braided universal enveloping bialgebras.

4.4 Vertex models

This section is included for historical completeness, to give some insight into how bialgebras arise naturally in the theory of exactly solvable lattice models in statistical mechanics. We will see many similarities with the computations of Section 4.1, with the added complexity now of a spectral parameter. The relevant bialgebras are infinitely generated.

The statistical mechanical models most directly related to our present exposition are the vertex models. These are built on a square lattice of M rows and N columns. Each bond takes values $1, \ldots, n$, and the partition function is the sum over all bond states of the products of the Boltzmann weights at each vertex. The weight at a vertex with adjacent bond states $ijkl$, as shown in Fig. 4.2, is $R^i{}_j{}^k{}_l$. We think of $R \in \mathrm{End}(V \otimes V)$, where $V = \mathbb{C}^n$.

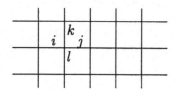

Fig. 4.2.   Vertex model on $M \times N$ lattice. State $ijkl$ has weight $R^i{}_j{}^k{}_l$.

Such a model is physical if R obeys certain unitarity, symmetry and positivity conditions. It is *exactly solvable* if R is $R(\lambda)$, i.e., depends on a parameter $\lambda \in \mathbb{C}$, the *spectral parameter* such that $R(\lambda)$ obeys the parametrised QYBE

$$R_{12}(\lambda)R_{13}(\lambda')R_{23}(\lambda'') = R_{23}(\lambda'')R_{13}(\lambda')R_{12}(\lambda),$$

for all $\lambda, \lambda', \lambda''$ such that $\lambda'' + \lambda = \lambda'$. In fact, a linear additive law is not strictly necessary. In solving the model, one typically imposes periodic boundary conditions on the lattice and the limit $M, N \to \infty$.

The main ingredient of exact solvability is to construct an infinite set of commuting operators. To construct these, consider

$$T(\lambda)^i{}_{jl_1l_2\cdots l_N}^{k_1k_2\cdots k_N} = R(\lambda)^i{}_{m_1}{}^{k_1}{}_{l_1}R(\lambda)^{m_1}{}_{m_2}{}^{k_2}{}_{l_2}\cdots R(\lambda)^{m_N-1}{}_j{}^{k_N}{}_{l_N}$$

as a multiindex matrix for each pair i, j and λ fixed, $T(\lambda)^i{}_j \in \mathrm{End}(V^N)$. We think of the collection $\{T(\lambda)^i{}_j\}$ as an $n \times n$ matrix $T(\lambda)$ with values in $\mathrm{End}(V^N)$. Let $\mathrm{Tr}\, T(\lambda) = T(\lambda)^i{}_i \in \mathrm{End}(V^N)$. This is called the *single-row transfer matrix with periodic boundary conditions*. The partition function for periodic boundary conditions is

$$Z_{R(\lambda)} = \mathrm{Tr}_{V^N}(\mathrm{Tr}\, T(\lambda))^M.$$

Proposition 4.4.1 *Suppose there exists a suitable λ for each λ', λ'' such that the above parametrised QYBE hold. Show that*

$$R(\lambda)^i{}_m{}^k{}_nT(\lambda')^m{}_jT(\lambda'')^n{}_l = T(\lambda'')^k{}_nT(\lambda')^i{}_mR(\lambda)^m{}_j{}^n{}_l.$$

This is the fundamental relation of quantum inverse scattering. *We conclude from this that*

$$[\mathrm{Tr}\, T(\lambda'), \mathrm{Tr}\, T(\lambda'')] = 0, \qquad \forall \lambda', \lambda''.$$

Proof: To see the fundamental relation, it will be convenient to adopt the notation R_{ij}, as above, and also $T_1 = T \otimes 1$, $T_2 = 1 \otimes T$ as elements of

$M_n(\text{End}(V^N)) \otimes M_n(\text{End}(V^N))$. Thus, we first prove

$$R(\lambda)_{12} T_1(\lambda') T_2(\lambda'') = T_2(\lambda'') T_1(\lambda') R(\lambda)_{12}.$$

Indeed, acting on $V \otimes V \otimes V^N$, we have $T_1 = R_{13} R_{14} \cdots R_{1N+2}$ and $T_2 = R_{23} R_{24} \cdots R_{2N+2}$. Then

$$
\begin{aligned}
& R(\lambda)_{12} R(\lambda')_{13} R(\lambda')_{14} \cdots R(\lambda')_{1N+2} R(\lambda'')_{23} R(\lambda'')_{24} \cdots R(\lambda'')_{2N+2} \\
& = R(\lambda)_{12} R(\lambda')_{13} R(\lambda'')_{23} R(\lambda')_{14} R(\lambda')_{15} \cdots \\
& \qquad \cdots R(\lambda')_{1N+2} R(\lambda'')_{24} R(\lambda'')_{25} \cdots R(\lambda'')_{2N+2} \\
& = R(\lambda'')_{23} R(\lambda')_{13} R(\lambda)_{12} R(\lambda')_{14} R(\lambda')_{15} \cdots \\
& \qquad \cdots R(\lambda')_{1N+2} R(\lambda'')_{24} R(\lambda'')_{25} \cdots R(\lambda'')_{2N+2} \\
& = \\
& \quad \vdots \\
& = R(\lambda'')_{23} R(\lambda')_{13} R(\lambda'')_{24} R(\lambda')_{14} \cdots R(\lambda'')_{2N+2} R(\lambda')_{1N+2} R(\lambda)_{12} \\
& = R(\lambda'')_{23} R(\lambda'')_{24} \cdots R(\lambda'')_{2N+2} R(\lambda')_{13} R(\lambda')_{14} \cdots R(\lambda')_{1N+2} R(\lambda)_{12},
\end{aligned}
$$

as required. For the first equality, we use that $R(\lambda'')_{23}$ commutes with $R(\lambda')_{14} \cdots R(\lambda')_{1N+2}$ since it acts on different copies of V, etc. We then use the parametrised QYBE, with a suitable numbering of the copies of V. Repeating these steps for $R(\lambda'')_{24}$, and so on, gives the penultimate line. Finally, we move some of the matrices to the right.

Now, on multiplying the fundamental relation above on the right by $R(\lambda)_{12}^{-1}$ and taking traces in the $\text{End}(V \otimes V)$, we have

$$
\begin{aligned}
\text{Tr}_{V \otimes V} T_2(\lambda'') T_1(\lambda') & = \text{Tr}_{V \otimes V}(R(\lambda)_{12} T_1(\lambda') T_2(\lambda'') R(\lambda)_{12}^{-1}) \\
& = \text{Tr}_{V \otimes V} T_1(\lambda') T_2(\lambda''),
\end{aligned}
$$

which gives the desired conclusion on evaluating the trace over $V \otimes V$. ∎

That the system is exactly solvable means, in practice, that the partition function can be expressed in a simple way in terms of the parameters of the model. This is typically in terms of products of rational functions, often of the type of interest in number theory.

As well as the statistical mechanical system, there is an associated one-dimensional quantum chain. This is a quantum system for a particle in space consisting of a chain of N sites. In the continuum limit as $N \to \infty$, one obtains a $1 + 1$ dimensional quantum integrable system. This has been called the transfer matrix method. A quantum integrable system is a quantisation of an integrable system of equations of motion in $1 + 1$ dimensions. A typical example is the nonlinear Schrödinger equation. One can also obtain conformal field theories as limits.

Example 4.4.2 *The solution of the parametrised QYBE, given by*

$$
R(\lambda) = \begin{pmatrix}
\frac{\mathrm{sn}(\lambda\mu+\mu)}{\mathrm{sn}(\mu)} & 0 & 0 & k\,\mathrm{sn}(\lambda\mu)\mathrm{sn}(\lambda\mu+\mu) \\
0 & \frac{\mathrm{sn}(\lambda\mu)}{\mathrm{sn}(\mu)} & 1 & 0 \\
0 & 1 & \frac{\mathrm{sn}(\lambda\mu)}{\mathrm{sn}(\mu)} & 0 \\
k\,\mathrm{sn}(\lambda\mu)\mathrm{sn}(\lambda\mu+\mu) & 0 & 0 & \frac{\mathrm{sn}(\lambda\mu+\mu)}{\mathrm{sn}(\mu)}
\end{pmatrix}
$$

defines a physically interesting model, the eight-vertex or XYZ model. Here, sn denotes a Jacobi elliptic function, dependent on a modulus parameter k. It is a ratio of two Jacobi theta-functions. $\mu \in \mathbb{C}$ is another parameter.

The term 'XYZ' derives from the existence of anisotropies in the associated quantum chain in all three space dimensions. The limit $k \to 0$ sends sn $\to$ sin, and $R(\lambda)$ then defines the *trigonometric six-vertex or XXZ model*, while in the limit $\mu \to 0$ we have $R(\lambda) = \lambda + P$, the rational six-vertex or XXX model. Also, for general k, μ, evaluation at the point $\lambda = 0$ again gives the solution $R = P$ of the matrix QYBE, as studied at the end of Section 4.1.

Looking back over the computation for Proposition 4.4.1, it is clear that the same conclusion would hold for the traces of *any* matrix of operators $\{T(\lambda)^i{}_j\}$ obeying the above fundamental relations. Different representations of these relations lead to different models, still exactly solvable. Thus it is natural to define the abstract algebra $A(\{R(\lambda)\})$ generated by 1 and $\{t(\lambda)^i{}_j; \lambda \in \mathbb{C}\}$ and the relations

$$
R(\lambda)_{12}\mathbf{t}_1(\lambda')\mathbf{t}_2(\lambda'') = \mathbf{t}_2(\lambda'')\mathbf{t}_1(\lambda')R(\lambda)_{12}
$$

for all $\lambda' = \lambda + \lambda''$. This is a bialgebra with coproduct

$$
\Delta t(\lambda)^i{}_j = t(\lambda)^i{}_k \otimes t(\lambda)^k{}_j.
$$

Taking the trace of any representation of this will give operators that commute at different values of λ. For example, $K \in M_n(\mathbb{C})$, such that $K \otimes K$ commutes with $R(\lambda)$ for all λ, is a representation. Similarly, we saw that the specific $T(\lambda) = R(\lambda)_{12} \cdots R(\lambda)_{1N+1}$ is a representation. Using the coproduct, one can show that the matrix $T_K(\lambda) = K_1 R(\lambda)_{12} \cdots R(\lambda)_{1N+1}$ is also a representation (it is the tensor product of K with T, in the sense of Chapter 9). So, we conclude that $[\mathrm{Tr}\, T_K(\lambda'), \mathrm{Tr}\, T_K(\lambda'')] = 0$. The corresponding model is one with twisted boundary conditions defined by K.

This is how bialgebras arose implicitly in the theory of exactly solvable vertex models or their associated integrable quantum systems.

4.5   Quantum linear algebra

To conclude this chapter, we move on from quantum matrices and matrix Lie algebras to quantum vector and quantum covector algebras. The idea is that these are algebras on which quantum matrices act and which are some kind of noncommutative–geometric analogue of $k(\mathbb{R}^n)$. This gives some very concrete examples of the abstract definitions of coactions and comodule algebras in Chapter 1.6.2. We will see that there are two points of view from which the theory can be developed. The first is that quantum vectors, quantum covectors and quantum matrices should all be examples of a general construction of quantum rectangular matrices. We develop this in Section 4.5.1. The theory is naturally bicovariant in that it mixes both left and right coactions in a natural way. A second line of development is to ask for all constructions to be covariant entirely from one side, say right-covariant under a background quantum group. We do this in Section 4.5.2. This second point of view will be picked up again in Chapter 10.2 where we show how to add quantum vectors and quantum covectors by means of a braided coaddition. The present sections cover only the algebraic structures and their comodule properties.

4.5.1   Bicovariant formulation

The first step for quantum linear algebra is to develop some popular notation for working with coproducts and coactions of matrix type. This will be far more convenient for the examples in this section than writing out the maps Δ, β in the usual way.

Lemma 4.5.1 *Let A be an algebra with n^2 matrix generators $\mathbf{t} = (t^i{}_j)$ and 1. Suppose that $\epsilon(t^i{}_j) = \delta^i{}_j$ extends multiplicatively to a map $\epsilon : A \to k$. Let $\Delta t^i{}_j = t^i{}_k \otimes t^k{}_j$ and let $\Delta(1) = 1 \otimes 1$. Then (A, ϵ, Δ) is a bialgebra iff the following holds: if $\mathbf{t}, \mathbf{t}'$ are two identical sets of generators of A, mutually commuting elementwise, then $\mathbf{t}'' = \mathbf{t}\mathbf{t}'$ is also a realisation of A (i.e. $t''^i{}_j = t^i{}_k t'^k{}_j$ also obey its relations). We can call such bialgebras quantum matrix bialgebras, and Δ is in matrix form.*

Proof: The requirement in the lemma is just saying that the Δ that we desire is an algebra homomorphism to $A \otimes A$ (i.e. $A \otimes A$ is a realisation of A), where $A \otimes A$ as an algebra is given by two mutually commuting copies of A, i.e. it is generated by $\mathbf{t} = \mathbf{t} \otimes 1$ and $\mathbf{t}' = 1 \otimes \mathbf{t}$ in $A \otimes A$. The main content of the notation is to omit the tensor product, distinguishing the elements of the second factor instead by the prime. When Δ, ϵ are defined, they automatically obey the relations of a coalgebra, as in Chapter 1.2

(the coassociativity and counity axioms). This follows from their particular form. Thus, $(\Delta \otimes \mathrm{id})\Delta t^i{}_j = (\Delta \otimes \mathrm{id})(t^i{}_k \otimes t^k{}_j) = t^i{}_l \otimes t^l{}_k \otimes t^k{}_j = (\mathrm{id} \otimes \Delta)(t^i{}_l \otimes t^l{}_j) = (\mathrm{id} \otimes \Delta)\Delta t^i{}_j$, and $(\epsilon \otimes \mathrm{id})\Delta t^i{}_j = (\epsilon \otimes \mathrm{id})(t^i{}_k \otimes t^k{}_j) = t^i{}_j$. Similarly on the other side. ∎

The quantum matrices $A(R)$ associated to an R-matrix solution of the QYBE are clearly of this form, and we have studied them extensively in Section 4.1. The lemma does nothing more than give us a nice way to talk about the coproduct as matrix multiplication. For example, the fact that the algebra $M_q(2)$ in Example 4.2.5 forms a matrix bialgebra means, in these terms, that if a, b, c, d and a', b', c', d' are two independent (mutually commuting) copies of $M_q(2)$, then

$$\begin{pmatrix} a'' & b'' \\ c'' & d'' \end{pmatrix} = \begin{pmatrix} a & b \\ c & d \end{pmatrix}\begin{pmatrix} a' & b' \\ c' & d' \end{pmatrix} \tag{4.27}$$

is also a realisation of $M_q(2)$. Since these are noncommuting algebras, we can imagine that they are represented by operators on some Hilbert space (for example). In the limit $q \to 1$, they become mutually commuting and can be viewed as having (in the representation) actual values. In this case, the value of $\mathbf{t''}$ is exactly the matrix product of the values of $\mathbf{t}, \mathbf{t'}$ as actual matrices. The quantum matrix preserves the form of this matrix product even when the generators become operators without definite values.

We can think about quantum vectors and quantum covectors in just the same way. Thus, rather than working directly with the vectors or covectors (column or row matrices), we think of the algebra $k(\mathbb{R}^n)$ consisting of functions on the set of vectors or covectors. In each case, the algebra is generated by the coordinate functions

$$x_i(X) = X_i, \quad v^i(W) = W^i, \tag{4.28}$$

where X and W are actual row and column vectors with entries X_i and W^i, respectively. Any actual row or column vector determines a representation of the corresponding abstract algebra generated by x_i or v^i, with these values. More generally then, we keep the abstract generators $\mathbf{x} = (x_i)$ and $\mathbf{v} = (v^i)$ and discard the usual underlying space. When the generators no longer commute, they can no longer have definite values as in (4.28), but we can still work with them anyway.

Lemma 4.5.2 *Let A be a matrix bialgebra (as above) and let V be an algebra with 1 and n generators $\mathbf{x} = (x_i)$ written as a row vector. Define $\beta(1) = 1 \otimes 1$ and $\beta(x_j) = x_i \otimes t^i{}_j$. Then β makes V a right comodule algebra iff the following holds: whenever $\mathbf{t}$ is a copy of the generators of A and $\mathbf{x}$ is a copy of the generators of V, commuting elementwise with the $\mathbf{t}$, then $\mathbf{x'} = \mathbf{xt}$ is a realisation of V. We call such algebras right quantum covector algebras.*

Proof: Here $t^i{}_j = 1 \otimes t^i{}_j$ and $x_i = x_i \otimes 1$ are the generators of the tensor product algebra $V \otimes A$ built from mutually commuting copies of A and V. The condition is just that the products $x'_j = x_i t^i{}_j$ are a realisation of V, i.e. that $\beta : V \to V \otimes A$ as defined is an algebra map. On the other hand, β as defined is already a right coaction from the form of its definition. This is because to be a coaction one needs $(\beta \otimes \mathrm{id})\beta = (\mathrm{id} \otimes \Delta)\beta$ and $(\mathrm{id} \otimes \epsilon)\beta = \mathrm{id}$, which we see automatically as $\beta(x_i) \otimes t^i{}_j = x_{i'} \otimes t^{i'}{}_i \otimes t^i{}_j = x_{i'} \otimes \Delta(t^{i'}{}_j)$ and $x_i \epsilon(t^i{}_j) = x_j$ due to the matrix form of Δ, ϵ. ∎

Lemma 4.5.3 *Let A be a matrix bialgebra as above and let V be an algebra with 1 and n generators $\mathbf{v} = (v^i)$ written as a column vector. Define $\beta(1) = 1 \otimes 1$ and $\beta(v^i) = t^i{}_j \otimes v^j$. Then β makes V a left comodule algebra iff the following holds: whenever $\mathbf{t}$ is a copy of the generators of A and $\mathbf{v}$ is a copy of the generators of V, commuting elementwise with the $\mathbf{t}$, then $\mathbf{v}' = \mathbf{t}\mathbf{v}$ is a realisation of V. We call such algebras left quantum vector algebras.*

Proof: This is strictly analogous to the preceding lemma. The point to note is that, because the indices of the generators v^i are up, or, equivalently, because we consider the generators as a column vector, the natural coaction of the matrix generators $\mathbf{t}$ is from the left and not from the right as before. ∎

Just as for the quantum matrices, there is a natural construction for these comodule algebras of vector and covector type for every R-matrix. More precisely, there is one of each for every R-matrix *and* a choice of a suitable normalisation constant λ. The allowed values of λ will be the eigenvalues of PR, where P is the permutation matrix. Equivalently, the constant λ should be such that $\lambda^{-1}PR$ has eigenvalue 1. Recall that every matrix PR obeys some minimal polynomial of the form

$$\prod_i (PR - \lambda_i) = 0. \tag{4.29}$$

This is called the *spectral decomposition* of the R-matrix, and λ must be taken from among these *skew-eigenvalues* λ_i. We do not assume that the λ_i are distinct, i.e. they could have some multiplicity. Generally speaking, the bigger the multiplicity, the more nontrivial will be the corresponding quantum vector and quantum covector algebras. The simplest nontriangular case is the Hecke one, where there are just two distinct λ_i. We saw at the end of Section 4.1 that most of the interesting low-dimensional R-matrices are, in fact, of this Hecke type.

Example 4.5.4 *We define* $V_R^*(\lambda, R)$ *to be the right quantum covector algebra with n generators* x_i *and 1, and relations*

$$\lambda x_j x_l = x_b x_a R^a{}_j{}^b{}_l, \quad i.e. \quad \lambda x_1 x_2 = x_2 x_1 R,$$

where $\mathbf{x} = (x_i)$ *as a row vector and* $\mathbf{x}_1 = \mathbf{x} \otimes 1$ *and* $\mathbf{x}_2 = 1 \otimes \mathbf{x}$ *refer to the position in* $M_n \otimes M_n$. *Writing* $\mathbf{t} = (t^i{}_j)$ *as an (algebra-valued) matrix, the assignment* $\mathbf{x}' = \mathbf{x}\mathbf{t}$ *makes* $V_R^*(\lambda, R)$ *into a right* $A(R)$-*comodule algebra.*

Proof: In our matrix notation, this is simply as follows. We check $\lambda x_1' x_2' = \lambda x_1 t_1 x_2 t_2 = x_2 x_1 R t_1 t_2 = x_2 x_1 t_2 t_1 R = x_2' x_1' R$, so the transformed covectors obey the same relations. We have used the assumption that the $\mathbf{x}, \mathbf{t}$ commute, as well as the relations in $V_R^*(\lambda, R)$ and the relations of $A(R)$. One can, of course, say this more formally in terms of the comodule $\beta(x_j) = x_i \otimes t^i{}_j$. $\blacksquare$

Example 4.5.5 *We define* $V_L(R, \lambda)$ *to be the left quantum vector algebra with n generators* v^i *and 1, and relations*

$$R^i{}_a{}^k{}_b v^a v^b = v^k v^i \lambda, \quad i.e. \quad R\mathbf{v}_1 \mathbf{v}_2 = \mathbf{v}_2 \mathbf{v}_1 \lambda,$$

where we write $\mathbf{v} = (v^i)$ *as a column vector and* $v_1 = v \otimes 1$ *etc. Then the assignment* $\mathbf{v}' = \mathbf{v}\mathbf{t}$ *makes* $V_L(R, \lambda)$ *into a left* $A(R)$-*comodule algebra.*

Proof: This is strictly analogous to the preceding construction with a left–right interchange. $\blacksquare$

The best-known example is the *quantum plane*, which is coacted upon by the quantum matrices $M_q(2)$ in Example 4.2.5. The R-matrix for this is Hecke, so there are two possible values of λ. The same applies for the eight-vertex solution in Section 4.1, as well as for the general $SL_q(n)$ solution, and their multiparameter and nonstandard variants.

Example 4.5.6 *Let R be the* sl_2 *R-matrix in Proposition 4.2.4. If* $\lambda = q^{\frac{1}{2}}$, *the quantum covectors and quantum vectors are*

$$V_R^*(\lambda, R): \quad yx = qxy, \qquad V_L(R, \lambda): \quad wv = qvw,$$

where $\mathbf{x} = (x \; y)$ *and* $\mathbf{v} = \begin{pmatrix} v \\ w \end{pmatrix}$. *The algebras can be identified, and we denote them by* $\mathbb{A}_q^{2|0}$. *If* $\lambda = -q^{-3/2}$ *and* $q^2 \neq -1$, *the quantum covectors and quantum vectors are*

$$V_R^*(\lambda, R): \quad \theta^2 = 0, \quad \vartheta^2 = 0, \quad \vartheta\theta = -q^{-1}\theta\vartheta,$$
$$V_L(R, \lambda): \quad \xi^2 = 0, \quad \eta^2 = 0, \quad \eta\xi = -q^{-1}\xi\eta,$$

where $\mathbf{x} = (\theta \; \vartheta)$ *and* $\mathbf{v} = \begin{pmatrix} \xi \\ \eta \end{pmatrix}$. *These algebras are denoted by* $A_{q^{-1}}^{0|2}$.

Proof: For R of the general form in Proposition 4.2.4, the only nonzero entries are $R^1{}_1{}^1{}_1$, $R^2{}_2{}^2{}_2$, $R^1{}_1{}^2{}_2$, $R^2{}_2{}^1{}_1$ and $R^1{}_2{}^2{}_1$. Hence the four equations for the covector algebra are $\lambda xx = xxR^1{}_1{}^1{}_1$, $\lambda yy = yyR^2{}_2{}^2{}_2$, $\lambda xy = yxR^1{}_1{}^2{}_2$ and $\lambda yx = xyR^2{}_2{}^1{}_1 + yxR^1{}_2{}^2{}_1$ if $\mathbf{x} = (x \; y)$, say. For $\lambda = q^{\frac{1}{2}}$, the first two are $xx = xx$, $yy = yy$; the third is the one stated; and the fourth is redundant. Similarly for the vectors. For the other value of λ, we have $xx = -q^2 xx$, etc. We have then labelled the generators in this case by symbols suggestive of 'fermionic' coordinates. The reason for this will appear in Chapter 10.2. They naturally have braid statistics which become fermionic in the limit $q \to 1$. ∎

We know from the above that the quantum matrices $A(R)$ coact on the quantum covector algebras by right multiplication and on the quantum vector algebras by left multiplication. For example, one can check explicitly that if x, y is a copy of $A_q^{2|0}$ as a right handed quantum covector and if a, b, c, d are an independent, i.e. mutually commuting, copy of $M_q(2)$, then

$$(x' \; y') = (x \; y) \begin{pmatrix} a & b \\ c & d \end{pmatrix} \tag{4.30}$$

is also a realisation of the right handed quantum covectors $A_q^{2|0}$. In the limit $q \to 1$, they become mutually commuting and can be viewed as having (in some representation) actual values. In this case, the value of $\mathbf{x}'$ is exactly the matrix transformation of the values of $\mathbf{x}$ by the values of $\mathbf{t}$ as an actual matrix. The quantum linear algebra preserves the form of this matrix transformation even when the generators become operators without definite values.

The same applies for the other choice of λ; i.e. the right handed quantum covectors $A_{q^{-1}}^{0|2}$ can be multiplied on the right by $M_q(2)$. One can also turn this around and see that the requirement that (4.30) preserves the quantum covector relations for both values of λ (both $A_q^{2|0}$ and $A_{q^{-1}}^{0|2}$ in the present case) forces the a, b, c, d to obey the quantum matrix relations of $M_q(2)$. Thus, not only does the quantum matrix act on the quantum planes, but in some sense it is generated by this transformation property. Note that, for this characterisation, one must consider both values of λ; otherwise, only some of the $M_q(2)$ relations will be obtained. The same principle applies generally as long as one considers the covector algebras for all the λ_i in the spectrum.

Likewise, one can check explicitly that, if v, w is a copy of $\mathbb{A}_q^{2|0}$ as a left handed quantum vector, and if a, b, c, d are a mutually commuting copy of $M_q(2)$, then

$$\begin{pmatrix} v' \\ w' \end{pmatrix} = \begin{pmatrix} a & b \\ c & d \end{pmatrix} \begin{pmatrix} v \\ w \end{pmatrix} \tag{4.31}$$

is also a left handed quantum vector. The same applies for the left quantum vector algebra $\mathbb{A}_{q^{-1}}^{0|2}$, and this property, for the two values of λ, again determines the relations of $M_q(2)$. The same principle applies generally, as long as one considers the vector algebras for all the allowed λ_i.

Example 4.5.7 *Let R be the sl_n R-matrix in Example 4.2.8. If $\lambda = q^{-\frac{1-n}{n}}$, the corresponding n-dimensional quantum covectors and quantum vectors are*

$$V_R^*(\lambda, R): \quad x_i x_j = q x_j x_i \text{ if } i > j,$$
$$V_L(R, \lambda): \quad v^i v^j = q v^j v^i \text{ if } i > j.$$

The algebras here can be identified, and are denoted by $\mathbb{A}_q^{n|0}$. If $\lambda = -q^{-\frac{1+n}{n}}$ and $q^2 \neq -1$, the corresponding quantum covectors and vectors are

$$V_R^*(\lambda, R): \quad \theta_i^2 = 0, \quad \theta_i \theta_j = -q^{-1}\theta_j \theta_i \quad \text{if } i > j,$$
$$V_L(R, \lambda): \quad \eta_i^2 = 0, \quad \eta_i \eta_j = -q^{-1}\eta_j \eta_i \quad \text{if } i > j,$$

where $\mathbf{x} = (\theta_i)$ and $\mathbf{v} = (\eta^i)$. The algebras in this case are denoted by $\mathbb{A}_{q^{-1}}^{0|n}$.

Proof: From the standard sl_n R-matrix in Example 4.2.8, we see that

$$R^i{}_j{}^k{}_l = q^{-\frac{1}{n}} \left(\delta^i{}_j \delta^k{}_l \begin{cases} q & \text{if } j = l \\ 1 & \text{if } j \neq l \end{cases} + (q - q^{-1})\delta^i{}_l \delta^k{}_j \ (\text{if } j > l) \right).$$

Hence, when $\lambda = q^{-\frac{1-n}{n}}$, we have

$$(j = l): \quad x_j x_j = x_j x_j, \qquad (j < l): \quad x_j x_l = q^{-1} x_l x_j,$$
$$(j > l): \quad x_j x_l = q^{-1} x_l x_j + (1 - q^{-2}) x_j x_l.$$

The second of these gives the equation stated, and the other two are redundant. Similarly for the vector algebras and for the other value of λ (where the first becomes $\theta_i^2 = -q^2 \theta_i^2$ and we assume $q^2 \neq -1$). ∎

The multiparameter R-matrices in Section 4.2 likewise have quantum planes (vectors and covectors). They are just the same, with q_{ij} in place of

q. Likewise, the nonstandard R-matrices in (4.17) have analogues. They are like the above, but with x_i 'fermionic' for those i where q is flipped over to $-q^{-1}$.

Example 4.5.8 *Let R be the two-parameter R-matrix in Example 4.2.12. The corresponding quantum covectors and quantum vectors are*

$$V_R^*(\lambda, R) = \mathbb{A}_p^{2|0} : \quad yx = pxy, \qquad V_L(R, \lambda) = \mathbb{A}_q^{2|0} : \quad wv = qvw,$$

for $\lambda = q$. Another choice, $\lambda = -p^{-1}$, gives

$$V_R^*(\lambda, R) = \mathbb{A}_{q^{-1}}^{0|2} : \quad \theta^2 = 0, \quad \vartheta^2 = 0, \quad \vartheta\theta = -q^{-1}\theta\vartheta,$$

$$V_L(R, \lambda) = \mathbb{A}_{p^{-1}}^{0|2} : \quad \xi^2 = 0, \quad \eta^2 = 0, \quad \eta\xi = -p^{-1}\xi\eta,$$

where $\mathbf{x} = (\theta \; \vartheta)$ and $\mathbf{v} = \begin{pmatrix} \xi \\ \eta \end{pmatrix}$.

Proof: The general form of R is the same as that in Example 4.5.6, with different values, p, in place of q in some places. For the second value of λ, we suppose that $pq \neq -1$. ∎

Example 4.5.9 *Let R be the Alexander–Conway R-matrix given in Example 4.2.13 and let $q^2 \neq -1$. The corresponding quantum covectors and quantum vectors are*

$$V_R^*(\lambda, R) : \quad yx = qxy, \quad y^2 = 0,$$
$$V_L(R, \lambda) : \quad wv = qvw, \quad w^2 = 0,$$

for $\lambda = q$. The algebras here can be identified, and are denoted by $\mathbb{A}_q^{1|1}$. Another choice, $\lambda = -q^{-1}$, gives the same results, with the roles of the 'fermionic' and 'bosonic' variables interchanged.

Proof: The general form of R is again the same as in Example 4.5.6, with some different values. This time, for the covectors we have $xx = xx$, as before, but $yy = -q^{-2}yy$, so that $y^2 = 0$. Thus, y behaves like a 'fermionic' coordinate. Similarly in the other cases. ∎

So far, we have introduced and computed some algebras for quantum vectors, covectors and matrices. Now we show how these various structures are special cases of a more general concept of quantum linear algebra. In usual linear algebra, one can multiply square matrices to obtain new square matrices, and we have seen the analogue of this in the form of a coproduct. One can also multiply a row vector by a matrix from the right, and a column vector by a matrix from the left, and we have seen the analogues of these in the form of right and left coactions. Another natural step in quantum linear algebra is to obtain a realisation of the quantum matrix algebra in the tensor product of a quantum vector with a quantum covector. This is a 'rank-one' quantum matrix.

Exercise 4.5.10 *Show that the assignment*

$$\mathbf{t} \mapsto \mathbf{vx} = \begin{pmatrix} v^1 x_1 & \cdots & v^1 x_n \\ \vdots & & \vdots \\ v^n x_1 & \cdots & v^n x_n \end{pmatrix}$$

is an algebra homomorphism $A(R) \to V_{\mathrm{L}}(R, \lambda) \otimes V_{\mathrm{R}}^*(\lambda, R)$. *Moreover, this assignment is covariant under the left and right coactions of* $A(R)$.

Solution: The relations for $V_{\mathrm{L}}(R, \lambda)$ are in Example 4.5.5. The $\mathbf{v}$ commute with the $\mathbf{x}$ in the tensor product algebra, so we have $R\mathbf{v}_1\mathbf{x}_1\mathbf{v}_2\mathbf{x}_2 = R\mathbf{v}_1\mathbf{v}_2\mathbf{x}_1\mathbf{x}_2 = \mathbf{v}_2\mathbf{v}_1\lambda\mathbf{x}_1\mathbf{x}_2 = \mathbf{v}_2\mathbf{v}_1\mathbf{x}_2\mathbf{x}_1 R = \mathbf{v}_2\mathbf{x}_2\mathbf{v}_1\mathbf{x}_1 R$, so that the $\mathbf{vx}$ realise (4.4). Next note that the coproduct of $A(R)$ can also be viewed as a left coaction $\beta_L = \Delta : A(R) \to A(R) \otimes A(R)$. Meanwhile, the tensor product algebra of the vectors and covectors also has a left coaction of $A(R)$ induced by the left coaction on $V_{\mathrm{L}}(R, \lambda)$. We show that our realisation map $\mathbf{t} \mapsto \mathbf{vx}$ is an intertwiner, i.e. it respects these left coactions. Thus, if we apply the coaction on $A(R)$ first, and then realise $A(R)$ as the tensor product, we have $\mathbf{t} \mapsto \mathbf{t} \otimes \mathbf{vx}$. But, if we apply the realisation first and then the left coaction on the quantum vectors, we have $\mathbf{t} \mapsto \mathbf{vx} \mapsto (\mathbf{t} \otimes \mathbf{v})\mathbf{x}$, which is just the same. Similarly, the coproduct of $A(R)$ can be viewed as a right coaction $\beta_R = \Delta$, and the assignment is likewise covariant under this, where the tensor product has the right coaction coming from the covectors.   ∎

   In this exercise, we see the appearance of the notion of a *bicomodule*. This is a vector space that is both a left and a right comodule, and the two coactions β_L, β_R commute in the sense

$$(\beta_L \otimes \mathrm{id})\beta_R = (\mathrm{id} \otimes \beta_R)\beta_L. \tag{4.32}$$

A *bicomodule algebra* is an algebra which is a bicomodule and both coactions are algebra homomorphisms. The coproduct of a bialgebra can always be viewed as a bicomodule structure, i.e. the left and right regular coactions from Chapter 1.6.2 always commute. This is just coassociativity of the coproduct. Since Δ is an algebra homomorphism, we have a bicomodule algebra. The tensor product of the quantum vectors and quantum covectors is also a bicomodule algebra, with the left coaction given by matrix multiplication from the left on the vectors, and the right coaction given by matrix multiplication from the right on the covectors.

We see in the exercise that the realisation of $A(R)$ in the tensor product algebra is fully compatible with these independent left and right coactions. One says that the construction is *bicovariant*.

Lemma 4.5.11 *Let H be a bialgebra or Hopf algebra. An H-bicomodule is nothing other than a right $H^{\mathrm{cop}} \otimes H$-comodule. An H-bicomodule algebra is nothing other than a right $H^{\mathrm{cop}} \otimes H$-comodule algebra.*

Proof: A left H-comodule can be viewed equivalently as a right coaction of H^{cop}. So, given a bicomodule V with commuting left and right coactions, as in (4.32), we define the $H^{\mathrm{cop}} \otimes H$-coaction by first applying β_R and then β_L viewed as a right coaction of H^{cop}. When β_L, β_R are algebra maps, then so is this combined coaction. In the converse direction, we use ϵ on H or H^{cop} to recover β_R, β_L, respectively. ∎

Thus, in the examples above, bicovariance under $A(R)$ could also be thought of as right-covariance under $A(R)^{\mathrm{cop}} \otimes A(R)$ if one wanted to reduce our point of view to the setting of right coactions in Chapter 1.6.2. Similarly, one could reduce our point of view to that of left coactions. On the other hand, for our matrix examples above, neither of these asymmetrical points of view is particularly natural. The most natural point of view is that of mutually commuting left and right-covariance, i.e. bicovariance under $A(R)$.

Finally, we put together all these various ideas about vectors, covectors and their transformation properties into a single systematic quantum linear calculus. The key idea here is that of a rectangular quantum matrix. For this we allow ourselves as data not one but two solutions of the QYBE. This notion is due to the author and M. Markl

Definition 4.5.12 *Let $R_1 \in M_m \otimes M_m$ and $R_2 \in M_n \otimes M_n$ be matrices, where the dimensions m, n can be different. We define the* rectangular quantum matrix algebra $A(R_1 : R_2)$ *to be generated by 1 and mn generators $\{t^\mu{}_i\}$ regarded as an $m \times n$ matrix of generators, with relations*

$$R_1{}^\mu{}_\alpha{}^\nu{}_\beta t^\alpha{}_j t^\beta{}_i = t^\nu{}_b t^\mu{}_a R_2{}^a{}_j{}^b{}_l, \quad \forall\, 1 \le \mu, \nu \le m,\ 1 \le j, l \le n,$$

$$\text{i.e.} \quad (R_1) \mathbf{t}_1 \mathbf{t}_2 = \mathbf{t}_2 \mathbf{t}_1 (R_2),$$

(4.33)

where summation is over $1 \le \alpha, \beta \le m$ and $1 \le a, b \le n$.

Here and below, it is not strictly required for the matrices R_i to obey the QYBE, though this is the main case of interest because it helps to ensure that the algebras have nontrivial representations, in much the same way as we saw for $A(R)$ in Section 4.1. Also, for the algebra to be nontrivial we require that there is at least one common eigenvalue in the spectral decompositions in the sense of (4.29) for the R_i.

Proposition 4.5.13 *For any R_1, R_2, R_3, we define the algebra homomorphisms*

$$\Delta_{R_1,R_2,R_3} : A(R_1 : R_3) \to A(R_1 : R_2) \otimes A(R_2 : R_3) \qquad (4.34)$$

by the matrix multiplication of the generators in the tensor product algebra. This is coassociative in the sense that, for any R_1, R_2, R_3, R_4, we have

$$(\text{id} \otimes \Delta_{R_2,R_3,R_4}) \circ \Delta_{R_1,R_2,R_4} = (\Delta_{R_1,R_2,R_3} \otimes \text{id}) \circ \Delta_{R_1,R_3,R_4} \qquad (4.35)$$

as a map $A(R_1 : R_4) \to A(R_1 : R_2) \otimes A(R_2 : R_3) \otimes A(R_3 : R_4)$.

Proof: The generators of $A(R_2 : R_3)$ are $\tau = \{\tau^i{}_J\}$ say, where $1 \le J \le p$ if $R_3 \in M_p \otimes M_p$. The generators of $A(R_1 : R_3)$ are, likewise, $\sigma = \{\sigma^\mu{}_J\}$, say, and the definition of Δ is, explicitly, $\Delta_{R_1,R_2,R_3}(\sigma^\mu{}_J) = t^\mu{}_i \otimes \tau^i{}_J$, where the sum is over $1 \le i \le n$ in the range appropriate for R_2. We write this compactly as $\sigma \mapsto t\tau$, where the rectangular matrix multiplication of the generators is understood. The proof that this is an algebra homomorphism is analogous to the solution of Exercise 4.5.10. Thus, $R_1 \sigma_1 \sigma_2 = R_1 t_1 \tau_1 t_2 \tau_2 = R_1 t_1 t_2 \tau_1 \tau_2 = t_2 t_1 R_2 \tau_1 \tau_2 = t_2 t_1 \tau_2 \tau_1 R_3 = \tau_2 t_2 t_1 \tau_1 R_3 = \sigma_2 \sigma_1 R_3$, as required. The coassociativity property (4.35) is clearly true on the generators, and both sides are algebra homomorphisms, so it holds for products of the generators also. ∎

The idea behind this coproduct-like structure is that the concept of the number of rows and columns of a usual matrix must be generalised by attaching an entire R-matrix to the rows, and another R-matrix to the columns. Just as one can multiply rectangular matrices when the columns of the first matrix matches the rows of the second, so the same principle applies to rectangular quantum matrices: one can comultiply them as shown when the column R-matrix of the first matches the row R-matrix of the second. On the other hand, whenever this matrix comultiplication is defined, it is coassociative. This is expressed by (4.35).

It is clear that $A(R : R) = A(R)$ reduces to the notion of a square quantum matrix, as studied in Section 4.1. We see that, more generally one could have rectangular quantum matrices where the number of rows and columns are different or, even if the number of rows and columns are the same, their flavour of R-matrices are different.

It is also clear that $A(R : \lambda) = V_L(R, \lambda)$ recovers the quantum vector algebra as $n \times 1$ rectangular quantum matrices. Here we regard λ as a trivial one-dimensional matrix. Likewise, $A(\lambda : R) = V_R^*(\lambda, R)$ recovers the quantum covector algebra as $1 \times n$ quantum matrices. Then the realisation in Exercise 4.5.10 is an example of quantum matrix multiplication. For example, the realisation of our standard example $M_q(2)$ in the quantum vectors $\mathbb{A}_q^{2|0}$ and quantum covectors $\mathbb{A}_q^{2|0}$ is

$$\begin{pmatrix} a & b \\ c & d \end{pmatrix} = \begin{pmatrix} v \\ w \end{pmatrix} \begin{pmatrix} x & y \end{pmatrix},$$

as one can verify explicitly using the relations in Examples 4.2.5 and 4.5.6.

Another example is the multiplication of a covector by a vector to give a scalar or 1×1 matrix. The quantum analogue of this inner product map is the element $x_i \otimes v^i \in V_R^*(\lambda, R) \otimes V_L(R, \lambda)$.

More generally, these outer and inner products of vectors and covectors are special cases of the rectangular matrix multiplication map

$$\Delta_{R_1, R_2, R_1} : A(R_1, R_1) \to A(R_1 : R_2) \otimes A(R_2 : R_1).$$

Likewise, all the left and right coactions above and their complicated properties are nothing other than further cases of the rectangular quantum matrix comultiplication map and its coassociativity property, namely

$$\beta_L = \Delta_{R_1, R_1, R_2} : A(R_1 : R_2) \to A(R_1 : R_1) \otimes A(R_1 : R_2),$$

$$\beta_R = \Delta_{R_1, R_2, R_2} : A(R_1 : R_2) \to A(R_1 : R_2) \otimes A(R_2 : R_2).$$

In general, $A(R_1 : R_2)$ is an $A(R_1 : R_1)$–$A(R_2 : R_2)$-bicomodule under these maps. Here the left coacting bialgebra and the right coacting bialgebra can be different. This provides the general point of view behind the construction of quantum vectors and quantum covectors in this section, and the quantum matrices in Section 4.1.

4.5.2 Covariant formulation

In this subsection, we consider briefly how to formulate vectors and covectors in such a way that there are not separate left and right coacting copies of $A(R)$, but rather a single copy under which everything is covariant. This is the usual way that one thinks about linear algebra being GL_n-covariant: if a covector transforms by a matrix $\mathbf{t}$, then a vector transforms by $\mathbf{t}^{-1}$ for the same matrix $\mathbf{t}$. This point of view is needed in any situation where we want to consider GL_n as acting on our vectors and covectors as a change of basis. Our strategy in the quantum case is to begin with the bicovariant picture above and let the left coactions be strictly related to the right coaction by means of an antipode, rather than being independent as was the case in the preceding subsection. To this end, we assume that $A(R)$ has been made into an honest Hopf algebra A, as discussed in Section 4.2. Note that one might expect that the quantum matrices themselves would transform in this case by conjugation as $\mathbf{t}^{-1}(\ )\mathbf{t}$; here we will explain a problem in the quantum case.

We will aim throughout for our constructions to be covariant under A in the sense of being right A-comodule algebras, as in Chapter 1.6.2. There is an equally good left handed theory too. In this case, the notion of quantum covector algebra in Lemma 4.5.2 and its example $V_R^*(\lambda, R)$ already fit the bill. To supplement this we need

Lemma 4.5.14 *Let A be a matrix Hopf algebra (as above but with an antipode S) and let V be an algebra with n generators $\mathbf{v} = (v^i)$ (written as a column vector) and 1. Define $\beta(1) = 1 \otimes 1$ and $\beta(v^j) = v^i \otimes St^j{}_i$. Then β makes V a comodule algebra iff the following holds: whenever $\mathbf{t}$ is a copy of the generators of A, and if $\mathbf{v}$ is a copy of the generators of V, commuting elementwise with the $\mathbf{t}$, then $\mathbf{v}' = \mathbf{t}^{-1}\mathbf{v}$ is a realisation of V. Here $\mathbf{t}^{-1} = S\mathbf{t}$, i.e. the matrix with entries $(St^i{}_j)$. We call such algebras right quantum vector algebras.*

Proof: Here $t^i{}_j = 1 \otimes t^i{}_j$ (as before), and $v^i = v^i \otimes 1$ are the generators of the tensor product algebra $V \otimes A$ built from mutually commuting copies of A and V. We use the fact that they mutually commute in the tensor product to write the $S\mathbf{t}$ on the left, even though it lives in the second factor of $V \otimes A$. The condition is just that the $v'^i = St^i{}_j v^j$ is a realisation of V, i.e. that $\beta : V \to V \otimes A$ as defined is an algebra map. Once again, β as defined is already a right coaction from the form of its definition. This is because $\beta(v^j) \otimes St^i{}_j = v^{j'} \otimes St^j{}_{j'} \otimes St^i{}_j = v^{j'} \otimes \Delta(St^i{}_{j'})$ and $v^i \epsilon(St^j{}_i) = v^j$ due to the matrix form of Δ, ϵ and, to the fact that S is an anticoalgebra map and $\epsilon \circ S = \epsilon$. ∎

Example 4.5.15 *The algebra $V_R(R, \lambda) = V_L(R, \lambda)^{\mathrm{op}} = V_L(R_{21}, \lambda)$ is a right quantum vector algebra. Explicitly, it is the algebra with n generators v^i and 1, and relations*

$$R^i{}_a{}^k{}_b v^b v^a = v^i v^k \lambda, \quad i.e. \quad R\mathbf{v}_2\mathbf{v}_1 = \mathbf{v}_1\mathbf{v}_2\lambda,$$

where we write $\mathbf{v} = (v^i)$ as a column vector and $\mathbf{t}^{-1} = (St^i{}_j)$ for the matrix inverse of $\mathbf{t}$ (with values in the respective algebras), and $v_1 = v \otimes 1$ etc. Then the assignment $\mathbf{v}' = \mathbf{t}^{-1}\mathbf{v}$ makes $V_R(R, \lambda)$ into a right A-comodule algebra.

Proof: The direct proof is similar to that of Example 4.5.5. In matrix notation, it is $\mathbf{v}'_1\mathbf{v}'_2\lambda = \mathbf{t}_1^{-1}\mathbf{t}_2^{-1}\mathbf{v}_1\mathbf{v}_2\lambda = \mathbf{t}_1^{-1}\mathbf{t}_2^{-1}R\mathbf{v}_2\mathbf{v}_1 = R\mathbf{t}_2^{-1}\mathbf{t}_1^{-1}\mathbf{v}_2\mathbf{v}_1 =$

$R\mathbf{v}'_2\mathbf{v}'_1$. We used the relations for $A(R)$ in a form obtained by applying the antipode S to the relations (4.4). We assume that A can be obtained from $A(R)$ (by imposing determinant-type relations or by inverting a determinant etc.) in a way consistent with the coaction. ∎

These algebras will be developed further in Chapter 10.2 in a slightly more general form as braided vectors and braided covectors. This connection with the braided theory is good only in the Hecke case, but includes the most popular examples, such as those computed above.

Next we consider the transformation property of quantum matrices in our right-covariant formulation. The one that we should aim for is characterised as follows.

Lemma 4.5.16 *Let A be a matrix Hopf algebra (as above) and let V be an algebra with n^2 generators $\mathbf{u} = (u^i{}_j)$ and 1. Define $\beta(1) = 1 \otimes 1$ and $\beta(u^i{}_j) = u^a{}_b \otimes (St^i{}_a)t^b{}_j$. Then β makes V a comodule algebra iff the following holds: whenever $\mathbf{t}$ is a copy of the generators of A, and if $\mathbf{u}$ is a copy of the generators of V, commuting with the $\mathbf{t}$, then $\mathbf{u}' = \mathbf{t}^{-1}\mathbf{u}\mathbf{t}$ is a realisation of V.*

Proof: Here $t^i{}_j = 1 \otimes t^i{}_j$ (as before), and $u^i{}_j = u^i{}_j \otimes 1$ are the generators of the tensor product algebra $V \otimes A$ built from mutually commuting copies of A and V. We again use the fact that they mutually commute in the tensor product to write the $St^i{}_a$ part on the left, even though it lives in the second factor of $V \otimes A$ along with the $t^b{}_j$ part. The condition is just that $u'^i{}_j = (St^i{}_a)u^a{}_b t^b{}_j$ is a realisation of V, i.e. that $\beta : V \to V \otimes A$ as defined is an algebra map. The map β as defined is already a right coaction because $\beta(u^a{}_b) \otimes (St^i{}_a)t^b{}_j = u^{a'}{}_{b'} \otimes (St^a{}_{a'})t^{b'}{}_b \otimes (St^i{}_a)t^b{}_j = u^{a'}{}_{b'} \otimes \Delta((St^i{}_{a'})t^{b'}{}_j)$, and $u^a{}_b \epsilon((St^i{}_a)t^b{}_j) = u^i{}_j$ due to Δ, ϵ being algebra homomorphisms and the arguments already given in the proofs of the two lemmas above. ∎

Our first observation concerning such comodule algebras is a negative one. We saw in Example 1.6.14 that every Hopf algebra coacts on itself by a right quantum adjoint coaction. Applying this to the matrix quantum group A, we see from its matrix comultiplication that this coaction is indeed $t^i{}_j \mapsto t^a{}_b \otimes (St^i{}_a)t^b{}_j$ as, desired. But, as soon as $A(R)$ is noncommutative, one finds that this is *not in general an algebra homomorphism*. So A coacts on itself or on $A(R)$, but this coaction does not give a comodule algebra.

On the other hand, we have given, in Section 4.3, a matrix description of quantum enveloping algebras in terms of certain braided matrices $B(R)$;

see Definition 4.3.1. Even though the quantum matrices A or $A(R)$ do not transform properly under conjugation, the braided ones do. From the point of view of Section 4.3, this is because the quantum enveloping algebras such as $U_q(g)$ are dual to the quantum matrix groups A. The left quantum adjoint action of the enveloping algebra on itself always respects its own algebra structure (see Example 1.6.3), and hence, by Proposition 1.6.11, its dualisation as a right coaction of A on the quantum enveloping algebra (the quantum coadjoint coaction in Example 1.6.15) likewise respects quantum enveloping algebra structure. So, while the adjoint coaction can fail to be a comodule algebra in the quantum case, the coadjoint coaction always works. Using the matrix description of quantum enveloping algebras from Section 4.3, it is not hard to see that this works at the level of $B(R)$ also. We have

Example 4.5.17 *The algebra $B(R)$ with generators $\mathbf{u} = (u^i{}_j)$ and relations as in (4.19) forms an A-comodule algebra under the assignment $\mathbf{u}' = \mathbf{t}^{-1}\mathbf{u}\mathbf{t}$.*

Proof: We give here a direct proof. We have $R_{21}\mathbf{t}_1^{-1}\mathbf{u}_1\mathbf{t}_1 R\mathbf{t}_2^{-1}\mathbf{u}_2\mathbf{t}_2 = R_{21}\mathbf{t}_1^{-1}\mathbf{u}_1\mathbf{t}_2^{-1}R\mathbf{t}_1\mathbf{u}_2\mathbf{t}_2 = \mathbf{t}_2^{-1}\mathbf{t}_1^{-1}R_{21}\mathbf{u}_1 R\mathbf{u}_2\mathbf{t}_1\mathbf{t}_2$. Here we used (4.4) in various forms and freely commuted $\mathbf{u}_1$ with $\mathbf{t}_2$, etc. (they live in different algebras and in different matrix spaces). Applying (4.19) to the result, we have, similarly, $\mathbf{t}_2^{-1}\mathbf{t}_1^{-1}\mathbf{u}_2 R_{21}\mathbf{u}_1 R\mathbf{t}_1\mathbf{t}_2 = \mathbf{t}_2^{-1}\mathbf{u}_2\mathbf{t}_2 R_{21}\mathbf{t}_1^{-1}\mathbf{u}_1\mathbf{t}_1 R$ so that the transformed $\mathbf{u}$ obey the same relations (4.19). ∎

This further justifies thinking of $B(R)$ as a kind of quantum matrix, like $A(R)$, but in a different (braided) version. It indeed has a (braided) matrix comultiplication and so, like $A(R)$, should be thought of as an analogue of the algebra of functions on M_n. This is in addition to its interpretation in Section 4.3 as describing quantum enveloping algebras in matrix form. These aspects, as well as analogues of Exercise 4.5.10 in this right-covariant formulation, must await the theory of braided groups in Chapter 10. We will also define there a coaddition for our square and rectangular quantum matrices, quantum vectors and quantum covectors in the case at least when R is Hecke.

4.5.3 Quantum automorphisms and diffeomorphisms

This is a miscellaneous subsection that is quite independent of the R-matrix theory above. Rather, it is a different source of noncommutative and noncocommutative bialgebras as the 'bialgebra of all transformations' of any given algebra. Nevertheless, one does obtain quantum matrices in the sense of Lemma 4.5.1 and one does include the linear transformations in the preceding subsections as part of the relevant bialgebra of all transformations.

Thus, just as every algebra has a group of all automorphisms, one has a similar concept of the 'maximal bialgebra' coacting on a given algebra. More precisely, we define a *comeasuring* of a unital algebra A by an algebra H as simply an algebra map $\beta : A \to A \otimes H$. The maximal object $\mathrm{Meas}(A)$ is called the *comeasuring algebra* of A and comes, by definition, with a comeasuring $\beta_U : A \to A \otimes \mathrm{Meas}(A)$ such that any comeasuring (H, β) is the image of this one in the sense that there exists a unique algebra map $\pi : \mathrm{Meas}(A) \to H$ with $\beta = (\mathrm{id} \otimes \pi) \circ \beta_U$.

Exercise 4.5.18 *If* $\mathrm{Meas}(A)$ *exists, show that it is a bialgebra and A is a* $\mathrm{Meas}(A)$*-comodule algebra with β_U as coaction.*

Solution: We need only note that $(\mathrm{Meas}(A)^{\otimes 2}, (\beta_U \otimes \mathrm{id}) \circ \beta_U)$ is also a comeasuring of A, hence we can conclude that there is a unique algebra map $\Delta : \mathrm{Meas}(A) \to \mathrm{Meas}(A)^{\otimes 2}$ connecting the comeasuring with β_U. Similarly, $(\mathrm{Meas}(A)^{\otimes 3}, (\beta_U \otimes \mathrm{id} \otimes \mathrm{id}) \circ (\beta_U \otimes \mathrm{id}) \circ \beta_U)$ is a comeasuring so there is an unique algebra map $\mathrm{Meas}(A) \to \mathrm{Meas}(A)^{\otimes 3}$ connecting this comeasuring with β_U. Both $(\Delta \otimes \mathrm{id}) \circ \Delta$ and $(\mathrm{id} \otimes \Delta) \circ \Delta$ can serve as this map and since it is unique, the two coincide with it and hence with each other. ∎

More nontrivial is to construct $\mathrm{Meas}(A)$ explicitly. Actually, the easiest way to construct it is to construct a slightly bigger object $\mathrm{Meas}_1(A)$ defined as 1 adjoined to the 'maximal' comeasuring object for nonunital algebras (which is a nonunital bialgebra by similar arguments to the above).

Proposition 4.5.19 *Let A be a finite-dimensional algebra with structure constants $c_{ij}{}^k$ in a basis $\{e_i\}$ Then $\mathrm{Meas}_1(A)$ is generated by 1 and indeterminates $t^i{}_j$ with the relations*

$$c_{ab}{}^k t^a{}_i t^b{}_j = t^k{}_a c_{ij}{}^a.$$

The coalgebra has the matrix form (a quantum matrix in the sense of Lemma 4.5.1) and the coaction is $\beta_U(e_j) = e_i \otimes t^i{}_j$.

Proof: The bialgebra and coaction properties follow by similar arguments to those above once we have proven the 'maximality'. We can also check these facts directly. For example, $\Delta(c_{ab}{}^k t^a{}_i t^b{}_j) = c_{ab}{}^k t^a{}_m t^b{}_n \otimes t^m{}_i t^n{}_j = t^k{}_a c_{mn}{}^a \otimes t^m{}_i t^n{}_j = t^k{}_b \otimes t^b{}_a c_{ij}{}^a = \Delta(c_{ij}{}^a t^k{}_a)$, while β_U being multiplicative is the defining relation of $\mathrm{Meas}_1(A)$. Now suppose H is an algebra and $\beta : A \to A \otimes H$ an algebra map. Let $\bar{t}^i{}_j \in H$ be defined by $\beta(e_j) = e_i \otimes \bar{t}^i{}_j$. Define $\pi(1) = 1$ and $\pi(t^i{}_j) = \bar{t}^i{}_j$. That this indeed extends as an algebra map is $e_k \otimes \pi(c_{ab}{}^k t^a{}_i t^b{}_j) = e_k \otimes c_{ab}{}^k \bar{t}^a{}_i \bar{t}^b{}_j = e_a e_b \otimes \bar{t}^a{}_i \bar{t}^b{}_j = \beta(e_i)\beta(e_j) = \beta(e_i e_j) = \beta(e_k)c_{ij}{}^k = e_k \otimes \pi(t^k{}_a c^a_{ij})$ as required, since $\{e_k\}$ are a basis.   ∎

Because of its universal nature, the construction is basis independent, i.e. if we change the basis the bialgebra is isomorphic after the corresponding change of variables.

Proposition 4.5.20 *If A in Proposition 4.5.19 has a unit, we chose a basis with $e_0 = 1$. Then $\mathrm{Meas}(A)$ is generated by 1, a vector $\mathbf{b} = (b_i)$ and a matrix $\mathbf{t} = (t^i{}_j)$ for $i, j \geq 1$, with the relations*

$$t^k{}_a c_{ij}{}^a = c_{ab}{}^k t^a{}_i t^b{}_j + b_i t^k{}_j + t^k{}_i b_j, \quad c_{ij}{}^0 + b_a c_{ij}{}^a = c_{ab}{}^0 t^a{}_i t^b{}_j + b_i b_j.$$

The coalgebra and coaction are of the matrix form for $\mathbf{t}$ and

$$\Delta b_j = b_i \otimes t^i{}_j + 1 \otimes b_j, \quad \epsilon b_i = 0, \quad \beta_U(e_j) = 1 \otimes b_j + e_i \otimes t^i{}_j.$$

Proof: By similar arguments to those in Proposition 4.5.19, the restriction to coactions that respect the unit of A corresponds to the quotient of $\mathrm{Meas}_1(A)$ by the further relations $t^i{}_0 = \delta^i{}_0$. The remaining generators $b_i = t^0{}_i$ and $t^i{}_j$ then have the relations stated as inherited from the relations of $\mathrm{Meas}_1(A)$.   ∎

Again, the construction depends only on the algebra; a change of basis that preserves $e_0 = 1$ gives an isomorphic bialgebra. Finally, there is a further quotient $\mathrm{Meas}_0(A)$ defined by setting $b_i = 0$. These are the transformations that not only preserve 1 but also the complement to the space spanned by 1. A change of basis $\{e_i\}$ of the complement gives an isomorphic bialgebra.

To get an idea of what these constructions mean from the point of view of noncommutative geometry, we consider (formally) the algebra $A = k[x]$ of polynomials in one variable. A 'diffeomorphism' in an algebraic setting is a map $k[x] \to k[x]$ induced by a polynomial $x \mapsto a(x) = a_0 + a_1 x + a_2 x^2 + \cdots$. If we compose two such transformations the composition law is

$$a \circ b(x) = \sum_j a(x^j)b_j = \sum_i x^i \sum_j \sum_{n_1 + \cdots + n_j = i} a_{n_1} \cdots a_{n_j} b_j$$

where $i, j, n_1 \in \mathbb{N}$. Notice that the inverse of a polynomial transformation cannot be expected to be a polynomial, i.e. one has only some kind of semigroup until one introduces topological completions. Also note that the composition law itself involves an infinite series. However, if we restrict to diffeomorphisms where $a_0 = 0$, i.e. which vanish at $x = 0$, we have only a finite number of terms as $n_1, \cdots, n_j \geq 1$ and hence an algebraic semigroup under composition. We have similar features for our comeasuring construction.

Example 4.5.21 $\mathrm{Meas}_0(k[x])$ *is the bialgebra generated by* 1 *and* t_i *with* $i \geq 1$ *with no relations (the free algebra) and the coalgebra*

$$\Delta t_i = \sum_j \sum_{n_1 + \cdots + n_j = i} t_{n_1} \cdots t_{n_j} \otimes t_j, \quad \epsilon t_i = \delta_{i,1},$$

where $n_1, \cdots, n_j \geq 1$. $\mathrm{Meas}(k[x])$ *is the free algebra generated similarly with an additional generator* t_0.

Proof: We formally apply Proposition 4.5.20 but one may then verify the required facts directly. The structure constants in the basis $e_i = x^i$ are

$$c_{ij}{}^k = \delta^k{}_{i+j}, \quad c_{ij}{}^0 = 0$$

for $i, j, k \geq 1$. Hence the algebra $\mathrm{Meas}(A)$ in Proposition 4.5.20 has relations

$$b_{i+j} = b_i b_j, \quad t^k{}_{i+j} = \sum_{m+n=k} t^m{}_i t^n{}_j + b_i t^k{}_j + t^k{}_i b_j$$

with all indices ≥ 1. These relations allow us to consider $t_0 \equiv b_1$ and $t_i \equiv t^i{}_1$ as the generators, obtaining the other generators by

$$t^i{}_j = \sum_{n_1 + \cdots + n_j = i} t_{n_1} \cdots t_{n_j}, \quad b_i = (t_0)^i,$$

where $i, j \geq 1$ but $n_1, \cdots, n_j \geq 0$. This is easily proven by induction. Putting this form into the above relations yields no further relations, i.e. the algebra is then free. We then obtain the coproduct from Proposition 4.5.20, but only as a formal expression (it involves an infinite sum). For $\mathrm{Meas}_0(k[x])$, however, we set $t_0 = 0$ and hence only have contributions in $t^i{}_j$ from $n_1, \cdots, n_j \geq 1$. Hence $t^i{}_j = 0$ for $j > i$. In this case our sums are all finite and we have an honest bialgebra. ∎

The coproduct here has the same form as the composition law for diffeomorphisms explained above, i.e. if we took the generators commuting we would have the coordinate algebra of the diffeomorphisms of the line (that fix the origin in the case of $\mathrm{Meas}_0(k[x])$). The comeasuring construction is a noncommutative version of this, in the present example the generators being totally noncommuting instead of commuting. We also have a formal coaction

$$\beta_U(x) = \sum_i x^i \otimes t_i$$

corresponding to the action of diffeomorphisms. It is possible to apply these ideas just as well to classical and quantum spaces as the the algebraic part of their 'quantum diffeomorphisms'. In higher dimensions one typically obtains nontrivial relations between the generators rather than totally free algebras. As an example, we can consider the quantum plane $A_q^{2|0}$ from Example 4.5.6.

Example 4.5.22 *The lowest order generators of* $\mathrm{Meas}_0(A_q^{2|0})$ *in the transformation*

$$\beta_U(x\ y) = (x\ y) \otimes \begin{pmatrix} a & b \\ c & d \end{pmatrix} + O(x^2, xy, y^2)$$

form a sub-bialgebra obeying the relations

$$dc = qcd, \quad ba = qab, \quad ad - da = q^{-1}bc - qcb$$

(half of the relations of the quantum matrices $M_q(2)$*) with the matrix coalgebra structure.*

Proof: One may compute $\mathrm{Meas}_0(A_q^{2|0})$ along the same lines as in Example 4.5.21. The structure constants in the basis $e_I = x^{i_0}y^{i_1}$ are

$$c_{IJ}{}^K = \delta^K{}_{I+J}q^{i_1 j_0},$$

where $I = (i_0, i_1)$ for $i_0, i_1 \geq 1$ is a multiindex and added componentwise. The relations for $t^I{}_J$ have a similar form as before, with some powers of q, and result in the fact that they can all be generated from $s_I = t^I{}_{(1,0)}$ and $t_I = t^I{}_{(0,1)}$. These are required to obey the residual relations

$$q \sum_{A+B=I} q^{a_1 b_0} s_A t_B = \sum_{A+B=I} q^{a_1 b_0} t_A s_B.$$

The lowest order generators in the formal coaction are

$$\begin{pmatrix} a & b \\ c & d \end{pmatrix} = \begin{pmatrix} s_{(1,0)} & t_{(1,0)} \\ s_{(0,1)} & t_{(0,1)} \end{pmatrix} = \begin{pmatrix} t^{(1,0)}{}_{(1,0)} & t^{(1,0)}{}_{(0,1)} \\ t^{(0,1)}{}_{(1,0)} & t^{(0,1)}{}_{(0,1)} \end{pmatrix}$$

and the relations among them then come out as stated. A shortcut to this computation if we only want these lowest order relations is to realise that the relations are nothing other than what is forced by the algebra $yx = qxy$ being respected by the transformation. We already know that $M_q(2)$ coacts in the required way from Example 4.5.6 and we just read the direct verification of that backwards to see exactly what relations are used.

We then have to argue that the higher order terms in the transformation do not affect the relations at this lowest level. ∎

Geometrically, the inclusion here corresponds to the differentiation of a diffeomorphism at the origin, i.e. truncation to its linear part. One can make a similar computation for the fermionic quantum plane $A_q^{0|2}$. This is a 4-dimensional algebra so we can more easily present the entire comeasuring bialgebra in this case. Indeed, the most general possible coaction necessarily has the form

$$\beta_U(\theta \; \vartheta) = 1 \otimes (b_1 \; b_2) + (\theta \; \vartheta) \otimes \begin{pmatrix} a & b \\ c & d \end{pmatrix} + \theta\vartheta \otimes (\alpha \; \beta)$$

for some generators as shown.

Example 4.5.23 *The comeasuring bialgebra* Meas$(A_q^{0|2})$ *has the form*

$$\mathbf{b} = (b_1 \; b_2 \; b_1 b_2), \quad \mathbf{t} = \begin{pmatrix} a & b & b_1 b + a b_2 \\ c & d & b_1 d + c b_2 \\ \alpha & \beta & ad - q^{-1} cb + b_1\beta + \alpha b_2 \end{pmatrix}$$

(giving the coalgebra) and the relations

$$b_1^2 = b_2^2 = 0, \quad \{b_1, b_2\}_q = 0, \quad \{b_1, a\} = \{b_1, c\} = \{b_2, b\} = \{b_2, d\} = 0,$$

$$\{b_1, b\}_q + \{a, b_2\}_q = 0, \; \{b_1, d\}_q + \{c, b_2\}_q = 0, \; ac - q^{-1} ca + \{b_1, \alpha\} = 0,$$

$$bd - q^{-1} db + \{b_2, \beta\} = 0, \; ad - da + \{b_1, \beta\}_q + \{\alpha, b_2\}_q = q^{-1} cb - qbc,$$

where $\{ \; , \; \}$ *denotes anticommutator and* $\{b_1, b_2\}_q \equiv b_1 b_2 + q b_2 b_1$, *etc.*

Proof: As basis we take $e_0 = 1$, $e_1 = \theta$, $e_2 = \vartheta$ and $e_3 = \theta\vartheta$. The only possibly-nonzero structure constants are $c_{0i}{}^j = \delta_i^j = c_{i0}{}^j$, $c_{12}{}^3 = 1$, $c_{21}{}^3 = -q^{-1}$. We then compute from Proposition 4.5.20, for example $c_{ij}{}^0 = 0$ for $i, j \neq 0$, so that $c_{12}{}^3 b_3 = b_1 b_2$ and $c_{21}{}^3 b_3 = b_2 b_1$ tell us b_3 and that $b_1 b_2 = -q b_2 b_1$, etc. Equivalently, we can take the ansatz above for β_U and write out all the relations required for this to be an algebra map. Comparison with the coaction in Proposition 4.5.20 gives the identification with the generators there. ∎

We see that the translation generators b_1, b_2 themselves form a fermionic quantum plane $A_q^{0|2}$. Setting them to zero gives Meas$_0(A_q^{0|2})$ as generated by a 2×2 quantum matrix with relations

$$ca = qac, \quad db = qbd, \quad ad - da = q^{-1} cb - qbc$$

(the other half of the $M_q(2)$ relations) and free α, β with no relations. Later on, in Chapter 10, we will see that $\theta = \mathrm{d}x$, $\vartheta = \mathrm{d}y$ generate a natural differential calculus on the bosonic quantum plane, in which case the above two examples identify $M_q(2)$ as the linear part of the 'diffeomorphisms' that preserve both its coordinate relations and its natural differential calculus.

As well as respecting a differential calculus, one can ask our 'diffeomorphisms' to respect various other structures of interest, giving quotients of $\mathrm{Meas}_0(A)$ etc. As a matter of fact, a look at the direct proof of the bialgebra structure in Proposition 4.5.19 shows that the same argument holds for *any* tensor $c_{i_1 \cdots i_n}{}^{j_1 \cdots j_m}$ in place of $c_{ij}{}^k$; there is a quantum matrix bialgebra generated by 1 and $t^i{}_j$ with the relations

$$c_{a_1 \cdots a_n}{}^{j_1 \cdots j_m} t^{a_1}{}_{i_1} \cdots t^{a_n}{}_{i_n} = t^{j_1}{}_{b_1} \cdots t^{j_m}{}_{b_m} c_{i_1 \cdots i_n}{}^{b_1 \cdots b_m}$$

and with the matrix coalgebra structure. For example, the quantum matrices $A(R)$ in Section 4.1 are just those that leave the tensor PR invariant in this sense. Moreover, we can always combine such invariance relations to obtain a matrix bialgebra leaving invariant any number of tensors of our choice.

For example, if A has a coproduct with structure constants $d_i{}^{jk}$ defined by $\Delta e_i = d_i{}^{jk} e_j \otimes e_k$, we can ask for this tensor to be preserved by the comeasurings, which translates as the additional relation

$$d_a{}^{jk} t^a{}_i = d_i{}^{ab} t^j{}_a t^k{}_b$$

to be added to those of $\mathrm{Meas}_1(A)$. Similarly for $\mathrm{Meas}(A)$ and $\mathrm{Meas}_0(A)$. In particular, we have mentioned at the end of Section 4.5.2 that we will later introduce a (braided) coaddition law on quantum planes such as $A_q^{2|0}$. At the moment we need to know only that it is a certain coalgebra structure and that this 'braided covector' algebra is dually paired with another copy of $A_q^{2|0}$ as a 'braided vector' algebra. This implies (after suitable normalisations) that the structure constants for this additive coalgebra have a similar but transposed form to those for the algebra of the quantum plane. Therefore, combining with the transpose of Example 4.5.22 and the upper triangular form of its higher order generators, we conclude that the quotient of $\mathrm{Meas}_0(A_q^{2|0})$ respecting the coaddition is *precisely* the bialgebra $M_q(2)$. This completes our picture of the latter as precisely the linear transformations within (or rather, for the coordinate algebra, a quotient of) a larger bialgebra of all transformations.

Finally, it should be stressed that we are not obliged to stick to geometrical examples like those above. We can simply put our favourite (and preferably finite-dimensional) algebra into Propositions 4.5.19 and 4.5.20 and obtain its comeasuring bialgebra as generalised 'quantum automorphisms'. For example, associated to the 2-dimensional 'complex number algebra' $k[i]$ with $i^2 + 1 = 0$ we obtain for Meas$_1$ a 2×2 quantum matrix bialgebra with relations

$$a^2 - c^2 = a = d^2 - b^2, \quad ac + ca = c = -(bd + db),$$

$$ab - cd = b = ba - dc, \quad ad + cb = d = bc + da$$

of which the trigonometric bialgebra in Example 1.6.20 is a natural quotient. For the algebra of functions on a finite set $\{i\}$ (with δ-function basis) one immediately obtains for Meas$_1$ a quantum matrix consisting of orthogonal families of projectors

$$t^i{}_j t^i{}_k = \delta_{j,k} t^i{}_j,$$

and so on. This is therefore an independent source of true noncommutative and noncocommutative bialgebras. It is not unrelated to the combinatoric bialgebras of the next chapter.

Notes for Chapter 4

The matrix bialgebras $A(R)$ had their origin in the quantum inverse scattering method and the theory of exactly solvable lattice models, as explained for vertex models in Section 4.4 in the parametrised case. This obviously leads to the version without a parameter by discarding the spectral parameter. For more of this physical background, see [27] and [81]. The use of these $A(R)$ to describe the dual of $U_q(g)$ is already in [22], but was studied more extensively in the work of L.D. Faddeev, N.Yu. Reshetikhin and L.A. Takhtajan [82]. The formulae for the R-matrices for these standard compact Lie group deformations quoted in Section 4.2 are taken from [82], to which we refer for details. These authors also isolated the elements $l^{\pm}$ in $U_q(g)$, which can be called the FRT description $U(R)$ of $U_q(g)$. For these standard quantum deformations, the universal R-matrix or quasitriangular structure for $U_q(g)$ was already known in principle from [22], and such things as q-determinants and the form of $l^{\pm}$ could be guessed by analogy with the underlying classical Lie algebra or Lie group.

A natural question arising from this approach was how to extend it to general R-matrices (obeying the QYBE), not necessarily related to any standard q-deformation. For example, it was not clear at first whether any generic R-matrix leads to a quasitriangular Hopf algebra with a universal R-matrix $\mathcal{R}$. The required general framework was developed by the author [1, 14, 83], on which Section 4.1 is based. Theorem 4.1.5 was obtained in these works in the form $\mathcal{R} : A(R) \to \tilde{U}(R)$; since $\tilde{U}(R)$ is dually paired with $A(R)$, the result can just as easily be formulated as a map $\mathcal{R} : A(R) \otimes A(R) \to k$, which is the more modern notion of a dual quasitriangular structure on a Hopf algebra. We have retained our original inductive proofs from [1], based on the 'partition functions' Z_R, this time giving full details and the more modern notation. The result was also elaborated in [84], following the above works.

Another problem when we want to begin with only the R-matrix is that, in general, we have no analogies with classical Lie groups, as we do in the FRT method, by which to guess q-determinants, the antipode, or the form of $\mathbf{l}^\pm$. These problems were solved [1, 14] by beginning with the abstract bialgebra $\tilde{U}(R)$ with $2n^2$ generators $\mathbf{L}^\pm$, and then adding the null relations to both $A(R)$ and $\tilde{U}(R)$ as explained in Lemma 4.1.10. This is the general method described in Sections 4.1 and 4.2. The notion of weak-antipode and its construction (or convolution-invertibility of the pairing) in Proposition 4.2.3 are due to the author [14]. In this way, we have an approach to matrix quantum groups and the corresponding quantum enveloping algebras depending only on the R-matrix [1, 14]. The nonstandard Example 4.1.11 is taken from [14].

These two approaches are related in Corollary 4.1.9, giving a method of obtaining the form of the $\mathbf{l}^\pm$ if the $\mathcal{R}$ is known. Example 4.2.10 for $U_q(sl_3)$ is taken from [83]. This relation was also studied by other authors, notably N. Burroughs.

The multiparameter $GL_{p,q}$ in Example 4.2.12 was studied in [85]. The nonstandard Example 4.2.13, which is related to (and arose from) the Alexander–Conway knot polynomial, was studied in [86]. The universal R-matrix for the corresponding quantum enveloping algebra in Example 4.2.14 was obtained [87] under the formal assumption that the K_i are exponentials of certain H_i. Thus,

$$K_1 = q^{\frac{H_1}{2}}, \quad K_2 = e^{\frac{i\pi}{2} H_2} q^{\frac{H_2}{2}},$$

$$\mathcal{R} = e^{-\frac{i\pi}{4} H_2 \otimes H_2} q^{\frac{1}{4}(H_1 \otimes H_1 - H_2 \otimes H_2)} (1 \otimes 1 + (1 - q^2) K_2 X_+ \otimes K_2^{-1} X_-).$$

Note that for a more precise theory one should work with dual quasitriangular Hopf algebras, using Theorem 4.1.5.

The classification of 4×4 R-matrices quoted at the end of Section 4.1 is due to J. Hietarinta, using computer methods [88]. The corresponding bialgebras $A(R)$ in the upper-triangular cases can be found in [89]. There are also plenty of 9×9 and higher R-matrices of interest. The dual quantum Heisenberg group related to the quantum enveloping algebra in Chapter 3.1 is of this type.

The approach to matrix quantum Lie algebras in Section 4.3 is due to the author [5, 34], where Propositions 4.3.2–4.3.6 can be found; it arose from the theory of braided groups, which we come to in Chapters 9.4 and 10. The importance of the generators $1 = 1^+ S1^-$ and the relations $R_{21} l_1 R l_2 = l_2 R_{21} l_1 R$ in the case of the standard quantum enveloping algebras $U_q(g)$ was long recognised by those working in quantum inverse scattering [82, 90], and this was one of the motivations behind [34]. Note, however, that our approach, based on Proposition 4.3.2 and braided matrices, works at the bialgebra level without requiring antipodes to begin with. There is also a connection between these matrix braided-Lie algebras and the theory of bicovariant differential calculus on quantum groups [91, 92], as noted subsequently in [93]. In this context, it is quite standard to consider as quantum vector fields not the action of elements h in the quantum enveloping algebra, but $h - \epsilon(h)$, which corresponds to the χ generators in Section 4.3. We have not tried to cover bicovariant differential calculus here, regarding the braided-Lie algebra as a more fundamental starting point [5, 34].

The theory of braided matrices $B(R)$ as quadratic algebras defined by $R_{21} \mathbf{u}_1 R \mathbf{u}_2 = \mathbf{u}_2 R_{21} \mathbf{u}_1 R$ in their own right, is due to the author in [94], where they were introduced as braided group variants of the usual quantum matrices $A(R)$ via a novel process of *transmutation*. Their ad-covariance in Example 4.5.17 and the basic example $BM_q(2)$ in Example 4.3.4 are taken from this 1990 work. The Hermitian $*$-structure in Proposition 4.3.7 is from [35] and the representation in Proposition 4.3.6 is from [5]. These braided matrices are part of a more general 'braided linear algebra' also developed by the author [95]. Section 4.5.2 is an introduction to this, with more to come in Chapters 7.4, 9.4 and 10. In fact, the standard matrix quantum groups G_q also transmute to braided versions, BG_q, as quotients of braided matrices $B(R)$. Hence, combining with Section 4.3, we arrive [5, 34] at the conclusion in Example 4.3.3 that the quantum enveloping algebras $U_q(g)$ can be viewed as matrix braided groups with relations in one to one correspondence (by transmutation) with the familiar q-determinant or other relations of the corresponding matrix groups. This is a self-duality result which is impossible at $q = 1$ and which is quite interesting for q-deformed physics; see the Notes to Chapter 10. Finally, we note that, like the relations of the quantum matrices in Section 4.5, the relations of the braided matrices $B(R)$ also have similarities with parametrised equations obeyed by the transfer matrices in certain 'reflection' models in quantum inverse scattering [96, 97]. Some of the algebraic properties of the braided matrices $B(R)$ were subsequently reiterated under this heading, following the braided groups work [34, 94, 95].

The matrix notation for matrix comultiplications and comodule algebras as in (4.27), (4.30) and (4.31) was stressed by Yu.I. Manin [98], who pioneered the view of matrix quantum groups as generated, in some sense, by linear these transformation properties. For example, the relations of $M_q(2)$ were obtained from the requirement to act on $\mathbb{A}_q^{2|0}$ as a covector and $\mathbb{A}_q^{0|2}$ as a vector. Our treatment in Section 4.5.1 differs slightly from this in that one does not need to mix vectors and covectors: either will do to characterise the quantum matrix provided we consider all values of λ. This point of view, which is especially important in the non-Hecke case, has been emphasised by A. Sudbery [99]; see also [100]. The standard Example 4.5.7 based on the sl_n R-matrix is often denoted by $\mathbb{C}_q^n$ when working over $\mathbb{C}$. For some R-matrices, such as the so_n ones, the corresponding quantum planes have a natural real form or *-structure, and can therefore be denoted more properly as $\mathbb{R}_q^n$. Again, the proper understanding of this must await our development of the theory of braided groups in Chapter 10.

These Zamolodchikov 'exchange' algebras $V_L(R, \lambda)$ and $V_R^*(\lambda, R)$ first arose in physics, in the context of quantum inverse scattering (the careful division into vector and covector types and their transformation properties came later). They were studied for nonstandard but Hecke-type R-matrices by D.I. Gurevich [101], who classified them according to their Poincaré series (this measures the growth of the dimension of the subspaces V_n spanned by n-fold products of the v^i). In this context, let us mention a common misconception in the physics literature. It is often stated that associativity of $V_L(R, \lambda)$, say, implies that R obeys the QYBE. Thus,

$$(\mathbf{v}_3\mathbf{v}_2)\mathbf{v}_1 = R_{23}\mathbf{v}_2(\mathbf{v}_3\mathbf{v}_1) = R_{23}R_{13}(\mathbf{v}_2\mathbf{v}_1)\mathbf{v}_3 = R_{23}R_{13}R_{12}\mathbf{v}_1\mathbf{v}_2\mathbf{v}_3,$$

$$\mathbf{v}_3(\mathbf{v}_2\mathbf{v}_1) = R_{12}(\mathbf{v}_3\mathbf{v}_1)\mathbf{v}_2 = R_{12}R_{13}\mathbf{v}_1(\mathbf{v}_3\mathbf{v}_2) = R_{12}R_{13}R_{23}\mathbf{v}_1\mathbf{v}_2\mathbf{v}_3$$

suggests the QYBE. However, equality of these expressions does not imply the QYBE since $\mathbf{v}_1\mathbf{v}_2\mathbf{v}_3$ may be trivial. The space spanned by $\{v^iv^jv^k\}$ can be much smaller than that spanned by $v^i \otimes v^j \otimes v^k$, and the two should not be confused. In fact, the Zamolodchikov algebras are, by definition, always associative, and the QYBE ensures that the ideal generated by their relations is not too big, hence that the resulting algebra is not too trivial.

The notion of rectangular quantum matrices $A(R_1 : R_2)$, and the consequent bicovariant approach given in Section 4.5.1, is due to the author and M. Markl [102], and studied further in [103]. It provides a powerful unifying point of view on the bialgebras $A(R)$ and the exchange algebras. The right-covariant picture in Section 4.5.2 is taken from the theory of braided linear algebra, as mentioned above. The right-covariant version of Exercise 4.5.10 can be found in [95]. It requires the notion of the braided tensor product of covariant algebras, which we shall come to in

Chapter 9.2.

The comeasuring bialgebra constructions and examples in Section 4.5.3 are due to the author in [13]. The starting point for Meas(A) in Exercise 4.5.18 is the dual of a more standard 'measuring bialgebra' in [7] which does not have an explicit formulation in terms of generators and relations. Also, the constructions in Examples 4.5.22 and 4.5.23 should not be confused with the abovementioned approach to $M_q(2)$ in [98] (which was limited to linear transformations) but proofs at lowest order have some points in common with that. Other results in [13] include conjunction with the R-matrix ideas; one may quotient $A(R)$ in Section 4.1 by the additional relations in Proposition 4.5.19 to obtain a dual quasitriangular bialgebra Meas$_1(A, R)$. This is particularly appropriate when A is compatible with the R-matrix in the sense that it forms a braided algebra (see Chapters 9 and 10), namely when

$$c_{ij}{}^a R^k{}_a{}^m{}_n = c_{ab}{}^k R^a{}_i{}^m{}_c R^b{}_j{}^c{}_n, \quad c_{jk}{}^a R^m{}_n{}^i{}_a = c_{ab}{}^i R^m{}_c{}^b{}_k R^c{}_n{}^a{}_j.$$

Also in [13] are braided group versions of these ideas. Meanwhile, a topological C^*-algebra setting for the simplest example of 'automorphisms' for finite sets was found independently in [104] and was also studied by S.L. Woronowicz for the two-point set. Finally, we mention that the notion of comeasuring also applies more generally to algebra maps $\beta : A \to B \otimes H$ between two algebras A, B. If their structure constants are $c_{ij}{}^k$ and $d_{IJ}{}^K$ respectively then the analogous Meas$_1(A, B)$ is the rectangular quantum matrix generated by $1, t^i{}_J$ with relations $c_{ab}{}^k t^a{}_I t^b{}_J = d_{IJ}{}^A t^k{}_A$, etc. These are noncommutative versions of the set of all maps between two spaces.

5

Quantum random walks and combinatorics

This is the first of two physically-oriented chapters in which we take a break from R-matrices and such topics and explore instead a completely different setting in which Hopf algebras arise naturally. This is the setting of logic, quantum mechanics and probability theory, in all of which contexts the coproduct Δ of the Hopf algebra plays the role of 'sharing out' possibilities. The theory has potential applications in computer science as well. In Boolean algebra and quantum mechanics, the product in the algebra corresponds in some sense to deduction of facts ('putting two and two together'), and supplementing this structure with a coproduct provides a kind of reverse of this process by sharing out possible explanations of a fact. This provides the general theme of the present chapter. The chapter can also be viewed as background material for the next chapter, where we present a number of concrete models of quantum-mechanical Hopf algebras of observables.

The simplest example of a combinatorial Hopf algebra is the shuffle algebra associated to an alphabet of symbols. Here the coproduct of a word is the formal sum of all pairs of words which *could be shuffled to give the original word*. Here, a shuffle of two words takes symbols from each of the words and interleaves them randomly while preserving the order of each word (this is exactly what you do when shuffling a pack of cards). Thus,

$$\Delta cat = \ \otimes cat + c \otimes at + ca \otimes t + ct \otimes a + at \otimes c + t \otimes ca + a \otimes ct + cat \otimes \ .$$

There are many more sophisticated variants associated to families of trees, to various kinds of graphs and to general distributive lattices. This is the topic of Section 5.1.

Another kind of example arises when we consider the Boolean algebra of subsets of a universal set Ω which happens to be a group. In this case the coproduct of a subset a is

$$\Delta a = \{(x, y) \mid xy \in a\} \subseteq \Omega \times \Omega,$$

the subset $\Delta a \subseteq \Omega \times \Omega$ consisting of all those things *which if multiplied would give something in* a. This is like a detective considering combinations of possible facts that would explain a given fact if they both held. This demonstrates our main theme in the chapter.

In Section 5.2, we pass from purely combinatorial questions of Boolean algebra, etc., to classical mechanics or classical probability theory. This means that we work with the commutative algebra of functions on a space rather than working with the Boolean algebra of subsets. The generalisation is a mild one (we just work with the characteristic functions of sets rather than with sets themselves), but enables us now to consider other classical observables (=random variables) and to formulate notions of probability. In this setting, a group structure on our underlying space allows us to make random walks: these come out very naturally and beautifully in terms of the coproduct Δ of the corresponding Hopf algebra. As a demonstration, we derive the diffusion equation for Brownian motion as the limit of a random walk on $\mathbb{R}$.

In these initial sections, the Hopf algebras considered are generally commutative or cocommutative. One can also write down noncommutative analogues as some kind of 'quantum combinatorics' or 'quantum random walk'. This is the topic of Section 5.3, where we show that, as well as being a purely mathematical generalisation of the previous section to the noncommutative case, these generalised random walks also have a nice physical interpretation in terms of quantum mechanics. We will not give any very physical examples of such systems (that is the task of Chapter 6), but we do makes some general remarks of interpretation. These general remarks are continued in Section 5.4, where we explore the notion of Hopf algebra duality in this context. Recall from Chapter 1 that the axioms of a Hopf algebra have an input–output symmetry given by interchanging the roles of product and coproduct. According to the theme of the present chapter, this corresponds to some kind of time-reversal; it is the topic of Section 5.4.

A word of warning is necessary about the style in Sections 5.2–5.4: probability theory and quantum mechanics are very well-developed mathematically and usually involve lots of analysis. There are several general texts from which the reader can learn about the formalism of sigma algebras, Borel sets, measures, Hilbert spaces, C^*-algebras, von Neumann algebras, etc. We will allude to some of these notions in order to maintain a traditional picture, but it is not our intention here to develop this standard theory. We work algebraically wherever possible.

5.1   Combinatorial Hopf algebras

The use of Hopf algebras in combinatorics is one of their traditional roles in the mathematics. Nevertheless, we include this material here because of its great potential for generalisation to the noncommutative noncocommutative (so-called quantum) case, and because of its rather deep nature.

Useful background for this section is the notion of a Boolean algebra. This is a set, $\mathcal{A}$, with two associative and commutative operations, which we denote by $\wedge$ and $\vee$, and a map $\bar{\ }$. The further abstract properties are

$$a \wedge 0 = 0, \quad a \wedge \bar{a} = 0, \quad a \wedge (b \vee c) = (a \wedge b) \vee (a \wedge c),$$
$$a \vee 1 = 1, \quad a \vee \bar{a} = 1, \quad a \vee (b \wedge c) = (a \vee b) \wedge (a \vee c), \tag{5.1}$$

for all $a, b, c \in \mathcal{A}$. Note that the axioms have a symmetry, or duality, given by interchanging $\wedge$ with $\vee$ and 1 with 0. De Morgan's theorem is the assertion that this symmetry is implemented by $\bar{\ }$. So,

$$\overline{(a \wedge b)} = \bar{a} \vee \bar{b}, \quad \overline{(a \vee b)} = \bar{a} \wedge \bar{b}, \quad \bar{1} = 0, \quad \bar{0} = 1. \tag{5.2}$$

One can take as a model the set of subsets of some universal set 1 with $\wedge$ the intersection and $\vee$ the union of subsets, and 0 the empty set. In this case, $\bar{a}$ is the complement of a.

Boolean algebras are equivalent to commutative rings, for which all elements are projections. The product for the ring is $\wedge$ and the addition is

$$a + b = (a \vee b) \wedge \overline{(a \wedge b)}. \tag{5.3}$$

Such rings are called *Stonean*. Since $a + a = 0$ for all a, these rings can be thought of as algebras over the field $\mathbb{Z}_{/2}$. The complementation in this ring language is $\bar{a} = 1 + a$. In these terms, then, a Boolean algebra is nothing other than a commutative algebra over $\mathbb{Z}_{/2}$ with the property that $a^2 = a$ for all a in the algebra. For example, the idempotents in any algebra over $\mathbb{Z}_{/2}$ form a Boolean algebra. We will return to the algebraic picture at the end of the section. For now, one can just think of subsets of a set Ω.

Lemma 5.1.1 *Let $\mathcal{A}$ be a Boolean algebra and let $A = k\mathcal{A}$ be the vector space spanned by elements of $\mathcal{A}$ over our ground field. Then*

$$\Delta a = \sum_{\substack{a_1 \wedge a_2 = 0 \\ a_1 \vee a_2 = a}} a_1 \otimes a_2 = \sum_{\substack{a_1 a_2 = 0 \\ a_1 + a_2 = a}} a_1 \otimes a_2; \quad \epsilon a = \begin{cases} 1 & \text{if } a = \emptyset \\ 0 & \text{otherwise} \end{cases}$$

defines a coalgebra, the Boolean coalgebra. *When $\mathcal{A}$ is the set of subsets of a universal set Ω, the coproduct expresses all possible splittings of a subset into disjoint pieces.*

Proof: We verify coassociativity by computing

$$(\Delta \otimes \mathrm{id})\Delta a = \sum_{\substack{a_1 \wedge a_2 = 0 \\ a_1 \vee a_2 = a}} \sum_{\substack{a_{11} \wedge a_{12} = 0 \\ a_{11} \vee a_{12} = a_1}} a_{11} \otimes a_{12} \otimes a_2 = \sum_{\substack{a_{11} \wedge a_{12} = 0,\ a_{11} \wedge a_2 = 0 \\ a_{12} \wedge a_2 = 0,\ a_{11} \vee a_{12} \vee a_2}} a_{11} \otimes a_{12} \otimes a_2,$$

where the sum over a_1, subject to the condition $a_{11} \vee a_{12} = a_1$, fixes a_1 once a_{11}, a_{12} are fixed (we change the order of summation). The condition $a_1 \vee a_2 = a$ then becomes $a_{11} \vee a_{12} \vee a_2 = a$, and the condition $a_1 \wedge a_2 = 0$ becomes $(a_{11} \vee a_{12}) \wedge a_2 = 0$. Using the distributive properties of $\wedge, \vee$ the latter is $(a_{11} \wedge a_2) \vee (a_{12} \wedge a_2) = 0$. But this expresses the two conditions $a_{11} \wedge a_2 = 0$ and $a_{12} \wedge a_2 = 0$. The computation of $(\mathrm{id} \otimes \Delta)\Delta a$ is entirely analogous, giving a sum $\sum a_1 \otimes a_{21} \otimes a_{22}$ which is exactly the same after a relabelling of the notation for the fragments. The counit clearly has the required property for a coalgebra. ∎

This construction clearly works for any *distributive lattice* (i.e. a structure where there are $\wedge, \vee$ operations that are distributive as above), but note that we are not asserting that this coalgebra makes kA into a bialgebra or Hopf algebra. On the other hand, we can obtain an ordinary bialgebra if we enlarge our point of view as follows. In the concrete case, instead of working with subsets $a \subseteq \Omega$, we work with what can be called *bags* of elements of Ω. A bag of elements is a collection like $\{x, x, y, x, z\}$, where $x, y, z \in \Omega$. The order does not matter. The difference between a bag and a subset is that for a bag we are allowed multiple copies of an element. Mathematically, a bag is the quotient of a subset of some $\Omega \times \Omega \times \cdots \times \Omega$ modulo an equivalence relation that identifies the copies of Ω. The number of copies here is allowed to be variable. The model is a bag of marbles of various colours but otherwise indistinguishable. The possible colours correspond to the elements of Ω.

One can take the union $\vee$ of bags in the obvious way, namely pool together the contents of the bags. Thus, $\{x, x, y\} \vee \{x, z\} = \{x, x, x, y, z\}$. We denote by $\emptyset$ the empty bag. In the reverse direction, however, there may be many ways to partition a given bag into two other bags. The notion of what we mean by distinct partitions (in order to be able to count them) is intuitively obvious but a little complicated mathematically: we consider temporarily that the repeated elements in our bag are distinct (we label them). Then we distribute the elements into two bags and afterwards forget the labelling inside each bag. The various placements of our labelled elements are considered distinct modulo the degeneracy caused by this last step. The model to keep in mind is opening and sharing out the contents of a bag of marbles. So long as you keep your eye on them, any duplicated marbles are distinguishable. Watching them, you put some into one bag of the partition and some into the other, i.e. you distinguish which of the distinguished marbles goes into which bag of the partition. Once the marbles are in their respective bags, you can no longer distinguish them within each bag. Two partitions are distinct *iff* you can tell them apart after this total procedure. In mathematical terms we lift the bag to a subset of $\Omega \times \cdots \times \Omega$, partition this subset as in Lemma 5.1.1 and project down the resulting two subsets to obtain two new bags. Some of the set partitions are then identified in this process.

Proposition 5.1.2 *Let A be the vector space spanned by the bags of elements of a set Ω and define the product as the union of bags, extended linearly. Then*

$$\Delta a = \sum_{partitions} a_1 \otimes a_2, \quad \epsilon a = \begin{cases} 1 & if\ a = \emptyset \\ 0 & otherwise \end{cases}$$

is a bialgebra, the binomial bialgebra. *Here the sum is over the distinct partitions of the bag a into bags a_1, a_2. There is also an antipode $Sa = (-1)^{|a|}a$ which makes A into a Hopf algebra. Here $|a|$ is the number of elements of a.*

Proof: It is easy to see that Δ is coassociative. The computation is as in Lemma 5.1.1, and in fact may be reduced to it if we work with the bag lifted to a subset of $\Omega \times \cdots \times \Omega$. To see that we have an algebra homomorphism, we first combine two bags a, b to give $a \vee b$ and then consider its possible partitions. We have to show that we have the same elements of $A \otimes A$ with the same multiplicities as first partitioning a, b separately and then combining the bags of the two partitions pairwise. To do this, we can consider each type $i \in \Omega$ separately since the elements of each type are unioned and partitioned independently of the other types. If there are n elements of this type in bag a, and m in bag b, then for Δab we have to share out $n + m$ elements, which we can do with r in one bag of the result (and $n + m - r$ in the other) in $\binom{n+m}{r}$ ways. On the other hand, if we share out the elements of this type in a, b separately, with s, say, going into the first bag of the result from a and $r - s$ going into this bag from b (and the rest going into the other bag of the result), there are $\binom{n}{s}\binom{m}{r-s}$ ways to do this since the partitionings of a, b are independent. Then we have to sum over the possible s. These two ways of reaching the final state of r in the first bag of the result and $n + m - r$ in the second are the same, due to the identity

$$\sum_{s=0}^{\min(r,n)} \binom{n}{s}\binom{m}{r-s} = \binom{n+m}{r}.$$

It is easy to check the counit and antipode. The antipode, however, is not very natural from a set-theoretic point of view, and for this reason is not usually emphasised in this context. ∎

This bialgebra takes its name from the counting needed in the proof, which can also be expressed as the following proposition.

Proposition 5.1.3 *If $|\Omega| = n$, then the binomial bialgebra in Proposition 5.1.2 can be identified with the commutative Hopf algebra $k[x_1, \ldots, x_n]$ of polynomials in n variables and coproduct $\Delta x_i = x_i \otimes 1 + 1 \otimes x_i$ extended to products as an algebra homomorphism. This is the usual commutative (Hopf) algebra of coordinate functions on the n-dimensional plane. The counit and antipode are $\epsilon x_i = 0$ and $S x_i = -x_i$.*

Proof: The elements of Ω can be identified with integers $1, \ldots, n$ say. A bag is then a vector $a = (a_i)_{i=1}^{i=n}$ with entries $a_i \in \mathbb{N}$, the number of times that i is in the bag. The union of bags is just the addition of the corresponding vectors. Associated to a vector we have the polynomial $\prod x_i^{a_i}$, so that the multiplication of the polynomials corresponds to the union of the bags. On the other hand, the coproduct in $k[x_1, \ldots, x_n]$ is

$$\Delta x_1^{a_1} \cdots x_n^{a_n} = (x_1 \otimes 1 + 1 \otimes x_1)^{a_1} \cdots (x_n \otimes 1 + 1 \otimes x_n)^{a_n}$$

$$= \sum_{b_i \leq a_i} \binom{a_1}{b_1} \cdots \binom{a_n}{b_n} x^{b_1} \cdots x^{b_n} \otimes x^{a_1 - b_1} \cdots x^{a_n - b_n}.$$

The binomial coefficients $\binom{a_i}{b_i}$ here are just the number of ways of sharing out the original a_i of the elements of type i in our bag into b_i and $a_i - b_i$ elements among the two bags, counted in the way explained for partitions. Hence it corresponds precisely to the coproduct in Proposition 5.1.2. ∎

Thus, from an algebraic point of view, the binomial bialgebra associated to the collection of bags of elements of a set Ω is nothing more than the algebra of polynomial functions in $|\Omega|$ independent variables, with the corresponding binomial coefficients reflecting the multiplicity in the coproduct in Proposition 5.1.2. This abstract way of thinking about binomial coefficients allows us to generalise them and their properties to a variety of combinatorial situations. We will see some of these later. Note that this question of the number of ways to partition a bag has a precise application in physics for the counting of possible states in a statistical mechanical system. For simplicity, consider Ω to be a singleton set (only one element or type). Then the binomial bialgebra is the algebra of functions in one variable x. If we ask how many ways there are to partition a bag of m indistinguishable such particles into n bags, say, to contain, a_i particles in the ith bag, the required number is precisely the multiplicity of $x^{a_1} \otimes x^{a_2} \otimes \cdots \otimes x^{a_n}$ in $\Delta^{n-1} x^m$. This is because the question of partitioning bags is coassociative, as explained in the proof of Proposition 5.1.2, and hence can be given by iterating the coproduct. But, by Proposition 5.1.3, we can equally well compute the required iterated coproduct in terms of $k[x]$. One obtains, of course, the multinomial coefficients

$$\Delta^{n-1} x^m = \sum_{a_1 + \cdots + a_n = m} \frac{m!}{a_1! a_2! \cdots a_n!} x^{a_1} \otimes \cdots \otimes x^{a_n}. \tag{5.4}$$

Indistinguishable particles counted and partitioned in this way are called *bosons*. Later, in Chapter 10.1, we will introduce the superplane Hopf algebras corresponding to counting *fermions*. The same applies to the quantum planes in Chapter 4.5 associated to R-matrices, which also have a linear coproduct, as we will see in Chapter 10.2, albeit not forming a usual Hopf algebra but a braided one. They correspond to partitioning of particles with braid statistics. In this general context, the logarithm of the number of partitions leading to a given configuration defines the *entropy* of the configuration.

Also generalising the above binomial algebra is the free tensor algebra, which was already encountered in Example 1.1.2. To see this, we take a more formal point of view and think of Ω as an alphabet of symbols. Instead of bags, we work with *words* $a = xxyxz$, etc., where x, y, z are from our alphabet. The only difference between a bag and a word is that for a bag we do not care about the order of the symbols. These words, of course, have a product given by concatenation, and a unit given by the empty word (they form a monoid). To explain the coproduct we need the notion of an *unshuffle* of two words. This precisely generalises the notion of partition to our linguistic setting. Thus, given two words a, b we define a *shuffle* $a \vee b$ to be a permutation of the combined word ab in such a way that the internal order of each word is preserved under the permutation. Thus, if x comes before y in the spelling of a then x also comes somewhere to the left of y in the shuffle. Likewise for b. Note that all repeated symbols are regarded as distinct from each other. As mentioned in the preamble above, we interleave the symbols of a with the symbols of b as if shuffling together a cut pack of cards. Now an unshuffle of a is a triple (a_1, a_2, σ), where the a_i are words and σ is a shuffle of $a_1 a_2$ resulting in a. This is the same operation as the notion of partition in

Proposition 5.1.2 except that all symbols are completely distinct in this enumeration.

Proposition 5.1.4 *Let Ω be a set and let A be the linear span of the collection of words on it. We equip A with an algebra structure given by the concatenation product, extended linearly. Then*

$$\Delta a = \sum_{\text{unshuffles}} a_1 \otimes a_2, \qquad \epsilon a = 0, \quad \text{if } a \neq \emptyset, \quad \epsilon \emptyset = 1$$

defines a bialgebra, the unshuffle *bialgebra. It can be identified with the free noncommutative algebra $k\langle x_1, \ldots x_n \rangle$ in $n = |\Omega|$ generators and coproduct $\Delta x_i = x_i \otimes 1 + 1 \otimes x_i$, extended to products as a bialgebra. There is also an antipode $Sa = (-1)^{|a|} \bar{a}$ which makes A into a Hopf algebra. Here $\bar{a}$ is the reversed word and $|a|$ is its length.*

Proof: The arguments are the same as in the proof of Proposition 5.1.3. The direct proof that the stated coproduct forms a bialgebra can easily be seen in examples, but quite tedious to formulate generally. On the other hand, it is clear that the required combinatorics is the same as for the coproduct of $k\langle x_1, \ldots, x_n \rangle$, assuming this is a bialgebra. So it is enough to verify this. It is clearly coassociative on the generators and extends freely as an algebra homomorphism to products of the generators (there are no relations to check). Moreover, if an algebra homomorphism is coassociative on two elements, it is easy to see that it is also coassociative on their product. Indeed, we have already encountered more complicated examples of Hopf algebras with generators in earlier chapters. The present one is abstractly the tensor algebra, as in Example 1.1.2, on the vector space with basis dual to $\{x_i\}$. The antipode corresponds algebraically to $Sx_i = -x_i$ extended as an antialgebra homomorphism. ∎

Unlike the binomial bialgebra, this example is noncommutative. It is still cocommutative. Generalising in a different direction, we come now to some noncocommutative combinatorial bialgebras or Hopf algebras. We return to Lemma 5.1.1 and describe a general class of *incidence coalgebras* containing the Boolean coalgebra. The idea is to generalise the notion of a set to that of a *poset*. A poset is a set equipped with a partial ordering $\leq$. This is a relation in the set such that not every two elements a, b in the set need be comparable, but $a \leq b$ and $b \leq a$ iff $a = b$, and for three elements, $a \leq b$ and $b \leq c$ implies $a \leq c$. For a, b in a poset, the set of elements between them is denoted by the interval $[a, b]$. Given two posets α, β, we define their product as the Cartesian product $\alpha \times \beta = \{(a, b) | a \in \alpha, \ b \in \beta\}$ equipped with the ordering $(a, b) \leq (a', b')$ iff $a \leq a'$ and $b \leq b'$. For a typical example, if Ω is a finite set, the set of all its subsets forms a poset with ordering given by inclusion, so $a \leq b$ iff $a \subseteq b$.

Lemma 5.1.5 *Let α be a poset such that all its intervals are finite sets. The vector space spanned by the intervals in α forms a coalgebra with*

$$\Delta[a,b] = \sum_{a \leq c \leq b} [a,c] \otimes [c,b], \qquad \epsilon[a,b] = \begin{cases} 1 & \text{if } a = b \\ 0 & \text{otherwise.} \end{cases}$$

This is the incidence coalgebra *associated with a poset.*

Proof: Coassociativity is immediate since

$$(\Delta \otimes \text{id})\Delta[a,b] = \sum_{a \leq c \leq b} \sum_{a \leq d \leq c} [a,d] \otimes [d,c] \otimes [c,b]$$

is the same as the sum over the region $a \leq d \leq b$ and $d \leq c \leq b$, which corresponds to $(\text{id} \otimes \Delta)\Delta$. The counity property is clear. ∎

Further coalgebras are obtained by quotienting this by any equivalence relation compatible with the partial ordering. For example, we let α be the poset of all subsets of a set Ω, partially ordered by inclusion, and take as equivalence relation $[a,b] \sim [c,d]$ if $b - a = d - c$ as sets. The equivalence classes in this case are just the difference sets, as subsets of Ω. In this case the reduced incidence coalgebra recovers to the Boolean coalgebra of Lemma 5.1.1. This is why it is a generalisation. In general, for any poset we can define the equivalence relation $[a,b] \sim [c,d]$ *iff* they are isomorphic as sets. This is a standard construction for reducing the incidence coalgebra to something simpler, which is often a bialgebra or Hopf algebra.

For completeness, we now outline the most famous of these *incidence bialgebras*, the *Faà di Bruno bialgebra*. (There are plenty of other variants associated to graphs, trees, etc.) Namely, let α denote the set of all set-theoretic partitions of the natural numbers $\mathbb{N}$ (excluding zero) into one infinite block and finitely many finite blocks. Here a set-theoretic partition means in the obvious sense, as in Lemma 5.1.1 (to split the set into disjoint subsets), but note now that on α there is a natural partial ordering: we say that $a \leq b$ if a can be obtained as a further partition of the pieces of b (a refinement). So α becomes a poset. To compute its reduced incidence coalgebra, we need to describe $[a,b]$ at least up to isomorphism. To do this, take a coarse-grained point of view in which the various blocks in the partition a are considered as distinct elements of a set $B(a)$, the set of blocks of a. To say that $a \leq b$ means that each block of the partition b breaks up into sub-blocks taken from a. In other words, the partitions $\geq a$ can be identified with meta-partitions of $B(a)$. We use the term meta-partitions for usual set-theoretic partitions of $B(a)$ in this coarse-grained point of view where the elements refer to entire blocks of our original partition a. When b is described in this way, we let λ_k be the number of meta-blocks of size k. For example,

$$a = \{\{1,2\}, \{3,5\}, \{4,6,\ldots\}\} \equiv \{\alpha, \beta, \gamma\}$$
$$\leq b = \{\{1,2,3,5\}, \{4,6,\ldots\}\} \equiv \{\{\alpha, \beta\}, \{\gamma\}\}$$

has $(\lambda_1, \lambda_2, \ldots) = (1, 1, 0, \ldots)$. This vector (λ_k) is a kind of index for the refinement $a \leq b$. Now consider $a \leq c \leq b$. Since $a \leq c$, c is a partition of $B(a)$ as explained above. But since $c \leq b$, it must be the case that each meta-block of c is a partition of a meta-block of b. For each meta-block in b of k elements, the possibilities for the corresponding part of c are described by Π_k, the set of partitions on k elements. Hence, $[a, b] \cong \Pi_1^{\lambda_1} \times \Pi_2^{\lambda_2} \times \cdots$. We see that, up to isomorphism, $[a, b]$ as a set is fully determined by the vector (λ_k). Hence the reduced incidence coalgebra, as a linear space, can be identified with the polynomials $k[x_1, x_2, \ldots]$ where $[a, b]$ corresponds to $x_1^{\lambda_1} x_2^{\lambda_2} \cdots$. The coalgebra structure in Lemma 5.1.5 comes out in this description as $\epsilon(x_n) = \delta^1{}_n$, and

$$\Delta x_n = \sum_{\lambda_1 + 2\lambda_2 + \cdots + n\lambda_n = n} \frac{n!}{\lambda_1! \cdots \lambda_n! (1!)^{\lambda_1} \cdots (n!)^{\lambda_n}} x_1^{\lambda_1} \cdots x_n^{\lambda_n} \otimes x_{\lambda_1 + \lambda_2 + \cdots + \lambda_n}.$$

$$(5.5)$$

For the product, we take the usual polynomial algebra in the $\{x_i\}$. The result then turns out to be a bialgebra. Unlike the binomial bialgebra above, we see that the Faà di Bruno bialgebra is noncocommutative. The construction has still further generalisations to matroids, trees, graphs and other structures, some of which are noncommutative as well as non-cocommutative. What we have seen is that questions of enumerating partitions and other combinatorial problems can be neatly encoded in the coproduct of relatively simple bialgebras. The model is the partitioning of indistinguishable particles (encoded by polynomials $k[x]$ in one variable), but the principle applies much more generally.

Among the further examples of combinatorial bialgebras, the ones associated to trees and graphs are especially important as tools for handling branching processes in symbol manipulation. For example, they have been used to obtain algorithms for the solving of nonlinear differential equations via computer algebra. We describe here only the simplest example, associated to rooted trees. A *rooted tree* is a tree with a single top node (the root). This branches into subtrees (the children of the node), each of which can branch again etc. Note that one can imagine that these trees are embedded disjointly in some space, but the position of the nodes in the space is not significant, only which nodes are connected to which others. One can multiply two rooted trees a and b to give a formal sum of rooted trees as follows. Remove the root from a so that it becomes a collection of trees which were the children of the root of a. Then graft

these children onto the nodes of b in all possible ways. If there are r
children from a and $n + 1$ nodes in b, then there are $(n + 1)^r$ possibilities.
The sum of all these possible graftings is ab. A worked example is shown
in Fig. 5.1(a). The rooted tree consisting of only a root and no children is
the identity element. One can also define a splitting of a rooted tree into
two rooted trees as follows: remove the root leaving the various children
subtrees, which are all considered distinct; divide this set of children into
two disjoint subsets (a usual set-theoretic partition), and glue each subset
of children back by adding a root for each. If there are r children in the
original rooted tree, there are 2^r possible ordered pairs of rooted trees for
the splittings.

Proposition 5.1.6 *Let A be the vector space spanned by the set of rooted
trees. This forms an algebra with product defined by the multiplication of
rooted trees as explained. The coalgebra*

$$\Delta a = \sum_{\text{splittings}} a_1 \otimes a_2, \qquad \epsilon(a) = \begin{cases} 1 & \text{if } a = \text{root only} \\ 0 & \text{otherwise} \end{cases}$$

*makes A into a bialgebra. Here the sum is over all splittings of the rooted
tree a into rooted trees a_1, a_2, as explained. There is also an antipode.*

Proof: This is illustrated in Fig. 5.1, which is both an example and a
template for the general proof. The product ab of two rooted trees a, b is
in part (a). The children of a have been made a bit smaller so that they
can easily be recognised and replaced by general ones. They are grafted
in all possible ways onto the nodes of b. Clearly, the internal structure
of the children is irrelevant here: the same form applies. The coproducts
of a, b are shown in part (b). The coproduct of the product ab is in part
(c). On the other hand, if one multiplies out Δa and Δb from part (b)
using the rules as in part (a), one obtains the same 18 terms. This is
an example, but clearly the same result holds for any internal structure
of the children. This proves the bialgebra property for the product of
a rooted tree with two children with a rooted tree with one child. The
general case is similar, but rather complicated to write out in detail. ∎

Because this bialgebra is cocommutative (and in fact a Hopf algebra),
we know from Chapter 1.5 that it has the flavour of a group algebra or
enveloping algebra. In the present case, one knows from general considerations that it is precisely the universal enveloping algebra, as in Example 1.5.7, of the Lie algebra $\mathcal{L}$ of primitive elements. These are the
elements of A for which the coproduct is linear. In the present case, it
is easy to see that $\mathcal{L}$ is the vector space spanned by the trees with exactly one child. We conclude that the space of rooted trees with one

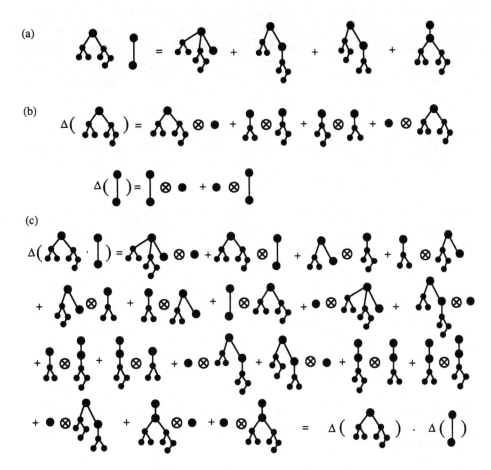

Fig. 5.1. Examples of (a) the product, (b) the coproduct and (c) the bialgebra property of rooted trees.

child forms a Lie algebra with product computed from the commutator of the multiplication above. However, since the allowed child is itself an arbitrary rooted tree, it is obvious that $\mathcal{L}$ is in one to one correspondence with A itself. So $A \cong U(A)$. These spaces are all graded by the number of nodes, and the dimension of the space spanned by the rooted trees of n nodes recovers the Cayley numbers. For ordered trees, one obtains Catalan numbers.

To conclude this section, we return to Boolean algebras of subsets and describe a different kind of bialgebra from the ones above; one that we shall generalise in later sections. We consider subsets of a set Ω which is a group or monoid. One bialgebra structure is obtained at once by noting that the set $\mathcal{A}(\Omega)$ of subsets inherits a monoid structure as well. The product of two subsets is their image in Ω under the product map. The unit is $\{e\}$. Then, by Example 1.5.2, we have a bialgebra on the vector space spanned by $\mathcal{A}$ over any field. The coproduct of a subset a is a linear combination of the subsets which multiply to a. More intrinsic from the point of view of set theory, however, is that the Boolean algebra $\mathcal{A}$ itself becomes a bialgebra or Hopf algebra.

Proposition 5.1.7 *Let $\mathcal{A}(\Omega)$ be the Boolean algebra of subsets of a finite group Ω. Then $\mathcal{A}$ is a Hopf algebra over $\mathbb{Z}/2$, with coproduct and counit*

$$\Delta a = \{(x,y)| \ xy \in a\} \subseteq \Omega \times \Omega, \quad \epsilon a = \begin{cases} 1 & \text{iff } e \in a \\ 0 & \text{otherwise,} \end{cases}$$

where e is the group identity element, and we identify $\mathcal{A}(\Omega) \otimes_{\mathbb{Z}/2} \mathcal{A}(\Omega) = \mathcal{A}(\Omega \times \Omega)$. The dual Hopf algebra is again $\mathcal{A}(\Omega)$, with elementwise product of subsets and diagonal coproduct.

Proof: This is really a case of Example 1.5.2 for functions on a group, in disguise. The reason is that subsets of a set Ω can be thought of just as easily as $\mathbb{Z}/2$-valued functions on Ω. The function corresponding to a subset a is the characteristic function

$$\chi_a(x) = \begin{cases} 1 & \text{if } x \in a \\ 0 & \text{otherwise.} \end{cases} \tag{5.6}$$

The usual pointwise product of these is exactly the intersection of the corresponding subsets, and their usual pointwise sum mod $\mathbb{Z}/2$ is the exclusive-or of the corresponding subsets,

$$\chi_a \chi_b = \chi_{a \wedge b}, \quad \chi_a + \chi_b = \chi_{a+b}. \tag{5.7}$$

The coproduct of χ_a from Example 1.5.2 is the function in $\Omega \times \Omega$ which is 1 precisely on the set of elements whose product is in a, and zero otherwise; i.e. the characteristic function of the set Δa as stated. The counit is equally clear. The dual Hopf algebra from Example 1.5.3 is more obvious: a $\mathbb{Z}/2$-linear combination of elements of Ω just means a subset. The product of subsets is their image under the product map. The coproduct is $\Delta a = \{(x,x)| \ x \in a\}$. ∎

In what follows we shall consider logic as the simplest 'theory of physics', with classical observables and quantum observables as generalisations. In classical mechanics we fix a field k and consider functions $k(\Omega)$. We can consider the Boolean algebra of characteristic functions $\mathcal{A}(\Omega)$ as contained in classical mechanics by specialising to the case over $\mathbb{Z}/2$. We can also consider the characteristic functions as embedded in a general $k(\Omega)$ as the set of functions with values $0, 1 \in k$. Their product corresponds to intersection of subsets and $\chi_a + \chi_b - \chi_a \chi_b$ to the union $a \vee b$, which provides the generalisation from questions of logic to questions of classical mechanics. The general point of view which underlies this is the following. We consider some structure that describes the possible states or configurations of the system – for example, the set Ω. We also have some dual structure describing the possible questions that can be asked about the system, the 'observables'. Dual means that there is a pairing or evaluation that evaluates an observable on a state to give a number, which we consider to be the answer to the corresponding question. For example, the observables could be the algebra of functions $k(\Omega)$, and the evaluation $f(x)$ considered to be the value of the observable f in a configuration $x \in \Omega$. Or the observables could be logical questions of membership of a subset, with value true or false. The characteristic functions are the logical observables viewed as classical ones. Their values are $0, 1$, and are considered to be the truth or falsity of a proposition as regards a particular configuration x. Other functions can have other ranges for their value, and together they make up the classical observables for the system.

With this algebraic way of thinking about physics in mind, we can ask: when is the algebra of observables a Hopf algebra, and what does it mean logically? Proposition 5.1.7 provides a typical example for the case when the algebra of observables is a Boolean algebra $\mathcal{A}$. The product is induced by the intersection of subsets, as explained in (5.7), i.e. the logical conjunction of propositions. What the coproduct $\Delta \mathcal{A} \to \mathcal{A} \otimes \mathcal{A}$ does is to supplement this deductive logic by 'inductive' laws, with Δa referring, in this example, to a proposition about pairs of configurations which would imply the proposition a after multiplying according to some given group law. One could view any Boolean algebra as an algebra over $\mathbb{Z}_{/2}$ and ask when it is, similarly, a bialgebra or Hopf algebra. Moreover, in our case, the logical Hopf algebra structure generalises to our more familiar Hopf algebra structure for $k(\Omega)$ from Example 1.5.2. Then the notion of 'induction' or 'cologic' generalises to the classical notion of a random walk or Markov process on Ω. This is the starting point for the next section. Later on, we will come to the quantum versions of these same considerations.

5.2 Classical random walks using Hopf algebras

In this section, we proceed from logic and pure combinatorics to applications of Hopf algebras in probability theory. In probability theory we work with a little more than subsets of a set. Namely, we have the notion of volume or area of a subset, and, moreover, we work with real-valued random variables rather than the characteristic functions corresponding to Boolean algebras, as in the preceding section. From a purely algebraic point of view, one can work over a general field k, but for the usual statistical interpretation we need to work over $\mathbb{R}$ and $\mathbb{C}$.

A *probability space* is a space Ω equipped with some density function $\rho(x)$ governing the probability density of the configuration being found in the region of each point x of Ω. More precisely, we are given here a measure $\mu(\mathrm{d}x) = \rho(x)\mathrm{d}x$ which defines the probability for the configuration to be found in the region $\mathrm{d}x$ (or in any other subset by integration over it).

In an algebraic language, we work with an algebra A of functions on Ω rather than with Ω itself. The elements of A are called *random variables*, and integration over Ω against any ρ gives the *expectation value* of random variable a. The density ρ in this context determines then a *state* of the system, in the sense that it assigns to each observable a in A a number

$$\phi(a) = \int \mathrm{d}x\, \rho(x) a(x), \qquad (5.8)$$

the *expectation value* of observable or random variable a in state ϕ.

In the end, then, we can just work directly with an algebra of observables A and a collection of states ϕ which directly assign to an observable a a number $\phi(a)$, which we call the expectation of a in state ϕ. These expectation values contain all the information that we actually need and make the notions of probability space and density redundant. If we do want to know the probability for the configuration to be in a subset α, we can just take the expectation value of the observable χ_α. This connects the algebraic picture with the usual probability density or measure picture. On the other hand, we can just work directly with observables and states and forget about these other concepts. This point of view will be especially useful in the quantum picture in later sections.

In the present section, we are going to demonstrate the utility of these algebraic ideas for ordinary probability theory by showing how Hopf algebras can be used to describe random walks and associated Markov processes. These need a group structure on the probability space, and their content is to make the kind of 'inference' mentioned at the end of Section 5.1. Even in simple and well-known cases, the use of Hopf algebras encodes the required computation in a very direct way. The reason for this is quite fundamental, as we have seen above.

5.2.1 Brownian motion on the real line

To be concrete, we develop our ideas first with the real line $\mathbb{R}$ as our probability space. A probability density function for us is then a pointwise positive function $\rho \in L^1(\mathbb{R})$ such that $\int \mathrm{d}x\rho(x) = 1$. We also allow atomic measures (δ-functions). We can use such a ρ to define a random walk as follows: at $x_0 = 0$, $x_{i+1} = x_i + X$, where X is chosen randomly with probability distribution ρ. The routine question to ask is: what is the probability distribution after n steps, i.e. of x_n? Clearly it is given by a probability density ρ^n (say) defined by

$$\rho^n(x) = \int_{y_1+\cdots+y_n=x} dy_1 \cdots dy_n \rho(y_1) \cdots \rho(y_n). \tag{5.9}$$

Note that in asking after the system at various steps, we are always asking the same mathematical question, namely the distribution of a random variable on the real line. Thus we take the point of view that the question, which we denote by X, is not changing with each step. Rather, it is the probability distribution of X which is changing with each step. This X is the abstract position observable of the random walk, i.e. 'where the particle is' (statistically speaking). Its expectation value and moments, etc. are changing with each time step. The probability distribution (5.9) immediately gives

$$\langle 1 \rangle_{\rho^n} = \langle 1 \rangle_\rho^n = 1, \quad \langle X \rangle_{\rho^n} = n\langle X \rangle_\rho,$$
$$\langle X^2 \rangle_{\rho^n} = n(\langle X^2 \rangle_\rho - \langle X \rangle_\rho^2) + n^2 \langle X \rangle_\rho^2, \tag{5.10}$$

where $\langle\ \rangle$ denotes expectation value in the state as marked. From this we see that the mean position is increasing with each step, as would be expected. Also, if we define $Y^{(n)} = \frac{1}{n}X$, then $\langle Y^{(n)} \rangle_{\rho^n} = \langle X \rangle_\rho$ is independent of n, while its variance $\langle Y^{(n)2} \rangle_{\rho^n} - \langle Y^{(n)} \rangle_{\rho^n}^2 = \frac{1}{n}(\langle X^2 \rangle_\rho - \langle X \rangle_\rho^2)$ decreases. The higher variances similarly decrease, so that the rescaled variables $Y^{(n)}$ become more and more sharply peaked. This is the basis of the central limit theorem.

This elementary computation can be done by means of Hopf algebras, as follows. We regard the abstract question X as approximated by an element of $L^\infty(\mathbb{R})$ (the Hopf–von Neumann algebra of bounded functions on the real line). Here we consider X as the coordinate function $X(x) = x$ and approximate it to lie in $L^\infty(\mathbb{R})$. Now, a probability distribution assigns to any power X^n, or more precisely any bounded function $a(X)$, an expectation value $\langle a(X) \rangle_\rho = \int dx \rho(x) a(x)$. This gives a positive linear functional ϕ on $L^\infty(\mathbb{R})$, corresponding to ρ via $\phi(a) = \langle a(X) \rangle_\rho$. (Positive means $\phi(a^*a) \geq 0$ for all a, and we assume also the condition $\phi(a) = \phi(a^*)$; these correspond to ρ real and positive.) The normalisation corresponds to $\phi(1) = 1$. Thus we can work directly with the $L^\infty(\mathbb{R})$ in place of random variables, and ϕ a (normal, unital) state on $L^\infty(\mathbb{R})$ in place of ρ. This is a familiar point of view in the context of classical mechanics (and quantum mechanics), where $L^\infty(\mathbb{R})$ is the algebra of observables of the system.

The coproduct structure on $L^\infty(\mathbb{R})$ is $(\Delta a)(x,y) = a(x+y)$, where $L^\infty(\mathbb{R}) \otimes L^\infty(\mathbb{R}) = L^\infty(\mathbb{R} \times \mathbb{R})$. It expresses the group law on $\mathbb{R}$ in algebraic terms, as we have seen in Example 1.5.2. For an approximation of the coordinate function, we will have approximately the additive comultiplication $(\Delta X)(x,y) = X(x+y) = x + y = (X \otimes 1 + 1 \otimes X)(x,y)$, or abstractly

$$\Delta X = X \otimes 1 + 1 \otimes X, \quad \epsilon X = 0. \tag{5.11}$$

Before going on to questions involving moments, we formally avoid our approximation problems by working explicitly with the algebra $\mathbb{C}[X]$ of polynomials in X, as an *algebraic model* of $L^\infty(\mathbb{R})$. We have already encountered this point of view in Chapter 4.5. Thus we work with the Hopf $*$-algebra generated by X with $X^* = X$ and the coproduct (5.11). There is also an antipode $SX = -X$. The formal axioms are in Definition 1.7.5.

The group law played a key role in (5.9), and now Δ plays the corresponding role in defining a new state ϕ^n. Indeed, the comultiplication on any Hopf algebra defines a multiplication in its dual, and it is this multiplication, *the convolution product*, that we need. Explicitly,

$$(\phi\psi)(a) = (\phi \otimes \psi)(\Delta a), \quad \text{e.g.} \quad \phi^n(a) = (\phi^{\otimes n})(\Delta^{n-1}a), \tag{5.12}$$

where Δ^{n-1} denotes Δ applied $n-1$ times. The abstract axioms of a Hopf $*$-algebra ensure that if ϕ is positive then ϕ^n is also positive, etc., i.e. it is also a state. Associativity of this convolution algebra is ensured by coassociativity of Δ, while ϵ is the identity state in the convolution algebra.

In these algebraic terms, the computation of (5.10) looks like

$$\langle 1 \rangle_{\rho^n} = \phi^n(1) = (\phi^{\otimes n})\Delta^{n-1}1 = (\phi \otimes \cdots \otimes \phi)(1 \otimes \cdots \otimes 1) = \phi(1)^n = 1,$$
$$\langle X \rangle_{\rho^n} = (\phi^{\otimes n})\Delta^{n-1}X = (\phi \otimes \cdots \otimes \phi)(X \otimes \cdots \otimes 1 + \cdots + 1 \otimes \cdots \otimes X)$$
$$= n\phi(X) = n\langle X \rangle_\rho,$$
$$\langle X^m \rangle_{\rho^n} = (\phi^{\otimes n})\Delta^{n-1}X^m$$

$$= (\phi \otimes \cdots \otimes \phi)\left(\sum_{i_1+\cdots+i_n=m} \binom{m}{i_1 \cdots i_n} X^{i_1} \otimes \cdots \otimes X^{i_n}\right)$$

$$= \sum \binom{m}{i_1 \cdots i_n} \phi(X^{i_1}) \cdots \phi(X^{i_n})$$

$$= \sum \binom{m}{i_1 \cdots i_n} \langle X^{i_1} \rangle_\rho \cdots \langle X^{i_n} \rangle_\rho.$$

Thus, we reproduce the results above in this new language. The computation in this form uses nothing other than the two equations (5.11) and (5.12). Not only does it bypass unpleasant convolution integrals as in (5.9), but it is conceptually rather cleaner as well. The conceptual picture is that the expectation value of X after n steps is simply the expectation value of $\Delta^{n-1}X$ in the tensor product system $\mathbb{C}[X]^{\otimes n}, \phi^{\otimes n}$ (or $L^{\infty}(\mathbb{R})^{\otimes n}, \phi^{\otimes n}$ in the topological setting). The tensor product system is the system for n independent random variables, each with the same distribution ϕ, and the iterated comultiplication embeds the algebra of observables for our one particle into this system. If we denote these co-ordinate functions that generate our n-fold tensor product by

$$X_i = 1 \otimes \cdots \otimes X \otimes \cdots \otimes 1 \qquad (5.13)$$

(embedded in the ith position), then the random variable X of our one particle embeds as $X_1 + X_2 + \cdots + X_n$. Regarding Δ^{n-1} as understood, we can simply write $X = X_1 + \cdots + X_n$. Here the X_i are n independent random variables.

To demonstrate the power of this formalism further, let us compute a specific example. In terms of ρ, we take an atomic measure with $\rho = p\delta_b + (1-p)\delta_{-b}$, i.e. peaked as δ-functions at $b, -b \in \mathbb{R}$. In terms of ϕ, we have

$$\phi(a) = (p\phi_b + (1-p)\phi_{-b})(a) = pa(b) + (1-p)a(-b), \qquad (5.14)$$

where $\phi_b(a) = a(b)$ is given by evaluation at b. We can also introduce the linear map $D(a) = a'(0)$. It is not a normalised state, but we can still view it as a linear functional and make the convolution product, etc., as above, i.e. we include it in the convolution algebra (5.12). Then the form of Δ in (5.11) gives

$$D^n(a) = (D \otimes \cdots \otimes D)(\Delta^{n-1}a) = \frac{\partial}{\partial x_1}\Big|_0 \cdots \frac{\partial}{\partial x_n}\Big|_0 a(x_1 + \cdots + x_n) = a^{(n)}(0).$$

Hence, in this convolution algebra, we have the expansion

$$\phi_b = \epsilon + bD + \frac{b^2}{2!}D^2 + \cdots \qquad (5.15)$$

or Taylor's theorem. This is a finite series when acting on polynomials. There are various analytic details for more general functions, which we suppress. We then compute the system after n steps, and its limit, as described by the states

$$\phi^n = \left(\epsilon + 2b\left(p - \frac{1}{2}\right)D + \frac{b^2}{2!}D^2 + \cdots\right)^n = \left(\epsilon + \frac{ct}{n}D + \frac{t\beta}{n}D^2 + \cdots\right)^n;$$

$$\phi^{\infty} = e^{t\beta D^2 + ctD}, \qquad (5.16)$$

where t, c, β are defined by $t = n\delta$, $\frac{b^2}{2} = \frac{t\beta}{n} = \beta\delta$ and $2b(p - \frac{1}{2}) = \frac{tc}{n} = \delta c$. We send $b \to 0$ and $n \to \infty$ (or $\delta \to 0$) with t, c, β fixed.

This limiting state describes our random walk after an infinite number of steps, with the step viewed as a step in time of size δ, which tends to zero, i.e. it is the continuous limit of the random walk. We can still evaluate our observables in this limit,

$$\phi^\infty(a) = (e^{t\beta D^2 + ctD})(a) = (e^{t\beta \frac{d}{dx}^2 + ct \frac{d}{dx}} a)(0), \qquad (5.17)$$

and we can also compute the distribution ρ^∞ corresponding to ϕ^∞. Formally, if ϕ is any state, we can recover the corresponding density by $\rho(x) = \phi(\delta_x)$ (the δ-function at x, which of course has to be approximated to lie in $L^\infty(\mathbb{R})$). We have

$$\rho^\infty(y) = \phi^\infty(\delta_y) = (e^{t\beta \frac{d}{dx}^2 + ct \frac{d}{dx}} \delta_y)|_{x=0} = (4\pi\beta t)^{-\frac{1}{2}} e^{-\frac{(y-ct)^2}{4\beta t}}, \qquad (5.18)$$

where the right hand side is the unique solution $G(y,t)$ of the diffusion equation

$$\frac{\partial}{\partial t}G(y,t) = \beta \frac{\partial^2}{\partial y^2}G(y,t) - c\frac{\partial}{\partial y}G(y,t). \qquad (5.19)$$

Another way to see that this must coincide with ρ^∞ is that the latter is characterised by $\int \rho^\infty(x)a(x) = e^{t\beta D^2 + ctD}(a)$ for all a: differentiating this $\frac{\partial}{\partial t}$, we have

$$\int \left(\frac{\partial}{\partial t}\rho^\infty\right)(x)a(x)\mathrm{d}x = ((\beta D^2 + cD)e^{t\beta D^2 + ctD})(a)$$

$$= \int \rho^\infty(x)\left(\beta\frac{\mathrm{d}^2}{\mathrm{d}x^2} + c\frac{\mathrm{d}}{\mathrm{d}x}\right)a(x)\mathrm{d}x$$

$$= \int \left(\left(\beta\frac{\mathrm{d}^2}{\mathrm{d}x^2} - c\frac{\mathrm{d}}{\mathrm{d}x}\right)\rho^\infty(x)\right)a(x)\mathrm{d}x,$$

where $(D \otimes \mathrm{id})\Delta a(X) = a'(X)$. Our derivation of (5.18) can be compared with more standard derivations based on Stirling's formula etc. In our case, we used only the well-known limit $(1 + \frac{Z}{n})^n \to e^Z$ in deriving the limiting state (5.16).

We have given these examples on $\mathbb{R}$. Clearly, we can have random walks on any probability space Ω which forms a group. Then $L^\infty(\Omega)$ becomes a Hopf–von Neumann algebra along the lines of Example 1.5.2, and we can work with this or a Hopf $*$-algebra as an algebraic model of it. The linear functional ϕ, corresponding to a measure on Ω, provides the probability distribution for a step of the random walk. We see that this can be iterated, and for suitable states can have a continuous limiting diffusion-type process.

5.2.2 *Markov processes*

Closely related to random walks are Markov processes. In the simplest setting, a Markov process describes a system with various possible classical states $i, j \in \Omega$. One can think of these as possible outcomes E of repeated trials in an experiment. We suppose that we know only the probabilities (p_i) for these possible outcomes, which we think of as a covector (row vector). This covector (p_i) is the *probability state* of the system, and we suppose, for this interpretation, that $p_i \geq 0$ and $\sum_i p_i = 1$. The system is now considered to undergo some kind of procedure such that the probabilities for the resulting state depend only linearly on the previous probabilities (p_i). From a statistical point of view, the assumption is that the probabilities for the outcomes of the experiment can be decomposed into conditional probabilities as

$$\text{Prob}(E = i \text{ at time } n + 1)$$
$$= \sum_j \text{Prob}(E = j \text{ at } n)\text{Prob}(E = i \text{ at the next step}|E = j).$$

This form of evolution or type of dependence on previous experiments is the defining property of a *Markov process*. We see that it consists of multiplication of the probability state covector by a matrix $T^i{}_j$ of conditional probabilities. The matrix T is characterised by the properties that (i) all its entries are ≥ 0 and (ii) each of its its rows adds up to 1. These ensure that pT is also a valid probability state. Matrices of this form are called *stochastic matrices*.

This language of Markov processes provides a second way of thinking about random walks. We suppose that our set Ω has its elements enumerated as $i = 0, 1, \ldots, m - 1$, say. So, $\Omega \subset \mathbb{N}$. A Markov process has independent increments if

$$T^i{}_j = T_{j-i} \tag{5.20}$$

for some m-tuple (T_i). These need to obey $T_i \geq 0$ and $\sum T_i = 1$, i.e. they are like the probability states themselves. The processes corresponding to the random walks in the preceding subsection are of this type.

Explicitly, let (ρ_i) be the probability distribution for the step of a random walk on Ω. The role of x above is played now by i. The probability state after n steps is the convolution as in (5.9). Thus, if $(p_i^{(n)})$ is the state after some number of steps, the state after one more step is

$$p_i^{(n+1)} = \sum_{j+k=i} p_j^{(n)} \rho_k = \sum_j p_j^{(n)} T^j{}_i, \qquad T^j{}_i = \rho_{i-j}, \tag{5.21}$$

which is of the form of a Markov process with independent increments. There are various boundary conditions one may impose here. If we think of Ω as a line, then we take $\rho_{i-j} = 0$ when $i - j$ is outside the range $0, \ldots, m - 1$. In this case, we may as well work with $\Omega = \mathbb{N}$ with p_i and ρ_i vanishing for $i \geq m$. Alternatively, we can shift our notation and think of $\Omega = \mathbb{Z}$. The corresponding random walk, in this case, is a discrete version of the random walk on $\mathbb{R}$ as above. Another alternative is to take periodic boundary conditions. Thus, we consider $i - j$ in (5.21) to be modulo m. In this case, the system can just as easily step round from $m - 1$ to 0 and $\Omega = \mathbb{Z}/m$, the cyclic group of order m. It is a discrete version of a random walk on the circle.

The corresponding algebraic construction is as follows. Let $A = \mathbb{C}(\Omega)$ be the Hopf algebra of functions on the group Ω, as in Example 1.5.2. For simplicity, we suppose that we are in the finite case such, as $\Omega = \mathbb{Z}/m$. (For the infinite case we would have to introduce some form of topology or a suitable algebraic model.) Then let δ_i be the δ-functions in A. They form a basis, and their coproduct is $\Delta \delta_i = \sum_{j+k=i} \delta_j \otimes \delta_k$. A state is a positive linear functional $A \to \mathbb{C}$, so these are determined by the row vectors $\phi(\delta_i) = \rho_i$. Their convolution according to (5.12) gives the states after $n + 1$ steps, just as above.

Also, we know in our algebraic language that the passage from any state ψ, corresponding to the probabilities $(p_i^{(n)})$, to the state for time $n + 1$ is nothing other than multiplication from the right by ϕ in the convolution algebra (5.12). We can consider this as an operator on the set of states, the *passive Markov transition operator* R_ϕ for the system:

$$R_\phi : A^* \to A^*, \qquad R_\phi(\psi) = (\psi \otimes \phi)\Delta. \tag{5.22}$$

This is just (5.21) in an abstract form and reproduces it when we bear in mind the form of Δ on the basis functions δ_i. It also defines a random walk and an associated Markov process for any group structure on Ω, not necessarily Abelian. It is expressed conveniently via the coproduct Δ.

These elementary computations demonstrate how Hopf algebras can be used in the theory of random walks and Markov processes. The reason is quite fundamental because the coproduct Δ encodes 'all possible ways of arriving' at a given outcome, as we have already seen in the combinatorial bialgebras and Hopf algebras of Section 5.1. We can also take continuous limits to obtain diffusion processes. Finally, these techniques conveniently convert from the random walk description to the description as a Markov process. We shall now formalise this use of Hopf algebras, and give some more examples. Note that, in the above, we did not need the probability distributions for all the steps of the walk to be the same; they could vary with the time n.

Definition 5.2.1 *A stationary random walk on a Hopf algebra A is a pair (A, ϕ), where A is a Hopf $*$-algebra or $*$-bialgebra and $\phi : A \to \mathbb{C}$ is a* state. *This is a normalised positive complex-linear functional, i.e. a linear functional such that $\phi(a^*a) \geq 0$ and $\overline{\phi(a)} = \phi(a^*)$ for all a, and $\phi(1) = 1$. The system, after n steps, is in state ϕ^n, where the product is in the convolution algebra. The right multiplication R_ϕ is the associated passive Markov transition operator. More generally, a nonstationary random walk is $(A, \phi_1, \phi_2, \ldots)$, where (ϕ_i) is a sequence of states describing the probability distribution for each step of the walk. The state of the system after n steps is now $\phi_1 \phi_2 \cdots \phi_n$.*

The model is $A = \mathbb{C}(\Omega)$, the algebra of suitable functions on a group or monoid Ω. The positivity requirement corresponds, in this case, to the state being given by integration against a normalised real positive function on Ω, the corresponding probability density function. Alternatively, we can use Exercise 1.5.5 to identify the states with weighted averages of the points in Ω themselves.

Example 5.2.2 *Let Ω be a finite group and let $A = \mathbb{C}(\Omega)$, as in Example 1.5.2. A state on A is an element $\phi \in \mathbb{C}\Omega$ of the form*

$$\phi = \sum_{u \in \Omega} \rho_u u, \qquad \rho_u \geq 0, \qquad \sum_{u \in \Omega} \rho_u = 1,$$

with value on A given by evaluation. The convolution product (5.12) in these terms is just the group product in Ω extended linearly. If $\sum_u p_u^{(n)} u$ is the state at time n, then the state at time $n+1$ for a random walk with step ϕ is

$$p_u^{(n+1)} = \sum_{v \in \Omega} p_v^{(n)} T^v{}_u, \qquad T^v{}_u = \rho_{v^{-1}u}.$$

Proof: We use Exercise 1.5.5 from Chapter 1 to identify A^* with the group algebra of Ω, with pairing given by evaluation. A basis of A^* is provided by the elements of Ω, and the product of basis elements is the group law of Ω. A general state must be a linear combination of these, subject to positivity and normalisation requirements. The requirement that $\phi(a^*) = \overline{\phi(a)}$ for the δ-functions δ_u, which form a basis of A, tells us that the coefficients ρ_u in ϕ must be real, while $\phi(a^*a) \geq 0$ for the δ-functions means that they must be positive. Finally, the normalisation $\phi(1) = 1$ tells us that they must sum to unity; i.e. ϕ is a *convex linear combination* of group elements. Conversely, it is clear that every convex linear combination gives a normalised state. If we take a Markov process point of view, then the operator of right multiplication on the states in this basis of group elements is $R_\phi(\sum_v p_v^{(n)} v) = (\sum_v p_v^{(n)} v)(\sum_u \rho_u u) = \sum_{u,v} p^{(n)} \rho_{v^{-1}u} vv^{-1}u = \sum_u (\sum_v p_v^{(n)} T^v{}_u) u$, as required. In effect, we are repeating the convolution algebra description of $\mathbb{C}\Omega$ in Example 1.5.4. ∎

Clearly this is a generalisation of (5.21) to the case where Ω is a general, possibly non-Abelian, group. On the other hand, we see that it is extremely easy to compute the evolution of the random walk using our algebraic method, even in this general case. We just multiply formal linear combinations in the group.

Example 5.2.3 *Let* $\Omega = \mathbb{Z}_{/3}$, *which we write multiplicatively as* $\Omega = \{e, u, u^2\}$. *The group algebra is* $\mathbb{C}[u]/(u^3 - 1)$. *A state is a convex linear combination of the elements of* Ω *as basis, i.e. a polynomial in* u *with real positive coefficients adding up to unity. We consider the random walk with probability* $\frac{1}{2}$ *to step up one or down in* $\mathbb{Z}_{/3}$, *i.e. probability distribution* $\rho = (0, \frac{1}{2}, \frac{1}{2})$. *The corresponding state is*

$$\phi = \frac{u + u^2}{2}.$$

Then the state after n steps is

$$\phi^n = \left(\frac{u + u^2}{2}\right)^n = \frac{1}{3}\left(1 - \left(-\frac{1}{2}\right)^{n-1}\right)e + \frac{1}{3}\left(1 - \left(-\frac{1}{2}\right)^n\right)(u + u^2).$$

Proof: According to Example 5.2.2, we just have to compute $\phi^n = (u + u^2)^n 2^{-n}$ for the state after n steps. To do this, let $c = \frac{u+u^2}{2}$. Then

$$c^2 = 4^{-1}(u + u^2)^2 = \frac{u^2 + u + 2}{4} = 2^{-1}(c + 1),$$

since $u^3 = 1$ in our group algebra. Then, on writing $c^n = a_n + b_n c$, say, we have $c^{n+1} = a_n c + b_n c^2 = \frac{1}{2}b_n + (a_n + \frac{1}{2}b_n)c$. This gives the recurrence relations $b_{n+1} = a_n + \frac{1}{2}b_n$ and $a_{n+1} = \frac{1}{2}b_n$ with $a_1 = 0, b_1 = 1$, which has the solution $a_n + b_n = 1$ and $a_n - \frac{1}{2}b_n = (-\frac{1}{2})^n$, yielding ϕ^n as shown. Note that we are performing an elementary calculation from trigonometry here, namely computing $\cos^n(\frac{2\pi}{3})$ in terms of $e^{i\frac{2\pi}{3}}$. The same applies to the case of general $\mathbb{Z}_{/n}$. Note also that, in this random walk, as $n \to \infty$, the state ϕ^n converges to the integral on the Hopf algebra $\int = \frac{1}{3}(e + u + u^2)$, as in Example 1.7.2. The evaluation of this on any observable $a \in C(\Omega)$ is

$$\langle a \rangle_{\phi\infty} = \int a = \frac{1}{3}\left(a(e) + a(u) + a(u^2)\right).$$

Of course, one can do these calculations using Fourier transforms, as well as in the form of a Markov process by writing out the Markov transition matrix corresponding to $(\rho_i) = (0, \frac{1}{2}, \frac{1}{2})$ in the basis $\{e, u, u^2\}$. This is

$$T = \frac{1}{2}\begin{pmatrix} 0 & 1 & 1 \\ 1 & 0 & 1 \\ 1 & 1 & 0 \end{pmatrix},$$

and writing out a recurrence relation for the powers T^n applied to the initial state $p^{(0)} = (1, 0, 0)$ is equivalent to the above algebraic computation. ∎

Example 5.2.4 *Let $\Omega = S_3$ be the permutation group on three elements. This has generators u, v and relations*

$$u^2 = e, \quad v^2 = e, \quad uvu = vuv.$$

We take $\{e, u, v, uv, vu, uvu\}$ as our basis of $\mathbb{C}\Omega$. A state on $A = \mathbb{C}(\Omega)$ is a convex linear combination of these. We compute the random walk associated to the state,

$$\phi = \frac{u + v}{2}, \quad i.e. \quad \rho = \left(0, \frac{1}{2}, \frac{1}{2}, 0, 0, 0\right).$$

The states after $2n$ and $2n + 1$ steps are

$$\phi^{2n} = \frac{1}{3}\left(1 - \left(-\frac{1}{2}\right)^{2n-1}\right) e + \frac{1}{3}\left(1 - \left(-\frac{1}{2}\right)^{2n}\right)(uv + vu),$$

$$\phi^{2n+1} = \frac{1}{3}\left(1 - \left(-\frac{1}{2}\right)^{2n+1}\right)(u + v) + \frac{1}{3}\left(1 - \left(-\frac{1}{2}\right)^{2n}\right) uvu.$$

Proof: We use the algebraic technique above. Note that u, v do not commute. Thus,

$$\left(\frac{u + v}{2}\right)^2 = \frac{u^2 + v^2 + uv + vu}{4} = \frac{1}{2}\left(1 + \frac{uv + vu}{2}\right) = \left(\frac{uv + vu}{2}\right)^2,$$

where $vuuv = e = uvvu$, so $vu = (uv)^{-1}$ and $uvuv = vuvv = vu$, so $(uv)^3 = e$. Hence, the even steps $\phi^{2n} = ((u + v)/2)^{2n}$ of the random walk are exactly the same as those of the $\mathbb{Z}_{/3}$ random walk in the preceding example, with uv in the role of u. The odd steps ϕ^{2n+1} are given by multiplying the even steps again by $(u + v)/2$, using the relations in the algebra. Thus, $(uv + vu)(u + v)/2 = uvu + (u + v)/2$ gives the second term of ϕ^{2n+1}, and combines also with the coefficient of $e(u + v)/2$ to give the first term as shown. The average of the odd and even states converges as $n \to \infty$ to the integral on $\mathbb{C}(S_3)$.

In this case, the Markov transition matrix in the above basis is

$$T = \frac{1}{3}\begin{pmatrix} 0 & 1 & 1 & 0 & 0 & 1 \\ 1 & 0 & 0 & 1 & 1 & 0 \\ 1 & 0 & 0 & 1 & 1 & 0 \\ 0 & 1 & 1 & 0 & 0 & 1 \\ 0 & 1 & 1 & 0 & 0 & 1 \\ 1 & 0 & 0 & 1 & 1 & 0 \end{pmatrix},$$

which also leads to the above solution on iterating and evaluation on $p^{(0)} = (1, 0, 0, 0, 0, 0)$.   ∎

Finally, from this algebraic point of view, in which the algebra A is emphasised, it is quite natural to think of the Markov transition operator 'actively' as an operator $A \to A$ inducing the above action on states. This is not usual in probability theory but is motivated by analogy with the correspondence between the so-called Schrödinger and Heisenberg pictures in quantum mechanics. In our case, the corresponding operator is clearly

$$R_\phi^* : A \to A, \qquad R_\phi^*(a) = (\mathrm{id} \otimes \phi) \Delta a, \qquad (5.23)$$

This R_ϕ^* is exactly the left action of A^* on A for any bialgebra A, as we have seen in Example 1.6.2. Working with this R_ϕ^* on A is equivalent to working with R_ϕ on A^*, so long as one keeps one point of view or the other. There is a subtlety, however, when one considers both views at the same time. It shows up when Δ is noncocommutative (so when the group structure on Ω is non-Abelian) and the walk is nonstationary; i.e. it is not evident in the above examples of random walks or Markov processes, but shows up in our more general setting when there is both 'curvature', in the sense of non-Abelian Ω, and a variation in time of the transition probabilities for the walk. If, at time n, we are in state ψ, then at time $n + 1$ we are in state $\psi \phi$. However, the corresponding 'evolution' of the random variables to achieve the same effect is

$$R_{\psi\phi}^*(a) = R_\psi^* \circ R_\phi^*(a),$$

which is conceptually in the wrong order for the active point of view. We should expect to first evolve A by an operator defined by ψ corresponding to the past evolution, and then by one corresponding to ϕ for the next step. The correct active description is as follows.

Proposition 5.2.5 *Let (A, ϕ) be a (stationary) random walk. The associated active Markov transition operator*

$$T_\phi : A \to A, \qquad T_\phi(a) = (\phi \otimes \mathrm{id}) \Delta a$$

is completely positive and $*$-preserving. Moreover, in terms of this operator, the expectation values after n steps are given by

$$\langle a \rangle_{\phi^n} = \phi^n(a) = \epsilon \circ (T_\phi)^n(a),$$

i.e. by evolution by T_ϕ for each step, followed by the counit. For a non-stationary random walk with varying states $\phi_1, \ldots, \phi_n$ at each step, the corresponding expectations after n steps are

$$\langle a \rangle_{\phi_1 \phi_2 \cdots \phi_n} = (\phi_1 \phi_2 \cdots \phi_n)(a) = \epsilon \circ T_{\phi_n} \cdots T_{\phi_1}(a).$$

Proof: For the expectations after n steps, we just use the elementary axioms of a bialgebra. The operator T is nothing other than the right action of A^* on A along the lines of Example 1.6.2 (but from the right). Note that $\phi = \epsilon \circ T_\phi$ (from the counity axioms) means that working 'actively' with T_ϕ as an operator in A is equivalent to working with states ϕ or R_ϕ. The fact that T is a right action is $T_{\phi\psi} = T_\psi \circ T_\phi$ and means that, for a nonstationary walk, our T for n-steps breaks up into repeated evolutions of A in the correct order as stated. The assertions of positivity and preservation of the $*$-structure assume that A is a Hopf $*$-algebra. Then, from Definition 1.7.5, we see that if $\phi(a^*) = \overline{\phi(a)}$ then

$$T_\phi(a^*) = a_{(1)}{}^*\phi(a_{(2)}{}^*) = a_{(1)}{}^*\overline{\phi(a_{(2)})} = T_\phi(a)^*,$$

so T_ϕ is $*$-preserving. The notion of complete positivity is a generalisation of the stochasticity requirement for Markov transition matrices, and says that T maps the positive elements of A (those of the form $a*a$) to positive ones, and, moreover, does the same when extended componentwise to the $*$-algebra $M_n(A)$ for any n. It is beyond our scope to prove this in detail here, but it follows from complete-positivity of Δ and our assumed positivity of ϕ. Any $*$-algebra homomorphism is evidently completely positive, and this applies to $\Delta : A \to A \otimes A$ with the tensor product $*$-structure on $A \otimes A$. ∎

It is easy enough to compute the active Markov transition operator in the setting of Example 5.2.2.

Exercise 5.2.6 *Let $A = \mathbb{C}(\Omega)$ for a finite group Ω and let $\{\delta_u\}$ be the basis of A given by Kronecker δ-functions. Then, a general element of A is $a = \sum_u a(u)\delta_u$. Show that, if we regard a as a column vector $(a(u))$, then the active Markov transition operator in the preceding proposition comes out as a matrix T_A on column vectors,*

$$T_\phi(a)(u) = \sum_{v \in \Omega} T_A{}^u{}_v\, a(v), \quad T_A{}^u{}_v = \rho_{vu^{-1}}.$$

Solution: We note that the coproduct in $\mathbb{C}(\Omega)$ in terms of the basis of δ-functions takes the form $\Delta\delta_v = \sum_u \delta_{vu^{-1}} \otimes \delta_u$. Hence $T_\phi(a) = \sum_v a(v)(\phi \otimes \mathrm{id})\Delta(\delta_v) = \sum_{u,v} a(v)\phi(\delta_{vu^{-1}})\delta_u = \sum_{u,v} a(v)\rho_{vu^{-1}}\delta_u$, as required. ∎

This exercise should be contrasted with the analogous computation for the matrix of R_ϕ^*, which must necessarily come out as our original Markov transition matrix T from Example 5.2.2, acting on the column vector a (because R_ϕ^* is the adjoint of the right action R_ϕ with matrix T). Thus, the matrices T and T_A coincide only if the coproduct is cocommutative, i.e. if the underlying group Ω in this setting is Abelian.

5.3   Quantum random walks

In the preceding section, we demonstrated how Hopf algebras and bialgebras can fruitfully be used to encode the essential structure of a random walk on a group. The resulting algebraic formulation in Definition 5.2.1 may be used to construct random walks in a wide variety of situations. In particular, all the combinatorial bialgebras in Section 5.1 have random walks on them. For example, a random walk on the Faà di Bruno bialgebra corresponds to a random walk on the set of partitions of the natural numbers.

Let us note now a remarkable feature of this algebraic formulation: there is no need in Definition 5.2.1 or Proposition 5.2.5 to assume that A is commutative. If A is commutative, then it has the flavour of the functions on a group Ω and we are essentially in the classical probabilistic picture that we have described in Section 5.2. Our task in the present section is to give the physical picture and an elementary example when A is noncommutative. The correct interpretation is provided not now by classical probability theory, but by quantum mechanics.

We recall very briefly some elements of quantum mechanics from an abstract algebraic point of view. First, there is some noncommutative *-algebra A of quantum observables. The elements of this are like the random variables above and are considered as 'questions' that one can ask about the system. The self-adjoint ones are considered to be the physical observables because their expectations will always be real numbers. The role of the characteristic functions χ_E is now played by projection operators, and their expectations lie in the range [0,1]. Secondly, the expectation values in this setting are given by evaluation in a state $\phi : A \to \mathbb{C}$ which should be a normalised positive linear functional. Thirdly, abstract quantum systems A, B have a tensor product: we take the tensor product algebra $A \otimes B$, which becomes a *-algebra with $(a \otimes b)^* = a^* \otimes b^*$. It should be thought of as the joint system with A, B independent subsystems. If ϕ, ψ are states on A, B respectively, then $\phi \otimes \psi$ is a state on $A \otimes B$. Its value $(\phi_1 \otimes \phi_2)(a \otimes b) = \phi_a(a)\phi_2(b)$ is the expectation for a, b treated independently. Finally, we can also consider the *conditional expectations* $\phi_1 \otimes \mathrm{id} : A \otimes B \to B$ and $\mathrm{id} \otimes \phi_2 : A \otimes B \to A$. They average out one of the subsystems and are related classically to the notion of conditional probabilities.

More familiar, perhaps, is the description of A realised concretely as an operator algebra acting on a Hilbert space, and, in this case, a typical state might be of the form

$$\phi = \sum_i \rho_i \langle \psi_i | \, | \psi_i \rangle, \qquad \rho_i \geq 0, \qquad \sum_i \rho_i = 1, \qquad (5.24)$$

where $|\psi_i\rangle$ are Hilbert space state vectors. When $A = B(\mathcal{H})$ (the bounded operators on a Hilbert space), one can show that the pure states, namely those that cannot be decomposed into nontrivial convex linear combinations of states, are exactly of the form $\langle\psi|\,|\psi\rangle$, where $|\psi\rangle \in \mathcal{H}$. A general state is then of the form (5.24), and the ρ_i are interpreted as probabilities or statistical weights for the general state. One needs to be more careful when $A \subset B(\mathcal{H})$.

As with the picture of probability density functions in Section 5.2.1, we could do away with this concrete realisation on Hilbert spaces as, in a certain sense, redundant; i.e. we can work directly with the abstract $*$-algebra of observables A and states on it. On the other hand, in the nice cases we can always reconstruct from the abstract algebra and a choice of vacuum the familiar concrete picture. This is the *Gelfand–Naimark–Segal (GNS) construction*, and we outline it briefly. Let $\mathcal{H} = A$ be a copy of A with A acting by left multiplication. Let $(a, b) = \phi_0(a^*b)$ be the inner product defined by a state ϕ_0. We kill the null vectors by passing to the quotient

$$\mathcal{H}_{\phi_0} = A/\{a \in A|\,(a, a) = 0\}, \tag{5.25}$$

which can now be completed to a Hilbert space with positive-definite inner product. The set of null states coincides with the vector space $\{b \in A|\,(a, b) = 0\ \forall a \in A\}$ because $|(a, b)|^2 \le |(a, a)||(b, b)|$ for any inner product space. This makes it clear that the action of A on itself by left multiplication descends to $\mathcal{H}_{\phi_0}$ (the vectors which are being set to zero form a left ideal of A). The vector in $\mathcal{H}_{\phi_0}$ corresponding to the identity element in A is denoted $|0\rangle$, the vacuum vector. Our original algebra A is now realised concretely as a $*$-subalgebra $A \subseteq B(\mathcal{H}_{\phi_0})$, and our state ϕ_0 is realised as the vacuum in the sense $\phi_0(a) = \langle 0|a|0\rangle$. This construction is the *Fock representation* of our abstract A with vacuum state ϕ_0. Finally, other states of interest in A can now be realised concretely in the form (5.24). Usually, of course, all these considerations are made in an infinite-dimensional setting with continuity assumptions etc., via the theory of von Neumann and C^* algebras. We concentrate, as usual, on the algebraic aspects of the system.

Thus, our abstract description of quantum mechanics as an abstract algebra and a state as a linear functional is fully equivalent to the usual picture of operators on a Hilbert space and Hilbert space state vectors. In fact, it is more precise than the usual picture because, by fixing an algebra of allowed observables A which need not be all of $B(\mathcal{H})$ (i.e. realising that not all the operators $B(\mathcal{H})$ may be observable in a given quantum system), one does not encounter some of the conceptual problems found in more naive points of view about quantum mechanics.

So far we have discussed what is called the kinematic structure of

quantum mechanics. Let us recall also that, in quantum mechanics, there
is a notion of time evolution given either as a unitary operator acting
on the Hilbert space vectors $|\psi\rangle_t = e^{\imath t H}|\psi\rangle$ (the Schrödinger picture) or
'actively' as the evolution

$$a_t = U_t a U_t^{-1}, \qquad U_t = e^{-\imath t H}, \tag{5.26}$$

The generator H is called the Hamiltonian of the system. There is a
relative minus sign between the two points of view, which we can dispense
with if we use only one or the other. In the 'active' picture we need only
the exponentiated U_t, acting by conjugation. More generally, we need a
one-parameter group of $*$-automorphisms of A under which the vacuum
should be invariant. In this case the GNS construction essentially forces
us back to the concrete form.

   This completes our lightning review of quantum mechanics. Now, we
have seen in Section 5.2 that when A has the additional structure of a
$*$-bialgebra or $*$-Hopf algebra, we can make random walks on it. We also
know from Section 1.7 that when A is a bialgebra then we have one natural
invariant linear functional $\int$, the integral on A. It is natural to use this
as the vacuum state of an algebraic GNS-type construction, and, indeed,
this is exactly how one can interpret the concrete realisation theorem
in Chapter 1.7. We have seen in Theorem 1.7.4 and Proposition 1.7.6
that $\int$ leads to a concrete description of A as operators acting by left-
multiplication on $\mathcal{H} = A$ and giving A realised concretely in $\mathrm{Lin}(\mathcal{H})$ as a
sub $*$-algebra. In our algebraic treatment we did not bother to divide by
the null-vectors to ensure that we have a positive-definite inner product,
though one can do this also as in the GNS construction.

Proposition 5.3.1 *Suppose A is a Hopf $*$-algebra and $A \subset \mathrm{Lin}(\mathcal{H})$ is the
left regular representation. Let ϕ be any state on A. Then the associated
active Markov transition operator takes the form*

$$T_\phi(a) = (\phi \otimes \mathrm{id})\left(W(a\triangleright \otimes 1)W^{-1}\right),$$

*where W is the unitary operator in $\mathrm{Lin}(\mathcal{H} \otimes \mathcal{H})$ from Theorem 1.7.4 and
Proposition 1.7.6. We call W the evolution operator of the joint system
$A \otimes A$.*

Proof: Here $a\triangleright$ denotes a acting in the left regular representation on $\mathcal{H}$.
W acts on $\mathcal{H} \otimes \mathcal{H}$, and we know from Theorem 1.7.4 that conjugation by
this is $\sum a_{(1)}\triangleright \otimes a_{(2)}\triangleright$. By definition, the action of ϕ on the first factor is
the same as $\phi(a_{(1)})$. The second factor $a_{(2)}$ acts by the left regular repre-
sentation $A \to A$. Note that this means that T_ϕ descends as an operator
to $\mathcal{H}_{\phi_0} \to \mathcal{H}_{\phi_0}$ if we want to work on any specific GNS representation,
such as the one associated to a right integral $\int$ in Proposition 1.7.6. ∎

Thus, our random walk, considered as an active Markov process, consists of the following. First, we embed $a \in A$ as $a \otimes 1 \in A \otimes A$. Here the first copy of A is the algebra of observables at time n and the second copy is the algebra of observables at time $n+1$ (i.e. one step further in time). Secondly, we evolve $a \otimes 1$ by the quantum evolution operator W of the joint system $A \otimes A$. Finally, we take a conditional expectation value in the first copy of A to leave us in the second copy at $n+1$. This represents averaging out or 'forgetting' the details of where the system might have been at the, now unobserved, time n. Thus, T_ϕ is the quantum step of a quantum random walk in a reasonably physical way according to the usual picture of quantum mechanics.

Remarkably, the data for this concrete interpretation agree with our data for a random walk given in Definition 5.2.1. One needs to fix a state ϕ_0 as a reference for the concrete realisation, while this or other states can be used for the random walk. This justifies calling Definition 5.2.1 (when A is noncommutative) a 'quantum random walk'. We now give a simple noncommutative example. In fact, it is quite hard to find a nontrivial truly noncommutative and noncocommutative Hopf algebra that really arises as the quantum algebra of observables of a familiar quantum system. The next chapter will be devoted to this. For the present, we will content ourselves with an elementary 'nonphysical' example, which at least has the merits of demonstrating the kinds of mathematical phenomena that can arise. This example is related to anyons in the setting of Chapter 10.1, and is also a variant of our random walk on the real line in Section 5.2.1 above.

For our Hopf algebra A, we take generators X, g, g^{-1} and relations, coproduct, counit, antipode and $*$-structure

$$gX = qXg, \quad gg^{-1} = 1 = g^{-1}g,$$

$$\Delta g = g \otimes g, \quad \Delta X = X \otimes g^{-1} + g \otimes X, \quad \epsilon g = 1, \quad \epsilon X = 0, \quad (5.27)$$

$$SX = -q^{-1}X, \quad Sg = g^{-1}, \quad X^* = X, \quad g^* = g, \quad q^* = q^{-1}.$$

This is a variant of Example 1.3.2, but note that q is a parameter of modulus 1 for the $*$-structure that we need. The elements $X^m g^i$ for $m \in \mathbb{N}, i \in \mathbb{Z}$ form a basis. The coproduct should be compared with (5.11). If ϕ is any state as above, the state ϕ^n after n steps is computed from (5.12) using the coproduct Δ. It is given by embedding X in $A^{\otimes n}$ via Δ^{n-1} and applying ϕ to each A. From (5.27), we have

$$\Delta^{n-1}g = g \otimes \cdots \otimes g,$$
$$\Delta^{n-1}X = X \otimes g^{-1} \otimes \cdots \otimes g^{-1}$$
$$+ g \otimes X \otimes g^{-1} \otimes \cdots \otimes g^{-1} + \cdots + g \otimes \cdots \otimes g \otimes X. \quad (5.28)$$

If we write the right hand expression as $\Delta^{n-1}X = X_1 + X_2 + \cdots + X_n$, where $X_1 = X \otimes g^{-1} \otimes \cdots \otimes g^{-1}$, $X_2 = g \otimes X \otimes g^{-1} \otimes \cdots \otimes g^{-1}$, etc., we see that the random variable describing the position after n steps is the sum of n random variables X_i embedded in $A^{\otimes n}$. They are not, however, independent in a usual sense. Instead,

$$X_i X_j = q^2 X_j X_i, \qquad i > j, \qquad (5.29)$$

from the relations in (5.27). We have also seen this algebra in Example 4.5.7 in another context. This q-independence shows up when we look at higher moments, where we need to compute $\Delta^{n-1}X^m$. We have already encountered the relevant structure in Lemma 3.2.2, and, using the same techniques as there, one finds

$$\Delta^{n-1}X^m = \sum_{i_1+\cdots+i_n=m} \frac{[m;q^2]!}{[i_1;q^2]!\cdots[i_n;q^2]!} X_1^{i_1} \cdots X_n^{i_n}, \quad [i;q^2] = \frac{1-q^{2i}}{1-q^2}$$

$$(5.30)$$

in view of the relations (5.29).

Proposition 5.3.2 *Consider a random walk on the Hopf *-algebra (5.27) with any state ϕ for the step of the random walk. The expectation values after n steps are*

$$\langle g \rangle_{\phi^n} = \langle g \rangle_\phi^n, \qquad \langle X \rangle_{\phi^n} = \langle X \rangle_\phi \frac{\langle g \rangle_\phi^n - \langle g^{-1} \rangle_\phi^n}{\langle g \rangle_\phi - \langle g^{-1} \rangle_\phi},$$

$$\langle X^2 \rangle_{\phi^n} = \langle X^2 \rangle_\phi \frac{\langle g^2 \rangle_\phi^n - \langle g^{-2} \rangle_\phi^n}{\langle g^2 \rangle_\phi - \langle g^{-2} \rangle_\phi}$$

$$+ [2;q^2] \langle Xg \rangle_\phi \langle g^{-1}X \rangle_\phi \frac{[n;\langle g^2 \rangle_\phi] - [n;\langle g^{-2} \rangle_\phi]}{\langle g^2 \rangle_\phi - \langle g^{-2} \rangle_\phi}.$$

These moments reduce to our previous results (5.10) for a random walk on $\mathbb{R}$ if we set $q \to 1$ and $g \to 1$ in a strong sense with $\langle g^{\pm 1} \rangle_\phi \sim \langle g \rangle_\phi^{\pm 1}$ and $\langle g^{\pm 2} \rangle_\phi \sim \langle g \rangle_\phi^{\pm 2}$.

Proof: The expectations of g and X follow at once from (5.28). For the expectation of X^2, we have to write out $\Delta X^2 = \sum_{i=1}^n X_i^2 + \sum_{i<j}[2;q^2]X_iX_j$ from (5.30) explicitly in terms of the X_i. For the first sum, we have $X^2 \otimes g^{-2} \otimes \cdots \otimes g^{-2} + g^2 \otimes X^2 \otimes g^{-2} \otimes \cdots \otimes g^{-2} + \cdots$, which gives the first term shown. For the second term, we note

$$X_iX_j = g^2 \otimes \cdots \otimes g^2 \otimes Xg \otimes 1 \otimes \cdots \otimes 1 \otimes g^{-1}X \otimes g^{-2} \otimes \cdots \otimes g^{-2},$$

where Xg is in the ith position and $g^{-1}X$ is in the jth. Applying $\phi^{\otimes n}$ gives a factor $\langle g^2 \rangle_\phi^{i-1} \langle g^{-2} \rangle_\phi^{n-j}$. Summing over j, we obtain, for the second term in $\langle X^2 \rangle_{\phi^n}$, the result $[2; q^2]\langle Xg \rangle_\phi \langle g^{-1}X \rangle_\phi$ times the expression

$$\sum_{i=1}^{n-1} \langle g^2 \rangle_\phi^{i-1} \frac{1 - \langle g^{-2} \rangle_\phi^{n-i}}{1 - \langle g^{-2} \rangle_\phi}$$

$$= \frac{1 - \langle g^2 \rangle_\phi^{n-1}}{(1 - \langle g^{-2} \rangle_\phi)(1 - \langle g^2 \rangle_\phi)} - \frac{\langle g^{-2} \rangle_\phi}{(1 - \langle g^{-2} \rangle_\phi)} \frac{(\langle g^2 \rangle_\phi^{n-1} - \langle g^{-2} \rangle_\phi^{n-1})}{(\langle g^2 \rangle_\phi - \langle g^{-2} \rangle_\phi)}$$

$$= (\langle g^2 \rangle_\phi - \langle g^{-2} \rangle_\phi)^{-1} \left(\frac{1 - \langle g^2 \rangle_\phi^n}{1 - \langle g^2 \rangle_\phi} - \frac{1 - \langle g^{-2} \rangle_\phi^n}{1 - \langle g^{-2} \rangle_\phi} \right),$$

after suitable reorganisation of the partial fractions. Note that, if we write $\langle g^{\pm 2} \rangle_\phi = 1 \pm 2\delta + O(\delta)$ and take $\delta \to 0$, then this expression tends to $\frac{n(n-1)}{2}$, as in Section 5.2   ∎

One can also compute the continuum limit of a random walk on this Hopf algebra once a suitable state is chosen. For example, one nonpositive (and therefore not very physical) choice q-deforming (5.14), and suitably scaled as we take the continuum-limit, gives a q-deformed version of the usual diffusion equation on $\mathbb{R}$. Many examples and their limits may be developed along similar lines for appropriate bialgebras or Hopf algebras and states on them.

5.4 Input–output symmetry and time-reversal

This section is strictly for fun and is of a philosophical character. No nonempty theorems will be proven, but rather we aim to describe a duality phenomenon of conceptual interest which is made possible when the algebra of observables underlying our random walk is a Hopf algebra. The uninterested reader can, and should, proceed directly to the next chapter. On the other hand, the ideas here are novel and could be interesting for theoretical physics, and perhaps also for computer science.

Note first of all that, in the previous sections of this chapter, we have recalled the algebraic structure of successively more advanced frameworks for models of physics, namely the sequence

{Boolean algebra : subsets of Ω}

$\subset$ {classical probability : $\mathbb{C}(\Omega)$}

$\subset$ {quantum mechanics : operator algebra A}.   (5.31)

The first embedding is via the characteristic functions. The second is via (5.24) in Section 5.3, which reduces to a probability state as in Section 5.2 if A is commutative and represented irreducibly; for then the Hilbert space collapses to one dimension and our picture of quantum mechanics collapses to our picture of classical probability.

In each of these cases, we explored the meaning of a coalgebra structure Δ on the relevant algebra as encoding some form of combinatorics or random walk. This corresponds to the role of Δ as expressing some kind of underlying group structure in a suitably general sense. In this section, we want to explore a different point of view on the coproduct, as restoring an input–output symmetry to our framework for physics. This is a new kind of duality in quantum mechanics and statistics which is made possible by the Hopf algebra structure. It was developed by the author.

The idea behind our considerations is as follows. Usually in quantum mechanics we have an algebra of observables A but only a set of states. Thus, there is a product $A \otimes A \to A$, but nothing in the other direction. On the other hand, when the algebra of observables A of our quantum system happens to be a bialgebra or Hopf algebra, then this symmetry is restored by the coproduct $\Delta : A \to A \otimes A$ (similarly for the unit and counit). In this case, then, we *can* multiply the states. Moreover, we have seen in Sections 5.2 and 5.3 one interpretation of this multiplication of states, namely as the convolution product (5.12) of a random walk. But, of course, our interpretation of multiplication of states is quite different from the multiplication in the algebra A. What then is the physical meaning or interpretation of this mathematical input–output symmetry provided by the presence of Δ?

The proposal which we shall explore in this section, and also in part of the next chapter, is that when we model a given system by means of an algebra of observables A and states in A^*, we are making an arbitrary distinction or 'polarisation' between some aspects which we consider to be the questions we could ask about the system and other aspects which we consider to be the reality or state that provides the answers to these questions. But, in fact (the proposal asserts), we cannot really tell which aspects are which. All we really know is the *pairing* between what we called the questions or observables, and what we called the states. Thus we consider that $\phi(a) = \langle a \rangle_\phi$ is the 'expectation value of observable a in state ϕ', but, in fact, all we really know are the numbers $\phi(a)$, which we could equally consider as $a(\phi)$. We have alluded to this point of view in general terms at the end of Section 5.1.

This is a new kind of duality principle, which is well established in modern mathematics, but whose physical implications in quantum mechanics have not been well explored. For a general quantum system the reversed interpretation in which $\phi \in A^*$ is regarded as an observable and $a \in A \subseteq A^{**}$ is regarded as a state does not progress very far because one has an algebra of observables, but only a set of states. It is exactly this possibility that opens up when A is a bialgebra or Hopf algebra.

Definition 5.4.1 *Let A be a quantum algebra of observables that happens to be a ∗-bialgebra or Hopf ∗-bialgebra. We define the* dual quantum system *to be the system with quantum algebra of observables A^* and states A, which we understand as A^{**}. We postulate as the* quantum Mach's principle *that this dual quantum system should have a physical interpretation as quantum mechanics which is equally as valid as our original interpretation of A.*

Note that this postulate is a far-from-empty constraint on whatever we propose as a systematic theory of physics. Not every quantum system can even begin to have this dual interpretation, and even when *A is* a bialgebra, A^* may look nothing like any quantum system with which we are familiar. The assertion of the postulate is that our theory of physics in this case cannot be complete, i.e. that we must add to it refinements and further structure until the reversed interpretation is equally possible. Put another way, there *are* physical models (presented in the next chapter) where such an observable–state symmetry does exist, i.e. where there is present a new kind of $\mathbb{Z}_{/2}$ symmetry in physics. It is proposed that where this symmetry does not exist in our model or theory of physics, the model is incomplete and needs to be enlarged. Unlike usual postulates of physics, this one concerns not any one theory and not any specific laws or experimental numbers, but rather it is a postulate about the semantic structure of any acceptable theory of physics. As such, it should be understood as a symmetry principle or tendency, not to be taken too exactly.

These are very broad considerations, and ones to which we shall return in Chapter 6.4 at the general level. Here we want to explore this possible duality in the specific context of quantum random walks. Because this duality is also related to some kind of 'input–output symmetry', it clearly also has something to do with the arrow of time. We begin then by recalling some facts about entropy in connection with random walks.

To begin with, in classical probability, the relative entropy between states ϕ, ψ on the algebra $\mathbb{C}(\Omega)$ is defined to be

$$S(\phi, \psi) = \sum_{i \in \Omega} (-\rho_i \log \rho_i + \rho_i \log r_i); \quad \forall \, \phi = \sum_i \rho_i i, \; \psi = \sum_j r_j j \quad (5.32)$$

and, in our conventions, is ≤ 0. Here $i, j \in \Omega$ are the pure states on $\mathbb{C}(\Omega)$ acting by evaluation (they are the δ-function distributions), and general mixed states are convex linear combinations of them. We used this point of view (which works for any set) in Example 5.2.2. We consider all entropy as relative entropy, although there can often be a natural reference state in the problem which we might take for granted in a given context. The classical entropy $S(\phi, \psi)$ measures the degree of impurity of

ϕ relative to ψ. It also has an information-theoretic interpretation, and, roughly speaking, $e^{\mathcal{S}(\phi,\psi)}$ is the probability per unit trial of the system appearing to be in state ϕ when it is in state ψ.

In quantum theory the relative entropy $\mathcal{S}(\phi,\psi) \leq 0$ is defined between two quantum states ϕ, ψ in a similar way. When the algebra A is that of bounded operators on a Hilbert space (so that states are of the form (5.24)), we can write similarly

$$\mathcal{S}(\phi,\psi) = \sum_{i,j}(-\rho_i \log \rho_i + \rho_i \log r_j)|\langle \phi_i | \psi_j \rangle|^2,$$

$$\forall \quad \phi = \sum_i \rho_i \langle \phi_i | \, | \phi_i \rangle, \quad \psi = \sum_j r_j \langle \psi_j | \, | \psi_j \rangle \tag{5.33}$$

This definition has a similar interpretation to that in the classical case and extends to abstract von Neumann and C^* algebras, as well as into more general algebraic situations. We do not need the full details of this here because we need only the following general fact, which holds for all these definitions: let $\mathcal{O} : A \to A$ be a completely positive operator in the sense that it maps positive elements to positive elements and continues to do so for all n when extended to $M_n(A) \to M_n(A)$. Then,

$$\mathcal{S}(\phi \circ \mathcal{O}, \psi \circ \mathcal{O}) \geq \mathcal{S}(\phi,\psi).$$

We take this as an axiom for any good formulation of entropy in an algebraic setting. It says that precomposing with $\mathcal{O}$ loses information and blurs the distinction between ϕ and ψ, hence their relative entropy increases. We have seen operators of this type in the theory of classical and quantum Markov processes.

Example 5.4.2 *Let Ω be a finite group and let $A = \mathbb{C}(\Omega)$. Let $\int$ be the integral on Ω in Example 1.7.2 and let ϵ be the counit. These are both classical states and*

$$\mathcal{S}\left(\epsilon, \int\right) = -\log|\Omega|$$

If ϕ is another classical state then $\mathcal{S}(\phi, \int) \geq \mathcal{S}(\epsilon, \int)$, with equality iff ϕ is pure.

Proof: This follows at once from the definition in (5.32). Here, $\int$ is a state with all its weights equal to $|\Omega|^{-1}$. For the second part, note that $-\rho_i \log \rho_i \geq 0$, since $0 \leq \rho_i \leq 1$, and is equal to zero only at $\phi_i = 0, 1$. Hence the sum of these terms is ≥ 0 with equality *iff* the state ϕ is pure. For the second term in (5.32), use that $\sum \rho_i = 1$. $\blacksquare$

Now let (A, ϕ) be a random walk; so the states are ϕ^n for successive n. There is obviously an arrow of time here since we are successively taking steps of the random walk. This arrow of time is correctly reflected by the entropy because R_ϕ^* and T_ϕ are completely positive, as mentioned for T_ϕ in Proposition 5.2.5, so we know at once that

$$0 \geq S(\phi^n, \phi^{n+1}) = S(\phi^{n-1} \circ R_\phi^*, \phi^n \circ R_\phi^*)$$
$$\geq S(\phi^{n-1}, \phi^n) \geq \cdots \geq S(\epsilon, \phi), \qquad (5.34)$$
$$0 \geq S(\phi^{n+1}, \int) = S(\phi^n \circ R_\phi^*, \int \circ R_\phi^*) \geq S(\phi^n, \int) \geq \cdots \geq S(\epsilon, \int).$$

The first sequence says that, as n grows, the states ϕ^n of our random walk change more and more slowly in the sense that the probability increases for the system to still appear in state ϕ^n when it is in the next ϕ^{n+1}. At least in the classical case, the lower bound is $S(\epsilon, \phi) = \log \rho_1$, and $e^{S(\epsilon, \phi)}$ is precisely the probability that the random walk stays where it is in one step, which is the probability after one step of the system still appearing to be in the previous state ϵ. The first sequence in (5.34) also says that the impurity of ϕ^n, which is generally less than that of ϕ^{n+1}, tends towards the latter. The second sequence says more, namely that, as the walk evolves, the states are converging in an entropic sense to the right integral $\int$ (which we assume for the sake of discussion to be positive). By this we mean only that the entropy of ϕ^n relative to $\int$ indeed increases with each step, and in this sense the state becomes more and more similar to $\int$. It also becomes more and more impure as its degree of impurity approaches the degree of impurity of $\int$ as claimed. This integral $\int$ represents a kind of 'maximal entropy' or 'maximally impure' state, and our random walk evolves towards it. In the classical case of functions on a finite group, the integral is precisely the maximally impure state consisting of the average of all the pure states with equal weight (this gives a probabilistic picture of Example 1.7.2). In proving the second sequence in (5.34), we used right-invariance of $\int$, as in Proposition 1.7.6 in Chapter 1. The same arguments and interpretations hold for a nonstationary random walk, but, in this case, we should distinguish between T_ϕ and R_ϕ^* and use the former in the first sequence. Clearly we have

$$0 \geq S(\phi_1 \phi_2 \cdots \phi_n, \phi_1 \phi_2 \cdots \phi_{n+1})$$
$$\geq S(\phi_2 \phi_3 \cdots \phi_n, \phi_2 \phi_3 \cdots \phi_{n+1}) \geq \cdots \geq S(\epsilon, \phi_{n+1})$$
$$0 \geq S(\phi_1 \phi_2 \cdots \phi_{n+1}, \int) \geq S(\phi_1 \phi_2 \cdots \phi_n, \int) \geq \cdots \geq S(\epsilon, \int).$$

The origin of the increase with each step of the entropy relative to $\int$ is that, whereas the evolution of the joint system by W in Proposition 5.3.1 is invertible, the conditional expectation given by evaluation in the state

ϕ for the step represents averaging or forgetting information. This is the intuitive reason (from the active point of view) that proceeding on the random walk is entropy increasing. Moreover, we demonstrated in Section 5.2.1 precisely how this discrete arrow of time can lead, in a suitable limit, to a more familiar continuous arrow of time.

We are now ready to have some fun with Hopf algebra duality in this context. Thus, if we have a random walk (A, ϕ) as above, but perversely take the dual point of view in which A is the 'algebra of states', then what process do we see? According to our postulate in Definition 5.4.1, we should have some equally physical interpretation of what is going on, though perhaps not as a random walk. Equally well, we can do a random walk in the dual point of view, and ask how does it appear from our usual point of view in which A is the algebra of observables of our quantum system?

A random walk in the dual A^* will be controlled by $a \in A$, regarded as an element of A^{**} in such a way as to give a normalised positive state on A^*. From our usual point of view, however, a is an element of the algebra of observables. The kind of elements turning up in this way can be characterised as follows:

Definition 5.4.3 *An element a in a Hopf $*$-algebra A is* copositive *if $\langle a, \phi^*\phi \rangle \geq 0$ for all $\phi \in A^*$. It is* normalised *if $\epsilon(a) = 1$ and* S-Hermitian *if $Sa = a^*$. A copositive element is usually taken with these additional properties as well.*

Clearly the theory of copositive elements is similar to the theory of states, indeed it is the theory of states on A^*. Thus, if a_i are copositive then so is any convex linear combination

$$a = \sum \rho_i a_i, \quad \rho_i \geq 0, \quad \sum \rho_i = 1.$$

A *pure copositive* element is a copositive element which cannot be further decomposed in this way. From Definition 1.7.5 it is clear that the requirement $\langle a^*, \phi \rangle = \overline{\langle a, \phi \rangle}$ for all $\phi \in A^*$ likewise translates into the S-Hermiticity condition. But, from the point of view of the algebra of observables, this S-Hermiticity condition is a notion that unifies the concepts of Hermitian, anti-Hermitian or unitary in usual quantum theory, depending on how the antipode S looks when acting in the element.

Example 5.4.4 *Let Ω be a finite group and let $A = \mathbb{C}(\Omega)$. Let $\hat{\Omega}$ denote the set of characters of Ω given by the traces of irreducible unitary representations and normalised so that $\chi(e) = 1$. The copositive normalised and S-Hermitian elements in A are of the form*

$$a = \sum_{\chi \in \hat{\Omega}} \rho_\chi \chi, \quad \rho_\chi \geq 0, \quad \sum_{\chi \in \hat{\Omega}} \rho_\chi = 1.$$

Proof: The dual A^* is $\mathbb{C}\Omega$, and its elements from Example 1.5.4 are of the form $\sum_u \phi_u u$, say. Let π be a unitary matrix representation of Ω with dimension d_π, and let $a = d_\pi^{-1}\mathrm{Tr}\,\pi$ be extended linearly to $\mathbb{C}\Omega$. Then a is normalised and S-Hermitian since $a(\phi^*) = d_\pi^{-1}\sum_u \bar{\phi}_u\mathrm{Tr}\,\pi(u^{-1}) = d_\pi^{-1}\sum_u \bar{\phi}_u\overline{\mathrm{Tr}\,\pi(u)} = \overline{a(\phi)}$ by π unitary. Moreover, a is copositive since

$$a(\phi^*\phi) = d_\pi^{-1}\sum_{u,v} \bar{\phi}_u\phi_v\mathrm{Tr}\,\pi(u^{-1})\pi(v)$$

$$= d_\pi^{-1}\sum_{i,j}\sum_{u,v} \bar{\phi}_u\phi_v\mathrm{Tr}\,\overline{\pi(u)^j{}_i}\pi(v)^j{}_i = d_\pi^{-1}\sum_{i,j}\left|\sum_u \phi_u\pi(u)^i{}_j\right|^2 \geq 0,$$

where we used that π is a unitary representation and wrote out the trace explicitly in an orthonormal basis. Conversely, given a copositive element, we view it as a state on $\mathbb{C}\Omega$. As such, it corresponds to a representation of $\mathbb{C}\Omega$ by the GNS construction. But nontrivial $*$-algebra representations of $\mathbb{C}\Omega$ correspond to unitary representations of Ω. Hence every normalised S-Hermitian copositive is of this form. Finally, every representation of Ω can be decomposed into irreducibles, and, corresponding to this, our copositive a decomposes into a convex linear combination of characters. ∎

Example 5.4.5 *Let Ω be a finite group. The integral element $\Lambda = \delta_e$ in $\mathbb{C}(\Omega)$ in the sense of Definition 1.7.1 is a copositive normalised and S-Hermitian element, and decomposes according to Example 5.4.4 as*

$$\Lambda = |\Omega|^{-1}\sum_{\chi\in\hat{\Omega}} d_\chi^2\chi,$$

where the sum is over the normalised characters and d_χ is the dimension of the corresponding representation.

Proof: The integral element Λ in $\mathbb{C}(\Omega)$ from Definition 1.7.1 is characterised by $a(u)\Lambda(u) = a(e)\Lambda(u)$ since the counit is $\epsilon(a) = a(e)$. This forces the normalised integral to be the Kronecker δ-function δ_e, centred at e. To see that it is of the form stated, it is enough to realise that δ_e is clearly constant on conjugacy classes, i.e. $\delta_e(u) = \delta_e(v^{-1}uv)$ for all $u, v \in \Omega$. Hence, it too is a sum of characters since these span the space of such functions. By the orthogonality of the characters in the form $|\Omega|^{-1}\sum_u \overline{\chi(u)}\chi'(u) = \delta_{\chi,\chi'}d_\chi^{-2}$, we can find the coefficients as stated.

Another way to see this result is to note that the integral Λ in $\mathbb{C}(\Omega)$ is the integral on $\mathbb{C}\Omega$, and, since $S^2 = \mathrm{id}$, we see from Proposition 1.7.3 that it is the normalised trace of the left regular representation π. But the left regular representation decomposes into $\sum_\chi d_\chi\pi_\chi$, where π_χ are the inequivalent irreducible representations, and it has dimension $|\Omega|$. Hence, its normalised trace is $|\Omega|^{-1} \sum_\chi d_\chi^2 \chi$. Note that we do not assume that Ω is Abelian. If this were the case, then $\mathbb{C}\Omega = \mathbb{C}(\hat{\Omega})$ from Exercise 1.5.5, and we would be in the classical situation given in Example 5.2.2, applied to $\hat{\Omega}$. ∎

Note that, at the Hopf algebra level, we are simply developing the theory of states on $\mathbb{C}\Omega$, but, unless Ω is Abelian, this Hopf algebra is not commutative and hence not of the familiar type where there is an interpretation of states in terms of classical probability distributions. In other words, a copositive classical observable in $\mathbb{C}(\Omega)$ appears from the dual point of view as a state on the dual algebra, but as such it is a quantum, and not a classical, state when Ω is non-Abelian. Meanwhile, from our usual point of view, it is an element of the classical observables (a random variable) and not a state at all.

One can continue in the spirit of our duality principle and make constructions for copositive elements which would appear from the dual view as standard constructions on states. For example, we define the *relative coentropy* $S^*(a,b)$ between two copositive observables a, b as the relative entropy of a, b as linear functionals on A^*. The heuristic interpretation of $S^*(a,b)$ is that $e^{S^*(a,b)}$ should be thought of as the probability per unit trial of mistaking the observable or random variable b for a when examined on random states of the system. In very general terms, it is the ratio of the impurity of a to that of b.

We now consider A equipped with a copositive element a. It induces a random walk in A^*, but what does this random walk correspond to from our usual point of view where a is in the algebra of observables? To be clear about the order, we should consider a nonstationary random walk in A^* with successive steps given by $a_1, a_2, \ldots$. From the dual point of view, we are successively convolving these 'states', but from our point of view in A we are simply multiplying them as elements of the algebra of observables. Hence, in the spirit of the duality principle of Definition 5.4.1, we should supplement our concepts to include such a process:

Definition 5.4.6 *A* creation process *is* $(A, a_1, a_2, \ldots)$, *where A is a Hopf *-algebra and a_i are normalised copositive S-Hermitian elements of A. The elements $1, a_1, a_2 a_1, \ldots$ are the successive stages of the creation process.*

We will justify this term both classically and quantum mechanically. Quantum mechanically, the successive products are precisely the way that states in Fock space are created from the vacuum. Thus, in any GNS representation (5.25) our so-called creation process projects to a sequence

$$|0\rangle, \ a_1|0\rangle, \ \ldots, \ a_n a_{n-1} \cdots a_1|0\rangle \tag{5.35}$$

of Hilbert space vectors. In the classical case, the role of the algebra of observables is played by $A = \mathbb{C}(\Omega)$ on a set Ω, and if we concentrate on the characteristic functions among these we are considering logic or Boolean algebra, as explained at the end of Section 5.1 and summarised in (5.31). Recall that the characteristic functions are the observables or random variables that specify membership of the subsets. We can use these if (for the sake of discussion) we do not worry about the copositivity requirements. Then, a creation process in this context is the sequence of subsets

$$\Omega, a_1, a_2 \wedge a_1, \ldots, a_n \wedge a_{n-1} \wedge \cdots \wedge a_1; \qquad a_i \subseteq \Omega \qquad (5.36)$$

corresponding to the product of their characteristic functions. The first step of the creation process in terms of characteristic functions is 1 (the identity function), the next step, $a_1.1$, specifies membership of one subset, $a_2.a_1.1$ further specifies an additional membership, etc. This models the way that classical concepts or observables are created as a series of specifications or conjunctions. In this classical case, of course, the order of the creation process does not matter, whereas for a general quantum system the order does matter.

Finally, our comments above about the entropy of a random walk now become comments about coentropy in a creation process. For example, the second sequence in (5.34) for the random walk from the dual point of view, appears from our point of view as

$$0 \geq \mathcal{S}^*(a^{n+1}, \Lambda) \geq \mathcal{S}^*(a^n, \Lambda) \geq \cdots \geq \mathcal{S}^*(1, \Lambda). \qquad (5.37)$$

Here the integral element is characterised by $a\Lambda = \epsilon(a)\Lambda$, and we assume that it is copositive and normalised etc. The coentropy increases when both arguments are multiplied by a copositive element because, from the dual point of view, this is just the action of precomposition of 'states' by R_a^* or T_a. Let us recall that these operations are the adjoint of right and left multiplication, respectively. The interpretation of (5.37) is that, for a typical copositive element a, the sequence $1, a, a^2, \ldots$ typically tends to Λ in a coentropic or probabilistic sense. In general terms, it becomes more and more impure as it approaches Λ in this coentropic sense. This is also true in the nonstationary case, with

$$\mathcal{S}^*(a_m a_{m-1} \cdots a_1, \Lambda) \geq \mathcal{S}^*(a_{m-1} a_{m-2} \cdots a_1, \Lambda) \geq \mathcal{S}^*(1, \Lambda).$$

We can easily see the phenomenon in the classical case of the algebra of functions on a group Ω as follows. In this case, the pure copositives are the characters, as we have seen in Example 5.4.4. The assertion is that if we take some typical (nonpure) copositive a and raise it to higher and higher powers then it 'converges' in some probabilistic sense to Λ, which is the δ-function at the identity. This is a typical phenomenon on $\mathbb{Z}_{/n}$ or (in the continuous setting) on $\mathbb{R}$ or S^1. Thus, for $\mathbb{R}$, the pure copositives are just the pure frequency waves $a_\omega = e^{\imath\omega(\ )}$. Our assertion that typical copositive elements give creation processes tending to the integral is, then, something well-known to engineers. For example, we can take the convex linear combination

$$a = \frac{e^{\imath\omega(\ )} + e^{-\imath\omega(\ )}}{2} = \cos\omega(\ ).$$

It is well-known that squaring such cosine waves introduces higher harmonics. Moreover, raising to higher and higher powers a^n makes the wave more and more impure as further harmonics are introduced, until we have something resembling the integral Λ. This is the maximally impure wave containing all frequencies as we have seen in Example 5.4.5. This is a simple physical interpretation of (5.37) in the classical case. Of course, the mathematical content in this group case is the same as the random walk on the dual (as calculated in Example 5.2.3 for $\mathbb{Z}_{/3}$), but our interpretation now as a result about multiplying observables is quite different.

If we do not worry about the copositivity condition, we can also give a heuristic picture of this general phenomenon in terms of membership of subsets as mentioned above. We consider a nonstationary creation process given by a family of subsets containing e (so that their characteristic functions $a_1, a_2, \ldots$ are correctly normalised). These indeed typically tend to $\{e\}$ (corresponding to Λ) as the only point definitely in their joint intersection. This is a second 'Boolean' interpretation of (5.37).

Finally, we take a closer look at the order of events when a random walk for us is considered to be a creation process from the dual point of view, and vice versa. Since the passage from our point of view to the dual one is always by the pairing, i.e. by dualising the product in A to get the coproduct in A^* etc., we will suppose, for the sake of discussion, that this continues to be the case after each step; i.e. we suppose that after n steps the dual process continues to be related to our original one by means of dualisation or adjoints. This is the observable–state duality principle that at any time an observable can equally well be thought of as a state from the dual point of view and vice versa. Thus, a nonstationary random walk from our point of view with successive steps $\phi_1, \phi_2, \ldots, \phi_n$ means, from our point of view, the successive quantum states

$$\epsilon, \ \phi_1, \ \phi_1\phi_2, \ \ldots, \ \phi_1 \cdots \phi_n.$$

But these same states appear from the dual point of view as observables or vectors

$$|\epsilon\rangle, \ \phi_1|\epsilon\rangle, \ \ldots, \ \phi_1 \cdots \phi_n|\epsilon\rangle,$$

where our counit provides the unit and hence the vacuum vector in any GNS construction for the dual. This is like a creation process from the dual point of view but with the steps taken *in the reverse order*. Thus, at each time n, the state appears as if it is created after n steps by a creation process with steps $\phi_n, \phi_{n-1}, \ldots, \phi_1$.

Likewise, if we have a sequence of copositive observables $a_1, a_2, \ldots, a_n$, which we use as a creation process with stages $1, a_1, a_2 a_1, \ldots, a_n \cdots a_1$ (or the corresponding Fock states as in (5.35) in any GNS representation) then these appear at each n from the dual point of view as states of a random walk with steps $a_n, a_{n-1}, \ldots, a_1$, i.e. in the reverse order.

Note that we are not saying precisely that the dual point of view has its arrow of time reversed relative to ours. For the sake of discussion, we have synchronised the 'events' in the two points of view. In the first 'event' the state/observable is 1, in the second it is ϕ_1, and in the third it is $\phi_1 \phi_2$, etc. But after n such 'events', what we see is a state to which we give one historical interpretation as a sequence of steps, while what is seen in the dual point of view is some observable (or Fock state) to which is given a historical interpretation as a creation process of a reversed sequence of steps.

On the other hand, there is clearly some kind of time-reversal going on in this dualisation, and it is interesting to speculate about the deeper significance of the above computations. Note that this phenomenon is not simply a matter of conventions but has a more fundamental origin in the fact that the adjoint of a series of maps, such as the composed T_{ϕ_i} in Proposition 5.2.5, is necessarily the composition of their adjoints but in the reversed order of application. Related to this, let us recall that in Theorem 1.7.4, both A and A^* can be realised concretely as operators on the same space. The coproduct of A (used by us in making our random walk) is obtained by embedding a as $a \triangleright \otimes \mathrm{id}$ and evolving with W, while the coproduct of A^* is obtained by embedding ϕ as $\mathrm{id} \otimes \phi \triangleright$ and evolving with W^{-1}. This again represents a kind of reversal between our view and the dual one.

Apart from this subtle reversal of order, we see that we have succeeded in identifying the dual process to a random walk as something equally familiar in quantum theory. When we allow both processes, our model is, in some sense, self-dual. The dual point of view has the same conceptual structure with the role of random walk and creation process interchanged. These considerations indicate, then, some of the philosophical or conceptual ideas that one can explore when the algebra of observables of a quantum system is a Hopf algebra. We still have to give some realistic examples of such quantum systems, and this we do in the next chapter.

Notes for Chapter 5

The use of coalgebras and bialgebras in combinatorics has a long history in the works of G.-C. Rota and others; see the review [105]. In fact, most applications in combinatorics concern coalgebras much as in Lemmas 5.1.1 and 5.1.5. This should not come as a surprise: coalgebras are about as common as algebras and hence arise naturally in many situations. They represent a point of view. One has coalgebras associated to subspaces of a linear space, to algorithms for iterative solutions of differential equations, to puzzles such as Latin squares, and much more. Rather more rare is the case when these form a bialgebra or Hopf algebra. That the coalgebra of partitions forms a bialgebra (the Faà di Bruno bialgebra) was proven by P. Doubilet [106].

The unshuffle bialgebra is even more standard [7]. We know from Proposition 5.1.4 that it is nothing other than the tensor algebra with no relations. Since this bialgebra is graded, one can define in a nice way its dual as built on the duals of each of its finite-dimensional spaces of given degree. In this case, the dual of the unshuffle bialgebra comes out as the *shuffle bialgebra*. It too consists of formal words in an alphabet Ω, but has a new (commutative) product given as the formal sum of all possible shuffles of one word into another word. Its coproduct, on the other hand, is noncommutative and consists of simply cutting the word into two in each possible place. For example,

$$xy \cdot z = xyz + xzy + zxy, \quad \Delta xyz = \otimes xyz + x \otimes yz + xy \otimes z + xyz \otimes .$$

Some applications to computer-science in connection with series-parallel problems have been proposed in [107] and elsewhere.

The bialgebra structure on families of trees in Proposition 5.1.6 is due to A. Dür [108] in a dual form and to R. Grossman and R.G. Larson [109] in the form given; the latter also studied labelled trees and applications to symbolic solving of partial differential equations. Some further generalisations to incidence bialgebras and Hopf algebras to families of graphs (and an elucidation of the antipode in such cases) can be found in [110]. The elementary $\mathbb{Z}/2$-Hopf algebra point of view on the Boolean algebra of subsets on a group in Proposition 5.1.7 features in work by the author.

Following this combinatorial work, it was natural to consider Hopf algebras as suitable for random walks and diffusion processes. This is a topic which comes under the heading of 'quantum probability'. The first general formulation to include noncommutative and noncocommutative Hopf algebras (as well as super ones) is in [111]. In this approach, however, the main topic of interest is not so much a discrete random walk but the structures that can arise in a suitable continuous limit. These are called *quantum stochastic processes*, and a fundamental question is how they may be represented on Fock space. This was done for general bialgebras by M. Schürmann [112], who constructed such a process for every bialgebra equipped with suitable data to define an exponential state. In the course of this (in the converse direction), Schürmann showed that, given a pre-Hilbert space $\mathcal{H}$ on which A acts and linear maps $D : A \to \mathbb{C}, \eta : A \to \mathcal{H}$ forming a 'cocycle' in the sense

$$a \triangleright \eta(b) = \eta(ab) - \eta(a)\epsilon(b), \quad \epsilon(a)D(b) - D(ab) + D(a)\epsilon(b) = -\langle \eta(a^*) | \eta(b) \rangle,$$

and if D is Hermitian in the sense $D(a^*) = \overline{D(a)}$, then it is *conditionally positive* and normalised in the sense

$$D((a - \epsilon(a))^*(a - \epsilon(a))) \geq 0, \qquad D(1) = 0.$$

This is not yet a state, but its exponential e^{tD} in the convolution algebra (5.12) is easily seen to be a state. We have seen an example of this in Section 5.2.1, where we built the pointwise-evaluation state ϕ_b in this way as the exponential of differentiation. The general theory provides a way of building states or positive linear functionals from infinitesimal data. Moreover, one can classify the possible D obeying the cocycle condition using cohomology.

Our more primitive treatment of random walks, rather than the associated diffusions or stochastic processes, is taken from the author's work [20], as is the noncommutative and noncocommutative toy example in Section 5.3. Its limiting q-diffusion process is also obtained in this work, and has some similarities with the stochastic process studied [112, 113] in the context of the Azema martingale. Other examples of random walks (such as on group duals) have been found by P. Biane [114]. Our elementary Examples 5.2.3 and 5.2.4 appear to be new. The active Markov transition operator and its representation in Proposition 5.3.1 is borrowed from the more general idea of a *dilation*, where the general form (though not in a bialgebra context) is a standard feature. Namely, when one has a completely positive (and in some sense 'irreversible') operator on an algebra A (like our T_ϕ), one can generally tensor product the algebra with another algebra such that the evolution on the joint system is unitary and hence reversible, while our original irreversible process is recovered from the dilated reversible one by projecting back with a conditional expectation.

Such ideas are useful in a broader context of quantum stochastics, not necessarily tied to bialgebras; a useful text is [115]. The formalism has also been applied to the problem of modelling continuous measurements in quantum mechanics. Here, the 'partial collapse' due to random measurements of the quantum system is modelled as a dissipative term in Schrödinger's equation. The evolution in this case tends to be irreversible. There are applications in quantum optics and in the theory of quantum systems coupled to heat baths.

The observable–state duality ideas in Section 5.4 were introduced by the author [116, 117, 118], and developed further in the context of random walks and their duals in [20]. Of course, we explored mainly conceptual questions here, but we will see rather more nontrivial examples and consequences in the next chapter. The basic question of when is the quantum algebra of observables of an actual quantum system a bialgebra or Hopf algebra was introduced in the author's Ph.D. thesis [116]. By contrast, the bialgebras and Hopf algebras in combinatorics and stochastic processes have so far been technical tools and not quantum-mechanical algebras of observables in this sense. Finally, for a general abstract approach to quantum theory (using C^* algebras and von Neumann algebras), see the text [119]. The GNS construction is standard now, but was pioneered by I.M. Gelfand, M.A. Naimark and I. Segal, as well as by G.W. Mackey.

6

Bicrossproduct Hopf algebras

This chapter is devoted to the description of a large class of noncommutative and noncocommutative Hopf algebras, called *bicrossproduct Hopf algebras*. Although less well-known than the familiar quantum groups of Chapters 3 and 4, they have the merit of arising genuinely as the quantum algebras of observables of quantum systems. They were introduced by the author in this context, as part of an algebraic approach to unifying quantum mechanics and gravity. This is the theme of the present chapter. Of course, these noncommutative and noncocommutative Hopf algebras are also interesting on purely mathematical grounds and can be fed into any of the modern quantum groups machinery.

The idea behind the construction of these bicrossproduct Hopf algebras is self-duality. Recall from Chapter 1 that the axioms of a Hopf algebra have a remarkable input–output symmetry. We have already seen some of the physical implications of this for a quantum algebra of observables in Chapter 5.4. Another more geometrical interpretation is that of putting quantum mechanics and gravity on an equal but mutually dual footing. The reason for this is that the quantum algebra of observables is a noncommutative version of the algebra of functions $C(X)$, where X is the classical phase-space. The degree of noncommutativity is controlled by $\hbar$, the physical Planck's constant. On the other hand, if this is a Hopf algebra, then we have the additional structure of a coproduct, which corresponds in the classical case to a group structure, as we know from Example 1.5.2. A noncocommutative coproduct corresponds, then, to a non-Abelian group structure on phase-space, but phrased in an algebraic way that makes sense for the quantum algebra of observables as well. But a non-Abelian group structure on phase-space means that the phase-space, as a Riemannian manifold, has curvature. In a physical setting, this should correspond also to curvature in the geometry of the underlying classical system of which X is the phase-space. In very general terms, the degree of noncocommutativity in the coproduct on the quantum algebra of observables measures curvature in the system and hence should be controlled by the gravitational coupling constant G. From this point of view, the search for a quantum algebra of observables which is a bialgebra or Hopf algebra is the search for a simple model in which both quantum effects and gravitational effects are unified into a single algebraic framework, and, moreover, in which they are broadly dual to each other.

In particular, if one believes in this philosophy, then a new and novel possibility opens up: suppose we have such a quantum system which is a Hopf algebra. Then we can, with some analysis in the infinite-dimensional case, form a dual Hopf algebra. It might be that this too is the quantum algebra of observables of some other quantum system. In this case, the product of our original system is the coproduct, and hence describes the geometry, of our dual system. Likewise, the coproduct or geometry of our original system is the product, and hence describes the quantum structure, of our dual system. In the best case, it might even be that this dual point of view is isomorphic to our original system with the roles of $\hbar$ and G reversed. This corresponds mathematically to the Hopf algebra being self-dual.

All of this is exhibited by the bicrossproduct Hopf algebras, providing their original motivation from physics. From the mathematical side, the idea behind the construction of this chapter is to combine the idea of cross product algebras from Proposition 1.6.6 with the idea of self-duality. Recall from Chapter 1.6 that these cross products arise when a Hopf algebra acts on an algebra. We begin in Section 6.1 by interpreting such cross products as quantisation, following established ideas in mathematical physics. Hence, we are led to ask: when is a cross product algebra a Hopf algebra? When is it self-dual or of self-dual form? The answer is to assume that the algebra which is acted upon is also a Hopf algebra and that it coacts back on the Hopf algebra which was acting. Then, by making the dual cross coproduct construction for the coalgebra, we will be forced to have something of manifestly self-dual form. If the action and coaction are suitably compatible, we will have a Hopf algebra. This is the idea behind the bicrossproducts, as we shall see in Section 6.2.

The simplest cases of bicrossproducts are associated to the following data. Suppose that a group factorises into two subgroups as a product GM. Then we will see that G acts on M and M acts back on G. This action and 'reaction' are just what it takes to get a bicrossproduct of the form

$$k(M) \!\blacktriangleright\!\!\triangleleft kG = \begin{cases} k(M) \!\rtimes\! kG & \text{as an algebra} \\ k(M) \!\blacktriangleright\!\!< kG & \text{as a coalgebra.} \end{cases}$$

It forms a Hopf algebra, and the dual Hopf algebra is $kM \!\blacktriangleright\!\!\triangleleft k(G)$, i.e. it

is of the same general form. Physically, for the first system, the group
M is position space and G is called the momentum group. When particles moving in M on orbits under G are quantised, they give this Hopf
algebra. Thus, in a Lie group and von Neumann algebra setting, we have
that the quantum algebras of observables for particles moving on certain
homogeneous spaces are indeed Hopf–von Neumann algebras. The dual
Hopf algebra then has just the same physical interpretation but with the
roles of M and G, i.e. of position and momentum, and the roles of the
action and reaction, interchanged.

Moreover, although this is a very natural way for the factorisation of
groups to arise, these same factorisation data are very common in mathematics and arise in a number of other contexts as well. We will see this
in Chapter 7 where, when generalised to the quantum group setting in
a more direct way, factorisation ideas lead to the quantum double Hopf
algebra. We will meet these ideas again at the Lie algebra and Lie group
level in Chapter 8, where they lead to classical integrable systems and
the theory of classical inverse scattering. The physical picture in these
applications is quite different from that in the present chapter. Likewise,
the resulting Hopf algebras are quite different.

In Section 6.3 we put these constructions into the framework of some
general extension theory for Hopf algebras. The general theory gives
slightly more than the bicrossproducts given by actions and coactions,
namely it allows the additional possibility of cocycles and dual-cocycles.

Finally, Section 6.4 returns to the deeper philosophical implications behind these constructions. One can argue that the self-duality phenomenon
demonstrated by bicrossproducts is not limited to Hopf algebras, which
are only the simplest algebraic structures that demonstrate a very general
duality between quantum and gravitational physics. From this point of
view, Einstein's equation, which equates gravitational and quantum parts
of physics, should ideally appear as some kind of self-duality condition.
This section is strictly for inspirational purposes. However, at least in the
author's view, we live in an era when both quantum physics and geometry
can be cast into a single algebraic language wherein duality ideas such as
these can and should be explored by physicists and mathematicians.

6.1 Quantisation on homogeneous spaces

We recall from Chapter 1.6 that when a group acts on an algebra by
automorphisms we can make a cross product algebra by this action. As
we saw in Proposition 1.6.6, this notion also makes sense for the action of a
general Hopf algebra or bialgebra. In this section, we develop the standard
picture of such cross products as quantisation, albeit in fairly algebraic

terms. This assumes an interest in geometry and quantum mechanics. The uninterested reader can just as easily proceed directly to the next section.

We shall consider, for our classical system, particles moving on homogeneous spaces. Thus, the *classical data* for us means (G, M, α), where, in the nicest case, M is a smooth manifold and G is a Lie group with Lie algebra g acting on M from the right by a map $\alpha : M \times G \to M$. In the best case, we suppose that the action is transitive and effective so that we can identify $M \cong_H \backslash G$, where $H \subset G$ is the isotropy group consisting of the elements of G that fix a point in M. Each element $\xi \in g$ generates a flow in M of the form

$$s(t) = \alpha_{e^{t\xi}}(s(0)), \qquad s(t) \in M, \quad t \in \mathbb{R}.$$

A fact from differential geometry is that, for G semisimple and H reductive, there is an indefinite metric on M such that these flows are the geodesics for this metric. The metric is induced by the Killing form and is Riemannian if G is compact. Otherwise it may be pseudo-Riemannian. This is the natural metric on a homogeneous space, as it appears on M.

Note that when describing the motion of the particle one needs to refer to the points of M, but one could equally well work with the algebra of functions $C^\infty(M)$, say. These are called the *classical position observables*. If the particle is at a point s in M, then this is equivalent to specifying all the values $f(s)$ for all $s \in M$. The various trajectories, or geodesics, are labelled by elements of the Lie algebra g, and so these are called the *classical momentum observables*. Another equivalent set-up is to work in terms of the phase-space $X = T^*M$, or rather with functions on it. This algebra $C^\infty(X)$ is often called the *classical algebra of observables*. But we prefer to work with the subset

$$g \otimes C^\infty(M) \subset C^\infty(X),$$

where $C^\infty(M) \subset C^\infty(X)$ via the pull-back of the projection $T^*M \to M$, and $g \subset C^\infty(X)$, via the differential map $\alpha_* : g \to \text{Vect}(M)$ and the inclusion $\text{Vect}(M) \subset C^\infty(X)$ which is given by evaluating a vector field with a covector. Explicitly, the vector fields $\alpha_*(\xi)$ induced by the action are defined by

$$(\alpha_* \xi)(f)(s) = \left.\frac{d}{dt}\right|_0 f(\alpha_{e^{t\xi}}(s)), \qquad \forall f \in C^\infty(M).$$

So, by working with the position and momentum observables as stated, we are picking out a natural subset of the algebra of observables. This consists of the functions on the phase-space that are linear in the fibre direction, where the precise choice is adapted to the action α. We only try to quantise this subset of observables. Such a restriction is needed in any approach to quantisation.

Now let us ask what it means to actually quantise this classical system. In physics, quantisation has two steps. First, one constructs a noncommutative *quantum algebra of observables*, and secondly one introduces a Hamiltonian to describe the dynamics. The first task is called the *quantum kinematics* and is the part that concerns us here. Usually, the quantum algebra of observables should be some *-subalgebra of operators on a Hilbert space, and it should, of course, contain the quantum counterparts of the classical observables of interest. In the present case we have identified the latter as the position and momentum observables above. We include them as subalgebras. These inclusions are called the *quantisation* maps, and we denote them by $\hat{}$, so we require

$$[\widehat{\xi}, \widehat{\eta}] = \widehat{[\xi, \eta]}, \quad \forall \xi, \eta \in g, \qquad \widehat{f}\widehat{h} = \widehat{fh}, \quad \forall f, h \in C^\infty(M). \qquad (6.1)$$

Finally, inside the quantum algebra of observables, we should have non-commutation relations between some of the observables. In our context we require the Heisenberg commutation relations between position and momentum:

$$[\widehat{\xi}, \widehat{f}] = (\widehat{\alpha_* \xi})(f), \qquad \alpha_* \xi = O(\hbar). \qquad (6.2)$$

This prescription follows a fairly conventional pattern except that we have written everything in a coordinate-invariant way. We assume that the action α has in it a parameter $\hbar$ such that the action becomes trivial as $\hbar \to 0$. In the physical interpretation, normalised vectors ψ of our Hilbert space determine the expectation values of our observables. So $\langle \psi | \widehat{f} | \psi \rangle$ is the expectation value of a position function $\widehat{f}$. From this point of view, the natural assumption, slightly weaker than the one above, would be that the various operators $\widehat{f}$ commute among themselves. Then any finite number of them can be simultaneously diagonalised. Hence, there are simultaneous eigenstates ψ_s, which we label by $s \in M$, such that the eigenvalue of $\widehat{f}$ is the classical value $f(s)$. The particle truly appears to be at s when it is in the state ψ_s, at least in so far as is detectable by the chosen position observables. To ensure this, we may as well assume that the classical position observables are embedded as a subalgebra. Then $\widehat{f}\widehat{h}\psi_s = \widehat{f}h(s)\psi_s = f(s)h(s)\psi_s = (\widehat{fh})(s)\psi_s = \widehat{fh}\psi_s$. This is the reason for our inclusion of the position observables $C^\infty(M)$ as a subalgebra of the quantum ones. Typically, in a real system, it is not possible to diagonalise all the f in this way: one has only approximate simultaneous eigenstates ψ_s, for which the particle can be said only to be near s.

More generally, if we fix some local coordinate chart, with coordinates $q = (q_i)$ valid in some open subset $U \subset M$, then we have the expectations $\langle \psi | \widehat{q_i} | \psi \rangle$ for any state ψ. If we are able to identify these expectations as $q_i(s)$ for some point $s \in U$, then s is the expectation value of the position.

However, if $\langle \psi | \hat{q}_i \psi \rangle$ lie outside the image of U, this interpretation breaks down, i.e. it is only local. Likewise, let $\{e_i\}$ be a basis for the Lie algebra g. Then the expectations $\langle \psi | \hat{e}_i \psi \rangle = p_i$ are the components of the expected momentum of the state. Here $p = (p_i) \in g^*$ is characterised abstractly as $\langle \psi | \hat{\xi} \psi \rangle = \xi(p)$ for all $\xi \in g$.

Finally, if K_{ij} is the Killing form of g with inverse K^{ij}, the operator $K^{ij} \hat{e}_i \hat{e}_j$ leads to a natural choice of Hamiltonian for a free particle. By definition, commutation with this operator gives the rate of change of the observables of the system (the Heisenberg formulation of quantum mechanics) and coincides to lowest order in $\hbar$ with the classical geodesic motion. At higher order the particle appears also to be under the influence of a gauge field. This is well-known, for the Hamiltonian corresponds here to motion on G which projects down to a metric + gauge field on $H \backslash G$; this is the usual point of view in Kaluza–Klein theory. Our choice is not the only possible one, but at least for this Hamiltonian the fact that the enveloping algebra $U(g)$ appears as a subalgebra of the quantum algebra of observables, and that K is ad-invariant, ensures that the momentum observables $\hat{e}_i$ are constant under this evolution. They commute with the free particle Hamiltonian and, moreover, via the Heisenberg commutation relations, implement infinitesimal translations in the group. This justifies the requirements above for our quantisation maps.

The construction of an operator algebra containing g and $C^\infty(M)$, embedded via maps $\hat{\ }$ as above, is the *quantisation problem* for our classical system. Another formulation (where the operators can be required to be bonded) is to work with the group G rather than the Lie algebra. Then we should work with the Heisenberg commutation relations in the group form

$$\widehat{e^{t\xi}} \, \hat{f} \, \widehat{e^{-t\xi}} = \widehat{\alpha_{e^{t\xi}}(f)}, \quad \alpha_{e^{t\xi}}(f)(s) = f(\alpha_{e^{t\xi}}(s)) \tag{6.3}$$

for all $e^{t\xi} \in G$ and $f \in C^\infty(M)$. The given right action of G on M induces the stated left action (again denoted by α) on the position observables. We can also generalise slightly and allow G to be locally compact rather than a Lie group at all. We call it the *momentum group*. For example, it could be finite. Likewise, we can allow M to be a locally compact space rather than a manifold, for example a finite set. Instead of smooth functions on M, we take for our position observables the continuous functions $C(M)$ vanishing at infinity (in a C^* algebra context) or $L^\infty(M)$ in a von Neumann algebra context if M is a measure space. This greater generality is quite convenient from the point of view of the quantum system. Finally, we do not really need that the action $\alpha : M \times G \to M$ is transitive etc. If not, then the system is not that of a particle moving on one homogeneous space, but describes a particle on M constrained to move on any one of the orbits of G in M. In the quantum picture, we do not have definite positions anyway and we may as well treat all the possible orbits together. In this looser setting then, our classical data is (G, M, α), where G is a group acting by α on M. The quantisation problem in this setting is to find an operator algebra and inclusions $\hat{}$ containing the position observables algebra and the momentum group, subject to the group-version of the Heisenberg commutation relations.

For the usual quantum-mechanical interpretation, the operator algebra should be a *-subalgebra of the bounded operators on a Hilbert space, and the quantisation maps should be *-algebra maps. So $\hat{}$ of an element of G should be unitary (or in the Lie algebra case $\hat{\xi}$ should be anti-Hermitian), the Hamiltonian mentioned above should be Hermitian, and the real-valued position observables should be Hermitian. The C^* algebra version of the above is called a *dynamical system*. However, as we explained in Chapter 5.3, one does not need any actual Hilbert space. Even in regular quantum mechanics, one can work more generally with the expectation values of the algebra of observables rather than with Hilbert space vectors. These expectation values are positive linear functionals on the algebra of observables and constitute the quantum states of the system from an algebraic point of view.

For our algebraic model, we will go one step further and work with everything finite, but over an arbitrary field k. In this case, we no longer have the notion of positivity and hence we have no direct probabilistic interpretation of the constructions. However, the purely algebraic quantisation problem is still well defined. So we have (G, M, α), where G is a finite group, M is a set and α is a right action. The latter defines equally well a left action of G on the algebra of functions $k(M)$. The algebraic quantisation problem is to find an algebra B and maps

$$k(M) \overset{\frown}{\rightarrow} B \overset{\frown}{\leftarrow} kG, \quad \widehat{u}\widehat{f}\widehat{u^{-1}} = \widehat{\alpha_u(f)}, \quad \forall u \in G, \ f \in k(M). \qquad (6.4)$$

We will see now that this algebraic problem is universally solved by the cross product construction that we have already encountered in Chapter 1.6.

Proposition 6.1.1 *Let G be a finite group acting from the right by α on a set M. Then the cross product algebra $k(M) \rtimes kG$ from Proposition 1.6.6, with its canonical inclusions of kG and $k(M)$, is a universal solution of the algebraic quantisation problem; i.e. it solves the problem, and if we have any other solution (6.4) then there is a unique algebra map*

$$\phi_B : k(M) \rtimes kG \to B$$

such that $1 \otimes u$ maps to $\hat{u}$ and $f \otimes 1$ maps to $\hat{f}$ for all $u \in G$ and $f \in k(M)$.

Proof: First, we check that $k(M)$ is indeed a left kG-module algebra. The action is by $(u \triangleright f)(s) = f(\alpha_u(s))$, and we have $(u \triangleright (fh))(s) = (fh)(\alpha_u(s))$ $= f(\alpha_u(s))h(\alpha_u(s)) = ((u \triangleright f)(u \triangleright h))(s)$ as required. Hence we can make the construction in Proposition 1.6.6. This is therefore built on $k(M) \otimes kG$ with product

$$(f \otimes u)(h \otimes v) = f(u \triangleright h) \otimes uv.$$

Another convenient description of the product is to take a basis of Kronecker δ-functions $\{\delta_s\}$ for $k(M)$ and the basis $\{\delta_s \otimes u\}$ for $k(M) \rtimes kG$. Then note that $u \triangleright \delta_t = \delta_{\alpha_{u^{-1}}(t)}$ and that the product of this with δ_s is zero unless it equals δ_s, i.e. unless $t = \alpha_u(s)$. Hence the product in this basis is

$$(\delta_s \otimes u)(\delta_t \otimes v) = \delta_{\alpha_u(s),t}\delta_s \otimes u, \quad \text{i.e.} \quad s \xrightarrow{u} t \xrightarrow{v} = \begin{cases} s \xrightarrow{uv} & \text{if } \alpha_u(s) = t \\ 0 & \text{otherwise}, \end{cases}$$

where the second expression is a compact shorthand notation in which we write $s \xrightarrow{u} = \delta_s \otimes u$. The unit is $\sum_s \delta_s \otimes e = \sum_s s \xrightarrow{1}$. The arrow notation expresses the idea that $u \mapsto \alpha_u(s)$ in M. Such transformations compose *iff* the end point of one arrow is the base of the next arrow. Otherwise the product is zero. Hence it is very easy to work with this cross product algebra.

Next, we know that $k(M) \otimes 1$ and $1 \otimes kG$ are natural subalgebras of the cross product, and we noted that one could equally well consider the cross product as generated by the two subalgebras with certain cross relations. These are now $uf = (u \triangleright f)u$, i.e. just (6.4). So from the proof of Proposition 1.6.6 we see that $k(M) \rtimes kG$ solves the quantisation problem, with the quantisation maps given by the obvious inclusions. That the cross product is freely generated by the subalgebras modulo these cross relations means that it is universal in the sense that any B also obeying these relations is a quotient of it. To put this a little more formally, suppose that we have a solution of the quantisation problem. Then let $\phi(f \otimes u) = \hat{f}\hat{u}$. This is a map from the cross product into B and obeys $\phi_B(f \otimes u)\phi_B(h \otimes v) = \hat{f}\hat{u}\hat{h}\hat{v} = \widehat{f u \triangleright h}\widehat{uv} = \phi_B(f(u \triangleright h) \otimes uv) = \phi_B((f \otimes u)(h \otimes v))$ as required. We also have $\phi_B(1 \otimes 1) = 1$ and that it restricts to $\hat{\ }$ on the two subalgebras as claimed. It is easy to see that this fixes ϕ_B.   ∎

It is also clear that this algebraic formulation of quantisation on homogeneous spaces works perfectly well if kG is replaced by an arbitrary bialgebra or Hopf algebra, and if A is replaced by an arbitrary H-module algebra. The formalisation of the universal property of cross products is then

Proposition 6.1.2 *Let A be a left H-module algebra and let $A \rtimes H$ be the associated cross product algebra as in Proposition 1.6.6, with $A \hookrightarrow$*

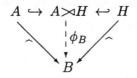

Fig. 6.1.   Universal property of the cross product algebra.

$A \bowtie H \hookleftarrow H$ *the standard inclusions. If we have an algebra* B *and maps* $A \overset{\frown}{\to} B \overset{\frown}{\leftarrow} H$ *such that*

$$\sum \widehat{h_{(1)}} \widehat{a} S\widehat{h_{(2)}} = \widehat{h \triangleright a}, \qquad \forall h \in H, a \in A$$

then there is a unique map $\phi_B : A \bowtie H \to B$ *such that the triangles in Fig. 6.1 commute. This is called the* universal property of cross products.

Proof: This is the general Hopf algebra version of the last example, and the proof is similar. We know from Proposition 1.6.6 that inside the cross product algebra we have the relations

$$\widehat{h}\widehat{a} = \sum \widehat{h_{(1)} \triangleright a}\,\widehat{h_{(2)}},$$

which are equivalent to the form shown when we have an antipode (if H is only a bialgebra then we should use this form of the relations rather than the form stated in the proposition). Now, given an algebra B and the maps $\widehat{}$, we define $\phi_B(a \otimes h) = \widehat{a}\widehat{h}$ and verify that $\phi_B((a \otimes h)(b \otimes g)) = \phi_B(a(h_{(1)} \triangleright b) \otimes h_{(2)}g) = \widehat{a}\widehat{h_{(1)} \triangleright b}\,\widehat{h_{(2)}}\widehat{g} = \widehat{a}\widehat{h}\widehat{b}\widehat{g} = \phi_B(a \otimes h)\phi_B(b \otimes g)$. So we have an algebra map $\phi_B : A \bowtie H \to B$. Its restrictions to the subalgebras A, H clearly coincide with the maps $\widehat{}$, and, moreover, this property fixes ϕ_B uniquely since a general element of $A \bowtie H$ is a sum of terms of the form $a \otimes h = (a \otimes 1)(1 \otimes h)$. ∎

This proposition is a trivial repackaging of Proposition 1.6.6, but it does nevertheless provide a powerful conceptual way of thinking about cross products. The universal property of the cross product determines it completely up to isomorphism (another universal object of this type would have to be isomorphic via the maps ϕ induced by the universal property of each). Another easy statement, more or less equivalent to the universal property, is the following.

Corollary 6.1.3 *Let* A *be an* H-*module algebra. The representations of the associated cross product* $A \bowtie H$ *are in one to one correspondence with* the *H-covariant A-modules. These are vector spaces on which both H and A are represented and for which the latter map, $A \otimes V \rightarrow V$, is an intertwiner for the action of H. Explicitly, we need actions $\triangleright$ of H, A such that*

$$h \triangleright (a \triangleright v) = \sum (h_{(1)} \triangleright a) \triangleright (h_{(2)} \triangleright v), \qquad \forall a \in A, \ h \in H, \ v \in V.$$

Proof: Recall from Chapter 1.6 and Exercise 1.6.8 that one can think of the condition that A is an H-module algebra as assuming that the product is an intertwiner or morphism in the category of H-modules. The condition in the proposition likewise asks that A acts on V, but this action is H-covariant in the same way (a morphism in the category of H-modules). We will study such topics more systematically in Chapter 9. From Exercise 1.6.8 we see that the required condition is the one stated. It is easy to see that it works, for if $A \rtimes H$ is represented on V, we can restrict or pull back via the standard inclusions of A, H so that both of these are represented on V. The cross relations imply that these two representations obey our covariance condition. In the other direction, it is clear that, given such actions on V, we can define $(a \otimes h) \triangleright v = a \triangleright (h \triangleright v)$ to be the corresponding representation of $A \rtimes H$.

The result is also clear if we define the cross product abstractly in terms of the universal property. In one direction we have the canonical inclusions as before. Conversely, if we have representations ρ of H, A on V, we consider them as maps $\rho : A \rightarrow \mathrm{Lin}(V)$ and $H \rightarrow \mathrm{Lin}(V)$. The intertwiner property

$$\rho(a)\rho(h)\rho(b)\rho(g) = \sum \rho(a(h_{(1)} \triangleright b))\rho(h_{(2)}g)$$

just says that we have a solution of the quantisation problem with $B = \mathrm{Lin}(V)$ and $\widehat{\ } = \rho$. Then we apply the universal property and get a representation $A \rtimes H \rightarrow \mathrm{Lin}(V)$. ∎

Example 6.1.4 *Let A be an H-module algebra. Then the cross product $A \rtimes H$ has a canonical representation on A itself, which we call the Schrödinger representation. It is*

$$(a \otimes h) \triangleright b = a(h \triangleright b), \qquad \forall a, b \in A, \ h \in H.$$

Proof: A acts on itself by the product map $A \otimes A \rightarrow A$. The module algebra assumption says that this map is an intertwiner for the action of H. Hence we have an example of the preceding corollary with $V = A$. One can check it directly as well. ∎

We call this the Schrödinger representation because, for the example $k(M) \rtimes kG$ above, the representation is on $k(M)$ itself. The action of the

$k(M)$ part of the cross product is by left multiplication. The action of kG is the one used in the cross product, which in this case is the one induced by the right action of G on M. This is just what we do in quantum mechanics. Note that for a quantum-mechanical picture we also need $*$-structures. All of the above is compatible with them if we have a unitary action in a suitable sense.

Proposition 6.1.5 *Let A be a left H-module algebra, where A is a $*$-algebra and H is a Hopf $*$-algebra over $\mathbb{C}$. Then the cross product is a $*$-algebra and the canonical inclusions are $*$-algebra maps iff*

$$(h \triangleright a)^* = (Sh)^* \triangleright a^*, \qquad \forall h \in H, \ a \in A.$$

In this case we say that the action is a $*$-action or unitary. Moreover, this $*$-algebra cross product is universal among solutions of the $*$-algebra quantisation problem for this data. Its $*$-representations are the H-covariant $*$-representations of A. The Schrödinger representation of Example 6.1.4 is an example.

Proof: The requirement that the standard inclusions are $*$-algebra maps is $(a \otimes 1)^* = a^* \otimes 1$ and $(1 \otimes h)^* = 1 \otimes h^*$, and this forces us to define

$$(a \otimes h)^* = ((a \otimes 1)(1 \otimes h))^* = (1 \otimes h^*)(a^* \otimes 1) = \sum h^*{}_{(1)} \triangleright a^* \otimes h^*{}_{(2)}$$

for the $*$-operation on $A \rtimes H$. By definition, this is an antialgebra map on the products shown. For a general product we have

$$\begin{aligned}
((a \otimes h)(b \otimes g))^* &= (a(h_{(1)} \triangleright b) \otimes h_{(2)} g)^* \\
&= (1 \otimes (h_{(2)} g)^*)((a(h_{(1)} \triangleright b))^* \otimes 1) \\
&= (1 \otimes g^*)(1 \otimes h_{(2)}{}^*)((h_{(1)} \triangleright b)^* \otimes 1)(a^* \otimes 1) \\
&= (1 \otimes g^*)(h_{(2)}{}^*{}_{(1)} \triangleright (h_{(1)} \triangleright b)^* \otimes h_{(2)}{}^*{}_{(2)})(a^* \otimes 1), \\
(b \otimes g)^*(a \otimes h)^* &= (1 \otimes g^*)(b^* \otimes 1)(1 \otimes h^*)(a^* \otimes 1) \\
&= (1 \otimes g^*)(b^* \otimes h^*)(a^* \otimes 1).
\end{aligned}$$

If the action is $*$-preserving as stated, then these expressions are equal because the middle term in the first case is $(h_{(2)}{}^*{}_{(1)} S^{-1} h_{(1)}{}^*) \triangleright b^* \otimes h_{(2)}{}^*{}_{(2)} = (h^*{}_{(2)} S^{-1} h^*{}_{(1)}) \triangleright b^* \otimes h^*{}_{(3)} = b^* \otimes h^*$ on using the properties of Hopf $*$-algebras. Conversely, if the equality of the two expressions holds for all $b^* \otimes h^*$, then we can apply it instead to $(S^{-1} h_{(1)} \triangleright b)^* \otimes h_{(2)}{}^*$ and hence deduce that $(S^{-1} h_{(1)} \triangleright b)^* \otimes h_{(2)}{}^* = h_{(2)(2)}{}^*{}_{(1)} \triangleright (h_{(2)(1)} S^{-1} h_{(1)} \triangleright b)^* \otimes h_{(2)(2)}{}^*{}_{(2)} = h^*{}_{(1)} \triangleright b^* \otimes h^*{}_{(2)}$. Now apply the map $* \otimes \epsilon \circ *$ to both sides to deduce that $(S^{-1} h) \triangleright b = (h^* \triangleright b^*)^*$, which is equivalent to the form stated. Let us also check

$$\begin{aligned}
*^2(a \otimes h) &= (1 \otimes h^*{}_{(2)}{}^*)((h^*{}_{(1)} \triangleright a^*)^* \otimes 1) = (1 \otimes h_{(2)})((Sh_{(1)}{}^*)^* \triangleright a \otimes 1) \\
&= (1 \otimes h_{(2)})(S^{-1} h_{(1)} \triangleright a \otimes 1) = (h_{(2)(1)} S^{-1} h_{(1)}) \triangleright a \otimes h_{(2)(2)} = a \otimes h,
\end{aligned}$$

as required for a ∗-structure.

Hence, $A \rtimes H$ solves the ∗-algebra quantisation problem with the standard inclusions as ∗-algebra maps. The general ∗-algebra quantisation problem is defined in the same way with $\hat{\ }$ assumed now to be ∗-algebra maps. It follows at once that the induced map ϕ_B for any other solution is also a ∗-algebra map since it is an algebra map and its restrictions to A, H are $\hat{\ }$. Finally, suppose that A, H are represented on a Hilbert space V in such a way that $\rho(a^*) = \rho(a)^\dagger$ and $\rho(h^*) = \rho(h)^\dagger$, where $\dagger$ is the ∗-operation on $\mathrm{Lin}(V)$ given by taking the adjoint operator. We have already encountered such ∗-representations in Chapter 3. We think of them abstractly as ∗-algebra maps $\rho : A \to \mathrm{Lin}(V)$ and $\rho : H \to \mathrm{Lin}(V)$. By the universal property, the induced map $A \rtimes H \to \mathrm{Lin}(V)$ is also a ∗-algebra map. ∎

All of this tells us that the cross product has all the properties that we need for a general solution of the quantisation problem in an algebraic setting. The same principle applies at the level of Lie groups or locally compact groups and C^* algebra or von Neumann algebra cross products, etc. This approach to quantisation is sometimes called *Mackey quantisation* and is quite popular in mathematical physics. We have seen that it also works perfectly well at the abstract Hopf algebra level also. Some canonical constructions are as follows.

Proposition 6.1.6 *Let H be a finite-dimensional Hopf algebra and let $w(H) = H \rtimes H^*$ be the cross product by the left coregular representation of H^* on H from Example 1.6.2. In this case, the Schrödinger representation is an isomorphism $H \rtimes H^* \cong \mathrm{Lin}(H)$ and, moreover, recovers the concrete realisations of $H \subseteq \mathrm{Lin}(H)$ and $H^* \subseteq \mathrm{Lin}(H)$ in Theorem 1.7.4.*

Proof: From Example 1.6.2 we know that we have H as an H^*-module algebra. Hence we have a cross product $H \rtimes H^*$. The action is $\phi \triangleright h = \sum h_{(1)} \langle \phi, h_{(2)} \rangle$ for $h \in H$ and $\phi \in H^*$. The explicit product structure from Proposition 1.6.6 is

$$(h \otimes \phi)(g \otimes \psi) = \sum h g_{(1)} \otimes \langle g_{(2)}, \phi_{(1)} \rangle \phi_{(2)} \psi.$$

From Example 6.1.5 above, we know that we have a canonical representation of $H \rtimes H^*$ on H itself. It is given by $\rho(h \otimes \phi)g = h g_{(1)} \langle \phi, g_{(2)} \rangle$, and we have to show that this map $\rho : H \rtimes H^* \to \mathrm{Lin}(H)$ is an isomorphism. Note that, from its definition, we see at once that it restricts to the realisations $H \subseteq \mathrm{Lin}(H)$ and $H^* \subseteq \mathrm{Lin}(H)$ already encountered in Chapter 1.7. To see the isomorphism, we construct the inverse map ρ^{-1}. Thus, let $A \in \mathrm{Lin}(H)$ and define

$$\rho^{-1}(A) = (A e_{a(2)}) S^{-1} e_{a(1)} \otimes f^a,$$

where $\{e_a\}$ is a basis of H and $\{f^a\}$ is the dual basis. We check

$$\rho(\rho^{-1}(A))g = (Ae_{a(2)})(S^{-1}e_{a(1)})g_{(1)}\langle f^a, g_{(2)}\rangle$$
$$= (Ag_{(2)(2)})(S^{-1}g_{(2)(1)})g_{(1)} = Ag,$$
$$\rho^{-1}(\rho(h\otimes\phi)) = (\rho(h\otimes\phi)e_{a(2)})(S^{-1}e_{a(1)})\otimes f^a$$
$$= he_{a(2)(1)}S^{-1}e_{a(1)}\langle\phi, e_{a(2)(2)}\rangle\otimes f^a = h\otimes\phi.$$

It is worth noting that the operators $W, W^{-1} \in \mathrm{Lin}(H)\otimes\mathrm{Lin}(H)$ in Theorem 1.7.4 are just the image under ρ of the canonical elements $(1\otimes f^a)\otimes(e_a\otimes 1)$ and $(1\otimes f^a)\otimes(Se_a\otimes 1)$ in $H{\rtimes}H^*\otimes H{\rtimes}H^*$. This provides a new proof of Theorem 1.7.4 by working in the algebra $H{\rtimes}H^*$, where it is immediate. ∎

We call this example $H{\rtimes}H^*$ the *Weyl algebra* of the Hopf algebra H because of the following example, which is an algebraic version of the usual Weyl algebra of quantum mechanics on a group.

Example 6.1.7 *The cross product $k(G){\rtimes}kG$ by the right action of G on itself is isomorphic to $\mathrm{Lin}(k(G))$ by the Schrödinger representation. It is the algebraic quantisation of a particle moving on G by translation. This example is called the* Weyl algebra *of the group.*

Proof: Here $\alpha_u(s) = su$ is the action by right multiplication. The induced left action is the left regular representation of G on $k(G)$. The Schrödinger representation is generated by this and by the action of $k(G)$ on itself by pointwise multiplication. We know from the preceding proposition that this representation fills out all of $\mathrm{Lin}(k(G))$. In other words, the claim is that every linear map $k(G) \to k(G)$ can be written as the action of a sum of products $f\otimes u$, where $f \in k(G)$ and $u \in G$. Over $\mathbb{C}$ we can make $\mathbb{C}G$ into a Hopf $*$-algebra by $u^* = u^{-1}$ and $\mathbb{C}(G)$ by $f^*(s) = \overline{f(s)}$ and make the copy of $k(G)$ that is acted upon into a Hilbert space via the integral via Proposition 1.7.6. Then the Schrödinger representation becomes an isomorphism of $*$-algebras. ∎

This example is our algebraic model of regular quantum mechanics on a group. When a group acts on itself by translation, the flows correspond to subgroups (and in the compact semisimple case, the associated metric is the usual Riemannian structure on the group). The most familiar case is

$$G = (\mathbb{R}, +); \qquad \alpha_u(s) = s + \hbar u. \tag{6.5}$$

This is not a finite group so the corresponding cross product requires some functional analysis (or a different algebraic model, see later); it is usually generated as a von Neumann algebra on the Hilbert space $L^2(\mathbb{R})$ by the action of pointwise multiplication and translation. An infinitesimal translation would mean differentiation which is unbounded, but finite translations are bounded operators, and this is why the group picture rather than a Lie algebra picture is useful here. Apart from this, we have the usual description of quantum mechanics in one dimension. If $q(s) = s$ is the coordinate function, the action is

$$(\hat{q}\psi)(s) = q(s)\psi(s), \quad \hat{u}\psi(s) = \psi(s + \hbar u), \qquad \forall \psi \in \mathcal{H}, \qquad (6.6)$$

and generates the group-version of the usual Heisenberg commutation relations in the cross product. Also, in this and similar cases, the algebra generated by the Schrödinger representation is essentially all of $B(\mathcal{H})$, which is the algebra of observables in this familiar case. We see that just the same features apply in our algebraic setting.

Another canonical construction is the semidirect product $H_{\mathrm{Ad}}{\rtimes}H$ by the quantum adjoint action in Example 1.6.3. We will be interested in this in Chapter 7 for the case when H is a factorisable Hopf algebra. Clearly, there are many other interesting examples as well. The quantum coadjoint action in Example 1.6.4, however, does not generally give a module algebra so we cannot use it to obtain a second general class of examples $H^*{}_{\mathrm{Ad}^*}{\rtimes}H$; but it does work if H is cocommutative, for, in this case, the coadjoint action does give a module algebra.

Example 6.1.8 $k(G)$ *is a* kG*-module algebra induced by the right adjoint action of* G *on itself (the coadjoint action). In this case,* $k(G){\rtimes}kG$ *is the algebraic quantisation of a particle constrained to move on conjugacy classes in* G*. The tensor product coalgebra makes it a Hopf algebra (nowadays called the* quantum double *of* G*).*

Proof: Unlike the previous Weyl algebras, where the action was transitive so that G had one orbit, we now have $\alpha_u(s) = u^{-1}su$ and there can be several orbits. By definition, they are the conjugacy classes in G. Our system quantises all of them simultaneously. On the other hand, this example has the interesting feature of being a Hopf algebra, albeit in a fairly trivial way. The coproduct is just the tensor product of the usual ones for $k(G)$ and kG. That these give a Hopf algebra will be proven in the next section as a case of Proposition 6.2.1, or of Example 6.2.12 with trivial β. One can check it directly as well. ∎

The Hopf algebra in this example is an important one, and we will study it further in Chapter 7. It is not usually thought of as a quantum algebra of observables, as we have presented it here, and is also in sharp contrast to the more familiar Weyl algebras. The more familiar Weyl algebras are never bialgebras or Hopf algebras, at least in the finite-dimensional case, because they are isomorphic to $\mathrm{Lin}(V)$ for some vector space V, and $\mathrm{Lin}(V)$ is never a bialgebra when V is nontrivial. To see this, suppose we have a counit $\epsilon : \mathrm{Lin}(V) \to k$. Since this is required to be an algebra map, its kernel is an ideal of $\mathrm{Lin}(V)$. But this algebra is simple and has no nontrivial ideals. Another way is to note that $\epsilon(A) = \mathrm{Tr}\, EA$ for some matrix E. That ϵ is an algebra map then quickly forces a contradiction. The infinite-dimensional case is more complicated, and, indeed, we have already seen a bialgebra in Example 1.3.2 which can be viewed as an infinite-dimensional Weyl algebra constructed via a duality pairing between two copies of $k\mathbb{N}$. In short, we can view cross products as quantisation on homogeneous spaces in some generalised form. In some cases the result is a Hopf algebra, and in some cases it is not.

Finally, although we have concentrated for simplicity on models based on finite groups, once we have the general Hopf algebraic formulation we can apply it also to other algebraic models of quantisation based on Lie algebras and Lie groups. We can replace the group algebra by the enveloping algebra from Example 1.5.7 and we can take for $k(G)$ the algebraic model of functions on G from Exercise 4.2.1. This gives a purely algebraic way of dealing with cases of interest in physics without involving too much functional analysis.

Thus, we saw in Exercise 1.6.5 that the notion of a module algebra A for $U(g)$ reduces to the idea that the Lie algebra g acts on A by derivations. Given such an action, the cross product $A {\rtimes} U(g)$ has the structure

$$(a \otimes \xi)(b \otimes \eta) = a(\xi {\triangleright} b) \otimes \eta + ab \otimes \xi\eta \qquad (6.7)$$

from Proposition 1.6.6. If we write $\hat{\xi} = 1 \otimes \xi$ and $\hat{a} = a \otimes 1$, then the cross relations in the cross product algebra are $\hat{\xi}\hat{a} = \widehat{\xi {\triangleright} a} + \hat{a}\hat{\xi}$, i.e. we have a (universal) solution of the quantisation problem (6.1)–(6.2) at the Lie algebra level. The Lie algebra version of the regular representation in Example 6.1.7 and the coadjoint representation in Example 6.1.8 are

Example 6.1.9 *Let $\rho : g \subseteq M_n$ be a Lie algebra and let $k(G)$ be the algebraic model of its Lie group as explained below Exercise 4.2.1, with matrix generators $t^i{}_j$. The vector fields for the action by right translation give a cross product Weyl algebra $k(G) {\rtimes} U(g)$. Explicitly,*

$$\xi {\triangleright} t^i{}_j = t^i{}_k \rho(\xi)^k{}_j, \quad [\hat{\xi}, \widehat{t^i{}_j}] = \widehat{t^i{}_k} \rho(\xi)^k{}_j, \quad i.e. \quad \xi {\triangleright} \mathbf{t} = \mathbf{t}\rho(\xi), \quad [\hat{\xi}, \hat{\mathbf{t}}] = \hat{\mathbf{t}}\rho(\xi).$$

Proof: We have already seen in Chapter 4.2 how to make an algebraic model $k(G)$, even without knowing the group G itself: we use ρ to define a pairing $\langle \mathbf{t}, \xi \rangle = \rho(\xi)$ between $k(M_n)$ and $U(g)$, and quotient by the null relations. In nice cases, the Lie algebra is associated to a group G and we can think of $t^i{}_j$ as its matrix coordinates, but this is not necessary for the construction. The duality pairing defines an action of $U(g)$ on $k(G)$, along the lines of Example 1.6.2, which comes out as stated in view of the matrix coproduct. One can also verify it directly. The resulting cross product comes out from Proposition 1.6.6, as explained above. Note that if we want $k(G)$ to be a Hopf algebra, we may need to add the inverse of the determinant function as well to our model, with cross relations fixed by those for t. We do not need any actual vector fields, but, in the geometrical setting, the Weyl algebra is the algebra of differential operators on G. It is not a Hopf algebra.   ∎

Example 6.1.10 *Let $\rho : g \subseteq M_n$ be as in the preceding example. The vector fields for the adjoint action give a cross product algebra $k(G) \rtimes U(g)$ with Lie algebra action and cross relations*

$$\xi \triangleright t^i{}_j = t^i{}_k \rho(\xi)^k{}_j - \rho(\xi)^i{}_k t^k{}_j, \quad [\hat{\xi}, \widehat{t^i{}_j}] = \widehat{t^i{}_k} \rho(\xi)^k{}_j - \rho(\xi)^i{}_k \widehat{t^k{}_j},$$

$$i.e. \quad \xi \triangleright t = [t, \rho(\xi)], \quad [\hat{\xi}, \hat{t}] = [\hat{t}, \rho(\xi)].$$

Proof: Again, we do not actually need the group G and actual vector fields in our algebraic model. We follow a version of Example 1.6.4 for the duality pairing as in Example 6.1.9. One can also verify the action directly. The cross product algebra then follows at once from Proposition 1.6.6. The cross product algebra in this case is a Hopf algebra, being a quantum double. In both examples, we do not need the pairing to be nondegenerate and we can construct $k(M_n) \rtimes U(g)$ in the same way.   ∎

Example 6.1.11 *Let $A = k[x]$ be the Hopf algebra of polynomials in one variable. Let $H = U(k)$, which we identify with $k[p]$, also the Hopf algebra of polynomials in one variable. Then*

$$p \triangleright f(x) = \hbar \frac{\partial}{\partial x} f(x), \quad [\hat{p}, \hat{x}] = \hbar$$

are the resulting action and cross product. The Lie algebra analogue of the Schrödinger representation gives the usual action on $\mathcal{H} = k[x]$ by multiplication and differentiation.

Proof: Here $k[x]$ is our algebraic model of the algebra of functions on $\mathbb{R}$, as generated by the coordinate function $x(s) = s$. We made a similar model in Chapter 5.2 in the context of random walks on $\mathbb{R}$. By contrast, $k[p]$ is our algebraic model of the momentum group $\mathbb{R}$ as an enveloping algebra of a Lie algebra. They are nevertheless isomorphic as Hopf algebras, both

having primitive coproduct $\Delta x = x \otimes 1 + 1 \otimes x$ and $\Delta p = p \otimes 1 + 1 \otimes p$. The rest is easy to check and is familiar from quantum mechanics. One can formally obtain the result as the infinitesimal version of (6.5)–(6.6) by taking the exponential map from $\mathbb{R}$ as the Lie algebra of $\mathbb{R}$ to $\mathbb{R}$ as the additive group, as the identity. So $u = tp$ is the parametrisation of group elements. To see that these points of view are consistent, one should keep in mind that $\widehat{u+v} = \hat{u}\hat{v}$ in the quantisation (6.6). Note that we have not put $\imath$ into our commutation relations here or in the above. In general, this means that when we work over $\mathbb{C}$, our Lie algebra basis vectors are represented by anti-Hermitian operators rather than by Hermitian ones.
∎

6.2 Bicrossproduct models

In the preceding section, we have formulated the quantisation of particles on orbits in algebraic terms, and we have seen that sometimes one can obtain a Hopf algebra for the quantum algebra of observables. The example of this phenomenon that we gave is of the following general form. The reader who is not yet comfortable with abstract Hopf algebra computations should proceed directly to the latter half of this section, where we give many more concrete examples. One can begin at Definition 6.2.10.

Proposition 6.2.1 *Let A, H be bialgebras and let A be an H-module algebra and an H-module coalgebra (one says that A is an H-module bialgebra). If H is cocommutative then the cross product $A \rtimes H$, with tensor product coalgebra structure, is a bialgebra. Likewise for Hopf algebras.*

Proof: We have a bialgebra with the tensor product coalgebra, since

$$\Delta((a \otimes h)(b \otimes g)) = \Delta\left(a(h_{(1)} \triangleright b) \otimes h_{(2)} g\right)$$
$$= a_{(1)}(h_{(1)} \triangleright b)_{(1)} \otimes h_{(2)(1)} \otimes a_{(2)}(h_{(1)} \triangleright b)_{(2)} \otimes h_{(2)(2)}$$
$$= a_{(1)}(h_{(1)(1)} \triangleright b_{(1)}) \otimes h_{(2)(1)} \otimes a_{(2)}(h_{(1)(2)} \triangleright b_{(2)}) \otimes h_{(2)(2)},$$
$$\Delta(a \otimes h)\Delta(b \otimes g) = (a_{(1)} \otimes h_{(1)})(b_{(1)} \otimes g_{(1)}) \otimes (a_{(2)} \otimes h_{(2)})(b_{(2)} \otimes g_{(2)})$$
$$= a_{(1)}(h_{(1)(1)} \triangleright b_{(1)}) \otimes h_{(1)(2)} g_{(1)} \otimes a_{(2)}(h_{(2)(1)} \triangleright b_{(2)}) \otimes h_{(2)(2)} g_{(2)},$$

where we have used the definition of $A \rtimes H$ from Proposition 1.6.6, and, for the third equality, the assumption that the A is a comodule algebra. We see that this only works for all such A if H is cocommutative. In this case, if H is a Hopf algebra then so is $A \rtimes H$. Its antipode is

$$S(a \otimes h) = (1 \otimes Sh)(Sa \otimes 1).$$

Both A and H are sub-Hopf algebras by the canonical inclusion. ∎

In this section we explain a natural generalisation of this result in which it is no longer assumed that H is cocommutative. Another generalisation to the case where H is quasitriangular will appear in Chapters 7.4 and 9.4 from the theory of braided groups. For the present work, the idea is the following: if we are to succeed and obtain a bialgebra or Hopf algebra of quantum observables, we will have a unified point of view in which the algebra is a quantum system and the coalgebra is some form of geometry. Hence, as mentioned in the preliminaries to this chapter, it is interesting to try to do this in a self-dual way. Thus, we should not only cross the product by an action of H on A, but we should also cross the coproduct by a coaction of A on H. This leads to the following general constructions, due to the author.

Theorem 6.2.2 *Let A, H be bialgebras, let A be a left H-module algebra and let H be a right A-comodule coalgebra by maps*

$$\alpha : H \otimes A \to A, \ \alpha(h \otimes a) = h \triangleright a, \quad \beta : H \to H \otimes A, \ \beta(h) = \sum h^{(\bar{1})} \otimes h^{(\bar{2})},$$

obeying the compatibility conditions

$$\epsilon(h \triangleright a) = \epsilon(h)\epsilon(a), \ \Delta(h \triangleright a) = \sum h_{(1)}{}^{(\bar{1})} \triangleright a_{(1)} \otimes h_{(1)}{}^{(\bar{2})}(h_{(2)} \triangleright a_{(2)}), \quad (6.8)$$

$$\beta(1) = 1 \otimes 1, \ \beta(gh) = \sum g_{(1)}{}^{(\bar{1})} h^{(\bar{1})} \otimes g_{(1)}{}^{(\bar{2})}(g_{(2)} \triangleright h^{(\bar{2})}), \quad (6.9)$$

$$\sum h_{(2)}{}^{(\bar{1})} \otimes (h_{(1)} \triangleright a) h_{(2)}{}^{(\bar{2})} = \sum h_{(1)}{}^{(\bar{1})} \otimes h_{(1)}{}^{(\bar{2})}(h_{(2)} \triangleright a), \quad (6.10)$$

for all $a, b \in A$ and $h, g \in H$. Then $A \rtimes H$ and $A \ltimes H$ form a bialgebra: the left–right bicrossproduct bialgebra associated to the compatible (co)actions and denoted by $A \blacktriangleright\!\!\triangleleft H$. If A, H are Hopf algebras then so is $A \blacktriangleright\!\!\triangleleft H$. Its antipode is

$$S(a \otimes h) = \sum (1 \otimes Sh^{(\bar{1})})(S(ah^{(\bar{2})}) \otimes 1).$$

Proof: To have a bialgebra, we need to verify that Δ and ϵ from the cross coproduct construction in Proposition 1.6.16 are algebra maps when we equip the same vector space $A \otimes H$ with the cross product algebra structure from Proposition 1.6.6. For the coproduct, we have

$$\Delta((a \otimes h)(b \otimes g)) = \Delta\left(a(h_{(1)} \triangleright b) \otimes h_{(2)}g\right)$$

$$= a_{(1)}(h_{(1)} \triangleright b)_{(1)} \otimes (h_{(2)}g)_{(1)}{}^{(\bar{1})} \otimes a_{(2)}(h_{(1)} \triangleright b)_{(2)}(h_{(2)}g)_{(1)}{}^{(\bar{2})} \otimes (h_{(2)}g)_{(2)}$$

$$= a_{(1)}(h_{(1)}{}^{(\bar{1})} \triangleright b_{(1)}) \otimes (h_{(3)}g_{(1)})^{(\bar{1})}$$

$$\otimes a_{(2)} h_{(1)}{}^{(\bar{2})}(h_{(2)} \triangleright b_{(2)})(h_{(3)}g_{(1)})^{(\bar{2})} \otimes h_{(4)}g_{(2)}$$

$$= a_{(1)}(h_{(1)}{}^{(\bar{1})} \triangleright b_{(1)}) \otimes h_{(3)(1)}{}^{(\bar{1})} g_{(1)}{}^{(\bar{1})}$$

$$\otimes a_{(2)} h_{(1)}{}^{(\bar{2})}(h_{(2)} \triangleright b_{(2)}) h_{(3)(1)}{}^{(\bar{2})}(h_{(3)(2)} \triangleright g_{(1)}{}^{(\bar{2})}) \otimes h_{(4)}g_{(2)}$$

$$= a_{(1)}({h_{(2)}}^{(\bar{1})} \triangleright b_{(1)}) \otimes {h_{(3)}}^{(\bar{1})} {g_{(1)}}^{(\bar{1})}$$
$$\otimes a_{(2)}(h_{(1)} \triangleright b_{(2)}) {h_{(2)}}^{(\bar{2})} {h_{(3)}}^{(\bar{2})} ({h_{(4)}} \triangleright {g_{(1)}}^{(\bar{2})}) \otimes h_{(5)} g_{(2)}$$
$$= a_{(1)}({h_{(2)}}^{(\bar{1})}{}_{(1)} \triangleright b_{(1)}) \otimes {h_{(2)}}^{(\bar{1})}{}_{(2)} {g_{(1)}}^{(\bar{1})}$$
$$\otimes a_{(2)}(h_{(1)} \triangleright b_{(2)}) {h_{(2)}}^{(\bar{2})} ({h_{(3)}} \triangleright {g_{(1)}}^{(\bar{2})}) \otimes h_{(4)} g_{(2)}$$
$$= a_{(1)}({h_{(1)}}^{(\bar{1})}{}_{(1)} \triangleright b_{(1)}) \otimes {h_{(1)}}^{(\bar{1})}{}_{(2)} {g_{(1)}}^{(\bar{1})}$$
$$\otimes a_{(2)}{h_{(1)}}^{(\bar{2})} (h_{(2)} \triangleright b_{(2)}) ({h_{(3)}} \triangleright {g_{(1)}}^{(\bar{2})}) \otimes h_{(4)} g_{(2)}$$
$$= a_{(1)}({h_{(1)}}^{(\bar{1})}{}_{(1)} \triangleright b_{(1)}) \otimes {h_{(1)}}^{(\bar{1})}{}_{(2)} {g_{(1)}}^{(\bar{1})}$$
$$\otimes a_{(2)}{h_{(1)}}^{(\bar{2})} (h_{(2)} \triangleright (b_{(2)}{g_{(1)}}^{(\bar{2})})) \otimes h_{(3)} g_{(2)}$$
$$= \left((a_{(1)} \otimes {h_{(1)}}^{(\bar{1})}) \otimes (a_{(2)}{h_{(1)}}^{(\bar{2})} \otimes h_{(2)}) \right) \left((b_{(1)} \otimes {g_{(1)}}^{(\bar{1})}) \otimes (b_{(2)}{g_{(1)}}^{(\bar{2})} \otimes g_{(2)}) \right)$$
$$= (\Delta(a \otimes h))(\Delta(b \otimes g)).$$

Here the first equality is the definition of the product and the second is that of the coproduct. The third uses (6.8), the fourth uses (6.9) and the fifth uses (6.10). The sixth uses the comodule coalgebra assumption (1.14), the sixth uses (6.10) again, the seventh uses the module algebra assumption (1.8) and the eighth recognises the product in $A \blacktriangleright\!\!\triangleleft H \otimes A \blacktriangleright\!\!\triangleleft H$, giving the result required. The proof for the counit is immediate. With more care, one can show that (6.8)–(6.10) are not only sufficient for a bialgebra but necessary too.

In the Hopf algebra case we verify the antipode thus:

$$\cdot(\mathrm{id} \otimes S)\Delta(a \otimes h) = (a_{(1)} \otimes {h_{(1)}}^{(\bar{1})})S(a_{(2)}{h_{(1)}}^{(\bar{2})} \otimes h_{(2)})$$
$$= (a_{(2)} \otimes {h_{(1)}}^{(\bar{1})})(1 \otimes S{h_{(2)}}^{(\bar{1})})(S(a_{(2)}{h_{(1)}}^{(\bar{2})}{h_{(2)}}^{(\bar{2})}) \otimes 1)$$
$$= (a_{(2)} \otimes {h_{(1)}}^{(\bar{1})} S{h_{(2)}}^{(\bar{1})})(S(a_{(2)}{h_{(1)}}^{(\bar{2})}{h_{(2)}}^{(\bar{2})}) \otimes 1)$$
$$= (a_{(2)} \otimes {h^{(\bar{1})}}_{(1)} S{h^{(\bar{1})}}_{(2)})(S(a_{(2)}h^{(\bar{2})}) \otimes 1)$$
$$= (a_{(2)} \otimes 1)(Sa_{(2)} \otimes 1)\epsilon(h) = \epsilon(a)\epsilon(h)$$

as required. We used associativity in the algebra $A \rtimes H$, and for the fourth equality we used the comodule coalgebra assumption (1.14). On the other side, we have

$$\cdot(S \otimes \mathrm{id})\Delta(a \otimes h) = (S(a_{(1)} \otimes {h_{(1)}}^{(\bar{1})}))(a_{(2)}{h_{(1)}}^{(\bar{2})} \otimes h_{(2)})$$
$$= (1 \otimes S{h_{(1)}}^{(\bar{1})(\bar{1})})(S(a_{(1)}{h_{(1)}}^{(\bar{1})(\bar{2})}) \otimes 1)(a_{(2)}{h_{(1)}}^{(\bar{2})} \otimes h_{(2)})$$
$$= (1 \otimes S{h_{(1)}}^{(\bar{1})(\bar{1})})((S{h_{(1)}}^{(\bar{1})(\bar{2})})(Sa_{(1)})a_{(2)}{h_{(1)}}^{(\bar{2})} \otimes h_{(2)})$$
$$= \epsilon(a)(1 \otimes S{h_{(1)}}^{(\bar{1})})((S{h_{(1)}}^{(\bar{2})}{}_{(1)}){h_{(1)}}^{(\bar{2})}{}_{(2)} \otimes h_{(2)})$$
$$= \epsilon(a)(1 \otimes S{h_{(1)}}^{(\bar{1})})(1 \otimes h_{(2)}) = \epsilon(a)\epsilon(h)$$

as required. We used associativity in the cross product algebra, and in the fourth equality we used the fact that we have a comodule. One can also show that if A, H have an invertible antipode then the antipode of

$A\bowtie H$ is also invertible.   ∎

There is a very similar construction with left and right interchanged. Thus,

Theorem 6.2.3 *Let A, H be bialgebras, let A be right H-module algebra and let H be a left A-comodule coalgebra by maps*

$$\alpha : A \otimes H \to A, \quad \alpha(a \otimes h) = a\triangleleft h, \quad \beta : H \to A \otimes H, \quad \beta(h) = \sum h^{(\bar{1})} \otimes h^{(\bar{2})},$$

obeying the compatibility conditions

$$\epsilon(a\triangleleft h) = \epsilon(a)\epsilon(h), \quad \Delta(a\triangleleft h) = \sum (a_{(1)}\triangleleft h_{(1)})h_{(2)}{}^{(\bar{1})} \otimes a_{(2)}\triangleleft h_{(2)}{}^{(\bar{2})}, \quad (6.11)$$

$$\beta(1) = 1 \otimes 1, \quad \beta(hg) = \sum (h^{(\bar{1})}\triangleleft g_{(1)})g_{(2)}{}^{(\bar{1})} \otimes h^{(\bar{2})}g_{(2)}{}^{(\bar{2})}, \quad (6.12)$$

$$\sum h_{(1)}{}^{(\bar{1})}(a\triangleleft h_{(2)}) \otimes h_{(1)}{}^{(\bar{2})} = \sum (a\triangleleft h_{(1)})h_{(2)}{}^{(\bar{1})} \otimes h_{(2)}{}^{(\bar{2})}, \quad (6.13)$$

for all $h, g \in H$ and $a, b \in A$. Then $H\ltimes A$ and $H\rtimes A$ form a bialgebra. It is the right–left bicrossproduct bialgebra and is denoted by $H\bowtie A$. If H, A are Hopf algebras then so is $H\bowtie A$. Its antipode is

$$S(h \otimes a) = \sum (1 \otimes S(h^{(\bar{1})}a))(Sh^{(\bar{2})} \otimes 1).$$

Proof: The proof is very similar to the above, and is in any case the original one appearing in the literature. It can be formally obtained from the one above by reflecting it in a mirror about a vertical axis and reversing the numbering of the coproduct and the numbering of the coaction. This is also formalised in the next lemma.   ∎

Lemma 6.2.4 *Let ${}^{op/op}$ applied to a bialgebra denote the bialgebra with opposite product and opposite coproduct. If A, H are bialgebras mutually (co)acting as in Theorem 6.2.3, then $H\bowtie A = \left((A^{op/op})\bowtie(H^{op/op})\right)^{op/op}$ Similarly in the Hopf algebra case.*

Proof: This is more or less self-evident from the remarks in Chapter 1.6 where we explained that the right handed cross product is a mirror image of the left handed one, and likewise that the left handed cross coproduct is the mirror image of the right handed one. The precise operation to go from one to the other is to reverse the lexicographical ordering of all expressions (hence the opposite algebras) and the ordering of the coproducts as well as the notation for the coaction and action. A brief inspection reveals that the further assumptions (6.11)–(6.13) are likewise the reverse of those stated in Theorem 6.2.2. Note that, in the Hopf algebra case, the op/op

Hopf algebra is anyway isomorphic to the original one via the antipode.
∎

We need both versions because ordinary Hopf algebra duality turns the left–right construction in Theorem 6.2.2 into one of the general right–left form in Theorem 6.2.3.

Proposition 6.2.5 *Let A, H be mutually coacting as in Theorem 6.2.2 and finite dimensional. Then $(A{\blacktriangleright\!\!\triangleleft}H)^* = A^*{\triangleright\!\!\blacktriangleleft}H^*$.*

Proof: The usual notion of Hopf algebra duality turns a left H-module algebra $\alpha = \triangleright : H \otimes A \to A$ into a left H^*-comodule coalgebra $\alpha^* : A^* \to H^* \otimes A^*$, as we note from Proposition 1.6.11 combined with Proposition 1.6.19. Similarly, a right A-comodule coalgebra $\beta : H \to H \otimes A$ dualises to a right A^*-module algebra $\beta^* = \triangleleft : H^* \otimes A^* \to H^*$. Explicitly, the relations between them are the usual adjunctions $\langle \alpha^*(\gamma), h \otimes a \rangle = \langle \gamma, h \triangleright a \rangle$ and $\langle \phi \triangleleft \gamma, h \rangle = \langle \beta(h), \phi \otimes \gamma \rangle$ for $\gamma \in A^*$ and $\phi \in H^*$. On the other hand, we have already remarked in Chapter 1.6 that, by such dualisation, the left cross product construction as in Proposition 1.6.6 becomes the left cross coproduct construction in Proposition 1.6.18. Diagrammatically, it corresponds to reflecting the proofs as diagrams about a horizontal axis. So α^* gives the data for a left cross coproduct. Similarly β^* becomes the data for a right cross product. Hence we identify the dual of $A{\blacktriangleright\!\!\triangleleft}H$ as being of the same form as in Theorem 6.2.3. The pairing is via the canonical parings between A, H and their duals. ∎

Hence we see that the dual Hopf algebra of a bicrossproduct is another bicrossproduct of the opposite handedness. The duality is via the usual maps. In view of the above lemma, an immediate variant is

Corollary 6.2.6 *Let H be a bialgebra or Hopf algebra with dual H^*. We define the categorical dual to be the bialgebra $H^\star = (H^*)^{\mathrm{op/op}}$. Then $(A{\blacktriangleright\!\!\triangleleft}H)^\star = H^\star{\blacktriangleright\!\!\triangleleft}A^\star$.*

Proof: This follows at once from Lemma 6.2.4 and the preceding proposition. Note that it is often natural to work with the categorical dual, basically whenever we want to stay in a left handed or a right handed setting for a particular construction. ∎

Finally, we know from Example 6.1.4 that the bicrossproducts have a canonical Schrödinger representation.

Proposition 6.2.7 *Let $A{\blacktriangleright\!\!\triangleleft} H$ be a left–right bicrossproduct bialgebra or Hopf algebra. Then the Schrödinger representation makes A into a left $A{\blacktriangleright\!\!\triangleleft} H$-module coalgebra, and its corresponding co-Schrödinger representation*

$$\phi{\triangleleft}(a \otimes h) = \sum \langle \phi_{(1)}, a \rangle \phi_{(2)} {\triangleleft} h,$$

for all $\phi \in A^*$, makes A^* into a right $A{\blacktriangleright\!\!\triangleleft} H$-module algebra. Here $(\phi{\triangleleft}h)(a) = \phi(h{\triangleright}a)$ is the dualisation of the left action used in the bicrossproduct. Similarly for right–left bicrossproducts.

Proof: From Example 6.1.4, we have an action $(a \otimes h){\triangleright}b = a(h{\triangleright}b)$ on $b \in A$. It obeys the comodule algebra property (1.9) from Chapter 1.6 since

$$\Delta((a \otimes h){\triangleright}b) = \Delta(a(h{\triangleright}b)) = (\Delta a)\Delta(h{\triangleright}b)$$
$$= a_{(1)}(h_{(1)}{}^{(\bar{1})}{\triangleright}b_{(1)}) \otimes a_{(2)}h_{(1)}{}^{(\bar{2})}(h_{(2)}{\triangleright}b_{(2)})$$
$$= (a_{(1)} \otimes h_{(1)}{}^{(\bar{1})}){\triangleright}b_{(1)} \otimes a_{(2)}(h_{(1)}{}^{(\bar{2})} \otimes h_{(2)}){\triangleright}b_{(2)},$$

where the third equality is (6.8). The result is the output of the coproduct of $A{\blacktriangleright\!\!\triangleleft} H$ from Theorem 6.2.2 acting on Δb, as required. As we know from Proposition 1.6.19, A^* is therefore a module algebra by dualisation. It is a combination of the left coregular representation and the dualisation of ${\triangleright}$, as stated. By symmetry, the same results hold for right–left bicrossproducts $H{\blacktriangleright\!\!\triangleleft} A$, with A becoming a right module coalgebra by $b{\triangleleft}(h \otimes a) = (b{\triangleleft}h)a$, and A^* a left module algebra by $(h \otimes a){\triangleright}\phi = h{\triangleright}\phi_{(1)}\langle\phi_{(2)}, a\rangle$; the proofs are strictly analogous. ∎

We now demonstrate the above theory with some examples. The first is a general construction for nontrivial bicrossproducts.

Example 6.2.8 *Let H be an arbitrary Hopf algebra. We view the right adjoint action in Example 1.6.9 as a left action of H^{op} on H, and the right adjoint coaction in Example 1.6.14 as a right coaction of H on H^{op} So,*

$$h{\triangleright}a = \sum (Sh_{(1)})ah_{(2)}, \quad \beta(h) = \sum h^{(\bar{1})} \otimes h^{(\bar{2})} = \sum h_{(2)} \otimes (Sh_{(1)})h_{(3)}$$

for all $a \in H$ and $h \in H^{\mathrm{op}}$. These form a left–right bicrossproduct $M(H) = H{\blacktriangleright\!\!\triangleleft} H^{\mathrm{op}}$ according to Theorem 6.2.2. Explicitly, its cross product and cross coproduct are built on $H \otimes H$ with structure

$$(a \otimes h)(b \otimes g) = \sum a(Sh_{(1)})bh_{(2)} \otimes gh_{(3)},$$
$$\Delta(a \otimes h) = \sum a_{(1)} \otimes h_{(2)} \otimes a_{(2)}(Sh_{(1)})h_{(3)} \otimes h_{(4)}.$$

If the antipode of H is invertible, then $M(H)$ has an antipode, which is also invertible. Likewise for a bicrossproduct $H \blacktriangleright\!\!\blacktriangleleft H^{\mathrm{cop}}$.

Proof: We already know from Example 1.6.9 that H has an adjoint action on itself making it into a right H-module algebra. Since H^{op} has the opposite algebra but the same coalgebra, H becomes a left H^{op}-module algebra as well. Again, since H^{op} has the same coalgebra, the right adjoint coaction of H on itself in Example 1.6.14 also makes H^{op} into a right H-comodule coalgebra. Hence we are in the setting of Theorem 6.2.2 and just have to verify the compatibility conditions (6.8)–(6.10). In this verification, as well as in the formulae stated, we are always working with the linear space H with all products written as usual products in H unless marked otherwise. We have

$$h_{(1)}{}^{(\bar{1})} \triangleright a_{(1)} \otimes h_{(1)}{}^{(\bar{2})}(h_{(2)} \triangleright a_{(2)}) = h_{(2)} \triangleright a_{(1)} \otimes (Sh_{(1)})h_{(3)}(h_{(4)} \triangleright a_{(2)})$$

$$= (Sh_{(2)(1)})a_{(1)}h_{(2)(2)} \otimes (Sh_{(1)})h_{(3)}(Sh_{(4)(1)})a_{(2)}h_{(4)(2)}$$

$$= (Sh_{(2)})a_{(1)}h_{(3)} \otimes (Sh_{(1)})a_{(2)}h_{(4)} = \Delta(h \triangleright a),$$

$$h_{(1)}{}^{(\bar{1})} \cdot_{\mathrm{op}} g^{(\bar{1})} \otimes h_{(1)}{}^{(\bar{2})}(h_{(2)} \triangleright g^{(\bar{2})})$$

$$= h_{(2)} \cdot_{\mathrm{op}} g_{(2)} \otimes (Sh_{(1)})h_{(3)}(h_{(4)} \triangleright ((Sg_{(1)})g_{(3)}))$$

$$= g_{(2)}h_{(2)} \otimes (Sh_{(1)})h_{(3)}(Sh_{(4)})(Sg_{(1)})g_{(3)}h_{(5)}$$

$$= (gh)_{(2)} \otimes (S(gh)_{(1)})(gh)_{(3)} = \beta(gh),$$

$$h_{(2)}{}^{(\bar{1})} \otimes (h_{(1)} \triangleright a)h_{(2)}{}^{(\bar{2})} = h_{(4)} \otimes (Sh_{(1)})ah_{(2)}(Sh_{(3)})h_{(5)}$$

$$= h_{(2)} \otimes (Sh_{(1)})ah_{(3)}$$

$$= h_{(2)} \otimes (Sh_{(1)})h_{(3)}(Sh_{(4)})ah_{(5)}$$

$$= h_{(1)}{}^{(\bar{1})} \otimes h_{(1)}{}^{(\bar{2})}(h_{(2)} \triangleright a),$$

where we simply evaluated the required properties in our example and used the basic properties of the antipode, etc. from Chapter 1. The other axioms in (6.8)–(6.10) are even easier. In the same way, one may at once compute the cross product algebra from Proposition 1.6.6 and the cross coproduct from Proposition 1.6.16 in terms of the action and coaction shown. One may likewise compute the antipode if H^{op} has an antipode. As we saw in Chapter 1, this is provided by the inverse of the antipode of H.

We can equally well view the right adjoint action from Example 1.6.9 as making H^{cop} into a right H-module algebra (since H^{cop} has the same algebra as H) and the right adjoint coaction from Example 1.6.14 as making H into a left H^{cop} comodule coalgebra. Thus, we can consider $a \triangleleft h = (Sh_{(1)})ah_{(2)}$ and $\beta(h) = (Sh_{(1)})h_{(3)} \otimes h_{(2)}$ for all $h \in H$ and $a \in H^{\mathrm{cop}}$, and make a right–left bicrossproduct $H \blacktriangleright\!\!\blacktriangleleft H^{\mathrm{cop}}$ according to

Theorem 6.2.3, with cross product and cross coproduct

$$(h \otimes a)(g \otimes b) = hg_{(1)} \otimes (Sg_{(2)})ag_{(3)}b,$$

$$\Delta(h \otimes a) = h_{(1)} \otimes (Sh_{(2)})h_{(4)}a_{(1)} \otimes h_{(3)} \otimes a_{(2)}$$

(6.14)

on $H \otimes H$. The proofs are strictly analogous. It should be clear from Proposition 6.2.5 that $H \bowtie H^{\mathrm{cop}} = (M(H^*))^*$ in the finite-dimensional case.   ∎

This example combines a Hopf algebra H with its 'mirror' H^{op}. It is a close cousin of the quantum double to be described in the next chapter, but has a rather simpler structure:

Proposition 6.2.9 *The bicrossproduct $M(H)$ is isomorphic to the tensor product $H \otimes H^{\mathrm{op}}$ as a bialgebra (and as a Hopf algebra when the antipode is invertible).*

Proof: It is clear that $H \subseteq M(H)$ as a sub-bialgebra by the canonical inclusion of a as $a \otimes 1$. However, the coproduct $\Delta : H \to H \otimes H$ also provides an embedding $H^{\mathrm{op}} \subseteq M(H)$. Indeed, combining these gives the linear map $\theta : H \otimes H^{\mathrm{op}} \to M(H)$ defined by $\theta(a \otimes h) = ah_{(1)} \otimes h_{(2)}$. It has inverse $\theta^{-1}(a \otimes h) = a(Sh_{(1)}) \otimes h_{(2)}$, and is therefore a linear isomorphism. It is a bialgebra map since

$$\theta(a \otimes h)\theta(b \otimes g) = (ah_{(1)} \otimes h_{(2)})(bg_{(1)} \otimes g_{(2)})$$
$$= ah_{(1)}(Sh_{(2)})bg_{(1)}h_{(4)} \otimes g_{(2)}h_{(3)} = \theta(ab \otimes gh),$$
$$\Delta\theta(a \otimes h) = a_{(1)}h_{(1)} \otimes h_{(4)} \otimes a_{(2)}h_{(2)}(Sh_{(3)})h_{(5)} \otimes h_{(6)}$$
$$= (\theta \otimes \theta)(a_{(1)} \otimes h_{(1)} \otimes a_{(2)} \otimes h_{(2)}),$$

using the elementary properties of the coproduct and antipode. The isomorphism of the counit and antipode (when it exists) is equally straightforward.   ∎

Now we come to some concrete examples of bicrossproducts associated to finite groups. We need a definition, which we shall also use in Chapter 7, of a pair of groups acting on each other in a compatible way.

Definition 6.2.10 *Two groups (G, M) form a right–left matched pair if there is a right action of G on M and a left action of M on G,*

$$\alpha : M \times G \to M, \quad \alpha_u(s) = s{\triangleleft}u, \qquad \beta : M \times G \to G, \quad \beta_s(u) = s{\triangleright}u,$$

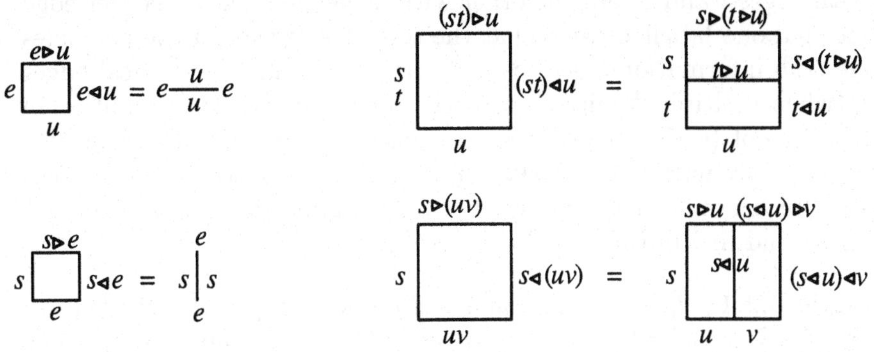

Fig. 6.2. The conditions for a matched pair of groups say that we can subdivide boxes whose boundaries are labelled by group elements according to the rule $s\,{}^{s\triangleright u}_{u}\square\,s\triangleleft u$.

obeying the conditions

$$\alpha_u(e) = e, \quad \alpha_u(st) = \alpha_{\beta_t(u)}(s)\alpha_u(t),$$
$$\beta_s(e) = e, \quad \beta_s(uv) = \beta_s(u)\beta_{\alpha_u(s)}(v). \tag{6.15}$$

In shorthand notation the full data here are

$$s\triangleleft e = s, \quad (s\triangleleft u)\triangleleft v = s\triangleleft(uv); \quad e\triangleleft u = e, \quad (st)\triangleleft u = (s\triangleleft(t\triangleright u))\,(t\triangleleft u);$$

$$e\triangleright u = u, \quad s\triangleright(t\triangleright u) = (st)\triangleright u; \quad s\triangleright e = e, \quad s\triangleright(uv) = (s\triangleright u)\,((s\triangleleft u)\triangleright v);$$

where the first two equations in each line say that we have an action, and the second two say that the action is almost by automorphisms, but twisted by the other action. It is also useful to employ a graphical notation which expresses these matching conditions as the ability to subdivide rectangles. Thus we adopt the notation that a square $s\,{}_{u}\square$ has on its top boundary u transformed by the action of s and on its right boundary s transformed by the action of u. This is a two-dimensional version of the arrow notation encountered in Proposition 6.1.2. Thus, $s\,{}_{u}\square = s\,{}^{s\triangleright u}_{u}\square\,s\triangleleft u$. In fact, one can see that labelling any two adjacent edges is enough to uniquely determine the other two edges to conform to this convention. We also adopt the convention that any group elements labelling the same edge are to be read as multiplied in the usual way for horizontal edges and downwards for vertical edges. In this notation, the conditions (6.15) become as shown in Fig. 6.2. At the top right the condition is that a box with edges st and u can be equally well viewed as a product of two boxes, one with edges t and u and the other with one edge s and the other edge the internal one labelled $t\triangleright u$. That the top edges agree on the two sides encodes the information that $\triangleright$ is an action, and that the vertical edges agree (when multiplied going downwards) expresses the last condition in the first line of (6.15). The condition concerning the identity element e says that a box with edge labelled by e can be collapsed to one of zero height, which notation is consistent with the gluing property. Likewise for the second line in Fig. 6.2.

Example 6.2.11 *If (G, M) is a matched pair of groups, then the induced action of kG on $k(M)$ and coaction of $k(M)$ on kG gives a left–right bicrossproduct Hopf algebra $k(M){\bowtie}kG$ as in Theorem 6.2.2. In the basis $\{\delta_s \otimes u\}$, the structure is explicitly*

$$(\delta_s \otimes u)(\delta_t \otimes v) = \delta_{\alpha_u(s),t}(\delta_s \otimes uv), \quad \Delta(\delta_s \otimes u) = \sum_{ab=s} \delta_a \otimes \beta_b(u) \otimes \delta_b \otimes u,$$

$$1 = \sum_s \delta_s \otimes e, \quad \epsilon(\delta_s \otimes u) = \delta_{s,e}, \quad S(\delta_s \otimes u) = \delta_{\alpha_u(s)^{-1}} \otimes \beta_s(u)^{-1}.$$

This structure is depicted in the graphical notation in Fig. 6.3(a).

Proof: We have seen in Proposition 6.1.1 that an action α of G on M gives the action $u\triangleright\delta_s = \delta_{\alpha_{u^{-1}}(s)}$ on $k(M)$ and results in the cross product algebra as stated. Now we have an additional 'back-reaction' β, and this induces a right coaction $\beta(u) = \sum_s \beta_s(u) \otimes \delta_s$ of $k(M)$ on kG. One readily sees that (6.15) ensure (6.8)–(6.9), while (6.10) is empty in the present setting. Hence we have a bicrossproduct Hopf algebra. The cross coproduct from Proposition 1.6.16 comes out immediately in the form stated.

This completes the proof. However, because of the importance of this example (and for readers who did not like the abstractions above), we will give now a direct proof, both a conventional one in the basis stated, and a pictorial one. The conventional proof that Δ as stated is an algebra homomorphism is

$$\Delta((\delta_s \otimes u)(\delta_t \otimes v)) = \delta_{\alpha_u(s),t}\Delta(\delta_s \otimes uv)$$

$$= \delta_{\alpha_u(s),t} \sum_{ab=s} \delta_a \otimes \beta_b(uv) \otimes \delta_b \otimes uv$$

$$= \delta_{\alpha_u(s),t} \sum_{ab=s} \delta_a \otimes \beta_b(u)\beta_{\alpha_u(b)}(v) \otimes \delta_b \otimes uv,$$

$$(\Delta\delta_s \otimes u)(\Delta\delta_t \otimes v) = \sum_{ab=s}\sum_{cd=t} (\delta_a \otimes \beta_b(u))(\delta_c \otimes \beta_d(v)) \otimes (\delta_b \otimes u)(\delta_d \otimes v)$$

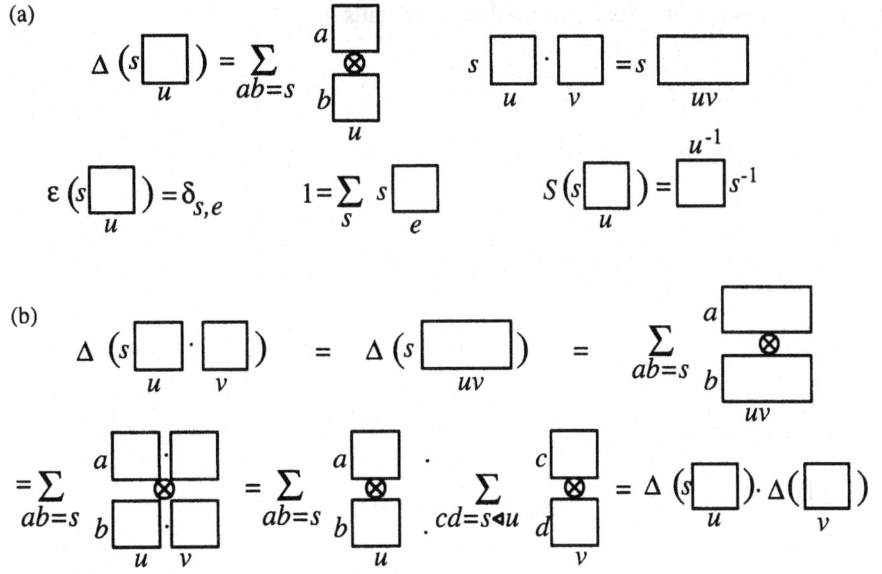

(a)

(b)

Fig. 6.3. (a) The Hopf algebra $k(M){\blacktriangleright\!\!\triangleleft}kG$ with basis elements $\delta_s \otimes u = s\,\square_u$ in graphical representation, and (b) proof that Δ is an algebra homomorphism.

$$= \sum_{ab=s}\sum_{cd=t} \delta_{\alpha_{\beta_b(u)}(a),c}\delta_a \otimes \beta_b(u)\beta_d(v) \otimes \delta_{\alpha_u(b),d}\delta_b \otimes uv$$

$$= \delta_{\alpha_{\beta_b(u)}(a)\alpha_u(b),t} \sum_{ab=s} \delta_a \otimes \beta_b(u)\beta_{\alpha_u(b)}(v) \otimes \delta_{\alpha_u(b),\alpha_u(b)}\delta_b \otimes uv.$$

In the penultimate line, in view of one of the Kronecker δ-functions, we replaced d by $\alpha_u(b)$. Likewise, the other Kronecker δ-function allows us to replace c by $\alpha_{\beta_b(u)}(a)$ in the constraint $cd = t$, which gives the δ-function at the front of the last expression. Comparing the two computations and a trivial one for the counit, we see that Δ, ϵ are algebra homomorphisms *iff* (6.15) hold.

The formula for the antipode comes from Theorem 6.2.2 with the above coaction induced by β as

$$S(\delta_s \otimes u) = \sum_t (1 \otimes \beta_t(u)^{-1})(S(\delta_s\delta_t) \otimes 1) = (1 \otimes \beta_s(u)^{-1})(\delta_{s^{-1}} \otimes 1)$$

$$= \delta_{\alpha_{\beta_s(u)}(s^{-1})} \otimes \beta_s(u)^{-1},$$

giving the form shown. Note here that, by considering the actions $\beta_s(uu^{-1})$ and $\alpha_u(s^{-1}s)$ etc., one has the useful identities

$$\beta_{\alpha_u(s)}(u^{-1}) = \beta_s(u)^{-1}, \quad \alpha_{\beta_s(u)}(s^{-1}) = \alpha_u(s)^{-1},$$

$$\text{i.e. } (s\triangleleft u)\triangleright u^{-1} = (s\triangleright u)^{-1}, \quad s^{-1}\triangleleft(s\triangleright u) = (s\triangleleft u)^{-1}.$$

$$(6.16)$$

Once one has the formula, the direct proof of the antipode is

$$\cdot(\mathrm{id}\otimes S)\Delta(\delta_s\otimes u) = \sum_{ab=s} (\delta_a\otimes\beta_b(u))(\delta_{\alpha_u(b)^{-1}}\otimes\beta_b(u)^{-1})$$

$$= \sum_{ab=s} \delta_{\alpha_{\beta_b(u)}(a),\alpha_u(b)^{-1}}\delta_a\otimes\beta_b(u)\beta_b(u)^{-1} = \sum_{ab=s} \delta_{\alpha_{\beta_b(u)}(a)\alpha_u(b),e}\delta_a\otimes e$$

$$= \sum_{ab=s} \delta_{\alpha_u(ab),e}\delta_a\otimes e = \delta_{s,e}\sum_a \delta_a\otimes e = \epsilon(\delta_s\otimes u),$$

$$\cdot(S\otimes\mathrm{id})\Delta(\delta_s\otimes u) = \sum_{ab=s} (\delta_{\alpha_{\beta_b(u)}(a)^{-1}}\otimes\beta_{ab}(u)^{-1})(\delta_b\otimes u)$$

$$= \sum_{ab=s} (\delta_{\alpha_{\beta_{ab}(u)}(a^{-1})}\otimes\beta_{ab}(u)^{-1})(\delta_b\otimes u)$$

$$= \sum_{ab=s} \delta_{a^{-1},b}\delta_{\alpha_{\beta_s(u)}(a^{-1})}\otimes\beta_s(u)^{-1}u$$

$$= \delta_{s,e}\sum_a \delta_{\alpha_u(a^{-1})}\otimes e = \epsilon(\delta_s\otimes u).$$

We used the matched pair conditions (6.15) and the identities (6.16).

The shorthand notation $\triangleright, \triangleleft$ is also quite convenient for these computations. Another way of doing these proofs is to use the graphical method shown in Fig. 6.3. Part (a) shows the Hopf algebra structure. Recall from our explanation of Fig. 6.2 that two adjacent edges of a box determine the other two. For example, the antipode shown is the same as the one used above in view of the identities (6.16). We add to this the convention that two boxes can be multiplied by gluing as shown if the vertical edges match in their values. Otherwise the product is zero. This is the product in $k(M)\blacktriangleright\!\!\blacktriangleleft kG$. This product is the same as the one in Proposition 6.1.1, where we used an arrow notation; the novel feature now is that, while the arrows there were one-dimensional objects, our boxes now are two-dimensional. We use the other dimension in a similar but dual way to make the coproduct. Thus, the coproduct of a box is the sum over labelled boxes such that when glued vertically they would give the labelled box we began with. This has a similar flavour to the combinatorial Hopf algebras in Chapter 5.1. Our convention is to read vertical expressions from top to bottom, so the upper box is the first output of the coproduct and the lower box is the second.

The proof that Δ is an algebra homomorphism is given in this language in Fig. 6.3(b). We compute Δ on a composite in the first line. The third equality decomposes each of the blocks of the coproduct into pieces. The subdivision picture of the matched pair conditions in Fig. 6.2 tells us exactly that this can be done in a way that the internal parallel edges in the fourth expression have matching values as needed for gluing. All the edges of all four boxes are fully determined by a, b, u, v according to the above conventions. In particular, the values c, d for the left edges of the right hand boxes must have product $cd = (ab) \triangleleft u$, which is the product of the right edges of the left hand boxes. This is shown in the fifth expression. But, written in this way, we have the product pairwise in two copies of the Hopf algebra of the outputs of Δ as required. The above proof of the antipode can likewise be cast in this form, as well as the easy proofs for the unit and counit.   ∎

Example 6.2.12 *If (G, M) is a matched pair of groups, then the induced action of kM on $k(G)$ and coaction of $k(G)$ on kM gives a right–left bicrossproduct Hopf algebra $kM {\blacktriangleright\!\!\triangleleft} k(G)$ as in Theorem 6.2.3. In the basis $\{s \otimes \delta_u\}$, the structure is explicitly*

$$(s \otimes \delta_u)(t \otimes \delta_v) = \delta_{u, \beta_t(v)} (st \otimes \delta_{uv}), \quad \Delta(s \otimes \delta_u) = \sum_{vw=u} s \otimes \delta_v \otimes \alpha_v(s) \otimes \delta_w,$$

$$1 = \sum_u e \otimes \delta_u, \quad \epsilon(s \otimes \delta_u) = \delta_{u,e}, \quad S(s \otimes \delta_u) = \alpha_u(s)^{-1} \otimes \delta_{\beta_s(u)^{-1}},$$

and is depicted in the graphical notation in Fig. 6.4. This Hopf algebra is dual to the preceding one by the standard evaluation of the δ-functions on group elements,

$$(k(M) {\blacktriangleright\!\!\triangleleft} kG)^* = kM {\blacktriangleright\!\!\triangleleft} k(G); \quad \langle s \underset{u}{\square}, t \underset{v}{\square} \rangle = \langle \delta_s \otimes u, t \otimes \delta_v \rangle = \delta_{s,t} \delta_{u,v}.$$

Proof: The proof that we have a Hopf algebra is strictly analogous by a left–right reversal of the one for the preceding example. This time, the right action α of G on M induces a left coaction α^* of $k(G)$ on kM, while the left action β of M on G induces a right action $\triangleleft$ of kM on $k(G)$. The formulae are

$$\alpha^*(s) = \sum_u \delta_u \otimes \alpha_u(s), \quad \delta_u \triangleleft s = \delta_{\beta_{s^{-1}}(u)}.$$

Putting this into Theorem 6.2.3 gives at once the formulae shown for the Hopf algebra structure. One may also verify the Hopf algebra axioms directly as we did for the preceding example. The dual pairing with the latter follows at once from Proposition 6.2.5 and is again easy to verify directly.

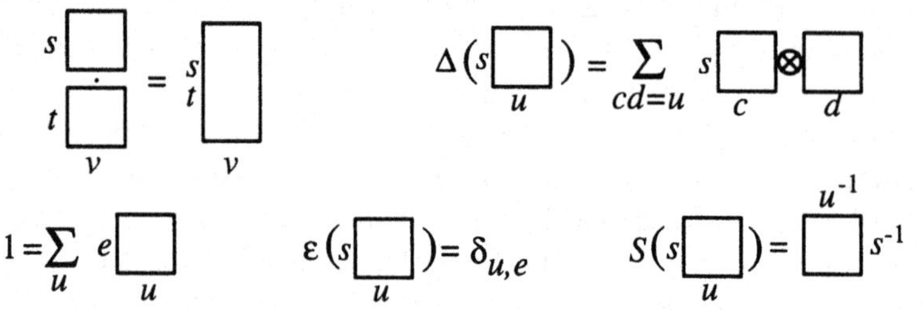

Fig. 6.4. The Hopf algebra $kM\bowtie k(G)$ with basis elements $s\otimes\delta_u = s\,\square_u$ in a graphical representation.

The corresponding graphical form of this Hopf algebra in Fig. 6.4 adopts the same conventions as explained above for the preceding example. This time, the boxes refer to the basis elements $\{s\otimes\delta_u\}$, but we work with them in the same way as before. The product in the algebra is represented by gluing, but now in the vertical axis (to be read from top to bottom) and is zero if the edges to be glued do not have the same group element. Likewise, we use the other left–right dimension in a similar but dual way, to express the coproduct as the sum of all labelled boxes which, when glued horizontally, would give the original one. ∎

Proposition 6.2.13 *The above bicrossproduct Hopf algebras $k(M)\bowtie kG$ and $kM\bowtie k(G)$ have $S^2 =$ id and integrals*

$$\int \delta_s \otimes u = \delta_{u,e}, \qquad \int s \otimes \delta_u = \delta_{s,e}.$$

Moreover, over $\mathbb{C}$ these bicrossproducts are mutually dual Hopf $*$-algebras with

$$(\delta_s \otimes u)^* = \delta_{\alpha_u(s)} \otimes u^{-1}, \qquad (s \otimes \delta_u)^* = s^{-1} \otimes \delta_{\beta_s(u)}.$$

The integral on each is a positive linear functional in the sense of Chapter 5.2.1 and self-adjoint when viewed as an element of the other Hopf $*$-algebra.

Proof: From the stated antipode in Example 6.2.11, it follows at once that $S^2 =$ id reduces to the identities

$$(s\triangleleft u)^{-1}\triangleleft(s\triangleright u)^{-1} = s^{-1}, \qquad (s\triangleleft u)^{-1}\triangleright(s\triangleright u)^{-1} = u^{-1}.$$

These, in turn, follow at once from (6.16). The integrals stated have the left-invariance property in Definition 1.7.1 since $(\mathrm{id}\otimes\int)\Delta(\delta_s\otimes u) =$

$\sum_{ab=s} \delta_a \otimes \beta_b(u) \int (\delta_b \otimes u) = \sum_a \delta_a \otimes \beta_{a^{-1}s}(u)\delta_{u,e} = (\sum_a \delta_a \otimes e) \int (\delta_s \otimes u)$.
The same map is also a right-invariant integral.

Next we work over $\mathbb{C}$. Then as an algebra we make the cross product into a $*$-algebra cross product (the action is automatically unitary in the required sense), so $\mathbb{C}(M)$ and $\mathbb{C}G$ are sub-$*$-algebras. Thus,

$$(\delta_s \otimes u)^* = (1 \otimes u^*)(\delta_s^* \otimes 1) = (1 \otimes u^{-1})(\delta_s \otimes 1)$$
$$= u^{-1} \triangleright \delta_s \otimes u^{-1} = \delta_{s \triangleleft u} \otimes u^{-1}.$$

This clearly obeys $*^2 = \mathrm{id}$, and hence, when extended antilinearly, defines a $*$-algebra. We check that it gives a Hopf $*$-algebra in the sense of Definition 1.7.5. Indeed,

$$(\Delta(\delta_s \otimes u))^{* \otimes *} = \sum_{ab=s} \delta_{a \triangleleft (b \triangleright u)} \otimes (b \triangleright u)^{-1} \otimes \delta_{b \triangleleft u} \otimes u^{-1}$$

$$= \sum_{a'b' = s \triangleleft u} \delta_{a'} \otimes b' \triangleright u^{-1} \otimes \delta_{b'} \otimes u^{-1} = \Delta((\delta_s \otimes u)^*),$$

where $b' = b \triangleleft u$ and $a' = a \triangleleft (b \triangleright u)$. Note that $a'b' = (ab) \triangleleft u$ and hence equals $s \triangleleft u$ iff $ab = s$. We also note that

$$(S \circ *)(\delta_s \otimes u) = S(\delta_{s \triangleleft u} \otimes u^{-1}) = \delta_{((s \triangleleft u) \triangleleft u^{-1})^{-1}} \otimes ((s \triangleleft u) \triangleright u^{-1})^{-1}$$
$$= \delta_{s^{-1}} \otimes s \triangleright u = (* \circ S)(\delta_s \otimes u).$$

We used (6.16) and a similar computation for the last equality. It follows at once that $(S \circ *)^2 = \mathrm{id}$. The $*$-structure for the dual Hopf algebra is obtained in the same way and comes out as stated. We check that the pairing is as Hopf $*$-algebras,

$$\langle (\delta_s \otimes u)^*, t \otimes \delta_v \rangle = \langle \delta_{s \triangleleft u} \otimes u^{-1}, t \otimes \delta_v \rangle = \delta_{t, s \triangleleft u} \delta_{u^{-1}, v}$$
$$= \delta_{s, t \triangleleft v} \delta_{u, v^{-1}} = \langle \delta_s \otimes u, t \triangleleft v \otimes \delta_{v^{-1}} \rangle = \langle \delta_s \otimes u, (S(t \otimes \delta_v))^* \rangle$$

as required.

Finally we look at the integrals and check the positivity required for them to be states. We write a general element of $\mathbb{C}(M) \blacktriangleright \!\!\!\triangleleft G$ as $h = \sum_{s,u} h_{s,u} \delta_s \otimes u$ and compute

$$\int h^* h = \int \sum_{s,u,t,v} \overline{h_{s,u}} h_{t,v} (\delta_{s \triangleleft u} \otimes u^{-1})(\delta_t \otimes v)$$

$$= \int \sum_{s,u,t,v} \overline{h_{s,u}} h_{t,v} (\delta_{s \triangleleft u} \otimes u^{-1}v)\delta_{s,t}$$

$$= \sum_{s,u,t,v} \overline{h_{s,u}} h_{t,v} \delta_{s,t} \delta_{u^{-1}v,e} = \sum_{s,u} |h_{s,u}|^2$$

so the integral is a state on our Hopf $*$-algebra. Similarly for the dual bicrossproduct. Each integral is a linear functional *on* the Hopf algebra and hence an integral *in* its dual. These integral elements are $\Lambda = \sum_u \delta_e \otimes u$

and $\Lambda = \sum_s s \otimes \delta_e$. They are each left and right integral elements and each obey $S\Lambda = \Lambda$ and $\Lambda^* = \Lambda$. This is relevant to the observable-state duality in Chapter 5.4 and Section 6.4 below.   ∎

Exercise 6.2.14 *Compute the fundamental operator W in Theorem 1.7.4 for the bicrossproduct Hopf algebra $k(M){\blacktriangleright\!\!\triangleleft}kG$ and show that it can be written in the form*

$$W = (V^{-1} \otimes 1)W_{k(M)}(V \otimes 1)(1 \otimes U^{-1})W_{kG}(1 \otimes U),$$

where $W_{k(M)}$ and W_{kG} are the fundamental operators for the factors of the bicrossproduct (acting in their relevant spaces) and the operators U, V are

$$V(\delta_s \otimes u) = \delta_s \otimes s{\triangleright}u, \qquad U(\delta_s \otimes u) = \delta_{s{\triangleleft}u} \otimes u.$$

Compute also the operator realisation in Theorem 1.7.4 for the bicross-product and its dual and check that, over $\mathbb{C}$, they become $*$-representations on the Hilbert space $\mathcal{H}$ defined by the integral in Proposition 1.7.6.

Solution: This is an elementary computation but useful when one wants to pass from finite groups to the Hopf–von Neumann operator algebra setting. For this reason we include it here as an exercise. It is clear that from Theorem 1.7.4, applied to each of kG and $k(M)$, we have $W_{kG}(u \otimes v) = u \otimes uv$ as a map $kG \otimes kG \to kG \otimes kG$ and $W_{k(M)}(\delta_s \otimes \delta_t) = \delta_{st^{-1}} \otimes \delta_t$ as a map $k(M) \otimes k(M) \to k(M) \otimes k(M)$. For the bicrossproduct Hopf algebra, the formula in Theorem 1.7.4 gives

$$\begin{aligned}
W(\delta_s \otimes u \otimes \delta_t \otimes v) &= \sum_{ab=s} \delta_a \otimes b{\triangleright}u \otimes (\delta_b \otimes u)(\delta_t \otimes v) \\
&= \sum_{ab=s} \delta_a \otimes b{\triangleright}u \otimes \delta_b \otimes uv\delta_{b{\triangleleft}u,t} \\
&= \delta_{s(t{\triangleleft}u^{-1})^{-1}} \otimes (t{\triangleleft}u^{-1}){\triangleright}u \otimes \delta_{t{\triangleleft}u^{-1}} \otimes uv,
\end{aligned}$$

which can be written as the product of operators as stated. As for the left regular and coregular action of the bicrossproduct and its dual, these are easily computed from the formulae for the coproduct in Example 6.2.11 as

$$\begin{aligned}
(\delta_s \otimes u){\triangleright}(\delta_t \otimes v) &= \delta_{s{\triangleleft}u,t}\delta_s \otimes uv, \\
(s \otimes \delta_u){\triangleright}(\delta_t \otimes v) &= \delta_{u,v}\delta_{ts^{-1}} \otimes s{\triangleright}v.
\end{aligned} \tag{6.17}$$

So, we see that both $k(M){\blacktriangleright\!\!\triangleleft}kG$ and its dual $kM{\triangleright\!\!\blacktriangleleft}k(G)$ are realised concretely as acting on the finite-dimensional vector space $\mathcal{H} = \{\delta_s \otimes u\}$.

Over $\mathbb{C}$, this becomes, from Proposition 1.7.6, a Hilbert space with inner product

$$(g, h) = \sum_{s,u} \overline{g_{s,u}} h_{s,u}, \quad \text{i.e.} \quad (\delta_s \otimes u, \delta_t \otimes v) = \delta_{s,t} \delta_{u,v}, \tag{6.18}$$

where the notation on the left is the same as in the proof of the preceding proposition. The computation is as in the proof of positivity of the integral. Next we check that the representations are unitary, i.e. $*$-representations. Thus,

$$((\delta_s \otimes u) \triangleright (\delta_t \otimes v), \delta_r \otimes w) = \delta_{s \triangleleft u, t} \delta_{s,r} \delta_{uv,w} = \delta_{(s \triangleleft u) \triangleleft u^{-1}, r} \delta_{s \triangleleft u, t} \delta_{u^{-1}w, v}$$

$$= \left(\delta_t \otimes v, (\delta_{s \triangleleft u} \otimes u^{-1}) \triangleright (\delta_r \otimes w) \right),$$

$$((s \otimes \delta_u) \triangleleft (\delta_t \otimes v), \delta_r \otimes w) = \delta_{u,v} \delta_{ts^{-1}, r} \delta_{s \triangleright v, w} = \delta_{s \triangleright u, w} \delta_{rs, t} \delta_{v, s^{-1} \triangleright w}$$

$$= \left(\delta_t \otimes v, (s^{-1} \otimes \delta_{s \triangleright u}) \triangleright (\delta_r \otimes w) \right),$$

using the formulae (6.17) and (6.18). This agrees with the $*$-structure on the bicrossproducts as computed in the last proposition. We leave it to the reader to verify in just the same way that W is unitary, i.e. to check that

$$(W(\delta_s \otimes u \otimes \delta_t \otimes v), \delta_{s'} \otimes u' \otimes \delta_{t'} \otimes v')$$

$$= \left(\delta_s \otimes u \otimes \delta_t \otimes v, W^{-1}(\delta_{s'} \otimes u' \otimes \delta_{t'} \otimes v') \right).$$

Indeed, $U, V, W_{\mathbb{C}G}$ and $W_{\mathbb{C}(M)}$ are all also unitary on their respective Hilbert spaces.   ∎

These bicrossproduct Hopf algebras have the interpretation as an algebraic version of a quantum algebra of observables as explained in Section 6.1, but have the added feature of being Hopf algebras. Moreover, Hopf algebra duality takes us to a construction of the same type but with the roles of the groups interchanged. They are also very interesting objects in their own right because of their two-dimensional nature. The product corresponds to one dimension and the coproduct naturally corresponds to the other dimension. Finally, let us note a much more classical result, not directly connected with our bicrossproduct Hopf algebras, but springing from the same input data. We will give its general Hopf algebra version in Chapter 7.

Proposition 6.2.15 *Given a matched pair of groups (G, M), we define the* double cross product group $G \bowtie M$ *as the set $G \times M$ with product, unit*

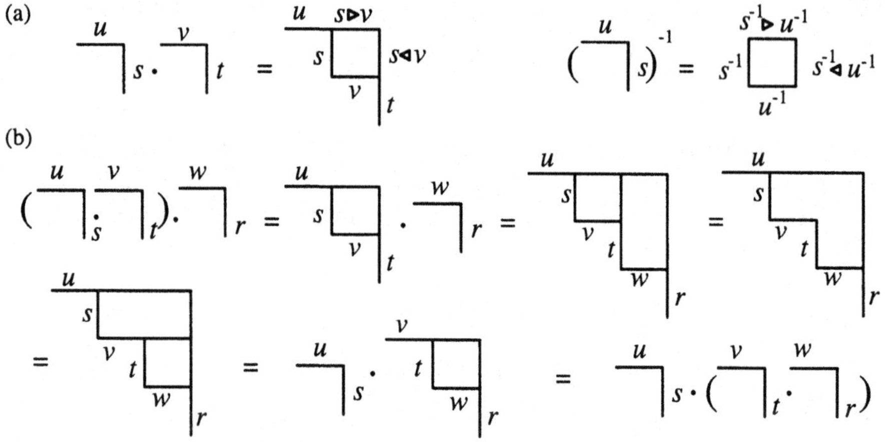

(a)

(b)

Fig. 6.5. Double cross product group $G\bowtie M$ with elements (u,s) represented graphically: (a) the product and inverse and (b) proof of associativity.

and inverse

$$(u,s)(v,t) = (u\beta_s(v), \alpha_v(s)t), \quad e = (e,e),$$

$$(u,s)^{-1} = (\beta_{s^{-1}}(u^{-1}), \alpha_{u^{-1}}(s^{-1})),$$

i.e. $(u,s)(v,t) = (u(s\triangleright v), (s\triangleleft v)t), \quad (u,s)^{-1} = (s^{-1}\triangleright u^{-1}, s^{-1}\triangleleft u^{-1}).$

The graphical form of these is shown in Fig. 6.5(a). Conversely, if a group $X = GM$ *in the sense that* $G \overset{i}{\hookrightarrow} X \overset{j}{\hookleftarrow} M$ *are two subgroups and the map* $G \times M \to X$ *given by multiplication in* X *is a bijection, then* (G,M) *are a matched pair and* $X \cong G\bowtie M$. *The required actions are recovered from*

$$j(s)i(u) = i(\beta_s(u))j(\alpha_u(s)).$$

Proof: In the first case we verify associativity directly. Thus,

$$((u,s)(v,t))\,(w,r) = (u\beta_s(v), \alpha_v(s)t)(w,r)$$
$$= (u\beta_s(v)\beta_{\alpha_v(s)t}(w), \alpha_w(\alpha_v(s)t)r)$$
$$= (u\beta_s(v)\beta_{\alpha_v(s)}(\beta_t(w)), \alpha_w(\alpha_v(s)t)r),$$
$$(u,s)\,((v,t)(w,r)) = (u,s)(v\beta_t(w), \alpha_w(t)z)$$
$$= (u\beta_s(v\beta_t(w)), \alpha_{v\beta_t(w)}(s)\alpha_w(t)r)$$
$$= (u\beta_s(v\beta_t(w)), \alpha_{\beta_t(w)}(\alpha_v(s))\alpha_w(t)r),$$

where, for the third equality, we used that β is a left action, and for the sixth that α is a right action. The two resulting expressions are equal in view of the matched pair conditions (6.15). One may verify the inverse easily in the same way, using the identities (6.16).

Conversely, given the factorisation assumption, we know that every element of X of the form $j(s)i(u)$ has a unique expression as the product of an element of G with an element of M, those elements being called $\alpha_u(s)$ and $\beta_s(u)$, respectively. One can verify that they obey the required conditions by putting the above proof of associativity into reverse gear.

Finally, we give in Fig. 6.5 the graphical version of the double cross product group. We write the ordered pairs as $(u, s) = \overset{u}{\daleth}s$. To multiply such elements, one should line them up and complete the picture with the square as shown. We recall that elements on the left and lower edges of the box force the elements on its upper and right edges as each moves through the box and is acted in so doing by the other. The result of the product is then read on the resulting bigger $\daleth$. Likewise, the inverse element is that induced on the upper and right edges from s^{-1} and u^{-1} on the left and lower ones.

Another way to think of this notation is to consider that an entire box represents an element in $X = G \bowtie M$ given by taking the product from the top left corner to the bottom right corner. Setting off in a clockwise direction gives $i(u)j(s) = (u, s)$, so this is the value of $\overset{u}{\daleth}s$. Likewise, the value of $s \overset{s \triangleright u}{\square} s \triangleleft u$ is $i(s \triangleright u)j(s \triangleleft u) = (s \triangleright u, s \triangleleft u)$, going clockwise. On the other hand, setting off in the anticlockwise direction gives the value $j(s)i(u)$, the two routes being equal in X.

In part (b) we show the proof of associativity in this notation. The fourth expression represents the fact that a box can be subdivided as we explained in Fig. 6.2 so that either the subdivision of the third expression or the subdivision of the fifth expression gives the same result, which we have therefore denoted by the L-shaped region. Its value is fully determined by the elements s, v, t, w on its boundary. ∎

The graphical technique here is very powerful, and underlies an interpretation of the matched pair conditions as a matched pair of (discrete) connections or gauge fields whose parallel transport takes one through the box either vertically or horizontally. This point of view underlies the construction of matched pairs of Lie groups based on Lie algebra data, as we shall see later, in Chapter 8.3. More useful in the finite group case is the characterisation in this proposition of a matched pair as corresponding to a group factorisation. In this context one has Lagrange's theorem and other counting arguments which help us to find suitable subgroups quite easily. Many examples of factorising groups are known to finite group theorists in connection with number theory and other contexts. If one of the subgroups is normal, then one of the actions is trivial, and our double cross product becomes an ordinary semidirect product of groups. This is more familiar but is not the generic situation when a group factorises. Here we content ourselves with a couple of examples (due to the author), relating to our theme of quantisation.

Example 6.2.16 *Let $G = T_+$ be the group of $n \times n$ upper-triangular matrices with 1 on the diagonal and let $M = T_-$ be the same but lower-triangular. To be finite, one can take the entries in a finite field. Let $\theta :$ $T_\pm \to T_\mp$ be the combined operation of inversion and matrix transposition. Then*

$$\alpha_u(s) = 1 + (s - 1)\theta(u), \quad \beta_s(u) = 1 + \theta(s)(u - 1)$$

obey the matching conditions in Definition 6.2.10. Hence we have a bi-crossproduct Hopf algebra $k(T_-) {\blacktriangleright\!\!\triangleleft} kT_+$. It is self-dual.

Proof: This is a matter of elementary computation. First, it is easy to see that α, β are well-defined as set maps $T_- \times T_+ \to T_\pm$. It is easy to see that they are, respectively, right and left actions, since θ is a group homomorphism. Thus, $\alpha_v(\alpha_u(s)) = 1 + (\alpha_u(s) - 1)\theta(v) = 1 + (s - 1)\theta(u)\theta(v) = \alpha_{uv}(s)$, and similarly for β.

Next, using the $\triangleleft, \triangleright$ shorthand for these actions we note the identities

$$s\theta(u) = (\theta(s \triangleright u))(s \triangleleft u), \quad \theta(s)u = (s \triangleright u)(\theta(s \triangleleft u)),$$

which follow on using the definition of θ as transpose-inverse. Thus, $(1 + \theta(s)(u-1))^{\text{tr}} s\theta(u) = (1 + (u^{\text{tr}} - 1)s^{-1})s\theta(u) = s\theta(u) - \theta(u) + u^{\text{tr}}\theta(u) = 1 + (s - 1)\theta(u)$, and similarly for the other identity. Using these, we have

$$\alpha_{\beta_t(u)}(s)\alpha_u(t) = (1 + (s - 1)\theta(t \triangleright u))(t \triangleleft u) = (1 + (st - 1)\theta(u)) = \alpha_u(st),$$

and likewise for β. Hence, (6.15) hold for the actions α, β as stated.

The dual Hopf algebra from Example 6.2.12 is $kT_- {\blacktriangleright\!\!\triangleleft} k(T_+)$, but $T_\pm \cong T_\mp$ via θ, and so one can expect the dual to be isomorphic after a change of conventions to our original $k(T_-) {\blacktriangleright\!\!\triangleleft} kT_+$. Explicitly, the self-duality pairing is

$$\langle \delta_s \otimes u, \delta_t \otimes v \rangle = \delta_{s,\theta(\beta_t(v))} \delta_{u,\theta(\alpha_v(t))}.$$

We can verify this directly once we note that

$$\theta(s \triangleright u) \triangleright \theta(s \triangleleft u) = \theta(s), \quad \theta(s \triangleright u) \triangleleft \theta(s \triangleleft u) = \theta(u), \quad (6.19)$$

which follows at once from the above identities. Thus, $\theta(s \triangleright u) \triangleright \theta(s \triangleleft u) = 1 + (s \triangleright u)(\theta(s \triangleleft u) - 1) = 1 + \theta(s)u - (1 + \theta(s)(u - 1)) = \theta(s)$, and similarly for the second half of (6.19). Then we compute

$$\langle \delta_s \otimes u \otimes \delta_t \otimes v, \Delta\delta_z \otimes w \rangle = \sum_{ab=z} \langle \delta_s \otimes u \otimes \delta_t \otimes v, \delta_a \otimes b{\triangleright}w \otimes \delta_b \otimes w \rangle$$

$$= \sum_{ab=z} \delta_s, \theta(a{\triangleright}(b{\triangleright}w))\delta_{u,\theta(a{\triangleleft}(b{\triangleright}w))}\delta_{t,\theta(b{\triangleright}w)}\delta_{v,\theta(b{\triangleleft}w)}$$

$$= \delta_{s,\theta(z{\triangleright}w)}\delta_{uv,\theta(z{\triangleleft}w)} \sum_{ab=z} \delta_{t,\theta(b{\triangleright}w)}\delta_{v,\theta(b{\triangleleft}w)} = \delta_{s,\theta(z{\triangleright}w)}\delta_{uv,\theta(z{\triangleleft}w)}\delta_{t,s{\triangleleft}u}$$

$$= \langle \delta_s \otimes uv \rangle \delta_{s{\triangleleft}u,t} = \langle (\delta_s \otimes u)(\delta_t \otimes v), \delta_z \otimes w \rangle.$$

In the third equality, we substituted $a{\triangleright}(b{\triangleright}w) = z{\triangleright}w$ and $\theta(a{\triangleleft}(b{\triangleright}w)) = \theta((ab){\triangleleft}w)\theta(b{\triangleleft}w)^{-1} = \theta(z{\triangleleft}w)v^{-1}$ in view of the other δ-function constraints. For the fourth equality, we used that $\theta(b{\triangleright}w) = \theta(b{\triangleright}w){\triangleleft}\theta(b{\triangleleft}w)v^{-1} = \theta(z{\triangleright}w){\triangleleft}\theta(z{\triangleleft}w)v^{-1} = s{\triangleleft}u$ in view of (6.19) and the other constraints. The pairing on the other side is another similar computation using the other half of (6.19).

Another way to see the self-duality, in view of the pairing in Example 6.2.12, is to establish the Hopf algebra isomorphism $k(T_-){\bowtie}kT_+ \to kT_-{\bowtie}k(T_+)$ given by $\delta_t \otimes v \mapsto \theta(t{\triangleright}v) \otimes \delta_{\theta(t{\triangleleft}v)}$, using (6.19) to establish that this is an algebra and coalgebra homomorphism. There are also variants of the above example with only upper- or only lower-triangular matrices and which are more obviously self-dual, whereas in the present conventions the dual is only isomorphic to the original bicrossproduct after using the group isomorphisms θ. ∎

Note that this example is similar to the Weyl algebra in Example 6.1.7 in that, for large matrices, the action α is like the right regular one. Unlike that case, however, it is a Hopf algebra. This phenomenon is clearest in the one-dimensional case. The possible matched pairs in this case can be completely classified.

Example 6.2.17 *Let $G = M = (\mathbb{R}, +)$ with its additive group structure. Then the general solution of the matching conditions in Definition 6.2.10 in a neighbourhood of the origin has two parameters $A, B \in \mathbb{R}$ and the form*

$$\alpha_u(s) = \frac{1}{B}\ln(1 + (e^{Bs} - 1)e^{-Au}), \quad \beta_s(u) = \frac{1}{A}\ln(1 + e^{-Bs}(e^{Au} - 1)).$$

Proof: The solution is of the form given in the preceding example and can be verified in the same way. It is only valid in a certain region

$$D_{A,B} = \{(u, s) \mid e^{Bs} + e^{Au} > 1\} \subseteq \mathbb{R} \times \mathbb{R} \tag{6.20}$$

containing a neighbourhood of the origin, where it is the general solution. This is essentially because one may think of the system (6.15) infinitesimally as a pair of cross-coupled first order partial differential equations, or equally as a single second order one, so that the general solution should have two free parameters. Indeed, the above solutions can be obtained directly by a detailed analysis of these partial differential equations. ∎

This is a deformation of the usual linear quantum mechanics action in (6.5) with $\hbar = -\frac{A}{B}$. We still have B available as an additional parameter. We will discuss its physical meaning in Section 6.4, where we relate it to the gravitational coupling constant. Suffice it to say that the solution is singular as we approach the origin from one side (according to the value of the parameters), having some of the features of a black-hole event horizon. This is true also for a general class of solutions coming from the Iwasawa decomposition, which we shall discuss in Chapter 8. For these, the solvable group M in the decomposition naturally sits inside Euclidean space, with the action on the whole Euclidean space being singular.

Finally, we give the algebraic version of the bicrossproduct Hopf algebra corresponding to this solution. Just as at the end of Section 6.1, we have only to note that, although we have emphasised finite group examples, the general Hopf algebra theorems, such as Theorem 6.2.2, apply just as well to enveloping algebras of Lie algebras.

Example 6.2.18 *We take $A = k[e^{-Bx}]$ and $H = U(k) = k[p]$ as in Example 6.1.11, except that we consider e^{-Bx} abstractly as an invertible group-like generator rather than x. These form a bicrossproduct with action and coaction*

$$p \triangleright e^{-Bx} = A(1 - e^{-Bx})e^{-Bx}, \quad \beta(p) = p \otimes e^{-Bx}.$$

The cross product algebra and cross coproduct coalgebra are

$$[p, e^{-Bx}] = A(1 - e^{-Bx})e^{-Bx}, \quad \Delta e^{-Bx} = e^{-Bx} \otimes e^{-Bx},$$

$$\Delta p = p \otimes e^{-Bx} + 1 \otimes p, \quad \epsilon e^{-Bx} = 1, \quad \epsilon p = 0,$$

$$S e^{-Bx} = e^{Bx}, \quad Sp = -pe^{Bx},$$

where we omit writing the quantisation maps ^.

Proof: Note first that, if we work formally over $\mathbb{C}[[B]]$ with x as the generator, and set $\hbar = -\frac{A}{B}$, then

$$p \triangleright x = \hbar(1 - e^{-Bx}), \quad [p, x] = \hbar(1 - e^{-Bx})$$

is the cross product algebra. This is the reason that we have written our generator in this suggestive way as e^{-Bx}. One can take either point of view. This is similar to the situation in Chapter 3 for quantum enveloping algebras. The stated action and coaction are obtained by computing the action and coaction at the group level for the group actions in Example 6.2.17, and then differentiating with respect to the group $G = \mathbb{R}$. This is because we are replacing the group algebra by the enveloping algebra $U(k) = k[p]$ just as in Example 6.1.11. As there, we keep our group $M = \mathbb{R}$ in the additive form but work with its coordinate function $x(s) = s$ or, more precisely, with the function $s \mapsto e^{-Bs}$ as the abstract generator. After obtaining these formulae, we then verify them directly in our algebraic picture. Thus we show that they extend uniquely to an action and coaction fulfilling the conditions of Theorem 6.2.2 for a bicrossproduct. It is probably more instructive to sketch it at the level of x, and we leave to the reader the task of rewriting it in terms of e^{-Bx}. First, that $\triangleright$ is a module algebra tells us that it acts on products as a derivation. Hence, $p \triangleright f(x) = \hbar(1 - e^{-Bx})\frac{d}{dx}f$. Secondly, the condition (6.9) tells us how β extends to products, namely

$$\beta(p^{n+1}) = (p \otimes e^{-Bx})\beta(p^n) + (\mathrm{id} \otimes p\triangleright)\beta(p^n),$$

which determines it completely. We recover, in fact, the formula for β in Example 6.2.17 from another point of view. Meanwhile, (6.10) is automatic, whereas (6.8) and the fact that β respects the coalgebra structure require some computations given these explicit forms for the action and coaction. The full proof here is quite lengthy and is essentially equivalent to the exponentiation theorem in Chapter 8.3 for matched pairs of Lie algebras. Putting the action and coaction into Theorem 6.2.2 then gives at once the structures as stated. We identify $e^{-Bx} \equiv e^{-Bx} \otimes 1$ and $p \equiv 1 \otimes p$ in $k[x] \blacktriangleright\!\!\triangleleft k[p]$. $\blacksquare$

This example is evidently a deformation by the parameter B of the usual quantum mechanics algebra of observables in Example 6.1.11. The difference is that now we have a Hopf algebra. Moreover, we know from the symmetry between α, β in Example 6.2.17 that this Hopf algebra is formally self-dual. Let us note also that its structure connects up with another more formal approach to quantisation or deformation by twisting in Chapter 2.3.

Proposition 6.2.19 *The bicrossproduct Hopf algebra $H = k[x]\blacktriangleright\!\!\triangleleft k[p]$ is triangular in the sense of Definition 2.1.12 and is twisting-equivalent to $U(b_+)$, where b_+ is the two-dimensional solvable Lie algebra. The triangular structure and twisting cocycle are*

$$\chi = e^{\frac{p \otimes x}{\hbar}}, \quad \mathcal{R} = e^{\frac{x \otimes p}{\hbar}} e^{-\frac{p \otimes x}{\hbar}}.$$

Proof: We first change variables to $X = e^{Bx} - 1$ so that $p \triangleright X = \hbar(1 - e^{-Bx})\frac{\partial}{\partial x}X = \hbar B(1 - e^{-Bx})e^{Bx} = \hbar BX$. In terms of this generator X, the bicrossproduct structure in Example 6.2.18 becomes

$$[p, X] = \hbar BX,$$

$$\Delta X = X \otimes 1 + 1 \otimes X + X \otimes X, \quad \Delta p = 1 \otimes p + p \otimes (1 + X)^{-1}.$$

The algebra here is just the enveloping algebra of the Lie algebra b_+ with generators p, X. Its coproduct is not that of an enveloping algebra. But since $[p \otimes x, p \otimes 1] = 0 = [p \otimes x, 1 \otimes X]$, while $[p \otimes x, 1 \otimes p] = -\hbar p \otimes X(1 + X)^{-1}$ and $[p \otimes x, X \otimes 1] = \hbar BX \otimes x$, we see at once that

$$\chi(X \otimes 1 + 1 \otimes X)\chi^{-1} = \Delta X, \quad \chi(p \otimes 1 + 1 \otimes p)\chi^{-1} = \Delta p.$$

This means that the coalgebra is the twisting as in Theorem 2.3.4 of $U(b_+)$. One can easily check the assertion for the counit as well. Moreover, since the latter has the trivial quasitriangular structure $1 \otimes 1$, we deduce that $\mathcal{R} = \chi_{21}\chi^{-1}$ is a quasitriangular structure for the bicrossproduct Hopf algebra. It is triangular in the sense $\mathcal{R}_{21}\mathcal{R} = 1$. ∎

By the self-duality, the above bicrossproduct Hopf algebra is therefore also dual quasitriangular and a twisting of the commutative algebra of functions on the group with Lie algebra b_+ by a 2-cocycle on it. The switching on of this 2-cocycle introduces the noncommutativity and explicitly achieves the quantisation.

6.3 Extension theory and cocycles

In this section we study cocycle versions of the cross product algebras and cross coproduct coalgebras of Chapter 1.6 and the conditions under which they fit together to form cocycle bicrossproduct bialgebras and Hopf algebras. We have already encountered cocycles in Chapter 2.3, and we use the same notions now in a slightly more general form in which the cocycles can have values in a module algebra or from a comodule coalgebra. We will need only 2-cocycles, and once again there is a non-Abelian cohomology $\mathcal{H}^2$ underlying the picture, as well as aspects of $\mathcal{H}^3$.

The standard context in which one usually encounters cocycles is in the theory of extensions of groups and Lie algebras. We will see that the cocycle version of bicrossproduct Hopf algebra construction solves the problem of extensions of Hopf algebras by other Hopf algebras, and includes the usual theory of extensions of groups and Lie algebras as a special case. We will give some examples of the formalism demonstrating this, including the important case of central extensions such as the Virasoro algebra. It should be noted, however, that the Hopf algebra theory has many more possibilities, the applications of which to physics remain largely unexplored.

We begin with the theory of cocycle cross products. For this we need data precisely generalising the formula (2.20) for a 2-cocycle on a bialgebra or Hopf algebra.

Definition 6.3.1 *A* cocycle left module algebra *for a bialgebra or Hopf algebra H is an algebra A, and linear maps $\alpha = \triangleright : H \otimes A \to A$ and $\chi : H \otimes H \to A$ such that*

$$h \triangleright 1 = \epsilon(h)1, \qquad h \triangleright (ab) = \sum (h_{(1)} \triangleright a)(h_{(2)} \triangleright b), \qquad (6.21)$$

$$1 \triangleright a = a$$
$$\qquad\qquad (6.22)$$
$$\sum h_{(1)} \triangleright (g_{(1)} \triangleright a)\chi(h_{(2)} \otimes g_{(2)}) = \sum \chi(h_{(1)} \otimes g_{(1)})((h_{(2)}g_{(2)}) \triangleright a),$$

$$\sum (h_{(1)} \triangleright \chi(g_{(1)} \otimes f_{(1)}))\, \chi(h_{(2)} \otimes g_{(2)}f_{(2)})$$
$$= \sum \chi(h_{(1)} \otimes g_{(1)})\chi(h_{(2)}g_{(2)} \otimes f), \qquad (6.23)$$
$$\chi(1 \otimes h) = \chi(h \otimes 1) = 1\epsilon(h),$$

for all $h, g, f \in H$ and $a, b \in A$. The map χ is a 2-cocycle on H with values in A and $\triangleright$ is a cocycle action.

The condition (6.21) is the same as for a module algebra, but the second condition (6.22) says that $\triangleright$ is not necessarily an action at all. Rather, it is an action only up to conjugation by the cocycle. Note the similarity with the notion of quasiassociativity (2.33) for a dual quasi-Hopf algebra in Chapter 2.4. The idea is the same. If H is cocommutative like a group algebra and A is commutative, then we are in the classical or so-called 'Abelian' situation. In this case, the cocycle cancels and $\triangleright$ is a usual action. This is why in the usual theory of cocycles on groups one studies cocycles with values in a G-module. By contrast, in the truly non-Abelian or quantum case, the 'quasi-action' and the cocycle do not decouple in this way. We are forced to consider the two together.

Proposition 6.3.2 *Let H be a bialgebra or Hopf algebra and let A be a cocycle left H-module algebra as defined. There is a cocycle cross product algebra $A_\chi {>\!\!\!\triangleleft} H$ built on $A \otimes H$ with product*

$$(a \otimes h)(b \otimes g) = \sum a(h_{(1)} \triangleright b)\chi(h_{(2)} \otimes g_{(1)}) \otimes h_{(3)}g_{(2)}$$

and the unit element $1 \otimes 1$.

Proof: This is a generalisation of Example 1.6.6, and its proof is similar. We verify associativity. Thus,

$$(a \otimes h)((b \otimes g)(c \otimes f)) = (a \otimes h)(b(g_{(1)} \triangleright c)\chi(g_{(2)} \otimes f_{(1)}) \otimes g_{(3)}f_{(2)})$$
$$= a\left(h_{(1)} \triangleright (b(g_{(1)} \triangleright c)\chi(g_{(2)} \otimes f_{(1)}))\right)\chi(h_{(2)} \otimes g_{(3)}f_{(2)}) \otimes h_{(3)}g_{(4)}f_{(3)}$$
$$= a(h_{(1)} \triangleright b)(h_{(2)} \triangleright (g_{(1)} \triangleright c))(h_{(3)} \triangleright \chi(g_{(2)} \otimes f_{(1)}))\chi(h_{(4)} \otimes g_{(3)}f_{(2)}) \otimes h_{(5)}g_{(4)}f_{(3)}$$
$$= a(h_{(1)} \triangleright b)(h_{(2)} \triangleright (g_{(1)} \triangleright c))\chi(h_{(3)} \otimes g_{(2)})\chi(h_{(4)}g_{(3)} \otimes f_{(1)}) \otimes h_{(5)}g_{(4)}f_{(2)}$$
$$= a(h_{(1)} \triangleright b)\chi(h_{(2)} \otimes g_{(1)})((h_{(3)}g_{(2)}) \triangleright c)\chi(h_{(4)}g_{(3)} \otimes f_{(1)}) \otimes h_{(5)}g_{(4)}f_{(2)}$$
$$= (a(h_{(1)} \triangleright b)\chi(h_{(2)} \otimes g_{(1)}) \otimes h_{(3)}g_{(2)})(c \otimes f) = ((a \otimes h)(b \otimes g))(c \otimes f).$$

Here the third equality uses (6.21), the fourth uses the 2-cocycle condition (6.23) and the fifth uses (6.22) for a cocycle action. Then we recognise the result. That $1 \otimes 1$ is the identity is easily verified. Note that, with a little more care, it is easy to reverse the arguments here: if we assume (6.21), then the other two conditions for a cocycle action are necessary as well as sufficient for associativity. ∎

Example 6.3.3 *Let G be a finite group and let A be an algebra. Then A is a cocycle G-module algebra if $\alpha = \triangleright : G \times A \to A$ and $\chi : G \times G \to A$ obey*

$$u \triangleright 1 = 1, \qquad u \triangleright (ab) = (u \triangleright a)(u \triangleright b),$$

$$1 \triangleright a = a, \quad (u \triangleright (v \triangleright a))\chi(u, v) = \chi(u, v)((uv) \triangleright a),$$

$$(u \triangleright \chi(v, w))\chi(u, vw) = \chi(u, v)\chi(uv, w), \quad \chi(u, e) = \chi(e, u) = 1,$$

for $u, v, w \in G$ and $a, b \in A$. The cocycle cross product algebra then takes the form

$$(a \otimes u)(b \otimes v) = a(u \triangleright b)\chi(u, v) \otimes uv.$$

If A is commutative, then $\alpha = \triangleright$ is a usual action and $\chi \in Z_\alpha^2(G, A)$ is a usual group 2-cocycle with respect to it.

Proof: We simply put $H = kG$ in the preceding example. We note that the decoupling of the notion of action and cocycle also works if A is noncommutative provided χ has its values in the centre of A. ∎

This justifies our cocycle terminology. Further justification comes from the following proposition.

Proposition 6.3.4 *Let A be an algebra, let H be a bialgebra and let $(\chi, \triangleright)$ be a left cocycle module algebra structure as defined in Definition 6.3.1.*

If $\gamma : H \to A$ is a convolution-invertible linear map with $\gamma(1) = 1$, then

$$\chi^\gamma(h \otimes g) = \sum \gamma(h_{(1)})(h_{(2)} \triangleright \gamma(g_{(1)}))\chi(h_{(3)} \otimes g_{(2)})\gamma^{-1}(h_{(4)}g_{(3)}),$$

$$h \triangleright^\gamma a = \sum \gamma(h_{(1)})(h_{(2)} \triangleright a)\gamma^{-1}(h_{(3)})$$

is also a left cocycle module algebra structure. We say that $(\chi, \triangleright)$ and $(\chi^\gamma, \triangleright^\gamma)$ are cohomologous and denote the equivalence classes of cocycle module algebra structures modulo such transformations by $\mathcal{H}^2(H, A)$, the non-Abelian 2-cohomology of H with values in A. A cocycle module algebra which is cohomologous to the trivial one is called a coboundary.

Proof: This is a straight verification from Definition 6.3.1. First, it is easy to see that $\triangleright^\gamma$ obeys (6.21) if $\triangleright$ does. Next we compute

$$h_{(1)} \triangleright^\gamma (g_{(1)} \triangleright^\gamma a)\chi^\gamma(h_{(2)} \otimes g_{(2)})$$

$$= \gamma(h_{(1)})\left(h_{(2)} \triangleright (\gamma(g_{(1)})(g_{(2)} \triangleright a)\gamma^{-1}(g_{(3)}))\right)\gamma^{-1}(h_{(3)})$$

$$\gamma(h_{(4)})(h_{(5)} \triangleright \gamma(g_{(4)}))\chi(h_{(6)} \otimes g_{(5)})\gamma^{-1}(h_{(7)}g_{(6)})$$

$$= \gamma(h_{(1)})(h_{(2)} \triangleright \gamma(g_{(1)}))(h_{(3)} \triangleright (g_{(2)} \triangleright a))(h_{(4)} \triangleright (\gamma^{-1}(g_{(3)})\gamma(g_{(4)})))$$

$$\chi(h_{(5)} \otimes g_{(5)})\gamma^{-1}(h_{(6)}g_{(6)})$$

$$= \gamma(h_{(1)})(h_{(2)} \triangleright \gamma(g_{(1)}))\chi(h_{(3)} \otimes g_{(2)})((h_{(4)}g_{(3)}) \triangleright a)\gamma^{-1}(h_{(5)}g_{(4)})$$

$$= \gamma(h_{(1)})(h_{(2)} \triangleright \gamma(g_{(1)}))\chi(h_{(3)} \otimes g_{(2)})\gamma^{-1}(h_{(4)}g_{(3)})$$

$$\gamma(h_{(5)}g_{(4)})((h_{(6)}g_{(5)}) \triangleright a)\gamma^{-1}(h_{(7)}g_{(6)})$$

$$= \chi^\gamma(h_{(1)} \otimes g_{(1)})((h_{(2)}g_{(2)}) \triangleright^\gamma a)$$

as required for (6.22). We used for the third equality the same property for $\chi, \triangleright$. Likewise, we verify

$$h_{(1)} \triangleright^\gamma \chi^\gamma(g_{(1)} \otimes f_{(1)})\chi^\gamma(h_{(2)} \otimes g_{(2)}f_{(2)})$$

$$= \gamma(h_{(1)})\left(h_{(2)} \triangleright \left(\gamma(g_{(1)})(g_{(2)} \triangleright \gamma(f_{(1)}))\chi(g_{(3)} \otimes f_{(2)})\gamma^{-1}(g_{(4)}f_{(3)})\right)\right)\gamma^{-1}(h_{(3)})$$

$$\gamma(h_{(4)})(h_{(5)} \triangleright \gamma(g_{(5)}f_{(4)}))\chi(h_{(6)} \otimes g_{(6)}f_{(5)})\gamma^{-1}(h_{(7)}g_{(7)}f_{(6)})$$

$$= \gamma(h_{(1)})(h_{(2)} \triangleright \gamma(g_{(1)}))(h_{(3)} \triangleright (g_{(2)} \triangleright \gamma(f_{(1)})))$$

$$(h_{(4)} \triangleright \chi(g_{(3)} \otimes f_{(2)}))\chi(h_{(5)} \otimes g_{(4)}f_{(3)})\gamma^{-1}(h_{(6)}g_{(5)}f_{(4)})$$

$$= \gamma(h_{(1)})(h_{(2)} \triangleright \gamma(g_{(1)}))(h_{(3)} \triangleright (g_{(2)} \triangleright \gamma(f_{(1)})))$$

$$\chi(h_{(4)} \otimes g_{(3)})\chi(h_{(5)}g_{(4)} \otimes f_{(2)})\gamma^{-1}(h_{(6)}g_{(5)}f_{(3)})$$

$$= \gamma(h_{(1)})(h_{(2)} \triangleright \gamma(g_{(1)}))\chi(h_{(3)} \otimes g_{(2)})$$

$$((h_{(4)}g_{(3)}) \triangleright \gamma(f_{(1)}))\chi(h_{(5)}g_{(4)} \otimes f_{(2)})\gamma^{-1}(h_{(6)}g_{(5)}f_{(3)})$$

$$= \chi^\gamma(h_{(1)} \otimes g_{(1)})\chi^\gamma(h_{(2)}g_{(2)} \otimes f)$$

as required. The second equality uses (6.21) and makes cancellations of γ, γ^{-1} as before. The third uses the 2-cocycle condition (6.23) for χ.

The fourth equality uses (6.22) for $\chi, \triangleright$. The properties with respect to the unit are clear provided $\gamma(1) = 1$. It is also easy to see that this transformation by γ gives a left action of the group of identity-preserving convolution-invertible maps $H \to A$, i.e. $(\chi^\mu)^\gamma = \chi^{\gamma\mu}$ and $(\triangleright^\mu)^\gamma = \triangleright^{\gamma\mu}$, where $\gamma\mu(h) = \gamma(h_{(1)})\mu(h_{(2)})$ is the convolution product of two such maps. Hence we have an equivalence relation given by this action, and the orbits are the cohomology classes in $\mathcal{H}^2(H, A)$.   ∎

This cohomology has many applications. We already know from the dual version of Proposition 2.3.5 that $\mathcal{H}^2(H, k)$ shows up when describing the inequivalent dual-twistings of H. We will see now a different and more general application whereby $\mathcal{H}^2(H, A)$ completely classifies our cocycle cross products. Note that H coacts on itself from the right and hence also on $A_\chi {\rtimes} H$, since this is built on $A \otimes H$ as a vector space. The coaction is $\beta = \mathrm{id} \otimes \Delta$, and it is easy to see that this always makes $A_\chi {\rtimes} H$ into a right H-comodule algebra. We will say that two cocycle cross products $A_\chi {\rtimes} H$ and $A_{\chi'} {\rtimes} H$ are *equivalent* or *regularly isomorphic* if they are isomorphic as H-comodule algebras under this right coaction. We require also that the isomorphism is the identity when restricted to the subalgebra A.

Proposition 6.3.5 *Two cocycle cross products are regularly isomorphic iff their corresponding cocycle module algebra structures are cohomologous, i.e. the equivalence classes of cocycle cross products $A_\chi {\rtimes} H$ are in one to one correspondence with elements of the non-Abelian cohomology $\mathcal{H}^2(H, A)$. In particular, $A_\chi {\rtimes} H$ is regularly isomorphic to the tensor product algebra $A \otimes H$ iff $(\chi, \triangleright)$ is a coboundary.*

Proof: We begin by showing that any algebra map $\tilde\gamma : A_{\chi'} {\rtimes} H \to A_\chi {\rtimes} H$ which is covariant under the right coaction of H and restricts to the identity on A, is necessarily of the form

$$\tilde\gamma(a \otimes h) = a\gamma(h_{(1)}) \otimes h_{(2)} \tag{6.24}$$

for a linear map $\gamma : H \to A$. We define $\gamma(h) = (\mathrm{id} \otimes \epsilon) \circ \tilde\gamma(1 \otimes h)$ and check that we have $a\gamma(h_{(1)}) \otimes h_{(2)} = a\tilde\gamma(1 \otimes h_{(1)})^{(1)}\epsilon(\tilde\gamma(1 \otimes h_{(2)})^{(2)}) \otimes h_{(2)} = a\tilde\gamma(1 \otimes h)^{(1)}\epsilon(\tilde\gamma(1 \otimes h)^{(2)}{}_{(1)}) \otimes \tilde\gamma(1 \otimes h)^{(2)}{}_{(2)} = (a \otimes 1)\tilde\gamma(1 \otimes h) = \tilde\gamma(a \otimes h)$, where $\tilde\gamma = \sum \tilde\gamma^{(1)} \otimes \tilde\gamma^{(2)}$ is an explicit notation for the output of $\tilde\gamma$ in $A \otimes H$. We used the assumption of covariance under the right coaction of H for the second equality. Conversely, it is clear that any linear map $\gamma : H \to A$ defines an H-comodule map $\tilde\gamma$. This is a generalisation of the idea behind the Markov transition operators in Chapter 5.3. Under this correspondence, the map $\tilde\gamma$ is an isomorphism *iff* γ is convolution-invertible. Here $\tilde\gamma^{-1}$ is provided in the same way as (6.24) with γ replaced by the convolution-inverse γ^{-1}.

Next, on evaluating the product in $A_\chi \rtimes H$ and the image of the product in $A_{\chi'} \rtimes H$ we have, respectively,

$$
\begin{aligned}
\tilde{\gamma}(a \otimes h)\tilde{\gamma}(b \otimes g) &= (a\gamma(h_{(1)}) \otimes h_{(2)}) \cdot (b\gamma(g_{(1)}) \otimes g_{(2)}) \\
&= a\gamma(h_{(1)})h_{(2)} \triangleright (b\gamma(g_{(1)}))\chi(h_{(3)} \otimes g_{(2)}) \otimes h_{(4)}g_{(3)} \\
&= a(h_{(1)} \triangleright^\gamma b)\chi^\gamma(h_{(2)} \otimes g_{(1)})\gamma(h_{(3)}g_{(2)}) \otimes h_{(4)}g_{(3)}, \\
\tilde{\gamma}((a \otimes h)(b \otimes g)) &= \tilde{\gamma}\left(a(h_{(1)} \triangleright' b)\chi'(h_{(2)} \otimes g_{(1)}) \otimes h_{(3)}g_{(2)}\right) \\
&= a(h_{(1)} \triangleright' b)\chi'(h_{(2)} \otimes g_{(1)})\gamma(h_{(3)}g_{(2)}) \otimes h_{(4)}g_{(3)}.
\end{aligned}
$$

Assuming these expressions coincide, we apply $\mathrm{id} \otimes \gamma^{-1}$ to both and multiply in A, to conclude that

$$
(h_{(1)} \triangleright^\gamma b)\chi^\gamma(h_{(2)} \otimes g) = (h_{(1)} \triangleright' b)\chi'(h_{(2)} \otimes g).
$$

Setting first $g = 1$ and then $b = 1$, we conclude that $(\chi', \triangleright') = (\chi^\gamma, \triangleright^\gamma)$ as required. Conversely, it is clear that if this equality does hold then $\tilde{\gamma}$ is an algebra homomorphism. It is also clear that $\tilde{\gamma}$ preserves the unit $1 \otimes 1$ *iff* $\gamma(1) = 1$. ∎

We turn now to a slightly more abstract point of view behind the above results. We define an *extension* of an algebra A by a bialgebra or Hopf algebra H as a right H-comodule algebra E such that A is its fixed point subalgebra under the coaction. Here, the fixed point subalgebra of a comodule algebra E is, by definition,

$$
E^H = \{e \in E| \; \beta(e) = e \otimes 1\} \subseteq E,
$$

where β is the coaction. This condition that β acts trivially is preserved by the product of E because β is an algebra homomorphism, so the fixed points indeed form a subalgebra. Thus we have an extension of A when it is given as the fixed point subalgebra of a comodule algebra E. Obviously, a morphism or extension-preserving map between two extensions E, E' of A should be defined as an algebra map $E \to E'$, which is compatible with the comodule structures (an intertwiner) and which is the identity when restricted to the subalgebra A.

This notion of extensions is an important idea in Galois theory, where one is interested in extensions of fields. It is also an important idea in noncommutative differential geometry and quantum group gauge theory. From both of these points of view it is natural to formulate the notion of a 'topologically trivial' extension as one for which there exists a convolution-invertible linear map $j : H \to E$ which is an intertwiner between the given coaction on E and the right regular coaction on H provided by its coproduct. Such an extension is said to be *cleft*. Without loss of generality, one can also assume that $j(1) = 1$.

Proposition 6.3.6 *A cocycle cross product by a Hopf algebra with a convolution-invertible cocycle is always a cleft extension. Conversely, every cleft extension of an algebra A by a Hopf algebra H is isomorphic to a convolution-invertible cocycle cross product. Hence, the inequivalent cleft extensions are in one to one correspondence with the non-Abelian convolution-invertible cohomology classes in $\mathcal{H}^2(H, A)$.*

Proof: It is easy to see that the cocycle cross product algebra $E = A_\chi{\rtimes}H$ is such an extension of A, with $i(a) = a \otimes 1$ as the identification of A as the fixed point subalgebra under the coaction $\beta = \mathrm{id} \otimes \Delta$ of H. Now consider the linear map $j : H \to E$ given by $j(h) = 1 \otimes h$. It obviously respects the coactions of H and it is convolution-invertible if χ is convolution-invertible. Here

$$j^{-1}(h) = \chi^{-1}(Sh_{(2)} \otimes h_{(3)}) \otimes Sh_{(1)}, \tag{6.25}$$

and one easily verifies that

$$
\begin{aligned}
j^{-1}(h_{(1)})(1 \otimes h_{(2)}) &= (\chi^{-1}(Sh_{(2)} \otimes h_{(3)}) \otimes Sh_{(1)})(1 \otimes h_{(4)}) \\
&= \chi^{-1}(Sh_{(3)} \otimes h_{(4)})\chi(Sh_{(2)} \otimes h_{(5)}) \otimes (Sh_{(1)})h_{(6)} \\
&= (\chi^{-1}\chi)(Sh_{(2)} \otimes h_{(3)}) \otimes (Sh_{(1)})h_{(4)} = 1 \otimes 1, \\
(1 \otimes h_{(1)})j^{-1}(h_{(2)}) &= (1 \otimes h_{(1)})(\chi^{-1}(Sh_{(3)} \otimes h_{(4)}) \otimes Sh_{(2)}) \\
&= (h_{(1)}{\triangleright}\chi^{-1}(Sh_{(6)} \otimes h_{(7)}))\chi(h_{(2)} \otimes Sh_{(5)}) \otimes h_{(3)}(Sh_{(4)}) \\
&= (h_{(1)}{\triangleright}\chi^{-1}(Sh_{(6)} \otimes h_{(7)}))\chi(h_{(2)} \otimes Sh_{(5)})\chi(h_{(3)}(Sh_{(4)}) \otimes h_{(8)}) \otimes 1 \\
&= (h_{(1)}{\triangleright}\chi^{-1}(Sh_{(6)} \otimes h_{(7)}))(h_{(2)}{\triangleright}\chi(Sh_{(5)} \otimes h_{(8)}))\chi(h_{(3)} \otimes (Sh_{(4)})h_{(9)}) \otimes 1 \\
&= (h_{(1)}{\triangleright}(\chi^{-1}\chi)(Sh_{(4)} \otimes h_{(5)}))\chi(h_{(2)} \otimes (Sh_{(3)})h_{(6)}) \otimes 1 = 1 \otimes 1
\end{aligned}
$$

as required. We used elementary properties of the antipode, and, for the penultimate equality, we used the 2-cocycle condition (6.23). Hence we have a cleft extension. Note also in this case that we could recover χ and $\triangleright$ from the product in $A_\chi{\rtimes}H$ by computing

$$
\begin{aligned}
j(h_{(1)})j(g_{(1)})j^{-1}(h_{(2)}g_{(2)}) &= (1 \otimes h_{(1)})(1 \otimes g_{(1)})j^{-1}(h_{(2)}g_{(2)}) \\
&= (\chi(h_{(1)} \otimes g_{(1)}) \otimes h_{(2)}g_{(2)})j^{-1}(h_{(3)}g_{(3)}) \\
&= (\chi(h_{(1)} \otimes g_{(1)}) \otimes 1)j(h_{(2)}g_{(2)})j^{-1}(h_{(3)}g_{(3)}) \\
&= \chi(h \otimes g) \otimes 1, \\
j(h_{(1)})(a \otimes 1)j^{-1}(h_{(2)}) &= ((h_{(1)}{\triangleright}a) \otimes h_{(2)})j^{-1}(h_{(3)}) = h{\triangleright}a \otimes 1.
\end{aligned}
$$

Conversely, given a cleft extension, we define $\chi : H \otimes H \to E$ and $\triangleright : H \otimes A \to E$ by

$$
\begin{aligned}
j(h_{(1)})j(g_{(1)})j^{-1}(h_{(2)}g_{(2)}) &= \chi(h \otimes g), \\
j(h_{(1)})i(a)j^{-1}(h_{(2)}) &= h{\triangleright}a,
\end{aligned}
\tag{6.26}
$$

where $i : A \hookrightarrow E$ explicitly denotes the inclusion of A in E as a fixed point subalgebra. It is clear that $(\chi, \triangleright)$ form a trivial cocycle in $\mathcal{H}^2(H, E)$ as the coboundary of j. On the other hand, H-covariance of j gives at once that the images of these maps $\chi, \triangleright$ are in the fixed point subalgebra, i.e. they are actually maps $H \otimes H \to A$ and $H \otimes A \to A$ as required. They still obey the cocycle conditions but are now viewed as an element of $\mathcal{H}^2(H, A)$. As such, they are not necessarily a coboundary since j itself need not map to A. We then build the cocycle cross product $A_\chi \rtimes H$ on the vector space $A \otimes H$. Finally, we verify that $\tilde{j} : A \otimes H \to E$ given by $\tilde{j}(a \otimes h) = i(a)j(h)$ is an isomorphism. The map $\tilde{j}$ is clearly covariant under the right coaction of H since j is, and it is an algebra homomorphism by virtue of the way that we have defined $\chi, \triangleright$. We have to prove that it is an isomorphism. To do this, we define

$$\tilde{j}^{-1}(e) = e^{(\bar{1})} j^{-1}(e^{(\bar{2})}{}_{(1)}) \otimes e^{(\bar{2})}{}_{(2)}, \qquad \forall e \in E,$$

where $\beta(e) = e^{(\bar{1})} \otimes e^{(\bar{2})}$ is our coaction of H on E. Again, covariance of j implies at once that j^{-1} is covariant in the sense

$$\beta \circ j^{-1} = (j^{-1} \otimes S) \circ \Delta,$$

involving the antipode of H. This is because $(j^{-1} \otimes S) \circ \Delta$ provides a left and right inverse to $(j \otimes \mathrm{id})\Delta$, and the latter is $\beta \circ j$ as our covariance assumption. Using this, it is easy to see that the image of $\tilde{j}^{-1} : E \to E \otimes H$ in fact lies in $A \otimes H$. One then easily verifies that it provides the inverse to $\tilde{j}$ as required.

For the proof of the assertion about cohomology, we have only to add to Proposition 6.3.5 the assumption that χ, χ' are convolution-invertible, which does not affect the proof. We note that this assumption is, in any case, quite natural in the cohomological picture, as we have seen already in Chapter 2.3 in the case of trivial A. One can say rather more in the cleft case. For example, every morphism between cleft extensions is an isomorphism. Indeed, $\gamma^{-1}(h) = (1 \otimes h_{(1)}) \tilde{\gamma}(j^{-1}(h_{(2)}))$ lies in the fixed point subalgebra of $A_\chi \rtimes H$ and hence defines a map $H \to A$, which one can verify is the convolution-inverse of γ. ∎

The noncommutative geometrical interpretation behind these notions is the following. Thus, one can think of a principal bundle or, more generally, a right G-space P as a base space $M = P/G$ 'extended' by the group G to give the total space P. The corresponding description in the dual language of algebras of functions $A = k(M)$, $E = k(P)$ and $H = k(G)$ is that H coacts on E from the right and that the fixed point subalgebra is A, i.e. an extension as above except that we do not require our algebras to be commutative. For an actual principal bundle one requires other conditions such as 'local triviality' which we have not encoded yet; we will return to these at the end of the section. On the other hand, the notion of a globally trivial bundle is clear, namely the existence of a global group coordinate $P \to G$ which, together with the projection $P \to M$ provides a global trivialisation $P \cong M \times G$ as right G-spaces. In the algebraic picture the role of global group coordinate is played by $j : E \to H$ and its existence (and that of its convolution inverse) ensures that that $E \cong A \otimes H$ as a right H-comodule and left A-module as proven above. Actually, we have proven that E is isomorphic to $A_\chi \rtimes H$ as algebras. In usual differential geometry one does not see the action and cocycle because one usually assumes that E is commutative and that j is an algebra map. The first of these assumptions forces any possible action $\triangleright$ to be trivial, and the second forces any possible cocycle χ to be trivial also. Hence, $E \cong A \otimes H$ as an algebra in this classical situation. This is why our non-Abelian cohomology does not enter the scene in the usual classical theory of trivial principal bundles. Also, in the physical picture of Section 6.1 where cross products are interpreted as quantisation, one assumes that the momentum group algebra $H = kG$ or enveloping algebra $U(g)$ is a subalgebra, which again forces any cocycle to be trivial. Physically, one expects the momentum to be realised without any kind of cocycle representation in the quantum system (unless there is some kind of anomaly). For this reason, one does not see the above cocycles in this conventional setting either. On the other hand, as soon as we go beyond these conventional settings, we have the possibility of nontrivial cocycles and correspondingly new quantum numbers provided by the cohomology classes in $\mathcal{H}^2(H, A)$. We can interpret them either from an extended quantisation point of view if H is a group or enveloping Hopf algebra or from a geometrical point of view if H is a coordinate Hopf algebra. This is another example of geometrical and quantum ideas being unified into one formalism.

This completes our introduction to the theory of extensions of algebras by Hopf algebras. For completeness, we give also the right handed and dual versions of some of the above constructions.

Proposition 6.3.7 *A cocycle right module algebra for a bialgebra or Hopf algebra H is an algebra A, and linear maps $\alpha = \triangleleft : A \otimes H \to A$ and $\chi : H \otimes H \to A$ such that*

$$1 \triangleleft h = 1\epsilon(h), \quad (ab) \triangleleft h = \sum (a \triangleleft h_{(1)})(b \triangleleft h_{(2)}), \qquad (6.27)$$

$$a \triangleleft 1 = a,$$

$$\sum \chi(h_{(1)} \otimes g_{(1)})((a \triangleleft h_{(2)}) \triangleleft g_{(2)}) = \sum (a \triangleleft (h_{(1)} g_{(1)})) \chi(h_{(2)} \otimes g_{(2)}), \qquad (6.28)$$

$$\sum \chi(h_{(1)}g_{(1)} \otimes f_{(1)})(\chi(h_{(2)} \otimes g_{(2)}) \triangleleft f_{(2)})$$
$$= \sum \chi(h \otimes g_{(1)}f_{(1)})\chi(g_{(2)} \otimes f_{(2)}), \qquad (6.29)$$
$$\chi(1 \otimes h) = \chi(h \otimes 1) = 1\epsilon(h),$$

for all $h, g, f \in H$ and $a, b \in A$. The map χ is called a right 2-cocycle on H with values in A and $\triangleleft$ is a right cocycle action. In this situation, there is a right cocycle cross product algebra $H{\ltimes}_\chi A$ on the vector space $H \otimes A$ with product

$$(h \otimes a)(g \otimes b) = \sum h_{(1)}g_{(1)} \otimes \chi(h_{(2)} \otimes g_{(2)})(a \triangleleft g_{(3)})b$$

and unit element $1 \otimes 1$.

Proof: This is strictly analogous to Definition 6.3.1 and Proposition 6.3.2. It is given precisely by a left–right reversal of the relevant formulae. ∎

The dualisation is equally easy. It only looks unfamiliar because of the need for an explicit notation.

Proposition 6.3.8 *A cocycle left comodule coalgebra for a bialgebra or Hopf algebra H is a coalgebra C and linear maps $\beta : C \to H \otimes C$ and $\psi : C \to H \otimes H$ such that*

$$(\mathrm{id} \otimes \epsilon) \circ \beta(c) = \epsilon(c),$$
$$\sum c^{(\bar{1})} \otimes c^{(\bar{2})}{}_{(1)} \otimes c^{(\bar{2})}{}_{(2)} = \sum c_{(1)}{}^{(\bar{1})} c_{(2)}{}^{(\bar{1})} \otimes c_{(1)}{}^{(\bar{2})} \otimes c_{(2)}{}^{(\bar{2})}, \qquad (6.30)$$

$$\sum ((\mathrm{id} \otimes \beta) \circ \beta(c_{(1)}))(\psi(c_{(2)}) \otimes 1)$$
$$= \sum (\psi(c_{(1)}) \otimes 1)((\Delta \otimes \mathrm{id}) \circ \beta(c_{(2)})), \qquad (6.31)$$
$$(\epsilon \otimes \mathrm{id}) \circ \beta(c) = c,$$

$$\sum ((\mathrm{id} \otimes \psi) \circ \beta(c_{(1)}))((\mathrm{id} \otimes \Delta) \circ \psi(c_{(2)}))$$
$$= \sum (\psi(c_{(1)}) \otimes 1)((\Delta \otimes \mathrm{id}) \circ \psi(c_{(2)})), \qquad (6.32)$$
$$(\epsilon \otimes \mathrm{id}) \circ \psi(c) = \epsilon(c) = (\mathrm{id} \otimes \epsilon) \circ \psi(c),$$

for all $c \in C$. In this situation there is a left cocycle cross product coalgebra $C^\psi {\blacktriangleright\!\!\triangleleft} H$ on the vector space $C \otimes H$ and coproduct

$$\Delta(c \otimes h) = \sum c_{(1)} \otimes c_{(2)}{}^{(\bar{1})} \psi(c_{(3)})^{(1)} h_{(1)} \otimes c_{(2)}{}^{(\bar{2})} \otimes \psi(c_{(3)})^{(2)} h_{(2)}$$

and counit $\epsilon(c \otimes h) = \epsilon(c)\epsilon(h)$. Here $\psi(c) = \sum \psi(c)^{(1)} \otimes \psi(c)^{(2)}$ is an explicit notation.

Proof: This is dual to Proposition 6.3.2, and the proof can be obtained by writing the construction there in terms of maps and then reversing arrows, as explained in the proof of Proposition 1.6.18 of which the present construction is a generalisation. The direct proof in the dual form follows in the same way as the proof of Proposition 1.6.16. In this case, as well as in the proof of the next theorem, it is helpful to work with the conditions (6.31) and (6.32) in the explicit notation. They are, respectively,

$$\sum((\mathrm{id} \otimes \beta) \circ \beta(c_{(1)}))(\psi(c_{(2)}) \otimes 1) = \sum \psi(c_{(1)})\Delta c_{(2)}{}^{(\bar{1})} \otimes c_{(2)}{}^{(\bar{2})},$$

$$\sum c_{(1)}{}^{(\bar{1})}\psi(c_{(2)})^{(1)} \otimes \psi(c_{(1)}{}^{(\bar{2})})\Delta\psi(c_{(2)})^{(2)} = \sum \psi(c_{(1)})\Delta\psi(c_{(2)})^{(1)} \otimes \psi(c_{(2)})^{(2)},$$

where $\psi(c) = \sum \psi(c)^{(1)} \otimes \psi(c)^{(2)}$, plus the conditions with respect to the counit. There is also a cocycle right cross coproduct $H \blacktriangleright\!\!\!< ^\psi C$ given by reflecting these formulae, and dual to Proposition 6.3.7.   ∎

The cohomological picture also holds in this cocycle comodule coalgebra setting. Thus, fix a coalgebra C and a bialgebra or Hopf algebra H. Then the group of convolution-invertible linear maps $\gamma : C \to H$ with $\epsilon \circ \gamma = \epsilon$ acts on the left cocycle comodule (ψ, β) by

$$\psi^\gamma(c) = \sum(\gamma(c_{(1)}) \otimes 1)((\mathrm{id} \otimes \gamma) \circ \beta(c_{(2)}))\psi(c_{(3)})(\Delta\gamma(c_{(4)})),$$

$$\beta^\gamma(c) = \sum(\gamma(c_{(1)}) \otimes 1)\beta(c_{(2)})(\gamma^{-1}(c_{(3)}) \otimes 1).$$

The equivalence classes under this are the elements of the non-Abelian cohomology $\mathcal{H}^2(C, H)$. We say that ψ is a 2-cocycle *for* H and valued from C. We see that $C = k$ reduces us to the cohomology space in Proposition 2.3.3 in the theory of twisting. By contrast, in the present setting, such cohomology spaces classify the coextensions of a coalgebra by a bialgebra or Hopf algebra H.

Now we are ready to take the cocycle cross product and cocycle cross coproduct together and ask when they form a bialgebra or Hopf algebra.

Theorem 6.3.9 *Let H and A be bialgebras or Hopf algebras and let $(\chi, \lhd)$, (ψ, β) make A a right cocycle H-module algebra and H a left cocycle A-comodule coalgebra as in Propositions 6.3.7 and 6.3.8, respectively. If, in addition, we have*

$$\epsilon(a \lhd h) = \epsilon(a)\epsilon(h),$$

$$\sum \psi(h_{(1)})\Delta(a \lhd h_{(2)}) = \sum \left((a_{(1)} \lhd h_{(1)})h_{(2)}{}^{(\bar{1})} \otimes a_{(2)} \lhd h_{(2)}{}^{(\bar{2})}\right)\psi(h_{(3)}),$$

$$\text{(6.33)}$$

$$\beta(1) = 1 \otimes 1,$$

$$\sum \beta(h_{(1)}g_{(1)})(\chi(h_{(2)} \otimes g_{(2)}) \otimes 1) \tag{6.34}$$
$$= \sum \chi(h_{(1)} \otimes g_{(1)})(h_{(2)}{}^{(\bar{1})} {\triangleleft} g_{(2)})g_{(3)}{}^{(\bar{1})} \otimes h_{(2)}{}^{(\bar{2})}g_{(3)}{}^{(\bar{2})},$$

$$\sum h_{(1)}{}^{(\bar{1})}(a {\triangleleft} h_{(2)}) \otimes h_{(1)}{}^{(\bar{2})} = \sum (a {\triangleleft} h_{(1)})h_{(2)}{}^{(\bar{1})} \otimes h_{(2)}{}^{(\bar{2})}, \tag{6.35}$$

and if (χ, ψ) obey a compatibility condition

$$\sum \psi(h_{(1)}g_{(1)})\Delta\chi(h_{(2)} \otimes g_{(2)})$$
$$= \sum \chi(h_{(1)} \otimes g_{(1)})(h_{(2)}{}^{(\bar{1})} {\triangleleft} g_{(2)})g_{(3)}{}^{(\bar{1})}(\psi(h_{(3)})^{(1)} {\triangleleft} g_{(4)})g_{(5)}{}^{(\bar{1})}$$
$$\otimes \chi(h_{(2)}{}^{(\bar{2})} \otimes g_{(3)}{}^{(\bar{2})})(\psi(h_{(3)})^{(2)} {\triangleleft} g_{(5)}{}^{(\bar{2})}))\psi(g_{(6)}) \tag{6.36}$$

and the conditions $\epsilon(\chi(h \otimes g)) = \epsilon(h)\epsilon(g)$, $\psi(1) = 1 \otimes 1$, then $H {\ltimes}_\chi A$ and $H^\psi {\rtimes} A$ form a bialgebra; the cocycle right–left bicrossproduct bialgebra, denoted by $H^\psi {\bowtie}_\chi A$.

Proof: This is a rather long computation to check the bialgebra axioms (and even longer for the antipode axioms in the Hopf algebra setting). We have to show that the cocycle cross coproduct Δ is a homomorphism for the cocycle cross product algebra. We concentrate on the products of elements where the constraint is nontrivial. The other cases are easy. Thus,

$$\Delta((1 \otimes a)(g \otimes 1)) = \Delta(g_{(1)} \otimes a {\triangleleft} g_{(2)})$$
$$= g_{(1)} \otimes g_{(2)}{}^{(\bar{1})}\psi(g_{(3)})^{(1)}(a {\triangleleft} g_{(4)})_{(1)} \otimes g_{(2)}{}^{(\bar{2})} \otimes \psi(g_{(3)})^{(2)}(a {\triangleleft} g_{(4)})_{(2)}$$
$$= g_{(1)} \otimes g_{(2)}{}^{(\bar{1})}(a_{(1)} {\triangleleft} g_{(3)})g_{(4)}{}^{(\bar{1})}\psi(g_{(5)})^{(1)} \otimes g_{(2)}{}^{(\bar{2})} \otimes (a_{(2)} {\triangleleft} g_{(4)}{}^{(\bar{2})})\psi(g_{(5)})^{(2)}$$
$$= g_{(1)} \otimes (a_{(1)} {\triangleleft} g_{(2)})g_{(3)}{}^{(\bar{1})}g_{(4)}{}^{(\bar{1})}\psi(g_{(5)})^{(1)} \otimes g_{(3)}{}^{(\bar{2})} \otimes (a_{(2)} {\triangleleft} g_{(4)}{}^{(\bar{2})})\psi(g_{(5)})^{(2)}$$
$$= g_{(1)} \otimes (a_{(1)} {\triangleleft} g_{(2)})g_{(3)}{}^{(\bar{1})}\psi(g_{(4)})^{(1)} \otimes g_{(3)}{}^{(\bar{2})}{}_{(1)} \otimes (a_{(2)} {\triangleleft} g_{(3)}{}^{(\bar{2})}{}_{(2)})\psi(g_{(4)})^{(2)}$$
$$= (1 \otimes a_{(1)} \otimes 1 \otimes a_{(2)}) \cdot (g_{(1)} \otimes g_{(2)}{}^{(\bar{1})}\psi(g_{(3)})^{(1)} \otimes g_{(2)}{}^{(\bar{2})} \otimes \psi(g_{(3)})^{(\bar{2})})$$
$$= (\Delta(1 \otimes a)) \cdot (\Delta(g \otimes 1))$$

as required. The second equality used (6.33), the third used (6.35) and the fifth equality used (6.28). In the last two lines, the product is in $H {\bowtie} A \otimes H {\bowtie} A$. Likewise, we verify

$$\Delta((h \otimes 1)(g \otimes 1)) = \Delta(h_{(1)}g_{(1)} \otimes \chi(h_{(2)} \otimes g_{(2)}))$$
$$= h_{(1)}g_{(1)} \otimes (h_{(2)}g_{(2)})^{(\bar{1})}\psi(h_{(3)}g_{(3)})^{(1)}\chi(h_{(4)} \otimes g_{(4)})_{(1)}$$

$$\otimes (h_{(2)}g_{(2)})^{(\bar{2})} \otimes \psi(h_{(3)}g_{(3)})^{(2)}\chi(h_{(4)} \otimes g_{(4)})_{(2)}$$

$$= h_{(1)}g_{(1)} \otimes (h_{(2)}g_{(2)})^{(\bar{1})}\chi(h_{(3)} \otimes g_{(3)})(h_{(4)}{}^{(\bar{1})} {\triangleleft} g_{(4)})$$

$$g_{(5)}{}^{(\bar{1})}(\psi(h_{(5)})^{(1)}{\triangleleft}g_{(6)})g_{(7)}{}^{(\bar{1})}\psi(g_{(8)})^{(1)} \otimes (h_{(2)}g_{(2)})^{(\bar{2})}$$

$$\otimes \chi(h_{(4)}{}^{(\bar{2})} \otimes g_{(5)}{}^{(\bar{2})})(\psi(h_{(5)})^{(2)}{\triangleleft}g_{(7)}{}^{(\bar{2})})\psi(g_{(8)})^{(2)}$$

$$= h_{(1)}g_{(1)} \otimes \chi(h_{(2)} \otimes g_{(2)})(h_{(3)}{}^{(\bar{1})}{\triangleleft}g_{(3)})g_{(4)}{}^{(\bar{1})}(h_{(4)}{}^{(\bar{1})}{\triangleleft}g_{(5)})g_{(6)}{}^{(\bar{1})}$$

$$(\psi(h_{(5)})^{(1)}{\triangleleft}g_{(7)})g_{(8)}{}^{(\bar{1})}\psi(g_{(9)})^{(1)} \otimes h_{(3)}{}^{(\bar{2})}g_{(4)}{}^{(\bar{2})}$$

$$\otimes \chi(h_{(4)}{}^{(\bar{2})} \otimes g_{(6)}{}^{(\bar{2})})(\psi(h_{(5)})^{(2)}{\triangleleft}g_{(8)}{}^{(\bar{2})})\psi(g_{(9)})^{(2)}$$

$$= h_{(1)}g_{(1)} \otimes \chi(h_{(2)} \otimes g_{(2)})(h_{(3)}{}^{(\bar{1})}{\triangleleft}g_{(3)})(h_{(4)}{}^{(\bar{1})}{\triangleleft}g_{(4)})g_{(5)}{}^{(\bar{1})}g_{(6)}{}^{(\bar{1})}$$

$$(\psi(h_{(5)})^{(1)}{\triangleleft}g_{(7)})g_{(8)}{}^{(\bar{1})}\psi(g_{(9)})^{(1)} \otimes h_{(3)}{}^{(\bar{2})}g_{(5)}{}^{(\bar{2})}$$

$$\otimes \chi(h_{(4)}{}^{(\bar{2})} \otimes g_{(6)}{}^{(\bar{2})})(\psi(h_{(5)})^{(2)}{\triangleleft}g_{(8)}{}^{(\bar{2})})\psi(g_{(9)})^{(2)}$$

$$= h_{(1)}g_{(1)} \otimes \chi(h_{(2)} \otimes g_{(2)})(h_{(3)}{}^{(\bar{1})}{\triangleleft}g_{(3)})(h_{(4)}{}^{(\bar{1})}{\triangleleft}g_{(4)})(\psi(h_{(5)})^{(1)}{\triangleleft}g_{(5)})$$

$$g_{(6)}{}^{(\bar{1})}g_{(7)}{}^{(\bar{1})}g_{(8)}{}^{(\bar{1})}\psi(g_{(9)})^{(1)} \otimes h_{(3)}{}^{(\bar{2})}g_{(6)}{}^{(\bar{2})}$$

$$\otimes \chi(h_{(4)}{}^{(\bar{2})} \otimes g_{(7)}{}^{(\bar{2})})(\psi(h_{(5)})^{(2)}{\triangleleft}g_{(8)}{}^{(\bar{2})})\psi(g_{(9)})^{(2)}$$

$$= h_{(1)}g_{(1)} \otimes \chi(h_{(2)} \otimes g_{(2)})((h_{(3)}{}^{(\bar{1})}h_{(4)}{}^{(\bar{1})}\psi(h_{(5)})^{(1)}){\triangleleft}g_{(3)})g_{(4)}{}^{(\bar{1})}g_{(5)}{}^{(\bar{1})}g_{(6)}{}^{(\bar{1})}$$

$$\psi(g_{(7)})^{(1)} \otimes h_{(3)}{}^{(\bar{2})}g_{(4)}{}^{(\bar{2})} \otimes \chi(h_{(4)}{}^{(\bar{2})} \otimes g_{(5)}{}^{(\bar{2})})(\psi(h_{(5)})^{(2)}{\triangleleft}g_{(6)}{}^{(\bar{2})})\psi(g_{(7)})^{(2)}$$

$$= h_{(1)}g_{(1)(1)} \otimes \chi(h_{(2)} \otimes g_{(1)(2)})((h_{(3)}{}^{(\bar{1})}\psi(h_{(4)})^{(1)}){\triangleleft}g_{(1)(3)})g_{(2)}{}^{(\bar{1})}\psi(g_{(3)})^{(1)}$$

$$\otimes h_{(3)}{}^{(\bar{2})}{}_{(1)}g_{(2)}{}^{(\bar{2})}{}_{(1)} \otimes \chi(h_{(3)}{}^{(\bar{2})}{}_{(2)} \otimes g_{(2)}{}^{(\bar{2})}{}_{(2)})(\psi(h_{(4)})^{(2)}{\triangleleft}g_{(2)}{}^{(\bar{2})}{}_{(3)})\psi(g_{(3)})^{(2)}$$

$$= \left(h_{(1)} \otimes h_{(2)}{}^{(\bar{1})}\psi(h_{(3)})^{(1)} \otimes h_{(2)}{}^{(\bar{2})} \otimes \psi(h_{(3)})^{(2)} \right)$$

$$\cdot \left(g_{(1)} \otimes g_{(2)}{}^{(\bar{1})}\psi(g_{(3)})^{(1)} \otimes g_{(2)}{}^{(\bar{2})} \otimes \psi(g_{(3)})^{(2)} \right)$$

$$= (\Delta(h \otimes 1)) \cdot (\Delta(g \otimes 1))$$

as required. We used (6.36) for the third equality, (6.34) for the fourth, (6.35) once for the fifth and twice for the sixth. For the seventh equality we used (6.27), and (6.30) was used for the eighth. In the last two expressions the product is in $H{\bowtie}A \otimes H{\bowtie}A$. That the counit is an algebra homomorphism is easy. This proof generalises those of Theorem 6.2.2 or 6.2.3 to the cocycle case. With more care one, can show that (6.33)–(6.36) are not only necessary for a bialgebra but sufficient too. There is also a left–right version $A_\chi{\bowtie}^\psi H$, given by reflecting these formulae. These constructions are due to the author.   ∎

Before turning to one or two examples, we note that, as for cocycle cross product algebras and cocycle cross coproduct coalgebras, there is an abstract extension theory point of view for the combined system. Recall that for an extension of algebras, now a right extension, we know that in the cleft case we have $E {\cong} H \otimes A$ as a right A-module and a left

H-comodule. In fact, this tensor product property of an extension, along with the Galois property mentioned above, is enough to show that the extension is cleft and hence of the cocycle cross product form as an algebra. Similar remarks apply in a dual formulation for the coalgebra structure. Hence, it is natural to make this tensor product property the starting point of the definition of an extension for bialgebras or Hopf algebras. Thus, in its simplest form, we define a *right–left extension* of a bialgebra algebra A by a bialgebra H to be a bialgebra E equipped with bialgebra maps

$$A \overset{i}{\hookrightarrow} E \overset{p}{\twoheadrightarrow} H \tag{6.37}$$

such that $E \cong H \otimes A$ as a left H-comodule and right A-module. Here $H \otimes A$ has the obvious left coaction of H via its coproduct and the obvious right action of A by its product. Likewise, E has a coaction $(p \otimes \mathrm{id})\Delta$ given by its coproduct followed by p, and by an action of A given by i followed by multiplication from the right in E.

Clearly, the cocycle bicrossproduct $H^\psi {\blacktriangleright\!\!\triangleleft}_\chi A$ obeys these conditions trivially and hence can be viewed as an extension. As before, not every extension is of this type, and we have to add more conditions (such as the Galois property and its dual form) to characterise the cocycle bicrossproducts fully in this way. Also, underlying this abstract extension theory is some kind of non-Abelian cohomology $\mathcal{H}^3(H, A)$ which remains to be understood. We content ourselves here with some examples and special cases.

Example 6.3.10 *Let G be a finite group and let A be a bialgebra or Hopf algebra. Let $\chi : G \times G \to A$, $\triangleleft : A \times G \to A$ be a cocycle right action, cf. Example 6.3.3, and let $\psi : G \to A \otimes A$ be a family of counital 2-cocycles in A as in (2.15). If*

$$\epsilon(a \triangleleft u) = \epsilon(a), \quad \psi(u)\Delta(a \triangleleft u) = ((\Delta a) \triangleleft (u \times u))\psi(u), \tag{6.38}$$

$$\psi(uv)\Delta\chi(u, v) = (\chi(u, v) \otimes \chi(u, v))(\psi(u) \triangleleft (v \times v))\psi(v), \tag{6.39}$$

$$\epsilon(\chi(u, v)) = 1, \quad \psi(e) = 1 \otimes 1,$$

for all $u, v \in G$ and $a \in A$, then we have a right–left bicrossproduct $kG^\psi {\blacktriangleright\!\!\triangleleft}_\chi A$. Explicitly, its bialgebra structure is given by

$$(u \otimes a)(v \otimes b) = uv \otimes \chi(u, v)(a \triangleleft v)b, \quad 1 = e \otimes 1,$$

$$\Delta(u \otimes a) = \sum u \otimes \psi(u)^{(1)} a_{(1)} \otimes u \otimes \psi(u)^{(2)} a_{(2)}, \quad \epsilon(u \otimes a) = \epsilon(a).$$

Proof: The conditions on $\chi, \triangleleft$ are the right handed version of those in Example 6.3.3; they correspond to a usual group cocycle on G with values in A if A is commutative. They can be taken from Proposition 6.3.7 as

$$\chi(u,v)((a \triangleleft u) \triangleleft v) = (a \triangleleft (uv))\chi(u,v),$$

$$\chi(uv,w)(\chi(u,v) \triangleleft w) = \chi(u,vw)\chi(v,w), \tag{6.40}$$

etc. This gives the algebra. The conditions on ψ are a special case of those in Proposition 6.3.8 and they give us a coalgebra. Note that, for each u, condition (6.32) reduces to the condition

$$\psi(u)_{23}(\mathrm{id} \otimes \Delta)\psi(u) = \psi(u)_{12}(\mathrm{id} \otimes \Delta)\psi(u) \tag{6.41}$$

and the counitarity condition in the notation of Chapter 2.3. For each u it is just the 2-cocycle condition $\psi(u) \in \mathcal{H}^2(k,A)$. We use these cocycles now, not to conjugate the coproduct as in Chapter 2.3, but merely to multiply it from the left as stated. Finally, conditions (6.33) and (6.36) for this algebra and coalgebra to form a bialgebra reduce to (6.38) and (6.39), while (6.34) and (6.35) are empty since kG is cocommutative and the coaction is trivial. Note also that if A is a Hopf algebra then it is possible (i) to show that $\chi(u,v)$ and $\psi(u)$ are everywhere invertible and (ii) to give the antipode of the cocycle bicrossproduct explicitly as

$$S(u \otimes a) = (1 \otimes Sa)(u^{-1} \otimes v(u^{-1})S\chi(u,u^{-1})),$$

$$v(u) = \sum \psi(u)^{(1)} S\psi(u)^{(2)}. \tag{6.42}$$

These ψ, like the 2-cocycles in Chapter 2.3, are generalisations of the quasitriangular structure, and v is similar to the associated element implementing the square of the antipode in Proposition 2.1.8.   ∎

The case where the coaction and ψ are trivial and A is a group algebra takes us into the classical situation.

Example 6.3.11 *Let G be a finite group and let M be a finite Abelian group on which G acts from the right by group automorphisms, and let $\chi :$ $G \times G \to M$ be a 2-cocycle in $Z_{\triangleleft}^2(G,M)$. This gives a bicrossproduct Hopf algebra $kG {\bowtie}_{\chi} kM$ with the tensor product coalgebra structure. It can be identified as kE where E is the associated group extension $M \hookrightarrow E \twoheadrightarrow G$. Explicitly, this can be realised as $E = G \times M$ with product and inverse*

$$(u,s)(v,t) = (uv, \chi(u,v)(s \triangleleft u)t),$$

$$(u,s)^{-1} = (u^{-1}, (s^{-1} \triangleleft u^{-1})\chi^{-1}(u,u^{-1})).$$

If the action is trivial then M lies in the centre of E, which is then called a central extension.

Proof: We use the preceding example, with $H = kG$, $A = kM$, ψ trivial and χ invertible so that we have a Hopf algebra bicrossproduct. Condition (6.38) becomes empty when we look at group-like elements $s \in M$, while condition (6.39) holds provided $\chi(u, v)$ for every u, v is group-like, i.e. an element of M. This puts our cocycle χ into the classical setting of a map $G \times G \to M$. The first of the cocycle conditions (6.40) holds for $\triangleleft$ a right action, while the second becomes that χ is a group cocycle in the usual sense with values in M as an Abelian group and G-module. The resulting Hopf algebra has the tensor product coalgebra structure in which $u \otimes s$ are group-like. We identify $u \otimes s$ with the elements (u, s) of the usual group extension E. This clearly contains M as a subgroup and projects to G as shown by forgetting the s coordinate. The extension is cleft, with $j(u) = (u, e)$ and inverse $j^{-1}(u) = (u, e)^{-1} = (u^{-1}, \chi^{-1}(u, u^{-1}))$. Finally, it is clear from the product shown that M is central in E *iff* the action is trivial. ∎

In this way we recover the theory of group extensions, at least in the finite case. More commonly in physics, we are interested in topological groups, and to recover this theory one could put the above into a C^* or Hopf–von Neumann context. For example, a common case is the central extension with $M = S^1$, the circle group, and $\chi : G \times G \to S^1$. Alternatively, there is also a purely algebraic version of the theory at the level of Lie algebras. We describe it briefly in the left handed version since we will need left handed Lie algebra cocycles again in Chapter 8.

In fact, we have already described left handed Lie algebra cocycles with trivial action in (2.22) in Chapter 2.3. When there is a left action, one adds to this coboundary formula the extra terms

$$\partial\chi(\xi_1, \ldots, \xi_{n+1}) = \cdots + \sum_{i=1}^{n+1}(-1)^{i+1}\xi_i \triangleright \chi(\xi_1, \ldots, \hat{\xi}_i, \ldots, \xi_{n+1}). \quad (6.43)$$

Here $\hat{\ }$ denotes omission and $\chi : g^{\otimes n} \to V$ is a totally antisymmetric linear map from tensor powers of the Lie algebra g to a vector space on which g acts. In particular, a left 1-cocycle and left 2-cocycle are, respectively, linear and antisymmetric maps obeying

$$\chi([\xi, \eta]) = \xi \triangleright \chi(\eta) - \eta \triangleright \chi(\xi), \quad (6.44)$$

$$\chi([\xi, \eta], \zeta) + \text{cyclic} = \xi \triangleright \chi(\eta, \zeta) + \text{cyclic}. \quad (6.45)$$

Given a 2-cocycle $\chi : g \otimes g \to V$, we view V as an Abelian Lie algebra on which $\triangleright$ is automatically an action by Lie algebra automorphisms. Applying the general theory above to $H = U(g)$ and $A = U(V)$ in a left handed version now gives for the bicrossproduct the enveloping algebra

$U(E)$, where E is the Lie algebra extension $V \hookrightarrow E \twoheadrightarrow g$ associated to the above data. Explicitly, this takes the form $E = V \oplus g$ with Lie bracket

$$[v \oplus \xi, w \oplus \eta] = (\xi \triangleright w - \eta \triangleright v + \chi(\xi, \eta)) \oplus [\xi, \eta].$$

If we identify $v = v \oplus 0$ and $\xi = 0 \oplus \xi$, then another way to write this is

$$[\xi, \eta]_E = [\xi, \eta] + \chi(\xi, \eta), \quad [\xi, v]_E = \xi \triangleright v, \quad [v, w]_E = 0.$$

The extension is central *iff* the action is trivial. A famous example is the Virasoro algebra, which is the central extension of $g = \text{diff}(S^1)$. In algebraic terms, this is $V = k$ and

$$g = \{L_n : [L_n, L_m] = (n - m)L_{n+m}\}, \quad \chi(L_n, L_m) = \lambda \delta_{n,-m} n(n^2 - 1)$$

for any value of λ. The extension consists of adding one more generator with zero bracket, and adding χ to the bracket of the L_n generators.

This completes our demonstration of how the general Hopf algebra theory recovers the conventional picture of extensions as special cases. We note, however, that there are plenty more novel possibilities in our general setting. Probably the simplest of these is to go back to Example 6.3.10 and let $G = \mathbb{Z}_{/2}$ be the group with two elements, $\{e, u\}$, say, and $u^2 = 1$. Then our family of 2-cocycles (6.41) becomes just one 2-cocycle for the bialgebra or Hopf algebra; our action $\triangleleft$ just becomes one automorphism T; and χ just becomes one element x, say.

Proposition 6.3.12 *Let A be a Hopf algebra and let $\psi \in A \otimes A$ be an invertible counital 2-cocycle as in (2.15) in Chapter 2.3. Recall from Theorem 2.3.4 that there is a twisted Hopf algebra A_ψ with the same algebra, and coproduct $\psi(\Delta\ )\psi^{-1}$. If $T : A \to A_\psi$ is a bialgebra map and x is an invertible element in A such that*

$$x^{-1}ax = T^2(a), \quad T(x) = x, \quad \Delta x = (x \otimes x)((T \otimes T)(\psi))\psi, \quad \epsilon(x) = 1,$$

then these data define a cocycle bicrossproduct Hopf algebra $k\mathbb{Z}_{/2}\ {}^{\psi}{\blacktriangleright\!\!\triangleleft}_x A$. It is generated by A as a sub-Hopf algebra and one additional generator w with

$$w^{-1}aw = T(a), \quad w^2 = x, \quad \Delta w = (w \otimes w)\psi, \quad \epsilon w = 1, \quad Sw = wU Sx,$$

where U is as in Theorem 2.3.4.

Proof: We analyse the requirements of Example 6.3.10 with $G = \mathbb{Z}_{/2}$ and obtain exactly the above data ψ, T, x as necessary and sufficient for a cocycle bicrossproduct of this general type. For a cocycle cross coproduct we need (6.41), but since G now has only two elements $\{e, u\}$, this requirement becomes $\psi(e) = 1 \otimes 1$ and $\psi(u) = \psi$, a 2-cocycle as in Chapter 2.3. Likewise, a right cocycle action of $\mathbb{Z}_{/2}$ simply means one linear map $T : A \to A$ and one element x in A with

$$a \triangleleft e = a, \quad a \triangleleft u = T(a), \quad \chi(e,e) = \chi(e,u) = \chi(u,e) = 1, \quad \chi(u,u) = x,$$

obeying (6.27), which reduces to $T(ab) = T(a)T(b)$ and the cocycle conditions (6.40) on the group. The first of the cocycle conditions has the content $\chi(u,u)((a\triangleleft u)\triangleleft u) = a\chi(u,u)$, i.e. the condition $x^{-1}ax = T^2(a)$ as stated. The second has as its content $\chi(u,u)\triangleleft u = \chi(u,u)$ since $u^2 = e$. This is the condition $T(x) = x$, and it is all that we need to have a $\mathbb{Z}_{/2}$ extension of the algebra. Next, for a bialgebra, the condition (6.38) becomes

$$\psi(\Delta \circ T(a)) = ((T \otimes T) \circ \Delta a)\psi, \quad \epsilon \circ T = \epsilon,$$

which, along with T an algebra map, can be written as the requirement that T is a bialgebra map $A \to A_\psi$. Finally, condition (6.39) has as its content the conditions on $\Delta x, \epsilon x$ as stated.

All this works if A is a bialgebra, and does not really require ψ, x to be invertible. In the Hopf algebra case, we do require ψ, x invertible and obtain a Hopf algebra for our bicrossproduct. The resulting Hopf algebra structure from Example 6.3.10 appears as stated, when we write $w = u \otimes 1$ and identify $a = 1 \otimes a$. ∎

Example 6.3.13 *We work formally over* $\mathbb{C}[[t]]$. *Let* $A = U_q(sl_2)$ *as in Chapter 3.2, let* $\psi = \mathcal{R}$ *be its quasitriangular structure, let* $x = \nu^{-1}$ *be the inverse of its ribbon element in the sense of Definition 2.1.10, and let* T *be the algebra automorphism*

$$T(H) = -H, \quad T(X_\pm) = -q^{\pm\frac{1}{2}}X_\mp.$$

These data obey the conditions in the preceding proposition. The resulting cocycle bicrossproduct Hopf algebra is generated by $U_q(sl_2)$ *and one element* w *adjoined with relations*

$$w^{-1}aw = T(a), \quad w^2 = \nu^{-1}, \quad \Delta w = (w \otimes w)\mathcal{R}, \quad \epsilon w = 1, \quad Sw = wq^{-H}.$$

This extended Hopf algebra is a variant of the quantum Weyl group *of* $U_q(sl_2)$, *and* T *is a variant of Lusztig's automorphism from Chapter 3.3.*

Proof: We already noted in Example 2.3.6 that the quasitriangular structure provides a simple example of a 2-cocycle for a Hopf algebra. The twisted coproduct is the opposite one. It is easy to verify that T is an algebra and anticoalgebra map and, hence, is an isomorphism between these two Hopf algebras as required. Moreover, in this example, $T^2 = \mathrm{id}$. Next, from the formula for $\mathcal{R}$ in Example 3.2.1, we see at once that $(T \otimes T)(\mathcal{R}) = \mathcal{R}_{21}$, so the conditions for x are

$$x \text{ central}, \quad \Delta x = (x \otimes x)\mathcal{R}_{21}\mathcal{R}, \quad T(x) = x.$$

The inverse of the ribbon element automatically obeys the first two of these. It remains to check the last one. We note that the ribbon element for $U_q(sl_2)$ exists and can be expressed as $\nu = uq^{-H} = vq^H$, where u, v are the elements in Proposition 2.1.8 implementing the square of the antipode. On the other hand, it is easy to see that $T \circ S = S^{-1} \circ T$ from the explicit form of S in Example 3.2.1. Hence, $T(\nu) = T(vq^H) = T(\mathcal{R}^{(1)})T(S\mathcal{R}^{(2)})T(q^H) = \mathcal{R}^{(2)}(S^{-1}\mathcal{R}^{(1)})q^{-H} = uq^{-H} = \nu$. This formula for the ribbon element also gives the antipode of w as stated. ∎

We have not said much here about the representation theory of cocycle cross products. This can be developed along the same lines as in Corollary 6.1.3 for cross products, and is described now by covariant cocycle representations of A. The fundamental idea is that when we have a cocycle obstruction to building a representation, we can neutralise the cocycle by passing to the extension. We have already seen this principle in some form in Proposition 6.3.6 where the cocycle is trivial when viewed in E. A similar principle holds for the cocycle cross coproducts and bicrossproducts and is evident in the last example, where the cocycle $\mathcal{R}$ appears cohomologically trivial in the bicrossproduct.

The need to extend by a cocycle to have reasonable representations is also familiar classically for the Virasoro algebra mentioned above and for affine Lie algebras. The same applies for the affine quantum groups. These are defined in just the same way as for $U_q(g)$ in Chapter 3.3 but with $d_i a_{ij}$ of affine type rather than positive definite. Given g, there is a natural extension to an affine Lie algebra $\widehat{g}$ with additional $H_0, X_{\pm 0}$ generators. For $\widehat{sl_2}$ the resulting relations are

$$q^{\frac{H_i}{2}} X_{\pm j} q^{-\frac{H_i}{2}} = q^{\pm(2\delta_{ij}-1)} X_{\pm j}, \quad [X_{\pm 0}^3, X_{\pm 1}] = [3] X_{\pm 0} [X_{\pm 0}, X_{\pm 1}] X_{\pm 0}.$$

The H_i mutually commute and the $[X_{\pm i}, X_{\mp j}]$ have their usual form as in Chapter 3.3, as do the coproducts. Note that $c = q^{H_0 H_1}$ is central and grouplike. Its value in irreducible representations is (an exponential of) the *central charge* or level of the representation. The quotient of $U_q(\widehat{sl_2})$ by $c = 1$ is the *loop quantum group* $U_q(Lsl_2)$. Strictly speaking, we should also add here an additional 'derivation' generator (which can be done) for this terminology to conform to the classical picture.

We note first that the vector spaces of these Hopf algebras are graded by the total degree of X_{+0} and X_{-0} when all expression are (say) ordered in such a way that the positive root generators are to the left of the negative ones (for example, the degree of $X_{+0}X_{+1}X_{+0}X_{-0}^2$ is $2 + 2 = 4$). This is because the q-Serre relations are homogeneous in $X_{\pm 0}$ while the relations between positive and negative roots can be used to reorder. Meanwhile, c, c^{-1} generate a sub-Hopf algebra $k\mathbb{Z} = k[c, c^{-1}]$.

Example 6.3.14 *The affine quantum group $U_q(\widehat{sl_2})$ is a cocycle central extension*

$$k\mathbb{Z} \hookrightarrow U_q(\widehat{sl_2}) \twoheadrightarrow U_q(Lsl_2)$$

and has the bicrossproduct form $U_q(\widehat{sl_2}) = k\mathbb{Z}_\chi{\blacktriangleright\!\!\blacktriangleleft}U_q(Lsl_2)$, where χ : $U_q(Lsl_2)^{\otimes 2} \to k\mathbb{Z}$ is a cocycle and the cross coproduct is by the grading viewed as a coaction $\beta : U_q(Lsl_2) \to U_q(Lsl_2) \otimes k\mathbb{Z}$.

Proof: The first step is to work with new generators $E_i = X_{+i}q^{\frac{H_i}{2}}$ and $F_i = q^{\frac{H_i}{2}}X_{-i}$ and $g = q^{H_1}$, $c = q^{H_0}q^{H_1}$ along the lines in Chapter 3.3. The structure of $U_q(\widehat{sl_2})$ becomes c central and

$$gE_0 = q^{-2}E_0g, \quad gE_1 = q^2E_1g, \quad gF_0 = q^2F_0g, \quad gF_1 = q^{-2}F_1g,$$

$$q^2E_0F_0 - F_0E_0 = \frac{g^{-2}c^2 - 1}{q - q^{-1}}, \quad q^2E_1F_1 - F_1E_1 = \frac{g^2 - 1}{q - q^{-1}},$$

$$\Delta g = g \otimes g, \quad \Delta E_0 = E_0 \otimes cg^{-1} + 1 \otimes E_0, \quad \Delta E_1 = E_1 \otimes g + 1 \otimes E_1,$$

$$\Delta c = c \otimes c, \quad \Delta F_0 = F_0 \otimes cg^{-1} + 1 \otimes F_0, \quad \Delta F_1 = F_1 \otimes g + 1 \otimes F_1,$$

along with the q-Serre relations of the same form as above in terms of our new generators E_i, F_i. We denote the corresponding generators in the quotient $U_q(Lsl_2)$ by q^h, e_i, f_i. They have the same form of structure as above with $c = 1$. It is clear that we have Hopf algebra maps as stated, where $k\mathbb{Z} = k[c, c^{-1}]$ is the sub-Hopf algebra generated by c, c^{-1}. By moving the c's to the left, we can clearly identify $U_q(\widehat{sl_2}) = k\mathbb{Z} \otimes U_q(Lsl_2)$ as linear spaces. This identification restricted to $U_q(Lsl_2)$ is the linear map $j : U_q(Lsl_2) \to U_q(\widehat{sl_2})$ which sends an expression in terms of q^h, e_i, f_i to the same expression with g, E_i, F_i. The grading $|\ |$ by the total e_0 and f_0 degree in expressions that are ordered with the e_i to the left defines a coaction $\beta(f) = f \otimes c^{|f|}$. Note that neither of these maps j, β are algebra maps. They are merely well-defined linear maps. It is easy to see however that $U_q(Lsl_2)$ becomes a right $k\mathbb{Z}$ comodule coalgebra, the latter because the coproduct preserves our above ordering. We are therefore in a position to make the cross coproduct coalgebra $k\mathbb{Z}{\blacktriangleright\!\!\blacktriangleleft}U_q(Lsl_2)$. For example,

$$\Delta(1 \otimes e_0) = (1 \otimes e_0) \otimes (c \otimes q^{-h}) + (1 \otimes 1) \otimes (1 \otimes e_0).$$

This is the coproduct of $U_q(\widehat{sl_2})$ on its identification with $k\mathbb{Z} \otimes U_q(Lsl_2)$. Meanwhile, for the algebra structure we show that j makes $k\mathbb{Z} \subset U_q(\widehat{sl_2})$ a cleft extension and hence by Proposition 6.3.6 a cocycle cross product $k\mathbb{Z}_\chi{\rtimes}U_q(Lsl_2)$. The action in the cross product is trivial since c is central. To show that it is a cleft extension and to construct χ we need to show that j has a convolution-inverse $j^{-1} : U_q(Lsl_2) \to U_q(\widehat{sl_2})$. We defer the proof of this to the next lemma, which provides also an explicit formula for this inverse and the resulting cocycle. Using the formulae in the lemma, the explicit form of χ on the lowest generators is

$$\chi(f_0 \otimes e_0) = j(f_{0(1)})j(e_{0(1)})(Sj((f_{0(2)}e_{0(2)})^{(\bar{1})}))(f_{0(2)}e_{0(2)})^{(\bar{2})}$$

$$= F_0 E_0 g^2 + F_0(Sj(q^{-h}e_0))c + E_0(Sj(f_0 q^{-h}))c + (Sj((f_0 e_0)^{(\bar{1})}))(f_0 e_0)^{(\bar{2})}$$

$$= F_0 E_0 g^2 + F_0 cg S E_0 + E_0 cg S F_0 + (Sj(q^2 e_0 f_0))c^2 + Sj(\frac{1 - g^{-2}}{q - q^{-1}})$$

$$= F_0 E_0 g^2 + F_0 cg S E_0 + E_0 cg S F_0 + c^2 S(q^2 E_0 F_0) + \frac{1 - g^2}{q - q^{-1}}$$

$$= E_0 cg S F_0 + c^2 S(F_0 E_0) + \frac{g^2 - c^2}{q - q^{-1}} + \frac{1 - g^2}{q - q^{-1}} = \frac{1 - c^2}{q - q^{-1}}$$

using the coproducts above, the definition of j and the algebra relations in $U_q(Lsl_2)$ and $U_q(\widehat{sl_2})$. We also used the antipode $SE_0 = -E_0 c^{-1}g$ etc., to cancel terms. One may verify that the cocycle cross product on $k\mathbb{Z} \otimes U_q(Lsl_2)$ in Proposition 6.3.2 indeed recovers the algebra structure of $U_q(\widehat{sl_2})$.   ■

The general Hopf algebra 'central extension' construction used here is a special case of a cocycle bicrossproduct in which the action is trivial and the dual cocycle is trivial. The coacting Hopf algebra A does not need to be commutative, but if it is then it is central, i.e. this is a natural generalisation of Example 6.3.11. The following lemma is the corresponding special case of the abstract picture of cocycle bicrossproducts as cleft and cocleft extensions.

Lemma 6.3.15 *Let* $A \hookrightarrow E \twoheadrightarrow H$ *be algebra maps between Hopf algebras with* $E = A \blacktriangleright\!\!\triangleleft H$ *as a coalgebra, the inclusion given by* $\otimes 1$ *and the surjection by* $\epsilon \otimes \mathrm{id}$. *If the inclusion* $j : H \to E$ *defined by* $1 \otimes$ *obeys* $aj(h) = a \otimes h = j(h)a$ *then* j *is convolution-invertible with*

$$j^{-1}(h) = \sum (Sj(h^{(\bar{1})}))h^{(\bar{2})}, \quad \forall h \in H$$

and $E = A_\chi \blacktriangleright\!\!\triangleleft H$, *a bicrossproduct with trivial action and the cocycle*

$$\chi(h \otimes g) = \sum j(h_{(1)})j(g_{(1)})(Sj((h_{(2)}g_{(2)})^{(\bar{1})}))(h_{(2)}g_{(2)})^{(\bar{2})}, \quad \forall h, g \in H.$$

Proof: We assume that H is a right A-comodule coalgebra so that we can form the right cross coproduct on $A \otimes H$ from Proposition 1.6.16. Then the inclusion of A and the projection to H are automatically Hopf algebra maps. There is also a coaction $E \to E \otimes H$ given by pushing out the coproduct of E and $A = E^H$, the fixed subalgebra under this. That is, we have an algebra extension of A by H. We show that it is a cleft extension. The required maps j, j^{-1} are defined as stated and we check

$$j^{-1}(h_{(1)})j(h_{(2)}) = (Sj(h_{(1)}{}^{(\bar{1})}))h_{(1)}{}^{(\bar{2})}j(h_{(2)})$$
$$= (S(1 \otimes h_{(1)}{}^{(\bar{1})}))(h_{(1)}{}^{(\bar{2})} \otimes h_{(2)}) = \epsilon(1 \otimes h) = \epsilon(h),$$

using the form of the coproduct of E. On the other side, we have

$$j(h_{(1)}) \; j^{-1}(h_{(2)}) = (1 \otimes h_{(1)})(S(1 \otimes h_{(2)}{}^{(\bar{1})}))h_{(2)}{}^{(\bar{2})}$$
$$= (1 \otimes h_{(1)}{}^{(\bar{1})})(S(1 \otimes h_{(2)}{}^{(\bar{1})}))(Sh_{(1)}{}^{(\bar{2})}{}_{(1)})h_{(1)}{}^{(\bar{2})}{}_{(2)}h_{(2)}{}^{(\bar{2})}$$
$$= (1 \otimes h_{(1)}{}^{(\bar{1})(\bar{1})})(S(1 \otimes h_{(2)}{}^{(\bar{1})}))(Sh_{(1)}{}^{(\bar{1})(\bar{2})})h_{(1)}{}^{(\bar{2})}h_{(2)}{}^{(\bar{2})}$$
$$= (1 \otimes h^{(\bar{1})}{}_{(1)}{}^{(\bar{1})})(S(1 \otimes h^{(\bar{1})}{}_{(2)}))(Sh^{(\bar{1})}{}_{(1)}{}^{(\bar{2})})h^{(\bar{2})}$$
$$= (1 \otimes h^{(\bar{1})}{}_{(1)}{}^{(\bar{1})})(S(h^{(\bar{1})}{}_{(1)}{}^{(\bar{2})} \otimes h^{(\bar{1})}{}_{(2)}))h^{(\bar{2})} = \epsilon(h^{(\bar{1})})h^{(\bar{2})} = \epsilon(h),$$

where the second equality inserts $(Sh_{(1)}{}^{(\bar{2})}{}_{(1)})h_{(1)}{}^{(\bar{2})}{}_{(2)}$, the third uses the coaction axiom, the fourth uses covariance of the coproduct of H under the coaction and the fifth that A is a sub-Hopf algebra. We can then use that S is the antipode in E to collapse the expression. Next, from the form of the coproduct it is clear that the map j intertwines the above coaction $E \to E \otimes H$ and the right regular coaction of H on itself. Hence all the conditions for a cleft extension are satisfied and we can apply Proposition 6.3.6. The cocycle bicrossproduct conditions in the (left–right version of) Proposition 6.3.9 must necessarily hold since we already know that E is a Hopf algebra. ∎

Finally, we conclude the section with some examples of more general extensions that are not cleft and hence not cocycle cross products. We return to the quantum principal bundle point of view touched upon briefly after Proposition 6.3.6. In this context we will denote the extension of an algebra A by P (rather than by E) to remind us that it plays the role of coordinate algebra of the total space of a principal bundle, except that we do not assume that it is commutative. Thus, P is required to be a right H-comodule algebra by a coaction β and $A = P^H$ is its fixed subalgebra, i.e. an extension. A *quantum principal bundle* is an extension for which 'local triviality' holds in the form of exactness of the sequence

$$0 \to P(\Omega^1 A)P \to \Omega^1 P \xrightarrow{\text{ver}} P \otimes \ker \epsilon \to 0.$$

Here

$$\text{ver} = (\cdot \otimes \text{id})\beta : P \otimes P \to P \otimes H$$

is built from the coaction and we restrict it to $\Omega^1 P \subset P \otimes P$, where $\Omega^1 P$ denotes the kernel of the product map of P. The latter is a construction which makes sense for any algebra and is called the *universal differential calculus* over the algebra. It comes with a map $\text{d} : P \to \Omega^1 P$ defined as

$d = 1 \otimes \mathrm{id} - \mathrm{id} \otimes 1$, obeying the Leibniz rule with respect to multiplication of '1-forms' in $\Omega^1 P$ by P from the left or right. Other choices of 'differential structure' are quotients of this one. From this point of view one should think of ver : $\Omega^1 P \to P \otimes H$ as the generator of the 'vertical vector fields' on the total space correponding classically to the group action of an element of the Lie algebra of G, except that we are viewing the latter as a Hopf algebra coaction. Its surjectivity corresponds to freeness of the action, while the other part of the exactness condition stated above specifies that its kernel or 'horizontal forms' (those annihilated by the vertical vector fields) is to coincide with $P(\Omega^1 A)P$, which are the forms from the base A 'pulled back' to forms on P. This is a key property in differential geometry which is usually proven using the axiom of local triviality of a bundle, and our exactness condition replaces that. One then finds that a theory of connections, associated bundles, covariant derivatives, etc., goes through at this level, which is to say 'quantum group gauge theory'. A connection, for example, is defined abstractly as an equivariant complement of $P(\Omega^1 A)P \subseteq \Omega^1 P$.

In the present case of the universal differential calculus is it possible to state the 'local triviality' exactness condition more algebraically. Indeed, since $P(\Omega^1 A)P = P(\mathrm{d}A)P$ is spanned by elements of the form $P(1 \otimes a - a \otimes 1)P$, specifying the kernel of ver and its surjectivity is equivalent to specifying that

$$\overline{\mathrm{ver}} : P \underset{A}{\otimes} P \to P \otimes H$$

is a bijection, where ver descends to $P \otimes_A P$ since the elements of A are fixed under the coaction β. This condition turns up also in the application of Hopf algebras to the theory of field extensions and is called the *Galois property*. So a Galois extension, where $\overline{\mathrm{ver}}$ is a bijection, is equivalent to a quantum principal bundle with the universal differential calculus.

A natural example of such a quantum principal bundle (beyond cleft extensions or trivial bundles) is a *quantum homogeneous space*. This is an extension $A = P^H$ forming a quantum principal bundle in which P is itself a Hopf algebra and the right coaction of H on P is given by pushing out the coproduct of P, $\beta = (\mathrm{id} \otimes \pi)\Delta$, for a Hopf algebra surjection $\pi : P \to H$. This is the analogue in our coordinate language of the idea of a quotient of a group by a subgroup and its associated bundle with fibre given by the subgroup. In our case we have

$$A \hookrightarrow P \overset{\pi}{\twoheadrightarrow} H$$

where the inclusion is an algebra map and π is a Hopf algebra map.

Example 6.3.16 *Let* $k\mathbb{Z} = k[g, g^{-1}]$ *with coproduct* $\Delta g = g \otimes g$ *and con-*

sider the Hopf algebra surjection

$$\pi : SL_q(2) \to k\mathbb{Z}, \quad \pi \begin{pmatrix} a & b \\ c & d \end{pmatrix} = \begin{pmatrix} g & 0 \\ 0 & g^{-1} \end{pmatrix}.$$

The standard *q-sphere is the fixed subalgebra* $S_q^2 = SL_q(2)^{k\mathbb{Z}}$ *under the coaction* $\beta = (\mathrm{id} \otimes \pi) \circ \Delta$ *of* $k\mathbb{Z}$. *It is the algebra generated by* 1 *and* $b_\pm, b_3$ *modulo the relations*

$$b_3^2 = b_3 + qb_-b_+, \quad b_\pm b_3 = q^{\pm 2}b_3 b_\pm + (1 - q^{\pm 2})b_\pm,$$

$$q^2 b_- b_+ = q^{-2} b_+ b_- + (q - q^{-1})(b_3 - 1),$$

and forms a quantum homogeneous space

$$S_q^2 \hookrightarrow SL_q(2) \twoheadrightarrow k\mathbb{Z}.$$

Proof: The coaction induced by π and the coproduct of the Hopf algebra $SL_q(2)$ in Proposition 4.2.6 is

$$\beta \begin{pmatrix} a & b \\ c & d \end{pmatrix} = \begin{pmatrix} a \otimes g & b \otimes g^{-1} \\ c \otimes g & d \otimes g^{-1} \end{pmatrix}.$$

It corresponds to a $\mathbb{Z}$-grading of $SL_q(2)$ according to the total degree of a, c minus the total degree of d, b in any expression. The elements $b_3 = ad$, $b_+ = cd$ and $b_- = ab$ are therefore invariant and clearly generate all the invariant combinations of generators. These generators of S_q^2 then inherit the relations shown from those of $SL_q(2)$ in Proposition 4.2.6. To prove that we have a quantum principal bundle, consider the elements $d \otimes a - qb \otimes c = Sa_{(1)} \otimes a_{(2)}$ and $a \otimes d - q^{-1}c \otimes b = Sd_{(1)} \otimes d_{(2)}$. Since every term in the right tensor factor has degree 1 in the first case and minus 1 in the second, it is clear that

$$\overline{\mathrm{ver}}(Sa_{(1)} \underset{S_q^2}{\otimes} a_{(2)}) = 1 \otimes g, \quad \overline{\mathrm{ver}}(Sd_{(1)} \underset{S_q^2}{\otimes} d_{(2)}) = 1 \otimes g^{-1}$$

by the antipode axioms. Similarly, by multiplicativity of the coproduct we can replace a by a^n to obtain $1 \otimes g^n$, etc. Thus we have

$$\overline{\mathrm{ver}}^{-1}(h \otimes g^n) = hSa^n_{(1)} \underset{S_q^2}{\otimes} a^n_{(2)}, \quad \overline{\mathrm{ver}}^{-1}(h \otimes g^{-n}) = hSd^n_{(1)} \underset{S_q^2}{\otimes} d^n_{(2)}$$

as the right inverse of $\overline{\mathrm{ver}}$. On the other side, let $h \in SL_q(2)$ have homogeneous degree $n \geq 0$, say. Then $\overline{\mathrm{ver}}(h' \otimes_{S_q^2} h) = h'h \otimes g^{|h|}$ and $\overline{\mathrm{ver}}^{-1}(h'h \otimes g^{|h|}) = h'hSa^n_{(1)} \otimes_{S_q^2} a^n_{(2)} = h' \otimes_{S_q^2} h(Sa^n_{(1)})a^n_{(2)} = h' \otimes_{S_q^2} h$ for all h'. This is because the terms in $Sa^n_{(1)}$ all have degree $-n$ so that the terms of $hSa^n_{(1)}$ lie in S_q^2. Similarly for $n \leq 0$. ∎

This treatment also works as ∗-algebras over $\mathbb{C}$. Then $S_q^2 = SU_q(2)^{\mathbb{C}\mathbb{Z}}$ inherits a ∗-structure from Example 4.2.16, namely

$$b_3^* = b_3, \quad b_+^* = -qb_-, \quad b_-^* = -q^{-1}b_+$$

when q is real. We think of $\mathbb{C}\mathbb{Z}$ as the coordinate ring of $U(1)$ with $g^* = g^{-1}$. Thus, this is the analogue of $S^2 = SU_2/U(1)$. Indeed, two of the relations of S_q^2 are 'q-commutativity' relations in the sense that they become that $b_3, b_\pm$ commute when $q = 1$. The remaining relation becomes, when $q = 1$,

$$x^2 + y^2 + z^2 = \frac{1}{4}$$

in terms of x, y, z defined by $b_\pm = \pm(x \pm \imath y)$ and $b_3 = z + \frac{1}{2}$. That is, these are complex coordinates for the sphere. Also, the associated bundle classically has a natural connection, the Dirac monopole. In our formulation, instead of a Lie-algebra valued one form it is a map $\omega :$ $\ker \epsilon \to \Omega^1 P$, given in the above example by

$$\omega(g^n - 1) = Sa^n{}_{(1)}da^n{}_{(2)}, \quad \omega(g^{-n} - 1) = Sd^n{}_{(1)}dd^n{}_{(2)}.$$

This indeed recovers the Dirac monopole on S^2 when computed for $q = 1$ with the usual coordinates.

6.4 Quantum–gravity and observable–state duality

In this section we want to explain the significance of the constructions of this chapter for the problem of unifying quantum mechanics and gravity, with the bicrossproduct Hopf algebras of Section 6.2 providing some concrete examples. There are two main things that Hopf algebras can tell us about this important problem.

First, as we have seen in many places, Hopf algebras are the simplest examples of noncommutative or quantum geometry in which the commutative algebra of functions on a space is replaced by a noncommutative one. We already know from Chapter 6.1 that our bicrossproducts, which as algebras are cross products, can be interpreted as the algebras of observables of quantum systems. Classically, such algebras of observables would be functions on a phase-space, so now in the quantum case we should think of them as functions on a quantum phase-space. They provide us then with concrete models in which to test out some ideas of quantum geometry in actual physical examples. In particular, we know that the simplest nontrivial geometries are based on group structures and this appears now in the form of a coproduct on our algebra of observables. This has been explained in the preamble to the present chapter and we demonstrate in detail how it works in the simplest cases. It could be

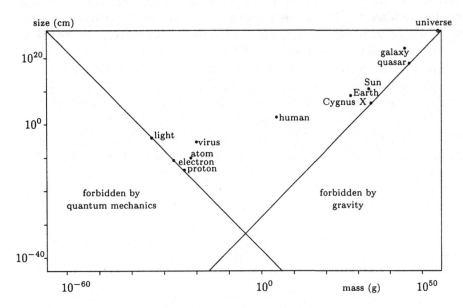

Fig. 6.6. Size v. mass of objects in the universe.

said that a large part of the problem of unifying quantum mechanics and gravity, or, in more general terms, of unifying the concepts of microscopic and macroscopic physics, is a matter of language. Quantum mechanics is usually formulated algebraically, whereas gravity is usually formulated more geometrically. Quantum geometry, with Hopf algebras as a basic example, addresses exactly this problem because it allows us to formulate both structures in the same algebraic language.

Secondly, once everything is converted to algebra in this way, one has to ask what genuinely new phenomena does this more unified picture of quantum mechanics and gravity allow? Here again, Hopf algebras and, in particular, our bicrossproduct models, provide examples of a genuinely new phenomenon. This is a kind of symmetry or isomorphism between quantum or microscopic phenomena and geometrical or macroscopic phenomena. There is a sense in which each represents the other and, moreover, a sense in which their roles can be interchanged. We have already touched upon this novel *principle of representation-theoretic self-duality* in Chapter 5.4, and we continue now with a description of its geometrical aspects. This approach to Planck-scale physics is due to the author; although the topic is necessarily speculative, it does demonstrate some of the important concepts that can explored using our algebraic techniques.

We begin by reminding the reader of the definition of the Planck scale.

This is the scale at which quantum and gravitational effects become equally dominant. The easiest way to explain this is to draw a plot, Fig. 6.6, of the mass and size of all objects in the universe. The striking feature is that everything lies in the V-shaped region in the middle. The region to the bottom left is forbidden by Heisenberg's uncertainty principle, i.e. by quantum effects. This says that $\Delta_x \Delta_p \geq \hbar$, where we take Δ_x as a measure of the size of an object and $\Delta_p \sim mc$ as a measure of the mass m. Here c is the speed of light. The region to the bottom right is also forbidden, but this time for gravitational reasons. If one tries to put too much mass into a given region, it forms a black hole, which only grows as more mass is put in. This slope is the line at which the gravitational energy in the spacetime around the object is comparable to its mass-energy, which is estimated by $\frac{Gmm}{\Delta_x} = mc^2$, where G is the gravitational coupling constant. The Planck scale is the intersection of these two lines and is the value in size and mass where both the quantum effects and the gravitational effects are strong. It comes out as

$$m_P = \left(\frac{\hbar c}{G}\right)^{\frac{1}{2}} = 2.177 \times 10^{-5} \text{ g}, \quad x_P = \left(\frac{\hbar G}{c^3}\right)^{\frac{1}{2}} = 1.616 \times 10^{-33} \text{ cm}.$$

It seems reasonable that phenomena should become simpler on these boundary lines and, conversely, more complicated as we approach the centre of the V-shaped region. It is significant that the range of human interaction is roughly in the centre: from a more conventional point of view, it reflects the complexity needed for life to form; from the more deconstructionist view (which we develop at the end of the section), it reflects the idea that we build physics around ourselves.

Now we turn to our bicrossproduct models. The simplest case is the one-dimensional bicrossproduct $\mathbb{C}[x] \blacktriangleright\!\!\triangleleft \mathbb{C}[p]$ from Example 6.2.18. What we know from Section 6.1 is that the general problem of introducing x, p relations such that conjugation by e^{tp} generates a given action α on the position coordinates, is solved by the cross product algebra defined by α. If we ask that the resulting quantum algebra remains a Hopf algebra (as is $\mathbb{C}[x] \otimes \mathbb{C}[p]$) of the same general type for the coalgebra, then we are forced to look for a bicrossproduct. One can also phrase this more formally as an extension problem $\mathbb{C}[x] \to E \to \mathbb{C}[p]$ according to Section 6.3. Then Example 6.2.17 tells us that the general solution to this problem has just two parameters A, B with action and reaction as stated there. In short, all possible quantisations of a particle in one dimension, for which the quantum algebra of observables remains a Hopf algebra of self-dual type, are classified by two parameters. We have identified the combination $\hbar = -\frac{A}{B}$ in Example 6.2.18. To see the meaning of the remaining parameter B we consider how the particle moves classically. Note that in our approach we keep p as a conserved momentum and keep the Hamiltonian $\frac{-p^2}{2m}$ so that the particle is in free-fall. The different motions of the particle are then controlled by changing the x, p commutation relations. It is more usual to keep the commutation relations fixed in a canonical form and to vary the Hamiltonian, and indeed we could reformulate things this way, except that the necessary change of variables would have to be singular. We know this because the usual quantum mechanics algebra in Example 6.1.11 is not a Hopf algebra, while our bicrossproduct one is. Recall also that in our conventions p is anti-Hermitian, i.e. $-\imath p$ is the physical momentum observable. Then

$$\frac{\mathrm{d}x}{\mathrm{d}t} = \frac{\imath}{\hbar}\left[-\frac{p^2}{2m}, x\right] = \left(\frac{-\imath p}{m}\right)(1 - e^{-Bx}) + O(\hbar), \qquad \frac{\mathrm{d}p}{\mathrm{d}t} = 0.$$

This is as operators, but in representations where the system behaves like a particle its classical trajectories will be of the form given here by the leading term. We identify $\frac{-\imath p}{m} = v_\infty$, the velocity at $x = \infty$. If we consider a particle falling in from infinity then we see that the particle approaches the origin $x = 0$ but does so more and more slowly. In fact, it takes an infinite amount of time to reach the origin, which therefore behaves in some ways like a black-hole event horizon. The present model is one dimensional but we can imagine that it is the radial part of some motion in spacetime, and can estimate the value of B on this basis. We find

$$B = \frac{c^2}{MG}, \quad \frac{\mathrm{d}x}{\mathrm{d}t} = v_\infty\left(1 - \frac{1}{\exp(\frac{c^2 x}{MG})}\right); \quad \text{cf.} \quad \frac{\mathrm{d}x}{\mathrm{d}t} = -c\left(1 - \frac{1}{1 + \frac{1}{2}\frac{c^2 x}{MG}}\right),$$

where the comparison is with an in-falling photon at radial distance x from the event horizon in the Schwarzschild black-hole solution of mass M. This analogy should not be pushed too far, since our present treatment is in nonrelativistic quantum mechanics, but it gives us at least one interpretation of the parameter B as being comparable to introducing the distortion in the geometry due to a gravitational mass M.

Another, more mathematical, way to reach the same conclusion is to take the limit $\hbar \to 0$. In this case, our algebra becomes commutative, but the coalgebra remains noncocommutative. In this case, $\mathbb{C}[x]\!\blacktriangleright\!\blacktriangleleft\mathbb{C}[p] \cong \mathbb{C}(X)$, where X is the group $\mathbb{R}\!\blacktriangleright\!\blacktriangleleft\mathbb{R}$ with group law

$$X = \{(s, u)\}, \quad (s, u)(t, v) = (s + t, ue^{-Bt} + v).$$

To see this, let $p(s, u) = u$ and $x(s, u) = s$ be the coordinate functions on X. They commute, and with coproducts determined from

$$(\Delta x)((s, u), (t, v)) = x((s, u)(t, v)) = s + t$$
$$= (x \otimes 1 + 1 \otimes x)((s, u), (t, v))$$
$$(\Delta p)((s, u), (t, v)) = p((s, u)(t, v)) = ue^{-Bt} + v$$

$$= (p \otimes e^{-Bx} + 1 \otimes p)((s, u), (t, v))$$

they generate our algebraic model $\mathbb{C}(X)$ of the functions on X. This is clearly the limit of $\mathbb{C}[x] \blacktriangleright\!\!\triangleleft \mathbb{C}[p]$. This group $\mathbb{R} \!\!\rtimes\!\! \mathbb{R}$ is therefore the underlying classical phase-space of our system. The noncocommutative coproduct equips it with a non-Abelian group structure. In geometrical terms, a non-Abelian group law corresponds to geometrical curvature. Note that since the group is not semisimple (it is solvable), its natural metric in the sense discussed in Section 6.1 is degenerate. So there are some subtleties here but the general principle is the same. In our case one can compute that the curvature on phase-space is of the order of B^2. For dynamical models with a reasonable degree of symmetry, one can expect that this should also be comparable to the curvature in position space. Comparing it with the curvature near a mass M, for example, would give the same estimate as above.

In summary, we see that the bicrossproduct model has two limits:

$$\mathbb{C}[x] \blacktriangleright\!\!\triangleleft \mathbb{C}[p] \quad \begin{array}{c} \overset{G \to 0}{\nearrow} \quad \mathbb{C}[x] \!\rtimes\! \mathbb{C}[p] \text{ usual quantum mechanics for } x > 0 \\[2mm] \underset{\hbar \to 0}{\searrow} \quad \mathbb{C}(X) \text{ usual curved geometry.} \end{array}$$

This illustrates our first goal of unifying a quantum system and a geometrical one. The most general unification within this framework allows only two free parameters, which we have identified in general terms as $\hbar, G$. Moreover, the existence of curvature in our geometrical system forces the dynamics of the quantum particle to be deformed from the usual one. This deformation forces, as we have seen, a structure not unlike a black-hole event horizon. With a little more care, one can estimate also how small $\hbar, G$ have to be in comparison to the other scales in the system. For example, we can take the two scales in the system as m, the mass of the quantum test-particle moving in the curved background and M, the active gravitational mass, which we estimate as causing comparable curvature. The flat-space quantum-mechanical picture is valid if $Bx \gg 1$. This says that we are far from the right hand slope in Fig. 6.6. If we consider a relativistic quantum particle, then its position is only defined up to its Compton wavelength $\frac{\hbar}{mc}$ (the left hand slope in Fig. 6.6), so that the condition that the system is not detectably different from usual flat-space quantum mechanics in the region $x > 0$ is estimated by the inequality

$$mM \ll m_P^2.$$

On the other hand, the system appears classical if $\hbar \ll \Delta p \Delta x$. This says that we are far from the left hand slope in Fig. 6.6. If we estimate $\Delta p \sim mc$ again and suppose that the smallest length scale of interest is the gravitational one $\frac{1}{B}$ (the right hand slope), then the condition that the quantum aspect of the algebra is not detectable is estimated by the inequality

$$mM \gg m_P^2.$$

Some of the general features here also hold for other bicrossproduct models associated to group factorisations. The action α describes, as in Section 6.1, the flow or 'metric' under which the quantum particle moves. But not every α admits a back-reaction β forming a matched pair in the sense (6.15) needed for the Hopf algebra property. This is a genuine constraint, which one can think of as an integrated form of a second order differential equation for α (or a pair of first order equations for α, β). Moreover, this constraint in the above examples, and also for more complicated examples in Chapter 8, does have qualitative similarities with the singularities forced by Einstein's equation for the metric in the presence of matter. From this point of view, the matching conditions (6.15) are some kind of toy version of Einstein's equation. For more complicated metrics, one would need to leave the class of Hopf algebras and construct more complicated quantum geometries, but exhibiting perhaps some of these same features.

Next we note that, by their construction, the bicrossproduct models are self-dual, as we know from the symmetry in Example 6.2.17 and the analysis of Example 6.2.12. To see this more precisely one has to introduce some functional analysis and work with Hopf–von Neumann algebras, or else proceed with formal power series. For our algebraic purposes, we construct directly the corresponding Hopf algebra $\mathbb{C}[\phi]\blacktriangleright\!\!\triangleleft\mathbb{C}[\psi]$, say, and show that it is dually paired. This time, the construction follows Theorem 6.2.3 with a right action and left coaction. These, and the resulting Hopf algebra, are

$$\psi\triangleleft\phi = \hbar^{-1}(1 - e^{-A\psi}), \quad \beta(\phi) = e^{-A\psi} \otimes \phi,$$

$$[\psi, \phi] = \hbar^{-1}(1 - e^{-A\psi}),$$

$$\Delta\phi = \phi \otimes 1 + e^{-A\psi} \otimes \phi, \quad \Delta\psi = \psi \otimes 1 + 1 \otimes \psi,$$

$$\epsilon\phi = \epsilon\psi = 0, \quad S\phi = -e^{A\psi}\phi, \quad S\psi = -\psi.$$

We have followed here exactly the same steps as in Example 6.2.18 but with the roles of the two groups in Example 6.2.17 or the roles of A, B interchanged. Next, we should think of $\mathbb{C}[\phi]$ as the universal enveloping algebra $U(\mathbb{R})$ and hence as being dually paired with the algebra $\mathbb{C}[x]$ in the usual way, namely by

$$\langle \phi^n, x^m \rangle = \delta^n{}_m n!, \quad \text{i.e.} \quad \langle \phi^n, f(x) \rangle = \left. \frac{\mathrm{d}^n}{\mathrm{d}x^n} \right|_0 f.$$

In the same way, $\mathbb{C}[p]$ is dually paired with $\mathbb{C}[\psi]$. Hence from Proposition 6.2.5, we conclude that $\mathbb{C}[\phi] \bowtie \mathbb{C}[\psi]$ is dually paired with $\mathbb{C}[x] \bowtie \mathbb{C}[p]$. Explicitly, it is

$$\langle \phi, : f(x,p) : \rangle = \left(\frac{\partial f}{\partial x} \right)(0,0), \quad \langle \psi, : f(x,p) : \rangle = \left(\frac{\partial f}{\partial p} \right)(0,0),$$

where $: f(x,p) := \sum f_{n,m} x^n p^m$ is the normal-ordered form of a function in the two variables x, p. For a strictly polynomial version, one should use $e^{-Bx}, e^{-A\psi}$ as abstract generators along the lines of Example 6.1.17. If we put in the physical interpretation of the parameters A, B above then we see that the natural dual of $\mathbb{C}[x] \bowtie_{\hbar,G} \mathbb{C}[p]$ is the bicrossproduct $\mathbb{C}[\phi] \bowtie_{\frac{1}{\hbar}, \frac{G}{\hbar}} \mathbb{C}[\psi]$ (keeping m, M, c^2 fixed, say).

Also, we recall from Chapter 5 that linear functionals on a quantum Hopf algebra of observables themselves form an 'algebra of states'. The physical states are the positive ones among them. In our model, then, the algebra of states is generated by the linear functionals ϕ, ψ. Moreover, we see that this algebra of states is exactly like a quantum system. It is natural to give ϕ the dimensions of inverse length, and ψ the dimensions of inverse momentum. Moreover, if we consider

$$x' = \hbar\psi, \quad p' = \hbar e^{A\psi} \phi,$$

we see that these have dimensions of length and momentum and have exactly the same Hopf algebra structure as $\mathbb{C}[x] \bowtie \mathbb{C}[p]$. Thus one could equally well regard this second Hopf algebra as the algebra of observables, with p' as momentum and x' as position. One would have the same picture as above in terms of geometry and quantum mechanics. The possibility of making such a reinterpretation of the same algebraic structures is our second and more radical theme demonstrated by the bicrossproduct models. In this reinterpretation, the roles of observables and states become reversed. Hence, the roles of noncommutativity in the algebra (of quantum origin) and noncocommutativity in the coalgebra (of geometrical origin as curvature on phase-space) become reversed. This is a new kind of symmetry principle, which one can propose as a speculative idea for the structure of Planck-scale physics.

In the remainder of this section, we put some of these considerations into a more general context which makes clear that they go much beyond Hopf algebras. After all, Hopf algebras are only the most simple quantum geometries, and we would like these ideas to have much wider applicability. This will also bring together many of the algebraic facts from earlier chapters. Note that it is not usual to question too much the physics behind the choice of an axiomatic structure, but this is exactly what we shall do. After all, axioms are 'physical observables' too and might have associated laws that govern their choice. The systematic way to address this kind of question is by means of category theory, which we shall not come to until Chapter 9. Suffice it to say that we consider types or categories of structures, as defined, for example, by a system of axioms. We also consider maps between categories, called *functors*. They give a way of realising one structure in terms of another. For example, the enveloping algebra construction in Example 1.5.7 is a functor from the category of Lie algebras to the category of Hopf algebras. Likewise, the group algebra Example 1.5.3 is a functor from finite groups to Hopf algebras.

Armed with this categorical point of view, we want to explore the following very simple idea about the nature of theoretical physics. Before doing so, it should be stressed that theoretical physics is not so much concerned with 'what' is observed but with the question 'why is it observed?'; i.e. it always seeks to explain structures in terms of still more fundamental structures or principles. This reductionist programme takes the naive view that there are indeed some fundamental laws of nature, of which our experiments and observations are representations. Thus it is supposed that something is absolutely true, and that something else measures or observes it. However, one of the themes throughout this book is that such evaluations should generally be thought of more symmetrically as a 'duality pairing' of one structure with another. An evaluation $f(x)$ can also be read $x(f)$, where f is an element of a dual structure. Since theoretical physics adopts the language of mathematics, such an 'observer-observed' reversed interpretation of the mathematical structure can always be forced, but will the dual interpretation also describe physics? The idea, which we have already discussed in Definition 5.4.1 and which we would like to elevate to a more general *principle of representation-theoretic self-duality*, is that the answer should be 'yes', i.e. that a fundamental theory of physics is incomplete unless such a role-reversal is possible. We can go further and hope to fully determine the (supposed) structure of fundamental laws of nature among all mathematical structures by this self-duality condition.

Such duality considerations are certainly evident in some form in the context of quantum theory and gravity. The situation is summarised to the left in Fig. 6.7. For example, Lie groups provide the simplest examples of Riemannian geometry, while the representations of similar Lie groups provide the quantum numbers of elementary particles in quantum theory. Thus, both quantum theory and non-Euclidean geometry are needed for a self-dual picture. Now, Hopf algebras precisely serve to unify these mu-

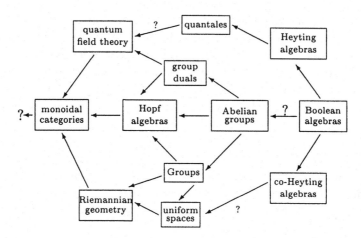

Fig. 6.7. Representation-theoretic approach to quantum-gravity. Self-dual structures are on the central axis.

tually dual structures. This was our motivation for the bicrossproduct models above. We have seen that imposing the duality requirement in the context of particles moving on homogeneous spaces forces the homogeneous spaces to come in pairs, and, moreover, constrains the allowed effective metric and dynamics in a way not too unlike Einstein's equations in this context.

If this principle is right, we should be able to use it as a guide to the required mathematical structure for the next more advanced theory of quantum mechanics and gravity. There are two steps, the first being to identify or invent the relevant self-dual category or axiom-system beyond Hopf algebras but with similar input–output symmetry or self-duality properties of the axioms. The second step would be to construct some concrete models in such a framework, as we did for Hopf algebras with the bicrossproduct models. The long-term aim is to have in this way a picture of the full Einstein's equation on a general metric (i.e. not only homogeneous ones) appearing as a self-duality constraint. Recall that Einstein's equation does indeed equate a geometrical object (the Einstein tensor which measures the curvature of spacetime) to a quantum-mechanical object (the vacuum expectation of the stress energy tensor which measures the matter content).

This is a long-term programme, but it provides a point of view in which Hopf algebras appear naturally as simple cases. Concerning what comes next, one proposal (also due to the author) is that the next most general category should be the category of all (small) monoidal categories C over a fixed one $\mathcal{V}$. A monoidal category is one with tensor products. We have already seen in Exercise 1.6.8 that the representations of a Hopf algebra indeed have a tensor product. We will see later, in Chapter 9.1, that one can consider such a category equipped with a functor $C \to V$ and show that there is then a dual one $C^\circ \to V$. Here C° is called the *dual monoidal category* and there is a functor $C \to {}^\circ(C^\circ)$ such that the double-dual is closely connected with the original. Monoidal categories are quite general objects, and many aspects of quantum field theory and Riemannian geometry can be encoded in terms of them. On the other hand, our prediction of these as the correct unifying structure is based on purely mathematical or philosophical considerations, and it is interesting that it is in tune with recent developments in theoretical physics on this front. Once such functors into monoidal categories are found, then the next task is to construct concrete examples, which would then model both a quantum theory from one point of view and a classical geometry from another point of view, simultaneously. This is the representation-theoretic approach to quantum-gravity.

We also note that, in addition to the 'end' of theoretical physics in the form of quantum-gravity, we can ask also about its 'birth'. We take the view that the simplest theories of physics are based on classical logic or, roughly speaking, Boolean algebras. It appears that the relevant duality here may be provided by complementation with Boolean algebras considered self-dual according to De Morgan's theorem. The situation is summarised on the right in Fig. 6.7. Going above the axis to Heyting algebras and beyond takes us into intuitionistic logic and ultimately into an axiomatic framework for quantum field theory. A Heyting algebra describes logic in which one drops the familiar 'law of the excluded middle' that either a proposition or its negation is true. This generalisation is also the essential feature of the logical structure of quantum mechanics. Dual to this is the notion of co-Heyting algebra and co-intuitionistic logic in which one drops the axiom that the intersection of a proposition and its negation is empty. It has been argued by F.W. Lawvere and his school that this intersection is like the 'boundary' of the proposition, and, hence, that these co-Heyting algebras are the 'birth' of geometry. Specifically, in a co-Heyting, algebra we define $\partial a = a \wedge {}^\sim a$, which then behaves as one would expect for the 'boundary' of a. It has the derivation property

$$\partial(a \wedge b) = (\partial a) \wedge b \vee a \wedge (\partial b),$$

where ${}^\sim a$ is the smallest element such that $a \vee {}^\sim a = 1$. The long-term programme at this end of physics is to develop this geometrical interpretation of co-intuitionistic logic further into the notion of metric spaces and ultimately into Riemannian or Lorentzian geometry. This ties up the present considerations with the combinatorial theme of Chapter 5.

In between these extremes, a well-known example of a self-dual category is that of (locally compact) Abelian groups. We have seen this already in

Corollary 1.5.6 for finite Abelian groups G. For every such group there is a dual one $\hat{G}$ of its representations. Especially important is that $\hat{\hat{G}} \cong G$. The position and momentum groups $\mathbb{R}^n$ in flat space are dual to each other in this way. Another example is for a particle on a circle, where the position is S^1 and the allowed momentum modes are labelled by $\hat{S}^1 = \mathbb{Z}$. Fourier transform interchanges the roles of the group and its dual.

This familiar setting gives a clue to the philosophical basis of the principle of representation-theoretic self-duality. For if some theorist thought that a group G was the 'true' structure underlying a law of physics and that $\hat{G}$ was its representations, a more experimentally-minded physicist might equally well consider $\hat{G}$ as the true object and G as its representations. Only in a self-dual category could both points of view be entertained. This says that the principle has its origins in the nature of the scientific method. If, for example, Einstein's equation and other laws of theoretical physics could be deduced from such a principle alone, we would have achieved a Kantian or Hegelian view of the nature of physical reality as a consequence of the choice to look at the world in a certain way.

Finally, let us note that, in the history of mathematics, and also of the physics with which we have drawn a parallel above, progress has not been directly from self-dual structures to self-dual structures on the central axis. Instead, there is a second principle of irreducibility or 'unificationism' which can take us off the self-dual axis. The point is that, as well as finding a self-dual category in which to work, the next step is to look, in particular, for self-dual objects in the category. This is the *strong form of the principle of representation-theoretic self-duality*, and its philosophical basis is that only in the self-dual case would the two points of view mentioned above be immediately isomorphic. Now, one way in which an object can always be made self-dual is to double it up, so, for example, $G \times \hat{G}$ is always self-dual. In the dynamical example it means that the conflict about which is real, position or momentum, can be resolved by taking both as real, i.e. adopting phase-space as the self-dual structure underlying our theory. A second principle of unificationism, however, now motivates us to not stop here but rather to deform $G \times \hat{G}$ to some other more unified structure G_1 in which the two factors interact nontrivially. This leads, for example, to the Heisenberg group as a deformation of the group $\mathbb{R}^n \times \widehat{\mathbb{R}^n}$. Such considerations will tend to take us off the self-dual axis. So, if G_1 is now a non-Abelian group, then its dual $\hat{G}_1$ is some other structure, and we are faced again with the problem of unifying the two. And so on. One could view this as a kind of 'engine' that drives the evolution of physics. Since the area below the self-dual axis is closely connected with geometry and that above is closely connected with quantum theory, the unification of the two is the kind of problem that we have considered. The bicrossproduct models solved this problem by embedding both groups and group duals into the category of Hopf algebras and making the unification there. This is the more categorical picture behind the bicrossproduct models of Section 6.2.

Notes for Chapter 6

The general theory of bicrossproduct Hopf algebras was introduced in the author's 1988 Ph.D. thesis [116], published in four papers [12, 18, 118, 120]. There were also two sequels [117, 14]. The present chapter is based heavily on these seven works.

In particular, the construction and study of general bicrossproduct Hopf algebras $H \blacktriangleright\!\!\triangleleft A$ in Theorem 6.2.2–Example 6.2.8 was introduced in [12]. Also developed in this work was a theory of double cross products of Hopf algebras, to be described in Chapter 7.2, and an extensive study of the bicrossproducts $k(M) \blacktriangleright\!\!\triangleleft kG$ associated to group factorisations in Examples 6.2.11, 6.2.12, 6.2.16 and 6.2.17. The further Proposition 6.2.13 and Exercise 6.2.14 are from [18], and the physical Example 6.2.18 is from [118]; both went somewhat further, to the Lie group and locally compact group setting. The diagrammatic description in Figs. 6.3 and 6.4 as gluing and ungluing of squares is from [46] by the author. Further works on bicrossproduct Hopf algebras associated to finite and Lie group factorisations are [121, 122, 123, 124] by the author and collaborators. The last of these shows that the quantum double (see Chapter 7) of a bicrossproduct is again a bicrossproduct for a certain 'doubled' group factorisation. It also classifies quantum differential calculi on bicrossproducts associated to finite group factorisations. Meanwhile, the subsequent observation in Proposition 6.2.19 is due to O. Ogievetsky.

A special 'Abelian' case of this bicrossproduct theory in which the coacting Hopf algebra is commutative and the other is cocommutative (so that (6.10) is empty in this case) was known previously from the work of W. Singer [125]. The construction based on finite group factorisation had also been noted previously in this context; see [126] by M. Takeuchi, where an interesting Hopf algebra is constructed in this way. We have retained in Section 6.2 our original motivation, examples and general theory from [12, 116], etc., which were somewhat more general and were in fact obtained independently of these works.

The group double cross products $G \bowtie M$ in Definition 6.2.10 and Proposition 6.2.15 has an even older history, at least for finite groups. It was introduced in several different contexts, including by the author from the physical motivation of action and 'reaction' (Mach's principle) [12]. The idea of factorisation of finite groups is quite standard, but not the result that the larger group can be built up from the factors via a double cross product by matching actions; we consider this, however, as only one application of a matched pair of actions. Probably the first non-trivial theorem about double cross product groups was the result [120] that every Lie algebra matched pair (under some technical restrictions) exponentiates to a Lie group matched pair; it will be the main topic of Chapter 8.3. Following [120], K. Mackenzie gave a useful generalisation to double groupoids [127]. The graphical notation in Fig. 6.2 is taken from this.

We have developed the chapter in left–right or right–left conventions for the action and reaction, as used in [12] and [14]. At the group level, it can also be convenient to use right–right conventions for the action α and reaction β, obeying

$$\bar{\alpha}_u(st) = \bar{\alpha}_u(s)\bar{\alpha}_{\bar{\beta}_{s(u)}}(t), \quad \bar{\beta}_s(uv) = \bar{\beta}_s(u)\bar{\beta}_{\bar{\alpha}_u(s)}(v)$$

as in [12] and [120] (where they were further converted into left actions via the group inverse, so that $\alpha_u(s)$ and $\beta_s(u)$ in these works denote $\bar{\alpha}_{u^{-1}}(s)$ and $\bar{\beta}_{s^{-1}}(u)$, respectively). They are related to our present conventions by

$$\bar{\alpha}_u(s) = \alpha_u(s^{-1})^{-1}, \quad \bar{\beta}_s(u) = \beta_{s^{-1}}(u).$$

Similarly in [18], where left–left conventions for group double cross products were used.

The construction of cocycle bicrossproducts $^\psi\!\bowtie_\chi$ in the general case in Theorem 6.3.9 is due to the author in [14]. The special 'Abelian' case where the coacting Hopf algebra is commutative and the other cocommutative is due to [125], where a cohomological picture of (ψ, χ) was given as a 3-cocycle in a double complex. The corresponding picture in the general case suggests an approach to non-Abelian 3-cohomology[14]. Examples of general 'non-Abelian' cocycle bicrossproducts were introduced by the author and Ya. S. Soibelman [71], including a construction of the so-called quantum Weyl group in the form $k\widetilde{W}^\psi\!\bowtie_\chi U_q(g)$. The partial quasitriangular structures, which play a role in the multiplicative formula (3.27) in Chapter 3.3, appear, from this point of view, as forming a cocycle ψ; see [71] for details. Example 6.3.13 is a simplified version for the case $U_q(sl_2)$. Example 6.3.10 is also taken from this work. The description of affine quantum groups $U_q(\hat{g})$ as cocycle central extensions $k\mathbb{Z}_\chi\!\bowtie U(Lg)$ is due to the author in [128], from which Example 6.3.14 and Lemma 6.3.15 are taken.

The cocycle cross product $\ltimes_\chi$ at the purely algebra level and the algebra part of the extension theory in the first half of Section 6.3 can be found in [129, 130] and elsewhere. The classification by cohomology spaces

$\mathcal{H}^2(H, A)$ in Proposition 6.3.4 appeared in [130] by Y. Doi; we have given a self-contained treatment. The case where A is commutative, and its cohomological picture, has an older history [48]. The cocycle cross co-product $^\psi\!\!\bowtie$ in Proposition 6.3.8 is from [14] as a step towards combining the two constructions to form a cocycle bicrossproduct. The connection of this non-Abelian cohomology with Drinfeld's twisting theory is due to the author [46]; see also [71].

The theory of quantum principal bundles at the end of Section 6.3 is due to T. Brzezinski and the author in [2], which also provided the corresponding 'gauge theory' of connections, associated bundles, etc. over a general algebra. The exactness condition or 'local triviality' arises naturally there. For the universal calculus it was later noted to be equivalent to the Galois condition which is part of the traditional theory of Hopf algebras and field extensions. Its role for describing certain quotient spaces was studied in [131] in a slightly different context, without the noncommutative geometrical picture of differentials, connections, etc. The paper [2] also gives the exactness condition for general non-universal differential calculi $\Omega^1(P)$; see also [132]. In this case one needs to choose a differential calculus $\Omega^1(H)$ on the Hopf algebra as well; see the Notes to Chapter 9. Example 6.3.16 for the q-monopole bundle over the standard q-sphere is due to [2] as well, along with the q-monopole itself. The q-monopole is studied further in [133], where its associated vector bundles are constructed as projective modules over S_q^2, thereby linking up with an older approach to noncommutative geometry based on the Serre-Swann theorem. The q-sphere algebra itself is a special case of the 2-parameter family of 'quantum spheres' introduced by P. Podleś in [134] in a C^*-algebra setting. Among recent works, the deformed monopole has been generalised to the full 2-parameter family using a generalisation of the gauge theory to the situation where H is merely a coalgebra, see [135] and related works. The generalisation of the gauge theory as well as the theory of cleft extensions and cocycle cross products to braided groups (see Chapters 9 and 10) is in [136]. Finally, we note that [137] extends the gauge theory to noncommutative Riemannian geometry via noncommutative analogues of frame bundles, metrics and metric-compatible covariant derivatives. It provides a working definition of a 'quantum Riemannian manifold' along these lines. It is significant that although our original physical motivation for cross products and extensions came from thinking of quantum systems, the same objects (and more general extensions) can be interpreted as noncommutative geometry as well.

Returning to our physical motivation in Section 6.1, the interpretation of cross products as quantisation was first emphasised in an equivalent form by G.W. Mackey and studied in detail by H.D. Doebner and J. Tolar [138] from the point of view of functional analysis. It is also

standard in physics, where momentum is the generator of translations
in the quantum algebra. The algebraic characterisation of quantisation
in terms of the universal property of cross products in Proposition 6.1.2
is from [12], where Examples 6.1.8 and 6.1.7 were the starting point as
quantum systems which were Hopf algebras and non-Hopf algebras, respectively. Proposition 6.1.6, which gives the Weyl algebra in general as
isomorphic to $\mathrm{Lin}(H)$, is taken from [20] by the author. It corresponds
in quantum mechanics to the idea that, in a certain sense, the position
and momentum operators generate all of $B(\mathcal{H})$ or all the compact operators (depending on the formulation), and neither of these is a bialgebra
or Hopf algebra in a natural way. This is why the requirement that the
algebra of observables should be a bialgebra or Hopf algebra imposes a
physical constraint on a quantum system. The introduction and study
of this constraint was the main goal in [116] and in the abovementioned
works by the author.

Although we have been interested in this chapter in physical examples, the general machinery which we have developed can be applied to
Hopf algebras coming from other contexts as well. For example, the cross
product $G_q{\times}U_q(g)$ by the left coregular representation can easily be computed using the FRT generators of Chapter 4, giving a q-deformation of
the Weyl algebra of quantum mechanics on a group in Example 6.1.7.
Example 6.1.8 also has a q-deformation, as a Hopf algebra $BG_q{\times}U_q(g)$
obtained via the theory of braided matrices [34, 35]. It turns out to be
isomorphic to the quantum double of $U_q(g)$, as we will see in Chapter 7.4.
It should be clear from this that the cross product and bicrossproduct
ideas in the present chapter are simply 'orthogonal' to the more popular
q-deformation ideas and not incompatible with them. They were developed independently.

The construction of actual physical models of this bicrossproduct type is
due to the author in [18] and [118]. The first step is the abovementioned
construction of Lie group matched pairs [120]. The second step is the
construction of corresponding *bicrossproduct Hopf–von Neumann algebras*
$L^\infty(M){\blacktriangleright\!\!\blacktriangleleft}\mathcal{M}(G)$, which were introduced in [18] for this purpose. They
were the first large class of noncommutative and noncocommutative Hopf–
von Neumann or Kac algebras. This work in functional analysis was then
taken up and developed further by S. Baaj and G. Skandalis [19] and
also by T. Yamanouchi [139]. These authors showed, among other things,
that, with a slightly more modern definition of Kac algebra (that need
not be involutive), a technical 'modularity' assumption in [18] could be
avoided. It has been beyond our scope to describe this functional analysis
here, and we refer to [18] and these works. We have limited ourselves to a
key step from [18] and [116], namely the construction of the fundamental
operator W for the finite group bicrossproduct models in Exercise 6.2.14.

See, however, Example 8.3.19, where we give a simplified version of these models at the level of matrix group coordinate algebras and enveloping algebras as in Example 6.1.8.

The more philosophical Section 6.4 is based on [117] and [118], where the *principle of representation-theoretic self-duality* was introduced as an algebraic approach to combining quantum mechanics and gravity. As far as I know, these works by the author were the first attempt of any kind to use Hopf algebras in connection with Planck-scale physics, and provided a new approach based on the idea of a single algebraic system with different limits connected by duality (a duality also connected with T-duality in string theory). Apart from these duality ideas, the other novel idea in the approach in [118] is that the 'Planck scale Hopf algebra' $\mathbb{C}[x] \blacktriangleright\!\!\triangleleft_{\hbar,G} \mathbb{C}[p]$ (when we put in the interpretation of the two parameters) is *the general solution* to the 1-dimensional extension problem $\mathbb{C}[x] \to E \to \mathbb{C}[p]$, i.e., we are led to discover $\hbar, G$ (or quantum and gravitation effects) as the two-parameters of possibilities to bring together x, p. For the same reason it should have a role in any unknown theory of quantum gravity (in some limited context) as part of its effective description, independent of its details; this is a practical application of noncommutative cohomology [118]. Among recent works, see [44] for quantum differential structures and Fourier theory on the Planck scale Hopf algebra. Meanwhile, further sequels of the duality ideas for category theorists appeared in [37] and [140]. Although of a conceptual nature, the duality perspective on the nature of physical reality presented here does lead one to new results. The important construction of the 'dual' $\mathcal{C}^\circ$ of a monoidal category was introduced in [37] and [140] on this basis; we cover it in detail in Chapter 9.1. It is also related to the quantum double algebra to be described in the next chapter, and is of independent interest for this reason as well. The remark on co-intuitionistic logic can be found in [141].

7

Quantum double and
double cross products

This chapter is all about one of the most celebrated of Hopf algebra constructions, the quantum double of V.G. Drinfeld. This construction associates to a Hopf algebra H a quasitriangular Hopf algebra $D(H)$. We have already studied the general theory of quasitriangular Hopf algebras or 'quantum groups' in Chapter 2, and we have given some examples in Chapter 3, including such standard quantum groups as $U_q(sl_2)$ and the more general $U_q(g)$. We will see now that these standard quantum groups are quotients of quantum doubles and that their all-important quasitriangular structure comes from this fact. This is the way in which the quasitriangular structures for these $U_q(g)$ were first found. Nowadays, the quantum double is also of independent interest, providing, as it does, a rich source of quantum groups.

The construction of the quantum double and its various properties is developed in Section 7.1 in an algebraic setting. We verify everything directly. The simplest case is the quantum double $D(G)$ of a group algebra kG, and this turns out to be the semidirect or cross product

$$D(G) = k(G) \rtimes kG$$

which was already encountered as a quantum algebra of observables in Chapter 6.1. On the other hand, the quantum double, like other quasi-triangular Hopf algebras, is usually interpreted physically as a quantum symmetry, and, from this point of view, even the simplest case $D(G)$ and cocycle extensions of it turn up in concrete models.

In Section 7.2, we come to a more general construction (due to the author) of *double cross products* of Hopf algebras. Given two Hopf algebras A, H which act on each other, we form a double cross product Hopf algebra

$$A \bowtie H = \begin{cases} \text{doubly crossed} & \text{as an algebra} \\ A \otimes H & \text{as a coalgebra.} \end{cases}$$

302

We will see that, whenever a Hopf algebra factorises into sub-Hopf algebras, the factors act on each other in the way required. Thus, we give the Hopf algebra version of the group factorisations already encountered in the preceding chapter. The general quantum double is an example, as

$$D(H) = H^{*\mathrm{op}} \bowtie H$$

by mutual coadjoint actions. Finally, it turns out that there is a general correspondence between the data for such double cross products and the data for bicrossproducts as in Chapter 6.2. We will also consider another way of viewing the quantum double, as a dual-twisting of $H^{*\mathrm{op}} \otimes H$ according to the machinery of Chapter 2.3.

By now, the reader knows that every general (categorical) construction based on Hopf algebras and associated notions of modules and comodules, etc., has a dual one. In our case it is the *double cross coproduct*

$$H \blacktriangleright\!\blacktriangleleft A = \begin{cases} H \otimes A & \text{as an algebra} \\ \text{doubly crossed} & \text{as a coalgebra.} \end{cases}$$

This is also to be found in Section 7.2. The dual of the quantum double (the quantum codouble) is an example. It is dual quasitriangular in the sense of Chapter 2.2.

We come, in the remaining two sections, to theorems about the quantum double $D(H)$ where H is already quasitriangular. The quantum double in this case is very special, and has both a double cross product structure as above and now a further map to a double cross coproduct

$$D(H) \to H \blacktriangleright\!\blacktriangleleft H.$$

This map becomes an isomorphism if H is factorisable. The result originates in the theory of quantum inverse scattering, but is also important for the interpretation of the quantum double as some form of 'complexification' of the quantum group H. This latter point of view underlies the modern interpretation of $U_q(su_2) \blacktriangleright\!\blacktriangleleft U_q(su_2)$ as the quantum Lorentz group $U_q(so(1,3))$. This is the topic of Section 7.3, where it provides our first exposure to the problem of q-deforming spacetime geometry.

Finally, in Section 7.4, we present a quite different kind of theorem, which asserts that when H is quasitriangular we have

$$D(H) \cong B \rtimes H,$$

where the symbol denotes a cross product structure as both an algebra and a coalgebra. Here B is not an ordinary Hopf algebra but rather a braided group. We will introduce this concept more formally in Chapters 9.4 and 10, needing for the present only the remarkable covariance properties associated with these objects. By working with these braided groups, we obtain a picture of $D(H)$ as an ordinary cross product exactly analogous to the classical situation for $D(G)$ mentioned above. This means, for example, that we can interpret the double in this case as the algebra of observables of a quantum system according to the point of view in Chapter 6.1. The codouble is also an ordinary cross product and hence a dual quantum system according to this philosophy. The cross product description of the quantum double is also useful in other contexts, such as ones where the double is a quantum symmetry. Moreover, this class of theorems is part of a general phenomenon whereby Hopf algebras that project onto other Hopf algebras are necessarily cross products of braided groups. This general theory is covered in Chapter 9.4. In the case of the quantum Lorentz group expressed in this new semidirect product form, the algebra B is a quotient of the 2×2 braided matrices encountered in Chapter 4.3. From the quantum system point of view on the cross product, however, it appears as the unit hyperboloid in q-deformed Minkowski space. We will study the latter further in Chapter 10.5. The dual role of a quantum or braided group, as both a symmetry and as geometrical coordinates, is a further manifestation of the generalised wave–particle duality which was explored in the preceding chapter.

7.1   Definition of $D(H)$

This section gives direct and elementary proofs of the construction of the quantum double $D(H)$ in the nicest setting where H is a finite-dimensional Hopf algebra. We also cover the basic properties and some elementary examples such as the quantum double $D(G)$ of a group algebra. We will see that the quantum double demonstrates much of the general theory of Hopf algebras and quasitriangular structures from Chapters 1 and 2. Although we concentrate in this section on the finite-dimensional case, the main formulae are written in a way that immediately generalises to the infinite-dimensional bialgebra setting by letting the role of H^* be played by a second bialgebra paired with H. One needs only that the pairing $\langle \, , \, \rangle$ is convolution-invertible, i.e. one can keep the abstract properties of $\langle \, , \, \rangle$ and $\langle \, , \, \rangle^{-1} = \langle S(\,), \, \rangle$ and use them in the formulae below. We come to these generalised quantum doubles in detail in Section 7.2, where they arise as examples of a general construction of Hopf algebra or bialgebra double cross products.

Theorem 7.1.1 *Let H be a finite-dimensional Hopf algebra. The quantum double $D(H)$, in a form containing $H, H^{*\mathrm{op}}$, is a quasitriangular Hopf algebra generated by these as sub-Hopf algebras with the cross relations and quasitriangular structure*

$$\sum \langle h_{(1)}, a_{(1)} \rangle h_{(2)} a_{(2)} = \sum a_{(1)} h_{(1)} \langle h_{(2)}, a_{(2)} \rangle, \quad \mathcal{R} = \sum_a f^a \otimes e_a,$$

where $h \in H$, $a \in H^*$ *and* $\{e_a\}$, *is a basis of* H *with* $\{f^a\}$ *a dual basis.* *This* $D(H)$ *can be realised on the vector space* $H^* \otimes H$ *with product*

$$(a \otimes h)(b \otimes g) = \sum b_{(2)} a \otimes h_{(2)} g \langle Sh_{(1)}, b_{(1)} \rangle \langle h_{(3)}, b_{(3)} \rangle$$

and tensor product unit, counit and coproduct; and likewise on the vector space $H \otimes H^*$ *with product*

$$(h \otimes a)(g \otimes b) = \sum h g_{(2)} \otimes b a_{(2)} \langle g_{(1)}, a_{(1)} \rangle \langle S g_{(3)}, a_{(3)} \rangle$$

and tensor product unit, counit and coproduct.

Proof: We give here a direct proof with the structure built explicitly on $H^* \otimes H$. It then becomes obvious that $h \equiv 1 \otimes h$ and $a \equiv a \otimes 1$ give the quantum double as generated by H and $H^{*\mathrm{op}}$ as stated. In the expressions, all products shown are the usual ones on H, H^*. First, we verify associativity. Thus,

$$((a \otimes h)(b \otimes g))(c \otimes f) = (b_{(2)} a \otimes h_{(2)} g)(c \otimes f)\langle Sh_{(1)}, b_{(1)}\rangle\langle h_{(3)}, b_{(3)}\rangle$$

$$= c_{(2)} b_{(2)} a \otimes h_{(3)} g_{(2)} f$$
$$\langle Sh_{(1)}, b_{(1)}\rangle\langle h_{(5)}, b_{(3)}\rangle\langle (Sg_{(1)})Sh_{(2)}, c_{(1)}\rangle\langle h_{(4)} g_{(3)}, c_{(3)}\rangle$$

$$= c_{(3)} b_{(2)} a \otimes h_{(3)} g_{(2)} f \langle Sh_{(1)}, b_{(1)}\rangle\langle h_{(5)}, b_{(3)}\rangle$$
$$\langle Sg_{(1)}, c_{(1)}\rangle\langle Sh_{(2)}, c_{(2)}\rangle)\langle h_{(4)}, c_{(4)}\rangle\langle g_{(3)}, c_{(5)}\rangle$$

$$= c_{(3)} b_{(2)} a \otimes h_{(2)} g_{(2)} f$$
$$\langle Sg_{(1)}, c_{(1)}\rangle\langle g_{(3)}, c_{(5)}\rangle\langle Sh_{(1)}, c_{(2)} b_{(1)}\rangle\langle h_{(3)}, c_{(4)} b_{(3)}\rangle$$

$$= (a \otimes h)(c_{(2)} b \otimes g_{(2)} f)\langle Sg_{(1)}, c_{(1)}\rangle\langle g_{(3)}, c_{(3)}\rangle = (a \otimes h)((b \otimes g)(c \otimes f)).$$

We used here the definition of the product and the elementary properties of the antipode and of the duality pairing between H and $H^{*\mathrm{op}}$. Next we verify that the tensor product coproduct is an algebra homomorphism:

$$\Delta(a \otimes h)\Delta(b \otimes g) = (a_{(1)} \otimes h_{(1)})(b_{(1)} \otimes g_{(1)}) \otimes (a_{(2)} \otimes h_{(2)})(b_{(2)} \otimes g_{(2)})$$

$$= b_{(2)} a_{(1)} \otimes h_{(2)} g_{(1)} \otimes b_{(5)} a_{(2)} \otimes h_{(5)} g_{(2)}$$
$$\langle Sh_{(1)}, b_{(1)}\rangle\langle h_{(3)}, b_{(3)}\rangle\langle Sh_{(4)}, b_{(4)}\rangle\langle h_{(6)}, b_{(6)}\rangle$$

$$= b_{(2)} a_{(1)} \otimes h_{(2)} g_{(1)} \otimes b_{(3)} a_{(2)} \otimes h_{(3)} g_{(2)}\langle Sh_{(1)}, b_{(1)}\rangle\langle h_{(4)}, b_{(4)}\rangle$$

$$= \Delta((a \otimes h)(b \otimes g)).$$

The unit and counit are easily checked and we have a bialgebra. Next, since we know from Exercise 1.3.3 that the antipode on $H^{*\mathrm{op}}$ is the inverse of the one on H^*, we know that the only possibility for the antipode is

$$S(a \otimes h) = (1 \otimes Sh)(S^{-1} a \otimes 1) = S^{-1} a_{(2)} \otimes Sh_{(2)}\langle h_{(1)}, a_{(1)}\rangle\langle Sh_{(3)}, a_{(3)}\rangle$$

if $H, H^{*\mathrm{op}}$ are to be sub-Hopf algebras. This is easily checked directly as

$$(S(a_{(1)} \otimes h_{(1)}))(a_{(2)} \otimes h_{(2)}) = (1 \otimes Sh_{(1)})(S^{-1} a_{(1)} \otimes 1)(a_{(2)} \otimes 1)(1 \otimes h_{(2)})$$

$$= \epsilon(h)\epsilon(a),$$

$$(a_{(1)} \otimes h_{(1)})S(a_{(2)} \otimes h_{(2)}) = (a_{(1)} \otimes h_{(1)})(1 \otimes Sh_{(2)})(S^{-1}a_{(2)} \otimes 1)$$
$$= \epsilon(h)\epsilon(a).$$

Next we make use of finite dimensionality to choose a basis and dual basis and write

$$\mathcal{R} = f^a \otimes 1 \otimes 1 \otimes e_a.$$

Then to verify (2.1) we need

$$(f^a{}_{(1)} \otimes 1) \otimes (f^a{}_{(2)} \otimes 1) \otimes (1 \otimes e_a) = (f^a \otimes 1) \otimes (f^b \otimes 1) \otimes (1 \otimes e_a e_b),$$

$$(f^a \otimes 1) \otimes (1 \otimes e_{a(1)}) \otimes (1 \otimes e_{a(2)}) = (f^b f^a \otimes 1) \otimes (1 \otimes e_b) \otimes (1 \otimes e_a),$$

which are easily verified by evaluating against general elements. Thus, evaluating against $a \in H^*$ in the third factor gives both sides of the first identity as $a_{(1)} \otimes 1 \otimes a_{(2)} \otimes 1 \otimes 1$. Here $f^a \langle e_a, a \rangle = a$ and $f^a \otimes f^b \langle e_a e_b, a \rangle = a_{(1)} \otimes a_{(2)}$. Similarly on evaluating against $h \in H$ for the second identity. To verify (2.2) we compute

$$\mathcal{R}\Delta(a \otimes h) = (f^a \otimes 1)(a_{(1)} \otimes h_{(1)}) \otimes (1 \otimes e_a)(a_{(2)} \otimes h_{(2)})$$
$$= a_{(1)} f^a \otimes h_{(1)} \otimes a_{(3)} \otimes e_{a(2)} h_{(2)} \langle Se_{a(1)}, a_{(2)} \rangle \langle e_{a(3)}, a_{(4)} \rangle,$$
$$(\tau \circ \Delta(a \otimes h))\mathcal{R} = (a_{(2)} \otimes h_{(2)})(f^a \otimes 1) \otimes (a_{(1)} \otimes h_{(1)})(1 \otimes e_a)$$
$$= f^a{}_{(2)} a_{(2)} \otimes h_{(3)} \otimes a_{(1)} \otimes h_{(1)} e_a \langle Sh_{(2)}, f^a{}_{(1)} \rangle \langle h_{(4)}, f^a{}_{(3)} \rangle.$$

Evaluating against a general element $g \in H$ in the first tensor factor gives the first expression as

$$\langle g, \mathcal{R}\Delta(a \otimes h) \rangle = \langle g_{(1)}, a_{(1)} \rangle h_{(1)} \otimes a_{(3)} \otimes g_{(3)} h_{(2)} \langle Sg_{(2)}, a_{(2)} \rangle \langle g_{(4)}, a_{(4)} \rangle$$
$$= h_{(1)} \otimes a_{(1)} \otimes g_{(1)} h_{(2)} \langle g_{(2)}, a_{(2)} \rangle,$$
$$= \langle g_{(2)}, a_{(2)} \rangle h_{(3)} \otimes a_{(1)} \otimes h_{(1)} e_a \langle (Sh_{(2)})g_{(1)} h_{(4)}, f^a \rangle$$
$$= \langle g, (\tau \circ \Delta(a \otimes h))\mathcal{R} \rangle$$

as required. Finally, we have to show that $\mathcal{R}$ is invertible. The easiest way to do this is to recall Lemma 2.1.2 and accordingly to define

$$\mathcal{R}^{-1} = S^{-1} f^a \otimes 1 \otimes 1 \otimes e_a.$$

That this is an inverse needs the identities

$$f^b S^{-1} f^a \otimes 1 \otimes 1 \otimes e_a e_b = 1 \otimes 1 \otimes 1 \otimes 1,$$

$$(S^{-1} f^a) f^b \otimes 1 \otimes 1 \otimes e_b e_a = 1 \otimes 1 \otimes 1 \otimes 1,$$

which become $a_{(2)} S^{-1} a_{(1)} = \epsilon(a)$ and $(S^{-1} a_{(2)}) a_{(1)} = \epsilon(a)$, respectively, when evaluated against $a \in H^*$ in the last tensor factor. The alternative realisation $h \equiv h \otimes 1$ and $a \equiv 1 \otimes a$ on $H \otimes H^*$ is similar. ∎

There is another version of this construction (Drinfeld's original form) containing H and $H^{*\text{cop}}$. This looks stranger in terms of H alone, but is equally good from the point of view of the pair of sub-Hopf algebras themselves.

Exercise 7.1.2 *The quantum double $D(H)$ in a form containing $H, H^{*\text{cop}}$ is a quasitriangular Hopf algebra generated by these as sub-Hopf algebras and the cross relations and quasitriangular structure*

$$\sum \langle h_{(1)}, a_{(2)} \rangle h_{(2)} a_{(1)} = \sum a_{(2)} h_{(1)} \langle h_{(2)}, a_{(1)} \rangle, \quad \mathcal{R} = \sum_a e_a \otimes f^a.$$

Show that it can be realised on the vector space $H \otimes H^*$ with

$$(h \otimes a)(g \otimes b) = \sum h g_{(2)} \otimes a_{(2)} b \langle g_{(1)}, S^{-1} a_{(3)} \rangle \langle g_{(3)}, a_{(1)} \rangle,$$

$$\Delta(h \otimes a) = \sum h_{(1)} \otimes a_{(2)} \otimes h_{(2)} \otimes a_{(1)}$$

and quasitriangular structure as above, and can likewise be realised on the vector space $H^* \otimes H$ with product and coproduct

$$(a \otimes h)(b \otimes g) = \sum a b_{(2)} \otimes h_{(2)} g \langle h_{(1)}, b_{(3)} \rangle \langle h_{(3)}, S^{-1} b_{(1)} \rangle,$$

$$\Delta(a \otimes h) = \sum a_{(2)} \otimes h_{(1)} \otimes a_{(1)} \otimes h_{(2)}.$$

Solution: This can be verified directly in the same manner as Theorem 7.1.1. Alternatively, we know from Exercise 1.3.3 that S provides an isomorphism between the opposite product and opposite coproduct Hopf algebras, so, in particular, the antipode of H^* viewed as a linear map $H^{*\text{op}} \to H^{*\text{cop}}$ provides an isomorphism from the $D(H)$ in Theorem 7.1.1 to the present form. The natural quasitriangular structure $\mathcal{R}$ here corresponds to the image of $\mathcal{R}_{21}^{-1}$ under this isomorphism. Later, we will give a unified point of view that includes both the forms here and the forms in Theorem 7.1.1 under a single construction. ∎

Hence there are several conventions for Drinfeld's quantum double. We will generally mean by $D(H)$ the first form in Theorem 7.1.1 as built on $H^* \otimes H$. This is the right handed quantum double in the form containing $H^{*\text{op}}$ and H.

Proposition 7.1.3 $D(H)$ *is factorisable in the sense of Definition 2.1.12.*

Proof: We compute the quantum inverse Killing form $Q : H \otimes H^* \to H^* \otimes H$ as

$$Q(h \otimes a) = \langle h \otimes a, (1 \otimes e_a)(f^b \otimes 1) \rangle (f^a \otimes 1)(1 \otimes e_b)$$

$$= \langle h, f^b{}_{(2)} \rangle \langle e_{a(2)}, a \rangle \langle Se_{a(1)}, f^b{}_{(1)} \rangle \langle e_{a(3)}, f^b{}_{(3)} \rangle f^a \otimes e_b$$
$$= f^a \otimes (Se_{a(1)}) h e_{a(3)} \langle e_{a(2)}, a \rangle = f^a a f^b \otimes (Se_a) h e_b.$$

This is easily seen to be surjective as required. Indeed, it is an isomorphism, with inverse provided by

$$Q^{-1}(a \otimes h) = e_{a(1)} h Se_{a(3)} \otimes f^a \langle a, e_{a(2)} \rangle = e_a h Se_b \otimes f^a a f^b.$$

∎

Proposition 7.1.4 *If H is a Hopf ∗-algebra, then $D(H)$ is also a Hopf ∗-algebra, with*

$$(a \otimes h)^* = (1 \otimes h^*)((S^2 a)^* \otimes 1),$$

and its quasitriangular structure $\mathcal{R}$ is antireal in the sense of Definition 2.1.15.

Proof: We first make $H^{*\mathrm{op}}$ into a Hopf ∗-algebra with $*^{\mathrm{op}} = S^{-2} \circ * = * \circ S^2$ and antipode S^{-1} (any even power in place of S^2 will do for $*^{\mathrm{op}}$, but this is the one that we need here). Since the quantum double is generated by these Hopf algebras, there is then a unique possibility for its ∗-structure such that the inclusions are ∗-algebra maps. Putting this into the relations in Theorem 7.1.1 and using the duality for Hopf ∗-algebras as in Definition 1.7.5, we see that this is indeed a ∗-algebra structure on $D(H)$:

$$[(a \otimes h)(b \otimes g)]^*$$
$$= (1 \otimes g^*)(1 \otimes h_{(2)}{}^*)((S^2 b_{(2)})^* \otimes 1)((S^2 a)^* \otimes 1)\overline{\langle Sh_{(1)}, b_{(1)} \rangle \langle h_{(3)}, b_{(3)} \rangle}$$
$$= (1 \otimes g^*)(1 \otimes h_{(2)}{}^*)((S^2 b_{(2)})^* \otimes 1)((S^2 a)^* \otimes 1)$$
$$\qquad \langle (Sh_{(1)})^*, (Sb_{(1)})^* \rangle \langle h_{(3)}{}^*, (Sb_{(3)})^* \rangle$$
$$= (1 \otimes g^*)(1 \otimes h^*{}_{(2)})((S^2 b)^*{}_{(2)} \otimes 1)((S^2 a)^* \otimes 1)$$
$$\qquad \langle h^*{}_{(1)}, (S^2 b)^*{}_{(1)} \rangle \langle h^*{}_{(3)}, S(S^2 b)^*{}_{(3)} \rangle$$
$$= (1 \otimes g^*)((S^2 b)^*{}_{(3)} \otimes h^*{}_{(3)})((S^2 a)^* \otimes 1)\langle Sh^*{}_{(2)}, (S^2 b)^*{}_{(2)} \rangle$$
$$\qquad \langle h^*{}_{(4)}, (S^2 b)^*{}_{(4)} \rangle \langle h^*{}_{(1)}, (S^2 b)^*{}_{(1)} \rangle \langle h^*{}_{(5)}, S(S^2 b)^*{}_{(5)} \rangle$$
$$= (1 \otimes g^*)((S^2 b)^* \otimes h^*)((S^2 a)^* \otimes 1)$$
$$= (1 \otimes g^*)((S^2 b)^* \otimes 1)(1 \otimes h^*)((S^2 a)^* \otimes 1) = (b \otimes g)^*(a \otimes h)^*.$$

For the fifth equality, we have used the duality between H^* and H and the antipode axioms. Moreover, $(S \circ *)(a \otimes h) = S[(1 \otimes h^*)((S^2 a)^* \otimes 1)] = S^{-3}(a^*) \otimes S(h^*)$ so that $(S \circ *)^2 = \mathrm{id}$ on $D(H)$. Equally well, we note that $* \circ S(a \otimes h) = S^{-1} \circ *(a \otimes h) = S^{-1}[(1 \otimes h^*)((S^2 a)^* \otimes 1)] = (Sa)^* \otimes (Sh)^*$. Since the coalgebra of the quantum double is the tensor product one, it is equally clear that it commutes with ∗ as it should. Hence we have a Hopf

*-algebra. For the second part, we find $\mathcal{R}^{*\otimes *} = (S^{-2}(f^{a*}) \otimes 1) \otimes (1 \otimes e_a{}^*)$ $= (S^{-1}f'^a \otimes 1) \otimes (1 \otimes e'_a) = (S \otimes \mathrm{id})\mathcal{R} = \mathcal{R}^{-1}$, where $e'_a = e_a{}^*$ is a new basis with dual basis $f'^a = S^{-1}(f^{a*})$. Hence the canonical quasitriangular structure of the quantum double is antireal. ∎

Proposition 7.1.5 $D(H)$ *is the dual-twisting in the sense of (2.23) by*

$$\chi(a \otimes h \otimes b \otimes g) = \epsilon(a)\langle Sh, b\rangle\epsilon(g),$$

$$\chi^{-1}(a \otimes h \otimes b \otimes g) = \epsilon(a)\langle h, b\rangle\epsilon(g).$$

Here χ is a convolution-invertible dual-cocycle on the tensor product Hopf algebra $H^{*\mathrm{op}} \otimes H$.

Proof: We verify (2.20) for χ on $H^{*\mathrm{op}} \otimes H$. Indeed,

$$\chi(b_{(1)} \otimes g_{(1)} \otimes c_{(1)} \otimes f_{(1)})\chi(a \otimes h \otimes (b_{(2)} \otimes g_{(2)})(c_{(2)} \otimes f_{(2)}))$$
$$= \epsilon(b_{(1)})\langle Sg_{(1)}, c_{(1)}\rangle\epsilon(f_{(1)})\epsilon(a)\langle Sh, c_{(2)}b_{(2)}\rangle\epsilon(g_{(2)}f_{(2)})$$
$$= \langle Sg, c_{(1)}\rangle\epsilon(a)\langle Sh, c_{(2)}b\rangle\epsilon(f),$$
$$\chi(a_{(1)} \otimes h_{(1)} \otimes b_{(1)} \otimes g_{(1)})\chi((a_{(2)} \otimes h_{(2)})(b_{(2)} \otimes g_{(2)}) \otimes c \otimes f)$$
$$= \epsilon(a_{(1)})\langle Sh_{(1)}, b_{(1)}\rangle\epsilon(g_{(1)})\epsilon(b_{(2)}a_{(2)})\langle Sh_{(2)}g_{(2)}, c\rangle\epsilon(f)$$
$$= \langle Sh_{(1)}, b\rangle\epsilon(a)\langle Sh_{(2)}g, c\rangle\epsilon(f),$$

which are the same from the properties of a pairing. Given a dual-cocycle on a bialgebra, we have a new bialgebra with twisted product

$$(a \otimes h) \cdot_\chi (b \otimes g)$$
$$= \chi(a_{(1)} \otimes h_{(1)} \otimes b_{(1)} \otimes g_{(1)})b_{(2)}a_{(2)} \otimes h_{(2)}g_{(2)}\chi^{-1}(a_{(3)} \otimes h_{(3)} \otimes b_{(3)} \otimes g_{(3)})$$
$$= \epsilon(a_{(1)})\langle Sh_{(1)}, b_{(1)}\rangle\epsilon(g_{(1)})b_{(2)}a_{(2)} \otimes h_{(2)}g_{(2)}\epsilon(a_{(3)})\langle h_{(3)}, b_{(3)}\rangle\epsilon(g_{(3)})$$
$$= \langle Sh_{(1)}, b_{(1)}\rangle b_{(2)}a \otimes h_{(2)}g\langle h_{(3)}, b_{(3)}\rangle.$$

One can check the antipode as well, using (2.25). ∎

Proposition 7.1.6 *A left $D(H)$-module is a vector space V which is both a left H-module by $\triangleright$ and a right H^*-module by $\triangleleft$, which are compatible in the sense*

$$\sum \langle h_{(1)}, a_{(1)}\rangle h_{(2)}\triangleright(v\triangleleft a_{(2)}) = \sum (h_{(1)}\triangleright v)\triangleleft a_{(1)}\langle h_{(2)}, a_{(2)}\rangle$$

for all $v \in V$, $a \in H^*$ and $h \in H$. The action of the quantum double is $(a \otimes h)\triangleright v = (h\triangleright v)\triangleleft a$.

Equivalently, a left $D(H)$-module is a vector space V which is both a left H-module by $\triangleright$ and a left H-comodule by $\beta(v) = \sum v^{(\bar{1})} \otimes v^{(\bar{2})}$ such that

$$\sum h_{(1)}v^{(\bar{1})} \otimes h_{(2)}\triangleright v^{(\bar{2})} = \sum (h_{(1)}\triangleright v)^{(\bar{1})}h_{(2)} \otimes (h_{(1)}\triangleright v)^{(\bar{2})}.$$

The action of the quantum double is $(a \otimes h) \triangleright v = (a \otimes \mathrm{id}) \circ \beta(h \triangleright v)$.

Proof: This is obvious from the relations in Theorem 7.1.1. To represent these on V means an action from the left of H and $H^{*\mathrm{op}}$ respecting the relations, the latter being the same thing as a right action of H^*. For the second part, we know from Proposition 1.6.11 that a right H^*-module corresponds to a left H-comodule. So we can equally well think of V as a left H-module and H-comodule in a compatible way. The relationship between the right action of H^* and the coaction $\beta(v) = v^{(\bar{1})} \otimes v^{(\bar{2})}$ is the usual dualisation $v \triangleleft a = \langle v^{(\bar{1})}, a \rangle v^{(\bar{2})}$, giving the compatibility as stated. This form has the advantage of giving the requirement for a $D(H)$-module entirely in terms of H. The modules in this second form are called *crossed H-modules*. The modules in the first form are called crossed $H - -H^*$-bimodules) and apply when we later generalise the quantum double to paired bialgebras. ∎

Example 7.1.7 *The adjoint action on $D(H)$ comes out as*

$$h \triangleright (a \otimes g) = \sum a_{(2)} \otimes h_{(2)} g S h_{(4)} \langle S h_{(1)}, a_{(1)} \rangle \langle h_{(3)}, a_{(3)} \rangle,$$

$$(b \otimes h) \triangleleft a = \sum (S^{-1} a_{(3)}) b a_{(1)} \otimes h_{(2)} \langle h_{(1)}, a_{(4)} \rangle \langle h_{(3)}, S^{-1} a_{(2)} \rangle.$$

Proof: We compute Ad from Example 1.6.3, for the case of $D(H)$. One needs only the formulae for its Hopf algebra structure as given in Theorem 7.1.1. Thus, $h \triangleright (a \otimes g) = (1 \otimes h_{(1)})(a \otimes g)(1 \otimes S h_{(1)})$ and $(b \otimes h) \triangleleft a = (a_{(1)} \otimes 1)(b \otimes h)(S^{-1} a_{(2)} \otimes 1)$ multiply out as stated. ∎

We know from Example 1.6.3 that $D(H)$ becomes a module algebra under this adjoint action. We should also expect that $D(H)$ on H itself along the lines of the quantum mechanical picture of cross products in Chapter 6.1 and its Schrödinger representation. We will see later, in Section 7.4, the precise sense in which the quantum double is a cross product.

Example 7.1.8 *The Schrödinger representation of $D(H)$ on H is defined as*

$$h \triangleright g = \sum h_{(1)} g S h_{(2)}, \qquad h \triangleleft a = \sum \langle a, h_{(1)} \rangle h_{(2)},$$

and makes H into a $D(H)$-module algebra.

Proof: We know that each of these are actions. The first is the left adjoint action Ad of H from Example 1.6.3, and the second is the right handed analogue L^* of the coregular representation, Example 1.6.2. We need only verify their compatibility in the sense of Proposition 7.1.6. The corresponding coaction is $\beta = \Delta$, and in this form the compatibility is

$$h_{(1)}g_{(1)} \otimes \mathrm{Ad}_{h_{(2)}}(g_{(2)}) = h_{(1)}g_{(1)} \otimes h_{(2)}g_{(2)}Sh_{(3)}$$
$$= h_{(1)}g_{(1)}(Sh_{(4)})h_{(5)} \otimes h_{(2)}g_{(2)}Sh_{(3)} = \mathrm{Ad}_{h_{(1)}}(g)_{(1)}h_{(2)} \otimes \mathrm{Ad}_{h_{(1)}}(g)_{(2)}$$

as required. Since H is a module algebra under both H and H^*, and the quantum double has the tensor product coalgebra structure, it follows that it is a module algebra under $D(H)$ as well. ∎

Clearly, the quantum double is a fertile testing ground for the general theory of Hopf algebras and quasitriangular Hopf algebras. Many of the other notions and ideas given in Chapters 1 and 2 can likewise be computed by the reader for $D(H)$. We turn now from the abstract quantum double for general H to some simple examples.

Example 7.1.9 *Let $H = kG$ be the group algebra of a finite group, as in Example 1.5.3. Then $D(G) = k(G) {\rtimes} kG$ is the cross product Hopf algebra by the coadjoint action in Example 6.1.8. In that context the Hopf algebra was the quantum algebra of observables of a quantum system. We learn now that it is quasitriangular. Explicitly, the full structure is*

$$(\delta_s \otimes u)(\delta_t \otimes v) = \delta_{u^{-1}su,t}\delta_t \otimes uv, \quad \Delta(\delta_s \otimes u) = \sum_{ab=s} \delta_a \otimes u \otimes \delta_b \otimes u,$$

$$\epsilon(\delta_s \otimes u) = \delta_{s,e}, \quad S(\delta_s \otimes u) = \delta_{u^{-1}s^{-1}u} \otimes u^{-1},$$

$$\mathcal{R} = \sum_{u \in G} \delta_u \otimes e \otimes 1 \otimes u.$$

Proof: This cross product feature is clear when H is cocommutative, for then the formula for the product of $D(H)$ in Theorem 7.1.1 can be written as a cross product, as in Proposition 1.6.6, by the coadjoint action from Example 1.6.4. In the present case, this is $(\delta_s \otimes u)(\delta_t \otimes v) = \sum_{abc=t} \delta_b \delta_s \otimes uv \delta_{a,u^{-1}} \delta_{c,u} = \delta_{u^{-1}su,t}\delta_t \otimes uv$, as in Proposition 6.1.1 with α the right adjoint action of G on itself. The full Hopf algebra structure here is also a case of Example 6.2.11, with β trivial, where the antipode and a useful diagrammatic notation are given. The new part is the quasitriangular structure from Theorem 7.1.1. It is also a nice exercise to verify the axioms (2.1) and (2.2) directly from these definitions. ∎

Example 7.1.10 *A left $D(G)$-module is a G-graded vector space on which G also acts in a compatible way according to*

$$|u \triangleright v| = u|v|u^{-1}, \qquad \forall v \in V, \ u \in G,$$

where $\triangleright$ is the action of G and $|\ |$ is the grading. Such a vector space is called a crossed G-module.

Proof: From Proposition 7.1.6, we know that a $D(G)$-module means a vector space V on which G and $k(G)$ (functions on G with pointwise product) act. We know from Example 1.6.7 that an action of $k(G)$ simply means a G-grading. The correspondence is $v \triangleleft a = va(|v|)$ on a vector v of degree $|v|$. The compatibility condition between the actions, which is needed for a $D(G)$-module, reduces to $a(u|v|)u \triangleright v = a(|u \triangleright v|u)$ for all $a \in k(G)$. This gives the stated compatibility condition between the action of G and the grading. Note that, more precisely, we should write $D(kG)$ and speak of crossed kG-modules.   ∎

A large class of $D(G)$-modules is provided by the crossed G-sets of J.H.C. Whitehead. This is a 'classic' construction in algebraic topology. By definition, a crossed G-set is a set M on which G acts, together with a map $\partial : M \to G$ such that $\partial(u \triangleright m) = u(\partial m)u^{-1}$ for all $u \in G, m \in M$. In this case the vector space kM with basis $\{m \in M\}$ clearly becomes a $D(G)$-module with $\triangleright$ extended linearly and degree $|m| = \partial m$. If M is a group, one usually demands that ∂ is a group homomorphism. In this case it is easy to see that the group algebra kM is a $D(G)$-module algebra, i.e. it is $D(G)$-covariant in the sense we have seen in Chapter 1.6.1. A further natural demand to make in this context is $(\partial m) \triangleright n = mnm^{-1}$ for all $m, n \in M$. Because of this background, the representations of $D(H)$ are called *crossed H-modules*. We study this category further in Chapter 9. For another kind of example, we have

Example 7.1.11 *Let* $H = \text{span}\{1, g, x, xg\}$ *denote the self-dual Hopf algebra in Example 2.1.7 with dual* $H^* = \text{span}\{1, f, y, yf\}$, *say, as in Example 2.2.6. Then,* $D(H)$ *is generated by* $\{1, g, x, f, y\}$ *with the additional relations*

$$fg = gf, \quad xy + yx = 1 - fg, \quad fx = -xf, \quad gy = -yg.$$

We can also consider the quotient of $D(H)$ *by the ideal generated by* $1 - fg$. *This is a quasitriangular Hopf algebra* $U = \text{span}\{1, g, x, y, xg, yg, xy, xyg\}$ *with relations, coalgebra antipode and quasitriangular structure*

$$x^2 = 0, \quad y^2 = 0, \quad g^2 = 1, \quad xg = -gx, \quad yg = -gy, \quad xy = -yx,$$

$$\Delta x = x \otimes 1 + g \otimes x, \quad \Delta y = y \otimes 1 + g \otimes y, \quad \epsilon g = 1, \quad \epsilon x = 0, \quad \epsilon y = 0,$$

$$Sg = g, \quad Sx = xg, \quad Sy = yg,$$

$$\mathcal{R} = 1 - 2p \otimes p + \alpha(x \otimes yg + 2xp \otimes yp); \quad p = \frac{1-g}{2}.$$

Proof: We described the dual of H in Example 2.2.6: it has the same structure as H, with the pairing as given there. For the double $D(H)$ we note that $H^{*\mathrm{op}}$ is also isomorphic to H, so the double is generated by two copies of the algebra with cross relations from Theorem 7.1.1. That these come out as stated is elementary and similar to the computation in Example 7.2.9 below. The natural quotient U involves identifying the 'Cartan elements' g, f, and provides one of the simplest finite-dimensional truly quasitriangular Hopf algebras. It has dimension eight, and is similar to $U_q(sl_2)$ in Chapter 3.4 at $q^2 = -1$. It is quasitriangular because U is the homomorphic image of the quasitriangular Hopf algebra $D(H)$. This is a typical application of the quantum double construction to generate new and interesting quasitriangular Hopf algebras. ∎

We conclude with some further general remarks about the quantum double. The first is to note that, in the proof of the bialgebra structure of $D(H)$ in Theorem 7.1.1, we do not really need that H is finite dimensional as long as we have the pairing $\langle\ ,\ \rangle$. Also, we do not really need an antipode on H, just a map that plays the role of the antipode evaluated against the pairing, i.e. a *weak antipode*, as in Proposition 4.2.3. Put another way, we just need to keep the formal properties of the maps $\langle\ ,\ \rangle$ and $\langle\ ,\ \rangle^{-1} = \langle S(\ ),\ \rangle$, the latter being characterised now as the convolution-inverse of the pairing by

$$\sum \langle h_{(1)}, a_{(1)}\rangle^{-1}\langle h_{(2)}, a_{(2)}\rangle = \epsilon(h)\epsilon(a) = \sum \langle h_{(1)}, a_{(1)}\rangle\langle h_{(2)}, a_{(2)}\rangle^{-1}.$$

One can easily see that it obeys

$$\langle h, ab\rangle^{-1} = \sum \langle h_{(2)}, a\rangle^{-1}\langle h_{(1)}, b\rangle^{-1},$$
$$\langle hg, a\rangle^{-1} = \sum \langle h, a_{(2)}\rangle^{-1}\langle g, a_{(1)}\rangle^{-1}, \tag{7.1}$$

instead of the usual pairing axioms. Then the two forms in Theorem 7.1.1 become bialgebras $A^{\mathrm{op}}\bowtie H$ and $H\bowtie A^{\mathrm{op}}$, constructed with $\langle\ ,\ \rangle$ in place of $\langle S(\ ),\ \rangle$. For example, the first has the product

$$(a \otimes h)(b \otimes g) = \sum b_{(2)}a \otimes h_{(2)}g\langle h_{(1)}, b_{(1)}\rangle^{-1}\langle h_{(3)}, b_{(3)}\rangle \tag{7.2}$$

on the vector space $A^{\mathrm{op}} \otimes H$. Note that the pairing here need not be nondegenerate, i.e. the construction is significantly more general than that above. We will come to it in Section 7.2 in the theory of double cross products, where it was first introduced (by the author). Proposition 7.1.5 likewise goes through just as easily, with

$$\chi(a \otimes h \otimes b \otimes g) = \epsilon(a)\langle h, b\rangle^{-1}\epsilon(g). \tag{7.3}$$

Similarly for the other main results above.

Finally, it is easy enough to dualise all these constructions in the usual way. We follow the same lines as in Chapter 2.2. Thus, $D(H)^*$ is a dual quasitriangular Hopf algebra built on $H \otimes H^*$ with tensor product algebra, and the coproducts

$$\Delta(h \otimes a) = \sum h_{(2)} \otimes f^a a_{(1)} f^b \otimes (Se_a) h_{(1)} e_b \otimes a_{(2)} \qquad (7.4)$$

as the duals of the first form in Theorem 7.1.1. Likewise for the second form built on $H^* \otimes H$. The codouble projects onto H^{cop}, H^* as Hopf algebras by the canonical projections provided by the counit. The antipode, dual quasitriangular structure and (in the Hopf $*$-algebra case) the $*$-structure are

$$S(h \otimes a) = e_a(S^{-1}h) Se_b \otimes f^a(Sa) f^b,$$
$$\mathcal{R}((h \otimes a) \otimes (g \otimes b)) = \epsilon(a)\epsilon(g)\langle h, b \rangle, \quad (h \otimes a)^* = h^* \otimes a^*. \qquad (7.5)$$

To see the last of these, one needs to use the pairing between a Hopf $*$-algebra and its dual in Definition 1.7.5 and $* \circ S$ on $D(H)$ as given in the proof of Proposition 7.1.4. Finally, the dual of Proposition 7.1.5 is that the codouble is the twisting in the sense of Theorem 2.3.4 of the Hopf algebra $H^{\mathrm{cop}} \otimes H^*$ by the 2-cocycle

$$\chi = \sum_a 1 \otimes f^a \otimes Se_a \otimes 1, \quad \chi^{-1} = \sum_a 1 \otimes f^a \otimes e_a \otimes 1. \qquad (7.6)$$

The direct proof that this gives an invertible 2-cocycle is similar to the proof in Theorem 7.1.1 that $\mathcal{R}$ there is quasitriangular. The various other properties of $D(H)$ can be dualised in just the same way.

7.2 Double cross product Hopf algebras

In this section, we include the quantum double in a more general construction of double cross products of Hopf algebras. This is due to the author and arose in a somewhat different context, namely in connection with bicrossproducts and the algebraic approach to quantum-gravity of Chapter 6. We will see the connection in Proposition 7.2.4. On the other hand, the notion of double cross products is natural and is applicable in any number of contexts. Whenever a Hopf algebra factorises in a natural way into sub-Hopf algebras, it is likely to be a double cross product. The examples include the quantum double of V.G. Drinfeld, and lead also to a natural generalisation of it based on a pairing or skew-pairing of bialgebras.

We have already encountered group double cross products $G \bowtie M$ in detail in Proposition 6.2.15, based on the notion of a matched pair of groups acting on each other in Definition 6.2.10. A usual semidirect or cross product of groups has one group acting on the other, while the double cross product has simultaneously an action α and 'reaction' β. As such, it is actually a more versatile and natural notion than the more well-known semidirect products. We now give the bialgebra or Hopf algebra version.

Definition 7.2.1 *Two bialgebras or Hopf algebras* (A, H) *form a right–left matched pair if H is a right A-module coalgebra and A is a left H-module coalgebra,*

$$\alpha : H \otimes A \to H, \quad \alpha(h \otimes a) = h \triangleleft a, \quad \beta : H \otimes A \to A, \quad \beta(h \otimes a) = h \triangleright a,$$

obeying the compatibility conditions

$$(hg) \triangleleft a = \sum (h \triangleleft (g_{(1)} \triangleright a_{(1)}))(g_{(2)} \triangleleft a_{(2)}), \quad 1 \triangleleft a = \epsilon(a), \tag{7.7}$$

$$h \triangleright (ab) = \sum (h_{(1)} \triangleright a_{(1)})((h_{(2)} \triangleleft a_{(2)}) \triangleright b), \quad h \triangleright 1 = \epsilon(h), \tag{7.8}$$

$$\sum h_{(1)} \triangleleft a_{(1)} \otimes h_{(2)} \triangleright a_{(2)} = \sum h_{(2)} \triangleleft a_{(2)} \otimes h_{(1)} \triangleright a_{(1)}. \tag{7.9}$$

Here the requirement that α, β are not only actions but give comodule algebras was explained in Chapter 1.6.1, and just requires that they be coalgebra maps.

Theorem 7.2.2 *Given a matched pair of bialgebras* (A, H), *we have a double cross product bialgebra $A \bowtie H$ built on the vector space $A \otimes H$ with product*

$$(a \otimes h)(b \otimes g) = \sum a(h_{(1)} \triangleright b_{(1)}) \otimes (h_{(2)} \triangleleft b_{(2)})g$$

and tensor product unit, counit and coproduct. Moreover, A and H are sub-bialgebras by the canonical inclusions, and $A \bowtie H$ is generated by these with the cross relations

$$ha = \sum (h_{(1)} \triangleright a_{(1)})(h_{(2)} \triangleleft a_{(2)}), \quad \forall h \in A, \, a \in A.$$

If A, H are Hopf algebras, then so is the double cross product with

$$S(a \otimes h) = (1 \otimes Sh)(Sa \otimes 1) = \sum (Sh_{(1)}) \triangleright (Sb_{(1)}) \otimes (Sh_{(2)}) \triangleleft (Sb_{(2)}).$$

Proof: We first verify associativity of the product, after which it is clear that $h \equiv 1 \otimes h$ and $a \equiv a \otimes 1$ realise the cross relations as stated. Thus,

$$(a \otimes h)\left((b \otimes g)(c \otimes f)\right) = (a \otimes h)\left(b(g_{(1)} \triangleright c_{(1)}) \otimes (g_{(2)} \triangleleft c_{(2)})f\right)$$
$$= a(h_{(1)} \triangleright (b_{(1)}(g_{(1)} \triangleright c_{(1)}))) \otimes (h_{(2)} \triangleleft b_{(2)}(g_{(2)} \triangleright c_{(2)}))(g_{(3)} \triangleleft c_{(3)})f$$
$$= a(h_{(1)} \triangleright b_{(1)})((h_{(2)} \triangleleft b_{(2)})g_{(1)} \triangleright c_{(1)}) \otimes (h_{(3)} \triangleleft b_{(3)}(g_{(2)} \triangleright c_{(2)}))(g_{(3)} \triangleleft c_{(3)})f$$
$$= a(h_{(1)} \triangleright b_{(1)})((h_{(2)} \triangleleft b_{(2)})g_{(1)} \triangleright c_{(1)}) \otimes (((h_{(3)} \triangleleft b_{(3)})g_{(2)}) \triangleleft c_{(2)}) f$$
$$= (a(h_{(1)} \triangleleft b_{(1)}) \otimes (h_{(2)} \triangleleft b_{(2)})g)(c \otimes f) = ((a \otimes h)(b \otimes g))(c \otimes f)$$

as required. We used the definitions and, for the second equality, that the actions are assumed to be coalgebra maps. The third equality is (7.8), while the fourth is (7.7) in reverse. We then use again that our actions are coalgebra maps and thereby recognise the result. The proof that the tensor product coproduct is an algebra homomorphism is as follows:

$$\Delta(a \otimes h)\Delta(b \otimes g) = (a_{(1)} \otimes h_{(1)})(b_{(1)} \otimes g_{(1)}) \otimes (a_{(2)} \otimes h_{(2)})(b_{(2)} \otimes g_{(2)})$$

$$= a_{(1)}(h_{(1)} \triangleright b_{(1)}) \otimes (h_{(2)} \triangleleft b_{(2)})g_{(1)} \otimes a_{(2)}(h_{(3)} \triangleright b_{(3)}) \otimes (h_{(4)} \triangleleft b_{(4)})g_{(2)}$$

$$= a_{(1)}(h_{(1)} \triangleright b_{(1)}) \otimes (h_{(3)} \triangleleft b_{(3)})g_{(1)} \otimes a_{(2)}(h_{(2)} \triangleright b_{(2)}) \otimes (h_{(4)} \triangleleft b_{(4)})g_{(2)}$$

$$= \Delta\left((a \otimes h)(b \otimes g)\right),$$

where we used (7.9) for the third equality. The unit, counit and antipode are more trivial, the last of these using the factorised form now that associativity is proved. With a little more care in the above proof, one easily sees that the converse is also true, i.e. that the product shown and tensor product coalgebra structure gives a bialgebra *iff* $\triangleright, \triangleleft$ make A, H a matched pair of bialgebras. ∎

Next we give a more abstract way of thinking about double cross products. The abstract definition we need is that a bialgebra or Hopf algebra X *factorises* as $X = AH$ if there are sub-bialgebras A, H such that the map $A \otimes H \to X$ given by the product in X is an isomorphism of vector spaces. This property clearly holds for $A \bowtie H$ by the canonical inclusions of A and H. We now also prove the converse.

Theorem 7.2.3 *If a bialgebra X factorises in the sense that there are sub-bialgebras*

$$A \overset{i}{\hookrightarrow} X \overset{j}{\hookleftarrow} H$$

such that $\cdot \circ (i \otimes j) : A \otimes H \to X$ *is an isomorphism of vector spaces, then A, H are a matched pair and $X \cong A \bowtie H$.*

Proof: Since $\cdot \circ (i \otimes j)$ is a linear isomorphism, we have a well-defined linear map $\Psi : H \otimes A \to A \otimes H$ defined by

$$j(h)i(a) = \cdot \circ (i \otimes j) \circ \Psi(h \otimes a).$$

Associativity in X, and the fact that i, j are algebra maps, tell us that

$$\cdot \circ (i \otimes j) \circ \Psi(hg \otimes a) = j(hg)i(a) = j(h)j(g)i(a)$$

$$= j(h) \cdot \circ (i \otimes j) \circ \Psi(g \otimes a) = \cdot \circ (i \otimes j) \circ (\mathrm{id} \otimes \cdot) \circ \Psi_{12} \circ \Psi_{23}(h \otimes g \otimes a),$$

$$\cdot \circ (i \otimes j) \circ \Psi(h \otimes ab) = j(h)i(ab) = j(h)i(a)i(b)$$

$$= \cdot \circ (i \otimes j) \circ \Psi(h \otimes a)i(b) = \cdot \circ (i \otimes j) \circ (\cdot \otimes \mathrm{id}) \circ \Psi_{23} \circ \Psi_{12}(h \otimes a \otimes b).$$

We conclude from this and, by a similar consideration of the identity element in X, that

$$\Psi \circ (\cdot \otimes \mathrm{id}) = (\mathrm{id} \otimes \cdot) \circ \Psi_{12} \circ \Psi_{23}, \quad \Psi(1 \otimes a) = a \otimes 1,$$

$$\Psi \circ (\mathrm{id} \otimes \cdot) = (\cdot \otimes \mathrm{id}) \circ \Psi_{23} \circ \Psi_{12}, \quad \Psi(h \otimes 1) = 1 \otimes h, \tag{7.10}$$

where the suffixes refer to the tensor factor on which Ψ acts. We see that the products in H and A, respectively, 'commute' with Ψ in a certain sense. This is the result at just the algebra level (we conclude that factorisations of algebras are of this form).

Next, we use the counits ϵ to define linear maps $\lhd : H \otimes A \to H$ and $\rhd : H \otimes A \to A$ by

$$\rhd = (\mathrm{id} \otimes \epsilon) \circ \Psi, \quad \lhd = (\epsilon \otimes \mathrm{id}) \circ \Psi. \tag{7.11}$$

Applying $\mathrm{id} \otimes \epsilon$ to the first line of (7.10) and $\epsilon \otimes \mathrm{id}$ to the second line, tells us that $\rhd$ is a left action and $\lhd$ is a right action as required. Applying $\epsilon \otimes \mathrm{id}$ to the first and $\mathrm{id} \otimes \epsilon$ to the second, tells us that

$$(hg) \lhd a = \cdot (h \lhd \Psi(g \otimes a)), \quad 1 \lhd a = \epsilon(a),$$

$$h \rhd (ab) = \cdot (\Psi(h \otimes a) \rhd b), \quad h \rhd 1 = \epsilon(h). \tag{7.12}$$

This is the result at the level of algebras equipped with homomorphisms ϵ (their factorisations are of this form).

Finally, we use that i, j are coalgebra maps to deduce that $\cdot \circ (i \otimes j) : A \otimes H \to X$ and $\cdot \circ (j \otimes i) : H \otimes A \to X$ are coalgebra maps. The inverse of $\cdot \circ (i \otimes j)$ is also a coalgebra map; hence, the ratio Ψ is also a coalgebra map; i.e. we conclude that

$$\Delta_{A \otimes H} \circ \Psi = (\Psi \otimes \Psi) \circ \Delta_{H \otimes A}, \quad (\epsilon \otimes \epsilon) \circ \Psi(h \otimes a) = \epsilon(a)\epsilon(h).$$

Now we apply $\mathrm{id} \otimes \epsilon \otimes \epsilon \otimes \mathrm{id}$, both sides of the first of these to conclude that

$$h_{(1)} \rhd a_{(1)} \otimes h_{(2)} \lhd a_{(2)} = \Psi(h \otimes a),$$

which means that our results above prove (7.7) and (7.8). Applying instead the map $\epsilon \otimes \mathrm{id} \otimes \mathrm{id} \otimes \epsilon$ gives, by contrast,

$$h_{(1)} \lhd a_{(1)} \otimes h_{(2)} \rhd a_{(2)} = \tau \circ \Psi(h \otimes a),$$

where τ is the usual transposition. This proves (7.9). Likewise, applying instead $\epsilon \otimes \mathrm{id} \otimes \epsilon \otimes \mathrm{id}$ gives that $\lhd$ is a coalgebra map, while applying $\mathrm{id} \otimes \epsilon \otimes \mathrm{id} \otimes \epsilon$ gives that $\rhd$ is a coalgebra map, too. Hence we have all the conditions needed for a matched pair in Definition 7.2.1. Finally, looking again at the equation for $j(h)i(a)$ above tells us now that $\cdot \circ (i \otimes j)$ becomes an isomorphism between the corresponding double cross product $A{\bowtie}H$ and our original bialgebra X.   ∎

Next we explain the connection with the bicrossproducts of Chapter 6.2. Recall that, from bialgebras mutually acting and coacting on each other, we obtained bicrossproduct bialgebras in Theorems 6.2.2 and 6.2.3. We know from Chapter 1.6 that actions and coactions are canonically related by dualisation. The associated bialgebras or Hopf algebras are quite different but the underlying data are in one to one correspondence.

Proposition 7.2.4 *If A, H are finite-dimensional bialgebras or Hopf algebras, they form a right–left matched pair with double cross product $A{\bowtie}H$ iff the corresponding action and coaction via dualisation give a bicrossproduct $H^{*}{\blacktriangleright\!\!\triangleleft}A$. Likewise, this is so iff dualising the other factor gives a bicrossproduct $H{\triangleright\!\!\blacktriangleleft}A^{*}$.*

Proof: In the first case, we dualise H so that A, as a left H-module coalgebra, corresponds to A, as a right H^{*}-comodule coalgebra, by Proposition 1.6.11. Meanwhile, H, as a right A-module coalgebra, corresponds to H^{*}, as a left A-module coalgebra, by Proposition 1.6.19. We recall the relevant conversions:

$$a^{(\bar{1})}\langle h, a^{(\bar{2})}\rangle = h{\triangleright}a, \quad \langle a{\triangleright}f, h\rangle = \langle f, h{\triangleleft}a\rangle, \quad \forall h \in H, \ a \in A, \ f \in H^{*},$$

where the left hand sides are the bicrossproduct data and the right hand sides are the double cross product data. We check the compatibility conditions in Theorem 6.2.2 as:

$$\langle \Delta(a{\triangleright}f), h \otimes g\rangle = \langle f, (hg){\triangleleft}a\rangle = \langle f_{(1)}, h{\triangleleft}(g_{(1)}{\triangleright}a_{(1)})\rangle\langle f_{(2)}, g_{(2)}{\triangleleft}a_{(2)}\rangle$$
$$= \langle (g_{(1)}{\triangleright}a_{(1)}){\triangleright}f_{(1)}, h\rangle\langle a_{(2)}{\triangleright}f_{(2)}, g_{(2)}\rangle$$
$$= \langle a_{(1)}{}^{(\bar{1})}{\triangleright}f_{(1)} \otimes a_{(1)}{}^{(\bar{2})}(a_{(2)}{\triangleright}f_{(2)}), h \otimes g\rangle,$$

which is the condition (6.8) if we are to build $H^{*}{\blacktriangleright\!\!\triangleleft}A$. This clearly also works the other way, to derive (7.7). In just the same way we have a precise correspondence for the other conditions for the bicrossproduct. Dualising A instead gives, in just the same way, the correspondence with the data for $H{\triangleright\!\!\blacktriangleleft}A^{*}$ in Theorem 6.2.3.   ∎

Example 7.2.5 *Let H be a finite-dimensional Hopf algebra. Then the quantum double $D(H) = H^{*\mathrm{op}}{\bowtie}H$ by the mutual left coadjoint actions Ad^{*} in Example 1.6.4.*

Proof: Let $\triangleright = \mathrm{Ad}^{*}$ be the left coadjoint action of H in Example 1.6.4, and let $\triangleleft = \mathrm{Ad}^{*}$ be the analogous left coadjoint action of H^{*} on H, which we view as a right action of $H^{*\mathrm{op}}$. We can easily compute that (7.7)–(7.9) hold so that $(H^{*\mathrm{op}}, H)$ form a matched pair of Hopf algebras. The product in the double cross product Hopf algebra is

$$(a \otimes h)(b \otimes g) = a \cdot_{\mathrm{op}} (h_{(1)} \triangleright b_{(1)}) \otimes (h_{(2)} \triangleleft b_{(2)}) g$$

$$= b_{(2)} a \langle h_{(1)}, (Sb_{(1)}) b_{(3)} \rangle \otimes \langle (Sh_{(2)}) h_{(4)}, b_{(4)} \rangle h_{(3)} g$$

$$= \langle h_{(1)}, Sb_{(1)} \rangle \langle h_{(2)}, b_{(3)} \rangle \langle Sh_{(3)}, b_{(4)} \rangle \langle h_{(5)}, b_{(5)} \rangle b_{(2)} a \otimes h_{(4)} g,$$

which collapses to the product for $D(H)$ in Theorem 7.1.1. In both cases, the unit and coproduct are the tensor product ones.   ∎

The corresponding bicrossproducts are $M(H^*)$ in Example 6.2.8 and its right–left version (6.14), respectively. We have already verified the compatibility conditions in the proof of these. So this gives an alternative way of constructing the quantum double, as corresponding to a bicross-product. Exactly the same results apply in suitable conventions to the other forms of $D(H)$ in Theorem 7.1.1 and Exercise 7.1.2. They are all double cross products by mutual quantum coadjoint actions of one form or another. Thus, $H \bowtie H^{*\mathrm{op}}$ uses the right coadjoint actions given by evaluating against the left adjoint coaction as in Example 1.6.17, the right action of H^* then being viewed as a left action of $H^{*\mathrm{op}}$. Also, we do not really require, for these constructions, that H is finite dimensional or even a Hopf algebra:

Example 7.2.6 *Let H, A be paired bialgebras as in Definition 1.4.3, with $\langle \ , \ \rangle$ convolution-invertible. There are mutual coadjoint actions*

$$h \triangleleft a = \sum h_{(2)} \langle h_{(1)}, a_{(1)} \rangle^{-1} \langle h_{(3)}, a_{(2)} \rangle,$$

$$h \triangleright a = \sum a_{(2)} \langle h_{(1)}, a_{(1)} \rangle^{-1} \langle h_{(2)}, a_{(3)} \rangle,$$

giving a double cross product bialgebra $A^{\mathrm{op}} \bowtie H$. Likewise, there are mutual coadjoint actions

$$a \triangleleft h = \sum a_{(2)} \langle h_{(1)}, a_{(1)} \rangle \langle h_{(2)}, a_{(3)} \rangle^{-1},$$

$$a \triangleright h = \sum h_{(2)} \langle h_{(1)}, a_{(1)} \rangle \langle h_{(3)}, a_{(2)} \rangle^{-1},$$

giving a double cross product bialgebra $H \bowtie A^{\mathrm{op}}$.

Proof: The first two actions are mutual left coadjoint actions of A, H associated to the pairing, and are verified as in Example 1.6.4 with $\langle S(\ ), \ \rangle$ replaced by $\langle \ , \ \rangle^{-1}$. The first of these is then viewed as a right action of A^{op} rather than as a left action of A. The resulting double cross product now has the structure quoted in (7.2). Likewise, the dualisation of the left adjoint coaction in Example 1.6.17 gives mutual right coadjoint actions of H, A on each other, which can be verified for any pairing. We view the second of these as a left action of A^{op}. The double cross product then has the structure $H \bowtie A^{\mathrm{op}}$.   ∎

These *generalised quantum doubles* are our first application of double cross products. Since the pairing need not be nondegenerate, this construction is very general and includes examples where the factors are far from dual to each other (and need not even be Hopf algebras). Note also that, purely as a matter of emphasis, it is sometimes useful to replace the symbol A^{op} here by some other bialgebra or Hopf algebra denoted by A. Of course, the same map $\langle \ , \ \rangle$ is no longer a duality pairing but a *skew-pairing*. By definition, a skew-pairing $\sigma : H \otimes A \to k$ between bialgebras H, A is nothing other than a pairing as in Definition 1.4.3 between H and A^{op}, i.e. a linear map obeying

$$\sigma(hg \otimes a) = \sum \sigma(h \otimes a_{(1)})\sigma(g \otimes a_{(2)}),$$

$$\sigma(h \otimes ab) = \sum \sigma(h_{(1)} \otimes b)\sigma(h_{(2)} \otimes a). \tag{7.13}$$

We have already encountered some similar formulae (2.6) in another context in Chapter 2.2. The convolution-inverse σ^{-1} exists if either H has an antipode or A has an inverse-antipode (cf. the proof of Lemma 2.2.2). Thus, our last construction could be written equally well, if we prefer, as:

Example 7.2.7 *Let A, H be two bialgebras and let $\sigma : H \otimes A \to k$ be a skew-pairing which is convolution-invertible. Then there are double cross product bialgebras $A \bowtie_\sigma H$ built on $A \otimes H$ with product*

$$(a \otimes h)(b \otimes g) = \sum ab_{(2)} \otimes h_{(2)}g\, \sigma^{-1}(h_{(1)} \otimes b_{(1)})\sigma(h_{(3)} \otimes b_{(3)})$$

with tensor product unit and counit. There is another isomorphic form $H_\sigma \bowtie A$ built on $H \otimes A$ and product

$$(h \otimes a)(g \otimes b) = \sum hg_{(2)} \otimes a_{(2)}b\sigma(a_{(1)} \otimes g_{(1)})\sigma^{-1}(a_{(3)} \otimes Sg_{(3)}).$$

Both forms contain A and H as sub-bialgebras, and are generated by these with cross relations

$$\sum \sigma(h_{(1)} \otimes a_{(1)})h_{(2)}a_{(2)} = \sum a_{(1)}h_{(1)}\sigma(h_{(2)} \otimes a_{(2)}).$$

Proof: What was called A^{op} in the preceding example is now called A, and what was called $\langle \ , \ \rangle$ in the various formulae is now called σ. So the actions in the first case are

$$h \triangleleft a = h_{(2)} \sigma^{-1}(h_{(1)} \otimes a_{(1)}) \sigma(h_{(3)} \otimes a_{(2)}),$$

$$h \triangleright a = a_{(2)} \sigma^{-1}(h_{(1)} \otimes a_{(1)}) \sigma(h_{(2)} \otimes a_{(3)}),$$

and so on. It is worth noting that, if σ is a convolution-invertible skew-pairing, then so is $\sigma_{21}^{-1} : A \otimes H \to k$, and

$$A \bowtie_\sigma H \cong H \bowtie_{\sigma_{21}^{-1}} A = H_\sigma \bowtie A,$$

which is why these generalised quantum doubles always come two at a time. It should be clear that the result is the same as Example 7.2.6, with a slightly different notation. ∎

If the factors are Hopf algebras, there is an antipode such that they become sub-Hopf algebras. Various other results from Section 7.1 likewise go through. For example, one can just replace $\langle\ ,\ \rangle$ by σ for the description of these double cross products as dual-twisting in (7.3). Another proposition which generalises is the quantum double for Hopf $*$-algebras. To explain this, we note that the axiom in Definition 1.7.5 which relates the $*$-structure on a Hopf $*$-algebra to that on its dual could be written equally well as

$$\overline{\langle h, a \rangle} = \langle h^*, a^{*\mathrm{op}} \rangle^{-1}, \quad a^{*\mathrm{op}} = (S^2 a)^*.$$

Hence Proposition 7.1.4 generalises to

Proposition 7.2.8 *Let H, A be two bialgebras which are skew-paired, and which are $*$-algebras such that $\overline{\sigma(h \otimes a)} = \sigma^{-1}(h^* \otimes a^*)$. Then $A \bowtie_\sigma H$ is a $*$-bialgebra with A, H as sub-$*$-bialgebras. It is a Hopf $*$-algebra if H, A are.*

Proof: We apply $*$ to the relations in Example 7.2.7, which become

$$\sigma^{-1}(h^*_{(1)} \otimes a^*_{(1)}) a^*_{(2)} h^*_{(2)} = h^*_{(1)} a^*_{(1)} \sigma^{-1}(h^*_{(2)} \otimes a^*_{(2)}).$$

These are the same relations again, for the elements h^*, a^*. Hence $*$ extends to the double cross product as a $*$-algebra. One can also say this more formally, as we did in Proposition 7.1.4. A $*$-bialgebra means that the coproduct and counit are $*$-algebra maps, which is easily verified. The rest of the details are as in Proposition 7.1.4. ∎

Now we compute a concrete example of the quantum double, namely the double of the very first nontrivial Hopf algebra that we encountered, namely Example 1.3.2 in Chapter 1.

Example 7.2.9 *Let $q^2 \in k$ with $q^2 \neq 0, 1$. Then the Hopf algebra H with generators X, g, g^{-1} and*

$$gX = q^2 Xg, \quad \Delta X = X \otimes 1 + g \otimes X, \quad \Delta g = g \otimes g,$$

$$\epsilon X = 0, \quad \epsilon g = 1, \quad Sg = g^{-1}, \quad SX = -g^{-1}X$$

is self-dual in the sense that it has a Hopf algebra pairing with itself given by

$$\langle g, g \rangle = q^2, \quad \langle g, X \rangle = 0, \quad \langle X, g \rangle = 0, \quad \langle X, X \rangle = \frac{1}{1 - q^{-2}}.$$

The generalised quantum double $D(H) = H^{\mathrm{op}} \bowtie H$ is a Hopf algebra extension

$$k\mathbb{Z} \hookrightarrow D(H) \twoheadrightarrow U_q(sl_2),$$

consisting of a version of $U_q(sl_2)$ extended by adjoining a central group-like generator.

Proof: We note that the use of q^2 (rather than another symbol such as q) is made here purely to tie up with the conventions of $U_q(sl_2)$ in Example 3.2.1. Likewise, the factor $\alpha = \frac{1}{1-q^{-2}}$ in the pairing is purely conventional since all the other equations in X are linear; it fixes the normalisation of X to match up with Example 3.2.1. The extension of the pairing from the generators to a map $H \otimes H \to k$ is determined according to the axioms in Definition 1.4.3, and can be verified directly by elementary means. In fact, we have already seen the necessary tools for this in Lemma 3.2.2, which can be used to compute ΔX^n, just as we did, for example, at the end of Chapter 5.3.

Here we will give a less direct but more conceptual 'geometrical' proof of the self-duality pairing. Recall that the Hopf algebra $k[x]$ of polynomial functions in one variable can equally well be regarded as the enveloping algebra of the one-dimensional Abelian Lie algebra k. We have already used this self-duality in Example 6.1.11 and the model in Chapter 6.4. Clearly, the same applies in higher dimensions. Our strategy is to prove the same result now for the Hopf algebra in algebra in Example 1.3.2, where the generators are q-commuting.

To this end, we define a right action of H on itself by

$$\phi(X, g) \triangleleft g = \phi(q^2 X, q^2 g), \quad \phi(X, g) \triangleleft X = \alpha \partial_q \phi(X, g),$$

where $\phi(X, g) = \sum \phi_{ab} X^a g^b$ is assumed to have powers of X to the left as shown, which we consider as 'normal-ordered'. The ∂_q is one of the usual

q-derivatives of functions in X extended trivially to H, namely

$$\partial_q X^m g^n = [m; q^2] X^{m-1} g^n; \quad [m; q^2] = \frac{1 - q^{2m}}{1 - q^2}.$$

We have already encountered similar q-derivatives at the end of Chapter 3.1, and we proceed in a similar way now. Since ∂_q lowers the degree by 1, it is clear that $(\phi \triangleleft g) \triangleleft X = q^2 (\phi \triangleleft X) \triangleleft g$ for all $\phi \in H$, so $\triangleleft$ indeed defines a right action of our Hopf algebra on itself. Moreover, one has

$$(\phi \psi) \triangleleft g = (\phi \triangleleft g)(\psi \triangleleft g), \quad (\phi \psi) \triangleleft X = (\phi \triangleleft X)\psi + (\phi \triangleleft g)(\psi \triangleleft X) \qquad (7.14)$$

for all normal-ordered polynomials $\phi(X, g), \psi(X, g)$. Note that the product $\phi \psi$ must be normal-ordered using the q-commutation relations between X, g before we can apply the definition of $\triangleleft$. The action of g on products is clear since the normal-ordering process commutes with scaling of the generators by q^2. The action of X is easily verified on monomials as

$$(X^m g^n X^k g^l) \triangleleft X = (q^{2nk} X^{m+k} g^{n+l}) \triangleleft X = q^{2nk} \alpha [m + k; q^2] X^{m+k-1} g^{n+l}$$
$$= q^{2nk} \left(\alpha [m; q^2] + q^{2m} \alpha [k; q^2] \right) X^{m+k-1} g^{n+l}$$
$$= \alpha [m; q^2] X^{m-1} g^n X^k g^l + q^{2(m+n)} X^m g^n \alpha [k; q^2] X^k g^l$$
$$= ((X^m g^n) \triangleleft X) X^k g^l + ((X^m g^n) \triangleleft g)((X^k g^l) \triangleleft X)$$

from the definition of $\triangleleft$ and the relations of H. This proves (7.14), which then implies that the product of H is covariant (a module algebra) under the action $\triangleleft$. This is because the generators X, g act in the required way according to (1.10), but, since the coproduct is an algebra homomorphism, any product of the generators must also act in the same way. This is a slight generalisation and variant of the more standard q-derivation property, at the end of Chapter 3.1.

On the other hand, the reader has met some similar actions $\triangleleft$ in Chapter 5.3 as the 'Markov transition operators' corresponding to the right coregular representation given by evaluating the coproduct against a pairing. We have turned this around by constructing $\triangleleft$ directly, and we are now in a position to read off a pairing. Namely, we define

$$\langle h, a \rangle \equiv \epsilon(h \triangleleft a) \quad \Rightarrow \quad \langle hg, a \rangle = \langle h, a_{(1)} \rangle \langle g, a_{(2)} \rangle$$

for $h, g, a \in H$, as an expression of our module-algebra property (7.14) of $\triangleleft$. This proves half of the axioms of a pairing in Definition 1.4.3. But since the resulting pairing,

$$\langle X^m g^n, X^k g^l \rangle = \epsilon \left(((X^k g^l) \triangleleft X^m) \triangleleft g^n \right) = \delta_{m,k} \alpha^m [m; q^2]! q^{2nl}, \qquad (7.15)$$

is symmetric in the two arguments of $\langle \, , \, \rangle$, we also deduce the other half, where the roles of a, h are interchanged. We also see from our construction of the pairing that $\langle h, ab \rangle = \langle h \triangleleft a, b \rangle$, so that $\triangleleft$ can then be recovered as the right coregular representation associated to the pairing.

We are now in the situation of Example 7.2.6, where $A = H$ is dually paired with H itself. It is easy to compute the associated quantum double Hopf algebras $D(H) = H^{\mathrm{op}} \bowtie H$ etc. as follows. We denote the generators of H^{op} by $\bar{X}, \bar{g}, \bar{g}^{-1}$, say, rather than by X, g, g^{-1} as before. They have the relations $\bar{X}\bar{g} = q^2 g\bar{X}$ and generate a copy of H^{op} inside the quantum double. The further cross relations are provided as in Theorem 7.1.1, and give

$$
\begin{aligned}
g\bar{g} &= \bar{g}g\langle Sg, g\rangle\langle g, g\rangle = \bar{g}g, \\
g\bar{X} &= \bar{X}_{(2)}g\langle Sg, X_{(1)}\rangle\langle g, X_{(2)}g\rangle = \bar{X}g\langle Sg, g\rangle\langle g, 1\rangle = \bar{X}gq^{-2}, \\
X\bar{g} &= \bar{g}X_{(2)}\langle SX_{(1)}, g\rangle\langle X_{(2)}\bar{g}, g\rangle = \bar{g}X\langle Sg, g\rangle\langle 1, g\rangle = \bar{g}Xq^{-2}, \\
X\bar{X} &= \bar{X}_{(2)}\langle SX, X_{(1)}\rangle\langle 1, X_{(2)}g\rangle \\
&\quad + \bar{X}_{(2)}X\langle Sg, X_{(1)}\rangle\langle 1, X_{(2)}g\rangle + \bar{X}_{(2)}g\langle Sg, X_{(1)}\rangle\langle X, X_{(2)}g\rangle \\
&= \langle -Sg, g\rangle\langle X, X\rangle + \bar{X}X\langle Sg, g\rangle + \bar{g}g\langle Sg, g\rangle\langle X, X\rangle \\
&= q^{-2}\bar{X}X + \frac{\bar{g}g - 1}{q^2 - 1}.
\end{aligned}
$$

This computes the quantum double of our Hopf algebra as $\bar{g}g^{-1}$ central and

$$
gX = q^2 Xg, \quad g\bar{X} = q^{-2}\bar{X}g, \quad X\bar{X} - q^{-2}\bar{X}X = \frac{\bar{g}g - 1}{q^2 - 1}, \tag{7.16}
$$

$$
\Delta X = X \otimes 1 + g \otimes X, \quad \Delta \bar{X} = \bar{X} \otimes 1 + \bar{g} \otimes \bar{X}
$$

with the corresponding counit and antipode. The generators $g, \bar{g}$ are group-like.

From this, it is easy to see that we have a Hopf algebra homomorphism $p : D(H) \to U_q(sl_2)$ defined by

$$
p(X) = F \equiv q^{-\frac{H}{2}} X_-, \quad p(\bar{X}) = \bar{E} \equiv q^{-\frac{H}{2}} X_+, \quad p(g) = p(\bar{g}) = q^{-H}, \tag{7.17}
$$

where $\bar{E}$ and F are similar to the generators in Chapter 3.3, and, together with $q^{\pm H}$, can be considered as generating a version of $U_q(sl_2)$. It is a sub-Hopf algebra of the usual form given in Example 3.2.1 since we do not use $q^{\pm \frac{H}{2}}$ themselves. With this form of $U_q(sl_2)$ understood, we have p as a surjection. Next, it is clear that the invertible element $C = \bar{g}g^{-1}$ generates $k\mathbb{Z} \subset D(H)$ as a commutative sub-Hopf algebra. Moreover, it is clear that $D(H) \cong U_q(sl_2) \otimes k\mathbb{Z}$ as vector spaces by $X^m \bar{X}^n g^k C^l \mapsto F^m \bar{E}^n q^{-Hk} \otimes C^l$. Finally, one verifies that this isomorphism respects multiplication by C from the right and the coaction of $U_q(sl_2)$ from the left in the manner required in (6.37) for a Hopf algebra extension in the setting of Chapter 6.3.

In fact, one can show that this extension is cleft and coclft, etc., and hence, by the theory of Chapter 6.3, it must be a cocycle bicrossproduct. In terms of the construction in Theorem 6.3.9, we find

$$D(H) \cong U_q(sl_2) \bowtie_\chi k\mathbb{Z}, \quad \beta \left(\left\{ \begin{matrix} \bar{E} \\ F \\ q^{-H} \end{matrix} \right\} \right) = \left\{ \begin{matrix} C \otimes \bar{E} \\ 1 \otimes F \\ 1 \otimes q^{-H} \end{matrix} \right., \quad \chi(\bar{E} \otimes F) = \frac{1 - C}{1 - q^{-2}},$$

with $\chi = \epsilon \otimes \epsilon$ on the other generators of $U_q(sl_2)$. The action $\lhd$ in the bicrossproduct and the cocycle ψ are both trivial. To see how this works, we recall from Proposition 6.3.7 that the cocycle bicrossproduct is built on the tensor product vector space, which we identify as $X = F \otimes 1$, $\bar{X} = \bar{E} \otimes 1$, $g = q^{-H} \otimes 1$ and $C \equiv 1 \otimes C$. It has the product in Proposition 6.3.7 as modified by the cocycle χ,

$$\bar{X}X = (\bar{E} \otimes 1)(F \otimes 1) = \bar{E}_{(1)} F_{(1)} \otimes \chi(\bar{E}_{(2)} \otimes F_{(2)})$$

$$= \bar{E}F \otimes 1 + g^2 \otimes \chi(\bar{E} \otimes F) = \bar{E}F \otimes 1 + g^2 \otimes \frac{1 - C}{1 - q^{-2}}$$

$$= q^2 F\bar{E} \otimes 1 + \frac{1 - g^2}{1 - q^{-2}} \otimes 1 + g^2 \otimes \frac{1 - C}{1 - q^{-2}}$$

$$= q^2 (F \otimes 1)(\bar{E} \otimes 1) + \frac{1 - g^2 \otimes C}{1 - q^{-2}} = q^2 X\bar{X} + \frac{1 - g^2 C}{1 - q^{-2}},$$

which are the relations (7.16) for the quantum double. The coalgebra structure is the cross coproduct from Proposition 1.6.18. From the coaction β we have

$$\Delta \bar{X} = \Delta(\bar{E} \otimes 1) = \bar{E} \otimes 1 \otimes 1 \otimes 1 + q^{-H} \otimes \beta(\bar{E}) \otimes 1 = \bar{X} \otimes 1 + gC \otimes \bar{X}$$

as required in (7.16). This completes our detailed description of $D(H)$. We see that, as an algebra, it is a central extension of $U_q(sl_2)$ by the cocycle χ, and, as a coalgebra, it is a cross coproduct.

Note that if we adjoin to our quantum double an element $q^{-c} = C^{\frac{1}{2}}$ then our cocycle χ becomes cohomologically trivial and the algebra of the quantum double becomes isomorphic to a tensor product. In this case, we can identify

$$X = q^{-\frac{H}{2}} X_-, \quad \bar{X} = q^{-\frac{H}{2}} X_+, \quad g = q^{-H} q^c, \quad \bar{g} = q^{-H} q^{-c} \qquad (7.18)$$

as an algebra isomorphism $U_q(sl_2) \otimes k\mathbb{Z}$, where $k\mathbb{Z}$ is generated now by q^{-c}. The coalgebra structure, however, remains a cross coproduct one. This time, we take the coaction of $k\mathbb{Z}$ as

$$\beta(F) = q^{-c} \otimes F, \quad \beta(\bar{E}) = q^c \otimes \bar{E},$$

and the cross coproduct from Proposition 1.6.18 then reproduces (7.16) as required. So, in this case, the extended quantum double is isomorphic to $U_q(sl_2) \bowtie k\mathbb{Z}$ as a Hopf algebra. Indeed, this coaction β respects the full bialgebra structure of $U_q(sl_2)$ (and corresponds to a useful $U(1)$ action on this quantum group); hence, we know from the dual of Proposition 6.2.1 that the cross coproduct in this case is necessarily a Hopf algebra.

Finally, we note that if we further assume that $q^{\pm H}$ have square roots then we have $U_q(sl_2)$ in its usual form in Example 3.2.1 rather than in the form above. The coproducts of $X_{\pm}$ in the quantum double come out as

$$\Delta X_{\pm} = X_{\pm} \otimes q^{\frac{H}{2}} + q^{-\frac{H}{2}} q^{\mp c} \otimes X_{\pm}. \tag{7.19}$$

To have these square roots, and hence to project onto $U_q(sl_2)$ in its usual form, one can repeat the above steps with our initial Hopf algebra H taken as generated by $X, g^{\pm\frac{1}{2}}$, say, and the corresponding formulae. The self-pairing is $\langle g^{\frac{1}{2}}, g^{\frac{1}{2}} \rangle = q^{\frac{1}{2}}$, where we take $q^{\frac{1}{2}}$ as our initial element of k rather than q^2. Of course, this distinction, and also the need for the cocycle χ, are both absent when, as is usually done, we work over formal power series $\mathbb{C}[[t]]$ and take H, c as generators. Also note, in passing, the similarity of this result with the structure of $U_q(\widehat{sl_2})$ in Example 6.3.14. ∎

The quantum double construction given here is a very powerful one and works just as well at roots of unity, etc. If one begins with a finite-dimensional version of the Hopf algebra H appropriate to a root of unity, the result is a version of the quasitriangular Hopf algebra $u_q(sl_2)$, reduced as in Chapter 3.4. Similarly for general $u_q(g)$. Note also that the natural generators from the point of view of these conventions for the double are similar to those in Chapter 3.3 (but slightly different), namely

$$\bar{E}_i = q_i^{-\frac{H_i}{2}} X_i, \quad F_i = q_i^{-\frac{H_i}{2}} X_{-i}, \quad g_i = q_i^{-H_i},$$

where the length $d_i = \frac{1}{2}(\alpha_i, \alpha_i) \in \mathbb{Z}$ of a root is used to define $q_i = q^{d_i}$. If a_{ij} is the symmetrisable Cartan matrix (so that $d_i a_{ij} = d_j a_{ji}$ and $a_{ii} = 2$), then the structure of $U_q(g)$ in this form is

$$F_i \bar{E}_j - q^{-a_{ij}} \bar{E}_j F_i = \delta_{ij} \frac{g_i^2 - 1}{q_i^2 - 1}, \quad g_i F_j g_i^{-1} = q_i^{a_{ij}} F_j, \quad g_i \bar{E}_j g_i^{-1} = q_i^{-a_{ij}} \bar{E}_j,$$

$$\Delta F_i = F_i \otimes 1 + g_i \otimes F_i, \quad \Delta \bar{E}_i = \bar{E}_i \otimes 1 + g_i \otimes \bar{E}_i$$

and q-Serre relations identical to those for $X_{\pm i}$. We take $H = U_q(b_-)$ as generated by $\{g_i, F_j\}$ and identify $U_q(b_-)^{\mathrm{op}} = U_q(b_+)$ as generated by $\{g_i, \bar{E}_j\}$. The identification is by $F_j \mapsto \bar{E}_j$. The self-duality pairing of $U_q(b_-)$ with itself is

$$\langle F_i, F_j \rangle = \delta_{ij} \frac{1}{1 - q_i^{-2}}, \quad \langle g_i, g_j \rangle = q_i^{a_{ij}},$$

and this allows one to construct $U_q(g)$ as a quotient of the quantum double $U_q(b_-)\bowtie U_q(b_+)$ according to Example 7.2.6. This pairing is also the same thing as a skew-pairing of $U_q(b_-)$ with $U_q(b_+)$ by

$$\sigma(F_i \otimes \bar{E}_j) = \delta_{ij}\frac{1}{1-q_i^{-2}}, \quad \sigma(g_i \otimes g_j) = q_i^{a_{ij}}$$

if one wants to use the language of Example 7.2.7.

Continuing in the $U_q(sl_2)$ case, we can also work over formal power series $\mathbb{C}[[t]]$, where $q = e^{\frac{t}{2}}$, and regard H as the generator to derive a power series form of the quasitriangular structure $\mathcal{R}$. The same applies to $U_q(g)$, using orthogonality of a PBW basis for $U_q(b_+)$ of the form in (3.32).

Proposition 7.2.10 *Working over formal power series $\mathbb{C}[[t]]$, the quasi-triangular structure of the quantum double in the previous example projects onto the quasitriangular structure of $U_q(sl_2)$ in Example 3.2.1. Moreover, the quantum double itself is isomorphic to a twisting according to Theorem 2.3.4 of the quasitriangular Hopf algebra $U_q(sl_2) \otimes U_{q^{-1}}(1)$ by cocycle $\chi = q^{-\frac{1}{2}c\otimes H}$. Here $U_{q^{-1}}(1)$ is the quantum line in Example 2.1.19 with c as generator.*

Proof: Recall that twisting in Theorem 2.3.4 consists of conjugating the coproduct by a 2-cocycle. Working over $\mathbb{C}[[t]]$, we can write $\chi = q^{-\frac{1}{2}c\otimes H}$ and easily see that it is a 2-cocycle for $U_q(sl_2) \otimes U_{q^{-1}}(1)$, where $U_{q^{-1}}(1)$ is the Hopf algebra with generator c and $\Delta c = c \otimes 1 + 1 \otimes c$. Then $[c \otimes H, (\ ) \otimes X_\pm] = \pm 2c(\ ) \otimes X_\pm$ tells us that the coproducts (7.19) are just the conjugation by χ of the usual coproduct of $U_q(sl_2)$. This identifies the Hopf algebra structure.

Next, we take a basis of $U_q(b_-)$ in the form $\{X^m(H-c)^n\}$. The pairing corresponding to $\langle g, g \rangle = q^2$ is

$$\langle (H-c)^m, (H-c)^n \rangle = \delta_{m,n}(\frac{4}{t})^m m!$$

since $q = e^{\frac{t}{2}}$, giving at once a dual basis for the self-pairing (7.15). The infinitesimal generator corresponding to $\bar{g}$ is $H + c$. Hence the quasitriangular structure in Theorem 7.1.1, understood now as a formal power series, is

$$\mathcal{R} = \sum_{a,b=0}^{\infty} (H+c)^b \bar{X}^a \otimes X^a(H-c)^b\frac{(1-q^{-2})^a(\frac{t}{4})^b}{[a;q^2]!b!}$$

$$= \sum_{a,b=0}^{\infty} (H+c)^b \bar{X}^a \otimes (H-c+2a)^b X^a\frac{(1-q^{-2})^a(\frac{t}{4})^b}{[a;q^2]!b!}$$

$$= q^{\frac{(H+c)\otimes(H-c)}{2}} \sum_{a=0}^{\infty} q^{Ha} q^{ca} \bar{X}^a \otimes X^a \frac{(1-q^{-2})^a}{[a;q^2]!}$$

$$= q^{\frac{1}{2}(H\otimes H - c\otimes c + c\otimes H - H\otimes c)} e_{q^{-2}}^{(1-q^{-2})q^c q^{\frac{H}{2}} X_+ \otimes q^{-\frac{H}{2}} X_-}$$

$$= q^{-\frac{1}{2}H\otimes c} \left(q^{\frac{1}{2}H\otimes H} e_{q^{-2}}^{(1-q^{-2})q^{\frac{H}{2}} X_+ \otimes q^{-\frac{H}{2}} X_-} q^{-\frac{1}{2}c\otimes c} \right) q^{\frac{1}{2}c\otimes H},$$

where the dual basis element is written as an element of $H^{\mathrm{op}} \subset D(H)$. For the second equality, we used the commutation relations between X, H. For the fourth equality, we used the relations again, in exponential form, to write

$$\frac{q^{Ha}(q^{-\frac{H}{2}} X_+)^a}{[a;q^2]!} = \frac{(q^{\frac{H}{2}} X_+)^a}{[a;q^{-2}]!},$$

and then recognised the $e_{q^{-2}}$ in Lemma 3.2.2. We see that $\mathcal{R}$ is exactly the twisting from Theorem 2.3.4 of the tensor product quasitriangular structure of $U_q(sl_2) \otimes U_{q^{-1}}(1)$, provided we take the element $q^{-\frac{1}{2}c\otimes c}$ for the quasitriangular structure of $U_{q^{-1}}(1)$, i.e. a version of the quantum line quasitriangular Hopf algebra in Example 2.1.19. ∎

The quantum double construction was how the quasitriangular structure of $U_q(sl_2)$ and other quantum enveloping algebras $U_q(g)$ was first constructed by Drinfeld. The dual basis in the general $U_q(g)$ case can be formulated using the quantum Weyl group, as mentioned in Chapter 3.3.

Exercise 7.2.11 *Rework Example 7.2.9 and Proposition 7.2.10 for the quantum double in the form $H \bowtie H^{\mathrm{cop}}$ with $H = U_q(b_+)$.*

Solution: This form is just as easy from our general machinery, and is more conventional in the context of $U_q(g)$. We begin with a new Hopf algebra H with generators $\{g, X\}$ and

$$gX = q^2 Xg, \quad \Delta X = X \otimes g + 1 \otimes X, \quad SX = -Xg^{-1},$$

which is self-dual with the self-pairing

$$\langle g, g \rangle = q^{-2}, \quad \langle X, X \rangle = \frac{1}{q - q^{-1}},$$

$$\Rightarrow \quad \langle g^m X^n, g^k X^l \rangle = q^{-2mk} \delta_{n,l} \frac{[n;q^{-2}]!}{(q-q^{-1})^n},$$

where $[n; q^{-2}]$ is as in Lemma 3.2.2. The proof is just the same as in Example 7.2.9 but with a left action $\triangleright$ based on right-derivatives rather than the more natural left-derivatives used there. Let $\bar{X}, \bar{g}$ be the generators of H viewed in H^{cop}. From the formulae in Exercise 7.1.2 (or from

Example 7.2.7), we compute the double $H \bowtie H^{\mathrm{cop}}$ as

$$\bar{g}g = g\bar{g}\langle g, g^{-1}\rangle\langle g, g\rangle = g\bar{g},$$

$$\bar{g}X = X_{(2)}\bar{g}\langle X_{(1)}, g^{-1}\rangle\langle X_{(3)}, g\rangle = X\bar{g}\langle 1, g^{-1}\rangle\langle g, g\rangle = X\bar{g}q^{-2},$$

$$\bar{X}g = g\overline{X_{(2)}}\langle g, S^{-1}X_{(3)}\rangle\langle g, X_{(1)}\rangle = g\bar{X}\langle g, g^{-1}\rangle\langle g, 1\rangle = g\bar{X}q^2,$$

$$\bar{X}X = X_{(2)}\bar{g}\langle X_{(1)}, g^{-1}\rangle\langle X_{(3)}, X\rangle + X_{(2)}\langle X_{(1)}, S^{-1}X\rangle\langle X_{(3)}, 1\rangle$$

$$+ X_{(2)}\bar{X}\langle X_{(1)}, g^{-1}\rangle\langle X_{(3)}, 1\rangle$$

$$= \bar{g}\langle X, X\rangle + g\langle X, -\bar{X}\rangle + X\bar{X} = X\bar{X} + \frac{\bar{g} - g}{q - q^{-1}}.$$

This computes the quantum double in this form as $g\bar{g}$ central and

$$gX = q^2 Xg, \quad g\bar{X} = q^{-2}\bar{X}g, \quad [X, \bar{X}] = \frac{g - \bar{g}}{q - q^{-1}},$$

$$\Delta X = X \otimes g + 1 \otimes X, \quad \Delta\bar{X} = \bar{X} \otimes 1 + \bar{g} \otimes \bar{X},$$

and $g, \bar{g}$ group-like. On the other hand, the mapping

$$p(X) = E \equiv X_+ q^{\frac{H}{2}}, \quad p(\bar{X}) = F \equiv q^{-\frac{H}{2}}X_-, \quad p(g) = q^H, \quad p(\bar{g}) = q^{-H}$$

gives us a homomorphism onto $U_q(sl_2)$ in a form generated by $E, F, q^{\pm H}$. Here E, F are the same as the preferred generators in Chapter 3.3. This time, $C = g^{-1}\bar{g}^{-1}$ generates a central sub-Hopf algebra $k\mathbb{Z}$, and once again we have the quantum double as a Hopf algebra central extension

$$D(H) \cong k\mathbb{Z}_\chi{\bowtie}U_q(sl_2), \quad \beta(F) = F \otimes C, \quad \chi(E \otimes F) = \frac{C - 1}{q - q^{-1}},$$

where we have given the nontrivial parts of the coaction and cocycle. The isomorphism of vector spaces here is $C^a\bar{X}^b X^c g^d \mapsto C^{a-b} \otimes F^b E^c q^{dH}$. In particular, $\bar{X} = C^{-1} \otimes F$, $X = 1 \otimes E$ and $g = 1 \otimes q^H$ is the required identification with a left–right version of the cocycle bicrossproduct construction in Theorem 6.3.9. The algebra is the left cocycle extension in Proposition 6.3.2 with $\triangleright$ trivial. The coalgebra is the right cross coproduct in Proposition 1.6.16.

As before, we have an ordinary tensor product algebra if we have suitable square roots or if we work over $\mathbb{C}[[t]]$ with $C = q^{-2c}$ and H, c as generators. In this case, we have the algebra of the quantum double as a tensor product $U_{q^{-1}}(1) \otimes U_q(sl_2)$ by the identification

$$X = X_+ q^{\frac{H}{2}}q^c, \quad \bar{X} = q^{-\frac{H}{2}}X_-, \quad g = q^H q^c, \quad \bar{g} = q^{-H}q^c.$$

The generator c enters into the coproducts in the quantum double in the same way as in (7.19). This means that the quantum double is a twisting of $U_{q^{-1}}(1) \otimes U_q(sl_2)$ by the same cocycle as in Proposition 7.2.10. Also, the self-pairing $\langle g, g \rangle = q^{-2}$ gives

$$\langle (H + c)^m, (H + c)^n \rangle = \delta_{m,n} \left(\frac{4}{t} \right)^m (-1)^m m!$$

from which we find the dual vectors to the $(H + c)^m$. The element $H + c$ is the infinitesimal generator of g, and it appears in H^{cop} as the infinitesimal generator $c - H$ of $\bar{g}$. Hence the quasitriangular structure from Exercise 7.1.2 is

$$\mathcal{R} = \sum_{a,b=0}^{\infty} (H + c)^a X^b \otimes (H - c)^a \bar{X}^b \frac{(q - q^{-1})^b (\frac{t}{4})^a}{[b; q^{-2}]! a!}$$

$$= q^{\frac{1}{2}(H+c) \otimes (H-c)} e_{q^{-2}}^{(q-q^{-1}) X_+ q^{\frac{H}{2}} q^c \otimes q^{-\frac{H}{2}} X_-}$$

$$= q^{-\frac{1}{2} H \otimes c} \left(q^{-\frac{1}{2} c \otimes c} q^{\frac{1}{2} H \otimes H} e_{q^{-2}}^{(1-q^{-2}) q^{\frac{H}{2}} X_+ \otimes q^{-\frac{H}{2}} X_-} \right) q^{\frac{1}{2} c \otimes H}.$$

This is the twisting from Theorem 2.3.4 of the tensor product quasitriangular structure on $U_{q^{-1}}(1) \otimes U_q(sl_2)$, where the quantum line has quasitriangular structure $q^{-\frac{1}{2} c \otimes c}$.

For the general case, with $U_q(g)$ given via the preferred generators E_i, F_i in (3.26) in Chapter 3.3, we take $H = U_q(b_+)$ as generated by $\{g_i, E_j\}$ and identify $U_q(b_+)^{\mathrm{cop}} = U_q(b_-)$ as generated by $\{g_i^{-1}, F_j\}$. The identification is by $E_j \mapsto F_j$ and $g_i \mapsto g_i^{-1}$. The self-duality pairing of $U_q(b_+)$ is

$$\langle E_i, E_j \rangle = \delta_{ij} \frac{1}{q_i - q_i^{-1}}, \quad \langle g_i, g_j \rangle = q_i^{-a_{ij}}; \quad \langle H_i, H_j \rangle = -\frac{2}{t} d_i a_{ij}.$$

One can also think of the pairing as a skew-pairing of $U_q(b_-)$ with $U_q(b_+)$ by

$$\sigma(F_i \otimes E_j) = \delta_{ij} \frac{1}{q_i - q_i^{-1}}, \quad \sigma(g_i \otimes g_j) = q_i^{a_{ij}}; \quad \sigma(H_i \otimes H_j) = \frac{2}{t} d_i a_{ij}.$$

We obtain once again the quasitriangular Hopf algebra $U_q(g)$ as a quotient of the quantum double $U_q(b_+) \bowtie U_q(b_-)$. ∎

Now we turn to a quite different class of examples, where the pairing is very far from nondegenerate. These generalised quantum doubles are also important in physics, and they will be studied further in the next section. For the moment, we emphasise that the next example provides an abstract picture of the bialgebra $\tilde{U}(R)$ already encountered in Example 4.1.6.

Example 7.2.12 *Let R be an invertible solution of the QYBE and let $A(R)$ be the associated bialgebra of Chapter 4.1. The skew-pairing of $A(R)$*

with itself provided by the dual quasitriangular structure in Theorem 4.1.5 gives actions

$$\mathbf{t}_1 \triangleleft \mathbf{s}_2 = R^{-1} \mathbf{t}_1 R, \quad \mathbf{t}_1 \triangleright \mathbf{s}_2 = R^{-1} \mathbf{s}_2 R$$

between two copies of $A(R)$ with generators $\mathbf{s}, \mathbf{t}$. The resulting double cross product $A(R) \bowtie A(R)$ has cross relations

$$R \mathbf{t}_1 \mathbf{s}_2 = \mathbf{s}_2 \mathbf{t}_1 R$$

and is the opposite bialgebra $\tilde{U}(R)^{\mathrm{op}}$ to the one in Example 4.1.6.

Proof: In the course of proving Theorem 4.1.5, we saw that the matrix R extends to a convolution-invertible map $\mathcal{R} : A(R) \otimes A(R) \to k$ obeying the conditions for a skew-pairing, since these are a subset of the conditions for a dual quasitriangular structure. So we are free to take $\langle \mathbf{t}_1, \mathbf{s}_2 \rangle = R$, $\langle \mathbf{t}_1, \mathbf{s}_2 \rangle^{-1} = R^{-1}$ as a pairing of $A(R)$ with $A(R)^{\mathrm{op}}$, or equivalently $\sigma(\mathbf{t}_1 \otimes \mathbf{s}_2) = R^{-1}$, $\sigma^{-1}(\mathbf{t}_1 \otimes \mathbf{s}_2) = R^{-1}$ as a skew-pairing of $A(R)$ with itself. Putting this into Examples 7.2.7 or 7.2.6, gives at once the actions $\triangleright, \triangleleft$ and cross relations as stated. ∎

We recall that our point of view in Chapter 4.1 was that $\tilde{U}(R)$ was itself some kind of generalised dual to $A(R)$, generalised in the sense that we had a pairing which was not expected to be nondegenerate. On the other hand, we can view Proposition 4.2.3 about the weak antipode as a computation that the pairing is convolution-invertible:

Example 7.2.13 *Let R be an invertible solution of the QYBE and let $\tilde{U}(R)$ with generators $\mathbf{L}^{\pm}$ be the bialgebra in Example 4.1.6. We denote the generators of $A(R)^{\mathrm{op}}$ by $\bar{\mathbf{t}}$, with the opposite relations to those in Example 4.1.3. The convolution-invertible pairing in Corollary 4.1.8 gives actions*

$$\mathbf{L}_2^+ \triangleleft \bar{\mathbf{t}}_1 = R^{-1} \mathbf{L}_2^+ R, \quad \mathbf{L}_1^- \triangleleft \bar{\mathbf{t}}_2 = R \mathbf{L}_1^- R^{-1},$$

$$\mathbf{L}_2^+ \triangleright \bar{\mathbf{t}}_1 = R^{-1} \bar{\mathbf{t}}_1 R, \quad \mathbf{L}_1^- \triangleright \bar{\mathbf{t}}_2 = R \bar{\mathbf{t}}_2 R^{-1}.$$

The resulting double cross product $A(R)^{\mathrm{op}} \bowtie \tilde{U}(R)$ has generators $\mathbf{L}^{\pm}, \bar{\mathbf{t}}$ and cross relations

$$R \mathbf{L}_2^+ \bar{\mathbf{t}}_1 = \bar{\mathbf{t}}_1 \mathbf{L}_2^+ R, \quad R \bar{\mathbf{t}}_2 \mathbf{L}_1^- = \mathbf{L}_1^- \bar{\mathbf{t}}_2 R.$$

Proof: The pairing is in Corollary 4.1.8, and its inverse is in Proposition 4.2.3 (where we emphasised a slightly different point of view of the pairing as a map $A(R) \to \tilde{U}(R)^*$). We put these formulae into Example 7.2.6 with $H = \tilde{U}(R)$ and $A = A(R)$. The product (7.2) then gives the cross relations as stated.   ∎

This example is therefore a weak or generalised version of the quantum double of $A(R)$, with $\tilde{U}(R)$ in the role of its dual. This was one of the original motivations behind the construction in Example 7.2.6. Note, on the other hand, that one can verify the actions in these two examples directly and hence construct the double cross products without necessarily going through the pairing or skew-pairing. From this point of view, it is easy to see that the same constructions can be iterated.

Example 7.2.14 *There are double cross product bialgebras* $A^{(m)}(R) = A^{(m-1)}(R) \bowtie A(R)$, *where* $A^{(m-1)}(R)$ *is a bialgebra with* $m - 1$ *matrix generators* $\{\mathbf{t}(i)\}$ *and relations*

$$R\mathbf{t}(i)_1\mathbf{t}(j)_2 = \mathbf{t}(j)_2\mathbf{t}(i)_1 R \quad \forall i \leq j,$$

and where the mutual actions are

$$\mathbf{t}(i)_1 \triangleleft \mathbf{t}_2 = R^{-1}\mathbf{t}_1(i)R, \quad \mathbf{t}(i)_1 \triangleright \mathbf{t}_2 = R^{-1}\mathbf{t}_2 R, \quad i = 1, \ldots, m - 1.$$

Proof: This is by induction. We verify first that the actions as stated extend consistently as a matched pair according to the conditions in Definition 7.2.1. At the lowest level, this is

$$\mathbf{t}(i)_1 \triangleleft R_{23}\mathbf{t}_2\mathbf{t}_3 = R_{23}R_{12}^{-1}R_{13}^{-1}\mathbf{t}(i)_1 R_{13}R_{12}$$
$$= R_{13}^{-1}R_{12}^{-1}\mathbf{t}(i)_1 R_{12}R_{13}R_{23} = \mathbf{t}(i)_1 \triangleleft \mathbf{t}_3\mathbf{t}_2 R_{23},$$

where the second equality uses the QYBE twice. That $\triangleright$ is an action is exactly similar for the relations of the form in $A^{(m-1)}(R)$. Extending according to the matching conditions (7.7), we have

$$R_{12}(\mathbf{t}(i)_1\mathbf{t}(j)_2) \triangleleft \mathbf{t}_3 = R_{12}(\mathbf{t}(i)_1 \triangleleft(\mathbf{t}(j)_2 \triangleright \mathbf{t}_3))(\mathbf{t}(j)_2 \triangleleft \mathbf{t}_3)$$
$$= R_{12}(\mathbf{t}(i)_1 \triangleleft R_{23}^{-1}\mathbf{t}_3 R_{23})R_{23}^{-1}\mathbf{t}(j)_2 R_{23} = R_{12}R_{23}^{-1}R_{13}^{-1}\mathbf{t}(i)_1\mathbf{t}(j)_2 R_{13}R_{23}$$
$$= R_{13}^{-1}R_{23}^{-1}R_{12}\mathbf{t}(i)_1\mathbf{t}(j)_2 R_{13}R_{23} = R_{13}^{-1}R_{23}^{-1}\mathbf{t}(j)_2\mathbf{t}(i)_1 R_{12}R_{13}R_{23}$$
$$= R_{13}^{-1}R_{23}^{-1}\mathbf{t}(j)_2 R_{23}\mathbf{t}(i)_1 R_{13}R_{12} = (\mathbf{t}(j)_2 \triangleleft R_{13}^{-1}\mathbf{t}_2 R_{13})R_{13}^{-1}\mathbf{t}(i)_1 R_{13}R_{12}$$
$$= (\mathbf{t}(j)_2 \triangleleft(\mathbf{t}(i)_1 \triangleright \mathbf{t}_3))(\mathbf{t}(i)_1 \triangleleft \mathbf{t}_3)R_{12} = (\mathbf{t}(j)_2\mathbf{t}(i)_1)R_{12} \triangleleft \mathbf{t}_3,$$

where we used the definitions of $\triangleright, \triangleleft$, the QYBE and the relations of $A^{(m-1)}(R)$, and then the QYBE and definitions in reverse. In the same way $\triangleright$ extends according to (7.8). One should then go back and check the actions $\triangleleft, \triangleright$ at order two and consistency of their extension, etc. A formal proof proceeds by induction on the degrees. That (7.9) holds is automatic from the definitions of $\vartriangleleft, \vartriangleright$. That these are module coalgebras is also auto-matic; thus, $(\Delta \mathbf{t}(i)_1) \vartriangleleft (\mathbf{t}_2 \otimes \mathbf{t}_2) = R^{-1} \mathbf{t}(i)_1 R \otimes R^{-1} \mathbf{t}(i)_1 R = \Delta(\mathbf{t}(i)_1 \vartriangleleft \mathbf{t}_2)$ and similarly for $\vartriangleright$. Finally, the relations of the double cross product from Theorem 7.2.2 are $\mathbf{t}(i)_1 \mathbf{t}_2 = (\mathbf{t}(i)_1 \vartriangleright \mathbf{t}_2)(\mathbf{t}(i)_1 \vartriangleleft \mathbf{t}_2) = R^{-1} \mathbf{t}_2 R R^{-1} \mathbf{t}(i)_1 R$, so that, setting $\mathbf{t}(m) = \mathbf{t}$, we identify the double cross product as $A^{(m)}(R)$. We note also that, if we write,

$$R(i) = \begin{cases} R & \text{for } i \leq 0 \\ R_{21}^{-1} & \text{for } i > 0, \end{cases}$$

then

$$R(i - j)_{12} R(i)_{13} R(j)_{23} = R(j)_{23} R(i)_{13} R(i - j)_{12} \qquad (7.20)$$

and $A^{(m)}(R)$ is the algebra with m matrix generators and relations

$$R(i - j) \mathbf{t}(i)_1 \mathbf{t}(j)_2 = \mathbf{t}(j)_2 \mathbf{t}(i)_1 R(i - j), \qquad \forall i, j = 1, \dots, m, \qquad (7.21)$$

i.e. the bialgebra associated to a solution of a 'parametrised QYBE' as in Chapter 4.4, but in a discretised version. ∎

Finally, it is easy to dualise all these constructions in the usual way. We follow the same lines as in Chapter 6.2.

Exercise 7.2.15 *Let H, A be bialgebras with H a right A-comodule alge-bra and A a left H-comodule algebra by*

$$\alpha : A \to A \otimes H, \quad \beta : H \to A \otimes H,$$

which are compatible in the sense

$$(\Delta \otimes \mathrm{id}) \circ \alpha(a) = \sum ((\mathrm{id} \otimes \beta) \circ \alpha(a_{(1)})) (1 \otimes \alpha(a_{(2)})), \qquad (7.22)$$

$$(\mathrm{id} \otimes \Delta) \circ \beta(h) = \sum (\beta(h_{(1)}) \otimes 1) ((\alpha \otimes \mathrm{id}) \circ \beta(h_{(2)})), \qquad (7.23)$$

$$\alpha(a)\beta(h) = \beta(h)\alpha(a). \qquad (7.24)$$

Show that there is a double cross coproduct bialgebra $H \bowtie A$ with tensor product algebra structure and counit, and

$$\Delta(h \otimes a) = \sum h_{(1)} \otimes \alpha(a_{(1)}) \beta(h_{(2)}) \otimes a_{(2)}.$$

If H, A are Hopf algebras then so is $H \bowtie A$.

Solution: This is strictly dual to Theorem 7.2.2. As usual, one should write all constructions as maps and then reverse the arrows. Another way to do this is to assume that H, A in Theorem 7.2.2 are finite dimensional and to compute the conditions there in terms of their duals H^*, A^*. The conversion is via $\langle \alpha(b), h \otimes a \rangle = \langle b, h \vartriangleleft a \rangle$ and $\langle \beta(g), h \otimes a \rangle = \langle g, h \vartriangleright a \rangle$. Af-terwards, we rename H^*, A^* as A, H and verify the resulting construction directly, following the dual ideas to those in the proof of Proposition 7.2.2. ∎

We note that the conditions (7.22) and (7.23) say that each coaction almost respects the coalgebra on which it acts, up to a correction by the other coaction. Clearly, one can develop the theory for double cross coproducts in the same way as for double cross products. They come equipped with bialgebra or Hopf algebra surjections

$$H \xleftarrow{p} H \blacktriangleright\!\!\blacktriangleleft A \xrightarrow{q} A, \qquad (7.25)$$

into which factors the double cross coproduct decomposes by $(p \otimes q) \circ \Delta$. Likewise, one has an important class of examples associated to the notion of a *copairing* or *skew-copairing* just dual to Definition 1.4.3 or (7.13). Two bialgebras H, A are *skew-copaired* if there is an element $\sigma \in A \otimes H$ such that

$$(\Delta \otimes \mathrm{id})\sigma = \sigma_{13}\sigma_{23}, \qquad (\mathrm{id} \otimes \Delta)\sigma = \sigma_{13}\sigma_{12}. \qquad (7.26)$$

We have seen similar equations (2.1) in another context in Chapter 2.1.

Exercise 7.2.16 *Let H, A be two bialgebras and let $\sigma \in A \otimes H$ be an invertible skew-copairing. Show that there is a double cross coproduct bialgebra $H \blacktriangleright\!\!\blacktriangleleft_\sigma A$ built on $H \otimes A$ with*

$$\Delta(h \otimes a) = \sigma_{23}^{-1}\Delta_{H \otimes A}(h \otimes a)\sigma_{23}, \quad S(h \otimes a) = \sigma_{21}(Sh \otimes Sa)\sigma_{21}^{-1}$$

with tensor product counit and algebra structure. Show that it is also the twisting as in Theorem 2.3.4 by $\chi = \sigma_{23}^{-1}$ and that it is a Hopf $*$-algebra if H, A are and if $\sigma^{*\otimes *} = \sigma^{-1}$. Show that there is a second isomorphic form $A_\sigma \blacktriangleright\!\!\blacktriangleleft H$ built on $A \otimes H$ with

$$\Delta(a \otimes h) = \sigma_{32}\Delta_{A \otimes H}(a \otimes h)\sigma_{32}^{-1}, \quad S(a \otimes h) = \sigma^{-1}(Sa \otimes Sh)\sigma.$$

Solution: Again, this is strictly dual to Example 7.2.7. It follows as an example of Exercise 7.2.15 as soon as we know that we have a matched pair of coactions α, β. For $H \blacktriangleright\!\!\blacktriangleleft_\sigma A$, the mutual coactions we use are

$$\alpha(a) = \sigma^{-1}(a \otimes 1)\sigma, \quad \beta(h) = \sigma^{-1}(1 \otimes h)\sigma,$$

which are clearly coactions from (7.26) applied to σ^{-1}, σ. They are manifestly algebra homomorphisms so H, A are comodule algebras for each other. The matching conditions (7.22)–(7.24) which we need from Exercise 7.2.15 are

$$(\Delta \otimes \mathrm{id})(\sigma^{-1}(a \otimes 1)\sigma)$$
$$= \sigma_{23}^{-1}\left(\sigma_{13}^{-1}(a_{(1)} \otimes 1 \otimes 1)\sigma_{13}\right)\sigma_{23}\sigma_{23}^{-1}(1 \otimes a_{(2)} \otimes 1)\sigma_{23},$$

$$(\mathrm{id} \otimes \Delta)(\sigma^{-1}(1 \otimes h)\sigma)$$
$$= \sigma_{12}^{-1}(1 \otimes h_{(1)} \otimes 1)\sigma_{12}\sigma_{12}^{-1}\left(\sigma_{23}^{-1}(1 \otimes 1 \otimes h_{(2)})\sigma_{23}\right)\sigma_{12},$$
$$\sigma^{-1}(a \otimes 1)\sigma\sigma^{-1}(1 \otimes h)\sigma = \sigma^{-1}(1 \otimes h)\sigma\sigma^{-1}(a \otimes 1)\sigma.$$

The first two follow at once from (7.26), while the third is automatic, so we have a double cross coproduct bialgebra. The numbers in these expressions refer as usual to the position in a tensor product. Alternatively, one may obtain the bialgebra as an example of twisting via Theorem 2.3.4 if one checks the cocycle conditions of $\chi = \sigma_{23}^{-1} \in H \otimes A \otimes H \otimes A$. Writing out the two sides of (2.15) we need

$$\sigma_{23}^{-1}(\Delta_{H \otimes A} \otimes \mathrm{id})\sigma_{23}^{-1} = \sigma_{45}^{-1}(\mathrm{id} \otimes \Delta_{H \otimes A})\sigma_{23}^{-1}.$$

Using (7.26), this becomes $\sigma_{23}^{-1}\sigma_{45}^{-1}\sigma_{25}^{-1} = \sigma_{45}^{-1}\sigma_{23}^{-1}\sigma_{25}^{-1}$, which holds automatically. This also gives the antipode when this is present for H, A. The element U in Theorem 2.3.4 is $(1 \otimes \sigma^{-(1)})(S\sigma^{-(2)} \otimes 1) = (S \otimes \mathrm{id})\sigma_{21}^{-1} = \sigma_{21}$. If H, A are $*$-algebras, we take the tensor product $*$-algebra structure. It is immediate from the form of the coproduct that we obtain a Hopf $*$-algebra if σ is unitary in the sense stated. These formulae are for the right handed form. The equivalent left handed form is $A_\sigma \bowtie H = A \bowtie_{\sigma_{21}^{-1}} H$, and is given by the same construction as above with A, H skew-copaired by σ_{21}^{-1}. ∎

In the same way, all the other results above may be dualised for these double cross coproducts. We see that the dual of the quantum double $D(H)^* = H^{\mathrm{cop}} \bowtie_\sigma H^*$ has the above structure with $\sigma = f^a \otimes e_a$. In this case, of course, there is also a dual quasitriangular structure (7.5).

7.3 Complexification of quantum groups

Until now, the bialgebras and Hopf algebras H, A above have been general. In this and the next section, we concentrate on the special features of the quantum double and its generalisations above in the case when H is quasitriangular or A is dual quasitriangular. We develop two specific themes to which this special case is significant. The one in the present section is to the problem of 'complexifying' quantum groups. An important example is the q-Lorentz group as a complexification of q-deformed su_2, with the quantum double as its quantum Iwasawa decomposition. A second theme is the problem of writing the quantum double as a semidirect or cross product and thereby giving it an interpretation as a quantum algebra of observables. This also connects with the complexification point of view, and is the topic of Section 7.4.

We have remarked already that a dual quasitriangular structure $\mathcal{R}$: $A \otimes A \to k$ on a bialgebra A is, in particular, a skew-pairing of the bialgebra with itself. Our Example 7.2.12 was already of this type. We now study this case in more detail. We know from Example 7.2.6 or Example 7.2.7 that we have $A \bowtie_{\mathcal{R}} A$ built on $A \otimes A$ with product

$$(a \otimes b)(c \otimes d) = \sum a c_{(2)} \otimes b_{(2)} d \mathcal{R}^{-1}(b_{(1)} \otimes c_{(1)}) \mathcal{R}(b_{(3)} \otimes c_{(3)}) \qquad (7.27)$$

and tensor product unit and coalgebra.

Proposition 7.3.1 *Let A be a dual quasitriangular bialgebra or Hopf algebra. Then the bialgebra or Hopf algebra $A \bowtie_{\mathcal{R}} A$ defined with skew-pairing by $\mathcal{R}$ has two dual quasitriangular structures,*

$$\mathcal{R}_L(a \otimes b \otimes c \otimes d)$$
$$= \sum \mathcal{R}^{-1}(d_{(1)} \otimes a_{(1)}) \mathcal{R}(a_{(2)} \otimes c_{(1)}) \mathcal{R}(b_{(1)} \otimes d_{(2)}) \mathcal{R}(b_{(2)} \otimes c_{(2)}),$$
$$\mathcal{R}_D(a \otimes b \otimes c \otimes d)$$
$$= \sum \mathcal{R}^{-1}(d_{(1)} \otimes a_{(1)}) \mathcal{R}^{-1}(c_{(1)} \otimes a_{(2)}) \mathcal{R}(b_{(1)} \otimes d_{(2)}) \mathcal{R}(b_{(2)} \otimes c_{(2)}),$$

for all $a, b, c, d \in A$. Here $\mathcal{R}_L$ restricts to $\mathcal{R}$ on each of the two factors of A, while $\mathcal{R}_D$ restricts to $\mathcal{R}_{21}^{-1}$ and $\mathcal{R}$, respectively.

Proof: The bialgebra structure is from Example 7.2.6 or Example 7.2.7 with the skew-pairing of A with itself provided by $\mathcal{R}$. So the actions and 2-cocycle (from the point of view in Proposition 7.1.5) are

$$b \triangleleft a = b_{(2)} \mathcal{R}^{-1}(b_{(1)} \otimes a_{(1)}) \mathcal{R}(b_{(3)} \otimes a_{(2)}),$$

$$b \triangleright a = a_{(2)} \mathcal{R}^{-1}(b_{(1)} \otimes a_{(1)}) \mathcal{R}(b_{(2)} \otimes a_{(3)}),$$

$$\chi(a \otimes b \otimes c \otimes d) = \epsilon(a) \mathcal{R}^{-1}(b \otimes c) \epsilon(d).$$

From the dual-twisting point of view, we start with $A \otimes A$ and twist its multiplication. But $A \otimes A$ has the tensor product dual quasitriangular structure, so from (2.24) we have a dual quasitriangular structure

$$\mathcal{R}_L(a \otimes b \otimes c \otimes d) = \chi(c_{(1)} \otimes d_{(1)} \otimes a_{(1)} \otimes b_{(1)}) \mathcal{R}(a_{(2)} \otimes c_{(2)})$$
$$\mathcal{R}(b_{(2)} \otimes d_{(2)}) \chi^{-1}(a_{(3)} \otimes b_{(3)} \otimes c_{(3)} \otimes d_{(3)}),$$

which computes as stated. Likewise, we could take the first copy of A in $A \otimes A$ as dual quasitriangular with $\mathcal{R}_{21}^{-1}$, so that this time

$$\mathcal{R}_D(a \otimes b \otimes c \otimes d) = \chi(c_{(1)} \otimes d_{(1)} \otimes a_{(1)} \otimes b_{(1)}) \mathcal{R}^{-1}(c_{(2)} \otimes a_{(2)})$$
$$\mathcal{R}(b_{(2)} \otimes d_{(2)}) \chi^{-1}(a_{(3)} \otimes b_{(3)} \otimes c_{(3)} \otimes d_{(3)}),$$

which computes as stated. There are altogether four natural dual quasitriangular structures on $A \otimes A$ according to whether we take $\mathcal{R}$ or $\mathcal{R}_{21}^{-1}$

for each, but the other two just give the inverse-transposes of $\mathcal{R}_L$ and $\mathcal{R}_D$. ∎

Next, if A is a Hopf $*$-algebra with $\mathcal{R}$ antireal then we already know from Proposition 7.2.8 that $(a \otimes b)^* = (1 \otimes b^*)(a^* \otimes 1)$ gives us a Hopf $*$-algebra structure on $A \bowtie_{\mathcal{R}} A$ with the two factors as sub-Hopf $*$-algebras. One can easily show by direct computation using the axioms (2.6) and (2.7) of a dual quasitriangular structure that $\mathcal{R}_L$ and $\mathcal{R}_D$ are both antireal in this case. This will be obvious also from the enveloping algebra point of view below. More special to the present setting is that there is also a natural $*$-structure if the dual quasitriangular structure of A is real.

Proposition 7.3.2 *In the situation of Proposition 7.3.1, if A is a real dual quasitriangular Hopf $*$-algebra, then*

$$(a \otimes b)^* = b^* \otimes a^*$$

makes $A \bowtie_{\mathcal{R}} A$ into a Hopf $*$-algebra. In this case we have $\mathcal{R}_D$ antireal and $\mathcal{R}_L$ real.

Proof: We verify first that we have a $*$-algebra for the product (7.27). Thus,

$$((a \otimes b)(c \otimes d))^* = d^* b^*_{(2)} \otimes c^*_{(2)} a^* \mathcal{R}^{-1}(c^*_{(1)} \otimes b^*_{(1)}) \mathcal{R}(c^*_{(3)} \otimes b^*_{(3)})$$
$$= (d^* \otimes c^*)(b^* \otimes a^*)$$

as required. Next, it is clear that $(* \otimes *) \circ \Delta = \Delta \circ *$ since the coproduct is the tensor product one. Finally, $* \circ S(a \otimes b) = (Sa \otimes 1)^*(1 \otimes Sb)^* = (1 \otimes * \circ Sa)(* \circ Sb \otimes 1) = (1 \otimes S^{-1}(a^*))(S^{-1}(b^*) \otimes 1) = S^{-1} \circ *(a \otimes b)$, so the conditions in Definition 1.7.5 are satisfied. Next,

$$\mathcal{R}_L((c \otimes d)^* \otimes (a \otimes b)^*) = \mathcal{R}_L(d^* \otimes c^* \otimes b^* \otimes a^*)$$
$$= \mathcal{R}^{-1}(a^*_{(1)} \otimes d^*_{(1)}) \mathcal{R}(d^*_{(2)} \otimes b^*_{(1)}) \mathcal{R}(c^*_{(1)} \otimes a^*_{(2)}) \mathcal{R}(c^*_{(2)} \otimes b^*_{(2)}),$$

which coincides with $\overline{\mathcal{R}_L(a \otimes b \otimes c \otimes d)}$ given that $\mathcal{R}$ is real. Likewise,

$$\mathcal{R}_D^{-1}((a \otimes b)^* \otimes (c \otimes d)^*) = \mathcal{R}_D^{-1}(b^* \otimes a^* \otimes d^* \otimes c^*)$$
$$= \mathcal{R}^{-1}(a^*_{(1)} \otimes d^*_{(1)}) \mathcal{R}(a^*_{(2)} \otimes c^*_{(1)}) \mathcal{R}(d^*_{(2)} \otimes b^*_{(1)}) \mathcal{R}(c^*_{(2)} \otimes b^*_{(2)}),$$

which coincides with $\overline{\mathcal{R}_D(a \otimes b \otimes c \otimes d)}$ for $\mathcal{R}$ real. This is a direct proof. It is also clear from the dual-twisting point of view in equation (2.26): $A \otimes A$ has the stated $*$-structure (as well as its more usual tensor product $*$-structure) with respect to which the twisting cocycle χ in the proof of Proposition 7.3.1 obeys the required reality condition. We use that $\mathcal{R}$ is real and invariant under $S \otimes S$. It is also clear that $U \circ S_{A \otimes A} = U$ in this case, so that W in (2.26) is trivial. Hence the same $*$ also works for

$A\bowtie_{\mathcal{R}}A$. We immediately conclude the reality and antireality of $\mathcal{R}_L, \mathcal{R}_D$ as well from this point of view. ∎

Matrix examples are provided whenever R is of real type in the sense of Definition 4.2.15 and sufficiently nice that one can quotient $A(R)$ to a dual quasitriangular Hopf algebra A.

Example 7.3.3 *Let $SU_q(2)$ be as in Proposition 4.2.4 with the R-matrix there and with the $*$-structure in Example 4.2.16. Then $SU_q(2)\bowtie SU_q(2)$, with generators $\mathbf{s}, \mathbf{t}$ as in Example 7.2.12, becomes a Hopf $*$-algebra with $t^i{}_j{}^* = Ss^j{}_i$, $s^i{}_j{}^* = St^j{}_i$ and has two dual quasitriangular structures*

$$\mathcal{R}_L(\mathbf{t}_1 \otimes \mathbf{s}_2 \otimes \mathbf{t}_3 \otimes \mathbf{s}_4) = R_{41}^{-1}R_{13}R_{24}R_{23},$$

$$\mathcal{R}_D(\mathbf{t}_1 \otimes \mathbf{s}_2 \otimes \mathbf{t}_3 \otimes \mathbf{s}_4) = R_{41}^{-1}R_{31}^{-1}R_{24}R_{23}.$$

Proof: We know from Example 4.2.16 that $SU_q(2)$ has a real dual quasitriangular Hopf $*$-algebra structure, and we use this now in the preceding proposition. The formulae for $\mathcal{R}_L, \mathcal{R}_D$ on the generators follow at once from Proposition 7.3.1 with the matrix comultiplications for $\mathbf{s}, \mathbf{t}$, and we use the corresponding compact notations. ∎

The same formulae apply in general for suitable R of real type. Alternatively, by working with $\mathbf{t}^\dagger = \mathbf{s}^{-1}$ and appropriate relations and $*$-structure $t^j{}_i{}^* = t^{\dagger i}{}_j$, one can eliminate the need for the antipode and work entirely at the bialgebra level. We will return to this example as a spinorial form of the q-Lorentz group of function algebra type in Chapter 10.

As usual, we also have the analogous theory for quasitriangular Hopf algebras such as the quantum enveloping algebras of Chapter 3. If H is a quasitriangular Hopf algebra, we use $\mathcal{R}$ itself as a skew-copairing in Exercise 7.2.16 to obtain a Hopf algebra $H\bowtie_{\mathcal{R}}H$ with tensor product algebra and counit, and

$$\Delta(h \otimes g) = \mathcal{R}_{23}^{-1}\Delta_{H \otimes H}(h \otimes g)\mathcal{R}_{23},$$
$$S(h \otimes g) = \mathcal{R}_{21}(Sh \otimes Sg)\mathcal{R}_{21}^{-1}. \tag{7.28}$$

We also know from Exercise 7.2.16 that in the Hopf $*$-algebra case, with $\mathcal{R}$ antireal, we have $(h \otimes g)^* = h^* \otimes g^*$ as a Hopf $*$-algebra structure. Both $\mathcal{R}_L$ and $\mathcal{R}_D$ in the next propositions are manifestly antireal with respect to it in this case.

Proposition 7.3.4 *Let H be a quasitriangular bialgebra. Then $H\bowtie_{\mathcal{R}}H$ with the skew-copairing given by $\mathcal{R}$ has two quasitriangular structures*

$$\mathcal{R}_L = \mathcal{R}_{41}^{-1}\mathcal{R}_{13}\mathcal{R}_{24}\mathcal{R}_{23}, \quad \mathcal{R}_D = \mathcal{R}_{41}^{-1}\mathcal{R}_{31}^{-1}\mathcal{R}_{24}\mathcal{R}_{23}.$$

and has a Hopf ∗-algebra structure

$$(h \otimes g)^* = \mathcal{R}_{21}(g^* \otimes h^*)\mathcal{R}_{21}^{-1}$$

when H is real quasitriangular. In this case $\mathcal{R}_L$ is real and $\mathcal{R}_D$ is antireal.

Proof: This is dual to Propositions 7.3.1 and 7.3.2. This time we know that $H \bowtie_{\mathcal{R}} H$ is the twisting of $H \otimes H$ by the 2-cocycle $\chi = \mathcal{R}_{23}^{-1}$ according to Theorem 2.3.4. Hence the tensor product quasitriangular structure $\mathcal{R}_{13}\mathcal{R}_{24}$ twists, according to that theorem, to $\mathcal{R}_{41}^{-1}(\mathcal{R}_{13}\mathcal{R}_{24})\mathcal{R}_{23}$, which is $\mathcal{R}_L$, while another quasitriangular structure on $H \otimes H$ is $\mathcal{R}_{31}^{-1}\mathcal{R}_{24}$, which twists to $\mathcal{R}_D$. The numbers here refer to the positions in $H^{\otimes 4}$.

Likewise, we initially take the ∗-structure $(h \otimes g)^* = g^* \otimes h^*$ on $H \otimes H$ and note that when $\mathcal{R}$ is real we have $\chi^{*\otimes*} = \mathcal{R}_{23}^{-1*\otimes*} = \mathcal{R}_{41}^{-1} = \chi_{21} = (S \otimes S)\chi_{21}$, which is the reality condition required in Proposition 2.3.7. Moreover, $S_{H \otimes H}U = U$, so that this ∗ on $H \otimes H$ conjugates to ∗ on $H \bowtie_{\mathcal{R}} H$ by just the same factor as for the antipode in (7.28). This is why the twisting of ∗ has just the same form as the twisting of S in this example. We also conclude the reality of $\mathcal{R}_L$ and the antireality of $\mathcal{R}_D$ from Proposition 2.3.7.

A direct proof of this proposition is also possible. The quasitriangularity follows routinely from that of H. Also, the stated ∗ is clearly an antilinear antialgebra map. Further,

$$
\begin{aligned}
(*\otimes*) &\circ \Delta(h \otimes g) \\
&= \mathcal{R}_{21}\mathcal{R}_{43}\left(\tau_{12}\tau_{34}((\mathcal{R}_{23}^{-1}(\Delta_{H \otimes H}(h \otimes g))\mathcal{R}_{23})^{*\otimes^4})\right)\mathcal{R}_{43}^{-1}\mathcal{R}_{21}^{-1} \\
&= \mathcal{R}_{21}\mathcal{R}_{43}\mathcal{R}_{41}(\Delta_{H \otimes H}(g^* \otimes h^*))\mathcal{R}_{41}^{-1}\mathcal{R}_{43}^{-1}\mathcal{R}_{21}^{-1} \\
&= \mathcal{R}_{23}^{-1}(\Delta_{H \otimes H}(\mathcal{R}_{21}(g^* \otimes h^*)\mathcal{R}_{21}^{-1}))\mathcal{R}_{23} = \Delta((h \otimes g)^*)
\end{aligned}
$$

as required, where we used $\tau_{12}\tau_{34} \circ (*\otimes*\otimes*\otimes*)(\mathcal{R}_{23}) = \tau_{12}\tau_{34}(\mathcal{R}_{32}) = \mathcal{R}_{41}$ for the second equality, and $\Delta_{H \otimes H}\mathcal{R}_{21} = \mathcal{R}_{23}\mathcal{R}_{21}\mathcal{R}_{43}\mathcal{R}_{41}$ from (2.2) for the third. Here τ denotes transposition. Also, $* \circ S(h \otimes g) = \mathcal{R}_{21}(\tau \circ (*\otimes*)(\mathcal{R}_{21}(Sh \otimes Sg)\mathcal{R}_{21}^{-1}))\mathcal{R}_{21}^{-1} = (Sh)^* \otimes (Sg)^*$ as $\mathcal{R}$ is real. Hence we have $(* \circ S)^2 = \text{id}$, as required for a Hopf ∗-algebra. Finally,

$$
\begin{aligned}
\mathcal{R}_D^{*\otimes*} &= \mathcal{R}_{21}\mathcal{R}_{43}(\tau_{12}\tau_{34}(\mathcal{R}_{32}\mathcal{R}_{42}\mathcal{R}_{13}^{-1}\mathcal{R}_{14}^{-1}))\mathcal{R}_{43}^{-1}\mathcal{R}_{21}^{-1} \\
&= \mathcal{R}_{21}\mathcal{R}_{43}\mathcal{R}_{41}\mathcal{R}_{31}\mathcal{R}_{24}^{-1}\mathcal{R}_{23}^{-1}\mathcal{R}_{43}^{-1}\mathcal{R}_{21}^{-1} = \mathcal{R}_{23}^{-1}\mathcal{R}_{24}^{-1}\mathcal{R}_{31}\mathcal{R}_{41} = \mathcal{R}_D^{-1},
\end{aligned}
$$

where we used the QYBE for $\mathcal{R}$ in Lemma 2.1.4 to move $\mathcal{R}_{43}^{-1}$ to the left until it cancelled, and similarly for $\mathcal{R}_{21}^{-1}$ until it cancelled. The analogous computation for $\mathcal{R}_L$ gives $\tau_{13}\tau_{24}(\mathcal{R}_L)$, which means that $\mathcal{R}_L$ is real. ∎

We also know from the general theory in Section 7.2 that there is a parallel left handed construction $H_{\mathcal{R}} \bowtie H$ given by a left–right reversal.

The corresponding formulae are

$$\Delta(h \otimes g) = \mathcal{R}_{32}\Delta_{H \otimes H}(h \otimes g)\mathcal{R}_{32}^{-1},$$

$$\mathcal{R}_L = \mathcal{R}_{14}\mathcal{R}_{13}\mathcal{R}_{24}\mathcal{R}_{32}^{-1}, \quad \mathcal{R}_D = \mathcal{R}_{14}\mathcal{R}_{13}\mathcal{R}_{42}^{-1}\mathcal{R}_{32}^{-1}, \tag{7.29}$$

$$S(h \otimes g) = \mathcal{R}^{-1}(Sh \otimes Sg)\mathcal{R}, \quad (h \otimes g)^* = \mathcal{R}^{-1}(g^* \otimes h^*)\mathcal{R},$$

which are the formulae above with $\mathcal{R}, \mathcal{R}_L$ and $\mathcal{R}_D$ replaced by their inverse-transposes and with the roles of the two factors interchanged. Both forms are useful as

$$H^* {\bowtie}_{\mathcal{R}} H^* = (H {\blacktriangleright\!\!\triangleleft}_{\mathcal{R}} H)^*, \quad H^* {\bowtie}_{\mathcal{R}} H^* = (H_{\mathcal{R}} {\triangleright\!\!\blacktriangleleft} H)^*,$$

where the second identification involves a left–right transposition of the factors and $(\ )^* = (\ )^{*\mathrm{op}/\mathrm{op}}$ is the categorical dual as in Corollary 6.2.6. It corresponds in diagrammatic terms to reflecting diagrammatic proofs in a mirror combined with the usual reversal of arrows.

Theorem 7.3.5 *If H is a finite-dimensional quasitriangular Hopf algebra, there are Hopf algebra maps*

$$H^{*\mathrm{op}} \xrightarrow{i} H_{\mathcal{R}} {\blacktriangleright\!\!\triangleleft} H \xleftarrow{j} H, \quad i(a) = (\mathrm{id} \otimes \mathrm{id} \otimes a)(\mathcal{R}_{13}\mathcal{R}_{32}^{-1}), \quad j = \Delta$$

defining a Hopf algebra homomorphism $\cdot \circ (i \otimes j) : D(H) \to H_{\mathcal{R}} {\blacktriangleright\!\!\triangleleft} H$. It maps the canonical quasitriangular structure of $D(H)$ to $\mathcal{R}_D$, and is an isomorphism if H is factorisable. Moreover, in the Hopf $*$-algebra cases, these are all $*$-algebra maps.

Proof: The coproduct j is always an algebra map to the tensor product coalgebra. Meanwhile, using (2.1) we have $i(ba) = (\mathrm{id} \otimes \mathrm{id} \otimes b \otimes a) \circ (\mathrm{id} \otimes \mathrm{id} \otimes \Delta)(\mathcal{R}_{13}\mathcal{R}_{32}^{-1}) = (\mathrm{id} \otimes \mathrm{id} \otimes b \otimes a)(\mathcal{R}_{14}\mathcal{R}_{42}^{-1}\mathcal{R}_{13}\mathcal{R}_{32}^{-1}) = i(a)i(b)$. These are coalgebra maps since

$$\Delta \circ j(h) = \mathcal{R}_{32}(h_{(1)} \otimes h_{(3)} \otimes h_{(2)} \otimes h_{(4)})\mathcal{R}_{32}^{-1} = (j \otimes j)(\Delta h),$$

$$\Delta \circ i(a) = \langle a, \mathcal{R}_{32} \left(\tau_{23} \circ (\Delta \otimes \Delta)(\mathcal{R}_{15}\mathcal{R}_{52}^{-1}) \right) \mathcal{R}_{32}^{-1} \rangle$$

$$= \langle \mathcal{R}_{32}\mathcal{R}_{15}\mathcal{R}_{35}\mathcal{R}_{52}^{-1}\mathcal{R}_{54}^{-1}\mathcal{R}_{32}^{-1}, a \rangle = \langle \mathcal{R}_{15}\mathcal{R}_{52}^{-1}\mathcal{R}_{35}\mathcal{R}_{32}\mathcal{R}_{54}^{-1}\mathcal{R}_{32}^{-1}, a \rangle$$

$$= i(a_{(1)}) \otimes i(a_{(2)}),$$

where the evaluation of a is in the fifth position. For j we used (2.2), while for i we used (2.1) and then the QYBE for $\mathcal{R}$ as in Lemma 2.1.4 to move $\mathcal{R}_{32}$ to the right and thereby cancel it. One can also do these computations in a completely explicit manner with $i(a) = \mathcal{R}^{(1)} \otimes \mathcal{R}^{-(2)} \langle \mathcal{R}^{(2)}\mathcal{R}^{-(1)}, a \rangle$. Next, $\cdot \circ (i \otimes j)$ is an algebra map from the quantum double in view of the identity

$$i(a_{(2)})j(h_{(2)})\langle Sh_{(1)}, a_{(1)}\rangle\langle h_{(3)}, a_{(3)}\rangle$$
$$= \mathcal{R}^{(1)}h_{(2)} \otimes \mathcal{R}^{-(2)}h_{(3)}\langle Sh_{(1)}, a_{(1)}\rangle\langle \mathcal{R}^{(2)}\mathcal{R}^{-(1)}, a_{(2)}\rangle\langle h_{(4)}, a_{(3)}\rangle$$
$$= \mathcal{R}^{(1)}h_{(2)} \otimes \mathcal{R}^{-(2)}h_{(3)}\langle (Sh_{(1)})\mathcal{R}^{(2)}\mathcal{R}^{-(1)}h_{(4)}, a\rangle = j(h)i(a),$$

where we use (2.2) to move $h_{(4)}$ to the left until it cancels with $Sh_{(1)}$. Since $\cdot \circ (i \otimes j)$ is manifestly a coalgebra map as a composition of coalgebra maps, we conclude that we have a Hopf algebra homomorphism as required.

If H is factorisable in the sense of Definition 2.1.12, then we have to show only that $\theta = \cdot \circ (i \otimes j)$ is a linear isomorphism to conclude from Theorem 7.2.3 that it is an isomorphism of Hopf algebras. Indeed, if the map $Q(a) = (a \otimes \mathrm{id})(\mathcal{R}_{21}\mathcal{R})$ is invertible then we define

$$\theta^{-1}(h \otimes g) = (Q^{-1}(hSg_{(1)}))_{(1)}\langle \mathcal{R}^{(1)}, (Q^{-1}(hSg_{(1)}))_{(2)}\rangle \otimes \mathcal{R}^{(2)}g_{(2)} \quad (7.30)$$

and verify that $\cdot \circ (i \otimes j)$ is injective in view of

$$\theta^{-1} \circ \cdot \circ (i \otimes j)(a \otimes h)$$
$$= (Q^{-1}(\mathcal{R}^{(1)}h_{(1)}S(\mathcal{R}^{-(2)}{}_{(1)}h_{(2)})))_{(1)}\langle \mathcal{R}^{(2)}\mathcal{R}^{-(1)}, a\rangle$$
$$\langle \mathcal{R}'^{(1)}, (Q^{-1}(\mathcal{R}^{(1)}h_{(1)}S(\mathcal{R}^{-(2)}{}_{(1)}h_{(2)})))_{(2)}\rangle \otimes \mathcal{R}'^{(2)}\mathcal{R}^{-(2)}{}_{(2)}h_{(3)}$$
$$= (Q^{-1}(\mathcal{R}^{(1)}S\mathcal{R}^{-(2)}))_{(1)}\langle \mathcal{R}^{(2)}\mathcal{R}^{-(1)}\mathcal{R}'^{-(1)}, a\rangle$$
$$\langle \mathcal{R}'^{(1)}, (Q^{-1}(\mathcal{R}^{(1)}S\mathcal{R}^{-(2)}))_{(2)}\rangle \otimes \mathcal{R}'^{(2)}\mathcal{R}'^{-(2)}h$$
$$= a_{(1)}\langle \mathcal{R}'^{-(1)}, a_{(3)}\rangle\langle \mathcal{R}'^{(1)}, a_{(2)}\rangle \otimes \mathcal{R}'^{(2)}\mathcal{R}'^{-(2)}h = a \otimes h,$$

where the first equality puts in the definitions (with primes to distinguish the copies of $\mathcal{R}$), the second uses (2.1) and also cancels $h_{(1)}Sh_{(2)}$, and the third uses Lemma 2.1.2 to recognise $Q^{-1}(Q(a_{(1)}))$. One may verify the inverse on the other side in a similar way if desired.

It is also instructive to map over the quasitriangular structure from Theorem 7.1.1 as $\cdot \circ (i \otimes j)(f^a \otimes 1 \otimes 1 \otimes e_a) = \mathcal{R}^{(1)} \otimes \mathcal{R}^{-(2)}\langle \mathcal{R}^{(2)}\mathcal{R}^{-(1)}, f^a\rangle \otimes e_{a(1)} \otimes e_{a(2)} = (\mathrm{id} \otimes \mathrm{id} \otimes \Delta)(\mathcal{R}_{13}\mathcal{R}_{32}^{-1}) = \mathcal{R}_D$.

Finally, we equip $H^{*\mathrm{op}}$ with the $*$-algebra structure $a^{*\mathrm{op}} = (S^2 a)^*$ in terms of the structure of H^*, as in Proposition 7.1.4. Then in the real quasitriangular case we have

$$(i(a))^* = \mathcal{R}^{-1}(\mathcal{R}^{-(2)*} \otimes \mathcal{R}^{(1)*})\mathcal{R}\langle \mathcal{R}^{-(1)*}\mathcal{R}^{(2)*}, (Sa)^*\rangle$$
$$= \langle \mathcal{R}_{12}^{-1}\mathcal{R}_{13}^{-1}\mathcal{R}_{32}\mathcal{R}_{12}, (Sa)^*\rangle = \langle \mathcal{R}_{32}\mathcal{R}_{13}^{-1}, (Sa)^*\rangle$$
$$= \langle \mathcal{R}_{32}\mathcal{R}_{13}^{-1}, S((S^2 a)^*)\rangle = i((S^2 a)^*)$$

as required. Here evaluation of $(Sa)^*$ is in the third position. The computation for the antireal case is easier, with the same trick for the last step to reach $i((S^2 a)^*)$. Meanwhile, $j(h)^* = \mathcal{R}^{-1}(\tau \circ (\Delta h)^{*\otimes *})\mathcal{R} = j(h^*)$ using (2.2) and that Δ respects $*$. Hence, in either case, we have i, j and $\cdot \circ (i \otimes j)$ as $*$-algebra maps. ∎

Similar results hold for the right handed form (7.28), and in a dual way for (7.27) and its left handed version. The natural maps are

$$H \xrightarrow{i} H \bowtie_{\mathcal{R}} H \xleftarrow{j} H^{*\mathrm{op}}, \quad i = \Delta, \quad j(a) = (\mathrm{id} \otimes \mathrm{id} \otimes a)(\mathcal{R}_{31}^{-1}\mathcal{R}_{23}),$$

$$H^{\mathrm{cop}} \xrightarrow{p} A_{\mathcal{R}} \bowtie A \xrightarrow{q} A, \quad p(a \otimes b) = (\mathrm{id} \otimes a \otimes b)(\mathcal{R}_{21}\mathcal{R}_{13}^{-1}), \quad q = \cdot, \quad (7.31)$$

$$A \xleftarrow{p} A \bowtie_{\mathcal{R}} A \xrightarrow{q} H^{\mathrm{cop}}, \quad p = \cdot, \quad q(a \otimes b) = (\mathrm{id} \otimes a \otimes b)(\mathcal{R}_{12}^{-1}\mathcal{R}_{31}),$$

where A is dual to H. Thus, in the factorisable case, we have the four versions

$$H_{\mathcal{R}} \bowtie H \cong D(H) = H^{*\mathrm{op}} \bowtie H, \qquad H \bowtie_{\mathcal{R}} H \cong H \bowtie H^{*\mathrm{op}},$$

$$H^{\mathrm{cop}} \bowtie A \cong A_{\mathcal{R}} \bowtie A, \qquad A \bowtie H^{\mathrm{cop}} \cong A \bowtie_{\mathcal{R}} A. \tag{7.32}$$

The proofs are strictly analogous to the proof of Theorem 7.3.5, by replacing $\mathcal{R}$ by $\mathcal{R}_{21}^{-1}$ for the right handed cases and by dualising for the lower line. For example, the inversion formula for the fourth isomorphism is

$$((p \otimes q) \circ \Delta)^{-1} (a \otimes h)$$
$$= \sum a_{(1)} S Q^{-1}(\mathcal{R}^{(1)}h)_{(1)} \otimes Q^{-1}(\mathcal{R}^{(1)}h)_{(2)} \langle \mathcal{R}^{(2)}, a_{(2)} \rangle, \tag{7.33}$$

using the inverse of the quantum Killing form map Q, as in (7.30).

The isomorphisms here have been stated for the finite-dimensional theory. This applies, for example, to the factorisable quasitriangular Hopf algebras $u_q(g)$ reduced at roots of unity as in Chapter 3.4. The theory is also useful in the infinite-dimensional cases such as $U_q(g)$ at generic q, regarded over formal power series in a deformation parameter. We need factorisability in a strict sense in which the map Q is invertible, which requires some form of completion. By redefining suitable generators, one can typically establish the isomorphisms at the algebraic level as well.

Example 7.3.6 *For $U_q(g)$ and its function algebra G_q in FRT form, we have homomorphisms $G_q^{\mathrm{op}} \xrightarrow{i} U_q(g)_{\mathcal{R}} \bowtie U_q(g) \xleftarrow{j} U_q(g)$ given by*

$$i(\bar{\mathbf{t}}) = 1^+ \otimes 1^-, \quad j(1^\pm) = 1^\pm \otimes 1^\pm,$$

providing a homomorphism from the quantum double of $U_q(g)$, in the from $G_q^{\mathrm{op}} \bowtie U_q(g)$ from Example 7.2.13, to $U_q(g)_{\mathcal{R}} \bowtie U_q(g)$. In a suitable setting, this is an isomorphism.

Proof: Here $\bar{\mathbf{t}}$ denotes the matrix generator of $A(R)^{\mathrm{op}}$, and $1^\pm$ denote those of $U_q(g)$, as in Chapter 4.3. Thus we take the quantum double in a form like that in Example 7.2.13 for the matrix relations, but at the Hopf algebra level. Putting in the matrix comultiplication, we have $i(\bar{\mathbf{t}}) = \mathcal{R}^{(1)}\langle \bar{\mathbf{t}}, \mathcal{R}^{(2)}\rangle \otimes \mathcal{R}^{-(2)}\langle \bar{\mathbf{t}}, \mathcal{R}^{-(1)}\rangle = \mathbf{l}^{+} \otimes \mathbf{l}^{-}$ from Corollary 4.1.9. Note that we identify $A(R)$ and $A(R)^{\mathrm{op}}$ as linear spaces by $\mathbf{t} = \bar{\mathbf{t}}$. It should be clear that the same computation holds for any matrix quantum group of this type, and that it provides an isomorphism $\theta = \cdot \circ (i \otimes j)$ when the quantum enveloping algebra is factorisable. Indeed, the inversion formula (7.30) tells us that $\theta(\bar{\mathbf{t}} \otimes S\mathbf{l}^{-}) = 1 \otimes 1$ and $\theta(\bar{t}^{a}{}_{j} \otimes S^{-1}l^{+i}{}_{a}) = S^{-1}l^{i}{}_{j} \otimes 1$, where $\mathbf{l} = \mathbf{l}^{+}S\mathbf{l}^{-}$ are the generators in Section 4.3. So we have an isomorphism of algebras if we identify generators in this way. ∎

Example 7.3.7 *In the matrix quantum group setting as above, we have homomorphisms $G_{q} \xleftarrow{p} G_{q}\bowtie_{\mathcal{R}}G_{q} \xrightarrow{q} U_{q}(g)^{\mathrm{cop}}$ given by*

$$p = \cdot, \quad q(\mathbf{s} \otimes 1) = S\mathbf{l}^{+}, \quad q(1 \otimes \mathbf{t}) = S\mathbf{l}^{-},$$

giving a homomorphism from $G_{q}\bowtie_{\mathcal{R}}G_{q}$ to the dual of the quantum double in the form $G_{q}\bowtie U_{q}(g)^{\mathrm{cop}}$.

Proof: We put in the matrix forms and use Corollary 4.1.9 to evaluate $\mathbf{t}$ against $\mathcal{R}$. The resulting homomorphism $\theta = (p \otimes q) \circ \Delta$ sends $\mathbf{s} \otimes 1$ to $\mathbf{t} \otimes S\mathbf{l}^{+}$ and $1 \otimes \mathbf{t}$ to $\mathbf{t} \otimes S\mathbf{l}^{-}$, which essentially provides an isomorphism in the factorisable case. This time the inversion formula (7.33) tells us that $\theta(S\mathbf{s} \otimes \mathbf{t}) = 1 \otimes 1$. The element with image $\mathbf{t} \otimes 1$ under θ is $\mathbf{t}SQ^{-1}(\mathbf{l}^{+})_{(1)} \otimes Q^{-1}(\mathbf{l}^{+})_{(2)}$, which again requires Q to be invertible as a map. ∎

For example, taking $H = U_{q}(su_{2})$ with its real quasitriangular structure in Example 3.2.3, we see that $U_{q}(su_{2})\bowtie_{\mathcal{R}}U_{q}(su_{2})$ is a Hopf *-algebra generated by two mutually commuting copies of $U_{q}(sl_{2})$ with a complicated coproduct, antipode and *-structure. One can compute the coproducts of its generators quite explicitly (they are not formal power series), by mapping over to the quantum double, taking the coproducts there, and mapping back. On the other hand, this form of the quantum double does bring out the reason that one could think of it as $U_{q}(so(1,3))$. Recall that the usual Lorentz group Lie algebra has six anti-Hermitian generators J_{i} (rotations) and K_{i} (boosts) with relations

$$[J_{i}, J_{j}] = \epsilon_{ijk}J_{k}, \quad [J_{i}, K_{j}] = \epsilon_{ijk}K_{k}, \quad [K_{i}, K_{j}] = -\epsilon_{ijk}J_{k},$$

where ϵ_{ijk} is the totally antisymmetric tensor. A convenient description over $\mathbb{C}$ is made possible by defining

$$M_{i} = \frac{1}{2}(J_{i} + \imath K_{i}), \quad N_{i} = \frac{1}{2}(J_{i} - \imath K_{i}),$$

which generate two mutually commuting and mutually adjoint copies of $U(sl_2)$ in the form

$$[M_i, M_j] = \epsilon_{ijk} M_k, \quad [N_i, N_j] = \epsilon_{ijk} N_k, \quad [M_i, N_j] = 0, \quad M_i^* = -N_i,$$

much as above. This is the interpretation of the two mutually commuting copies of $U_q(sl_2)$ with a $*$-structure that interchanges the two factors.

Moreover, another form of the usual Lorentz group is based on the Iwasawa decomposition, whereby a Lorentz transformation is uniquely factorised (in the sense of Definition 6.2.10) into a spatial rotation as an element of SU_2 and a boost as an element of a solvable group $SU_2^{*\mathrm{op}}$. We will see this in detail in Chapter 8.3. At the Lie algebra level there is a linear splitting $so(1,3) = su_2^{*\mathrm{op}} \oplus su_2$, and the general theory is that every complexification of a compact semisimple Lie algebra (in this case $so(1,3) = sl(2, \mathbb{C})$ as a six-dimensional real Lie algebra) has a similar splitting. From this point of view, the Hopf algebra factorisation theorem in Theorem 7.3.5 is exactly a Hopf algebra version of the Iwasawa decomposition of $H \bowtie H$ as the 'complexification' of H into two factors $H^{*\mathrm{op}} \bowtie H$.

In the application of these constructions as defining a q-Lorentz group, we see that we have two forms. In one form we have $U_q(su_2) \bowtie U_q(su_2)$, consisting of two mutually adjoint and commuting copies of $U_q(su_2)$, while in the second form we have the quantum double $U_q(su_2)^{*\mathrm{op}} \bowtie U_q(su_2)$, consisting of the physical rotations and

$$U_q(su_2)^{*\mathrm{op}} \equiv U_q(su_2^{*\mathrm{op}}).$$

Here, we view the left hand side as, by definition, the q-deformation of the enveloping algebra of $su_2^{*\mathrm{op}}$. This kind of reinterpretation is called a *quantum–geometry transformation*. We will see it in more detail at the end of the next section. This then makes precise the sense in which the formulation of the q-Lorentz group as the quantum double of $U_q(su_2)$ is a q-deformation of the classical Iwasawa decomposition.

7.4 Cross product structure of quantum doubles

In this section, we study the quantum double from another point of view (also due to the author) which is not directly connected with the double cross products and double cross coproducts above. Rather, we find that, in the case when our Hopf algebra H is quasitriangular, its quantum double $D(H)$ is, in fact, isomorphic to an ordinary cross product and cross coproduct. This therefore has some advantages over the double cross product description of Example 7.2.6, which we needed for the general case. This time, the physical application or interpretation is a different

one from the one in the last section. Indeed, we know from Chapter 6.1
that *-algebra cross products can be interpreted as quantum algebras of
observables of dynamical systems. As a result, both the quantum double
and its dual $D(H)^*$ have such quantum-mechanical interpretations when
H is quasitriangular. Another application of such cross product theorems
is in quantum group gauge theory, as we have mentioned in Chapter 6.3,
so the cross product structure is important also in this context and gives
a third interpretation of the quantum double as a quantum frame bundle
of some form.

We have seen the semidirect or cross product phenomenon already in
the group case in Example 7.1.9, and earlier in Example 6.1.8, and noted
that the double cross product degenerates into this cross product form
when H is cocommutative. Since a quasitriangular Hopf algebra is almost
cocommutative – up to $\mathcal{R}$ – we should expect a similar theorem in this
case with suitable modification by $\mathcal{R}$. This is the mathematical content
of this section. We will give a fairly self-contained point of view here,
but, in fact, all the results in this section arise as an application of the
theory of braided groups and their bosonisation, which we will come to
in Chapters 9.4 and 10. Suffice it to say that the key difference between
the cross products as algebras and coalgebras in this section and the
bicrossproduct theory in Chapter 6.2 is that here there is only one Hopf
algebra which is both acting and coacting – there is no 'back-reaction'
of the (co)acted-upon object. The price that we pay for this is that the
(co)acted-upon object B is no ordinary bialgebra or Hopf algebra (its
algebra and coalgebra do not fit together in the usual way), but is a
braided one. We begin with the purely algebraic results which are useful
in other contexts also, and then discuss the abovementioned quantum-
mechanical interpretation.

The only part of the braided group theory that we need here is the idea
that braided groups are fully covariant under a Hopf algebra. This Hopf
algebra plays the role of $\mathbb{Z}/2$ in the theory of superalgebras, where everything is $\mathbb{Z}/2$-covariant. We have already seen this for braided matrices in
Chapter 4.5.2, where the covariance is under the corresponding quantum
matrices. We now build such a covariant object associated to any Hopf
algebra. We already know that the product of a Hopf algebra is always
covariant under its own adjoint action from Example 1.6.3 or coadjoint
coaction from Example 1.6.15. Similarly, the coproduct of a Hopf algebra is always covariant under its adjoint coaction from Example 1.6.14 or
coadjoint action from Example 1.6.4. On the other hand, we do not in
general have covariance of the coproduct under the adjoint action or the
product under the coadjoint action. Instead, one finds

Theorem 7.4.1 *Let A be a dual quasitriangular Hopf algebra. Then A*

has a second associative product

$$a \underline{\cdot} b = \sum a_{(2)} b_{(3)} \mathcal{R}(a_{(3)} \otimes S b_{(1)}) \mathcal{R}(a_{(1)} \otimes b_{(2)}),$$

which we call the covariantised *or* braided product *of A. We write* $\underline{A} = (A, \underline{\cdot})$ *for A with this new product. It is an A-comodule algebra under the adjoint coaction* Ad *in Example 1.6.14. If A is a Hopf *-algebra with* $\mathcal{R}$ *real, then there is also a transmuted *-structure* $\underline{*} = * \circ S$.

Proof: This construction originates in the transmutation theory of Chapter 9.4, where a more conceptual proof will be given. Here we outline a direct proof of associativity and covariance as a complement to those considerations. We use the definition of $\underline{\cdot}$ and the axioms (2.6) to break down multiplications in the argument of $\mathcal{R}$ as

$$
\begin{aligned}
a \underline{\cdot} (b \underline{\cdot} c) &= a_{(2)} (b \underline{\cdot} c)_{(3)} \mathcal{R}(a_{(3)} \otimes S(b \underline{\cdot} c)_{(1)}) \mathcal{R}(a_{(1)} \otimes (b \underline{\cdot} c)_{(2)}) \\
&= a_{(2)} b_{(4)} c_{(5)} \mathcal{R}(a_{(3)} \otimes S(b_{(2)} c_{(3)})) \\
&\qquad \mathcal{R}(a_{(1)} \otimes b_{(3)} c_{(4)}) \mathcal{R}(b_{(5)} \otimes S c_{(1)}) \mathcal{R}(b_{(1)} \otimes c_{(2)}) \\
&= a_{(3)} b_{(4)} c_{(5)} \mathcal{R}(b_{(5)} \otimes S c_{(1)}) \mathcal{R}(a_{(4)} \otimes S b_{(2)}) \mathcal{R}(a_{(5)} \otimes S c_{(3)}) \\
&\qquad \mathcal{R}(a_{(2)} \otimes b_{(3)}) \mathcal{R}(a_{(1)} \otimes c_{(4)}) \mathcal{R}(b_{(1)} \otimes c_{(2)}) \\
&= a_{(3)} b_{(4)} c_{(5)} \mathcal{R}(b_{(5)} \otimes S c_{(1)}) \mathcal{R}(a_{(4)} \otimes S b_{(2)}) \mathcal{R}(a_{(5)} \otimes S c_{(3)}) \\
&\qquad \mathcal{R}(S a_{(2)} \otimes S b_{(3)}) \mathcal{R}(S a_{(1)} \otimes S c_{(4)}) \mathcal{R}(S b_{(1)} \otimes S c_{(2)}), \\
(a \underline{\cdot} b) \underline{\cdot} c &= (a \underline{\cdot} b)_{(2)} c_{(3)} \mathcal{R}((a \underline{\cdot} b)_{(3)} \otimes S c_{(1)}) \mathcal{R}((a \underline{\cdot} b)_{(1)} \otimes c_{(2)}) \\
&= a_{(3)} b_{(4)} c_{(3)} \mathcal{R}(a_{(4)} b_{(5)} \otimes S c_{(1)}) \\
&\qquad \mathcal{R}(a_{(2)} b_{(3)} \otimes c_{(2)}) \mathcal{R}(a_{(5)} \otimes S b_{(1)}) \mathcal{R}(a_{(1)} \otimes b_{(2)}) \\
&= a_{(3)} b_{(4)} c_{(5)} \mathcal{R}(b_{(5)} \otimes S c_{(1)}) \mathcal{R}(a_{(4)} \otimes S c_{(2)}) \mathcal{R}(a_{(5)} \otimes S b_{(1)}) \\
&\qquad \mathcal{R}(a_{(2)} \otimes c_{(3)}) \mathcal{R}(a_{(1)} \otimes b_{(2)}) \mathcal{R}(b_{(3)} \otimes c_{(4)}).
\end{aligned}
$$

One then regroups and uses the QYBE in Lemma 2.2.3 twice to see that these expressions are equal. It is easy to see that the unit is unchanged. Moreover, this new product makes $\underline{A}$ an A-comodule algebra since

$$
\begin{aligned}
\beta(a \underline{\cdot} b) &= a_{(3)} b_{(4)} \otimes (S(a_{(2)} b_{(3)})) a_{(4)} b_{(5)} \mathcal{R}(a_{(5)} \otimes S b_{(1)}) \mathcal{R}(a_{(1)} \otimes b_{(2)}) \\
&= a_{(3)} b_{(4)} \otimes (S(b_{(2)} a_{(1)})) a_{(4)} b_{(5)} \mathcal{R}(a_{(5)} \otimes S b_{(1)}) \mathcal{R}(a_{(2)} \otimes b_{(3)}) \\
&= a_{(3)} b_{(4)} \otimes (S a_{(1)}) a_{(5)} (S b_{(1)}) b_{(5)} \mathcal{R}(a_{(4)} \otimes S b_{(2)}) \mathcal{R}(a_{(2)} \otimes b_{(3)}) \\
&= a_{(2)} \underline{\cdot} b_{(2)} \otimes (S a_{(1)}) a_{(3)} (S b_{(1)}) b_{(3)} = \beta(a) \beta(b).
\end{aligned}
$$

We used $\beta = $ Ad and axiom (2.7) to obtain the second equality, and again to obtain the third.

Finally, in the real Hopf *-algebra case, we have

$$
\begin{aligned}
(a \underline{\cdot} b)^{\underline{*}} &= * \circ S(a \underline{\cdot} b) = (S a_{(2)})^* (S b_{(2)})^* \overline{\mathcal{R}((S a_{(1)}) a_{(3)} \otimes S b_{(1)})} \\
&= a_{(2)}{}^{\underline{*}} b_{(2)}{}^{\underline{*}} \mathcal{R}((S b_{(1)})^* \otimes (S(S a_{(3)})^*)(S a_{(1)})^*)
\end{aligned}
$$

$$= a_{(2)}{}^{*}b_{(2)}{}^{*}\mathcal{R}(b_{(1)}{}^{*} \otimes (Sa_{(3)}{}^{*})a_{(1)}{}^{*})$$
$$= a^{*}{}_{(2)}b^{*}{}_{(1)}\mathcal{R}(b^{*}{}_{(2)} \otimes (Sa^{*}{}_{(1)})a^{*}{}_{(3)})$$
$$= a^{*}{}_{(2)}b^{*}{}_{(1)}\mathcal{R}(b^{*}{}_{(2)} \otimes a^{*}{}_{(3)})\mathcal{R}(b^{*}{}_{(3)} \otimes Sa^{*}{}_{(1)})$$
$$= \mathcal{R}(b^{*}{}_{(1)} \otimes a^{*}{}_{(2)})b^{*}{}_{(2)}a^{*}{}_{(3)}\mathcal{R}(b^{*}{}_{(3)} \otimes Sa^{*}{}_{(1)})$$
$$= b^{*}{}_{(2)}a^{*}{}_{(2)}\mathcal{R}((Sb^{*}{}_{(1)})b^{*}{}_{(3)} \otimes Sa^{*}{}_{(1)}) = b^{*}{}_{\cdot}a^{*}{}_{\cdot},$$

using the elementary properties of $\mathcal{R}$.

We also note that, in this covariantised algebra, one has the identity

$$\mathcal{R}(a_{(1)} \otimes b_{(1)})b_{(2)}{}_{\cdot}\mathcal{R}(b_{(3)} \otimes a_{(2)})a_{(3)}$$

$$= a_{(1)}{}_{\cdot}\mathcal{R}(a_{(2)} \otimes b_{(1)})b_{(2)}\mathcal{R}(b_{(3)} \otimes a_{(3)})$$

(7.34)

by direct computation and the elementary properties of $\mathcal{R}$. We will derive this more conceptually in Chapter 9.4 as a braided commutativity condition. ∎

This construction also works for any bialgebra mapping to A: it too can be transmuted via similar formulae in such a way that the new product is A-covariant. The formulae are the same, but first mapping to A before applying $\mathcal{R}$. As usual, we have an analogous theory for quasitriangular Hopf algebras or bialgebras to which they map.

Theorem 7.4.2 *Let H be a quasitriangular Hopf algebra and let $\triangleright$ be the quantum adjoint action as in Example 1.6.3. Then H has a second coalgebra structure*

$$\underline{\Delta}h = \sum h_{(1)}S\mathcal{R}^{(2)} \otimes \mathcal{R}^{(1)}\triangleright h_{(2)},$$

which we call the covariantised or braided coproduct of H. We write $\underline{H} = (H, \underline{\Delta})$ for H with this new coproduct. It is an H-module coalgebra under the adjoint action. If H is a Hopf $*$-algebra and $\mathcal{R}$, real, then the $*$ operation obeys $(* \otimes *) \circ \underline{\Delta} = \tau \circ \underline{\Delta} \circ *$.

Proof: This is strictly dual to the preceding proposition and also originates in Chapter 9.4. Equivalently, its proofs are given by writing out the preceding proofs as diagrams, reflecting in a mirror and reversing arrows. If we write all maps pointing downwards, then it just means turning the diagrammatic proofs upside down. Note that the covariantisation procedures here, and in the preceding theorem, are clearly invertible. Thus, $\underline{\Delta}h = h_{(1)}\mathcal{R}^{(2)} \otimes \mathcal{R}^{(1)}\triangleright h_{(2)}$ recovers the original coproduct from $\underline{\Delta}h \equiv h_{(1)} \otimes h_{(2)}$. ∎

There is much more structure in these theorems, some of which we will see later in Chapters 9 and 10. Here we mention that $\underline{A}$ with its unchanged coalgebra and $\underline{H}$ with its unchanged algebra both have co-variantised antipodes $\underline{S}$ which are fully covariant and commute with the appropriate $*$-structure in the Hopf $*$-algebra case. One also has

Proposition 7.4.3 *Let H be a quasitriangular Hopf algebra with dual A. The quantum inverse Killing form $Q : A \to H$ maps the covariantised product of A to the product of H and the coproduct of A to the covariantised coproduct of H. If H is real quasitriangular, then Q becomes a $*$-algebra homomorphism for the relevant $*$-structures as stated above.*

Proof: Using the covariantised product from Theorem 7.4.1, we have

$$Q(a \underline{\cdot} b)$$
$$= \mathcal{R}^{(1)} \mathcal{R}'^{(2)} \langle a_{(2)} b_{(2)}, \mathcal{R}^{(2)} \mathcal{R}'^{(1)} \rangle \langle a_{(1)}, \mathcal{R}''^{(1)} \rangle \langle a_{(3)}, \mathcal{R}'''^{(1)} \rangle \langle b_{(1)}, (S\mathcal{R}'''^{(2)}) \mathcal{R}''^{(2)} \rangle$$
$$= \mathcal{R}^{(1)} \mathcal{R}'^{(2)} \langle a \otimes b, \mathcal{R}''^{(1)} \mathcal{R}^{(2)}{}_{(1)} \mathcal{R}'^{(1)}{}_{(1)} \mathcal{R}'''^{(1)} \otimes (S\mathcal{R}'''^{(2)}) \mathcal{R}''^{(2)} \mathcal{R}^{(2)}{}_{(2)} \mathcal{R}'^{(1)}{}_{(2)} \rangle.$$

Using the axioms for $\mathcal{R}$, this is the evaluation with $a \otimes b$ on the last two factors of $X^{(1)} \otimes X^{(2)} \mathcal{R}'''^{(1)} \otimes (S\mathcal{R}'''^{(2)}) X^{(3)}$, where $X = \mathcal{R}_{23} \mathcal{R}_{13} \mathcal{R}_{12} \mathcal{R}_{21} \mathcal{R}_{31} = \mathcal{R}_{12} \mathcal{R}_{13} \mathcal{R}_{23} \mathcal{R}_{21} \mathcal{R}_{31}$ by the QYBE for $\mathcal{R}$. Writing $\mathcal{R}_{13} = (\mathrm{id} \otimes S)(\mathcal{R}_{13}^{-1})$, combining the elements on which S acts and using the QYBE again for them, we obtain (after cancellations) the element $\mathcal{R}_{12} \mathcal{R}_{21} \mathcal{R}_{13} \mathcal{R}_{31}$. The pairing of this with $a \otimes b$ is just $Q(a)Q(b)$ as required. Next we compute with the braided coproduct to give $\underline{\Delta} Q(a)$ as

$$\mathcal{R}^{(1)}{}_{(1)} \mathcal{R}'^{(2)}{}_{(1)} \mathcal{R}''^{(2)} S\mathcal{R}'''^{(2)} \otimes \mathcal{R}'''^{(1)} \mathcal{R}^{(1)}{}_{(2)} \mathcal{R}'^{(2)}{}_{(2)} \mathcal{R}''^{(1)} \langle a, \mathcal{R}^{(2)} \mathcal{R}'^{(1)} \rangle$$

or evaluation of a in the third position on $X^{(1)} S\mathcal{R}'''^{(2)} \otimes \mathcal{R}'''^{(1)} X^{(2)} \otimes X^{(3)}$ where $X = \mathcal{R}_{13} \mathcal{R}_{23} \mathcal{R}_{32} \mathcal{R}_{31} \mathcal{R}_{21} = \mathcal{R}_{13} \mathcal{R}_{23} \mathcal{R}_{21} \mathcal{R}_{31} \mathcal{R}_{32}$ by the QYBE. Writing $\mathcal{R}_{31} = (S \otimes \mathrm{id})(\mathcal{R}_{31}^{-1})$, combining the arguments of S, using the QYBE again and cancelling, now gives the element $\mathcal{R}_{13} \mathcal{R}_{31} \mathcal{R}_{23} \mathcal{R}_{32}$. The pairing of this with a gives $(Q \otimes Q) \circ \underline{\Delta} a$ as required.

Finally, in the Hopf $*$-algebra case, one has $Q(a^{\underline{*}}) = \langle (Sa)^*, Q^{(1)} \rangle Q^{(2)} = \overline{\langle a, Q^{(1)*} \rangle Q^{(2)**}} = (Q(a))^*$ for the real case in view of the remarks in Definition 2.1.15.   ■

We now need the following lemma, which asserts that if a quasitriangular Hopf algebra acts on something covariantly, then it also coacts covariantly.

Lemma 7.4.4 *Let H be a quasitriangular Hopf algebra. If B is a left H-module (algebra, coalgebra) then $\beta = \mathcal{R}_{21} \triangleright(\ )$ makes it automatically a left H-comodule (algebra, coalgebra). The original action and this induced coaction makes B into a $D(H)$-module (algebra, coalgebra) in the crossed module form in Proposition 7.1.6.*

Proof: We just need the axioms (2.1), which are those of a skew-copairing as in (7.26) of H with itself. Thus the result as stated can be viewed as a variant of Proposition 1.6.11 for the self-dual case. A direct proof is easy: if B is an H-module, then β as stated is clearly a comodule in view of the second half of (2.1). If the action respects the algebra of B, then $\beta(ab) = \mathcal{R}_{21}\triangleright(ab) = \mathcal{R}^{(2)} \otimes (\mathcal{R}^{(1)}{}_{(1)}\triangleright a)(\mathcal{R}^{(1)}{}_{(2)}\triangleright b) = \beta(a)\beta(b)$ in view of the first half of (2.1). Likewise if B is a comodule algebra.

For completeness, we check that the original action and the induced coaction here obey the condition in Proposition 7.1.6 if (2.2) holds. The condition there is $h_{(1)}\mathcal{R}^{(2)} \otimes (h_{(2)}\mathcal{R}^{(1)})\triangleright v = \mathcal{R}^{(2)}h_{(2)} \otimes (\mathcal{R}^{(1)}h_{(1)})\triangleright v$ for all v, h. So this is where we fully use that $\mathcal{R}$ is a quasitriangular structure. Thus, the automatic coaction is such as to amplify every H-module (co)algebra to one of $D(H)$ or, more generally, to a crossed H-module. ∎

What this lemma means is that, if H acts on an algebra and coalgebra covariantly, we can make a cross product algebra $\rtimes$ as in Proposition 1.6.6, and, by the automatic coaction, we can make a cross coproduct coalgebra $\blacktriangleright\!\!\triangleleft$ as in Proposition 1.6.18. In nice cases, one might expect that the result is a Hopf algebra. This is exactly the situation for our covariantised $\underline{A}$ and $\underline{H}$ above.

Theorem 7.4.5 *Let H be a finite-dimensional quasitriangular Hopf algebra with dual A. Then the cross product and cross coproduct $\underline{A}\!\bowtie\!H$ by the left coadjoint action of H from Example 1.6.4 and its induced coaction from Lemma 7.4.4 is a Hopf algebra. The maps*

$$H^{*\mathrm{op}}\overset{i}{\hookrightarrow}\underline{A}\!\bowtie\!H\overset{j}{\hookleftarrow}H,$$

$$i(a) = \sum a_{(1)}\langle \mathcal{R}^{-(1)}, a_{(2)}\rangle \otimes \mathcal{R}^{-(2)}, \quad j(h) = 1 \otimes h,$$

provide an isomorphism

$$\cdot \circ (i \otimes j)(a \otimes h) = \sum a_{(1)}\langle \mathcal{R}^{-(1)}, a_{(2)}\rangle \otimes \mathcal{R}^{-(2)}h : \quad D(H)\cong\underline{A}\!\bowtie\!H.$$

The algebra structure is a $*$-algebra cross product in the sense of Proposition 6.1.5 if H is a real quasitriangular Hopf $*$-algebra.

Proof: The product and coproduct structures are the standard ones from Propositions 1.6.6 and 1.6.18, respectively, with the product of $\underline{A}$ in Theorem 7.4.1 and its usual coproduct. If one wants to see how the cross (co)product looks in terms of the structures of H and A, the formulae are

$$(a \otimes h)(b \otimes g) = a_{(2)}b_{(3)}\langle h_{(1)}, (Sb_{(2)})b_{(4)}\rangle\mathcal{R}((Sa_{(1)})a_{(3)} \otimes Sb_{(1)}) \otimes h_{(2)}g,$$

$$\Delta(a \otimes h) = a_{(1)} \otimes \mathcal{R}^{(2)}h_{(1)} \otimes \mathcal{R}^{(1)}\triangleright a_{(2)} \otimes h_{(2)}, \tag{7.35}$$

$$S(a \otimes h) = (1 \otimes Sh)(S^{-1}a_{(3)} \otimes \mathcal{R}^{(2)})\langle \mathcal{R}^{(1)}, a_{(2)}Sa_{(4)}\rangle u(a_{(5)})\mathcal{R}(a_{(6)} \otimes a_{(1)}),$$

where the functional u is from Proposition 2.2.4. One can verify the antipode directly. Also, it is easy to see from the form of product and coproduct that j is a bialgebra map. For i we compute directly

$$i(a)i(b) = (a_{(1)} \otimes \mathcal{R}^{-(2)})(b_{(1)} \otimes \mathcal{R}'^{-(2)})\langle \mathcal{R}^{-(1)}, a_{(2)}\rangle\langle \mathcal{R}'^{-(1)}, b_{(2)}\rangle$$
$$= a_{(1)} \underline{\cdot}(\mathcal{R}^{-(2)} \triangleright b_{(1)}) \otimes \mathcal{R}''^{-(2)}\mathcal{R}'^{-(2)}\langle \mathcal{R}^{-(1)}\mathcal{R}''^{-(1)}, a_{(2)}\rangle\langle \mathcal{R}'^{-(1)}, b_{(2)}\rangle$$
$$= a_{(1)} \underline{\cdot} b_{(2)}\mathcal{R}^{-1}(a_{(2)} \otimes (Sb_{(1)})b_{(3)}) \otimes \mathcal{R}''^{-(2)}\mathcal{R}'^{-(2)}\langle \mathcal{R}''^{-(1)}, a_{(3)}\rangle\langle \mathcal{R}'^{-(1)}, b_{(4)}\rangle$$
$$= b_{(1)}a_{(1)} \otimes \mathcal{R}''^{-(2)}\mathcal{R}'^{-(2)}\langle \mathcal{R}''^{-(1)}, a_{(2)}\rangle\langle \mathcal{R}'^{-(1)}, b_{(2)}\rangle = i(ba),$$

where the second equality multiplies out in the cross product algebra according to Proposition 1.6.6, the third puts in the form of the coadjoint action, while the fourth puts in the covariantised product from Theorem 7.4.1 and cancels $\mathcal{R}$ and $\mathcal{R}^{-1}$. To do this, we also used (2.7). For the coalgebra structure, we have

$$\Delta i(a) = a_{(1)} \otimes \mathcal{R}^{(2)}\mathcal{R}^{-(2)}{}_{(1)} \otimes \mathcal{R}^{(1)} \triangleright a_{(2)} \otimes \mathcal{R}^{-(2)}{}_{(2)}\langle \mathcal{R}^{-(1)}, a_{(3)}\rangle$$
$$= a_{(1)} \otimes \mathcal{R}^{(2)}\mathcal{R}^{-(2)} \otimes a_{(3)} \otimes \mathcal{R}'^{-(2)}\langle \mathcal{R}^{(1)}, (Sa_{(2)})a_{(4)}\rangle\langle \mathcal{R}^{-(1)}\mathcal{R}'^{-(1)}, a_{(5)}\rangle$$
$$= a_{(1)} \otimes \mathcal{R}^{(2)} \otimes a_{(3)} \otimes \mathcal{R}'^{-(2)}\langle S\mathcal{R}^{(1)}, a_{(2)}\rangle\langle \mathcal{R}'^{-(1)}, a_{(4)}\rangle = (i \otimes i) \circ \Delta a,$$

where we use the cross coproduct from Proposition 1.6.18 in the first equality, (2.1) and the definition of the coadjoint action for the second, and then combine some evaluations against $a_{(4)}$ to cancel $\mathcal{R}'$ and $\mathcal{R}^{-1}$.

Next, unlike the proof of Theorem 7.3.5, it is immediate that the map $\cdot \circ (i \otimes j)$ has inverse $a \otimes h \mapsto a_{(1)}\langle \mathcal{R}^{(1)}, a_{(2)}\rangle \otimes \mathcal{R}^{(2)}h$. So, by Theorem 7.2.3, we conclude that our Hopf algebra is isomorphic to a double cross product $H^{*\mathrm{op}} \bowtie H$. One easily sees, by looking at $j(h)i(a)$, that it is $D(H)$.

Finally, in the $*$-algebra case, we assume that $\mathcal{R}$ is real so that we can use the covariantised $*$-structure for $\underline{A}$ in Theorem 7.4.1. Then $(h \triangleright a)^{\underline{*}} = (Sa_{(2)})^*\langle h, (Sa_{(1)})a_{(3)}\rangle = (Sa)^*{}_{(2)}\langle (Sh)^*{}_{(2)}, (Sa)^*{}_{(3)}\rangle\langle (Sh)^*{}_{(1)}, S(Sa)^*{}_{(1)}\rangle = (Sh)^* \triangleright (a^{\underline{*}})$, as required in Proposition 6.1.5. Here $\underline{A}$ and H are sub-$*$-algebras. Note that the $*$-algebra structure here does not generally give a Hopf $*$-algebra (and does not correspond to the one for $D(H)$ in Proposition 7.1.4). ∎

Theorem 7.4.6 *Let H be a quasitriangular Hopf algebra. Then the cross product and cross coproduct $\underline{H} \bowtie H$ by the left adjoint action and its induced coaction is a Hopf algebra. The maps*

$$H \overset{p}{\leftarrow} \underline{H} \bowtie H \overset{q}{\to} H, \quad p = \cdot, \quad q = \epsilon \otimes \mathrm{id},$$

provide an isomorphism

$$(p \otimes q) \circ \Delta(h \otimes g) = \sum hg_{(1)} \otimes g_{(2)} : \quad \underline{H} \bowtie H \cong H_{\mathcal{R}} \bowtie H$$

to (7.29). The algebra structure is a $*$-algebra cross product if H is a Hopf $*$-algebra.

Proof: The algebra structure here is the cross product from Proposition 1.6.6 by the adjoint action Ad. It is always a ∗-algebra cross product in the Hopf ∗-algebra case, as already noted in Chapter 6.1. We put the covariantised coproduct in Theorem 7.4.2 into Proposition 1.6.18 for the cross coproduct structure. If one wants the formulae in terms of H, they are

$$(1 \otimes h)(g \otimes 1) = h_{(1)} g S h_{(2)} \otimes h_{(3)},$$

$$\Delta(h \otimes g) = h_{(1)} S \mathcal{R}^{(2)}{}_{(1)} \otimes \mathcal{R}^{(2)}{}_{(2)} g_{(1)} \otimes \mathcal{R}^{(1)} \triangleright h_{(2)} \otimes g_{(2)}, \tag{7.36}$$

$$S(h \otimes g) = (1 \otimes Sg)(S((S\mathcal{R}^{(2)}{}_{(2)})(\mathcal{R}^{(1)} \triangleright h)) \otimes S\mathcal{R}^{(2)}{}_{(1)}).$$

That q is a bialgebra homomorphism is easy. That p an algebra homomorphism comes out as $p((1 \otimes h)(g \otimes 1)) = p(h_{(1)} \triangleright g \otimes h_{(2)}) = (h_{(1)} \triangleright g) h_{(2)} = h_{(1)} g(S h_{(2)}) h_{(3)} = hg$. That it respects the coalgebra follows at once from the form of $\Delta(h \otimes g)$. Also from the form of Δ we see at once the formula for $(p \otimes q) \circ \Delta$, which is clearly a linear isomorphism. The dual form of Theorem 7.2.3 then tells us that $\underline{H} \bowtie H$ is a double cross coproduct, which we recognise as stated. ∎

In the factorisable case, these two theorems become the same under the isomorphism Q, but in general they are distinct. This applies even for the important class of matrix examples, to which we now turn. Thus, it is clear from (7.34) that the covariantised version of the quantum matrices $A(R)$ as in Chapter 4.1 is nothing other than the braided matrices $B(R)$ in Chapter 4.3. We just put the matrix comultiplication of t into the formulae in Theorem 7.4.1 to give

$$u^i{}_j = t^i{}_j,$$

$$u^i{}_j u^k{}_l = t^a{}_b t^d{}_l R^i{}_a{}^c{}_d \tilde{R}^b{}_j{}^k{}_c,$$

$$u^i{}_j u^k{}_l u^m{}_n = t^d{}_b t^s{}_u t^z{}_n R^i{}_a{}^p{}_q R^a{}_d{}^w{}_y \tilde{R}^b{}_c{}^v{}_w \tilde{R}^c{}_j{}^k{}_p R^q{}_s{}^y{}_z \tilde{R}^u{}_l{}^m{}_v, \tag{7.37}$$

$$\mathbf{u} = \mathbf{t},$$

$$R_{12}^{-1} \mathbf{u}_1 R_{12} \mathbf{u}_2 = \mathbf{t}_1 \mathbf{t}_2,$$

$$R_{23}^{-1} R_{13}^{-1} R_{12}^{-1} \mathbf{u}_1 R_{12} \mathbf{u}_2 R_{13} R_{23} \mathbf{u}_3 = \mathbf{t}_1 \mathbf{t}_2 \mathbf{t}_3,$$

etc., where we write **u** for **t** with this new product. Their relations from (7.34) recover those in Definition 4.3.1. This is historically how the braided matrices were introduced. The same applies after adding the additional relations for a quantum group G_q, which appear as some corresponding relations on the transmuted or covariantised side as a braided group $BG_q = \underline{G}_q$. We will study such braided groups in their own right in Chapter 10. Meanwhile, Theorem 7.4.2 converts $U_q(g)$ to a braided group $BU_q(g) = \underline{U}_q(g)$ with the same algebra but a transmuted or covariantised coproduct. The map Q in Proposition 7.4.3, which is surjective in the factorisable case, recovers the map $Q(\mathbf{u}) = 1^+ S 1^-$ in Example 4.3.3. The usual FRT generators $1^\pm$ are useful if one wants to work with the ordinary coproduct of $U_q(g)$, while the braided matrix generator $\mathbf{u}$ (or rather, its image $1 = Q(\mathbf{u})$) is useful when one wants to work with the covariantised coproduct from Theorem 7.4.2. In these standard factorisable cases, the map Q is essentially an isomorphism (of braided groups), i.e. we can just work with $BU_q(g) = BG_q$ identified in a suitable sense (as explained below Example 4.3.3). We do this to avoid repeating essentially the same formulae for the two situations.

In the factorisable case then, we have only one theorem, with two interpretations, depending on whether we regard the braided matrices as versions of the quantum function algebra with covariantised product and matrix coproduct $\underline{\Delta}\mathbf{u} = \mathbf{u} \otimes \mathbf{u}$, or as versions of the quantum enveloping algebras with the usual algebra and the covariantised coproduct $\underline{\Delta}1 = 1 \otimes 1$. We proceed with $\mathbf{u} = 1$ identified so that the reader can make either interpretation.

Example 7.4.7 *The cross product $BU_q(g) {>\!\!\!\triangleleft} U_q(g)$ by the quantum adjoint action, as in Theorem 7.4.5, is isomorphic to the quantum double in the form $U_q(g)_{\mathcal{R}}{\bowtie}U_q(g)$ from Example 7.3.6. Similarly, $BG_q {>\!\!\!\triangleleft} U_q(g)$ by the quantum adjoint coaction, as in Theorem 7.4.6, is isomorphic to the quantum double $G_q^{op}{\bowtie}U_q(g)$ from Example 7.2.13. Both cross products are generated by $\mathbf{u}, 1^\pm$ (as explained above) with cross relations, coproduct and antipode*

$$R 1_2^+ \mathbf{u}_1 = \mathbf{u}_1 R 1_2^+, \quad R_{21}^{-1} 1_2^- \mathbf{u}_1 = \mathbf{u}_1 R_{21}^{-1} 1_2^-,$$

$$\Delta\mathbf{u} = \mathbf{u}\mathcal{R}^{(2)} \otimes \mathcal{R}^{(1)}{\triangleright}\mathbf{u}, \quad S\mathbf{u} = ((\mathbf{u}\mathcal{R}^{(1)}){\triangleright}\mathbf{u}^{-1}) S\mathcal{R}^{(2)},$$

where ${\triangleright}$ is the quantum adjoint action. Here the $1^\pm$ appear as a sub-Hopf algebra. For the standard quantum enveloping algebras with q real, we have a $*$-algebra cross product with $u^i{}_j{}^* = u^j{}_i$ and $1^{\pm i}{}_j{}^* = S 1^{\mp j}{}_i$.

Proof: This is a special case of either Theorem 7.4.5 or Theorem 7.4.6 depending on how we regard the braided matrices $B(R)$. The quantum adjoint action was given in (4.21) in the proof of Proposition 4.3.5, leading at once to the relations shown. The coproduct is just that stated in Theorem 7.4.5 with $\mathbf{u} \equiv \mathbf{u} \otimes 1$, etc., while $\mathcal{R}$ lives in the other (acting) copy of $U_q(g)$ as generated by $1^\pm \equiv 1 \otimes 1^\pm$. The antipode is harder and comes from the bosonisation theory in Chapter 9.4; here u is the element in Proposition 2.1.8. Note that, on using the axioms for $\mathcal{R}$, one can also compute the coproduct further as

$$\Delta\mathbf{u} = (\mathbf{u}\mathcal{R}^{(2)} \otimes \mathcal{R}^{(1)}\mathbf{u})\mathcal{R}_{21}^{-1}. \tag{7.38}$$

The associated Hopf algebra maps are

$$U_q(g) \overset{p}{\leftarrow} BU_q(g) {\rtimes} U_q(g) \overset{q}{\rightarrow} U_q(g), \quad G_q^{\mathrm{op}} \overset{i}{\rightarrow} BG_q {\rtimes} U_q(g) \overset{j}{\leftarrow} U_q(g),$$

$$p(\mathbf{u} \otimes 1) = 1^+ S 1^-, \quad p(1 \otimes 1^\pm) = 1^\pm, \quad q = \epsilon \otimes \mathrm{id}, \tag{7.39}$$

$$i(\bar{\mathbf{t}}) = \mathbf{u} \otimes 1^-, \quad j(1^\pm) = 1 \otimes 1^\pm,$$

and provide the required isomorphisms $(p \otimes q) \circ \Delta(\mathbf{u} \otimes 1) = 1^+ S 1^- \otimes 1$, $(p \otimes q) \circ \Delta(1 \otimes 1^\pm) = 1^\pm \otimes 1^\pm$ and $\cdot \circ (i \otimes j)$. This works for general matrix quantum groups associated to R-matrices. For the standard deformations, the two results coinicide by factorisability. Thus

$$U_q(g)_{\mathcal{R}}{\bowtie} U_q(g) {\cong} BU_q(g){\rtimes} U_q(g) \equiv BG_q {\rtimes} U_q(g) {\cong} G_q^{\mathrm{op}}{\bowtie} U_q(g) \tag{7.40}$$

are all forms of the quantum double of $U_q(g)$.   ∎

For example, Theorem 7.4.6 tells us that

$$U_q(su_2){\rtimes} U_q(su_2){\cong} U_q(su_2)_{\mathcal{R}}{\bowtie} U_q(su_2), \tag{7.41}$$

which gives a useful new form of the q-Lorentz group Hopf algebra, as an ordinary cross product. The isomorphism holds generally without needing factorisability. The description of $U_q(su_2)$ in terms of $\mathbf{u}$ needed here was given in Example 4.3.4.

On the other hand, from the point of view of Theorem 7.4.5, we have an interesting geometrical interpretation of the quantum double as quantum mechanics in the general sense of cross products as explained in Chapter 6.1. Here, H in Theorem 7.4.5 plays the role of (quantum) momentum group, and the covariantised $\underline{A}$ play the role of functions on a space. So, from this point of view, we would prefer to write the content of (7.41) as

$$D(U_q(su_2)){\cong} BSL_q(2){\rtimes} U_q(su_2), \tag{7.42}$$

where we regard $BSL_q(2)$ as the covariantised or braided version of the functions on some space, which is the position space of our quantum system. So this is our algebra of position observables in the language of Chapter 6.1. It is described in terms of the same $\mathbf{u}$ as above with the relations as given in Example 4.3.4, but the point of view is different. This isomorphism (7.42) is quite general, and does not make use of factorisability. In this way, we view the resulting cross product (7.42) as a quantum system along the lines of a q-deformed version of Example 6.1.9.

In order to describe the geometrical interpretation of the position space further, we recall that $BSL_q(2)$ is given by beginning with the algebra of 2×2 braided Hermitian matrices $BM_q(2)$, as explained in Chapter 4.3, and adding the determinant relation. So we begin with the geometrical picture of these. Since ordinary 2×2 Hermitian matrices provide a useful definition of Minkowski spacetime, it is clear that $BM_q(2)$ should be viewed likewise as a q-deformed version of the algebra of coordinate functions on this. Crucial for this identification is the correct $*$-structure

$$\begin{pmatrix} a^* & b^* \\ c^* & d^* \end{pmatrix} = \begin{pmatrix} a & c \\ b & d \end{pmatrix}, \tag{7.43}$$

which is the one from Theorem 7.4.1. It is in sharp contrast to the situation for the usual quantum matrices $M_q(2)$ before covariantisation. The self-adjoint generators

$$t = \frac{qd + q^{-1}a}{2}, \quad x = \frac{b + c}{2}, \quad y = \frac{b - c}{2i}, \quad z = \frac{d - a}{2} \tag{7.44}$$

provide, up to scaling, some natural spacetime coordinates. Here t (the time direction) is central, while the braided determinant $\underline{\det}(\mathbf{u})$ is

$$ad - q^2 cb = \frac{4q^2}{(q^2 + 1)^2} t^2 - q^2 x^2 - q^2 y^2 - \frac{2(q^4 + 1)q^2}{(q^2 + 1)^2} z^2 + 2q \left(\frac{q^2 - 1}{q^2 + 1} \right)^2 tz \tag{7.45}$$

and plays the role of the metric on this q-Minkowski space.

We will explore this q-Minkowski space and its invariance under the q-Lorentz group in Chapter 10.5. For the present, we see that setting the braided determinant equal to unity, as we do for $BSL_q(2)$, means that our position observables are a q-deformation of a hyperboloid or 3-sphere in Minkowski space. This $*$-algebra can be denoted BS_q^3 for this reason. Moreover, the action of $U_q(su_2)$ on the position observables comes out from (4.21) as

$$X_+ \triangleright \begin{pmatrix} a & b \\ c & d \end{pmatrix} = \begin{pmatrix} -q^{\frac{3}{2}}c & -q^{\frac{1}{2}}(d - a) \\ 0 & q^{-\frac{1}{2}}c \end{pmatrix} \rightarrow \left[\begin{pmatrix} a & b \\ c & d \end{pmatrix}, \begin{pmatrix} 0 & 1 \\ 0 & 0 \end{pmatrix} \right],$$

$$X_- \triangleright \begin{pmatrix} a & b \\ c & d \end{pmatrix} = \begin{pmatrix} q^{\frac{1}{2}}b & 0 \\ q^{-\frac{1}{2}}(d - a) & -q^{-\frac{3}{2}}b \end{pmatrix} \rightarrow \left[\begin{pmatrix} a & b \\ c & d \end{pmatrix}, \begin{pmatrix} 0 & 0 \\ 1 & 0 \end{pmatrix} \right], \tag{7.46}$$

$$H \triangleright \begin{pmatrix} a & b \\ c & d \end{pmatrix} = \begin{pmatrix} 0 & -2b \\ 2c & 0 \end{pmatrix} \rightarrow \left[\begin{pmatrix} a & b \\ c & d \end{pmatrix}, \begin{pmatrix} 1 & 0 \\ 0 & -1 \end{pmatrix} \right],$$

where the limits are as $q \rightarrow 1$. These make clear that our action is a q-deformation of the one induced on the generators of the ring of coordinate functions by the adjoint action. So $U_q(su_2)$ has precisely the role of q-deformed angular momentum. One can see that it keeps the time direction fixed (because it is central and hence invariant under the quantum adjoint action) and that it acts on the x, y, z coordinates like a rotation.

The quantisation of this system as a $*$-cross product is then isomorphic to the quantum double $D(U_q(su_2))$. Thus, the quantum double should be viewed as a q-deformed version of quantum motion on spheres. Of course, in the q-deformed setting, there are neither actual points nor actual orbits in the usual sense. Moreover, one has to quantise the spheres together as a foliation of a hyperboloid in Minkowski space (rather than the more obvious setting of spheres in Euclidean space) for this interpretation to work in the q-deformed case.

Note also that q here is considered quite orthogonal to the process of quantisation, and need not be related to any physical $\hbar$. Thus in our interpretation, we have two independent processes, which mutually commute:

$$
\begin{array}{ccc}
C(S^3_{\text{Mink}}) \otimes su_2 & \xrightarrow{\text{deformation}} & BS^3_q \otimes U_q(su_2) \\[2mm]
\text{quantisation} \downarrow & & \downarrow \text{quantisation} \qquad (7.47) \\[2mm]
C(S^3_{\text{Mink}}) {\rtimes} U(su_2) & \xrightarrow{\text{deformation}} & BS^3_q {\rtimes} U_q(su_2).
\end{array}
$$

From a structural point of view, we had to use the covariantised or braided version when q-deforming here in order to maintain the cross product structure of the quantisation. We have seen that these braided versions are naturally tied to a $*$-structure corresponding to the signature of Minkowski space.

Next we turn to the dual constructions. Recall from Chapters 5.4 and 6.4 that one of the striking questions raised by an algebra of observables being a Hopf algebra was the possibility of a symmetry between observables and states, in which the roles of these are interchanged by Hopf algebra duality. So, is the dual of the quantum double also the algebra of observables of some other quantum system, dual to the first? The answer, from a mathematical point of view, is potentially 'yes' because both the algebra and coalgebra were crossed, so the dual Hopf algebra is also a cross product. To study this, we use the conventions for a right handed cross product by a Hopf algebra A. Of course, there is a more usual dual in which left handed cross coproducts dualise to left handed cross products, as explained in Chapter 1.6.2. We prefer here to make also a left–right reversal appropriate to the categorical dual. Then the condition for a $*$-algebra cross product analogous to Proposition 6.1.5 is

$$
(b{\triangleleft}a)^* = b^* {\triangleleft} (Sa)^* \qquad (7.48)
$$

for all $a \in A$ and all elements b on which it acts. We begin with the dual of Lemma 7.4.4 as

Lemma 7.4.8 *Let A be a dual quasitriangular Hopf algebra. If B is a right A-comodule (algebra, coalgebra) then*

$$b \triangleleft a = \sum b^{(\bar{1})} \mathcal{R}(b^{(\bar{2})} \otimes a)$$

makes it automatically a right A-module (algebra, coalgebra). The original coaction and this induced action make B a right $D(H)$-module (algebra, coalgebra) in the (right) crossed module form along the lines of Proposition 7.1.6.

Proof: This is a strictly dual and right handed version of Lemma 7.4.4 and follows at once from (2.6). A right $D(H)$-module, in the crossed module form in Proposition 7.1.6, means a right H-module and H-comodule which are compatible in the sense $b^{(\bar{1})} \triangleleft h_{(1)} \otimes b^{(\bar{2})} h_{(2)} = (b \triangleleft h_{(2)})^{(\bar{1})} \otimes h_{(1)} (b \triangleleft h_{(2)})^{(\bar{2})}$. ∎

Theorem 7.4.9 *Let H be a finite-dimensional quasitriangular Hopf algebra with dual A. Then the cross coproduct and cross product $A \blacktriangleright\!\!\triangleleft H$ by the right coadjoint coaction of A and its induced action from Lemma 7.4.8 is a Hopf algebra. The maps*

$$A \xleftarrow{p} A \blacktriangleright\!\!\triangleleft H \xrightarrow{q} H^{\mathrm{cop}},$$

$$p(a \otimes h) = \mathrm{id} \otimes \epsilon, \quad q(a \otimes h) = \sum \langle \mathcal{R}^{-(2)}, a \rangle \mathcal{R}^{-(1)} h,$$

provide an isomorphism

$$(p \otimes q) \circ \Delta(a \otimes h) = \sum a_{(1)} \otimes \langle \mathcal{R}^{-(2)}, a_{(2)} \rangle \mathcal{R}^{-(1)} h : \quad A \blacktriangleright\!\!\triangleleft H \cong A \blacktriangleright\!\!\blacktriangleleft H^{\mathrm{cop}}$$

to the dual of the quantum double of H. The algebra structure is a $*$-algebra cross product if H is a real quasitriangular Hopf $*$-algebra.

Proof: The general Hopf algebra construction is strictly analogous to that in Theorem 7.4.5 in a dual and right handed form. The coaction is $\beta(h) = e_a h S e_b \otimes f^a f^b$, where $\{e_a\}$ is a basis of H and $\{f^a\}$ is a dual basis (as in Proposition 7.1.3). This is such that evaluation against an element of H is the usual adjoint action in Example 1.6.3. We use it to make a cross coproduct according to Proposition 1.6.16. The induced action from Lemma 7.4.8 comes out as

$$h \triangleleft a = \mathcal{R}^{(1)} \triangleright h \langle \mathcal{R}^{(2)}, a \rangle,$$

where $\triangleright$ is the usual left adjoint action in Example 1.6.3. We make the cross product as stated via Proposition 1.6.10. If one wants the Hopf algebra structure explicitly in terms of A and H, it is

$$(a \otimes h)(b \otimes h) = ab_{(1)} \otimes \mathcal{R}^{(1)} \triangleright h \langle \mathcal{R}^{(2)}, b_{(2)} \rangle,$$

$$\Delta(a \otimes h) = a_{(1)} \otimes e_a h_{(1)}(S\mathcal{R}^{(2)})(Se_b) \otimes a_{(2)} f^a f^b \otimes \mathcal{R}^{(1)} \triangleright h_{(2)}, \qquad (7.49)$$

$$S(a \otimes h) = \left(S(f^a f^b) \otimes S^{-1}(\mathcal{R}^{(2)} e_a h(Se_b) u \mathcal{R}^{(1)}) \right)(Sa \otimes 1),$$

where the element u is from Proposition 2.1.8.

The maps p, q are verified in the dual way to the proof of Theorem 7.4.5. Not dual, however, is the question of $*$-structure for the cross product algebra (we did not consider the coalgebra $*$-structure in Theorem 7.4.5, which corresponds now to the algebra). For this we check

$$(h \triangleleft a)^* = \overline{\langle \mathcal{R}^{(2)}, a \rangle}(\mathcal{R}^{(1)} \triangleright h)^* = \langle (S\mathcal{R}^{(2)})^*, a^* \rangle (S\mathcal{R}^{(1)})^* \triangleright h^*$$

$$= \langle \mathcal{R}^{(2)*}, a^* \rangle \mathcal{R}^{(1)*} \triangleright h^* = \langle \mathcal{R}^{(2)}, a^* \rangle (S\mathcal{R}^{(1)}) \triangleright h^*$$

$$= \langle \mathcal{R}^{(2)}, S^{-1}(a^*) \rangle \mathcal{R}^{(1)} \triangleright h^* = h^* \triangleleft S^{-1}(a^*) = h^* \triangleleft (Sa)^*$$

as required. Hence, if A is antireal dual quasitriangular, then $A {\blacktriangleright\!\!\blacktriangleleft} \underline{H}$ is a $*$-algebra cross product. ∎

Theorem 7.4.10 *Let A is a dual quasitriangular Hopf algebra. Then the cross coproduct and cross product $A {\blacktriangleright\!\!\blacktriangleleft} \underline{A}$ by the right adjoint coaction and its induced action is a Hopf algebra. The maps*

$$A \overset{i}{\hookrightarrow} A {\blacktriangleright\!\!\blacktriangleleft} \underline{A} \overset{j}{\hookleftarrow} A, \quad i(a) = a \otimes 1, \quad j = \Delta,$$

provide an isomorphism

$$\cdot \circ (i \otimes j)(a \otimes b) = \sum ab_{(1)} \otimes b_{(2)} : \qquad A {\bowtie}_{\mathcal{R}} A \cong A {\blacktriangleright\!\!\blacktriangleleft} \underline{A}.$$

Proof: This is dual to Theorem 7.4.6 and is obtained by dualising the proof there and changing from left to right handed conventions. The right adjoint coaction in Example 1.6.14 is used for the cross coproduct. The induced action is $b \triangleleft a = a_{(2)} \mathcal{R}((Sb_{(1)})b_{(3)} \otimes a)$ from Lemma 7.4.8 and is used for the cross product. If one wants to know the resulting Hopf algebra structure of $A {\blacktriangleright\!\!\blacktriangleleft} \underline{A}$ explicitly in terms of A, it is

$$(a \otimes c)(b \otimes d) = ab_{(1)} \otimes c_{(3)} d_{(2)} \mathcal{R}((Sc_{(2)})c_{(4)} \otimes Sd_{(1)}) \mathcal{R}((Sc_{(1)})c_{(5)} \otimes b_{(2)}),$$

$$\Delta(a \otimes c) = a_{(1)} \otimes c_{(2)} \otimes a_{(2)}(Sc_{(1)})c_{(3)} \otimes c_{(4)}, \qquad (7.50)$$

$$S(a \otimes c) = u(c_{(3)}) \mathcal{R}(c_{(5)} \otimes c_{(2)}) (S((Sc_{(1)})c_{(6)}) \otimes Sc_{(4)}) (Sa \otimes 1),$$

where u is from Proposition 2.2.4. The maps i, j can be verified directly to be Hopf algebra maps. ∎

Example 7.4.11 *The cross product $G_q \bowtie\!\!\!\!\!< BG_q$ induced by the quantum adjoint coaction, as in Theorem 7.4.9, is isomorphic to the quantum co-double in the form $G_q \bowtie_\mathcal{R} G_q$ from Example 7.3.7. Similarly, $G_q \bowtie\!\!\!\!\!< BU_q(g)$ induced by the quantum coadjoint coaction, as in Theorem 7.4.10, is isomorphic to the quantum codouble $G_q \bowtie U_q(g)^{\mathrm{cop}}$. Both cross products are generated by the standard matrix generators $\mathbf{t}, \mathbf{u}$ with cross relations, coproduct and antipode*

$$\mathbf{u}_1 \mathbf{t}_2 = \mathbf{t}_2 R^{-1} \mathbf{u}_1 R, \quad \Delta \mathbf{u} = \mathbf{u} \otimes \mathbf{t}^{-1}(\ )\mathbf{t} \mathbf{u}, \quad S\mathbf{u} = \mathbf{u}^{-1} \cdot S\left(\mathbf{t}^{-1}(\ )\mathbf{t}\right),$$

where $\mathbf{t}^{-1}(\ )\mathbf{t}$ has a space for the matrix indices of $\mathbf{u}$ or $\mathbf{u}^{-1}$ to the left to be inserted. Here $\mathbf{t}$ appears as a sub-Hopf algebra.

Proof: This is a special case of either Theorem 7.4.9 or Theorem 7.4.10 according to one's point of view on the braided matrices. The coaction is the standard adjoint one $\beta(\mathbf{u}) = \mathbf{t}^{-1}\mathbf{u}\mathbf{t}$ from Example 1.6.14 for the matrix form of $\mathbf{t}$. The matrix form and Corollary 4.1.9 give the action from Theorem 7.4.9 as

$$\mathbf{u}_1 \triangleleft \mathbf{t}_2 = \mathbf{l}_2^+ \triangleright \mathbf{u}_1 = R^{-1} \mathbf{u}_1 R.$$

Then, from Propositions 1.6.10 and 1.6.16, we obtain the cross product and coproduct structures as stated. The antipode is most easily obtained from (9.51) in Chapter 9.4. The explicit formulae are $\Delta u^i{}_j = u^a{}_b \otimes (St^i{}_a)t^b{}_c u^c{}_j$ and $Su^i{}_j = (\underline{S}u^a{}_b)S((St^i{}_a)t^b{}_j)$, where $\underline{S}\mathbf{u} = \mathbf{u}^{-1}$. The maps in Theorems 7.4.9 and 7.4.10 likewise come out as

$$G_q \overset{i}{\hookrightarrow} G_q \bowtie\!\!\!\!\!< BG_q \overset{j}{\hookleftarrow} G_q, \quad G_q \overset{p}{\hookleftarrow} G_q \bowtie\!\!\!\!\!< BU_q(g) \overset{q}{\hookrightarrow} U_q(g)^{\mathrm{cop}},$$

$$i(\mathbf{s}) = \mathbf{t} \otimes 1, \quad j(\mathbf{t}) = \mathbf{t} \otimes \mathbf{u}, \tag{7.51}$$

$$p = \mathrm{id} \otimes \epsilon, \quad q(\mathbf{t} \otimes 1) = S\mathbf{l}^+, \quad q(1 \otimes \mathbf{u}) = \mathbf{l}^+ S\mathbf{l}^-,$$

and provide isomorphisms $\cdot \circ (i \otimes j)$ and $(p \otimes q) \circ \Delta(\mathbf{t} \otimes 1) = \mathbf{t} \otimes S\mathbf{l}^+$, $(p \otimes q) \circ \Delta(1 \otimes \mathbf{u}) = 1 \otimes \mathbf{u}$. This works for general matrix quantum groups based on R-matrices. For the standard deformations, the two results coincide by factorisability. Thus,

$$G_q \bowtie_\mathcal{R} G_q \cong G_q \bowtie\!\!\!\!\!< BG_q \equiv G_q \bowtie\!\!\!\!\!< U_q(g) \cong G_q \bowtie U_q(g)^{\mathrm{cop}} \tag{7.52}$$

are all forms of the dual of the quantum double of $U_q(g)$.   ∎

This emphasises the cross product structure of the dual of the quantum double and of the complexified quantum function algebras in Section 7.3. For another point of view, we regard the braided matrices $\mathbf{u}$ as the covariantised version of the quantum function algebra and hence the coordinate functions on some kind of (braided) space. Then we can interpret these cross products as quantisations according to Chapter 6.1. To be concrete, we assume $|q| = 1$ and again take for our position observables the braided matrices $BSL_q(2)$ but with a new $*$-structure,

$$\begin{pmatrix} a^* & b^* \\ c^* & d^* \end{pmatrix} = \begin{pmatrix} a & q^2 b \\ q^2 c & q^2 d + (1 - q^2)a \end{pmatrix}, \tag{7.53}$$

in contrast to the Hermitian $*$-structure above. This $*$-structure corresponds, from the point of view of an enveloping algebra, to $U_q(sl(2,\mathbb{R}))$ in Example 3.2.5, so one could denote this as the braided matrix algebra $BSL_q(2,\mathbb{R})$. It has an action from the right of the quantum matrices $\mathbf{t}$, for which we take the new $*$-structure

$$\begin{pmatrix} t^1{}_1{}^* & t^1{}_2{}^* \\ t^2{}_1{}^* & t^2{}_2{}^* \end{pmatrix} = \begin{pmatrix} t^1{}_1 & qt^1{}_2 \\ q^{-1}t^2{}_1 & t^2{}_2 \end{pmatrix}. \tag{7.54}$$

This corresponds to Example 3.2.5 as the dual Hopf $*$-algebra, and we denote it by $SL_q(2,\mathbb{R})$. The general picture is that, this time, the R-matrix obeys $\overline{R} = R^{-1}$, so the usual real-type $*$-structures is not appropriate.

We are in the position of Theorem 7.4.9 and hence we have a (right handed) $*$-algebra cross product dual as a Hopf algebra to the quantum double model above. We can think of it as a new quantum system in the sense of Chapter 6.1. This time, q is required to be of modulus 1 rather than real as before. The position observables from this point of view are $BSL_q(2,\mathbb{R})$, and they correspond this time to a particle constrained to a sphere in $\mathbb{R}^{2,2}$. The momentum (quantum group) is based on an interpretation $SL_q(2,\mathbb{R}) \cong U_q(su_2^*)$, along similar lines to that at the end of Section 7.3. There is more than one way to make this interpretation precise, but here is at least one. We define

$$q^\xi = t^1{}_1, \quad q^\eta = t^2{}_2, \quad \zeta = \frac{t^1{}_2}{q - q^{-1}}, \quad \chi = \frac{t^2{}_1}{q - q^{-1}},$$

where we suppose that q is generic; then the algebra relations become

$$[\chi, \xi] = \chi = [\eta, \chi], \quad [\zeta, \xi] = \zeta = [\eta, \zeta], \quad [\chi, \zeta] = 0,$$

$$q^\xi q^\eta = 1 + (q - q^{-1})^2 q^{-1} \zeta \chi,$$

while its usual matrix coproduct becomes a q-deformation of the usual linear one for enveloping algebras. The $*$-structure becomes

$$\xi^* = -\xi, \quad \eta^* = -\eta, \quad \chi^* = -q^{-1}\chi, \quad \zeta^* = -q\zeta.$$

Proceeding formally, we suppose that $q = e^t$ (where t in our case is imaginary) and deduce from the Campell–Baker–Hausdorff formula applied to these equations that

$$[\xi, \eta] = O(t), \quad \eta = -\xi + O(t).$$

The scaling of the generators is critical here, and means that (with scaling as stated) we have a deformation of the Lie algebra

$$[\chi, \xi] = \chi, \quad [\zeta, \xi] = \zeta, \quad [\chi, \zeta] = 0, \tag{7.55}$$

with its usual linear coproduct. The Lie algebra here is su_2^*, which will be studied further in Chapter 8. This motivates our physical interpretation of this *-algebra cross product.

Finally, the right action of this q-momentum group on the coordinate generators of $BSL_q(2, \mathbb{R})$ is

$$\begin{pmatrix} a & b \\ c & d \end{pmatrix} \triangleleft \xi = \begin{pmatrix} 0 & -b \\ c & 0 \end{pmatrix}, \quad (\ ) \triangleleft \zeta = 0, \quad \begin{pmatrix} a & b \\ c & d \end{pmatrix} \triangleleft \chi = \begin{pmatrix} -qc & -(d-a) \\ 0 & q^{-1}c \end{pmatrix}.$$

In the limit $q \to 1$, we see that ξ, χ, ζ become anti-self-adjoint and have a *-representation on the algebra of functions on $SL(2, R)$.

This is the system whose quantisation as a *-algebra cross product is the dual of the quantum double of $U_q(sl(2, \mathbb{R}))$, at least for generic q. Thus, we have for the dual of the quantum double in this case the interpretation

$$\begin{array}{ccc}
su_2^* \otimes \mathbb{C}(SL(2, \mathbb{R})) & \xrightarrow{\text{deformation}} & U_q(su_2^*) \otimes BSL_q(2, \mathbb{R}) \\
\\
\text{quantisation} \downarrow & & \downarrow \text{quantisation} \qquad (7.56) \\
\\
U(su_2^*) {\ltimes} \mathbb{C}(SL(2, \mathbb{R})) & \xrightarrow{\text{deformation}} & U_q(su_2^*) {\ltimes} BSL_q(2, \mathbb{R}).
\end{array}$$

The cross product quantisation of this system is then the dual Hopf algebra to the quantum double and hence dual to the system (7.47). This possibility of a dual interpretation is an interesting principle that we have already discussed in Chapters 5.4 and 6.4 above.

Notes for Chapter 7

The quantum double construction $D(H)$ is due to V.G. Drinfeld [22], where it was introduced by generators and relations built from the structure constants of H in a basis. It was subsequently studied by the author [12, 116, 14, 142, 143], from where much of the material in this chapter is taken. The slightly more abstract form in Theorem 7.1.1 in which we build the double explicitly on the vector space $H^* \otimes H$ is from [12], where elementary Hopf-algebraic proofs of Drinfeld's results were given. Also introduced in this paper was the description as a double cross product $H {\bowtie} H^{*\mathrm{cop}}$ by mutual coadjoint actions. The alternative conventions with $H^{*\mathrm{op}} {\bowtie} H$ were introduced in [14] and [143] as a way to avoid the use of the inverse antipode of H. The generalised quantum double (7.2), which works for dually paired bialgebras, is due to the author [14], where the notion of convolution-invertible pairing or *weak antipode* was introduced for just this purpose.

The representation theory given in Proposition 7.1.6 and Lemma 7.4.4 is from [143] by the author, where the interpretation of J.H.C. Whitehead's notion of crossed G-sets [144] as modules of $D(G)$ in Example 7.1.10 was also introduced. The generalisation to a category of vector spaces on which a Hopf algebra acts and coacts in a compatible way had been proposed [145] as a category of 'crossed bimodules', but was shown in [143] to be the category of representations of Drinfeld's quantum double (in a form appropriate for the infinite-dimensional case), i.e. not a new braided category (see Chapter 9). The $*$-algebra structure on the quantum double in Proposition 7.1.4 is from [35], Example 7.1.11 is from [146] and the factorisability property is from [33]. The expression of the quantum double as a dual-twisting in Proposition 7.1.5 is from [147].

The general theory of Hopf algebra double cross products $\bowtie$ in Section 7.2 is due to author. The theory was introduced in 1987 [12] in connection with the bicrossproduct models described in Chapter 6.2. Definition 7.2.1, Theorem 7.2.2 and Proposition 7.2.4 (which connects the two constructions) and Example 7.2.5 are taken from this work. Theorem 7.2.3 has been strengthened slightly [148] from the original form [12]. Note that the first half of the proof of Theorem 7.2.3 covered the theory of algebra factorisations $B\bowtie_\Psi C$ as well. Drinfeld's quantum double provided an important example in the form $D(H) = H\bowtie H^{*\mathrm{cop}}$ etc. of the double cross product theory [12]. This form of the quantum double has subsequently been useful, even in the finite-dimensional case [143, 149].

The introduction of generalised quantum doubles was one of the first applications of the double cross product theory, along the lines of Example 7.2.6; we have emphasised the pairing and its convolution inverse, $\langle\ ,\ \rangle^{-1}$, as maps $H \otimes A \to k$ rather than the original form $H \to A^*$ in [14]. These generalised quantum doubles go significantly beyond the usual quantum double because the pairing need not be nondegenerate, i.e. the two factors in the double cross product can be far from strictly dual. The examples $A(R)\bowtie A(R)$ etc. of generalised quantum doubles in Examples 7.2.12–7.2.14 were introduced in [14] on this basis. Example 7.2.7 represents a subsequent trend to write the pairing in the generalised quantum double as a skew-pairing of H with A^{op}. This is, of course, a purely cosmetic shift of notation and not a new construction.

The celebrated application of the quantum double of $U_q(b_+)$ as projecting onto $U_q(sl_2)$ in Example 7.2.9, Proposition 7.2.10 and Exercise 7.2.11, is due to Drinfeld [22], where the general $U_q(g)$ case is also covered. An explicit form for the resulting quasitriangular structures appeared later (see Chapter 3 for details). The structural descriptions $D(U_q(b_-)) = U_q(sl_2)\bowtie_\chi k\mathbb{Z}$ as a central extension of $U_q(sl_2)$ in Example 7.2.9, and as a twisting of $U_q(sl_2) \otimes U_{q^{-1}}(1)$ in Proposition 7.2.10, seem to be more recent observations of the author [148]. Our approach to constructing the self-pairing is also unconventional and has the advantage that the nondegeneracy of the pairing is clear, at least when q is generic. This is because $\langle \phi(X, g), X^m g^n \rangle = \alpha^m (\partial_q^m \phi)(0, q^{2n})$ vanishing for all m, n implies that ϕ vanishes too. Some interesting finite quantum doubles are computed in [123], namely the quantum doubles of the bicrossproduct Hopf algebras from Chapter 6.2.

The theory of double cross coproducts $\bowtie$ in Exercise 7.2.15 is just the dual theory to that of double cross products from [12], and is not usually emphasised for this reason. It is taken from [148]. A special case is the generalised quantum codouble $H \bowtie_\sigma A$ in Exercise 7.2.16, which generalises the *dual* of Drinfeld's quantum double. This particular example recovers a previous 'twisted square' construction obtained by N.Yu. Reshetikhin and M.A. Semenov-Tian-Shansky [33] from another point of view; these authors were generalising the quantum double of a factorisable Hopf algebra (see Section 7.3); the generalisation of the quantum double from this twisted square point of view, and of the dual of the quantum double from the point of view of a double cross coproduct, are two independent routes to the construction in Exercise 7.2.16.

Section 7.3 is devoted to these twisted squares $H \bowtie_{\mathcal{R}} H$. Theorem 7.3.5 is due to [33], where the maps i, j and $\cdot \circ (i \otimes j)$ are given and the latter asserted to be an isomorphism to the quantum double in the factorisable case. We could not find any indication of the proof of invertibility of $\circ (i \otimes j)$ in [33], and have instead obtained (7.30) using braided group methods (from the semidirect product forms of the quantum double in Section 7.4). Also, the term 'twisted square' appears to antedate the general theory of twisting of Hopf algebras as we have presented it in Chapter 2.3. However, it is obvious enough that (7.28) is a particular example, and we have therefore presented the material accordingly.

The use of this twisted square construction (or its dual) as complexification appeared in [150] by the author. Proposition 7.3.2, constructing $A \bowtie_{\mathcal{R}} A$ as a Hopf ∗-algebra, and its duality with the twisted square, are taken from this work. The dual quasitriangular structure $\mathcal{R}_D$ was also given explicitly, while the second one, $\mathcal{R}_L$, is obtained in just the same way. It was used in [151], and elsewhere. The enveloping algebra version in Proposition 7.3.4 just converts these results to the twisted square via the duality from [150]. Also explained in this work was the isomorphism of $A \bowtie A$ with the dual of the quantum double in the general factorisable case, as well as its use in the matrix setting to recover the $A(R) \bowtie A(R)$ construction from [14]. The special case of the complexification of $SU_q(2)$ as q-Lorentz group and its isomorphism to the quantum double appeared in [152]. The ∗-structure for this particular example, and the use of the

R-matrix form $A(R) \bowtie A(R)$ for it, appeared independently in [153]. The identification of these two approaches and their further identification as the dual of the twisted square $U_q(su_2) \bowtie U_q(su_2)$ is due to the author [150]. This provided, thereby, a description of the q-Lorentz enveloping algebra as a twisting of two copies of $U_q(su_2)$. One can also base the q-Lorentz quantum group on multiindex R-matrices $\mathbf{R}_L$ [151], or $\mathbf{R}_{II}$ [153], so that it takes the familiar form of a matrix quantum group as in Chapter 4.1. This was developed by U. Meyer [151], where $\mathbf{R}_L$ is also used to introduce an addition law on q-Minkowski space, as we will see in Chapter 10.5.

Section 7.4 on cross product theorems for the quantum double is due to the author [34, 35, 143, 154]; one of which [35] developed the interpretation of the quantum double as q-quantum mechanics of a particle on a unit hyperboloid in q-Minkowski space. This work (from mid-1992) was one of first in which the braided matrices $BM_q(2)$ from [94] were put to serious use as a definition of q-Minkowski space; see Chapter 10 for a full discussion. On the mathematical side, [143] also showed that for a finite-dimensional Hopf algebra, quasitriangular structures are in one to one correspondence with Hopf algebra projections $D(H) \to H$ which cover the canonical inclusion. This then ensures a general cross product and cross coproduct or 'biproduct' [155] form. The further picture behind these results must await the bosonisation theory of braided groups in Chapter 9.4. The required covariantisation constructions $\underline{A}, \underline{H}$ in Theorems 7.4.1 and 7.4.2 are part of this braided group theory [38, 146, 156]. Note, also, that the natural cross product $*$-structure on a bosonisation does not form a usual Hopf $*$-algebra, but something weaker [47]. The examples $B(R)$ were obtained [94] by covariantisation of $A(R)$, as we have already seen in Chapter 4. See the Notes to Chapters 4, 9 and 10 for further details on all of these topics.

8

Lie bialgebras and
Poisson brackets

This chapter is devoted to the infinitesimal or semiclassical structures underlying the theory of quantum groups. The infinitesimal notion of a group is that of a Lie algebra: we follow the same line now for a Hopf algebra or quantum group. Another point of view which we consider comes from classical and quantum mechanics: we can think of a Hopf algebra as a noncommutative deformation of the commutative algebra of functions $\mathbb{C}(G)$ on a group G, recovered as the limit $t \to 0$, say, of a deformation parameter t. In this case one can imagine an order-by-order expansion in powers of t. Since t controls the degree of noncommutativity, it plays a role mathematically analogous to that of Planck's constant $\hbar$ in some approaches to quantisation. Recall that, in quantum mechanics, the algebra of observables is a noncommutative version of the classical algebra of functions on phase-space. The data governing the lowest order of deformation of this are contained in the semiclassical theory. They consists in our case of geometrical structures such as Poisson brackets on the group G.

In fact, we have already seen Hopf algebras in Chapter 6 (the bicross-product models) which really are quantisations, while for other more well-known cases, such as the quantum function algebras of Chapter 4, we follow Drinfeld and take this view more as a mathematical analogy. Nevertheless, it is a very powerful analogy, and it leads to a rich theory of Poisson brackets and associated data on group manifolds. While classical mechanics is an old and well-established subject, the special properties of classical mechanics on group manifolds were not studied very systematically until recent years: it is exactly this situation to which the semiclassical theory of quantum groups leads us. Such systems, as opposed to general classical mechanical systems, are in a certain sense 'completely integrable', and are rather interesting for this reason.

Because the data for classical mechanics and quantisation are purely geometrical, there is plenty of scope in this chapter for tangent bundles, symplectic structures and such topics. Some familiarity with classical differential geometry will be required for this reason in the later sections, but, for the most part, we limit ourselves by taking a fairly algebraic line wherever possible. We concentrate in the first two sections on giving the semiclassical versions of the various Hopf algebraic results already covered in previous chapters. This is our theme. Thus, the semiclassical or infinitesimal notion of a Hopf algebra is a *Lie bialgebra* consisting of (g, δ), where g is a Lie algebra and $\delta : g \to g \otimes g$ is the *cobracket* obeying axioms dual to those of a Lie algebra. It is also compatible with the Lie algebra structure. Of special interest are quasitriangular Lie bialgebras, where $\delta = \partial r$ and $r \in g \otimes g$ is a solution of the classical Yang–Baxter equation. These notions are all due to V.G. Drinfeld, who also introduced the double Lie bialgebra $D(g)$ as an example. It is the semiclassical analogue of the quantum double in Chapter 7. These results form the topics of Sections 8.1 and 8.2.

We also give in Section 8.3 the Lie algebra version of the notion of a matched pair of groups in Chapter 6, as a pair of Lie algebras acting on each other. These give *double cross sum Lie algebras*, with Drinfeld's double understood now as $g^{*\mathrm{op}} \bowtie g$ by mutual coadjoint actions. We also show that, subject to some completeness conditions, every Lie algebra matched pair exponentiates to a group one. Both results are due to the author in connection with the theory of bicrossproduct Hopf algebras. We also formulate these results at the Lie bialgebra level (matched pairs of Lie bialgebras). In Section 8.4 we show how to exponentiate the results about Lie bialgebras to the group level as results about Poisson–Lie groups. This makes contact with Hamiltonian mechanics and classical integrable systems.

The reader may reasonably wonder why we have left the semiclassical theory till Chapter 8 of the book. Usually in physics, one begins with the more geometrical and accessible theory of classical mechanics and then tries to quantise it. The reason for our reversing the order here is an important one, both for the mathematics of Hopf algebras and from the physical point of view. The theory of Hopf algebras has indeed its own intrinsic structure and motivations, such as themes of duality and other themes of braiding to come in the next two chapters. The semiclassical theory is only one small part of this, and one should not imagine that the theory of Hopf algebras is completely determined as 'quantising' such structures. Rather, one can argue that it is the other way around, and that much of the beautiful theory of Lie algebras etc. is a remnant of a deeper theory and deeper principles which apply to Hopf algebras. The same holds more generally in physics; it can be argued that, to our best knowledge, the real world is a quantum one governed by its own deeper principles (which we would like to elucidate), and that the familiar features of classical mechanics are only what remain in one special limit in which $\hbar \to 0$. If quantum groups teach us anything about quantum physics, it is that it has its own intrinsic mathematical structure and principles which are not only the deformation or 'quantisation' of familiar classical concepts.

8.1   Lie bialgebras and the CYBE

We have seen in earlier chapters two ways in which a general Hopf algebra could be regarded as generalising or deforming a group. Thus the matrix quantum groups of Chapter 4 are ones of 'function algebra type', i.e. like the functions on a group G. Meanwhile, the quantum enveloping algebras, such as $U_q(g)$ in Chapter 3, are more like the enveloping algebra of a Lie algebra or the group algebra of its associated group. Corresponding to these two points of view are two ways to develop the semiclassical theory. We begin with the latter framework and defer the former to Section 8.4.

Thus, we think of a group algebra or enveloping algebra as our model. Then the correct infinitesimalisation of the product structure is just the notion of a Lie algebra itself. In any associative algebra, one can consider the commutator $[A, B] = AB - BA$ and see that it obeys the Jacobi identity. For another example, the primitive elements in any Hopf algebra (those for which the coproduct is additive) are closed under this bracket, giving an associated Lie algebra.

Dualising this idea, we have a similar notion for the coalgebra. Thus, given any coalgebra with coproduct Δ, we can study the properties of the *cocommutator* $\delta = \Delta - \tau \circ \Delta$, where τ is transposition. It obeys the *co-Jacobi identity* and forms an example of a *Lie coalgebra*.

Of course, in a bialgebra or Hopf algebra we have both ideas together and they should be compatible in some way. These considerations suggest the following definition of a Lie bialgebra. We work, for convenience, over a field k of characteristic zero.

Definition 8.1.1 *A* Lie bialgebra *is* $(g, [\ , \ ], \delta)$, *where* g *is a Lie algebra over* k *and* $\delta : g \to g \otimes g$ *obeys*

$$\xi_{[1]} \otimes \xi_{[2]} = -\xi_{[2]} \otimes \xi_{[1]}, \tag{8.1}$$

$$(\mathrm{id} \otimes \delta) \circ \delta\xi + \mathrm{cyclic} = 0, \tag{8.2}$$

$$\delta([\xi, \eta]) = \mathrm{ad}_\xi(\delta(\eta)) - \mathrm{ad}_\eta(\delta(\xi)), \tag{8.3}$$

i.e.

$$\delta([\xi, \eta]) = \sum [\xi, \eta_{[1]}] \otimes \eta_{[2]} + \eta_{[1]} \otimes [\xi, \eta_{[2]}] - [\eta, \xi_{[1]}] \otimes \xi_{[2]} + \xi_{[1]} \otimes [\eta, \xi_{[2]}],$$

where ad *is the Lie algebra adjoint action extended to tensor products as a derivation, and* $\delta\xi = \sum \xi_{[1]} \otimes \xi_{[2]}$ *is a shorthand notation.*

The first two axioms here are those of a Lie coalgebra and are such that g^* becomes a Lie algebra by dualising δ. The second condition is an infinitesimal compatibility condition, and says, in cohomological terms, that

$$\delta \in Z^1_{\mathrm{ad}}(g, g \otimes g), \quad \text{i.e.} \quad \partial \delta = 0 \tag{8.4}$$

in the sense of the Lie algebra cohomology coboundary operator (2.22) in Chapter 2.3 or Chapter 6.3, where we allowed for a nontrivial action such as we need here. A map between Lie bialgebras is a linear map respecting the brackets (a Lie algebra homomorphism) and the cobrackets (a Lie coalgebra homomorphism).

Proposition 8.1.2 *If* $(g, [\ ,\ ], \delta)$ *is a finite-dimensional Lie bialgebra, then so is* g^* *by dualisation according to*

$$\langle [\phi, \psi], \xi \rangle = \langle \phi \otimes \psi, \delta \xi \rangle, \quad \langle \delta \phi, \xi \otimes \eta \rangle = \langle \phi, [\xi, \eta] \rangle$$

for all ϕ, ψ *in* g^* *and* ξ, η *in* g. *More generally, two Lie bialgebras are said to be* dually paired *if their Lie brackets and Lie cobrackets are related in this way.*

Proof: As remarked, from the cobracket we define the dual map $[\ ,\ ]_{g^*}$: $(g \otimes g)^* \to g^*$. Since $g^* \otimes g^* \subseteq (g \otimes g)^*$ by the obvious inclusion, we can restrict it to a map $g^* \otimes g^* \to g^*$. This works even if g is infinite dimensional, and we clearly obtain a Lie algebra. On the other hand, we can also dualise the bracket to a map $\delta_{g^*} : g^* \to (g \otimes g)^*$ as stated, and in the finite-dimensional case this becomes a map $g^* \to g^* \otimes g^*$. It clearly obeys the co-Jacobi identity if $[\ ,\ ]$ on g obeys the Jacobi identity. We verify the compatibility axiom as

$$\begin{aligned}
\langle \delta([\phi, \psi]), \xi \otimes \eta \rangle &= \langle [\phi, \psi], [\xi, \eta] \rangle = \langle \phi \otimes \psi, \delta([\xi, \eta]) \rangle \\
&= \langle \phi \otimes \psi, [\xi, \eta_{[1]}] \otimes \eta_{[2]} + \eta_{[1]} \otimes [\xi, \eta_{[2]}] \rangle - (\xi \leftrightarrow \eta) \\
&= \langle \phi_{[1]} \otimes [\phi_{[2]}, \psi] + \psi_{[1]} \otimes [\phi, \psi_{[2]}], \xi \otimes \eta - \eta \otimes \xi \rangle \\
&= \langle [\phi, \psi_{[1]}] \otimes \psi_{[2]} + \psi_{[1]} \otimes [\phi, \psi_{[2]}] - (\phi \leftrightarrow \psi), \xi \otimes \eta \rangle
\end{aligned}$$

for all ϕ, ψ in g^* and ξ, η in g. We used the definition of the bracket and cobracket of g^*, and for the third equality we used the cocycle property (8.3) for (g, δ). ∎

This means that Lie bialgebras have the 'self-dual' nature of Hopf algebras, and is a second motivation for their definition. One can likewise work through various other elementary constructions from Chapters 1 and 2 in this Lie bialgebra setting, such as (co)actions of Lie bialgebras on algebras and Lie algebras. Especially, we recall that we had the notion of a quasitriangular Hopf algebra in Chapter 2.1. Its Lie bialgebra version arises quite naturally from the following consideration: since δ is a 1-cocycle, as explained above, a natural class of *coboundary Lie bialgebras* consists of Lie bialgebras for which

$$\delta = \partial r, \quad \text{i.e.} \quad \delta\xi = \text{ad}_\xi(r) \equiv \sum [\xi, r^{[1]}] \otimes r^{[2]} + r^{[1]} \otimes [\xi, r^{[2]}] \qquad (8.5)$$

for some $r = \sum r^{[1]} \otimes r^{[2]} \in g \otimes g$. Here $r \in C^0_{\text{ad}}(g, g \otimes g)$ is a 1-cochain and ∂ is the appropriate Lie algebra coboundary operator. So the Lie cobracket in this class is cohomologically trivial. This does not mean that the Lie bialgebra itself is trivial.

Proposition 8.1.3 $(g, [\,,\,], r)$ *defines a coboundary Lie bialgebra iff*

$$\text{ad}_\xi([\![r, r]\!]) = 0, \quad \text{ad}_\xi(r + r_{21}) = 0,$$

for all $\xi \in g$, where ad *is the usual Lie algebra adjoint action extended to products in the usual way (as a derivation) and where*

$$[\![r, s]\!] = [r_{12}, s_{13}] + [r_{12}, s_{23}] + [r_{13}, s_{23}],$$

i.e.

$$[\![r, s]\!] = \sum [r^{[1]}, s^{[1]}] \otimes r^{[2]} \otimes s^{[2]} + r^{[1]} \otimes [r^{[2]}, s^{[1]}] \otimes s^{[2]} + r^{[1]} \otimes s^{[1]} \otimes [r^{[2]}, s^{[2]}].$$

Proof: Since $\delta = \partial r$, $d^2 = 0$ implies that $\delta \in Z^1_{\text{ad}}(g, g \otimes g)$, so this condition is automatic. For the antisymmetry condition, we need

$$\sum [\xi, r^{[1]}] \otimes r^{[2]} + r^{[1]} \otimes [\xi, r^{[2]}] + r^{[2]} \otimes [\xi, r^{[1]}] + [\xi, r^{[2]}] \otimes r^{[1]} = 0,$$

which is just that $r + r_{21}$ is ad-invariant as stated. Next, assuming this, we see that $\delta = \partial r = \partial r_-$ if $r = r_- + r_+$, where r_- is the antisymmetric part of r and r_+, the symmetric part. Keeping this in mind, we compute the condition needed for the co-Jacobi identity as

$(1 \otimes \delta)\delta\xi + \text{cyclic}$
$\quad = [\xi, r_-^{[1]}] \otimes [r_-^{[2]}, r_-'^{[1]}] \otimes r_-'^{[2]} + [\xi, r_-^{[1]}] \otimes r_-'^{[1]} \otimes [r_-^{[2]}, r_-'^{[2]}]$
$\quad\quad + [[\xi, r_-^{[2]}], r_-'^{[1]}] \otimes r_-'^{[2]} \otimes r_-^{[1]} + [[\xi, r_-^{[2]}], r_-'^{[2]}] \otimes r_-^{[1]} \otimes r_-'^{[1]} + \text{cyclic}$
$\quad = ([\xi, \,] \otimes \text{id}) [\![r_-, r_-]\!] + \text{cyclic}$
$\quad = ([\xi, \,] \otimes \text{id}) [\![r_-, r_-]\!] + (\text{id} \otimes [\xi, \,] \otimes \text{id}) [\![r_-, r_-]\!] + (\text{id} \otimes [\xi, \,]) [\![r_-, r_-]\!]$
$\quad = \text{ad}_\xi([\![r_-, r_-]\!]),$

where, for the first equality, the terms in the cyclic average involving ξ in the second and third positions in $g \otimes g \otimes g$ were rotated into the first factor. r'_- denotes another independent copy of r_-. For the second equality, the primed and unprimed copies were interchanged in the third term. We the used antisymmetry of r_- in the third and fourth terms, and the Jacobi identity in g. But since r_- is antisymmetric, $[\![r_-, r_-]\!]$ is already invariant under the permutations, which gives the third equality. Thus, the co-Jacobi identity is precisely that $[\![r_-, r_-]\!]$ be ad-invariant. The next lemma then ensures that this is the same as requiring that $[\![r, r]\!]$ be ad-invariant, as stated. $\blacksquare$

Lemma 8.1.4 *Let $r = r_- + r_+$ in $g \otimes g$, where r_- is antisymmetric, and r_+ is symmetric and ad-invariant. Then $[\![r, r]\!] = [\![r_-, r_-]\!] + [\![r_+, r_+]\!]$ and $[\![r_+, r_+]\!]$ is ad-invariant.*

Proof: This is because the cross terms $[\![r_\pm, r_\mp]\!]$ vanish; for example,

$$[r_-^{[1]}, r_+^{[1]}] \otimes r_-^{[2]} \otimes r_+^{[2]} + r_-^{[1]} \otimes [r_-^{[2]}, r_+^{[1]}] \otimes r_+^{[2]}$$
$$+ r_-^{[1]} \otimes r_+^{[1]} \otimes [r_-^{[2]}, r_+^{[2]}] + [r_+^{[1]}, r_-^{[1]}] \otimes r_+^{[2]} \otimes r_-^{[2]}$$
$$+ r_+^{[1]} \otimes [r_+^{[2]}, r_-^{[1]}] \otimes r_-^{[2]} + r_+^{[1]} \otimes r_-^{[1]} \otimes [r_+^{[2]}, r_-^{[2]}] = 0$$

by the symmetries of $r_\pm$ and ad-invariance of r_+. For the second part, we have

$$[\![r_+, r_+]\!] = r_+^{[1]} \otimes [r_+^{[2]}, r_+'^{[1]}] \otimes r_+'^{[2]}$$

because r_+ is ad-invariant. As an element of $g \otimes g \otimes g$, this is ad-invariant since

$$[\xi, r_+^{[1]}] \otimes [r_+^{[2]}, r_+'^{[1]}] \otimes r_+'^{[2]} + r_+^{[1]} \otimes [\xi, [r_+^{[2]}, r_+'^{[1]}]] \otimes r_+'^{[2]}$$
$$+ r_+^{[1]} \otimes [r_+^{[2]}, r_+'^{[1]}] \otimes [\xi, r_+'^{[2]}] = 0,$$

for all $\xi \in g$. We used the Jacobi identity on the centre term and then the ad-invariance of r_+. $\blacksquare$

We have emphasised an explicit notation in these proofs, but one may also use a compact notation where numerical suffixes denote the position in a tensor product of copies of g. Thus, $[\![r, s]\!]$ was also given in this way in Proposition 8.1.3. In the same notation, the definition $\delta = \partial r$ implies useful identities such as

$$(\mathrm{id} \otimes \delta)r = [r_{12}, r_{23}] + [r_{13}, r_{23}], \quad (\delta \otimes \mathrm{id})r = [r_{13}, r_{12}] + [r_{23}, r_{12}], \quad (8.6)$$

which can then be used to give a more compact proof of the above results. Also note that, in practice, one does not need the full generality of a coboundary Lie bialgebra, but various special cases. Following the terminology for quantum groups, one has:

Definition 8.1.5 *A coboundary Lie bialgebra* $(g, [\ ,\ ], r)$ *is called* quasi-triangular *if* $[\![r, r]\!] = 0$. *It is called* triangular *if* $[\![r, r]\!] = 0$ *and* $r_{21} = -r$. *It is called* factorisable *if* $[\![r, r]\!] = 0$ *and* $(r + r_{21}) : g^* \to g$ *is a linear surjection.*

The equation $[\![r, r]\!] = 0$ is called the *classical Yang–Baxter equation* (CYBE). It is quite natural if we think of r as an infinitesimal version of the quasitriangular structure or universal R-matrix as in Chapter 2.1. Given (8.6), the CYBE is equivalent to either of the conditions

$$(\mathrm{id} \otimes \delta)r = [r_{13}, r_{12}], \quad (\delta \otimes \mathrm{id})r = [r_{13}, r_{23}]. \tag{8.7}$$

The 'triangular' case corresponds to $\mathcal{R}$ triangular, while the 'factorisable' case is the other extreme and corresponds to $\mathcal{R}$ factorisable.

Lemma 8.1.6 *Let* $(g, [\ ,\ ], r)$ *be a coboundary Lie bialgebra. Then* $r \in g \otimes g$ *obeys* $[\![r, r]\!] = 0$ *iff* $r_1 : g^* \to g$ *defined by* $r_1(\phi) = \sum \langle \phi, r^{[1]} \rangle r^{[2]}$ *is a Lie algebra homomorphism. Likewise, iff it is a Lie coalgebra anti-homomorphism. Likewise iff* $r_2 : g^* \to g$ *defined by* $r_2(\phi) = \sum r^{[1]} \langle \phi, r^{[2]} \rangle$ *is a Lie algebra antihomomorphism or Lie coalgebra homomorphism. The quasitriangularity of* r *is also equivalent to*

$$r_-([\phi, \psi]) = [r_-(\phi), r_-(\psi)] + [r_+(\phi), r_+(\psi)],$$

where $r_\pm = \frac{1}{2}(r_1 \pm r_2)$ *and where* r_+ *obeys*

$$r_+([\phi, \psi]) = [r_1(\phi), r_+(\psi)] - [r_1(\psi), r_+(\phi)] - 2[r_+(\phi), r_+(\psi)]$$
$$= [r_-(\phi), r_+(\psi)] + [r_+(\phi), r_-(\psi)].$$

Proof: Here g^* has the Lie algebra structure given by dualising $\delta = \partial r$ according to Proposition 8.1.2, so

$$\langle [\phi, \psi], r'^{[1]} \rangle r'^{[2]} - \langle \phi, r^{[1]} \rangle \langle \psi, r'^{[1]} \rangle [r^{[2]}, r'^{[2]}]$$
$$= \langle \phi \otimes \psi \otimes \mathrm{id}, \delta r'^{[1]} \otimes r'^{[2]} - r^{[1]} \otimes r'^{[1]} \otimes [r^{[2]}, r'^{[2]}] \rangle$$
$$= \langle \phi \otimes \psi \otimes \mathrm{id}, \mathrm{ad}_{r^{[1]}}(r') \otimes r^{[2]} - r^{[1]} \otimes r'^{[1]} \otimes [r^{[2]}, r'^{[2]}] \rangle$$
$$= -\langle \phi \otimes \psi \otimes \mathrm{id}, [\![r, r]\!] \rangle.$$

The vanishing of this for all ϕ, ψ is the CYBE condition $[\![r, r]\!] = 0$. The computation for r_2 is strictly analogous. One can also characterise the CYBE in terms of the Lie coalgebra. Thus, for r_2 we compute

$$\delta(r_2(\phi)) - (r_2 \otimes r_2)(\delta(\phi)) = \delta r^{[1]} \langle \phi, r^{[2]} \rangle - r^{[1]} \otimes r'^{[1]} \langle \delta \phi, r^{[2]} \otimes r'^{[2]} \rangle$$
$$= ([r'^{[1]}, r^{[1]}] \otimes r^{[2]} + r^{[1]} \otimes [r'^{[1]}, r^{[2]}]) \langle \phi, r'^{[2]} \rangle - r^{[1]} \otimes r'^{[1]} \langle \phi, [r^{[2]}, r'^{[2]}] \rangle$$
$$= -\langle \mathrm{id} \otimes \phi, [\![r, r]\!] \rangle$$

so that $[\![r, r]\!] = 0$ is also equivalent to r_2 a Lie coalgebra map (defined in the obvious way as commuting with δ). Similarly for the equivalence with

r_1 a Lie anticoalgebra map, i.e. a Lie coalgebra map with the opposite Lie cobracket on one side. For the last part, we compute

$$r_+([\phi, \psi]) = \langle \delta r_+{}^{[1]}, \phi \otimes \psi \rangle r_+{}^{[2]}$$
$$= \langle [r_+{}^{[1]}, r^{[1]}] \otimes r^{[2]} + r^{[1]} \otimes [r_+{}^{[1]}, r^{[2]}], \phi \otimes \psi \rangle r_+{}^{[2]}$$
$$= [r_2(\psi), r_+(\phi)] + [r_1(\phi), r_+(\psi)]$$
$$= [r_1(\phi), r_+(\psi)] - [r_1(\psi), r_+(\phi)] + 2[r_+(\psi), r_+(\phi)],$$

where we used the definitions of the bracket in g^* in terms of δ on g and ad-invariance of r_+ under $\mathrm{ad}_{r_1(\phi)}$ and $\mathrm{ad}_{r_2(\psi)}$. We then use $r_- = r_1 - r_+$ for the quasitriangularity condition in terms of r_-. ∎

This is the infinitesimal version of Exercise 2.1.5, and it tells us precisely when a coboundary Lie bialgebra is quasitriangular. It is also useful in calculations with Lie bialgebras, and will play an important role in later sections. One can also use Lemma 8.1.6 as an easy way to construct quasitriangular Lie bialgebras, as follows. First, it is easy to see that the Lie algebra structure on g^* in these terms is just

$$[\phi, \psi] = \mathrm{ad}^*_{r_1(\phi)}(\psi) + \mathrm{ad}^*_{r_2(\psi)}(\phi), \quad \langle \mathrm{ad}^*_\xi(\phi), \eta \rangle = \langle \phi, -\mathrm{ad}_\xi(\eta) \rangle, \quad (8.8)$$

where ad^* is the left coadjoint action of g on g^*. So one can start with a Lie algebra g and find a Lie bracket on g^* of the form (8.8), where $r_1, -r_2 : g^* \to g$ are Lie algebra homomorphisms and such that $r_1 + r_2$ is an intertwiner between the coadjoint and adjoint actions. The result is necessarily a quasitriangular Lie bialgebra by Lemma 8.1.6.

The last part of Lemma 8.1.6 tells us how to proceed in terms of the symmetric and antisymmetric parts of r. If r obeys the CYBE, then the antisymmetric part r_- will not generally do so, as we have seen already in Lemma 8.1.4. It can, nevertheless, be useful to focus on this skew-symmetric part. In particular, if we are only interested in the finite-dimensional factorisable case, then $2r_+ : g^* \to g$ is a linear isomorphism, and we can focus on the linear map

$$r_M = r_- \circ (2r_+)^{-1} : g \to g.$$

In terms of this, the last part of Lemma 8.1.6 tells us that the CYBE, or quasitriangularity condition, becomes

$$r_M([\xi, \eta]_{r_M}) = [r_M(\xi), r_M(\eta)] + \frac{1}{4}[\xi, \eta],$$
$$[\xi, \eta]_{r_M} \equiv [\xi, r_M(\eta)] - [\eta, r_M(\xi)], \quad (8.9)$$

which is the so-called *modified classical Yang–Baxter equation (MCYBE)*.

Probably the most important setting for these constructions is when g is a complex semisimple Lie algebra. First, one knows then that $H^1_{\mathrm{ad}}(g, g \otimes g)$

is trivial and hence that all such Lie bialgebras are coboundary. Secondly, one has a canonical ad-invariant invertible bilinear form K (the Killing form) and its inverse K^{-1} provides a natural choice for the symmetric part r_+. Also, by changing this symmetric part to a different multiple of K^{-1}, we do not change the Lie cobracket, but (from Lemma 8.1.4) we do add an ad-invariant part to $[\![r, r]\!]$. Typically, one could render $[\![r, r]\!] = 0$ in this way, which is why the concept of quasitriangularity is quite natural in this setting. Demanding $[\![r, r]\!] = 0$ generally fixes the symmetric part. If it is zero, we have a triangular Lie bialgebra, while, if it is a nonzero multiple of K^{-1}, then we have a factorisable one (and these are the only possibilities when g is simple). If we are interested only in the latter case, one can fix $2r_+ = K^{-1}$ from the start and solve (8.9) for r_-.

Finally, for our last piece of general theory, we give the analogue of the twisting construction of Chapter 2.3.

Theorem 8.1.7 *Let (g, δ) be a Lie bialgebra. If χ obeys the condition*

$$\mathrm{ad}_\xi((\mathrm{id} \otimes \delta)\chi + \mathrm{cyclic} + [\![\chi, \chi]\!]) = 0, \quad \mathrm{ad}_\xi(\chi + \chi_{21}) = 0, \qquad (8.10)$$

for all $\xi \in g$, then

$$\delta_\chi = \delta + \partial\chi, \quad \textit{i.e.} \quad \delta_\chi\xi = \delta\xi + \mathrm{ad}_\xi(\chi),$$

is also a Lie bialgebra.

Proof: This is a generalisation of Proposition 8.1.3, on which we build. As in the proof there, the requirement that δ_χ remains antisymmetric in its output tells us that $\chi + \chi_{21}$ needs to be ad-invariant. Then $\partial\chi = \partial\chi_-$, where χ_- is the antisymmetric part of χ. We verify the co-Jacobi identity as

$(\mathrm{id} \otimes \delta_\chi)\delta_\chi\xi + \mathrm{cyclic}$

$\quad = (\mathrm{id} \otimes \delta)\delta\xi + (\mathrm{id} \otimes \partial\chi)\delta\xi + ((\mathrm{id} \otimes \delta)\partial\chi)\xi + ((\mathrm{id} \otimes \partial\chi)\partial\chi)\xi + \mathrm{cyclic}$

$\quad = \xi_{[1]} \otimes [\xi_{[2]}, \chi^{[1]}] \otimes \chi^{[2]} + \xi_{[1]} \otimes \chi^{[1]} \otimes [\xi_{[2]}, \chi^{[2]}] + \chi^{[1]} \otimes \delta[\xi, \chi^{[2]}]$

$\quad\quad + [\xi, \chi^{[1]}] \otimes \delta\chi^{[2]} + \mathrm{cyclic} + \mathrm{ad}_\xi([\![\chi, \chi]\!]),$

where the second equality uses the fact that δ already obeys the co-Jacobi identity to cancel $(\mathrm{id} \otimes \delta)\delta$ along with the calculation in Proposition 8.1.3 to compute $(\mathrm{id} \otimes \partial\chi)\partial\chi$. The remaining four terms are from the definition of $\partial\chi$. We then use the cocycle axiom (8.3) to compute these four terms as

$\xi_{[1]} \otimes [\xi_{[2]}, \chi^{[1]}] \otimes \chi^{[2]} + \xi_{[1]} \otimes \chi^{[1]} \otimes [\xi_{[2]}, \chi^{[2]}] + \chi^{[1]} \otimes [\xi, \chi^{[2]}_{[1]}] \otimes \chi^{[2]}_{[2]}$

$\quad + \chi^{[1]} \otimes \chi^{[2]}_{[1]} \otimes [\xi, \chi^{[2]}_{[2]}] - \xi_{[2]} \otimes \chi^{[1]} \otimes [\chi^{[2]}, \xi_{[1]}] - \xi_{[1]} \otimes [\chi^{[2]}, \xi_{[2]}] \otimes \chi^{[1]}$

$\quad + [\xi, \chi^{[1]}] \otimes \delta\chi^{[2]} + \mathrm{cyclic} = \mathrm{ad}_\xi((\mathrm{id} \otimes \delta)\chi + \mathrm{cyclic}),$

where we also used the fact that the expressions are being cyclically averaged to rotate the $-$ terms to the right and left, respectively, to give the terms shown. We then used the fact that the symmetric part of χ is ad-invariant to cancel, giving just three terms under the cyclic average, which we recognised as $\mathrm{ad}_\xi((\mathrm{id}\otimes\delta)\chi)$. This gives the condition for the co-Jacobi identity. The proof of the cocycle axiom (8.3) for δ_χ is easy. We have

$$\delta_\chi([\xi,\eta]) = \delta([\xi,\eta]) + \mathrm{ad}_{[\xi,\eta]}\chi$$
$$= \mathrm{ad}_\xi(\delta(\eta)) - \mathrm{ad}_\eta(\delta(\xi)) + \mathrm{ad}_\xi\mathrm{ad}_\eta\chi - \mathrm{ad}_\eta\mathrm{ad}_\xi\chi$$
$$= \mathrm{ad}_\xi(\delta_\chi(\eta)) - \mathrm{ad}_\eta(\delta_\chi(\xi)),$$

because δ obeys the cocycle condition and ad is a Lie algebra action (in this case on $g\otimes g$). ∎

Note that, in particular, if $(g,[\ ,\ ],r)$ is a quasitriangular Lie bialgebra and $\chi\in g\otimes g$ obeys

$$[\![r,\chi]\!] + [\![\chi,r]\!] + [\![\chi,\chi]\!] = 0, \quad \mathrm{ad}_\xi(\chi+\chi_{21}) = 0,$$

then $(g,[\ ,\ ],r+\chi)$ is also a quasitriangular Lie bialgebra. This is also clear from Proposition 8.1.3 and the definition of quasitriangularity. We see, for example, that every quasitriangular Lie bialgebra is the twisting of the zero cobracket by $\chi=r$ in the sense of Theorem 8.1.7. The triangular case is likewise a twisting of the zero cobracket, with this time $\chi=r$ antisymmetric. This case has a standard cohomological interpretation, as the next exercise demonstrates.

Exercise 8.1.8 *An easy way to satisfy the twisting requirements in Theorem 8.1.7 is to require*

$$(\mathrm{id}\otimes\delta)\chi + \mathrm{cyclic} + [\![\chi,\chi]\!] = 0, \quad \chi+\chi_{21} = 0.$$

Show that if we view $\chi\in g\otimes g$ as a linear functional in $(g^*\otimes g^*)^*$, then these conditions are exactly that

$$\partial_{g^*}\chi + [\chi,\chi] = 0, \quad \chi\in C^2(g^*,k),$$

where ∂_{g^*} is the usual Lie algebra coboundary, acting on χ as a 2-cochain on g^*.

Hence, or otherwise, obtain the following dual-twisting theorem for any Lie bialgebra: if $(g,[\ ,\ ],\delta)$ is a Lie bialgebra and $\chi\in C^2(g,k)$ obeys

$$\partial\chi(\phi\otimes\psi\otimes\varphi) + \sum\chi(\phi_{[1]}\otimes\psi)\chi(\phi_{[2]}\otimes\varphi)$$
$$+ \sum\chi(\psi_{[1]}\otimes\varphi)\chi(\psi_{[2]}\otimes\phi) + \sum\chi(\varphi_{[1]}\otimes\phi)\chi(\varphi_{[2]}\otimes\psi) = 0$$

then so is $(g, [\ ,\ ]_\chi, \delta)$ *with*

$$[\phi, \psi]_\chi = [\phi, \psi] + \sum \phi_{[1]}\chi(\phi_{[2]} \otimes \psi) - \sum \psi_{[1]}\chi(\psi_{[2]} \otimes \phi).$$

Solution: A usual Lie algebra cochain means an antisymmetric linear function on our Lie algebra, which is one of the conditions for χ. The Lie algebra in question is g^*. The usual Lie algebra coboundary operator was given in (2.22), and appears now as

$$(\partial\chi)(\phi \otimes \psi \otimes \varphi) = \chi([\phi, \psi] \otimes \varphi) + \text{cyclic},$$

where '+cyclic' means to add the same term with $\phi \to \psi \to \varphi \to \phi$ and again the same term with two such rotations. It is obvious that $(\partial\chi)(\phi \otimes \psi \otimes \varphi) = \langle\phi \otimes \psi \otimes \varphi, (\text{id} \otimes \delta)\chi + \text{cyclic}\rangle$ under the pairing in Proposition 8.1.2. Likewise, it is clear that

$$\langle\phi \otimes \psi \otimes \varphi, [\![\chi, \chi]\!]\rangle = \chi(\phi_{[1]} \otimes \psi)\chi(\phi_{[2]} \otimes \varphi) + \text{cyclic},$$

giving the conditions stated. Finally, the twisting of the Lie cobracket on g in Theorem 8.1.7 dualises to the formulae stated. On the other hand, these formulae and their direct proof all make sense for any Lie bialgebra g, not necessarily dual to our original g and not necessarily finite dimensional. ∎

We see that theorems for Lie bialgebras can be dualised in just the same way as for Hopf algebras. The dual of the full Theorem 8.1.7 (with the conditions holding only up to the coadjoint action) can also be formulated in the same way. Moreover, it is clear that a *dual quasitriangular Lie bialgebra* is a Lie bialgebra equipped with a linear map $r : g \otimes g \to k$ such that

$$[\xi, \eta] = \sum \xi_{[1]}r(\xi_{[2]} \otimes \eta) + \eta_{[1]}r(\xi \otimes \eta_{[2]}), \tag{8.11}$$

$$\sum r(\xi \otimes \eta_{[1]})r(\eta_{[2]} \otimes \zeta) + r(\xi_{[1]} \otimes \eta)r(\xi_{[2]} \otimes \zeta) + r(\xi \otimes \zeta_{[1]})r(\eta \otimes \zeta_{[2]}), \tag{8.12}$$

with the symmetric part of r assumed to be ad-invariant. The special case where r is antisymmetric is the *dual triangular* case and is the dual-twisting, in the sense of the second part of Exercise 8.1.8, of g with the zero bracket and $\chi = r$. Also, just as we concluded (8.7) in the quasitriangular theory, we conclude now the dual formulae

$$r([\xi, \eta] \otimes \zeta) = \sum r(\xi \otimes \zeta_{[1]})r(\eta \otimes \zeta_{[2]}),$$
$$r(\xi \otimes [\eta, \zeta]) = \sum r(\xi_{[1]} \otimes \zeta)r(\xi_{[2]} \otimes \eta), \tag{8.13}$$

for all $\xi, \eta, \zeta \in g$. We will not need this dual theory much; we return to it only briefly, in Exercise 8.3.10 below.

This completes the basic general theory of Lie bialgebras. Before giving examples of Lie bialgebras and quasitriangular Lie bialgebras, we explain the motivation which was given above a little more formally. Thus, a *quantised enveloping algebra* means a Hopf algebra over $\mathbb{C}[[t]]$ generated by a vector space g, with relations and coproduct of the form

$$\xi\eta - \eta\xi = [\xi,\eta] + O(t), \quad (\Delta - \tau \circ \Delta)\xi = t\delta\xi + O(t^2),$$

where t is a formal deformation parameter and τ is the usual transposition map. Here $[\ ,\ ] : g \otimes g \to g$ is assumed antisymmetric in its input and $\delta : g \to g \otimes g$ is assumed antisymmetric in its output. One can easily see that $(g, \{\ ,\ \}, \delta)$ is necessarily a Lie bialgebra. Moreover, if the Hopf algebra has a quasitriangular structure of the form

$$\mathcal{R} = 1 + tr + O(t^2),$$

then $\delta = \partial r$, and our Lie bialgebra is quasitriangular. Twisting is also compatible with this analysis, and it leads to a theorem of uniqueness up to twisting for triangular quantisations of $U(g)$. A similar, but more complicated, result applies also for the standard quasitriangular quantisations $U_q(g)$; these are twistings of $U(g)$ as a quasi-Hopf algebra, as mentioned in Chapter 2.4.

Example 8.1.9 *Let $b_+ = \mathrm{span}\{H, X\}$ be the two-dimensional Lie algebra with $[H, X] = 2X$. It forms a Lie bialgebra with*

$$\delta(H) = 0, \quad \delta(X) = \frac{1}{2}(X \otimes H - H \otimes X).$$

Proof: We check the co-Jacobi identity (8.2) and cocycle axiom (8.3) as

$$(\mathrm{id} \otimes \delta) \circ \delta X + \mathrm{cyclic} = \frac{1}{2}(X \otimes X \otimes H - X \otimes H \otimes X + H \otimes X \otimes X$$
$$-X \otimes X \otimes H + X \otimes H \otimes X - H \otimes X \otimes X) = 0,$$

$$\delta([H, X]) = \delta(2X) = X \otimes H - H \otimes X,$$
$$\mathrm{ad}_H(\delta(X)) - \mathrm{ad}_X(\delta(H)) = \frac{1}{2}([H, X] \otimes H - H \otimes [H, X]),$$

as required. ∎

This Lie bialgebra is self-dual. It is a sub-Lie bialgebra (in the obvious sense) of the following important example.

Example 8.1.10 *Let $sl_2 = \mathrm{span}\{H, X_\pm\}$ be the Lie algebra sl_2 with $[H, X_\pm] = \pm 2X_\pm$ and $[X_+, X_-] = H$. This forms a quasitriangular Lie* *bialgebra with*

$$\delta(H) = 0, \quad \delta(X_\pm) = \frac{1}{2}(X_\pm \otimes H - H \otimes X_\pm), \quad r = X_+ \otimes X_- + \frac{1}{4}H \otimes H.$$

Proof: We already know the co-Jacobi identity from the preceding example, and also part of the cocycle axiom for a Lie bialgebra. The remaining part of the cocycle axiom is

$$\text{ad}_{X_+}(\delta(X_-)) - \text{ad}_{X_-}(\delta(X_+))$$
$$= \frac{1}{2}([X_+, H] \otimes X_- - [X_+, X_-] \otimes H + H \otimes [X_+, X_-] - X_- \otimes [X_+, H]$$
$$- [X_-, H] \otimes X_+ + [X_-, X_+] \otimes H - H \otimes [X_-, X_+] + X_+ \otimes [X_-, H])$$
$$= 0 = \delta(H),$$

as required. This example is also simple enough to verify directly that r as stated obeys the CYBE and that $\delta = \partial r$, in which case the Lie bialgebra axioms are assured by Proposition 8.1.3. For the CYBE we have

$$[\![r, r]\!] = \frac{1}{4}[H, X_+] \otimes H \otimes X_- + \frac{1}{4}[X_+, H] \otimes X_- \otimes H$$
$$+ X_+ \otimes [X_-, X_+] \otimes X_- + \frac{1}{4}H \otimes [H, X_+] \otimes X_- + \frac{1}{4}X_+ \otimes [X_-, H] \otimes H$$
$$+ \frac{1}{4}X_+ \otimes H \otimes [X_-, H] + \frac{1}{4}H \otimes X_+ \otimes [H, X_-] = 0,$$

given the relations of sl_2. Also, we have that

$$r + r_{21} = X_+ \otimes X_- + X_- \otimes X_+ + \frac{1}{2}H \otimes H \tag{8.14}$$

is ad-invariant as required for antisymmetry. For example,

$$\text{ad}_{X_+}(r + r_{21}) = [X_+, X_-] \otimes X_+ + \frac{1}{2}[X_+, H] \otimes H$$
$$+ X_+ \otimes [X_+, X_-] + \frac{1}{2}H \otimes [X_+, H] = 0,$$

given the relations of sl_2. Similarly for ad_H and ad_{X_-}. ∎

This Lie bialgebra is not self-dual. Letting $\{\phi, \psi_\pm\}$ denote the dual basis to $\{H, X_\pm\}$, we find the dual Lie bialgebra from Proposition 8.1.2 as the following.

Example 8.1.11 *Let $sl_2^* = \text{span}\{\phi, \psi_\pm\}$ be the Lie algebra with relations*

$$[\psi_\pm, \phi] = \frac{1}{2}\psi_\pm, \quad [\psi_+, \psi_-] = 0,$$
$$\delta\psi_\pm = \pm 2(\phi \otimes \psi_\pm - \psi_\pm \otimes \phi), \quad \delta\phi = \psi_+ \otimes \psi_- - \psi_- \otimes \psi_+.$$

Proof: For a direct proof that this is a Lie bialgebra, we already know part of the co-Jacobi identity and part of the cocycle axiom from the b_+ example. The remaining co-Jacobi identity is

$$(\mathrm{id} \otimes \delta)\delta\phi = 2(-\psi_+ \otimes \phi + \psi_+ \otimes \psi_- \otimes \phi - \psi_- \otimes \phi \otimes \psi_+ + \psi_- \otimes \psi_+ \otimes \phi$$
$$-\psi_- \otimes \psi_+ \otimes \phi + \psi_- \otimes \phi \otimes \psi_+ - \phi \otimes \psi_+ \otimes \psi_- + \phi \otimes \psi_- \otimes \psi_+$$
$$-\phi \otimes \psi_- \otimes \psi_+ + \phi \otimes \psi_+ \otimes \psi_- - \psi_+ \otimes \psi_- \otimes \phi + \psi_+ \otimes \phi \otimes \psi_-) = 0,$$

while the remaining cocycle axiom is checked by

$$\mathrm{ad}_{\psi_+}(\delta(\psi_-)) - \mathrm{ad}_{\psi_-}(\delta(\psi_+))$$
$$= 2(-[\psi_+, \phi] \otimes \psi_- + \psi_- \otimes [\psi_+, \phi] - [\psi_-, \phi] \otimes \psi_+ + \psi_+ \otimes [\psi_-, \phi]) = 0$$

using the relations.

Alternatively, we can show that this is the structure induced on the vector space spanned by $\{\phi, \psi_\pm\}$, as the dual basis to the basis for sl_2 in the preceding example. In this case, the Lie bialgebra axioms follow from Proposition 8.1.2. Thus, the nonzero pairings are

$$\langle \phi, H \rangle = 1, \quad \langle \psi_+, X_+ \rangle = 1, \quad \langle \psi_-, X_- \rangle = 1,$$

and we can verify the required duality. For example,

$$\langle \delta\psi_+, H \otimes X_+ \rangle = 2\langle \phi \otimes \psi_+ - \psi_+ \otimes \phi, H \otimes X_+ \rangle = 2 = \langle \psi_+, [H, X_+] \rangle,$$
$$\langle \delta\phi, X_+ \otimes X_- \rangle = \langle \psi_+ \otimes \psi_- - \psi_- \otimes \psi_+, X_+ \otimes X_- \rangle = 1 = \langle \phi, H \rangle,$$
$$\langle \psi_+ \otimes \phi, \delta X_+ \rangle = \langle \psi_+ \otimes \phi, \frac{1}{2}(X_+ \otimes H - H \otimes X_+) \rangle = \frac{1}{2}\langle \psi_+, X_+ \rangle \langle \phi, H \rangle$$
$$= \frac{1}{2} = \langle \frac{1}{2}\psi_+, X_+ \rangle = \langle [\psi_+, \phi], X_+ \rangle,$$

and similarly for ψ_-. Other pairings which give zero are more easily checked. ∎

Some of these examples are quite easily handled in a tensor notation. Just as one can introduce structure constants for a Lie algebra, one can do likewise for a Lie cobracket. Dualisation and some other constructions are then quite easy in this tensor notation.

Exercise 8.1.12 *Let* $g = \mathrm{span}\{e_i\}$ *be a finite-dimensional Lie bialgebra with this basis. Introduce structure constants by*

$$[e_i, e_j] = e_a c^a{}_{ij}, \quad \delta e_i = e_a \otimes e_b d^{ab}{}_i.$$

Show that the axioms of a Lie bialgebra are

$$c^i{}_{jk} = -c^i{}_{kj}, \quad d^{ij}{}_k = -d^{ji}{}_k,$$

$$c^l{}_{ia}c^a{}_{jk} + c^l{}_{ja}c^a{}_{ki} + c^l{}_{ka}c^a{}_{ij} = 0,$$

$$d^{ij}{}_a d^{ak}{}_l + d^{jk}{}_a d^{ai}{}_l + d^{ki}{}_a d^{aj}{}_l = 0,$$

$$d^{ij}{}_a c^a{}_{kl} = c^i{}_{ka}d^{aj}{}_l - c^i{}_{la}d^{aj}{}_k - c^j{}_{ka}d^{ai}{}_l + c^j{}_{la}d^{ai}{}_k,$$

(8.15)

*for the antisymmetry, Jacobi identity, co-Jacobi identity and cocycle ax-
ioms. Show that the dual-Lie bialgebra is built on the dual basis* $g^* = \{f^i\}$,
say, with

$$\langle f^i, e_j \rangle = \delta^i{}_j, \quad [f^i, f^j] = d^{ij}{}_a f^a, \quad \delta f^i = c^i{}_{ab} f^a \otimes f^b.$$

Equivalently, with the dual basis denoted by $g^* = \{e_i^*\}$, *the structure constants* c^*, d^* *are given via*

$$\langle e_i^*, e_j \rangle = \delta_{ij}, \quad c^{*i}{}_{jk} = d^{jk}{}_i, \quad d^{*ij}{}_k = c^k{}_{ij}.$$

Solution: This is a matter of writing out Definition 8.1.1 and Proposi-
tion 8.1.2 in a basis. Thus $\delta([e_i, e_j]) = \delta(e_a)c^a{}_{ij} = e_c \otimes e_d d^{cd}{}_a c^a{}_{ij}$, etc.
In this way we arrive immediately at (8.15) for the various conditions.
On the other hand, if we write c^*, d^* for the structure constants of g^* as
stated, it is easy to see that the second line of (8.15) for c^* is the third line
for d, and that the third line for d^* is the second line for c. The fourth
line of (8.15) is self-dual in this sense. This gives the tensorial proof of
Proposition 8.1.2. ∎

For completeness, we note also that, for a quasitriangular Lie bialgebra
in this tensorial notation, we need $r = e_a \otimes e_b r^{ab}$ such that

$$c^j{}_{ia}r^{ak} + c^k{}_{ia}r^{ja} + c^k{}_{ia}r^{aj} + c^j{}_{ia}r^{ka} = 0,$$

$$[\![r, r]\!]^{ijk} \equiv c^i{}_{ab}r^{aj}r^{bk} + c^j{}_{ab}r^{ia}r^{bk} + c^k{}_{ab}r^{ia}r^{jb} = 0.$$

(8.16)

Given such a tensor r, we have the Lie cobracket structure constants

$$d^{ij}{}_k = c^i{}_{ka}r^{aj} + c^j{}_{ka}r^{ia}.$$

(8.17)

Note also that a *real form* of a Lie algebra over $\mathbb{C}$ is a choice of basis in
which the structure constants are real; in this case, we have a Lie algebra
over $\mathbb{R}$. For Lie bialgebras over $\mathbb{C}$, we have this possibility for the Lie coalgebra also. So a *real–real form* of a Lie bialgebra g over $\mathbb{C}$ is a choice of
basis such that both the structure constants c, d are real. We regard the
real span in this basis as a Lie bialgebra u over $\mathbb{R}$. The dual basis gives

the dual real–real form u^* of g^*. This is the situation for Examples 8.1.10 and 8.1.11, which have real forms $sl(2, \mathbb{R})$ and $sl(2, \mathbb{R})^*$, respectively. We also have the tensor r real for $sl(2, \mathbb{R})$. Its quantisation is the quasitriangular Hopf $*$-algebra $U_q(sl(2, \mathbb{R}))$ in Example 3.2.5 with t imaginary, with the $*$-structure corresponding to the view that the elements of u (the real span of our basis of g) should be appear anti-Hermitian in $*$-representations.

Another natural possibility is a *half-real form* u of g, defined as the real span in a basis where c is real and d is imaginary. The basis $e_i^* = -\imath f^i$, where $\{f^i\}$ is the dual basis, defines the dual half-real form u^* of g^*. Its structure constants are

$$c^{*i}{}_{jk} = -\imath d^{jk}{}_i, \quad d^{*ij}{}_k = \imath c^k{}_{ij}. \tag{8.18}$$

In this context, the quantisation as above is a Hopf $*$-algebra with t real. In this setting the natural possibilities for r in the quasitriangular case are $\bar{r} = r_{21}$ (real-type) or r imaginary (antireal type). The following example gives the half-real forms su_2 and su_2^*.

Example 8.1.13 *Working over* $k = \mathbb{C}$, *the Lie algebras* sl_2 *and* sl_2^* *have dual bases*

$$e_1 = -\frac{\imath}{2}(X_+ + X_-), \quad e_2 = -\frac{1}{2}(X_+ - X_-), \quad e_3 = -\frac{\imath}{2}H,$$
$$f^1 = \imath(\psi_+ + \psi_-), \quad f^2 = -(\psi_+ - \psi_-), \quad f^3 = 2\imath\phi.$$

The structure constants and r-matrix in this basis are

$$c^i{}_{jk} = \epsilon^i{}_{jk}, \quad d^{ij}{}_k = \imath(\delta^i{}_k \delta^j{}_3 - \delta^j{}_k \delta^i{}_3), \quad r^{ij} = -\delta^{ij} + \imath \epsilon^{ij}{}_3,$$

where ϵ *is the totally antisymmetric tensor with* $\epsilon^1{}_{23} = \epsilon^{12}{}_3 = 1$.

Writing $\xi = e_i \xi^i$ *and* $\phi = \phi_i(-\imath f^i)$ *for 3-vectors* $\vec{\xi} = (\xi^i)$, $\vec{\phi} = (\phi_i)$, *we have*

$$[\vec{\xi}, \vec{\eta}] = \vec{\xi} \times \vec{\eta}, \quad \delta(\vec{\xi}) = \imath \vec{\xi} \wedge \vec{e_3}, \quad r = -\mathrm{id} + \imath * \vec{e_3},$$
$$[\vec{\phi}, \vec{\psi}] = \vec{e_3} \times (\vec{\phi} \times \vec{\psi}), \quad \delta(\vec{\phi}) = \imath * \vec{\phi},$$

where $\times$ *is the usual vector product,* $\wedge$ *is the outer product,* $*$ *is the Hodge* $*$-*operator and* $\vec{e_3}$ *is the basis vector in the third component. These Lie bialgebras are the half-real forms* su_2 *of* sl_2 *and* su_2^* *of* sl_2^*, *respectively.*

Proof: In the usual representation, the basis for sl_2 here is the anti-Hermitian one, $e_i = -\frac{\imath}{2}\sigma_i$, where σ_i are the Pauli matrices. Hence, or otherwise, it is clear that the Lie algebra structure constants in this basis are those of the real form su_2 or $so(3)$. One can easily check that

$\langle f^i, e_j \rangle = \delta^i{}_j$ given the duality pairing in Example 8.1.11, so we again have a dual basis. Then

$$\delta(e_1) = -\frac{\imath}{4}(X_+ \otimes H - H \otimes X_+ + X_- \otimes H - H \otimes X_-)$$
$$= \imath(e_1 \otimes e_3 - e_3 \otimes e_1),$$
$$\delta(e_2) = -\frac{1}{4}(X_+ \otimes H - H \otimes X_+ - X_- \otimes H + H \otimes X_-)$$
$$= \imath(e_2 \otimes e_3 - e_3 \otimes e_2),$$
$$\delta(e_3) = 0,$$

which can then be written compactly as stated. In the vectorial form, we use the usual vector product, outer product and Hodge $*$-operator

$$(\xi \times \eta)^i = \epsilon^i{}_{jk}\xi^j\eta^k, \quad (\xi \wedge \eta)^{ij} = \xi^i\eta^j - \eta^i\xi^j, \quad (*\xi)^{ij} = \epsilon^{ij}{}_a\xi^a,$$

respectively, to express the structure of su_2.

The structure of sl_2^* is then obtained in the dual basis $\{f^i\}$ according to the preceding example; it has the roles of the structure constants c, d reversed. In this dual basis, the Lie algebra is not real, but the cobracket is real. For the vectorial picture we choose a new basis, $\{-\imath f^i\}$, and have the real form su_2^* as stated. It is a solvable real Lie algebra if we view it over $\mathbb{R}$. The cobracket is defined now by the Hodge $*$-operator times $\imath$, i.e. it is again not real. So both these Lie bialgebras are not defined over $\mathbb{R}$ as bialgebras. They are half-real forms in the sense above. ∎

We conclude with a general class of examples associated to complex semisimple Lie algebras, with sl_2 as the simplest example. One can either verify this construction directly, much as we did for sl_2, or prove it from the general theory in the next section.

Proposition 8.1.14 *Let g be a semisimple Lie algebra over $\mathbb{C}$ in the notation of Chapter 3.3. Then,*

$$\delta(H_i) = 0, \quad \delta(X_{\pm i}) = \frac{d_i}{2}(X_{\pm i} \otimes H_i - H_i \otimes X_{\pm i}),$$
$$r = \sum_{\alpha \in \Delta^+} d_\alpha X_\alpha \otimes X_{-\alpha} + \frac{1}{2}\sum_{ij} d_i(a^{-1})_{ij}H_i \otimes H_j$$

is a quasitriangular Lie bialgebra. Here r is the Drinfeld–Jimbo *solution of the CYBE and $d_\alpha = \frac{1}{2}(\alpha, \alpha)$.*

Proof: As with the definition of $U_q(g)$ in Chapter 3.3, it is enough to give the Lie cobracket only on the simple root vectors since these generate the Lie algebra. To verify the Lie bialgebra axioms directly, we note first that the structure in each sl_2 subalgebra (for each i) is as in Example 8.1.10, so the co-Jacobi identity part of the cocycle axiom holds for these generators. The Lie cobracket extends according to the cocycle axiom (8.3), in which case one can deduce the co-Jacobi identity on the $X_{\pm\alpha}$ too. The general form of the Lie cobracket is

$$\delta(X_{\pm\alpha}) = \frac{d_\alpha}{2} X_\alpha \wedge H_\alpha + \sum_{\beta+\gamma=\alpha} c^{\pm\alpha}_{\pm\beta,\pm\gamma} X_{\pm\beta} \wedge X_{\pm\gamma}, \qquad (8.19)$$

where the sum is over positive roots adding up to α and the c are constants. We write $x \wedge y \equiv x \otimes y - y \otimes x$ for brevity. Thus, assuming (8.19) as induction hypothesis,

$$\delta([X_i, X_\alpha]) = \frac{d_{\alpha+\alpha_i}}{2}[X_i, X_\alpha] \wedge H_{\alpha+\alpha_i} + \alpha(d_i H_i) X_i \wedge X_\alpha$$
$$+ \sum_{\beta+\gamma=\alpha} c^\alpha_{\beta,\gamma}[X_i, X_\beta] \wedge X_\gamma + \sum_{\beta+\gamma=\alpha} c^\alpha_{\beta,\gamma} X_\beta \wedge [X_i, X_\gamma]$$

if $\alpha + \alpha_i$ is a positive root, using the identities

$$[d_\alpha H_\alpha, X_i] = \alpha(d_i H_i) X_i, \quad [d_i H_i, X_\alpha] = \alpha(d_i H_i) X_\alpha.$$

One may then build up the $c^\alpha_{\beta,\gamma}$ explicitly. Similarly for the negative roots. One can also verify that the extension of δ is consistent with the Serre relations (3.20) if we apply δ to both sides.

Alternatively, one can verify that r as stated obeys the CYBE and hence deduce the above from Proposition 8.1.3. For this one needs some elementary facts about the X_α from the theory of complex semisimple Lie algebras. We take these in the standard Chevalley form, where $[X_\alpha, X_{-\alpha}] = H_\alpha$ and $[X_\alpha, X_\beta] = N_{\alpha,\beta} X_{\alpha+\beta}$ when $\alpha + \beta$ is a root, and where the $N_{\alpha,\beta}$ are integers with $N_{-\alpha,-\beta} = -N_{\alpha,\beta}$. An abstract proof follows also from the classical double construction in the next section. ∎

Proposition 8.1.15 *The complex semisimple quasitriangular Lie bialgebra in the preceding proposition is factorisable and has a compact half-real form* u *with* r *of real type. The dual half-real form* u* *is noncompact and solvable.*

Proof: The half-real form u can be defined in the same manner as in Example 8.1.10 for the su_2 case for each simple root. Thus we let u be the real Lie algebra generated by the sub-Lie algebras

$$x_i = -\frac{\imath}{2}(X_i + X_{-i}), \quad y_i = -\frac{1}{2}(X_i - X_{-i}), \quad z_i = -\frac{\imath}{2}H_i.$$

The structure constants for each i are real by Example 8.1.10. One can compute the Lie brackets between them for $i \neq j$ and see that their structure constants are also real, as are those between new root vectors that are generated. The new part is to consider now the Lie cobracket. This has imaginary structure constants on the above generators as we know from Example 8.1.10 for each i. But if $\delta(\xi), \delta(\eta)$ have imaginary coefficients in our basis, then the cocycle axiom (8.3) tells us that $\delta([\xi, \eta])$ does also, since the Lie brackets introduce only real coefficients. Hence our basis is a half-real form in the sense explained in Section 8.1.

If one wants to work explicitly with the full basis rather than generators, the corresponding real basis is provided by

$$x_\alpha = -\frac{\imath}{2}(X_\alpha + X_{-\alpha}), \quad y_\alpha = -\frac{1}{2}(X_\alpha - X_{-\alpha}), \quad z_i = -\frac{\imath}{2}H_i,$$

with Lie brackets

$$[x_\alpha, y_\alpha] = z_\alpha, \quad [y_\alpha, z_\beta] = \frac{1}{2}a_{\beta\alpha}x_\alpha, \quad [z_\alpha, x_\beta] = \frac{1}{2}a_{\alpha\beta}y_\beta,$$

$$[x_\alpha, x_\beta] = \frac{1}{2}(N_{\alpha,\beta}y_{\alpha+\beta} + N_{\alpha,-\beta}y_{\alpha-\beta}),$$

$$[x_\alpha, y_\beta] = \frac{1}{2}(N_{\alpha,\beta}x_{\alpha+\beta} - N_{\alpha,-\beta}x_{\alpha-\beta}),$$

where $(\alpha, \beta) = d_\alpha a_{\alpha\beta} = d_\beta a_{\beta\alpha}$ and where $\alpha \neq \beta$ for the last two relations. We only need consider $\alpha \in \Delta^+$. The coproduct is

$$\delta x_\alpha = \imath(x_\alpha \otimes z_\alpha - z_\alpha \otimes x_\alpha), \quad \delta y_\alpha = \imath(y_\alpha \otimes z_\alpha - z_\alpha \otimes y_\alpha), \quad \delta z_\alpha = 0.$$

Next, we note that the Killing form K of a Lie algebra is defined via the trace in the adjoint representation. Hence, in the basis of a real form u, the Killing form is a real symmetric matrix. The real form is called *compact* if K is negative definite. It means that the associated Lie group is compact. In our basis above, the Killing form and its inverse come out as follows; the x, y, z directions are mutually orthogonal, and

$$K(z_i, z_j) = -\frac{1}{4}a_{ji}d_i^{-1}, \quad K(x_\alpha, x_\beta) = -\frac{1}{2}\delta_{\alpha,\beta} = K(y_\alpha, y_\beta),$$

$$K^{-1} = -4\sum_{ij} d_i(a^{-1})_{ij}z_i \otimes z_j - 2\sum_{\alpha \in \Delta^+} d_\alpha(x_\alpha \otimes x_\alpha + y_\alpha \otimes y_\alpha) = 2r_+.$$

Here we regarded $K : g \to g^*$ and inverted it. The inverse K^{-1} can then be regarded as an element of $g \otimes g$, which we see coincides with $2r_+$ in Proposition 8.1.14. This tells us that g is factorisable. We also see that $r_+ \in u \otimes u$, so its tensor in this basis is real. By contrast,

$$r_- = \imath \sum_{\alpha \in \Delta^+} d_\alpha(x_\alpha \otimes y_\alpha - y_\alpha \otimes x_\alpha),$$

so that the tensor for this antisymmetric part in our basis is imaginary. Hence, r is of real type in the sense explained above.

Next we compute u^*. By further diagonalising the Killing form (or otherwise), one can see that the dual basis to the x, y, z basis above is a real basis for the vector space $K(u)$. The basis $e^\star$ has a factor of $\imath$ relative to this dual basis. Hence,

$$u^\star = \imath K(u).$$

This gives a convenient description for the dual half-real form defined by (8.18). Let $\bar{\ } = \imath K(\ )$. Then its Lie algebra is defined by

$$[\bar{\xi}, \bar{\eta}] = [r_-(\bar{\xi}), \eta]^- - [r_-(\bar{\eta}), \xi]^-$$

using (8.8) and the ad-invariance of K, where we regard r_- as a map as in Lemma 8.1.6. In our case, we obtain

$$r_-(\bar{x}_\alpha) = \frac{1}{2} y_\alpha, \quad r_-(\bar{y}_\alpha) = -\frac{1}{2} x_\alpha, \quad r_-(\bar{z}_i) = 0,$$

from which the Lie bracket of $u^\star$ can be immediately computed. In particular, note that $[u^\star, u^\star] \subsetneq u^\star$ since the $\bar{z}_i$ are not in the image of the bracket. Considering higher and higher Lie brackets, one easily sees that $u^\star$ is solvable. It follows from general Lie algebra theory that it is also noncompact. ∎

8.2 Double Lie bialgebra

In this section, we give the infinitesimal or semiclassical counterparts of the results of Chapter 7 for the quantum double, i.e. we now study the classical double Lie bialgebra. This is built on the vector space $g^\star \oplus g$, at least when g is a finite-dimensional Lie bialgebra. We assume this now for convenience, but, as with the theory of the quantum double, it is easy to extend to the infinite-dimensional case by assuming a pair of Lie bialgebras with a duality pairing.

Proposition 8.2.1 *Let g be a finite-dimensional Lie bialgebra with dual $g^\star$ as in Proposition 8.1.2. There is a quasitriangular Lie bialgebra, $D(g)$, the* classical double *of g built on $g^\star \oplus g$ as a vector space, with*

$$[\phi \oplus \xi, \psi \oplus \eta] = ([\psi, \phi] + \sum \psi_{[1]} \langle \psi_{[2]}, \xi \rangle - \phi_{[1]} \langle \phi_{[2]}, \eta \rangle)$$

$$\oplus ([\xi, \eta] + \sum \xi_{[1]} \langle \psi, \xi_{[2]} \rangle - \eta_{[1]} \langle \phi, \eta_{[2]} \rangle),$$

$$\delta(\phi \oplus \xi) = \sum (\phi_{[1]} \oplus 0) \otimes (\phi_{[2]} \oplus 0) + \sum (0 \oplus \xi_{[1]}) \otimes (0 \oplus \xi_{[2]}),$$

$$r = \sum_a (f^a \oplus 0) \otimes (0 \oplus e_a).$$

Here $g^{\star\text{op}}, g$ appear as sub-Lie bialgebras, where $(\ )^{\text{op}}$ denotes the opposite (negated) Lie bracket. The set $\{e_a\}$ is a basis of g and $\{f^a\}$, a dual basis.

Proof: We will see later a more general construction of Lie algebra double cross sums, of which the Lie bialgebra structure here is an example. For the present, we give a direct proof. Recall that, for the Hopf algebra cross products in Chapter 1.6 or the quantum double in Chapter 7.1, one can either build the structure explicitly on the tensor product vector space built from the two factors, or give it implicitly in terms of cross relations. The same applies here: we can either build the double Lie bialgebra on the direct sum vector space $g^* \oplus g$ as stated, or note that every element of the direct sum has a unique decomposition into a vector in g^* and a vector in g, viewed in the canonical way in $g^* \oplus g$ and added there. Thus,

$$\phi \oplus \xi = \phi \oplus 0 + 0 \oplus \xi \equiv \phi + \xi,$$

where we identify $\phi \equiv \phi \oplus 0$ and $\xi \equiv 0 \oplus \xi$. In this notation, the double Lie bialgebra is

$$[\phi, \psi]_{D(g)} = [\psi, \phi], \quad [\xi, \eta]_{D(g)} = [\xi, \eta],$$

$$[\xi, \phi]_{D(g)} = \phi_{[1]}\langle \phi_{[2]}, \xi \rangle + \xi_{[1]}\langle \phi, \xi_{[2]} \rangle,$$

$$\delta_{D(g)}\phi = \delta\phi, \quad \delta_{D(g)}\xi = \delta\xi, \quad r = f^a \otimes e_a,$$

where the right hand sides are in terms of the structure of g, g^*. It is easy to verify that this obeys the Jacobi identity, so that we indeed have a Lie algebra. The main case to check is the cross bracket. We have

$$
\begin{aligned}
[\xi, [\phi, \psi]_{D(g)}]_{D(g)} &= -[\phi, \psi]_{[1]}\langle [\phi, \psi]_{[2]}, \xi \rangle - \xi_{[1]}\langle [\phi, \psi], \xi_{[2]} \rangle \\
&= -[\phi, \psi_{[1]}]\langle \psi_{[2]}, \xi \rangle - \psi_{[1]}\langle [\phi, \psi_{[2]}], \xi \rangle - (\phi \leftrightarrow \psi) \\
&\quad - \langle (\mathrm{id} \otimes \delta)\delta\xi, \mathrm{id} \otimes \phi \otimes \psi \rangle \\
&= -[\phi, \psi_{[1]}]\langle \psi_{[2]}, \xi \rangle - \psi_{[1]}\langle \phi, \xi_{[1]} \rangle \langle \psi_{[2]}, \xi_{[2]} \rangle - (\phi \leftrightarrow \psi) \\
&\quad - \langle (\mathrm{id} \otimes \delta)\delta\xi, \mathrm{id} \otimes \phi \otimes \psi \rangle,
\end{aligned}
$$

using the definitions and the cocycle axiom (8.3). On the other hand,

$$
\begin{aligned}
[[\xi, \phi]_{D(g)}, \psi]_{D(g)} &= -[\phi_{[1]}, \psi]\langle \phi_{[2]}, \xi \rangle + [\xi_{[1]}, \psi]_{D(g)}\langle \phi, \xi_{[2]} \rangle \\
&= -[\phi_{[1]}, \psi]\langle \phi_{[2]}, \xi \rangle + \psi_{[1]}\langle \psi_{[2]}, \xi_{[1]} \rangle \langle \phi, \xi_{[2]} \rangle \\
&\quad + \langle (\delta \otimes \mathrm{id})\delta\xi, \mathrm{id} \otimes \psi \otimes \phi \rangle,
\end{aligned}
$$

so that this, minus a similar expression with $\psi \leftrightarrow \phi$, equals the one above in view of the co-Jacobi identity for g. Also, r obeys the CYBE because

$$
\begin{aligned}
[\![r, r]\!] &= [f^a, f^b]_{D(g)} \otimes e_a \otimes e_b + f^a \otimes [e_a, f^b]_{D(g)} \otimes e_b \\
&\quad + f^a \otimes f^b \otimes [e_a, e_b]_{D(g)} \\
&= -[f^a, f^b] \otimes e_a \otimes e_b + f^a \otimes f^b_{[1]}\langle f^b_{[2]}, e_a \rangle \otimes e_b \\
&\quad + f^a \otimes e_{a[1]}\langle f^b, e_{a[2]} \rangle \otimes e_b + f^a \otimes f^b \otimes [e_a, e_b] = 0,
\end{aligned}
$$

using the useful identities

$$f^a \otimes e_{a[1]} \langle \phi, e_{a[2]} \rangle = [f^a, \phi] \otimes e_a, \quad f^a{}_{[1]} \langle f^a{}_{[2]}, \xi \rangle \otimes e_a = f^a \otimes [e_a, \xi] \quad (8.20)$$

for all $\phi \in g^*$ and $\xi \in g$. These are immediately seen by evaluating against $\mathrm{id} \otimes \psi$ and $\eta \otimes \mathrm{id}$, respectively, and using the duality pairing in Proposition 8.1.2. We also use the identities (8.20) to compute

$$\begin{aligned}
\mathrm{ad}_\phi(r) &= [\phi, f^a]_{D(g)} \otimes e_a + f^a \otimes [\phi, e_a]_{D(g)} \\
&= [f^a, \phi] \otimes e_a - f^a \otimes \phi_{[1]} \langle \phi_{[2]}, e_a \rangle - f^a \otimes e_{a[1]} \langle \phi, e_{a[2]} \rangle = \delta\phi, \\
\mathrm{ad}_\xi(r) &= [\xi, f^a]_{D(g)} \otimes e_a + f^a \otimes [\xi, e_a]_{D(g)} \\
&= f^a{}_{[1]} \langle f^a{}_{[2]}, \xi \rangle \otimes e_a + \xi_{[1]} \langle f^a, \xi_{[2]} \rangle \otimes e_a + f^a \otimes [\xi, e_a] = \delta\xi,
\end{aligned}$$

so that $\delta = \partial r$. In the same way (or since the result here is manifestly anti-symmetric), we see also that $\mathrm{ad}_\phi(r + r_{21}) = 0 = \mathrm{ad}_\xi(r + r_{21})$, as required in Proposition 8.1.3. Hence, we have a quasitriangular Lie bialgebra. ∎

The symmetric part r_+ is necessarily ad-invariant, as we know from Section 8.1. As for every quasitriangular Lie bialgebra, if this is invertible, it provides a natural symmetric bilinear form.

Proposition 8.2.2 *The classical double $D(g)$ of any Lie bialgebra is fac-torisable in the sense of Definition 8.1.5. The inverse of $2r_+ : D(g)^* \to D(g)$ is an ad-invariant map which we regard as an ad-invariant sym-metric bilinear form*

$$K(\phi \oplus \xi, \psi \oplus \eta) = \langle \phi, \eta \rangle + \langle \psi, \xi \rangle$$

on $D(g)$.

Proof: Clearly, $2r_+(\xi \oplus \phi) = 2\langle \xi \oplus \phi, r_+{}^{[1]} \rangle r_+{}^{[2]} = \phi \oplus \xi$ since $\{e_a\}$ and $\{f^a\}$ in Proposition 8.2.1 are dual bases. The pairing between $D(g)^*$ and $D(g)$ is

$$\langle \xi \oplus \phi, \psi \oplus \eta \rangle = \langle \phi, \eta \rangle + \langle \psi, \xi \rangle.$$

We have seen that $2r_+$ is ad-invariant as an element of $D(g) \otimes D(g)$, hence it is also equivariant as a map, where $D(g)^*$ has the coadjoint action. Hence, its inverse $D(g) \to D(g)^*$ is also equivariant and can be regarded as evaluation against an ad-invariant element $K \in D(g)^* \otimes D(g)^*$ as stated. This completes the proof. If one wanted to check directly that K is indeed ad-invariant, this is

$$\begin{aligned}
K([\phi \oplus \xi, \psi \oplus \eta], \varphi \oplus \zeta) &+ (\psi \oplus \eta \leftrightarrow \varphi \oplus \zeta) \\
= \langle [\psi, \phi], \zeta \rangle &+ \langle \delta\psi, \zeta \otimes \xi \rangle - \langle \delta\phi, \zeta \otimes \eta \rangle \\
&+ \langle \varphi, [\xi, \eta] \rangle + \langle \varphi \otimes \psi, \delta\xi \rangle - \langle \varphi \otimes \phi, \delta\eta \rangle + (\psi \oplus \eta \leftrightarrow \varphi \oplus \zeta)
\end{aligned}$$

$$= \langle [\psi, \phi], \zeta \rangle + \langle \psi, [\zeta, \xi] \rangle - \langle \phi, [\zeta, \eta] \rangle$$
$$+ \langle \varphi, [\xi, \eta] \rangle + \langle [\varphi, \psi], \xi \rangle - \langle [\varphi, \phi], \eta \rangle + (\psi \oplus \eta \leftrightarrow \varphi \oplus \zeta) = 0$$

by antisymmetry of the Lie brackets.   ∎

Example 8.2.3 *Let* $b_- = \mathrm{span}\{H, X\}$ *be the two-dimensional Lie alge-bra with relations* $[H, X] = -2X$ *equipped with a Lie bialgebra structure of the same form as in Example 8.1.9. Then* $D(b_-)$ *is an extension*

$$0 \to k \to D(b_-) \to sl_2 \to 0,$$

where sl_2 *is the Lie bialgebra from Example 8.1.10. Moreover,* $D(b_-) = (sl_2 \oplus u(1))_\chi$, *as twisted by cocycle*

$$\chi = \frac{1}{4}(c \otimes H - H \otimes c).$$

Here $u(1)$ *is the one-dimensional quasitriangular Lie bialgebra with gen-erator* c *and* $r = -\frac{1}{4} c \otimes c$.

Proof: We take $\delta(H) = 0$ and $\delta(X) = \frac{1}{2}(X \otimes H - H \otimes X)$ so that b_- is similar to b_+ in Example 8.1.9. (It has the opposite Lie algebra structure.) From that example, we know that this is a self-dual Lie bialgebra with self-pairing $\langle H, H \rangle = 4$, $\langle X, X \rangle = 1$ and $\langle H, X \rangle = 0$. Hence we can also identify $b_+ = b_-^{\mathrm{op}} = b_-^{*\mathrm{op}}$ as Lie bialgebras. To avoid confusion when we compute the double, we now write the generators more explicitly as $b_\pm = \mathrm{span}\{H_\pm, X_\pm\}$ when we intend H, X to be viewed in one or other Lie bialgebra. The double $D(b_-)$ in Proposition 8.2.1 is then built from $b_\pm$ as sub-Lie bialgebras and the Lie bracket

$$[X_-, X_+] = X_{+[1]}\langle X_{+[2]}, X_- \rangle + X_{-[1]}\langle X_+, X_{-[2]} \rangle$$
$$= -\frac{1}{2}H_+\langle X_+, X_- \rangle - \frac{1}{2}H_-\langle X_+, X_- \rangle = -H_-,$$

$$[X_-, H_+] = \frac{1}{2}X_-\langle H_+, H_- \rangle = 2X_-, \quad [H_-, H_+] = 0,$$

$$[H_-, X_+] = \frac{1}{2}X_+\langle H_+, H_- \rangle = 2X_+,$$

between them. If we now write $H_\pm = H \pm c$ then $[c, H] = 0$, $[c, X_\pm] = 0$, while $H, X_\pm$ generate the Lie algebra sl_2 in Example 8.1.10. The Lie coproduct is the tensor product one

$$\delta(X_\pm) = \frac{1}{2}(X_\pm \otimes (H \pm c) - (H \pm c) \otimes X_\pm), \quad \delta(H) = 0 = \delta(c),$$

inherited from $b_\pm$. Setting $c = 0$ projects onto sl_2 as a Lie bialgebra.

Also, the quasitriangular structure from Proposition 8.2.1 is evidently $r = \frac{1}{4}H_+ \otimes H_- + X_+ \otimes X_-$ according to the duality pairing above, which projects to that of sl_2 by setting $c = 0$. If we do not make the projection, we have

$$r = r_{sl_2} - \frac{1}{4}c \otimes c + \frac{1}{4}(c \otimes H - H \otimes c),$$

which is the twisting according to Theorem 8.1.7 by the stated χ of $sl_2 \oplus u(1)$. ∎

This is the analogue of Examples 7.2.9 and 7.2.10. Clearly, the same construction works for general complex semisimple Lie algebras g. We can work algebraically, using only the Cartan matrix and other data, as explained in Chapter 3.3. Thus b_- is the Lie algebra spanned by the Cartan elements and the negative root vectors. It is not quasitriangular but has Lie cobracket as in Proposition 8.1.14. b_+ is similar, and is spanned by the Cartan elements and the positive roots. Moreover, we identify $b_+ = b_-^{op}$ and $b_-^* = b_-$ by the self-pairing

$$\langle X_{-i}, X_{-j} \rangle = \frac{\delta_{ij}}{d_i}, \quad \langle H_i, H_j \rangle = \frac{2a_{ij}}{d_i},$$

and the double projects $D(b_-) \to g$. This is one way to prove Proposition 8.1.14 using our abstract techniques. We outline a second way at the end of this section which is based on the Iwasawa decomposition.

The dual Lie bialgebra to the Lie bialgebra double is the *Lie bialgebra codouble* $D(g)^*$. It is built on $g \oplus g^*$ with the direct sum Lie algebra structure and a complicated Lie cobracket, which one can work out from Proposition 8.1.2. The semiclassical analogue of the dual of Proposition 7.1.5 is

Proposition 8.2.4 *Let g be a finite-dimensional Lie bialgebra. The co-double $D(g)^*$ is the twisting as in Theorem 8.1.7 of the direct sum Lie bialgebra $g^{\text{cop}} \oplus g^*$ by*

$$\chi = \sum (0 \oplus f^a) \otimes (e_a \oplus 0) - (e_a \oplus 0) \otimes (0 \oplus f^a),$$

where $\{e_a\}$ is a basis of g and $\{f^a\}$, a dual basis. Here g^{cop} denotes the opposite (negated) cobracket.

Proof: The direct sum Lie algebra $g \oplus g^*$ means

$$[\xi \oplus \phi, \eta \oplus \psi] = [\xi, \eta] \oplus [\phi, \psi],$$

or equivalently that g, g^* are Lie subalgebras with $[\xi, \phi] = 0$ for the Lie bracket between them. The duality pairing of the codouble with the double is

$$\langle \xi \oplus \phi, \psi \oplus \eta \rangle = \langle \phi, \eta \rangle + \langle \psi, \xi \rangle,$$

and, using this, one easily computes the Lie cobracket of the codouble from the Lie bracket of the double as

$$\langle \delta(\xi \oplus \phi), (\psi \oplus \eta) \otimes (\varphi \oplus \zeta) \rangle = \langle \xi \otimes \phi, [\psi \oplus \eta, \varphi \oplus \zeta] \rangle$$
$$= \langle -\delta\xi, \psi \otimes \varphi \rangle + \langle [\xi, \eta], \varphi \rangle - \langle [\xi, \zeta], \psi \rangle$$
$$+ \langle \delta\phi, \eta \otimes \zeta \rangle + \langle [\phi, \varphi], \eta \rangle - \langle [\phi, \psi], \zeta \rangle$$
$$= \langle \delta_{g^{\mathrm{cop}} \oplus g^*}(\xi \oplus \phi) + \mathrm{ad}_{\xi \oplus \phi}(\chi), (\psi \oplus \eta) \otimes (\varphi \oplus \zeta) \rangle$$

for all $\phi, \psi, \varphi \in g^*$ and $\xi, \eta, \zeta \in g$. From the penultimate expression, one can write the Lie cobracket on the Lie subalgebras as

$$\delta_{D(g)^*}(\xi) = -\delta\xi - \mathrm{ad}_\xi + \tau \circ \mathrm{ad}_\xi, \quad \delta_{D(g)^*}(\phi) = \delta\phi - \mathrm{ad}_\phi + \tau \circ \mathrm{ad}_\phi,$$

where we regard $\mathrm{ad}_\xi : g \to g$ and $\mathrm{ad}_\phi : g^* \to g^*$ as elements of $g \otimes g^*$ in the natural way. On the other hand, the expression involving χ writes the Lie cobracket as a twisting. Here the Lie cobracket on $g^{\mathrm{cop}} \oplus g^*$ is $\delta_{g^{\mathrm{cop}} \oplus g^*}(\xi) = -\delta\xi$ and $\delta_{g^{\mathrm{cop}} \oplus g^*}(\phi) = \delta\phi$, and we evaluate $\mathrm{ad}_{\xi \oplus \phi}(\chi)$ using the direct sum Lie algebra structure.

To show that this is indeed a twisting, we verify the conditions in Theorem 8.1.7, in fact in the stronger form in Exercise 8.1.8. Writing

$$\chi = f^a \otimes e_a - e_a \otimes f^a,$$

with the inclusion into the direct sum Lie algebra understood, we compute

$$[\![\chi, \chi]\!] = [f^a, f^b] \otimes e_a \otimes e_b + [e_a, e_b] \otimes f^a \otimes f^b - f^a \otimes [e_a, e_b] \otimes f^b$$
$$- e_a \otimes [f^a, f^b] \otimes e_b + f^a \otimes f^b \otimes [e_a, e_b] + e_a \otimes e_b \otimes [f^a, f^b]$$
$$= [f^a, f^b] \otimes e_a \otimes e_b + [e_a, e_b] \otimes f^a \otimes f^b + \mathrm{cyclic}$$
$$= f^a \otimes \delta e_a + e_a \otimes \delta f^a + \mathrm{cyclic} = -(\mathrm{id} \otimes \delta_{g^{\mathrm{cop}} \oplus g^*})\chi + \mathrm{cyclic},$$

as required.   ∎

This completes the basic general theory of the classical double and its dual. Now we come to structure theorems for $D(g)$ in the case when g is quasitriangular. The first is the semiclassical analogue of Proposition 7.3.4 and Theorem 7.3.5, and is relevant to the theory of classical inverse scattering. The second is the analogue of the main result of Chapter 7.4, and comes from the theory of braided groups.

Theorem 8.2.5 *Let* $(g, [\ ,\ ], r)$ *be a quasitriangular Lie bialgebra. There is a quasitriangular Lie bialgebra* $g_r {\bowtie} g$ *given by twisting* $g \oplus g$ *by*

$$\chi = \sum (r^{[1]} \oplus 0) \otimes (0 \oplus r^{[2]}) - (0 \oplus r^{[2]}) \otimes (r^{[1]} \oplus 0).$$

Moreover, in the finite-dimensional case, there is a homomorphism θ : $D(g) \to g_r \bowtie g$ of Lie bialgebras given by

$$\theta(\phi \oplus \xi) = (r_2(\phi) + \xi) \oplus (-r_1(\phi) + \xi),$$

which is an isomorphism when g is factorisable. Here, r_1, r_2 are linear maps as in Lemma 8.1.6.

Proof: We construct this Lie bialgebra by twisting, using Theorem 8.1.7. Later on, we give a double cross sum construction $\bowtie$ for Lie bialgebras, with the present construction an example of the dual of this (hence the notation). As a Lie algebra, we take $g \oplus g$ with elements ϕ, ψ etc. in the left copy of g, ξ, η etc. in the right copy and $[\xi, \phi] = 0$ between them. To be clear, we denote the two copies by g_L, g_R, say. Note that we can view $r \in g \otimes g$ in four ways, namely

$$r_{LL} \in g_L \otimes g_L, \quad r_{LR} \in g_L \otimes g_R, \quad r_{RL} \in g_R \otimes g_L, \quad r_{RR} \in g_R \otimes g_R,$$

and in this notation we define

$$r_{g \bowtie g} = r_{LL} + r_{RR} + r_{LR} - \tau(r_{LR}).$$

One can check directly that it obeys the CYBE given that r does, and hence gives a Lie bialgebra by Proposition 8.1.3. One can also take a slightly more efficient twisting point of view and proceed via Theorem 8.1.7 or Exercise 8.1.8. First, $g \oplus g$ is clearly itself a quasitriangular Lie bialgebra with the direct sum quasitriangular structure, and we write

$$r_{g \bowtie g} = r_{g \oplus g} + \chi, \quad r_{g \oplus g} = r_{LL} + r_{RR}, \quad \chi = r_{LR} - \tau(r_{LR}).$$

We compute

$$\begin{aligned}
\llbracket \chi, \chi \rrbracket &= [r_{12}, r_{13}]_{LRR} + [r_{21}, r_{31}]_{RLL} - [r_{12}, r_{32}]_{LRL} \\
&\quad -[r_{21}, r_{23}]_{RLR} + [r_{13}, r_{23}]_{LLR} + [r_{31}, r_{32}]_{RRL} \\
&= [r_{12}, r_{13}]_{LRR} + [r_{13}, r_{23}]_{LLR} + \text{cyclic},
\end{aligned}$$

where the zero bracket between the L, R copies eliminates many of the terms. We then recognise the remaining terms as elements of $g_L \otimes g_R \otimes g_R$ and $g_L \otimes g_L \otimes g_R$, and their cyclic rotations. The numerical suffix notation is as in Proposition 8.1.3 and refers to the position of g in a tensor product, be it g_L or g_R. We likewise compute

$$\begin{aligned}
(\mathrm{id} \otimes \delta_{g \oplus g})\chi &= (\mathrm{id} \otimes \delta_R)r_{LR} - (\mathrm{id} \otimes \delta_L)\tau(r_{LR}) \\
&= [r_{12}, r_{23}]_{LRR} + [r_{13}, r_{23}]_{LRR} - [r_{21}, r_{13}]_{RLL} - [r_{31}, r_{23}]_{RLL}
\end{aligned}$$

since the Lie cobrackets δ_L, δ_R on g_L, g_R are given by r_{LL} and r_{RR}, respectively. Adding these computations, we find

$$(\mathrm{id} \otimes \delta_{g \oplus g})\chi + \text{cyclic} + \llbracket \chi, \chi \rrbracket = \llbracket r, r \rrbracket_{LRR} + \llbracket r, r \rrbracket_{LLR} + \text{cyclic} = 0$$

as required. Hence we apply Theorem 8.1.7 (in the strong form in Exercise 8.1.8) to construct $g \bowtie g$ as a quasitriangular Lie bialgebra.

This twisting construction also gives us a second and different quasitriangular structure by starting with $-r_{21}$ for the quasitriangular structure of g_R. This gives the same Lie bialgebra since $\mathrm{ad}_\xi(r) = \mathrm{ad}_\xi(-r_{21})$ by ad-invariance, so $(g, r) \oplus (g, -r_{21})$ is just as good a starting point for twisting by χ. In this case,

$$r^D_{g \bowtie g} = r_{LL} - \tau(r_{RR}) + r_{LR} - \tau(r_{LR})$$

is the second quasitriangular structure on the same Lie bialgebra. This is the semiclassical analogue of Proposition 7.3.4 in the variant appropriate to $g_r \bowtie g$.

Finally, we construct the map θ from the double Lie algebra. Since this is generated by $g^{*\mathrm{op}}, g$, it is enough to give Lie algebra homomorphisms

$$g^{*\mathrm{op}} \xrightarrow{i} g_r \bowtie g \xleftarrow{j} g$$

such that together they respect the cross relations in the double. We define these as

$$i(\phi) = r_2(\phi) \oplus (-r_1(\phi)), \quad j(\xi) = \xi \oplus \xi,$$

where j is clearly a Lie algebra homomorphism since $[\xi \oplus \xi, \eta \oplus \eta] = [\xi, \eta] \oplus [\xi, \eta]$ in the direct sum Lie algebra. The map i is equally clearly a Lie algebra homomorphism once we recall that r_2 and $-r_1$ are Lie algebra homomorphisms $g^{*\mathrm{op}} \to g$ from Lemma 8.1.6. Then we define $\theta(\phi \oplus \xi) = i(\phi) + j(\xi)$ and verify that

$$\begin{aligned}
\theta([\xi, \phi]) &= i(\phi_{[1]})\langle \phi_{[2]}, \xi \rangle + j(\xi_{[1]})\langle \phi, \xi_{[2]} \rangle \\
&= (r^{[1]}\langle \phi, [r^{[2]}, \xi] \rangle + \langle \mathrm{ad}_\xi(r), \mathrm{id} \otimes \phi \rangle) \\
&\quad \oplus (-r^{[2]}\langle \phi, [r^{[1]}, \xi] \rangle + \langle \mathrm{ad}_\xi(-r_{21}), \mathrm{id} \otimes \phi \rangle) \\
&= [\xi, r_2(\phi)] \oplus [\xi, -r_1(\phi)] = [\theta(\xi), \theta(\phi)]
\end{aligned}$$

as required. This gives θ as a Lie algebra homomorphism $D(g) \to g \bowtie g$. It is easy to see that it is an isomorphism if the symmetric part r_+ is invertible: the inverse is

$$\theta^{-1}(\eta \oplus \xi) = \frac{1}{2} r_+^{-1}(\eta - \xi) \oplus \left(\frac{1}{2} r_1(r_+^{-1}(\eta - \xi)) + \xi \right).$$

Moreover, the quasitriangular structure of $D(g)$ maps over as

$$(\theta \otimes \theta)(f^a \otimes e_a) = (r_2(f^a) \oplus -r_1(f^a)) \otimes (e_a \oplus e_a) = r^D_{g \bowtie g}$$

on writing out the four terms. Since the Lie bialgebra structure of both $D(g)$ and $g \bowtie g$ are determined by the Lie algebra and the quasitriangular structure, it follows that θ is a Lie bialgebra homomorphism. ∎

For the next theorem, we recall the notion of a *semidirect sum* of Lie algebras. Thus, if α is an action of a Lie algebra g by automorphisms on a Lie algebra m, the semidirect sum $m \rtimes g$ is built on $m \oplus g$ with

$$[\phi \oplus \xi, \psi \oplus \eta] = ([\phi, \psi] + \alpha_\xi(\psi) - \alpha_\eta(\phi)) \oplus [\xi, \eta] \qquad (8.21)$$

for $\phi \oplus \xi, \psi \oplus \eta \in m \rtimes g$. Equivalently, we can write $\phi \equiv \phi \oplus 0$ and $\xi \equiv 0 \oplus \xi$ so that g, m are Lie subalgebras and

$$[\xi, \phi] = \alpha_\xi(\phi)$$

is the bracket between them in the semidirect sum Lie algebra.

Theorem 8.2.6 *Let $(g, [\ ,\ ], r)$ be a quasitriangular Lie bialgebra. Then there is a homomorphism $\theta : D(g) \to g \rtimes g$ of Lie algebras, where the semidirect sum is by the adjoint action of g on itself. It is given by*

$$\theta(\phi \oplus \xi) = 2r_+(\phi) \oplus (\xi - r_1(\phi))$$

and is an isomorphism in the factorisable case. Here, r_1, r_+ are as in Lemma 8.1.6.

Proof: This time we define Lie algebra homomorphisms,

$$g^{*\mathrm{op}} \xrightarrow{i} g \rtimes g \xleftarrow{j} g,$$

$$i(\phi) = 2r_+(\phi) \oplus (-r_1(\phi)), \quad j(\xi) = 0 \oplus \xi,$$

where i is indeed a Lie algebra homomorphism, as

$$
\begin{aligned}
i([\phi, \psi]_{g^{*\mathrm{op}}}) &= 2r_+([\psi, \phi]) \oplus [r_1(\phi), r_1(\psi)] \\
&= ([2r_+(\phi), 2r_+(\psi)] + [-r_1(\phi), 2r_+(\psi)] \\
&\quad -[-r_1(\psi), 2r_+(\phi)]) \oplus [r_1(\phi), r_1(\psi)] \\
&= [2r_+(\phi) \oplus (-r_1(\phi)), 2r_+(\psi) \oplus (-r_1(\psi))] \\
&= [i(\phi), i(\psi)].
\end{aligned}
$$

We used Lemma 8.1.6 for r_1, r_+, and for the last step we used the semidirect sum Lie bracket (8.21) with $\alpha = \mathrm{ad}$. We then define $\theta(\phi \oplus \xi) = i(\phi) + j(\xi)$ and check the cross bracket

$$
\begin{aligned}
\theta([\xi, \phi]) &= i(\phi_{[1]})\langle \phi_{[2]}, \xi \rangle + j(\xi_{[1]})\langle \phi, \xi_{[2]} \rangle \\
&= 2r_+^{[2]}\langle \phi, [r_+^{[1]}, \xi] \rangle \oplus (-r^{[2]}\langle \phi, [r^{[1]}, \xi] \rangle + \langle \mathrm{ad}_\xi(-r_{21}), \mathrm{id} \otimes \phi \rangle) \\
&= [\xi, 2r_+(\phi)] \oplus [\xi, -r_1(\phi)] = [0 \oplus \xi, 2r_+(\phi) \oplus (-r_1(\phi))] = [\theta(\xi), \theta(\phi)]
\end{aligned}
$$

as required. We used ad-invariance of r_+ and the semidirect sum Lie algebra from (8.21). Finally, if r_+ is invertible, then so is θ, with inverse

$$\theta^{-1}(\eta \oplus \xi) = \frac{1}{2}r_+^{-1}(\eta) \oplus \frac{1}{2}r_1(r_+^{-1}(\eta)) + \xi$$

as one may easily verify.   ■

Unwinding these two isomorphisms with $D(g)$, one also obtains a Lie algebra isomorphism

$$g \bowtie g \cong g_r \bowtie g, \quad \xi \oplus \eta \mapsto (\xi + \eta) \oplus \eta.$$

This is the semiclassical analogue of Theorem 7.4.6, and it holds directly between these Lie algebras even if they are infinite dimensional or if r_+ is not invertible. It means, in particular, that $g \bowtie g$ inherits a quasitriangular Lie bialgebra structure. One can develop in this way the semiclassical analogues of the various remaining results in Chapter 7.3 and 7.4. We limit ourselves here to the question of a natural real form of the classical double and its role as complexification. This was one of the important themes in the Hopf algebra theory of Chapter 7.3.

Proposition 8.2.7 *Let g be a quasitriangular Lie bialgebra and let $D(g)$ be its double. There is a surjection of quasitriangular Lie bialgebras*

$$p : D(g) \to g, \quad p(\phi \oplus \xi) = r_2(\phi) + \xi.$$

Moreover, if $u = \{e_i\}$ is a half-real form of g, then $D(u) = \{e_i^*, e_i\}$ is a half-real form of $D(g)$. If the quasitriangular structure is of real type and factorisable, then p becomes an isomorphism $D(u) \cong g$ of real Lie algebras.

Proof: The first part works over any field and follows at once by composing the map $\theta : D(g) \to g_r \bowtie g$ in Theorem 8.2.5 with the surjection $g_r \bowtie g \to g$ sending $\xi \oplus \eta \to \xi$. It is clear also that $(p \otimes p)(f^a \otimes e_a) = r_2(f^a) \otimes e_a = r$, with the result that the quasitriangular structure of $D(g)$ maps onto that of g.

For the second part of the proposition, we work over $\mathbb{C}$ and suppose that $\{e_i\}$ is a half-real form as explained in Section 8.1. The Lie algebra structure constants are real and therefore define a real Lie algebra u, but the Lie cobracket structure constants are imaginary. The dual half-real form $u^* = \{e_i^*\}$ was described in (8.18); we define $D(u)$ as the real Lie algebra built on the real vector space $u^* \oplus u$ and observe that is a half-real form of $D(g)$. Indeed, the Lie algebra structure constants of $D(g)$ are provided by those of g^{*op} and g (both real in this basis) and those of the cross relations

$$[e_i, e_j^*] = e_{j\,[1]}^* \langle e_{j\,[2]}^*, e_i \rangle + e_{i\,[1]} \langle e_j^*, e_{i\,[2]} \rangle = e_a^* c^j{}_{ai} + e_a(-\imath d^{aj}{}_i), \tag{8.22}$$

which are also real in this basis as the d are imaginary. The Lie cobracket is the direct sum one and has imaginary structure constants in this basis because u, u^* do. The quasitriangular structure of the double is $\imath e_a^* \otimes e_a$

in this basis. Hence, the corresponding tensor is imaginary (an antireal quasitriangular structure).

Finally, we consider both of these aspects together. Thus we suppose that u is a half-real form and has a real quasitriangular structure r, so $\bar{r} = r_{21}$. This means that the symmetric part r_+ is real and the antisymmetric part r_- is imaginary. Now, $r_2 = r_+ - r_-$ so

$$p(e_i^\star \oplus e_j) = r_2(-\imath f^i) + e_j = -\imath e_a r^{ai} + e_j = -\imath e_a r_+^{ai} + \imath e_a r_-^{ia} + e_j.$$

From this, we see that $p(\phi \oplus \xi)$ has $r_+(\phi)$ as its imaginary part and $-r_-(\phi) + \xi$ as its real part if $\phi \oplus \xi \in u^\star \oplus u$. In other words, p maps the real form $D(u)$ into a complex element of g. If we suppose further that r_+ is invertible (the factorisable case), then we can invert this map by

$$p^{-1}(\xi + \imath\eta) = \imath r_+^{-1}(\eta) \oplus \xi + \imath r_-(r_+^{-1}(\eta)),$$

for all $\xi, \eta \in u$. Now, writing elements of g as complex linear combinations $\{\xi + \imath\eta;\ \xi, \eta \in u\}$ is precisely the description of g as the *complexification* of the real Lie algebra u. It has a Lie bracket which is the same as that of u, extended by linearity to complex linear combinations. As a Lie algebra over $\mathbb{R}$, this complexification has twice the dimension of u. We see that it is isomorphic to the half-real form $D(u)$ of the double. ∎

Since $D(g)$ is, in fact, isomorphic to $g_r \bowtie g$ in the factorisable case, we can go further and compute the composite map

$$\theta \circ p^{-1} : g \hookrightarrow g_r \bowtie g, \quad \xi + \imath\eta \mapsto (\xi + \imath\eta) \oplus (\xi - \imath\eta)$$

at once from p^{-1} above and θ in Theorem 8.2.4. For example, we recover the standard embedding

$$sl(2, \mathbb{C}) \hookrightarrow sl(2, \mathbb{C})_r \bowtie sl(2, \mathbb{C})$$

used in physics whereby a generator of $sl(2, \mathbb{C})$ (the Lorentz group Lie algebra) is realised in two commuting copies of sl_2. This was explained at the end of Chapter 7.3 as one of the motivations for the Hopf algebra constructions there.

Also, just as the double $D(g)$ is generated by $g^{\star\mathrm{op}}$ and g with certain cross relations, so $D(u)$ is generated as a real Lie algebra by $u^{\star\mathrm{op}}$ and u with certain cross relations. Indeed, every element of the vector space of $D(u)$ can be written uniquely as a part in $u^{\star\mathrm{op}}$ and a part in u. One writes $D(u) = u^{\star\mathrm{op}} \bowtie u$. We will study such decompositions of Lie algebras quite systematically in the next section, where they arise as the infinitesimal version of the double cross products of Chapter 6.2. Such decompositions are also an important part of the standard theory of Lie algebras. We obtain one of these immediately as an application of Proposition 8.2.7.

Corollary 8.2.8 *Let g be a complex semisimple Lie algebra with the quasitriangular Lie bialgebra structure in Proposition 8.1.14 and let u be its compact half-real form in Proposition 8.1.15. Then the complexification $u_\mathbb{C} = g$ regarded as a real Lie algebra decomposes into $u^{\star op} \oplus u$, where u, $u^{\star op}$ are Lie subalgebras and $u^{\star op}$ is solvable (the Iwasawa decomposition).*

Proof: We already know from Proposition 8.1.15 that we are in the situation of Proposition 8.2.7. So $p : D(u) \cong g$ is an isomorphism of real Lie algebras. We gave r_- explicitly in Proposition 8.1.15, and $r_+ = \frac{1}{2}K^{-1}$, so $r_2 = r_+ - r_-$ is explicitly

$$r_2(\vec{x}_\alpha) = \frac{\imath}{2}x_\alpha - \frac{1}{2}y_\alpha = \imath r_2(\vec{y}_\alpha), \quad r_2(\vec{z}_i) = \frac{\imath}{2}z_i,$$

and gives the embedding $u^{\star op} \to g$. The compact real form u is also included in g as the real span of the x, y, z basis. ∎

This gives a Yang–Baxter theoretic proof of the *Iwasawa decomposition* in the theory of complex semisimple Lie algebras, obtained now via the Drinfeld–Jimbo solution of the CYBE. Alternatively, if one already knows the Iwasawa decomposition in terms of the roots etc., one can deduce the form of r in Proposition 8.1.14.

Example 8.2.9 *There is an isomorphism $sl(2, \mathbb{C}) \cong D(su_2)$, as generated by the real Lie bialgebras su_2 and $su_2^{\star op}$ from Example 8.1.13 and cross relations*

$$[\vec{\xi}, \vec{\phi}]_{su_2^{\star op}} = \vec{\xi} \times \vec{\phi}, \quad [\vec{\xi}, \vec{\phi}]_{su_2} = \vec{\phi} \times (\vec{\xi} \times \vec{e_3})$$

for the parts lying in $su_2^{\star op}$ and su_2, respectively.

Proof: The general cross relations for $D(u)$ in (8.22) become

$$[e_i, e_j^\star] = e_a^\star \epsilon^a{}_{ij} + e_i \delta_{j3} - e_3 \delta_{ij}$$

in the setting of Example 8.1.13. This then looks as stated in terms of the real-valued vectors $\vec{\xi}, \vec{\phi}$ describing general elements of $su_2, su_2^{\star op}$ by $\xi = e_i\xi^i$ and $\phi = \phi_i e_i^\star$. The map p in Proposition 8.2.7 is

$$p(\phi \oplus \xi) = r_2(\phi) + \xi = \phi_i r_2(e_i^\star) + e_i\xi^i = \phi_i\omega^i + \left(-\frac{\imath}{2}\sigma_i\right)\xi^i,$$

where σ_i are the usual Pauli matrices and

$$\omega^i = r_2(e_i^\star) = -i r_2(f^i) = -\frac{1}{2}\sigma_a r^{ai} = \frac{1}{2}(\sigma_i + \imath\epsilon_{ia3}\sigma_a)$$

using the structure constants in Example 8.1.10. Explicitly,

$$\omega^1 = \begin{pmatrix} 0 & 1 \\ 0 & 0 \end{pmatrix}, \quad \omega^2 = \begin{pmatrix} 0 & -\imath \\ 0 & 0 \end{pmatrix}, \quad \omega^3 = \begin{pmatrix} \frac{1}{2} & 0 \\ 0 & -\frac{1}{2} \end{pmatrix}.$$

These ω^i are a concrete matrix realisation of the Lie algebra $su_2^{\star\text{op}}$, as they must be since r_2 is a Lie algebra homomorphism $sl_2^{\star\text{op}} \to sl_2$ from Lemma 8.1.6. We have

$$su_2^{\star\text{op}} = \{\begin{pmatrix} x & z \\ 0 & -x \end{pmatrix}; \quad x \in \mathbb{R}, \ z \in \mathbb{C}\}$$

as spanned by these ω^i. This is a solvable Lie algebra. Moreover, it is clear from the above that every complex linear combination of $-\frac{\imath}{2}\sigma_i$, i.e. every element of $sl(2, \mathbb{C})$, can be written uniquely in terms of $\omega^i, -\frac{\imath}{2}\sigma_i$ with real coefficients. Hence $sl(2, \mathbb{C}) = su_2^{\star\text{op}} \oplus su_2$ as real vector spaces. ∎

This example gives a description of the Lorentz group Lie algebra as real 3-vectors $\vec{\xi}$ for rotation and certain other real 3-vectors $\vec{\phi}$ governing the boosts.

8.3 Matched pairs of Lie algebras and their exponentiation

In this section, we give the semiclassical or infinitesimal version of the double cross product of groups in Chapter 6.2 double cross products Hopf algebras in Chapter 7.2. We will also give Lie bialgebra versions of the associated bicrossproduct Hopf algebras. The latter are interesting as the algebras of observables of quantum systems, in which context it is natural to look for examples based on Lie groups for the position and momentum. We can construct them by starting at the Lie algebra level and then exponentiating to the group level. This is one of the original motivations for Lie algebra double cross sums and is the main goal of the section. The theory is due to the author. On the way, we will also tie up the constructions with Drinfeld's Lie bialgebra theory, with the classical double $D(g)$ as an example. A further connection with classical inverse scattering, Poisson brackets etc., is the topic of the next section.

Definition 8.3.1 *Two Lie algebras* (g, m) *form a* right–left matched pair *if there is a right action of* g *on* m *and a left action of* m *on* g,

$$\alpha : m \otimes g \to m, \quad \alpha_\xi(\phi) = \phi \triangleleft \xi, \quad \beta : m \otimes g \to g, \quad \beta_\phi(\xi) = \phi \triangleright \xi,$$

obeying the conditions

$$\alpha_\xi([\phi, \psi]) = [\alpha_\xi(\phi), \psi] + [\phi, \alpha_\xi(\psi)] - \alpha_{\beta_\phi(\xi)}(\psi) + \alpha_{\beta_\psi(\xi)}(\phi),$$

$$\beta_\phi([\xi, \eta]) = [\beta_\phi(\xi), \eta] + [\xi, \beta_\phi(\eta)] - \beta_{\alpha_\eta(\phi)}(\xi) + \beta_{\alpha_\xi(\phi)}(\eta). \tag{8.23}$$

In shorthand notation, the full data here are

$$[\phi, \psi] \triangleright \xi = \phi \triangleright (\psi \triangleright \xi) - \psi \triangleright (\phi \triangleright \xi), \qquad \phi \triangleleft [\xi, \eta] = (\phi \triangleleft \xi) \triangleleft \eta - (\phi \triangleleft \eta) \triangleleft \xi,$$

$$\phi \triangleright [\xi, \eta] = [\phi \triangleright \xi, \eta] + [\xi, \phi \triangleright \eta] + (\phi \triangleleft \xi) \triangleright \eta - (\phi \triangleleft \eta) \triangleright \xi,$$

$$[\phi, \psi] \triangleleft \xi = [\phi \triangleleft \xi, \psi] + [\phi, \psi \triangleleft \xi] + \phi \triangleleft (\psi \triangleright \xi) - \psi \triangleleft (\phi \triangleright \xi).$$

This is the analogue of Definition 6.2.10. Just as group matched pairs are in correspondence with group factorisations, we have for the analogue of Proposition 6.2.15 the following.

Proposition 8.3.2 *Given a matched pair of Lie algebras* (g, m), *we define the* double cross sum $g \bowtie m$ *as the vector space* $g \oplus m$ *with the Lie bracket*

$$[\xi \oplus \phi, \eta \oplus \psi] = ([\xi, \eta] + \beta_\phi(\eta) - \beta_\psi(\xi)) \oplus ([\phi, \psi] + \alpha_\eta(\phi) - \alpha_\xi(\psi)),$$

i.e. $\quad [\xi \oplus \phi, \eta \oplus \psi] = ([\xi, \eta] + \phi \triangleright \eta - \psi \triangleright \xi) \oplus ([\phi, \psi] + \phi \triangleleft \eta - \psi \triangleleft \xi)).$

Conversely, if a Lie algebra $\Xi \cong g \oplus m$ *in the sense that* $g \xrightarrow{i} \Xi \xleftarrow{j} m$ *are Lie subalgebras and every element of* Ξ *decomposes uniquely into an element of* g *and an element of* m, *then* (g, m) *are a matched pair and* $\Xi \cong g \bowtie m$. *The required actions are recovered from*

$$[j(\phi), i(\xi)] = i(\phi \triangleright \xi) + j(\phi \triangleleft \xi).$$

Proof: The double cross sum Lie algebra contains g, m as Lie subalgebras, but with the cross relations

$$[\phi, \xi] = \phi \triangleright \xi + \phi \triangleleft \xi,$$

where the first term on the right lies in g and the second term lies in m. We suppress the maps i, j. To prove the Jacobi identity, we concentrate on the mixed case

$$[\phi, [\xi, \eta]] = \phi \triangleright [\xi, \eta] + \phi \triangleleft [\xi, \eta],$$
$$[[\phi, \xi], \eta] = [\phi \triangleright \xi + \phi \triangleleft \xi, \eta] = [\phi \triangleright \xi, \eta] + (\phi \triangleleft \xi) \triangleright \eta + (\phi \triangleleft \xi) \triangleleft \eta.$$

The second expression, minus a similar one with $\xi \leftrightarrow \eta$, equals the first expression exactly because of the second half of (8.23) for the part in g, and because ◁ is a Lie algebra action for the part in m. The other mixed Jacobi identity $[[\phi, \psi], \xi] = [[\phi, \xi], \psi] + [\phi, [\psi, \xi]]$ is strictly analogous using the first half of (8.23) and that ▷ is an action. Conversely, if the cross relations have this general form, and if every element decomposes uniquely into a part in g and a part in m, then the above computations can be pushed backwards to conclude that ▷, ◁ obey the matched pair conditions. ∎

Note that if ▷ is a left action of a Lie algebra then −▷ is a right action, etc. It is easy to see that if $(g, m, \triangleleft, \triangleright)$ is a right–left matched pair then so is $(m, g, -\triangleright, -\triangleleft)$, and $g{\bowtie}m \cong m{\bowtie}g$ in that they are both Lie algebras containing g, m with the same cross relations. The only difference is that, in one case, we choose to build the Lie algebra on the vector space $g \oplus m$, and in the other case we choose to build the Lie algebra on $m \oplus g$.

Next we give the analogue of Examples 6.2.11 and 6.2.12, which associated Hopf algebras to matched pairs of groups. This time we naturally associate a Lie bialgebra to a matched pair of Lie algebras.

Proposition 8.3.3 *If* (g, m) *form a matched pair of Lie algebras, let* $\alpha^* = \triangleright : g \otimes m^* \to m^*$ *denote the adjoint of the action* α *and let* $\{e_a\}$ *be a basis of* m *with dual basis* $\{f^a\}$. *There is a bicross sum Lie bialgebra* $m^* {\blacktriangleright\!\!\triangleleft} g$ *built on* $m^* \oplus g$ *with*

$$[f \oplus \xi, h \oplus \eta] = (\xi \triangleright h - \eta \triangleright f) \oplus [\xi, \eta],$$

$$\delta(f \oplus \xi) = \sum_a (0 \oplus e_a \triangleright \xi) \otimes (f^a \oplus 0) - (f^a \oplus 0) \otimes (0 \oplus e_a \triangleright \xi)$$

$$+ \sum (f_{[1]} \oplus 0) \otimes (f_{[2]} \oplus 0),$$

for all $f \oplus \xi, h \oplus \eta \in m^* {\blacktriangleright\!\!\triangleleft} g$, *where* $\delta(f) = \sum f_{[1]} \otimes f_{[2]}$ *is the Lie coalgebra given by dualisation of the Lie bracket of* m. *Similarly for* $m {\bowtie} g^*$.

Proof: The Lie algebra is the semidirect product of m^* with zero Lie bracket by the left action of g. The Lie coalgebra structure is similar but dual: g has the zero cobracket and is coacted upon by m^*, which has the Lie cobracket dual as in Proposition 8.1.2 to the Lie bracket on m. If we identify $\xi \equiv 0 \oplus \xi$, $f \equiv f \oplus 0$ as usual, then the Lie bialgebra structure is that m^*, g are Lie subalgebras and

$$[\xi, f] = \xi \triangleright f, \quad \delta f = \langle [e_a, e_b], f \rangle f^a \otimes f^b, \quad \delta \xi = e_a \triangleright \xi \otimes f^a - f^a \otimes e_a \triangleright \xi.$$

The co-Jacobi identity on f is already ensured since m^* is a Lie subbialgebra. On ξ it follows easily from ▷ an action. For the cocycle axiom

(8.3), the mixed case is

$$
\begin{aligned}
\delta([\xi, f]) &= \delta(\xi \triangleright f) = \langle [e_a, e_b], \xi \triangleright f \rangle f^a \otimes f^b = \langle [e_a, e_b] \triangleleft \xi, f \rangle f^a \otimes f^b \\
&= (\langle [e_a \triangleleft \xi, e_b], f \rangle + \langle [e_a, e_b \triangleleft \xi], f \rangle \\
&\quad + \langle e_a \triangleleft (e_b \triangleright \xi), f \rangle - \langle e_a \triangleleft (e_b \triangleright \xi), f \rangle) f^a \otimes f^b \\
&= \xi \triangleright \delta f + (e_b \triangleright \xi) \triangleright f \otimes f^b - f^a \otimes (e_a \triangleright \xi) \triangleright f \\
&= [\xi, f_{[1]}] \otimes f_{[2]} + f_{[1]} \otimes [\xi, f_{[2]}] - [f, e_b \triangleright \xi] \otimes f^b + f^a \otimes [f, e_a \triangleright \xi]
\end{aligned}
$$

as required. We used the definition of δf by dualisation for the second equality, the definition of $\alpha^* = \triangleright$ for the third, the first half of the matched pair condition (8.23) for the fourth and then worked backwards from the definitions. The other nontrivial case of the cocycle axiom (8.3) is

$$
\begin{aligned}
\delta([\xi, \eta]) &= (\mathrm{id} - \tau)(e_a \triangleright [\xi, \eta] \otimes f^a) \\
&= (\mathrm{id} - \tau)([e_a \triangleright \xi, \eta] \otimes f^a + (e_a \triangleleft \xi) \triangleright \eta \otimes f^a) - (\xi \leftrightarrow \eta) \\
&= (\mathrm{id} - \tau)([\xi, e_a \triangleright \eta] \otimes f^a + e_a \triangleright \eta \otimes \xi \triangleright f^a) - (\xi \leftrightarrow \eta) \\
&= (\mathrm{id} - \tau)([\xi, e_a \triangleright \eta] \otimes f^a + e_a \triangleright \eta \otimes [\xi, f^a]) - (\xi \leftrightarrow \eta)
\end{aligned}
$$

as required. We used the definition of the Lie cobracket on g, the second half of (8.23), the definition of α^* as adjoint to $\triangleleft$ and the Lie cobracket on g in reverse.   ∎

The double cross sum Lie algebra could also be viewed (like any Lie algebra) as a Lie bialgebra with zero cobracket. So both of the preceding two constructions can be viewed as producing Lie bialgebras. Indeed, they are both special cases of general Lie bialgebra double cross sums (cf. Theorem 7.2.3) and Lie bialgebra bicross sums (cf. Theorems 6.2.2 and 6.2.3), respectively. We outline this briefly.

The first step is to formalise the infinitesimal notions of Lie bialgebra actions and coactions along the lines given in Chapter 1.6 for Hopf algebras. These notions are (fairly obviously) the following, according to the general principles of infinitesimalisation outlined in Section 8.1. We have already met in Chapter 1.6 the concept of a *left Lie algebra module* or representation space. This is a vector space m, say, on which g acts by $\triangleright$ such that

$$
\xi \triangleright (\eta \triangleright \phi) - \eta \triangleright (\xi \triangleright \phi) = [\xi, \eta] \triangleright \phi, \tag{8.24}
$$

for all $\xi, \eta \in g$ and $\phi \in m$. A *left g-module Lie algebra* is a Lie algebra m on which g acts by automorphisms in the sense

$$
\xi \triangleright [\phi, \psi] = [\xi \triangleright \phi, \psi] + [\phi, \xi \triangleright \psi], \tag{8.25}
$$

for all $\phi, \psi \in m$, and $\xi \in g$. We have already encountered this notion in Section 8.2, and we know that we can form a semidirect sum Lie algebra

$m \bowtie g$. In the same spirit, a *left g-module Lie coalgebra* is when m is a Lie coalgebra with cobracket δ, say, and

$$\delta(\xi \triangleright \phi) = \xi \triangleright \delta \phi, \tag{8.26}$$

where the action of ξ on the right is by extension to $m \otimes m$ as a derivation. Similarly for right modules. Dualising these notions, one has the following: if g is a Lie coalgebra, then a *right Lie comodule* is a vector space m and a map $\beta : m \to m \otimes g$ such that

$$(\mathrm{id} \otimes \delta) \circ \beta = (\mathrm{id} \otimes (\mathrm{id} - \tau)) \circ (\beta \otimes \mathrm{id}) \circ \beta, \tag{8.27}$$

where τ is the usual permutation or transposition map. We say β is a right *Lie coaction* of the Lie coalgebra g. A *right g-comodule Lie coalgebra* is a Lie coalgebra m on which g coacts such that

$$(\delta \otimes \mathrm{id}) \circ \beta = ((\mathrm{id} - \tau) \otimes \mathrm{id}) \circ (\mathrm{id} \otimes \beta) \circ \delta. \tag{8.28}$$

In this case there is a semidirect sum Lie coalgebra $g \blacktriangleright\!\!< m$ built on $g \oplus m$. Finally, a *right g-comodule Lie algebra* is a Lie algebra m on which g coacts such that

$$\beta([\phi, \psi]) = \mathrm{ad}_\phi(\beta(\psi)) - \mathrm{ad}_\psi(\beta(\phi)), \tag{8.29}$$

for all $\phi, \psi \in m$. Here, $\mathrm{ad}_\phi, \mathrm{ad}_\psi$ act trivially on g and by the adjoint action on m. The left handed theory is strictly analogous.

Proposition 8.3.4 *Let (g, m) be a matched pair of Lie algebras as above, with both g and m now Lie bialgebras and with $\triangleright, \triangleleft$ making g a left m-module Lie coalgebra and m a right g-module Lie coalgebra, such that*

$$\phi \triangleleft (\delta \xi) + (\delta \phi) \triangleright \xi = 0, \tag{8.30}$$

$$i.e. \quad \phi \triangleleft \sum \xi_{[1]} \otimes \xi_{[2]} + \sum \phi_{[1]} \otimes \phi_{[2]} \triangleright \xi = 0,$$

for all $\xi \in g$, $\phi \in m$. Then the direct sum Lie coalgebra structure makes $g \bowtie m$ into a Lie bialgebra, the double cross sum Lie bialgebra.

Proof: The Lie algebra structure is as in Proposition 8.3.2. The Lie cobracket is the one inherited from g, m as Lie sub-bialgebras. Hence the co-Jacobi and cocycle axioms hold on each of these separately. We have only to check the cocycle axiom (8.3) in the mixed case as

$$\delta([\phi, \xi]) = \delta(\phi \triangleright \xi + \phi \triangleleft \xi) = \delta_g(\phi \triangleright \xi) + \delta_m(\phi \triangleleft \xi) = \phi \triangleright \delta_g(\xi) + \delta_m(\phi) \triangleleft \xi,$$

$$\mathrm{ad}_\phi(\delta(\xi)) - \mathrm{ad}_\xi(\delta(\phi))$$

$$= \phi \triangleright \xi_{[1]} \otimes \xi_{[2]} + \phi \triangleleft \xi_{[1]} \otimes \xi_{[2]} + \xi_{[1]} \otimes \phi \triangleright \xi_{[2]} + \xi_{[1]} \otimes \phi \triangleleft \xi_{[2]}$$

$$+ \phi_{[1]} \triangleright \xi \otimes \phi_{[2]} + \phi_{[1]} \triangleleft \xi \otimes \phi_{[2]} + \phi_{[1]} \otimes \phi_{[2]} \triangleright \xi + \phi_{[1]} \otimes \phi_{[2]} \triangleleft \xi,$$

using the definition of the $[\phi, \xi]$ and using, for the third equality, that the actions are module Lie coalgebras in the sense of (8.26) and its right handed version. The actions extend to tensor products as a derivation, so $\phi \triangleright \delta \xi = \phi \triangleright \xi_{[1]} \otimes \xi_{[2]} + \xi_{[1]} \otimes \phi \triangleright \xi_{[2]}$, accounting for four of the eight terms in the last expression. The remaining four terms cancel pairwise in view of (8.30). Hence the two expressions are equal.					∎

Proposition 8.3.5 *Let* m, g *be Lie bialgebras, let* m *be a left* g-*module Lie algebra and let* g *be a right* m-*comodule Lie coalgebra by maps*

$$\alpha : g \otimes m \to m, \quad \alpha_\xi(f) = \xi \triangleright f, \quad \beta : g \to g \otimes m, \quad \beta(\xi) = \sum \xi^{[1]} \otimes \xi^{[2]},$$

obeying the compatibility conditions

$$\delta(\xi \triangleright f) = \xi \triangleright \delta(f) + (\mathrm{id} - \tau)(\sum \xi^{[1]} \triangleright f \otimes \xi^{[2]}), \tag{8.31}$$

$$\beta([\xi, \eta]) = \xi \triangleright \beta(\eta) - \eta \triangleright \beta(\xi), \tag{8.32}$$

$$\sum \xi_{[1]} \otimes \xi_{[2]} \triangleright f + \sum \xi^{[1]} \otimes [\xi^{[2]}, f] = 0, \tag{8.33}$$

where the action on the output of β *in (8.32) is by the adjoint action of* g *on itself and, in addition, the action* $\triangleright$ *of* g *on* m. *Then* $m \rtimes g$ *and* $m \ltimes g$ *form a Lie bialgebra, the left–right bicross sum Lie bialgebra* $m \bowtie g$.

Proof: The Lie algebra is the semidirect sum (8.21) on $m \oplus g$. So, identifying $f \equiv f \oplus 0$ and $\xi \equiv 0 \oplus \xi$, we have m, g as Lie subalgebras. The mixed relations and the cobracket are

$$[\xi, f] = \xi \triangleright f, \quad \delta_{m \bowtie g}(\xi) = \delta(\xi) + (\mathrm{id} - \tau) \circ \beta(\xi), \quad \delta_{m \bowtie g}(f) = \delta(f).$$

The Lie coalgebra is a semidirect cobracket, and it is easy to see that it obeys the co-Jacobi identity given that β is a comodule Lie coalgebra structure. It is clear that m is a Lie sub-bialgebra. We have, in fact,

$$m \hookrightarrow m \bowtie g \to g$$

as maps of Lie bialgebras (a Lie bialgebra extension) by the canonical inclusion and the projection which sends m to zero. To see that $m \bowtie g$ is a Lie bialgebra in the first place, we have only to check the cocycle axiom (8.3). We have

$$\delta([\xi, f]) = \delta(\xi \triangleright f) = \xi \triangleright \delta(f) + (\mathrm{id} - \tau)(\xi^{[1]} \triangleright f \otimes \xi^{[2]}),$$

$$\mathrm{ad}_\xi(\delta(f)) - \mathrm{ad}_f(\delta_{m \bowtie g}(\xi)) = \xi \triangleright \delta(f) + \xi_{[1]} \triangleright f \otimes \xi_{[2]} + \xi_{[1]} \otimes \xi_{[2]} \triangleright f$$

$$+ (\mathrm{id} - \tau)(\xi^{[1]} \triangleright f \otimes \xi^{[2]} - \xi^{[1]} \otimes [f, \xi^{[2]}])$$

on writing out the definitions and using $[\xi, f] = \xi \triangleright f$. The second equality is (8.31). Four of the terms in the last expression cancel pairwise using (8.33), so that the two expressions are equal. Also, we have

$$
\begin{aligned}
\delta m_{\blacktriangleright \triangleleft} g([\xi, \eta]) &= \delta([\xi, \eta]) + (\mathrm{id} - \tau) \circ \beta([\xi, \eta]) \\
&= \mathrm{ad}_\xi(\delta(\eta)) + (\mathrm{id} - \tau)(\xi \triangleright \beta(\eta)) - (\xi \leftrightarrow \eta) \\
&= \mathrm{ad}_\xi(\delta m_{\blacktriangleright \triangleleft} g(\eta)) - \mathrm{ad}_\eta(\delta m_{\blacktriangleright \triangleleft} g(\xi)),
\end{aligned}
$$

where $\xi \triangleright \beta(\eta) = [\xi, \eta^{[1]}] \otimes \eta^{[2]} + \eta^{[1]} \otimes \xi \triangleright \eta^{[2]}$.   ∎

There is a strictly analogous right–left bicross sum Lie bialgebra $g_{\blacktriangleright \triangleleft} m$ if m is a right g-module Lie algebra and g is a left m-comodule Lie coalgebra. Moreover, one can easily see that, in the finite-dimensional case, we have $(g_{\blacktriangleright \triangleleft} m)^* \cong g^* {}_{\blacktriangleright \triangleleft} m^*$ as the analogue of Proposition 6.2.5. Also, the data for these constructions are all in one to one correspondence. This is the analogue of Proposition 7.2.4.

Proposition 8.3.6 *Finite-dimensional Lie bialgebras* (g, m) *form a right–left matched pair of Lie bialgebras with double cross sum Lie bialgebra* $g_{\bowtie} m$ *iff* $m^* {}_{\blacktriangleright \triangleleft} g$ *is a bicross sum by dualising the relevant actions. Likewise, iff* $m_{\blacktriangleright \triangleleft} g^*$ *is a bicross sum by dualisation.*

Proof: The formulae for dualisation in the first case were already given in Proposition 8.3.3: the left action of g on m^* and the right coaction of m^* on g are defined by

$$
\langle \xi \triangleright f, \phi \rangle = \langle f, \phi \triangleleft \xi \rangle, \quad \beta(\xi) = e_a \triangleright \xi \otimes f^a,
$$

for all $\xi \in g$, $f \in m^*$ and $\phi \in m$, where $\{e_a\}$ is a basis of m with dual basis $\{f^a\}$. The formulae for dualisation to the second bicross sum $m_{\blacktriangleright \triangleleft} g^*$ are similar: the right action of m on g^* and the left coaction of g^* on m are defined by

$$
\langle f \triangleleft \phi, \xi \rangle = \langle f, \phi \triangleright \xi \rangle, \quad \beta(\phi) = f^a \otimes \phi \triangleleft e_a,
$$

for all $\phi \in m$, $f \in g^*$ and $\xi \in g$, where $\{e_a\}$ is now a basis of g with dual basis $\{f^a\}$.   ∎

One could clearly go on to develop all the theory of Chapters 6 and 7 in this infinitesimal setting: the theory of Lie bialgebra extensions, cocycles and dual-cocycles etc. Here we content ourselves with tying up with some of the theory of Drinfeld's quasitriangular Lie bialgebras already encountered in Sections 8.1 and 8.2. The above theory of matched pairs and bicross sums arose in a very different physical context, as we know from Chapter 6, but the classical double $D(g)$ provides the point of contact.

Example 8.3.7 *Suppose that* g, g^* *are finite-dimensional Lie algebras. Then* (g^{*op}, g) *form a matched pair of Lie algebras by the mutual coadjoint actions of each on the other iff* g *is a Lie bialgebra. In this case they are a matched pair of Lie bialgebras and* $g^{*op} \bowtie g$ *and* $g \bowtie g^{*op}$ *are double cross sum Lie bialgebras, isomorphic to* $D(g)$ *in Proposition 8.2.1. Moreover,* $g \blacktriangleright\!\!\triangleleft g^{cop}$ *by the mutual adjoint action and adjoint coaction is a bicross sum Lie bialgebra isomorphic to* $g \oplus g^{cop}$. *Similarly for* $g^{cop} \blacktriangleright\!\!\triangleleft g$.

Proof: Assuming we have a Lie bialgebra, we take mutual actions

$$\xi \triangleleft \phi = \xi_{[1]} \langle \xi_{[2]}, \phi \rangle, \quad \xi \triangleright \phi = \phi_{[1]} \langle \xi, \phi_{[2]} \rangle,$$

for $\xi \in g$ and $\phi \in g^{*op}$, and verify that they obey the matched pair conditions in Definition 8.3.1 and Proposition 8.3.4. These are the mutual coadjoint actions, equivalent to the definition of ad^* mentioned already in (8.8). We verify the first half of (8.23) as

$$[\xi, \eta] \triangleleft \phi = [\xi, \eta]_{[1]} \langle \phi, [\xi, \eta]_{[2]} \rangle$$
$$= [\xi, \eta_{[1]}] \langle \eta_{[2]}, \phi \rangle + \eta_{[1]} \langle [\xi, \eta_{[2]}], \phi \rangle - (\xi \leftrightarrow \eta)$$
$$= [\xi, \eta \triangleleft \phi] - \eta_{[1]} \langle \eta_{[2]}, \xi \triangleright \phi \rangle - (\xi \leftrightarrow \eta) = [\xi, \eta \triangleleft \phi] - \eta \triangleleft (\xi \triangleright \phi) - (\xi \leftrightarrow \eta),$$

using the definitions of $\triangleleft, \triangleright$ and the cocycle axiom (8.3). The proof for the second half of (8.23) is similar, remembering to use the Lie bracket $[\phi, \psi]^{op} = [\psi, \phi]$ in terms of the usual Lie bracket of g^*. It is clear that this proof is reversible; i.e. if we have a Lie algebra and Lie coalgebra structure on g (or a Lie algebra structure on g^*) then the matching of the coadjoint actions implies the cocycle axiom (8.3) for a Lie bialgebra. The remaining condition (8.30) is automatic as

$$\xi \triangleleft (\delta \phi) + (\delta \xi) \triangleright \phi = \xi_{[1]} \langle \xi_{[2]}, \phi_{[1]} \rangle \otimes \phi_{[2]} + \xi_{[1]} \otimes \phi_{[1]} \langle \xi_{[2]}, \phi_{[2]} \rangle = 0,$$

using the definitions of $\triangleright, \triangleleft$ and antisymmetry of $\delta \phi$.

The other variant $g \bowtie g^{*op}$ is the same, with $\triangleright \leftrightarrow - \triangleleft$ as explained after Proposition 8.3.2. So, in this case,

$$\phi \triangleright \xi = \langle \xi_{[1]}, \phi \rangle \xi_{[2]}, \quad \phi \triangleleft \xi = \langle \xi, \phi_{[1]} \rangle \phi_{[2]},$$

for $\xi \in g$ and $\phi \in g^{*op}$. It is just as easy to verify the construction in this form.

For the last part, we use Proposition 8.3.6 to conclude that we have $g \blacktriangleright\!\!\triangleleft g^{cop}$ as a bicross sum. The right action of g on g^{cop} and the left coaction of g^{cop} on g are

$$\eta \triangleleft \xi = [\eta, \xi], \quad \beta(\xi) = -\delta(\xi),$$

for $\xi \in g$ and $\eta \in g^{cop}$, in terms of the bracket and cobracket of g. It is also easy enough to see directly that these maps obey a right–left version of (8.31)–(8.33). The first two of these reduce to the cocycle axiom (8.33), while the last is automatic. The data for this bicross sum are equivalent to the data for a coadjoint matched pair; on the other hand, it is easy to see that $g \bowtie g^{cop} \to g \oplus g^{cop}$ by $\xi \oplus \eta \mapsto \xi \oplus (\xi + \eta)$ is an isomorphism of Lie bialgebras. This is the semiclassical analogue of Example 6.2.8 and Proposition 6.2.9. The other variant based on (g, g^{*op}) gives $g^{cop} \bowtie g$. Note that this example should not be confused with (g, g^{*op}) as a matched pair of Lie bialgebras with zero Lie cobracket on the same Lie algebras g, g^{*op}, in which case $g^{cop} \bowtie g$ would be given by Proposition 8.3.3. ∎

   This gives the abstract structure of these complicated objects from Section 8.2 in terms of canonical ones on g, and also characterises a Lie bialgebra as equivalent to such a *coadjoint matched pair*. The latter point of view on Lie bialgebras is useful in the infinite dimensional case. The theory of matched pairs is also useful when we want to work with real forms as Lie algebras over $\mathbb{R}$. Recall from Section 8.1 that every complex semisimple Lie algebra has a natural compact real form u and another associated noncompact solvable real Lie algebra $u^{\star}$.

Example 8.3.8 *Let u be a half-real form of a complex Lie bialgebra g. Then $u^{\star op} \bowtie u$ and $u \bowtie u^{\star op}$ are double cross sums, isomorphic to the half-real form $D(u)$ of $D(g)$ in Proposition 8.2.7 (and to the complexification of u in the semisimple case).*

Proof: The point is that the Lie algebra here is a double cross sum working entirely over $\mathbb{R}$. We need not complicate the situation by bringing in $\imath$ at this stage. The calculation was already done in the context of defining $D(u)$ in (8.22), and, as there, we find the actions for the second form as

$$ e_i \triangleleft e_j^{\star} = e_a c^{\star i}{}_{aj}, \quad e_i \triangleright e_j^{\star} = e_a^{\star} c^j{}_{ai} $$

in our real bases $\{e_i\}$ for u and $\{e_i^{\star}\}$ for $u^{\star}$. Once the Lie algebra is constructed, there is a Lie bialgebra structure on $u, u^{\star op}$ and hence on $u^{\star op} \bowtie u$. The coalgebra structure constants are imaginary in our basis because the structure constants for $u^{\star op}, u$ are. ∎

   Thus, the theory of double cross sums of Lie algebras clarifies the structure of $D(u)$ as a Lie algebra over $\mathbb{R}$. It is not simply Drinfeld's double Lie bialgebra of u working over $\mathbb{R}$, not least because u is not a Lie bialgebra over $\mathbb{R}$.

Example 8.3.9 $D(su_2) = so(1,3)$ *in Example 8.2.9 can be given explicitly as double cross sums*

$$su_2^{\star \mathrm{op}} \bowtie su_2 : \quad \vec{\xi} \triangleleft \vec{\phi} = \vec{\phi} \times (\vec{\xi} \times \vec{e_3}), \quad \vec{\xi} \triangleright \vec{\phi} = \vec{\xi} \times \vec{\phi},$$

$$su_2 \bowtie su_2^{\star \mathrm{op}} : \quad \vec{\phi} \triangleright \vec{\xi} = \vec{\phi} \times (\vec{e_3} \times \vec{\xi}), \quad \vec{\phi} \triangleleft \vec{\xi} = \vec{\phi} \times \vec{\xi}.$$

Proof: We put the structure constants in Example 8.1.13 into Example 8.3.8 for the first version, and use the $\triangleright \leftrightarrow - \triangleleft$ symmetry for the second version. The $\times$ are usual vector space cross products, and it is easy enough to verify (8.23) directly using familiar vector calculus. The double cross sums in this case are a description of the Lorentz group Lie algebra. ∎

Another advantage of the double cross sum construction is that it works well in the infinite-dimensional setting. The point is that, in the infinite-dimensional case, it is not pleasant to work with Lie bialgebras because of the tensor product. But we have seen that we can work equivalently with coadjoint matched pairs, a notion which is perfectly clear in the infinite-dimensional case and suitable for adding topological considerations, though we do not do so here.

Exercise 8.3.10 *Let g, m be dually paired Lie bialgebras in the sense of the formulae in Proposition 8.1.2, where $\langle \ , \ \rangle : g \otimes m \to k$ is given and neither Lie bialgebra need be finite dimensional. Check that the first part of Example 8.3.7 still goes through in this setting, giving a Lie bialgebra $m^{\mathrm{op}} \bowtie g$, the generalised double Lie bialgebra. Reformulate this as $m \bowtie_\sigma g$, where g, m are skew-copaired by $\sigma : g \otimes m \to k$. Check that $g \bowtie_r g$ is an example whenever g is a dual quasitriangular Lie bialgebra.*

Solution: This is entirely straightforward from the double cross sum point of view. It is clear that if we keep the properties $\langle \xi, [\phi, \psi] \rangle = \langle \delta \xi, \phi \otimes \psi \rangle$ etc. for a pairing, then $\triangleright, \triangleleft$ in the proof of Example 8.3.7 are still well-defined and the proof of (8.23) and (8.30) still go through. The role of g^* there is now played by m. On the other hand, we do not even require that the pairing be nondegenerate, i.e. the Lie bialgebras g, m could be very far from actually being dual to each other.

The second part is a cosmetic shift: a *skew-pairing* between Lie bialgebras is a linear map $\sigma : g \otimes m \to k$ such that

$$\sigma([\xi, \eta], \phi) = \sum \sigma(\xi, \phi_{[1]}) \sigma(\eta, \phi_{[2]}),$$

$$\sigma(\xi, [\phi, \psi]) = \sum \sigma(\xi_{[1]}, \psi) \sigma(\xi_{[2]}, \phi),$$

(8.34)

for all $\xi, \eta \in g$ and $\phi, \psi \in m$. This is nothing other than a pairing of g, m^{op}. So the role of m^{op} above is now played by m. It is the infinitesimal version of (7.13) in Chapter 7.2. Note that $-\sigma_{21} : m \otimes g \to k$ is also a skew-pairing and gives an isomorphic generalised double $g_\sigma \bowtie m \equiv g \bowtie_{-\sigma_{21}} m \cong m \bowtie_\sigma g$. Finally, it is clear from (8.13) that a dual quasitriangular structure on g gives, in particular, a skew-pairing of g with itself. Hence we have an example $g \bowtie_r g$ of a generalised double. We also have $g_r \bowtie g$, and we have already encountered the dual construction to this in Section 8.2.   ∎

This gives the infinitesimal version of the generalised quantum double examples in Chapter 7.2. One can also compute the bicross sums $g \bowtie_r g^*$ etc., which correspond to the same data via Proposition 8.3.6 if g is finite dimensional. Also, the double construction $D(g)$ for Lie bialgebras is related to the following general construction for matched pairs.

Theorem 8.3.11 *Let $(g, m, \triangleleft, \triangleright)$ be a Lie algebra matched pair. Then $(g^{\mathrm{op}} \oplus m, g \bowtie m)$ is also a Lie algebra matched pair, where $g \bowtie m$ acts by the left adjoint action and $g^{\mathrm{op}} \oplus m$ acts by $-\triangleleft$ for the action of g^{op} and $-\triangleright$ for the action of m.*

Proof: First, $g \bowtie m$ acts on itself by its (right) adjoint action, and since $g^{\mathrm{op}} \oplus m$ has the same vector space, it acts in this too. The action and the matching 'back-reaction' of $g^{\mathrm{op}} \oplus m$ are

$$(\xi \oplus \phi) \triangleright (\eta \oplus \psi) = ([\xi, \eta] + \phi \triangleright \eta - \psi \triangleright \xi) \oplus ([\phi, \psi] + \phi \triangleleft \eta - \psi \triangleleft \xi),$$

$$(\xi \oplus \phi) \triangleleft (\eta \oplus \psi) = (-\psi \triangleright \xi) \oplus (-\phi \triangleleft \eta),$$

for all $(\xi \oplus \phi) \in g \bowtie m$ and $(\eta \oplus \psi) \in g^{\mathrm{op}} \oplus m$. The second of these is indeed a right action because $\triangleleft$ is a right action of g, hence $-\triangleleft$ of g^{op}, and $\triangleright$ is a left action of m, hence $-\triangleright$ a right action. Because the actions $(\triangleleft, \triangleright)$ are matching, it is not hard to guess that these induced actions should also be matching. This is immediately verified from the definitions inserted into (8.23) for these Lie algebras. Thus,

$$(\xi \oplus \phi) \triangleright [\eta \oplus \psi, \zeta \oplus \varphi]_{g^{\mathrm{op}} \oplus m} = (\xi \oplus \phi) \triangleright ([\zeta, \eta] \oplus [\psi, \varphi])$$
$$= ([\xi, [\zeta, \eta]] + \phi \triangleright [\zeta, \eta] - [\psi, \varphi] \triangleright \xi) \oplus ([\phi, [\psi, \varphi]] + \phi \triangleleft [\zeta, \eta] - [\psi, \varphi] \triangleleft \xi)$$

from the definitions. Computing this using the Jacobi identity and the matched pair conditions (8.23) gives 14 terms. On the other hand, computing $[(\xi \oplus \phi) \triangleright (\eta \oplus \psi), \zeta \oplus \varphi]_{g^{\mathrm{op}} \oplus m}$ etc. for the right hand side of the second part of (8.23) for these Lie algebras gives the same 14 terms after cancellation. One has to prove the first half of (8.23) in the same explicit way.   ∎

This theorem generalises the classical double to the double of a general matched pair of Lie algebras. The following exercise generalises the fact that the double is always quasitriangular. This in turn makes contact with constructions in the theory of classical integrable systems.

Exercise 8.3.12 *If $g \bowtie m$ is a Lie algebra double cross sum, show that $r_M(\xi \oplus \phi) = \frac{1}{2}(-\xi \oplus \phi)$ obeys the modified Yang–Baxter equation (8.9). Conclude that if $g \bowtie m$ has a symmetric ad-invariant element $K^{-1} \in (g \bowtie m)^{\otimes^2}$ (e.g. if it is complex semisimple) then it becomes a quasitriangular Lie bialgebra with $r = (\mathrm{id} \otimes r_M)K^{-1} + \frac{1}{2}K^{-1}$. Derive the quasitriangular structure of the double in Proposition 8.2.1.*

Solution: This result is usually approached another way and with emphasis on r_M, corresponding to r_- as explained when deriving the MCYBE in Section 8.1. Actually, we can work just as well in terms of $\pi = r_M + \frac{1}{2}$: $g \to g$ obeying the equivalent equations

$$\pi([\xi, \eta]_\pi) = [\pi(\xi), \pi(\eta)], \quad [\xi, \eta]_\pi = [\pi(\xi), \eta] + [\xi, \pi(\eta)] - [\xi, \eta]. \quad (8.35)$$

This corresponds more directly to r rather than to r_- and is just the CYBE in terms of the map $\pi = r \circ (2r_+)^{-1} : g \to g$. If g has on it an invertible symmetric ad-invariant bilinear map K, then we regard it as a map $g \to g^*$ and take its inverse. We conclude from Lemma 8.1.6 that $r_1 = \pi \circ K^{-1} : g^* \to g$ defines a quasitriangular structure on g in the form of a map. If we want to work only with r as an element of $g \otimes g$ and not as a map, it is just

$$r = (\mathrm{id} \otimes \pi)(K^{-1}),$$

where we regard $K^{-1} \in g \otimes g$. Actually, we need only assume this element K^{-1} and (8.35) to check directly that r is a quasitriangular structure for g.

We now apply this point of view to the Lie algebra $g \bowtie m$. Since we know its explicit double cross sum description in Proposition 8.3.2, we can just define $\pi(\xi \oplus \phi) = 0 \oplus \phi$ (projection onto the second component of the direct sum). Then

$$\begin{aligned}
[\xi \oplus \phi, \eta \oplus \psi]_\pi &= [\xi \oplus \phi, 0 \oplus \psi] - [\eta \oplus \psi, 0 \oplus \phi] - [\xi \oplus \phi, \eta \oplus \psi] \\
&= (-\psi \triangleright \xi) \oplus ([\phi, \psi] - \xi \triangleright \psi) + \phi \triangleright \eta \oplus ([\phi, \psi] + \phi \triangleleft \eta) \\
&\quad + (-[\xi, \eta] - \phi \triangleright \eta + \psi \triangleright \xi) \oplus (-[\phi, \psi] - \phi \triangleleft \eta + \psi \triangleleft \xi) \\
&= [\eta, \xi] \oplus [\phi, \psi],
\end{aligned}$$

for all $\xi, \eta \in g$ and $\phi, \psi \in m$. This is clearly a second Lie algebra structure on $g \bowtie m$ (namely $g^{\mathrm{op}} \oplus m$), and, equally clearly, $\pi : g^{\mathrm{op}} \oplus m \to g \bowtie m$ is a Lie algebra homomorphism as required in (8.35).

To give one example, starting with $D(g)$ constructed in Example 8.3.7 as a double cross sum Lie algebra $g^{*op}\bowtie g$, we can verify directly that K^{-1} as stated in Proposition 8.2.2 is ad-invariant. Then $(\mathrm{id} \otimes \pi)(K^{-1}) = (f^a \oplus 0) \otimes (0 \oplus e_a)$ as in Proposition 8.2.1. Here $\{e_a\}$ is a basis of g and $\{f^a\}$, a dual basis.   ∎

So far, we have concentrated on infinitesimal versions of our favourite Hopf algebra constructions. However, exponentiating in a more conventional way, we also obtain Lie group constructions. In the case of the bicross sums, we will now prove that every Lie algebra bicross sum of the form in Proposition 8.3.3, subject to some technical completeness conditions, exponentiates to a Lie group matched pair. This then gives a Hopf algebra bicrossproduct, constructed as in Chapter 6.2, of which $m^* \blacktriangleright\!\!\!\triangleleft g$ is the infinitesimal part. The existence of the Lie group matched pair is easy to see from some general geometric considerations. Its explicit construction is another matter and we take it up now.

The first step of the exponentiation is quite elementary: the Lie algebras g, m in our matched pair exponentiate in a natural way to connected simply connected Lie groups G, M, say. The easiest way to do this is to choose a faithful linear representation and exponentiate the corresponding matrices. Since our matched pair comes with linear actions g on m and m on g, it is natural to try to use these: if they are not faithful, then it means that we may as well have used quotient Lie algebras of g or m where the representations are faithful. In this nicest (reduced) case at least, we construct the action of G on m from the right and M on g from the left, as the corresponding matrices. We continue to denote these actions by $\triangleleft$ and $\triangleright$, respectively. In more complicated (such as infinite-dimensional) cases, this first step may need some care, but in principle it is the whole point of Lie algebras that Lie algebra actions generally extend to ones of the associated group.

Example 8.3.13 *Let (su_2, su_2^{*op}) be the matched pair in Example 8.3.9. The Lie group SU_2^{*op} is*

$$SU_2^{*op} = \{\vec{s} \in \mathbb{R}^3; \quad s_3 > -1\}, \quad \vec{s}\vec{t} = \vec{s} + (s_3 + 1)\vec{t}.$$

The actions of SU_2 on su_2^{*op} and SU_2^{*op} on su_2 are

$$\vec{\phi} \triangleleft u = \mathrm{Rot}_{u^{-1}}(\vec{\phi}), \quad \vec{s} \triangleright \vec{\xi} = \vec{\xi} + \frac{\vec{s}}{s_3 + 1} \times (\vec{e_3} \times \vec{\xi}),$$

where Rot is the usual 3-vector rotation associated to an element of SU_2.

Proof: The action of SU_2 is clear from the infinitesimal action in Example 8.3.9. The standard rotation here can be realised easily by writing $\vec{\phi}$

as an element $-\frac{1}{2}\sigma_i\phi^i$ of su_2, in which case $\mathrm{Rot}_u = u(\ )u^{-1} = \mathrm{Ad}_u$, the usual adjoint action of SU_2 on its Lie algebra. For the second part we write the action of $su_2^{*\mathrm{op}}$ in Example 8.3.9 as a matrix

$$\vec{\phi}\triangleright\vec{\xi} = \begin{pmatrix} -\phi_3 & 0 & 0 \\ 0 & -\phi_3 & 0 \\ \phi_1 & \phi_2 & 0 \end{pmatrix} \begin{pmatrix} \xi_1 \\ \xi_3 \\ \xi_3 \end{pmatrix}$$

and exponentiate it as:

$$e^{\phi} = \begin{pmatrix} e^{-\phi_3} & 0 & 0 \\ 0 & e^{-\phi_3} & 0 \\ \phi_1\frac{1-e^{-\phi_3}}{\phi_3} & \phi_2\frac{1-e^{-\phi_3}}{\phi_3} & 1 \end{pmatrix} = (s_3+1)^{-1}\begin{pmatrix} 1 & 0 & 0 \\ 0 & 1 & 0 \\ s_1 & s_2 & s_3+1 \end{pmatrix},$$

where $s_i = \phi_i(e^{\phi_3} - 1)/\phi_3$. This gives a description of $SU_2^{*\mathrm{op}}$ as certain matrices acting on 3-vectors $\vec{\xi}$. Matrix multiplication then gives the group law as stated in terms of $\vec{s}$. We identify the group as an open subset of $\mathbb{R}^3$ with a deformed (noncommutative) version of the usual addition on $\mathbb{R}^3$. Note that inversion in these coordinates is $s^{-1} = -\vec{s}/(s_3+1)$.   ∎

To proceed further, we need some elementary differential geometry on Lie groups. The main ingredient is the operation of left (or right) translation. Let $L_u : G \to G$ denote the map $L_u(v) = uv$. Its differential L_{u*} at $v \in G$ is a map $T_vG \to T_{uv}G$ between fibres of the tangent bundle. In particular, $L_{u*} : g \equiv T_eG \to T_uG$ gives a tangent vector at each point $u \in G$ associated to any Lie algebra element. This is the *left-invariant vector field*

$$\tilde{\xi}(u) = L_{u*}(\xi), \quad \tilde{\xi}(f)(u) = \frac{\mathrm{d}}{\mathrm{d}\tau}\Big|_0 f(ue^{\tau\xi}), \quad \forall f \in C^{\infty}(G), \qquad (8.36)$$

where $\tilde{\xi}$ is defined by its evaluation on all test functions f. One can check that $\tilde{\xi}$, as an operator from functions to functions, intertwines the action of G induced by left translation. One also has the left *Maurer–Cartan form* ω^L on G as the g-valued 1-form whose evaluation with $\tilde{\xi}$ at any point is ξ, for any $\xi \in g$. So,

$$\langle \omega^L, \tilde{\xi}\rangle = \xi, \quad \mathrm{d}\omega^L + \frac{1}{2}[\omega^L, \omega^L] = 0,$$

where the [,] denotes the Lie bracket in the Lie algebra g in which ω^L takes its values. There are also right-invariant vector fields based on the operator $R_u(v) = vu$ and a corresponding right handed Maurer–Cartan form on G.

For our right–left matched pairs, we have two groups G, M: we will generally use the left handed Maurer–Cartan form ω^L on G and a right handed Maurer–Cartan form ω^R when working on M. The latter is explicitly defined by

$$\tilde{\phi}(s) = R_{s*}(\phi), \quad \tilde{\phi}(f)(s) = \left.\frac{\mathrm{d}}{\mathrm{d}\tau}\right|_0 f(e^{\tau\phi}s), \quad \langle\omega^R, \tilde{\phi}\rangle = \phi,$$

for $f \in C^\infty(M)$ and all $\phi \in m$. Working with Maurer–Cartan forms is equivalent to (but more compact than) writing all equations evaluated explicitly on generic left- or right-invariant vector fields.

Finally, we recall from Chapter 6.1 that, more generally, whenever a Lie group acts on a manifold, we likewise have vector fields whose flows generate the action. Thus, if we did succeed in forming a group matched pair (G, M, α, β), then the differential of the action α of G on M from the right would be a vector field $\alpha_{*\xi}$ on M for each $\xi \in g$. Likewise, the differential of the action β of M on G from the left would be vector fields $\beta_{*\phi}$ on G for all $\phi \in m$. Again, we can refer the tangent vectors at each point back to the image of Lie-algebra valued functions a_ξ, b_ϕ defined by

$$\alpha_{*\xi}(s) \equiv L_{s*}(a_\xi(s)), \quad \beta_{*\phi}(u) \equiv R_{u*}(b_\phi(u)).$$

Moreover, by differentiating the conditions (6.15) of our supposed group matched pair, we deduce

$$\alpha_{*\xi}(st) = L_{s*}(\alpha_{*\xi}(t)) + R_{t*}(\alpha_{*t\triangleright\xi}(s)), \quad \alpha_{*\xi}(e) = 0,$$
$$\beta_{*\phi}(uv) = R_{v*}(\beta_{*\phi}(u)) + L_{u*}(\beta_{*\phi\triangleleft u}(v)), \quad \beta_{*\phi}(e) = 0, \tag{8.37}$$

or, in terms of $a : M \times g \to m$ and $b : m \times G \to g$, the equations

$$a_\xi(st) = \mathrm{Ad}_{t^{-1}}(a_{t\triangleright\xi}(s)) + a_\xi(t), \quad a_\xi(e) = 0,$$
$$b_\phi(uv) = b_\phi(u) + \mathrm{Ad}_u(b_{\phi\triangleleft u}(v)), \quad b_\phi(e) = 0, \tag{8.38}$$

i.e. $a \in Z^1_{\triangleright^* \otimes \mathrm{Ad}_R}(M, g^* \otimes m)$, $\quad b \in Z^1_{\mathrm{Ad} \otimes \triangleleft^*}(G, g \otimes m^*)$,

where we understand these equations exactly as group cocycles with values in the stated linear spaces. In the first case, Ad_R is the right adjoint action of M on m and $\triangleright^*$ is the right action of M on g^*, given by dualising its action on g. We regard a as a map $M \to g^* \otimes m$. In the second case, Ad is the (usual) left adjoint action of G on g and $\triangleleft^*$ is the left action of G on m^*, given by dualising its right action on m. The general definition of a group cocycle with values in a left-module has already been encountered in Example 6.3.3. The definition of right-module cocycles needed for a is strictly analogous. We aim to push this line of calculation backwards.

Proposition 8.3.14 *Let $(g, m, \triangleleft, \triangleright)$ be a Lie algebra matched pair with associated groups G acting on m and M acting on g. There are unique* *group cocycles a, b in the sense of (8.38) with differential at the identity given by $\lhd, \rhd$, respectively.*

Proof: We differentiate the desired (8.38) one more time to obtain

$$\langle da_\xi, \tilde{\phi} \rangle(s) \equiv \tilde{\phi}(a_\xi)(s) \equiv \left.\frac{d}{d\tau}\right|_0 a_\xi(e^{\tau\phi}s) = \mathrm{Ad}_{s^{-1}}(\phi\lhd(s\rhd\xi)), \quad a_\xi(e) = 0,$$

$$\langle db_\phi, \tilde{\xi} \rangle(u) \equiv \tilde{\xi}(b_\phi)(u) \equiv \left.\frac{d}{d\tau}\right|_0 b_\phi(ue^{\tau\xi}) = \mathrm{Ad}_u((\phi\lhd u)\rhd\xi), \quad b_\phi(e) = 0,$$

or, in terms of Maurer–Cartan forms on M and G, respectively,

$$da_\xi = \mathrm{Ad}_{s^{-1}}(\omega^R\lhd(s\rhd\xi)), \quad db_\phi = \mathrm{Ad}_u((\phi\lhd u)\rhd\omega^L), \tag{8.39}$$

with the same boundary conditions. In all these equations we treat a, b as functions on the relevant group with values in the representation spaces $g^* \otimes m$ etc. On the other hand,

$$d(\mathrm{Ad}_u((\phi\lhd u)\rhd\omega^L))$$
$$= (d\mathrm{Ad}_u)((\phi\lhd u)\rhd\omega^L) + \mathrm{Ad}_u((d(\phi\lhd u))\rhd\omega^L) + \mathrm{Ad}_u((\phi\lhd u)\rhd d\omega^L)$$
$$= \mathrm{Ad}_u\left(\mathrm{ad}_{\omega^L}((\phi\lhd u)\rhd\omega^L) + (((\phi\lhd u)\lhd\omega^L)\rhd\omega_L) - \frac{1}{2}(\phi\lhd u)\rhd[\omega^L, \omega^L]\right) = 0,$$

on using the second half of the matched pair axiom (8.23) to compute $(\phi\lhd u)\rhd[\omega^L, \omega^L]$. Note that in any linear representation $\lhd$ one can think of $\phi\lhd u$ as a (vector-valued) function. Then $d(\phi\lhd u) = (\phi\lhd u)\lhd\omega^L$. The same principle was used for Ad_u. One can say informally that $\omega^L = u^{-1}du = -(du^{-1})u$ in linear representations. The proof of $d(\mathrm{Ad}_{s^{-1}}(\omega^R\lhd(s\rhd\xi))) = 0$ is strictly analogous, using $\omega^R = (ds)s^{-1} = -sds^{-1}$ as far as $\mathrm{Ad}_{s^{-1}}$ and $s\rhd\xi$ are concerned. Since $H^1(M) = H^1(G) = 0$, we conclude the existence of a_ξ, b_ϕ as solutions to (8.39), i.e. these equations are integrable. Finally, they imply at once the desired (8.38), as we can expect since (8.39) were obtained by their differentiation. Indeed, let $f(v) = b_\phi(uv) - b_\phi(u) - \mathrm{Ad}_u(b_{\phi\lhd u}(v))$. Then $df(v) = db_\phi(uv) - \mathrm{Ad}_u(db_{\phi\lhd u}(v)) = 0$ from (8.39). Hence $f \equiv 0$. Similarly for the first half of (8.38). ∎

In fact, these steps can be summarised in cohomological terms: the Lie algebra matched pair equations (8.23) in terms of Lie algebra cohomology are

$$\lhd \in Z^1_{\rhd^* \otimes -\mathrm{ad}}(m, g^* \otimes m), \quad \rhd \in Z^1_{\mathrm{ad} \otimes \lhd^*}(g, g \otimes m^*).$$

We have already used special cases of Lie algebra cocycles in Section 8.1, while the general formula (for the left handed theory) is in (6.44) in Chapter 6.3. On the other hand, a general fact about Lie algebras is that Lie algebra cocycles always exponentiate to cocycles on the associated connected simply connected groups.

Example 8.3.15 *For the Lie algebra matched pair* (su_2, su_2^{*op}) *in Examples 8.3.9 and 8.3.13 we have the matching group cocycles*

$$a_{\vec{\xi}}(\vec{s}) = (s_3 + 1)^{-1}(\vec{s} - \frac{1}{2}\vec{e}_3\frac{s^2}{s_3 + 1}) \times \vec{\xi}, \quad b_{\vec{\phi}}(u) = \phi \times (\vec{e}_3 - \mathrm{Rot}_u(\vec{e}_3)).$$

Proof: These are obtained by solving the equations (8.39) on M and G. For b_ϕ we use the actions from Examples 8.3.9 and 8.3.13. Using that $\mathrm{Ad}_u = \mathrm{Rot}_u$ and that $\times$ is rotationally-invariant, we have to solve

$$\frac{d}{d\tau}\Big|_0 b_\phi(ue^{\tau\xi}) = \phi \times \mathrm{Rot}_u(\vec{e}_3 \times \xi), \quad \forall \xi.$$

It is clear that b_ϕ as stated does this. It is unique for the given boundary conditions.

For the a_ξ, we compute the adjoint action from the matrix description in Example 8.3.13 as

$$\mathrm{Ad}_{\vec{s}}(\phi) = L_{\vec{s}*} \circ R_{\vec{s}^{-1}*}(\phi) = \vec{\phi}(s_3 + 1) - \phi_3\vec{s}.$$

One can also obtain it as $\mathrm{Ad}_s = L_{s*} \circ R_{s^{-1}*}$ in any other coordinate system. In the $\vec{s}$ coordinates, the operators L_{s*} and R_{s*} are

$$L_{\vec{s}*}(\phi) = \frac{d}{d\tau}\Big|_0 \left(\vec{s} + (s_3 + 1)\vec{\phi}\left(\frac{e^{\tau\phi_3} - 1}{\phi_3}\right)\right) = \vec{\phi}(s_3 + 1),$$

$$R_{\vec{s}*}(\phi) = \frac{d}{d\tau}\Big|_0 \left(\vec{\phi}\left(\frac{e^{\tau\phi_3} - 1}{\phi_3}\right) + e^{\tau\phi_3}\vec{s}\right) = \vec{\phi} + \phi_3\vec{s},$$

giving the same formula for $\mathrm{Ad}_{\vec{s}}$. Note that Ad acts on the Lie algebra itself, unlike L_{s*}, R_{s*}, which have their outputs in other fibres of the tangent bundle. Using the actions from Examples 8.3.9 and 8.3.13, and our calculation for $\mathrm{Ad}_{\vec{s}^{-1}}$, we need to solve

$$\frac{d}{d\tau}\Big|_0 a_\xi(e^{\tau\phi}s) = \frac{\phi \times \xi}{s_3 + 1} - e_3 \times \xi\frac{\phi \cdot s}{(s_3 + 1)^2}, \quad \forall \phi.$$

It is easy to see that a_ξ as stated does this and is the unique solution for the given boundary conditions.

Once a_ξ, b_ϕ are found from these differential equations, it is of course possible to verify explicitly that they obey the cocycle conditions (8.38), as they must by our constructions. This is easy for b, while for a it takes some tedious 3-vector calculus. ∎

Next, we need the following lemma.

Lemma 8.3.16 *Let* $(g, m, \lhd, \rhd)$ *be a right–left matched pair of Lie algebras, let* G, M *be the associated groups and let* a_ξ, b_ϕ *be as constructed* *above.* **Then**

$$[\phi, \psi] \triangleleft u = [\phi \triangleleft u, \psi \triangleleft u] + (\phi \triangleleft b_\psi) \triangleleft u - (\psi \triangleleft b_\phi) \triangleleft u,$$

$$s \triangleright [\xi, \eta] = [s \triangleright \xi, s \triangleright \eta] + s \triangleright (a_\xi \triangleright \eta) - s \triangleright (a_\eta \triangleright \xi),$$

for all $\phi, \psi \in m$, $\xi, \eta \in g$, $u \in G$ and $s \in M$.

Proof: For each $\xi, \eta \in g$, we consider the vector-valued function $f(s) = s \triangleright [\xi, \eta] - [s \triangleright \xi, s \triangleright \eta] + s \triangleright (a_\xi \triangleright \eta) - s \triangleright (a_\eta \triangleright \xi)$ and compute that $df = \omega^R \triangleright f$. This follows from $d(s \triangleright (\ )) = \omega^R \triangleright (s \triangleright (\ ))$, as above, and da_ξ from the first half of (8.38). Since $f(e) = 0$, we conclude that $f(s) = 0$ for all s. Similarly for the first half of the lemma.   ∎

Using this lemma, one can show that the vector fields $\widetilde{a_\xi}(s) = L_{s*}(a_\xi(s))$ and $\widetilde{b_\phi}(u) = R_{u*}(b_\phi(u))$ obey

$$[\widetilde{a_\xi}, \widetilde{a_\eta}] = \widetilde{a_{[\xi,\eta]}}, \quad [\widetilde{b_\phi}, \widetilde{b_\psi}] = -\widetilde{b_{[\phi,\psi]}}$$

under the Lie bracket of vector fields. This says that we have an action of the Lie algebra g on M and the Lie algebra m on G in terms of these vector fields. The vector fields of course obey (8.37) by our earlier construction.

Finally, to obtain our Lie group matched pair, we just have to exponentiate these vector fields; i.e. we have to find group actions $\alpha = \triangleleft, \beta = \triangleright$ such that $\alpha_{*\xi} = \widetilde{a_\xi}$ and $\beta_{*\phi} = \widetilde{b_\phi}$. This means solving some first order nonlinear equations for paths $s(\tau) \in M$ and $u(\tau)$ in G such that

$$\begin{aligned}
\frac{ds(\tau)}{d\tau} &= L_{s(\tau)*}(a_\xi(s(\tau))), \quad s(0) = s, \\
\frac{du(\tau)}{d\tau} &= R_{u(\tau)*}(b_\phi(u(\tau))), \quad u(0) = u.
\end{aligned} \tag{8.40}$$

These can be solved separately, and the result is $s(1) = s \triangleleft u$ if $u = e^\xi$ and $u(1) = s \triangleright u$ if $s = e^\phi$. By our construction, these $(G, M, \triangleleft, \triangleright)$ will then obey the conditions for a group matched pair in the sense of Definition 6.2.10.

The general theory of when vector fields for Lie algebra actions exponentiate to group actions requires some further analysis. One needs basically that the relevant vector fields are complete, as always happens, for example, if the underlying manifold is compact. On the other hand, compactness is too restrictive to assume for both groups of the matched pair if we want to cover many of the examples of interest. In fact, we will finish the general theory in another way: by showing that each of the group actions α or β can be constructed in terms of the other. Hence, if one exists, so does the other. This cross-coupled formulation can also be used to view the problem as a single second order system rather than solving separately.

Theorem 8.3.17 *Let* (g, m) *be a Lie algebra matched pair and let* a, b *be the associated group cocycles on* G, M *as constructed above. Suppose that* b *exponentiates to a group action* $\triangleright$ *of* M *on* G. *Then let*

$$A_u(s) = w^R \triangleleft (s \triangleright u), \qquad s \triangleleft u = \overleftarrow{P} e^{\int_e^s A_u},$$

where the $\{A_u\}$ *are a family of zero-curvature gauge fields (connections) on a trivial* M*-bundle over* M *and* $\overleftarrow{P}$ *denotes parallel transport from the right. Likewise, if* a *exponentiates to an action* $\triangleleft$ *of* G *on* M *then let*

$$B_s(u) = (s \triangleleft u) \triangleright w^L, \qquad s \triangleright u = \overrightarrow{P} e^{\int_e^u B_s},$$

where $\{B_s\}$ *is a family of zero-curvature connections on a trivial* G*-bundle over* G *and* $\overrightarrow{P}$ *denotes the usual parallel transport. In either case,* $(G, M, \triangleleft, \triangleright)$ *is a matched pair of Lie groups.*

Proof: The parallel transport or path-ordered exponential symbols are the usual ones in physics, with $\overrightarrow{P} e^{\int}$ probably the more conventional. It means, in a concrete representation of the relevant groups, that we take a path $u(\tau)$ from $u(0) = e$ to $u(1) = u$ and compute the limit of $\prod_i (1 + (\tau_{i+1} - \tau_i)\langle \frac{du}{d\tau}, B_s \rangle (u(\tau_i)))$ taken in the order with increasing i from left to right. More formally, $\beta(\tau) = \overrightarrow{P} e^{\int B_s}$ along the path $u(\tau)$ is the unique solution of the equations

$$\beta(0) = e, \qquad L_{\beta(\tau)^{-1}*}\left(\frac{d\beta(\tau)}{d\tau}\right) = \langle \frac{du(\tau)}{d\tau}, B_s \rangle (u(\tau)).$$

In geometrical terms, one can consider $G \times G \to G$ in a trivial way as a principal G-bundle, projecting onto the left hand factor, say. Projection to the other factor provides a global trivialisation and an associated trivial connection. Any other g-valued 1-form on the base manifold can be viewed as a connection by adding to the trivial one. Its parallel transport lifts a path $u(\tau)$ on the base to a path $(u(\tau), \beta(\tau))$.

In the present case, however, all the connections B_s obey the equation $dB_s + \frac{1}{2}[B_s, B_s] = 0$ on G, i.e. they have zero curvature. Thus,

$$dB_s = (s \triangleleft u) \triangleright dw^L + (d(s \triangleleft u)) \triangleright w^L$$
$$= -\frac{1}{2}(s \triangleleft u) \triangleright [w^L, w^L] + (s \triangleleft u) \triangleright (a_{w^L}(s \triangleleft u) \triangleright w^L) = -\frac{1}{2}[B_s, B_s],$$

where the second equality was (as usual) the Maurer–Cartan equations and the differential (on G) of the action (in this case of G on m). The third equality is Lemma 8.3.16 above. Hence, in the present case, the parallel transport depends only on the endpoint u of the path. Hence it defines a map $M \times G \to G$, which we call $\triangleright$.

This is the second half of the theorem. The first half is strictly analogous, starting this time with $\triangleright$ known and used to define connections A_u. This time we construct by parallel transport a map $\alpha = \triangleleft$. It is a path-ordered exponential along a path $s(\tau)$ in M, taken this time in increasing order from right to left. Equivalently, it is the solution to

$$\alpha(0) = e, \quad R_{\alpha(\tau)^{-1}*}\left(\frac{d\alpha(\tau)}{d\tau}\right) = \langle\frac{ds(\tau)}{d\tau}, A_u\rangle(s(\tau)).$$

Because the ordering is unconventional, the appropriate zero-curvature condition is also slightly unconventional, namely $A_u \overleftarrow{d} + \frac{1}{2}[A_u, A_u] = 0$, where d acts as usual but is viewed as coming from the right.

Finally, it is easy to see from this point of view that $\triangleright, \triangleleft$ so constructed indeed obey the matched pair conditions given in Definition 6.2.10. Of course, we should expect this since (8.37) are obeyed by $\tilde{a}, \tilde{b}$, and were just the differentials of these desired conditions. Indeed, the difference between the left and right hand sides of the group matched pair conditions each obey a suitable first order differential equation. Holding near the identity then implies that the differences vanish identically.   ∎

The parallel transport point of view expresses the differential cocycle conditions (8.38) in exponentiated form in a manner suggested by the diagrammatic form of the matched pair conditions in Chapter 6.2. We arrive at the following conclusion: that every Lie algebra matched pair exponentiates to cocycles a, b; if the corresponding vector fields for one of these is complete, then the Lie algebra matched pair exponentiates to a Lie group matched pair. These exponentiations on the associated connected simply connected Lie groups are uniquely determined by the Lie algebra matched pair data.

Example 8.3.18 $(SU_2, SU_2^{\star op})$, *as in Example 8.3.13, form a right–left matched pair of groups in the sense of Definition 6.2.10 with mutual actions*

$$\alpha_u(\vec{s}) = \mathrm{Rot}_{u^{-1}}\left(\vec{s} - \frac{1}{2}\vec{e}_3\frac{s^2}{s_3 + 1}\right) + \frac{1}{2}\vec{e}_3\frac{s^2}{s_3 + 1},$$

$$\beta_{\vec{s}}\begin{pmatrix} x & y \\ -\bar{y} & \bar{x} \end{pmatrix} = \frac{\begin{pmatrix} (s_3 + 1)x - (s_1 - \imath s_2)\bar{y} & y \\ -\bar{y} & (s_3 + 1)\bar{x} - (s_1 + \imath s_2)y \end{pmatrix}}{\sqrt{|(s_3 + 1)x - (s_1 - \imath s_2)\bar{y}|^2 + |y|^2}},$$

where $| \ |$ *is the usual complex norm and* $|x|^2 + |y|^2 = 1$. *The double cross product* $SU_2 \bowtie SU_2^{\star op}$ *is the real six-dimensional Lie group* $SL(2, \mathbb{C})$

covering the usual Lorentz group.

Proof: The vector fields we have to exponentiate in the first case are obtained from Example 8.3.15 as

$$\widetilde{a_\xi}(\vec{s}) = L_{\vec{s}*}(a_\xi) = \left(\vec{s} - \frac{1}{2}\vec{e_3} \frac{s^2}{s_3 + 1} \right) \times \vec{\xi}.$$

Thus, we have to solve for a path $\vec{s}(\tau)$ such that

$$\frac{d\vec{s}(\tau)}{d\tau} = \left(\vec{s}(\tau) - \frac{1}{2}\vec{e_3} \frac{s(\tau)^2}{s(\tau)_3 + 1} \right) \times \vec{\xi}, \quad \vec{s}(0) = \vec{s}.$$

This is immediate once we note that $s(\tau)^2/(s(\tau)_3 + 1)$ is constant and hence equal to its initial value $s^2/(s_3 + 1)$. We then define $\alpha_u(\vec{s}) = \vec{s}(1)$, where $u = e^\xi$.

For the vector fields in the second case, we prefer to use the matrix coordinates of SU_2 as $u = \begin{pmatrix} x & y \\ -\bar{y} & \bar{x} \end{pmatrix}$, say, where $x, y \in \mathbb{C}$ and $|x|^2 + |y|^2 = 1$. If we also use the usual matrix description of su_2 as Hermitian matrices, then $R_{u*} = (\ )u$, i.e. just multiplication from the right. We could also have used matrix coordinates for the vector fields $\widetilde{a_\xi}$. Proceeding from Example 8.3.15, we also write $\phi, \xi, \vec{e_3}$ as elements of su_2 using the standard Pauli matrix basis $\{\frac{1}{2}\sigma_i\}$. Then $\times$ becomes the usual matrix commutator and $\text{Rot}_u = u(\ )u^{-1}$. We write

$$\phi = \frac{-\imath}{2} \begin{pmatrix} \phi_3 & \phi_- \\ \phi_+ & -\phi_3 \end{pmatrix}, \quad \phi_\pm \equiv \phi_1 \pm \imath\phi_2, \quad e_3 - ue_3u^{-1} = -\imath \begin{pmatrix} y\bar{y} & xy \\ \bar{x}\bar{y} & -y\bar{y} \end{pmatrix},$$

and therefore we have to find a path $u(\tau)$ such that

$$\frac{du(\tau)}{d\tau} = \widetilde{b_\phi}(u(\tau)) = [\phi, e_3 - u(\tau)e_3u(\tau)^{-1}]u(\tau), \quad u(0) = u,$$

or, more explicitly,

$$\frac{d}{d\tau}\begin{pmatrix} x(\tau) \\ y(\tau) \end{pmatrix} = \frac{1}{2}\begin{pmatrix} 2xy\bar{y}\phi_3 + x^2y\phi_+ + (x\bar{x} - 2)\bar{y}\phi_- \\ -2x\bar{x}y\phi_3 + xy^2\phi_+ + \bar{x}y\bar{y}\phi_- \end{pmatrix}, \quad \begin{pmatrix} x(0) \\ y(0) \end{pmatrix} = \begin{pmatrix} x \\ y \end{pmatrix}.$$

We observe that $y/\bar{y}$ is a constant of motion, as is $x\bar{x} + y\bar{y}$ (which remains equal to unity). Finding also

$$\frac{d}{d\tau}\frac{x(\tau)}{\bar{y}(\tau)} = \phi_3 \frac{x(\tau)}{\bar{y}(\tau)} - \phi_-$$

enables us to solve the system. We then define $\beta_{\vec{s}}(u) = u(1)$, where $\vec{s} = e^\phi = \vec{\phi}(e^{\phi_3} - 1)/\phi_3$ from Example 8.3.13.

We conclude from the general theory above that these $\alpha = \triangleleft, \beta = \triangleright$ give a Lie group matched pair. From (the Lie version of) Proposition 6.2.15, we have also a double cross product group. It is connected and simply connected and has Lie algebra $su_2 \bowtie su_2^{\star \text{op}}$, which we know from Example 8.2.9 to be the Iwasawa decomposition of $sl(2, \mathbb{C})$ as a real Lie algebra. ∎

The principle that every action has a 'reaction' has been one of our main themes here and in Chapter 6.2. We have seen matched pairs of Lie algebras, groups, Hopf algebras (in Chapter 7.3) and now *matched pairs of families of gauge fields* in Theorem 8.3.17. One can obviously take this principle much further. Whereas historically, in mathematics, semidirect or cross products are more well-known, in physics it can be argued that double cross products etc. are much more natural. In the above example, the orbits of SU_2 are spheres

$$\mathcal{O}_\lambda = \{\vec{s}, \quad s_1^2 + s_2^2 + (s_3 - \lambda)^2 = \lambda^2 + 2\lambda\}, \quad \lambda > 0$$

in $\mathbb{R}^3$, but not concentrically nested. They fill out the region $\{s_3 > -1\}$. This distortion of usual rotations allow the back-reaction in the form of the region $\{s_3 > -1\}$ acting back on SU_2 by β. The back-reaction has two kinds of orbits, points and disks, in both cases labelled by an angle θ,

$$\mathcal{O}_\theta^{(1)} = \{\begin{pmatrix} e^{i\theta} & 0 \\ 0 & e^{-i\theta} \end{pmatrix}\}, \quad \mathcal{O}_\theta^{(2)} = \{\begin{pmatrix} x & y \\ -\bar{y} & \bar{x} \end{pmatrix}, \quad |x|^2 + |y|^2 = 1, \quad \frac{y}{\bar{y}} = e^{2i\theta}\}.$$

This is the system with the roles of acting and acted-upon object interchanged.

We are ready now to explain the construction of bicrossproduct Hopf algebras based on matched pairs of Lie algebras (the original motivation for the above work). We have explained the construction for finite groups in Chapter 6.2. For locally compact topological groups, including the Lie groups in the preceding example, one can proceed using the theory of Hopf–von Neumann algebras. Alternatively, in the Lie group setting, one also has the purely algebraic possibility of constructing bicrossproducts of the form $\mathbb{C}(M) \blacktriangleright\!\!\!\triangleleft U(g)$, where $\mathbb{C}(M)$ is an algebraic model of the functions on M and $U(g)$ is the enveloping algebra of g. Our cocycle a above defines an infinitesimal action of g on M and hence, by evaluation of the corresponding vector fields, a left action of the Lie algebra g on $\mathbb{C}(M)$,

$$(\xi \triangleright X)(s) = \widetilde{a_\xi}(X)(s) = \frac{d}{d\tau}\Big|_0 X(se^{\tau a_\xi(s)})$$

for all $X \in \mathbb{C}(M)$ and $\xi \in g$. The algebra of $\mathbb{C}(M) \blacktriangleright\!\!\!\triangleleft U(g)$ is the cross product as given in Chapter 1.6 by this action. Its coalgebra, on the other hand, is the cross coproduct given by a right coaction

$$\beta : g \to g \otimes \mathbb{C}(M), \quad \beta(\xi)(s) = s \triangleright \xi,$$

for all $\xi \in g$ and $s \in M$, extended to products of the generators of $U(g)$ according to the bicrossproduct conditions (6.8) and (6.9) in Theorem 6.2.2.

Example 8.3.19 *The Lie group SU_2^{*op} has a Hopf algebra of functions $\mathbb{C}(SU_2^{*op})$ with generators $\{X_i, (X_3+1)^{-1}\}$ and*

$$[X_i, X_j] = 0, \quad \Delta X_i = X_i \otimes 1 + (X_3+1) \otimes X_i, \quad \epsilon X_i = 0, \quad SX_i = -\frac{X_i}{X_3+1}.$$

The Lie group matched pair in Example 8.3.18 then leads to a bicross-product Hopf $*$-algebra $\mathbb{C}(SU_2^{*op}) \blacktriangleright\!\!\blacktriangleleft U(su_2)$ with

$$[e_i, e_j] = \epsilon_{ijk} e_k, \quad [e_i, X_j] = \epsilon_{ijk} X_k - \frac{1}{2}\epsilon_{ij3}\frac{X^2}{X_3+1}, \quad \epsilon e_i = 0,$$

$$\Delta e_i = e_i \otimes \frac{1}{X_3+1} + e_3 \otimes \frac{X_i}{X_3+1} + 1 \otimes e_i, \quad Se_i = e_3 X_i - e_i(X_3+1),$$

where e_i are the $U(su_2)$ generators. The $*$-structure is $e_i^* = -e_i$ and $X_i^* = X_i$.

Proof: The algebraic analogue of the algebra of functions on SU_2^{*op} is worked out in the usual way by thinking of the X_i as the coordinate functions $X_i(s) = s_i$ in the description of Example 8.3.13, and computing the coproduct as corresponding in the usual way to the underlying group law there. The vector fields $\widetilde{a_\xi}$ come from Example 8.3.15, while $s \triangleright \xi$ comes from Example 8.3.13. This gives the action and coaction

$$e_i \triangleright X_j = \epsilon_{ijk}\left(X_k - \frac{1}{2}\delta_{k3}\frac{X^2}{X_3+1}\right), \quad \beta(e_i) = e_i \otimes \frac{1}{X_3+1} + e_3 \otimes \frac{X_i}{X_3+1}$$

in our usual basis $\{e_i\}$ for su_2. This action and coaction form a bicross-product system in the sense of (6.8) and (6.9), with A our algebra of functions generated by $\{X_i\}$ and $H = U(su_2)$. The resulting bicrossproduct Hopf algebra has the relations and coproduct as stated, constructed in the usual semidirect way that we know from Chapter 1.6. The $*$-structure comes from the general theory in Chapter 6.1, giving, in fact, a Hopf $*$-algebra. Note that we do not need to know the full exponentiation in Example 8.3.18 because we give the resulting Hopf algebra structure only on the generators.   ∎

The Hopf $*$-algebra in this example is an algebraic model of the quantum algebra of observables of a system consisting of particles moving on orbits (the nonconcentrically nested spheres) in the region $\{s_3 > -1\}$. The dual Hopf algebra is the quantum algebra of observables of the system with the roles of position and momentum reversed, i.e. particles moving on orbits in SU_2. This is the setting we know from Chapter 6.

On the other hand, the 'semidirect product' form of the construction with $U(su_2)$ acting on an algebra with three generators means that one could obviously also think of this Hopf algebra as a deformation of the enveloping algebra of the 3-dimensional 'Poincaré' group of translations and rotations on $\mathbb{R}^3$. Before describing this explicitly we trivially switch to some other generators (equally natural in Example 8.3.19), namely

$$X_i = \lambda p_i, \quad X_3 + 1 = e^{\lambda p_3},$$

for $i = 1, 2$, where we also take the opportunity to introduce a scaling parameter for the new generators. (For an algebraic treatment one should nevertheless consider $e^{\lambda p_3}$ as the true generator rather than p_3 in what follows.) The cross relations and coalgebra in Example 8.3.19 become

$$[e_i, p_3] = \epsilon_{i3k} p_k, \quad [e_3, p_j] = \epsilon_{3jk} p_k, \quad [e_3, p_3] = 0$$

$$[e_i, p_j] = \epsilon_{ij3} e^{-\lambda p_3} \left(\frac{e^{2\lambda p_3} - 1}{2\lambda} - \frac{\lambda}{2} \sum_{i=1,2} p_i^2 \right)$$

$$\Delta p_i = p_i \otimes 1 + e^{\lambda p_3} \otimes p_i, \quad \Delta e_i = e_i \otimes e^{-\lambda p_3} + 1 \otimes e_i + \lambda e_3 \otimes p_i e^{-\lambda p_3}$$

and $\epsilon e_i = \epsilon p_i = 0$, for $i, j = 1, 2$. The coalgebra for p_3, e_3 has the primitive linear form. Meanwhile, we know from Proposition 8.3.3 that the same matched pair data $(su_2, su_2^{*\text{op}})$ that leads to the bicrossproduct Hopf algebra also gives a bicross sum Lie algebra.

Example 8.3.20 *The bicrossproduct Hopf algebra* $\mathbb{C}(SU_2^{*\text{op}}) \blacktriangleright\!\!\triangleleft U(su_2)$ *in Example 8.3.19 is a deformation of* $U(\mathbb{R}^3 \blacktriangleright\!\!\triangleleft su_2)$, *where the bicross sum* $\mathbb{R}^3 \blacktriangleright\!\!\triangleleft so_3$ *from Proposition 8.3.3 is the Lie algebra of translations* $\vec{p}$ *and rotations* $\vec{e}$ *equipped with Lie cobracket* $\delta p_3 = \delta e_3 = 0$ *and, for* $i = 1, 2$,

$$\delta p_i = p_3 \otimes p_i - p_i \otimes p_3, \quad \delta e_i = e_3 \otimes p_i - p_i \otimes e_3 + p_3 \otimes e_i - e_i \otimes p_3.$$

Proof: This is easily read off from the Hopf algebra structure in the form above, with $(\text{id} - \tau) \circ \Delta = \lambda \delta + O(\lambda^2)$, etc. On the other hand, it necessarily has the bicross sum form in Proposition 8.3.3 given by applying Proposition 8.3.3 to the matched pair of Lie algebras $(su_2, su_2^{*\text{op}})$ underlying our construction of the Hopf algebra. This is $su_2^{\text{cop}} \blacktriangleright\!\!\triangleleft su_2$, where su_2 has its usual Lie bracket (and the zero Lie cobracket) and $su_2^{\text{cop}} = \mathbb{R}^3$ has the zero Lie bracket (and minus the usual Lie cobracket of su_2). Since the actions in the matched pair were originally obtained as coadjoint ones, the action and Lie coaction in the bicross sum are also provided by the same data. Explicitly,

$$e_i \triangleright p_j = \epsilon_{ijk} p_k, \quad \beta(e_3) = 0, \quad \beta(e_i) = e_3 \otimes p_i - e_i \otimes p_3.$$

■

Also, Proposition 6.2.7 from the theory of bicrossproducts in Chapter 6 provides a canonical action on $U(su_2^{*op})$ as a module algebra, so the latter becomes a deformation of the coordinates of $\mathbb{R}^3$ in this point of view. This is one of the simplest examples of noncommutative geometry in which a classical enveloping algebra is regarded 'up side down' as the coordinate ring of some deformed space. We denote the generators of $U(su_2^{*op})$ dual to the components of $\vec{p}$ by the components of $\vec{x}$, so

$$[x_3, x_i] = \lambda x_i, \quad [x_i, x_j] = 0, \quad i, j = 1, 2$$

(as in Example 8.1.13 in a different notation). We write a general element of the enveloping algebra as $: f(x_1, x_2, x_3):$ for an ordinary function f, where the normal ordering puts x_3 to the right.

Example 8.3.21 $U(su_2^{*op})$ *is a right* $\mathbb{C}(SU_2^{*op}){\blacktriangleright\!\!\triangleleft}U(su_2)$*-module algebra by Proposition 6.2.7. Explicitly, for all* i, j,

$$: f(x_1, x_2, x_3): {\triangleleft}p_i =: \frac{\partial}{\partial x_i} f(x_1, x_2, x_3):, \quad x_i{\triangleleft}e_j = \epsilon_{ijk}x_k$$

Proof: We need only compute the action on the generators x_i since the extension to products is via the coproduct (as a module algebra). The action of p_i according to Proposition 6.2.7 is the right coregular one, $x_i{\triangleleft}p_j = \langle x_i, p_j\rangle = \delta_{ij} = \frac{\partial}{\partial x_j}x_i$. The form of the coproduct of the p_i is sufficiently simple that this form extends as differentiation provided we keep the normal ordering. We have seen this already in Chapter 6.4, for example. The action of e_i is the dual of the action (the cross relation) in the bicrossproduct Hopf algebra, as defined by $\langle x_i{\triangleleft}e_j, f(p_1, p_2, p_3)\rangle = \langle x_i, e_j{\triangleright}f(p_1, p_2, p_3)\rangle$. But only the lowest order in p_i in $e_j{\triangleright}f$ will contribute, which is also the lowest order in λ. Hence the action at this level is a usual rotation. It is rather more complicated on products of the x_i according to the deformed coproduct of the e_i. ∎

Finally, by scaling the e_i generators rather than the p_i, we can instead view $\mathbb{C}(SU_2^{*op}){\blacktriangleright\!\!\triangleleft}U(su_2)$ as a deformation of the classical coordinate algebra of a group with Lie bialgebra dual to the one in Example 8.3.20. This is $su_2^{*op}{\blacktriangleright\!\!\triangleleft}\mathbb{R}^3$, where $\mathbb{R}^3$ stands for su_2^* with the zero Lie bracket but its usual Lie cobracket, and su_2^{*op} has its usual Lie bracket and zero Lie cobracket. Explicitly, the cross relations and Lie cobracket of the bicross sum are

$$[x_i, f_j] = \delta_{i3}f_j - f_i\delta_{j3}, \quad \delta f_i = \epsilon_{ijk}f_j \otimes f_k, \quad \delta x_i = \epsilon_{ijk}(f_j \otimes x_k - x_k \otimes f_j),$$

where $\{f_i\}$ are a dual basis to the $\{e_i\}$. The group $SU_2^{*op}{\blacktriangleright\!\!\triangleleft}\mathbb{R}^3$ of this Lie bialgebra is the classical phase space of which the bicrossproduct Hopf algebra in Example 8.3.19 is a quantisation from the point of view of Poisson geometry. We turn to this topic next.

8.4   Poisson–Lie groups

In this section we return to the theory of Lie bialgebras from Sections 8.1 and 8.2 and consider how it too exponentiates from the Lie algebra level to the Lie group level. This aspect of the theory has applications to Hamiltonian mechanics on group manifolds, in the form of systems that are completely integrable. Its infinite-dimensional version also makes contact with the classical inverse scattering method in the theory of solitons, though we shall not develop this here. Of interest to us is that we describe also quantum group function algebras, as in Chapter 4, as deformations (though not really quantisations) of classical data on the original Lie group. The theory of matched pairs of Lie algebras from the last section, arising as we have seen in quite a different context, also makes an appearance as a way to solve such systems.

The basis of Hamiltonian mechanics is the notion of *Poisson bracket* structure on a manifold M, say. This means a bilinear map $\{\ ,\ \}$: $C^\infty(M) \otimes C^\infty(M) \to C^\infty(M)$ such that

$$\{f, g\} = -\{g, f\}, \quad \{f, gh\} = \{f, g\}h + g\{f, h\},$$

$$\{f, \{g, h\}\} + \{h, \{f, g\}\} + \{g, \{h, f\}\} = 0,$$

hold for all $f, g, h \in C^\infty(M)$. This means a Lie algebra structure on $\mathbb{C}^\infty(M)$, but one which also respects its pointwise product in the sense that $\{f,\ \}$ (and hence $\{\ , f\}$) are derivations, i.e. vector fields. This derivation condition means that $\{\ ,\ \}$ is described by a 2-tensor field γ on M in the form

$$\{f, g\} = \langle \gamma, \mathrm{d}f \otimes \mathrm{d}g \rangle = \gamma^{ij}\partial_i(f)\partial_j(g).$$

In terms of this tensor field, the remaining conditions for a Poisson bracket are

$$\gamma^{ij} = -\gamma^{ji}, \quad \gamma^{ia}\partial_a\gamma^{jk} + \gamma^{ja}\partial_a\gamma^{ki} + \gamma^{ka}\partial_a\gamma^{ij} = 0 \qquad (8.41)$$

for antisymmetry and the Jacobi identity. This is an easy calculation provided that one notes that $\gamma^{ia}\gamma^{jk}$ + cyclic in (ijk) vanishes by antisymmetry. A *Poisson manifold* means a manifold equipped with such a 2-tensor field or Poisson bracket. A map between Poisson manifolds just means a map θ between manifolds which respects their Poisson brackets in the sense $\{f \circ \theta, g \circ \theta\} = \{f, g\} \circ \theta$ for all f, g. The product of Poisson manifolds is also a Poisson manifold by adding the Poisson brackets or their corresponding 2-tensors in the obvious way. Note that, traditionally, mathematicians have concentrated on Poisson brackets where the 2-tensor γ is everywhere an invertible matrix. In this case, one says that the structure is *symplectic*, and $\omega = \gamma^{-1}$ is called the associated symplectic 2-form. In terms of ω, the condition for the Jacobi identity is just that $d\omega = 0$ in the usual sense of a closed form on a manifold. One of the lessons of the theory of Lie bialgebras, however, is that one should not focus too soon on this special invertible case. Many Poisson brackets of interest are not invertible. On the other hand, one can generally fill out the manifold with a family of submanifolds, called the *symplectic leaves*, where on each leaf the Poisson bracket is invertible.

We will be concerned here only with Poisson manifolds which are Lie groups. In this context it is natural to demand the condition that the product map $G \times G \to G$ respects the Poisson brackets (a Poisson map), where $G \times G$ has the direct product structure. Explicitly, this means

$$\{f, g\}(uv) = \{f \circ L_v, g \circ L_u\}(v) + \{f \circ R_v, g \circ R_v\}(u),$$

$$\text{i.e.} \quad \gamma(uv) = L_{u*}(\gamma(v)) + R_{v*}(\gamma(u)) \tag{8.42}$$

in terms of γ. Here L_{u*} and R_{v*} act on both vectorial indices of γ. We say that G is a *Poisson–Lie group* when it is equipped with a Poisson bracket respecting the group product in this way. This notion is due to V.G. Drinfeld.

On a group manifold, it is natural to write our 2-tensor field γ as given by right-translation of a Lie-algebra valued function on the group. We write

$$\gamma(u) \equiv R_{u*}(D(u)), \quad D : G \to g \otimes g,$$

where g is the Lie algebra of G, and ask how our conditions look in terms of this. After a bit of work, the result for antisymmetry and the Jacobi identity are

$$\tau(D(u)) = -D(u), \tag{8.43}$$

$$\sum D(u)^{[1]} \otimes \widetilde{D(u)}^{[2]}(D)(u) - \frac{1}{2} \sum D(u)^{[1]} \otimes \mathrm{ad}_{D(u)^{[2]}}(D(u))$$

$$+ \text{ cyclic} = 0. \tag{8.44}$$

Here, $D(u) = \sum D(u)^{[1]} \otimes D(u)^{[2]}$ and $\sim$ denotes extension as a right-invariant vector field by R_{u*}. This vector field then acts on D as a $g \otimes g$-valued function. In the second term, ad acts on $g \otimes g$ as a derivation. The equation has values in $g \otimes g \otimes g$, and we add the two cyclic rotations of the values. Although (8.44) looks formidable, it is easy enough to derive from (8.41) in matrix coordinates $\{t^i{}_j\}$ associated to a linear representation, say. It can also be derived more invariantly. Equations (8.43)–(8.44) characterise a Poisson bracket on our group manifold.

For a Poisson–Lie group, we further require the condition

$$D(uv) = \mathrm{Ad}_u(D(v)) + D(u), \quad \text{i.e.} \quad D \in Z^1_{\mathrm{Ad}}(G, g \otimes g), \tag{8.45}$$

where Ad is the action of G on $g \otimes g$ in the usual way (on both factors of the tensor product), and we use the usual group cohomology as we have already encountered it in Chapter 6.3. This is clear from (8.42) and the definition of D in terms of γ.

Theorem 8.4.1 *If (g, δ) is a Lie bialgebra, then the associated connected and simply connected Lie group G is a Poisson–Lie group by exponentiating δ from a Lie algebra to a Lie group cocycle. Conversely, if (G, D) is a Poisson–Lie group, then its Lie algebra is a Lie bialgebra by differentiating D at the identity.*

Proof: We have already performed much more complex cocycle exponentiations in Proposition 8.3.14. The strategy is the same (and assured by general theory). Thus, we differentiate the desired group cocycle condition (8.45) to obtain the differential equation

$$\frac{\mathrm{d}}{\mathrm{d}\tau}\bigg|_0 D(ue^{\tau\xi}) \equiv \tilde{\xi}(D)(u) = \mathrm{Ad}_u(\delta\xi), \quad D(e) = 0,$$

where $\tilde{\xi}$ is here the left-invariant vector field corresponding to $\xi \in g$. Equivalently, in terms of Maurer–Cartan forms, we can write the equation for D as

$$\mathrm{d}D = \mathrm{Ad}_u(\delta\omega^L).$$

As before, the right hand side is closed since

$$\mathrm{d}(\mathrm{Ad}_u(\delta\omega^L)) = \mathrm{Ad}_u \mathrm{ad}_{\omega_L}(\delta\omega^L) + \mathrm{Ad}_u \delta \mathrm{d}\omega^L = 0,$$

using the Maurer–Cartan equations and the 1-cocycle condition (8.3) to expand $\delta([\omega^L, \omega^L])$. Hence, by $H^1(G) = 0$, we know that we have a solution D to our problem. It is unique for the boundary conditions stated. After finding it, we let $f(v) = D(uv) - D(u) - \mathrm{Ad}_u(D(v))$, and we see at once that $\mathrm{d}f = 0$, hence $f \equiv 0$. We conclude (8.45).

To see that we also have (8.44) is somewhat harder. We first prove the useful lemma that if $D : G \to g \otimes g$ obeys the group cocycle condition (8.45) then

$$\mathrm{Ad}_u \circ \delta \circ \mathrm{Ad}_{u^{-1}}(\xi) = \frac{\mathrm{d}}{\mathrm{d}\tau}\bigg|_0 D(u\mathrm{Ad}_{u^{-1}}(e^{\tau\xi})) = \frac{\mathrm{d}}{\mathrm{d}\tau}\bigg|_0 D(e^{\tau\xi}u)$$

$$= \frac{\mathrm{d}}{\mathrm{d}\tau}\bigg|_0 (\mathrm{Ad}_{e^{\tau\xi}}(D(u)) + D(e^{\tau\xi})) = \mathrm{ad}_\xi(D(u)) + \delta\xi, \qquad (8.46)$$

for all $u \in G$ and $\xi \in g$. This is half-way between the Lie algebra 1-cocycle condition (8.3) and our group one.

This expression is also the right-invariant vector field generated by ξ, evaluated on D. In the light of this, the remaining condition (8.44) becomes

$$\frac{1}{2}\sum D(u)^{[1]} \otimes \mathrm{ad}_{D(u)^{[2]}}(D(u)) + (\mathrm{id} \otimes \delta)D(u) + \text{cyclic} = 0. \qquad (8.47)$$

To prove this for our particular D, we consider the left hand side as a function $f(u)$ and compute its differential as

$$\begin{aligned}
\mathrm{d}f(u) &= \frac{1}{2}\mathrm{Ad}_u(\omega^L{}_{[1]}) \otimes \mathrm{ad}_{\mathrm{Ad}_u(\omega^L{}_{[2]})}(D(u)) \\
&\quad + \frac{1}{2}D(u)^{[1]} \otimes \mathrm{ad}_{D(u)^{[2]}}(\mathrm{Ad}_u \circ \delta\omega^L) \\
&\quad + \mathrm{Ad}_u(\omega^L{}_{[1]}) \otimes \delta \circ \mathrm{Ad}_u(\omega^L{}_{[2]}) + \text{cyclic} \\
&= \mathrm{Ad}_u(\omega^L{}_{[1]}) \otimes \mathrm{ad}_{\mathrm{Ad}_u(\omega^L{}_{[2]})}(D(u)) \\
&\quad + \mathrm{Ad}_u(\omega^L{}_{[1]}) \otimes \delta \circ \mathrm{Ad}_u(\omega^L{}_{[2]}) + \text{cyclic} \\
&= \mathrm{Ad}_u(\omega^L{}_{[1]}) \otimes \mathrm{Ad}_u \circ \delta \circ \mathrm{Ad}_{u^{-1}} \circ \mathrm{Ad}_u(\omega^L{}_{[2]}) + \text{cyclic} \\
&= \mathrm{Ad}_u((\mathrm{id} \otimes \delta) \circ \delta(\omega^L) + \text{cyclic}) = 0
\end{aligned}$$

just because δ obeys the co-Jacobi identity. The first equality here is $\mathrm{d}D$ from the above. The second occurs because the second term here equals the first under the cyclic average (using antisymmetry of D and δ). The third then uses (8.46) to recognise the result. Since $f(e) = 0$, we conclude (8.47) as required.

In the reverse direction we just define $\delta(\xi) = \frac{\mathrm{d}}{\mathrm{d}\tau}\big|_0 D(e^{\tau\xi})$, and we can easily check that it gives a Lie bialgebra. ∎

This tells us that Lie bialgebras on g are in strict correspondence with Poisson–Lie group structures on the associated group. In just the same way, it is mathematically a trivial exercise to map over *all* our Lie bialgebra constructions to corresponding ones for the theory of Poisson–Lie groups. Rather than repeat all of Sections 8.1 and 8.2 in this way, we highlight here just one or two of the main theorems. The reader can apply the same principle to the other constructions.

Exercise 8.4.2 *If $(g, r, [\,,\,])$ is a quasitriangular Lie bialgebra, show that the corresponding Poisson–Lie structure is given by*

$$D(u) = \mathrm{Ad}_u(r) - r, \qquad \gamma(u) = L_{u*}(r) - R_{u*}(r)$$

on the corresponding Lie group G. Deduce that the functions $I(G)$, which are constant on conjugacy classes, have zero Poisson-bracket among themselves, and that, if $H \in I(G)$ is taken as Hamiltonian, then the corresponding equations of motion are

$$\frac{\mathrm{d}f}{\mathrm{d}\tau} = (L_{u*} - R_{u*})(r_1(L_{u*}^* \mathrm{d}H)) f.$$

Solution: In this case, the Lie cobracket is $\delta\xi = \mathrm{ad}_\xi(r)$ and we have to solve $\mathrm{d}D = \mathrm{Ad}_u \circ \mathrm{ad}_{\omega^L}(r)$, but the right hand side is $\mathrm{d}\mathrm{Ad}_u(r)$ by the property $\mathrm{d}u = u\omega^L$ in a linear representation. So $\mathrm{Ad}_u(r)$ is a solution of the differential equations, though not of the boundary conditions. Since $D(e) = 0$, we see that D as stated is the solution.

If H is an ad-invariant function on G, then $L_{u*}(\xi) = R_{u*}(\xi)$ when acting on it, for any $\xi \in g$. So we deduce at once that

$$\dot{f}(u) = \{H, f\}(u) = (L_{u*}(r^{[1]})H)(L_{u*}(r^{[2]})f) - (R_{u*}(r^{[1]})H)(R_{u*}(r^{[2]})f)$$
$$= (L_{u*}(r^{[1]})H)(L_{u*}(r^{[2]})f - R_{u*}(r^{[2]})f)$$
$$= \langle r^{[1]}, L_{u*}^*(\mathrm{d}H)\rangle(L_{u*}(r^{[2]})f - R_{u*}(r^{[2]})f),$$

in terms of the adjoint operator $L_{u*}^* : T_u^* G \to T_e^* G = g^*$ defined by $L_{u*}(\xi)H = \langle\xi, L_{u*}^* \mathrm{d}H\rangle$ for all $\xi \in g$. We see that $\dot{f}$ has the form stated. From an active transformation point of view, this means that the flows for this Hamiltonian system are particularly simple: we just integrate the left-invariant minus right-invariant vector field generated by the Lie-algebra valued function $r_1(L_{u*}^* \mathrm{d}H)$. It is clear that if $f \in I(G)$ also, then $\{H, f\} = 0$. ∎

We know that some useful coordinates on a group are often given by the matrix elements $t^i{}_j$ defined by a suitable linear representation. This was also the starting point of Chapter 4. In this case, we have

$$\{t^i{}_j, t^k{}_l\} = t^i{}_a t^k{}_b r^a{}_j{}^b{}_l - r^i{}_a{}^k{}_b t^a{}_j t^b{}_l, \quad \text{i.e.} \quad \{\mathbf{t}_1, \mathbf{t}_2\} = [\mathbf{t}_1 \mathbf{t}_2, r_{12}]$$

in the compact notation where numerical suffices refer to the position in a matrix tensor product. Here, $r \in M_n \otimes M_n$ is the matrix of the quasitriangular structure in the linear representation. This is immediate once we recall that the left-invariant and right-invariant vector fields in such linear coordinates are

$$L_*(\xi) = t^i{}_a \xi^a{}_j \frac{\partial}{\partial t^i{}_j}, \quad R_*(\xi) = \xi^i{}_a t^a{}_j \frac{\partial}{\partial t^i{}_j},$$

where $\xi^i{}_j$ is the matrix of $\xi \in g$ in the linear representation. Also, if R is a solution of the QYBE with associated matrix quantum group $A(R)$ or its quotient, as in Chapter 4, and $R = 1 + tr + O(t^2)$ as matrices, then it is clear that the relations of the quantum group are

$$[\mathbf{t}_1, \mathbf{t}_2] = t\{\mathbf{t}_1, \mathbf{t}_2\} + O(t^2),$$

where t is the deformation parameter. This shows the role of Poisson brackets as prescribing the lowest order deformation governing the relations of the quantum group if its R-matrix is a deformation of the identity. Also, the existence of a large class of functions with mutually zero Poisson bracket is like the exact solvability for vertex models in Chapter 4.4, although it is considered here in the simplest setting without a spectral parameter.

Example 8.4.3 *Let su_2 have the quasitriangular Lie bialgebra structure in Example 8.1.13. The associated Poisson–Lie bracket on SU_2 with matrix coordinates $\begin{pmatrix} a & b \\ c & d \end{pmatrix}$ is*

$$\{a, b\} = -\frac{1}{2}ab, \quad \{a, c\} = -\frac{1}{2}ac, \quad \{a, d\} = -cb,$$

$$\{b, c\} = 0, \quad \{b, d\} = -\frac{1}{2}bd, \quad \{c, d\} = -\frac{1}{2}cd,$$

where the algebra is commutative and $ad - bc = 1$. This agrees with the lowest order commutators in $SU_q(2)$ in Example 4.2.5.

Proof: The matrix of the quasitriangular structure from Example 8.1.13 is

$$r = -\frac{1}{4}\sigma_i \otimes \sigma_j r^{ij} = \frac{1}{4}\begin{pmatrix} 1 & 0 & 0 & 0 \\ 0 & -1 & 4 & 0 \\ 0 & 0 & -1 & 0 \\ 0 & 0 & 0 & 1 \end{pmatrix},$$

where $r^i{}_j{}^k{}_l$ has rows ik and columns jl. This gives the formulae shown as an immediate example of Exercise 8.4.2 in the matrix form. We can take into account the reality properties by writing the coordinates as $\begin{pmatrix} x & y \\ -\bar{y} & \bar{x} \end{pmatrix}$ and $|x|^2 + |y|^2 = 1$ as we did in Example 8.3.18. Then, $x = x_1 + \imath x_2$, $y = x_3 + \imath x_4$ for real x_i gives $SU_2 = S^3$ in real coordinates. In this case, we write

$$\{x_1, x_2\} = \frac{\imath}{2}(x_3^2 + x_4^2), \quad \{x_1, x_3\} = -\frac{\imath}{2}x_2 x_3, \quad \text{etc.}$$

Also, let us recall that the Hopf algebra $M_q(2)$ from Chapter 4 is of the corresponding R-matrix form. That the q-determinant of $M_q(2)$ is central corresponds semiclassically to $ad - bc$ having zero Poisson bracket with the other generators, which we know since our Poisson bracket descends to functions on SU_2. ∎

Next, since every Lie bialgebra has a dual one, working over $\mathbb{R}$ or over $\mathbb{C}$ with half-real forms, we have a corresponding *dual Poisson–Lie group* $G^\star$ to a Poisson–Lie group G. We dualise the Lie bialgebra and (say) take its associated connected and simply connected Lie group, with Poisson bracket from Theorem 8.4.1.

Example 8.4.4 *Let $su_2^{\star\mathrm{op}}$ be the Lie bialgebra in Example 8.1.13 (taken with the opposite bracket). The associated Poisson bracket on $SU_2^{\star\mathrm{op}}$ in Example 8.3.13 with coordinates $\{X_i\}$ is*

$$\{X_i, X_j\} = \imath \epsilon_{ijk}(X_3 + 1)X_k - \frac{\imath}{2}\epsilon_{ij3}X^2.$$

Proof: Here the Lie bialgebra is not quasitriangular so we need the full force of Theorem 8.4.1. We have to solve the equation

$$\frac{\mathrm{d}}{\mathrm{d}\tau}\bigg|_0 D(\vec{s}e^{\tau\vec{\phi}}) \equiv L_{\vec{s}*}(\vec{\phi})(D) = \mathrm{Ad}_{\vec{s}}(\delta\vec{\phi}),$$

where $\delta\phi$ is from Example 8.1.13. We use our usual basis $\{e_i^*\}$. Taking $L\vec{s}*$ and $\mathrm{Ad}_{\vec{s}}$ from the proof of Example 8.3.15, we have the differential equation

$$\frac{\partial}{\partial s_i}D^{jk} = \imath(s_3 + 1)\epsilon_{ijk} - \imath\epsilon_{ij3}s_k + \imath\epsilon_{ik3}s_j,$$

which can be solved easily by applying the Hodge $*$-operator to both sides to obtain a vector equation. The unique solution of the system for the boundary condition $D(0) = 0$ is

$$D^{ij}(\vec{s}) = \imath\epsilon_{ija}s_a + \frac{\imath}{2}s^2\epsilon_{ij3}, \quad \text{i.e.} \quad D = \imath * \left(\vec{s} + \frac{1}{2}s^2\vec{e}_3\right),$$

$$\gamma^{ij}(\vec{s}) = \imath\epsilon_{ija}(s_3 + 1)s_a - \frac{\imath}{2}s^2\epsilon_{ij3}, \quad \text{i.e.} \quad \gamma = \imath * \left((s_3 + 1)\vec{s} - \frac{1}{2}s^2\vec{e}_3\right),$$

where we used $R_{\vec{s}*}$ from the proof of Example 8.3.15 to compute γ. This gives the Poisson bracket, which we state more formally by distinguishing between the points in the group that are labelled by $\vec{s}$ and the coordinate functions $X_i(\vec{s}) = s_i$. ∎

In both these examples, we have a Poisson–Lie group with imaginary Poisson bracket in a real coordinate system on the Lie group. This is characterised by $\overline{\{f, g\}} = \{\bar{g}, \bar{f}\}$. One can, of course, multiply by $-\imath$ for more conventional real Poisson brackets. In addition, we also have an algebraic model in which we work abstractly with a Hopf algebra

$\mathbb{C}(SU_2)$ or $\mathbb{C}(SU_2^{*\mathrm{op}})$ and the generators as shown. In this case, we have to work with the axioms of a Poisson–Lie group algebraically, i.e. the notion of a *Poisson–Hopf algebra*. It means a Hopf algebra A and a map $\{\,,\,\} : A \otimes A \to A$ obeying the axioms of a Poisson bracket, and such that the additional condition

$$\Delta\{a, b\} = \sum a_{(1)}b_{(1)} \otimes \{a_{(2)}, b_{(2)}\} + \{a_{(1)}, b_{(1)}\} \otimes b_{(2)}a_{(2)} \qquad (8.48)$$

holds. This is just (8.45) with the group multiplication expressed now as the coproduct. One can easily check that the preceding two examples are commutative Hopf algebras of this type. With the ordering as in (8.48), one also has that any Hopf algebra is a Poisson–Hopf algebra with $\{\,,\,\}$ given by the commutator. In the Hopf $*$-algebra case, it is natural to require $\{f, g\}^* = \{g^*, f^*\}$ as well.

Another important construction in the preceding section was that of a double cross sum Lie bialgebra. Now we have

Proposition 8.4.5 *If (g, m) is a matched pair of Lie bialgebras such that $g \bowtie m$ is a Lie bialgebra, as in Proposition 8.3.4, then the associated double cross product group $G \bowtie M$ from Theorem 8.3.17 is a Poisson–Lie group with the direct product Poisson–Lie structure. Moreover, the actions $\lhd, \rhd$ of the two factors on each other are Poisson maps.*

Proof: We have constructed the double cross product Lie group $G \bowtie M$ in Section 8.3 (at least locally), and we add to this now the assumption that both G, M are Poisson, which we know corresponds to g, m Lie bialgebras. The matching condition (8.30), which we assume now, ensures that $g \bowtie m$ is a Lie bialgebra with the direct sum Lie cobracket. Hence, by Theorem 8.4.1, our double cross product group is Poisson–Lie with the direct product Poisson bracket

$$\{f, g\}(u, s) = \{f(u, \ ), g(u, \ )\}_M(s) + \{f(\ , s), g(\ , s)\}_G(u)$$

as we deduce at once from $(u, s) = (u, e)(e, s)$. The corresponding structure maps are

$$D(u, s) = \mathrm{Ad}_{(u,e)}D^M(s) + D^G(u), \quad \gamma(u, s) = L_{(u,e)*}\gamma^M(s) + R_{(e,s)*}\gamma^G(u),$$

where one can check directly that D solves the differential equation in Theorem 8.4.1 in terms of δ on $g \bowtie m$. For the explicit form of D, we compute

$$\mathrm{Ad}_{(u,e)}(0 \oplus \phi) = \left.\frac{\mathrm{d}}{\mathrm{d}\tau}\right|_0 (u, e)(e, e^{\tau\phi})(u^{-1}, e) = \left.\frac{\mathrm{d}}{\mathrm{d}\tau}\right|_0 (u(e^{\tau\phi} \rhd u^{-1}), e^{\tau\phi} \lhd u^{-1})$$

$$= \mathrm{Ad}_u(b_\phi(u^{-1})) \oplus \phi \lhd u^{-1} = (-b_{\phi \lhd u^{-1}}(u)) \oplus \phi \lhd u^{-1},$$

using the product in $G\bowtie M$ from Proposition 6.2.15 and, for the last equality, the cocycle condition (8.38) for b. This then allows one to compute $D(u,s)$ explicitly as an element of $g\bowtie m \otimes g\bowtie m$. On the other hand, we have

$$R_{(u,s)*}(\xi \oplus \phi) = R_{(e,s)*}(\widetilde{\xi}(u) + \widetilde{b_\phi}(u)) + L_{(u,e)*}(\widetilde{\phi\triangleleft u}(s)),$$

which is obtained similarly by differentiating, where $\sim$ denotes extension as right-invariant vector fields on G or M, viewed on $G \times M$. This also involves b and cancels its occurrence in D when computing the tensor $\gamma(u,s)$ from D. The latter then comes out the same as the one corresponding to the direct product Poisson bracket on $G \times M$.

Also, recall from Proposition 8.3.2 that the actions of the Lie algebra matched pair were required to respect the Lie cobracket on which they act. After exponentiating these actions to the group level one can check that they then respect the Poisson brackets of the Lie group on which they act.   ∎

Some important examples are $D(u) = u^{*op}\bowtie u$ when u is a half-real form of a complex Lie-bialgebra, and $g\bowtie g$ whenever g is dual quasitriangular. The corresponding Poisson–Lie groups are $U^{*op}\bowtie U$ as a complexification of U, $G\bowtie G$, etc. whenever the completeness conditions for the exponentiation theorems in the last section go through (for example, when one of the factors is compact). The last part of the proposition says that the Poisson bracket of each group is in some sense covariant under the other group (although not in the usual manner of a Hamiltonian action). An easy corollary (see below) from the theory of coadjoint matched pairs is that the orbits under the action $\beta = \triangleright$ of U^{*op} are the symplectic leaves for the Poisson–Lie structure on U (and vice versa). The different kinds of symplectic leaves in the corresponding U are labelled by the Weyl group W. The action of U^{*op} which preserves them is called the *dressing action*.

As an application we demonstrate how some of the results in this section can be used to integrate or 'exactly solve' an important class of Hamiltonian systems. We consider the Poisson–Lie structure on $G\bowtie M$ obtained from Exercise 8.4.2 in the case of the quasitriangular structure r obtained on $g\bowtie m$ from Exercise 8.3.12.

Exercise 8.4.6 *Suppose that the double cross sum Lie algebra $g\bowtie m$ has an ad-invariant element K^{-1}, as in Exercise 8.3.12, and let $G\bowtie M$ denote the associated Poisson–Lie group induced by the quasitriangular structure described there. Let $H \in I(G\bowtie M)$ be a choice of smooth Hamiltonian function. Show that the equations of motion on the group manifold $G\bowtie M$ are solved by*

$$(u_\tau, s_\tau) = \mathcal{S}_\tau(u_0, s_0)\mathcal{S}_\tau^{-1} = \mathcal{U}_\tau^{-1}(u_0, s_0)\mathcal{U}_\tau,$$

where $\mathcal{U}, \mathcal{S}$ *are paths in* G, M *determined by*

$$e^{-\tau h} = \mathcal{U}_\tau \mathcal{S}_\tau; \quad h = K^{-1}(L^*_{(u_0,s_0)*}\mathrm{d}H) \in g\bowtie m$$

in the double cross product group $G\bowtie M$. *Finding* $\mathcal{U}, \mathcal{S}$ *is called the factorisation problem, and we see that it solves the system.*

Solution: We fix the arbitrary initial point (u_0, s_0) throughout. Then $\mathrm{d}H$ evaluated at (u_0, s_0) is translated back by L^* to $T^*_{(e,e)}G\bowtie M = (g\bowtie m)^*$. Then K^{-1} turns this into the Lie algebra element h. We exponentiate this in the group $G\bowtie M$. This is easy enough to do in practice, for example in a linear representation. But, since every element of the double cross product group factorises uniquely into an element in G and an element in M, we obtain the paths $\mathcal{U}, \mathcal{S}$. Writing their tangents in terms of Lie algebra elements $\xi(\tau), \phi(\tau)$, where

$$\frac{\mathrm{d}\mathcal{U}}{\mathrm{d}\tau} \equiv R_{\mathcal{U}*}(\xi(\tau)), \quad \frac{\mathrm{d}\mathcal{S}}{\mathrm{d}\tau} \equiv L_{\mathcal{S}*}(\phi(\tau)),$$

and differentiating the factorisation equation, we see that

$$-\mathrm{Ad}_{\mathcal{S}_\tau}(h) = \mathrm{Ad}_{\mathcal{U}_\tau^{-1}}(\xi(\tau)) + \mathrm{Ad}_{\mathcal{S}_\tau}(\phi(\tau)).$$

The first term on the right lies in g, and the second lies in m. On the other hand, the quasitriangular structure for $g\bowtie m$ from Exercise 8.3.12 has the form $r_1 = \pi \circ K^{-1}$, where π denotes projection onto the m part. Hence,

$$r_1(L^*_{(u_\tau,s_\tau)*}\mathrm{d}H) = \pi \circ K^{-1} \circ \mathrm{Ad}^*_{\mathcal{S}_\tau}(L^*_{(u_0,s_0)*}\mathrm{d}H)$$
$$= \pi \circ \mathrm{Ad}_{\mathcal{S}_\tau}(h) = -\mathrm{Ad}_{\mathcal{S}_\tau}(\phi(\tau)),$$

where we consider the path $(u_\tau, s_\tau) = \mathrm{Ad}_{\mathcal{S}_\tau}(u_0, s_0)$ and use that H is ad-invariant and then that K^{-1} is ad-covariant. Note that (u_0, s_0) commutes with $e^{-\tau h}$ again by this ad-covariance, which gives the second form stated for our path. We see that our path has tangent vector

$$\frac{\mathrm{d}}{\mathrm{d}\tau}(u_\tau, s_\tau) = \mathrm{Ad}_{\mathcal{S}_\tau}\left((R_{(u_0,s_0)*} - L_{(u_0,s_0)*})(\phi(\tau))\right)$$
$$= (L_{(u_\tau,s_\tau)*} - R_{(u_\tau,s_\tau)*})\left(r_1(L^*_{(u_\tau,s_\tau)*}\mathrm{d}H)\right),$$

which is the tangent vector we have to integrate if we want to solve the Hamiltonian system in Exercise 8.4.2 applied now to the Poisson–Lie group $G\bowtie M$. We see that our path solves this Hamiltonian system with the initial point (u_0, s_0) and general Hamiltonian function H. It induces the required $\dot{f}$ from the point of view of classical observables on G if we think of these as evolving rather than as the points themselves. Note also that, when our double cross product group is given explicitly by actions ◁, ▷ as constructed in Section 8.3, we obtain (u_τ, s_τ) explicitly in terms of these and the decomposition of h.

∎

This is a geometric version of the Adler–Kostant–Symes theorem and is the key construction behind the theory of *classical inverse scattering*. It is more usually stated in terms of orbits in $(g \rtimes m)^*$ and applied in the theory of solitons in the context of infinite-dimensional loop–group or current algebras. An important setting is provided by the Gauss decomposition into upper- and lower-triangular matrices.

There are also double cross bracket Poisson–Lie groups $G \bowtie M$ (on the group $G \times M$) and bicrossproduct Poisson–Lie groups $M \bowtie G$, etc., developed in the same way by exponentiating the Lie bialgebra theory in Sections 8.2 and 8.3. We content ourselves here with the part of the theory relevant to the bicrossproduct quantum systems also associated to a matched pair.

Proposition 8.4.7 *Let* (g, m) *be a matched pair of Lie algebras. The Poisson–Lie group* $M \bowtie g^*$ *associated to the bicross sum* $m \bowtie g^*$ *in Proposition 8.3.3 is the semidirect product* $M \ltimes g^*$ *(where* g^* *is regarded as an Abelian group) equipped with Poisson tensor*

$$\gamma = \partial^i \otimes \alpha_{*e_i} - \alpha_{*e_i} \otimes \partial^i + \gamma_{KK}$$

where $\alpha_{*\xi}$ *is the vector field for the action of* g *on* M *from Proposition 8.3.14,* ∂^i *is differentiation on* g^* *and* γ_{KK} *defines the Kirrillov–Kostant bracket on* g^*. *Explicitly,*

$$\{\xi, \eta\} = [\xi, \eta], \quad \{f, g\} = 0, \quad \{\xi, f\} = \alpha_{*\xi}(f)$$

for all ξ, η *linear functions on* g^* *and* f, g *functions on* M. *Similarly for the Poisson bracket on* $m^* \bowtie G$ *using the vector fields* β_*.

Proof: We assume that the matched pair has been exponentiated as in Proposition 8.3.14 so that M is the connected and simply connected Lie group associated to m. The action of M on g (exponentiating that of m) dualises to a right action ◁ on g^*. The group product of $M \ltimes g^*$ is then $(s, \phi)(t, \psi) = (st, \phi ◁ t + \psi)$. Meanwhile, the Lie coalgebra $m \bowtie g^*$ is given according to Proposition 8.3.3 (with the roles of m, g interchanged) as

$$\delta(x \oplus \phi) = \delta\phi + f^i \otimes x ◁ e_i - x ◁ e_i \otimes f^i,$$

where δ on the right refers to the Lie coalgebra structure on g^* defined by the Lie algebra g. Here $x \in m$ and $e_i \in g$ acts by the right action in the matched pair. As explained below Proposition 8.3.14, we can regard the left Lie coaction $\alpha(x) = f^i \otimes x ◁ e_i$ (say) as a right handed cocycle in $Z^1_{\triangleleft \otimes \mathrm{ad}_R}(m, g^* \otimes m)$. To find the Poisson–Lie bracket we prefer to use a right handed version of Theorem 8.4.1 where $\gamma = L_* \bar{D}$ and $\bar{D}$ is a right handed cocycle on the Poisson–Lie group with differential δ at the identity. In our case it means $\bar{D}(s, \phi)$ with values in $(m \bowtie g^*)^{\otimes 2}$ obeying

$$\bar{D}((s, \phi)(t, \psi)) = \mathrm{Ad}_{(t,\psi)^{-1}} \bar{D}(s, \phi) + \bar{D}(t, \phi).$$

Clearly, $\bar{D}(e, \phi) = \delta\phi$ because Ad on the Abelian group g^* is trivial. Let $\bar{D}(s) \equiv \bar{D}(s, 0)$. It has to solve $\bar{D}(st) = \mathrm{Ad}_{t^{-1}} \bar{D}(s) + \bar{D}(t)$ where $\mathrm{Ad}_{t^{-1}}(x \oplus \phi) = \mathrm{Ad}_{t^{-1}}(x) \oplus \phi \triangleleft t$ on each copy of $m \bowtie g^*$. The initial condition is determined by $\delta x = (\mathrm{id} - \tau)\alpha(x)$ and this form is preserved by the above cocycle equation as a differential equation on M, i.e. $\bar{D}(s) = (\mathrm{id} - \tau)a(s)$ for some $a(s)$ required to obey the same cocycle condition, i.e. a right cocycle $a \in Z^1_{\triangleleft \otimes \mathrm{Ad}_R}(M, g^* \otimes m)$. Since a has differential α at $e \in M$, it is precisely the function constructed in Proposition 8.3.14 in the exponentiation of a matched pair of Lie algebras. Finally, we put these results together using the cocycle condition for $\bar{D}(s, \phi)$. Thus,

$$\bar{D}(s, \phi) = \bar{D}((s, 0)(e, \phi)) = \mathrm{Ad}_{(e, -\phi)} \bar{D}(s) + \delta\phi$$

and hence we have $\gamma(s, \phi) = L_{*(s,\phi)} \bar{D}(s, \phi) = L_{*s} R_{*\phi} \bar{D}(s) + \gamma_{KK}(\phi) = R_{*\phi} f^i \otimes \alpha_{*e_i} - \alpha_{*e_i} \otimes R_{*\phi} f^i + \gamma_{KK}$. Here the vector field on g^* generated by $f^i \in g^*$ is denoted by ∂^i (the usual partial differential in the direction of f^i) and $\alpha_{*\xi} = L_{*\xi} a$ is the vector field corresponding to a in Proposition 8.3.14. An explicit formula for the Kirrillov–Kostant bracket is $\gamma_{KK}(\phi) = \langle \phi, [e_i, e_j] \rangle \partial^i \otimes \partial^j$. The formulae on linear functions on g^* and functions on M are immediate. The proof for $m^* \bowtie G$ is strictly analogous, using this time exactly the left handed versions (as stated) of Proposition 8.3.3 and Theorem 8.4.1. One has

$$\gamma(\xi, u) = \beta_{*e_i} \otimes \partial^i - \partial^i \otimes \beta_{*e_i} + \gamma_{KK}(\xi),$$

where $\xi \in m^*$, $u \in G$ and γ_{KK} is the Kirrillov–Kostant bracket on m^*. Here $\{e_i\}$ is a basis of m^* with differentials $\{\partial^i\}$. ∎

The Poisson–Lie group $M \bowtie g^*$ is the classical phase space underlying the bicrossproduct Hopf algebras $\mathbb{C}(M) \bowtie U(g)$ when these are constructed as in Chapter 6.2 and Section 8.3. Compare with the quantisation prescription in Chapter 6.1. Dually, the same Hopf algebra is a deformation of the enveloping algebra of $m^* \bowtie g$ since the Poisson–Lie group $m^* \bowtie G$ is the phase space of the dual quantum system $U(m) \bowtie \mathbb{C}(G)$ given by the dual Hopf algebra. In both cases one inserts scaling parameters to express the quantisation or deformation. Once again, important examples are provided by coadjoint matched pairs $(U, U^{*\mathrm{op}})$. A concrete example of the resulting Hopf algebra was already given as Example 8.3.19.

Exercise 8.4.8 *For any Lie bialgebra g with associated coadjoint matched pair (G, G^{*op}), show that the Poisson–Lie brackets on G, G^{*op} are*

$$\gamma^G(u) = \beta_{*f^i}(u) \otimes R_{u*}e_i, \quad \gamma^{G^{*op}}(s) = L_{s*}f^i \otimes \alpha_{*e_i}(s)$$

where $\{e_i\}$ is a basis of g and $\{f^i\}$ a dual basis. Conclude that the Hamiltonian vector fields $X_f = \{f, \} = (df \otimes id)\gamma$ for a function f on G or G^{*op}, respectively, obey

$$\beta_{*\phi}(f) = \langle \phi, \omega^R(X_f) \rangle, \quad \alpha_{*\xi}(f) = -\langle \xi, \omega^L(X_f) \rangle$$

for $\phi \in g^*$, $\xi \in g$ and the relevant Maurer–Cartan forms on G, G^{*op} Hence conclude that the Poisson bracket of functions f, g on $g^{cop} \bowtie G$ is

$$\{f, g\} = \omega^R(X_f) \cdot \partial g - \omega^R(X_g) \cdot \partial f + (\partial g \otimes \partial f) \cdot \delta$$

where $\xi \cdot \partial$ is differentiation in the direction ξ on g^{cop}, δ is the given Lie cobracket on g and X_f is the Hamiltonian vector field on G associated to f as a function on G (with its g^{cop} variable held fixed).

Solution: The first part is a trivial observation that we could have made at the start of the section: for any Lie bialgebra g the exponentiation of δ to $D \in Z^1_{Ad}(G, g \otimes g)$ (for any Poisson–Lie group) is a special case of Proposition 8.3.14 in the case where the matched pair is taken to be the coadjoint one (g, g^{*op}). For in this case the action $\triangleright$ of $m = g^{*op}$ is the coadjoint one and hence it is just δ when regarded as a cocycle $g \to g \otimes g^{cop}$ in $Z^1_{ad}(g, g \otimes g^{cop})$. So Proposition 8.3.14 already solved this as $D(u) = b(u) = b_{f^i}(u) \otimes e_i$ there. Then $\gamma = R_*D = R_*b = \beta_*$ as stated, where $\beta_{*\phi}$ is the vector field for the coadjoint action of $\phi \in g^{*op}$ on G. The inversion formula for β_* in terms of X_f follows at once by the (right handed) Maurer–Cartan form property. The last part puts this elementary observation into Proposition 8.4.7 along with an explicit formula for γ_{KK} on $g^{cop} = (g^{*op})^*$. There are identical results for G^{*op} with a left–right reversal. ∎

The first part of the exercise is the reason that the symplectic leaves for the Poisson–Lie group G are the same as the oribits under G^*: the functions giving zero Hamiltonian vector fields are precisely the functions constant on orbits. The second part of the exercise exhibits the 'semidirect bracket' structure $g^{cop} \bowtie G$ on the group $g^{cop} \rtimes G$ as built from those of G and g^{cop}.

Notes for Chapter 8

Section 8.1 is essentially due to V.G. Drinfeld [22, 157]. He introduced the notion of Lie bialgebras and quasitriangular Lie bialgebras. The r-matrix solution of the CYBE for g semisimple is due to Drinfeld and to M. Jimbo [158]. The half-real forms $u, u^\star$ and the vectorial picture of the su_2 and $su_2^\star$ are taken from [120] by the author. Theorem 8.1.7 on the twisting of Lie bialgebras is presented as the semiclassical analogue of Theorem 2.3.4; it is implicit in [22]. Lie coalgebras themselves (without the bialgebra axiom) have an earlier history; see [159], where a PBW theorem is obtained.

In terms of applications, the most interesting class of examples for physics are infinite-dimensional ones, in which context the CYBE becomes the 'parametrised' CYBE. Results on the classification of the solutions of both the parametrised case and the 'constant' (without parameter) case were obtained by A.A. Belavin and Drinfeld [160]. Some further results are due to A. Stolin [161, 162]. Among infinite-dimensional solutions, there are ones associated to Kac–Moody Lie algebras [22], and solutions on the Lie algebra $diff(S^1)$ introduced by E. Beggs and the author [163].

The classical double Lie bialgebra $D(g)$ in Proposition 8.2.1 is due to Drinfeld [157], being defined as preserving the canonical bilinear form in Proposition 8.2.2. The 'constructive approach' as a Lie bialgebra double cross sum $g^{\star\mathrm{op}}\bowtie g$ by mutual coadjoint actions is due to the author [12]. Proposition 8.2.4 on the dual of $D(g)$ as a twisting appears to be new. Theorem 8.2.5 for g factorisable is due to M.A. Semenov-Tian-Shansky [164]; see also [33]. We have presented it as a semiclassical analogue $g\blacktriangleright\blacktriangleleft g$ of a double cross coproduct (see Theorem 7.3.5), now with two quasitriangular structures. Theorem 8.2.6 on $D(g)$ as a semidirect sum is due to the author [34]. The paper [164] also considered aspects of the Iwasawa decomposition, while the general Iwasawa-type theory in Proposition 8.2.7, and the in-depth analysis of half-real forms $u, u^\star$ for the Drinfeld–Jimbo case in Proposition 8.1.15 and Corollary 8.2.8 are taken from [120] by the author.

Section 8.3 on the general theory of matched pairs of Lie algebras and Lie groups was introduced in the author's 1988 Ph.D. thesis [116] and published in [12] and [120], on which this section is based. The double cross sum Lie algebras $g\bowtie m$ in Proposition 8.3.2 and their connection with splittings of Lie algebras are in [12]. The notion of a *coadjoint matched pair* and its equivalence with the axioms of a Lie bialgebra in Example 8.3.7 were also introduced by the author in this work, with $D(g)$ as the associated double cross sum $g^{\star\mathrm{op}}\bowtie g$. The double cross sum theory is more general, however, than Drinfeld's classical double (for which the splittings are the so-called 'Manin triples') because, in the general double cross sum theory, m, g need not have the same dimension and the actions need not be strictly coadjoint ones. This greater generality is also important for the infinite-dimensional theory. Thus, $diff(S^1)$ was constructed in [163] actually as half of a topological matched pair of coadjoint type. In this setting, it was possible (for the first time) to give a family of classical r-matrices r_b for the complexification of $diff(S^1)$, namely as maps [163]

$$r_b(\phi)(z) = -\frac{\imath}{4\sqrt{b(z)}} \int_\gamma \frac{\phi(w)}{w\sqrt{b(w)}} dw \qquad (8.49)$$

defined by contour integration in the complex plane. For the real-analytic case, there turned out [163] to be a subtle winding-number obstruction which had not previously been expected.

Theorem 8.3.17 about the exponentiation of matched pairs of Lie algebras appeared in the second paper [120] from the author's thesis, as a main result. This work gave a new Yang–Baxter theoretic proof of the Iwasawa decomposition of the complexification of a compact Lie group, as well as full details of the nontrivial Example 8.3.18 on $SU_2^{*op}\bowtie SU_2$. Some of the Lie algebra double cross sum constructions (but not, for example, the main exponentiation Theorem 8.3.17) were also obtained independently by Y. Kosmann-Schwarzbach and F. Magri [165] in the context of integrable systems and Poisson geometry. See also the subsequent work [166]. The doubled Lie algebra matched pair $(g^{op} \oplus m, g\bowtie m)$ in Theorem 8.3.11 and the bicross sum Lie bialgebras $m\blacktriangleright\!\!\triangleleft g$ in Propositions 8.3.5 and 8.3.6 are of more recent interest [122]. We have, in the same way, groups $(G^{op} \times M, G\bowtie M)$ forming a matched pair, a general theory of bicrossproducts $M\blacktriangleright\!\!\triangleleft G$, and so on.

The original physical motivation for the double cross sum theory was in connection with quantum algebras of observables and Mach's principle that every action has a back-reaction. It then led to Lie-group versions of the bicrossproduct Hopf (von Neumann) algebras of Chapter 6 [18], including $\mathbb{C}(SU_2^{*op})\blacktriangleright\!\!\triangleleft U(su_2)$ in Example 8.3.19. We have presented a simplified algebraic treatment [122] by suppressing the functional analysis. This example, when viewed as a deformed 'Poincaré' algebra in three dimensions (see Example 8.2.20) was generalised in [121] to the 3+1-dimensional Poincaré algebra $U(so_{1,3})\blacktriangleright\!\!\triangleleft \mathbb{C}(\mathbb{R}\ltimes\mathbb{R}^3)$. Explicitly, this has [121] the usual translation and Lorentz generators $[p_\mu, p_\nu] = 0$, $[M_i, M_j] = \epsilon_{ijk}M_k$, $[N_i, N_j] = -\epsilon_{ijk}M_k$, $[M_i, N_j] = \epsilon_{ijk}N_k$ and the cross relations

$$[p_0, M_i] = 0, \quad [p_i, M_j] = \epsilon_{ijk}p_k, \quad [p_0, N_i] = -p_i,$$

$$[p_i, N_j] = -\delta_{ij}(\frac{1 - e^{-2\lambda p_0}}{2\lambda} + \frac{\lambda}{2}\vec{p}^2) + \lambda p_i p_j,$$

$$\Delta N_i = N_i \otimes 1 + e^{-\lambda p_0} \otimes N_i + \lambda\epsilon_{ijk}p_j \otimes M_k, \quad \Delta p_i = p_i \otimes 1 + e^{-\lambda p_0} \otimes p_i$$

and p_0, M_i primitive. This was also shown in [121] to be (nontrivially) isomorphic to a rather more complicated κ-deformed Poincaré algebra which had been found in [167] by contraction from $U_q(so_{3,2})$. Here $\lambda =$

κ^{-1} in the present conventions. The paper [121] also established (for the first time) the algebra $[t, x_i] = \lambda x_i$ as the appropriate deformed Minkowski spacetime on which the κ-deformed Poincaré algebra acts covariantly (hitherto only noncovariant actions of it on usual Minkowski spacetime had been considered). This algebra was already known in the 3-dimensional case, indeed $[x_3, x_i] = x_i$ is the Lie algebra su_2^{*op} in [120] and its enveloping algebra is necessarily the dual of the translation sector of $C(SU_2^{*op}) \bowtie U(su_2)$. It is part of the dual quantum system $U(su_2^{*op}) \bowtie \mathbb{C}(SU_2)$ according to the observable-state duality in Chapter 6.4. The 3+1-dimensional bicrossproduct formulation of [121] similarly provided the coordinate ring of the κ-deformed Poincaré algebra as another bicrossproduct dual to it, allowing a link with hitherto unconnected other approaches to deformed Poincaré algebras. We note also that enveloping algebras of the above type were first proposed as noncommutative spacetime in [168] in the context of regularising divergences in quantum field theory.

This bicrossproduct theory arose independently of the Yang–Baxter equation and inverse scattering, which are more directly tied up with the theory of Poisson–Lie groups, Poisson actions, etc. We give in Section 8.4 only some of the key ingredients leading up to the latter topic. The main Theorem 8.4.1 that Lie bialgebras exponentiate to Poisson–Lie groups is due to Drinfeld [157]. Proposition 8.4.5 about matched pairs of Poisson–Lie groups appears to be new, while the case of the double $U^{*op} \bowtie U$ recovers ideas from the theory of classical inverse scattering. The Adler–Kostant–Symes theorem in Exercise 8.4.6 was originally due to these gentlemen, and developed in a more modern form in [169, 170, 171]; see the text [172]. Our formulation based on matched pairs of Lie algebras and double cross product groups is, in principle, more general than the standard approach based on Lax pairs and symplectic geometry. The examples $M \bowtie g^*$ in Proposition 8.4.7 underly the corresponding bicrossproduct Hopf algebras $\mathbb{C}(M) \bowtie U(g)$ but our explicit construction of them by exponentiation is new. Exercise 8.4.8 is a corollary, the first part (which gives the symplectic leaves) being immediate from the point of view of coadjoint matched pairs in [12, 120] (where the role of δ is replaced by the coadjoint action).

9

Representation theory

After the technical work of the preceding three chapters, we now take a break and return to more conceptual issues, i.e. to the general theory of Hopf algebras and quasitriangular structures. We have already seen early on, in Exercise 1.6.8, that if a Hopf algebra is represented in vector spaces V, W then it is also naturally represented in $V \otimes W$. So the representations of the Hopf algebra have among themselves a tensor product operation $\otimes$. This is a key property of group representations, and is just as important for Hopf algebras. We have already used it several times in previous chapters in the course of our Hopf algebra constructions. In this chapter, we want to study this phenomenon quite systematically and more conceptually using the language of *category theory*. It should be possible to come to this chapter directly after Chapters 1 and 2, viewing the intermediate ones as providing examples and applications. For readers who want to skim this chapter, the key calculation, which also contains the essence of the entire chapter, is Theorem 9.2.4.

Category theory itself is often considered to be a hard subject by physicists, and for this reason we will try to be informal and nontechnical as much as possible. In truth, a category C just means a collection of objects (in our case they will be representations), and a specification of what it is to be an allowed map or *morphism* between any two objects. One can also have a notion of maps or *functors* between categories. For example, in our case, the tensor product operation $\otimes : C \times C \to C$ will be a functor. It makes the category of representations of any bialgebra or Hopf algebra into a *monoidal category* $(C, \otimes)$. This is the main point of Section 9.1. We also give some theorems about general monoidal categories (not necessarily coming from a bialgebra or Hopf algebra), such as the construction of a *dual monoidal category* C°.

In Section 9.2, we specialise to the case of quasitriangular bialgebras or Hopf algebras, as in Chapter 2. The quasitriangular structure or 'universal R-matrix' is our new ingredient, and it turns out to be just what is needed to ensure isomorphisms $\Psi : V \otimes W \cong W \otimes V$ between any two representations in a coherent way. We take such maps Ψ (the braiding or quasisymmetry) for granted in the case of group representations, where they are just the usual permutation at the level of the underlying vector spaces. However, in the general quantum group case, they are more complicated since $\Psi^2 \neq \mathrm{id}$. The category of representations becomes in this way a *braided monoidal* or *quasitensor* $(\mathcal{C}, \otimes, \Psi)$.

So far this works at the level of bialgebras. In Section 9.3, we assume that we really have a Hopf algebra (with an antipode). Just as group-inversion allows one to define conjugate representations, so the antipode is just what it takes to define a conjugate V^* of any Hopf-algebra representation. In this case, the category of representations is *rigid*. Related to this, one has a notion of the space of tensor operators or 'linear maps' $\underline{\mathrm{Hom}}(V, W)$ between representations, which is itself a representation.

This formalisation of the various abstract properties of the representations of a Hopf algebra is useful for at least two reasons, both of which we develop as themes in the chapter. The first reason, covered in Section 9.2, is that many of the constructions in previous chapters can now be understood very naturally in these categorical terms, enabling one to think conceptually about them rather than simply writing down formulae. In particular, we develop a categorical way of thinking about module algebras or quantum group covariant systems. We have already noted in Chapter 1.6 that a module algebra just means an algebra which is an object in the category of representations of a quantum group. So each quantum group generates a category or 'universe' in which its covariant algebras live. We will see that, in this universe, the covariant algebras naturally acquire *braid statistics*. For example, superalgebras (with Bose–Fermi statistics) are nothing other than algebras living in the category or universe generated by a nontrivial quantum group $\mathbb{Z}'/_2$. So two conceptually quite different ideas, that of nontrivial statistics and that of covariance under a group, are unified in the notion of quantum group covariance. A concrete spin-off is the notion of *braided tensor product* of covariant algebras. We will need it extensively in Chapter 10, where it leads into the theory of braided groups.

A second reason that our formalisation is useful is that if we really have understood the full structure underlying the representations of a given Hopf algebra, we should be able to reconstruct that Hopf algebra entirely from its representations. This is our goal in Section 9.4, where we show that basically every braided monoidal category for which the objects can be identified in a strict way with vector spaces is equivalent to the representations of some quantum group. We also cover the quasi-Hopf algebra case in this section. As well as being a test of the thoroughness of our understanding, it is actually rather useful to be able to pass backwards and forwards in this way between working with a Hopf algebra and working with its representations. This is actually the main principle of Fourier theory, where we know that many constructions for groups can appear very simple in terms of their representations (e.g. in terms of plane waves) but very complicated in terms of the original group, and vice versa. The results of this chapter precisely allow one to follow the same principle for quantum groups. It is also the case that, in some physical contexts (such as low-dimensional quantum field theory), one frequently knows a lot about the representations of the system (the states) and their tensor products, etc. without being sure of the quantum group or other generalised symmetry object behind them.

The reconstruction theorems in this section follow, in fact, the same pattern as the Fourier convolution theorem for finite Abelian groups in Corollary 1.5.6, except that they are much more general in virtue of being phrased categorically. Recall that if G is a finite Abelian group and $\hat{G}$ is its group of nontrivial one-dimensional representations, then one can recover (the group algebra of) G by taking the functions on $\hat{G}$ and making them into a Hopf algebra. This Hopf algebra of functions on $\hat{G}$ is then isomorphic (the Fourier convolution theorem) to the group algebra of G. Our strategy in Section 9.4 is just the same, namely we take certain functions on our given category $\mathcal{C}$ and make them into a Hopf algebra.

Another application, covered along the way in Section 9.3, is the notion of *categorical dimension* or *rank* of a representation. This is a natural and obvious construction from the point of view of the abstract braided category structure, yet recovers nicely such things as the q-numbers $[n] = \frac{q^n - q^{-n}}{q - q^{-1}}$ familiar from examples such as $U_q(sl_2)$ and the more general $U_q(g)$. It also makes the connection between quantum groups and the theory of knot-invariants.

9.1 Categories, functors and monoidal products

In this section we give a fairly self-contained introduction to the elements of category theory that we shall need. Hopf algebras enter here only as providing a class of examples at the end of the section. The subject is actually very easy, provided one glosses over certain technical issues of axiomatic set theory. At this nontechnical level, every mathematician or theoretical physicist is a category theorist, for it just means being clear about the objects that one is dealing with and what are the allowed transformations or maps between them, which we all do at least at an intuitive level.

Formally, a category $\mathcal{C}$ is a collection of *objects* $V, W, Z, U, \ldots$, and a set

Mor(V, W) of *morphisms* for each V, W. The sets Mor(V, W), Mor(Z, U) are disjoint unless $V = Z$ and $W = U$. There should also be specified a composition operation $\circ$ with properties analogous to the composition of maps. Thus, if $\phi \in$ Mor(V, W), $\psi \in$ Mor(W, Z), then there should be an element $\psi \circ \phi$ in Mor(V, Z), and, for any three morphisms for which $\circ$ is defined, we should have associativity of this composition $\circ$. Further, every set Mor(V, V) should contain an identity element id$_V$ such that $\phi \circ$ id $= \phi$, id $\circ \phi = \phi$ for any morphism for which $\circ$ is defined. A morphism $\phi \in$ Mor(V, W) is called an *isomorphism* if there exists a morphism $\phi^{-1} \in$ Mor(W, V) such that $\phi \circ \phi^{-1} \in$ Mor(W, W) and $\phi^{-1} \circ \phi \in$ Mor(V, V) are identity morphisms.

This is a bit abstract because we have not said that the collection of objects is a set. To do so would be too restrictive, whereas to gloss over the distinction in too naive a fashion can lead to paradoxes. Indeed, specifying or building up a universe of sets and classes or collections of sets, etc., is nontrivial to do consistently. This is the task of axiomatic set theory, and we assume in what follows that it has been done in some reasonable way. Also, we have not said that each of our objects V, W, etc. are themselves built from sets. In practice, they will indeed each be given concretely as sets equipped with further structure, but we are not tied to this. In the concrete case, a morphism $\phi : V \to W$ can be given as a map between the sets underlying each of the objects and restricted in some way that respects their additional structures. But, once again, we are not tied to this concrete case either. That is the reason why we avoid saying that $\mathcal{C}$ is a set of objects. But we still write $V, W \in \mathcal{C}$ as a natural extension of our set theory notation. Likewise, we avoid saying that a morphism $\phi \in$ Mor(V, W) is a map, but we still write $\phi : V \to W$ as a natural notation.

Technically, all our categories will be equivalent to essentially small ones. A small category is one where all the objects are indeed subsets of some universal set, constructed beforehand. Essentially small means that our categories can be approximated by small ones in a certain sense. Effectively, we do not need to worry about the technical points above, provided we say things in the correct formal way so that they can be understood precisely, when required.

We will also need the notion of a (covariant) *functor* $F : \mathcal{C} \to \mathcal{V}$ between categories $\mathcal{C}, \mathcal{V}$, say. As the notation suggests, this means a 'map' between the two categories which respects their structure. Thus, for every $V \in \mathcal{C}$, we specify an object $F(V) \in \mathcal{V}$, and, for every morphism $\phi : V \to W$, we specify a morphism $F(\phi) : F(V) \to F(W)$, such that $F(\phi \circ \psi) = F(\phi) \circ F(\psi)$ for any two morphisms that can be composed. It is sometimes useful to consider also the notion of a *contravariant functor*, which is defined instead as sending $\phi : V \to W$ to $F(\phi) : F(W) \to F(V)$ such that $F(\phi \circ \psi) = F(\psi) \circ F(\phi)$. We have mentioned these obvious concepts already in the discussion of Chapter 6.4. Obviously two categories are isomorphic if there exist mutually inverse functors between them.

Less obvious perhaps is the notion of a *natural transformation* $\theta : F \to G$ or $\theta \in \mathrm{Nat}(F, G)$ between two functors $F, G : \mathcal{C} \to \mathcal{V}$. This means, in fact, an entire collection $\{\theta_V | \ V \in \mathcal{C}\}$, where each $\theta_V : F(V) \to G(V)$ is a morphism of $\mathcal{V}$ and such that, for any morphism $\phi \in \mathrm{Mor}(V, W)$ in $\mathcal{C}$, we have

$$\theta_W \circ F(\phi) = G(\phi) \circ \theta_V. \tag{9.1}$$

There is a similar formula (with a reversal) if F, G are contravariant. This definition says that the collection $\{\theta_V\}$ is coherent or *functorial*. The natural transformation θ is called a *natural isomorphism* or *natural equivalence of functors* if each θ_V is an isomorphism. At least in the small case, the functors $\mathcal{C} \to \mathcal{V}$ are the objects of a new category $[\mathcal{C}, \mathcal{V}]$ with morphisms given by natural transformations. Also, for the recursively minded, it should be clear that the concept of a category itself defines a category, Cat. Its objects are (say) small categories $\mathcal{C}, \mathcal{V}$ and its morphisms $\mathrm{Mor}(\mathcal{C}, \mathcal{V})$ are the functors $\mathcal{C} \to \mathcal{V}$.

Natural transformations look a bit forbidding but are quite ubiquitous. Many constructions in physics are natural just because they respect certain functors: they can often be formulated literally as natural transformations. Also, the notion of two categories being equivalent needs this concept. We say that $\mathcal{C}$ is equivalent to $\mathcal{V}$ if there exist functors $F : \mathcal{C} \to \mathcal{V}$ and $G : \mathcal{V} \to \mathcal{C}$ such that $G \circ F$ and $F \circ G$ are naturally equivalent to the identity functors $\mathcal{C} \to \mathcal{C}$ and $\mathcal{V} \to \mathcal{V}$, respectively. This is a bit weaker than saying that two categories are isomorphic.

Finally, we give both a geometric and an algebraic picture to help us think about categories. The geometrical picture consists in ignoring the above warnings and thinking of the objects of our category as 'points' in a set. We can think of the morphisms as arrows or 'paths' connecting the points. Then a functor maps the points and paths of one category over to points and paths of the other. One can also think of a functor $F : \mathcal{C} \to \mathcal{V}$ as defining a kind of fibre bundle over $\mathcal{C}$ and connection (or gauge field) on it. The fibre over each 'point' V in $\mathcal{C}$ is $F(V)$, and the parallel transport along each 'path' ϕ in $\mathcal{C}$ is $F(\phi)$. Another fibre bundle is to have fibre $\mathrm{Mor}(F(V), F(V))$ over each point, which works better because each fibre is a set. In this case we can speak of 'sections' of this bundle: they are just functions θ which have a value $\theta_V \in \mathrm{Mor}(F(V), F(V))$ at each point V in $\mathcal{C}$. This time the functor F defines parallel transport $F(\phi) \circ (\ ) \circ F(\phi)^{-1}$ along the 'path' ϕ in $\mathcal{C}$, where we assume for the sake of our picture that $F(\phi)$ is invertible. Then a natural transformation $\theta \in \mathrm{Nat}(F, F)$ is just a section of this fibre bundle which is flat or covariantly constant under this parallel transport. This is just the content of the coherence condition (9.1), except of course that we moved $F(\phi)^{-1}$ to $F(\phi)$ on the other side in order not to assume that it is invertible. The same remarks apply generally for $\theta \in \mathrm{Nat}(F, G)$ between two functors. The fibre over V is $\mathrm{Mor}(F(V), G(V))$ and the parallel transport is formally $G(\phi) \circ (\ ) \circ F(\phi)^{-1}$.

So we can think of $\mathrm{Nat}(F, F)$ as certain 'covariantly constant functions' on $\mathcal{C}$ with values in endomorphisms. It is obvious that we can 'pointwise multiply' such coherent functions and that the identity natural transformation is the identity for this. So we have an algebra $\mathrm{Nat}(F, F)$ (or at least a unital semigroup) associated to any nice category. We will need it later.

Of course, the category Set of sets has all these features. Objects are sets (e.g. subsets of some given universal set), morphisms are set maps, etc. Another example is the category Vec of vector spaces. The objects are vector spaces, the morphisms are the linear maps, etc. Any algebra also gives a natural category and functor.

Example 9.1.1 *Let A be a unital algebra and let $\mathcal{C} = {}_A\mathcal{M}$ be the category of A-modules. The objects are vector spaces on which A acts. The morphisms are linear maps that commute with (intertwine) the action of A. An example of a functor is the functor ${}_A\mathcal{M} \to$ Vec that assigns to each representation its underlying vector space (i.e. throws away the action of A). This is called the forgetful functor. The natural transformations of the forgetful functor are in correspondence with the elements of the algebra A itself.*

Proof: One can easily see that our axioms of a category are satisfied. This example is concrete in that all objects are concrete sets with structure, and all morphisms are maps that respect the structure. The forgetful functor is $F(V) = V$, where the first V is as an A-module and the second is just as a plain vector space. It also sends $F(\phi) = \phi$. For the last part, if $a \in A$ then one can define $\theta_V(v) = a \triangleright v$ and check that the coherence or naturality condition (9.1) reduces to the definition of morphisms, namely maps that commute with the action of $a \in A$. Conversely, given a natural transformation $\theta \in \mathrm{Nat}(F, F)$, consider $V = A$ as the left regular representation (by multiplication) and define $a = \theta_A(1)$, where 1 is the identity element. We now check that these two constructions are mutually inverse. Thus, if we start with a and define θ by the first construction, then the corresponding element of A by the second construction is $\theta_A(1) = a.1 = a$, so we recover a. In the other direction, if we start with θ and define $a = \theta_A(1)$, and then the corresponding natural transformation by the first construction is $\theta_V(v) = (\theta_V \circ F(\phi_v))(1) = (F(\phi_v) \circ \theta_A)(1) = a \triangleright v$,

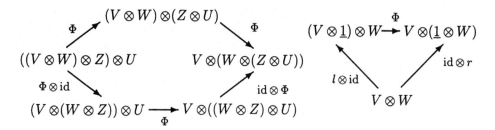

Fig. 9.1.   The pentagon condition for $\otimes$ ensures associativity and the triangle condition ensures compatibility with the unit object.

where $\phi_v : A \to V$ defined by $\phi_v(a) = a \triangleright v$ is a morphism from the left regular representation to V since $\triangleright$ is an action. We used (9.1) and that F is the forgetful functor. So we recover θ. Hence $\text{Nat}(F, F)$ and A are in one to one correspondence. The algebra structure mentioned above on $\text{Nat}(F, F)$ just corresponds to that of A in this example, as is clear since $\triangleright$ is an action.   ∎

So far, our remarks have been very general. From now on, we consider not general categories, but categories equipped with some kind of product, which we denote by $\otimes$. This need not have anything to do with tensor products of vector spaces, but this is our model.

Definition 9.1.2 *A* monoidal category *is* $(\mathcal{C}, \otimes, \underline{1}, \Phi, l, r)$, *where* $\mathcal{C}$ *is a category and* $\otimes : \mathcal{C} \times \mathcal{C} \to \mathcal{C}$ *is a functor which is associative in the sense that there is a natural equivalence* $\Phi : (\ \otimes\ ) \otimes \to \otimes (\ \otimes\ )$, *i.e. there are given functorial isomorphisms*

$$\Phi_{V,W,Z} : (V \otimes W) \otimes Z \cong V \otimes (W \otimes Z), \quad \forall V, W, Z \in \mathcal{C}$$

obeying the pentagon condition *in Fig. 9.1. We also require a unit object* $\underline{1}$ *and natural equivalences between the functors* $(\ ) \otimes \underline{1}$, $\underline{1} \otimes (\ )$ *and the identity functor* $\mathcal{C} \to \mathcal{C}$, *i.e. there should be given functorial isomorphisms* $l_V : V \cong V \otimes \underline{1}$ *and* $r_V : V \cong \underline{1} \otimes V$, *obeying the* triangle condition *in Fig. 9.1.*

The pentagon condition shown arises because there are two ways to go from $((V \otimes W) \otimes Z) \otimes U$ to $V \otimes (W \otimes (Z \otimes U))$ by applying Φ repeatedly. We demand for consistency that these two ways coincide. The remarkable fact is that, once this is demanded for all four objects, then all other consistency problems of this nature are automatically solved as well. This means, in practice, that we can just omit brackets and write expressions such as $V \otimes W \otimes Z \otimes U$ quite freely. There will be several ways to fill in the brackets and Φ in our expressions, but all the different ways will

$$(F(V) \otimes F(W)) \otimes F(Z) \xrightarrow{\ c \otimes \mathrm{id}\ } F(V \otimes W) \otimes F(Z) \xrightarrow{\ c\ } F((V \otimes W) \otimes Z)$$

$$\Big\downarrow \Phi \qquad\qquad\qquad\qquad\qquad\qquad\qquad\qquad \Big\downarrow F(\Phi)$$

$$F(V) \otimes (F(W) \otimes F(Z)) \xrightarrow{\ \mathrm{id} \otimes c\ } F(V) \otimes F(W \otimes Z) \xrightarrow{\ c\ } F(V \otimes (W \otimes Z))$$

Fig. 9.2. A monoidal functor F respects the relevant Φ maps.

coincide. The reader can easily check this consistency to low order. The formal proof is due to S. Mac Lane. The maps l, r associated to the unit object also take care of themselves once the consistency condition stated is satisfied.

Also, if $\mathcal{C}, \mathcal{V}$ are both monoidal, then we say that $F : \mathcal{C} \to \mathcal{V}$ is a *monoidal functor* if it respects the monoidal products in the sense that the two functors $F^2, F \circ \otimes : \mathcal{C} \times \mathcal{C} \to \mathcal{V}$ are naturally equivalent, where $F^2(V, W) = F(V) \otimes F(W)$. Thus, a monoidal functor is a functor that comes with functorial isomorphisms $c_{V,W} : F(V) \otimes F(W) \cong F(V \otimes W)$ such that the condition in Fig. 9.2 holds. We also require

$$F(\underline{1}) = \underline{1}, \quad c_{\underline{1},V} \circ l_{F(V)} = F(l_V), \quad c_{V,\underline{1}} \circ r_{F(V)} = F(r_V)$$

for compatibility with the unit object.

The category Set of sets is a monoidal category with $\otimes = \times$, the direct product of sets. The category Vec of vector spaces is monoidal with $\otimes$ the usual tensor product. In both cases, the isomorphisms Φ are the obvious ones. The unit objects are the singleton set and the field k, respectively.

Example 9.1.3 *Let H be a bialgebra or Hopf algebra and let $_H\mathcal{M}$ be the category of algebra representations in Example 9.1.1. Then $\otimes$, defined (using module notation) as in Exercise 1.6.8 by $h \triangleright (v \otimes w) = \sum h_{(1)} \triangleright v \otimes h_{(2)} \triangleright w$ using the coproduct, makes $_H\mathcal{M}$ into a monoidal category. The forgetful functor $_H\mathcal{M} \to$ Vec is a monoidal functor.*

Proof: We have already checked in Exercise 1.6.8 that $V \otimes W$ is a representation of H if V, W are. We just use the coproduct to split an element of H as stated. We have to check now that the other technical conditions are also satisfied. We first check that $\otimes$ is a functor: to be a functor, it must also map a morphism in $_H\mathcal{M} \times _H\mathcal{M}$ (i.e. a pair of intertwiners (ϕ, ψ)) to a morphism $\phi \otimes \psi$ in $_H\mathcal{M}$. In our case, $\phi \otimes \psi$ is defined to be the linear map $\phi \otimes \psi$. It is easy to see that this is indeed an intertwiner and that this assignment is compatible with $\circ$. Next, define the isomorphism Φ to be the same as the usual one at the level of the underlying vector spaces. So $\Phi_{V,W,Z}((v \otimes w) \otimes z) = v \otimes (w \otimes z)$ for all elements v, w, z

in their respective spaces. It is an intertwiner or morphism since

$$h{\triangleright}((v \otimes w) \otimes z) = h_{(1)(1)}{\triangleright}v \otimes h_{(1)(2)}{\triangleright}w \otimes h_{(2)}{\triangleright}z$$
$$= h_{(1)}{\triangleright}v \otimes h_{(2)(1)}{\triangleright}w \otimes h_{(2)(2)}{\triangleright}z = h{\triangleright}(v \otimes (w \otimes z))$$

by coassociativity of Δ. It is clear that Φ is a functorial isomorphism and obeys the pentagon condition just because these hold for the category Vec of vector spaces. The unit object is the trivial representation $\underline{1} = k$ made possible by the counit in H, $h{\triangleright}\lambda = \epsilon(h)\lambda$, $\forall h \in H, \lambda \in k$. It is easy to see that it has the desired properties under $\otimes$ using the axioms of the counit and the obvious definitions $l = (\ ) \otimes 1$ and $r = 1 \otimes (\ )$. Finally, since the forgetful functor just forgets the action of H while Φ is the same as for vector spaces, it is clear that the forgetful functor is monoidal with $c = \mathrm{id}$. ∎

Example 9.1.4 *Let (H, Φ) be a quasibialgebra or quasi-Hopf algebra in the sense of Chapter 2.4. Then $_H\mathcal{M}$ is a monoidal category with $\otimes$ as before and with Φ given by the action of ϕ followed by the usual associativity isomorphism for vector spaces. Explicitly,*

$$\Phi_{V,W,Z}((v \otimes w) \otimes z) = \sum \phi^{(1)}{\triangleright}v \otimes (\phi^{(2)}{\triangleright}w \otimes \phi^{(3)}{\triangleright}z).$$

The forgetful functor is monoidal iff H is twisting-equivalent to an ordinary bialgebra or Hopf algebra, i.e. iff ϕ is a coboundary.

Proof: This is just like the preceding example, except that we should be careful since we do not have coassociativity. So $h{\triangleright}((v \otimes w) \otimes z)) = (\Delta h){\triangleright}((v \otimes w) \otimes z) = ((\Delta \otimes \mathrm{id}) \circ \Delta h){\triangleright}(v \otimes w \otimes z)$, where the multiple actions are also denoted by $\triangleright$. Similarly for the other bracketing. So

$$\Phi(h{\triangleright}((v \otimes w) \otimes z)) = \phi((\Delta \otimes \mathrm{id}) \circ \Delta h){\triangleright}(v \otimes w \otimes z)$$
$$= ((\mathrm{id} \otimes \Delta) \circ \Delta h)\phi{\triangleright}(v \otimes w \otimes z) = h{\triangleright}\Phi((v \otimes w) \otimes z)$$

in view of axiom (2.29). It is then easy to check that the 3-cocycle axiom for ϕ precisely corresponds to the pentagon condition in Fig. 9.1 for Φ. Also, the maps Φ are isomorphisms because ϕ is invertible, and functorial because they are all defined 'uniformly' by the action of an element of $H \otimes H \otimes H$: they commute with any maps that commute with the action of H, i.e. with any morphisms. Similarly, we can define functorial isomorphisms $c_{V,W}(v \otimes w) = \chi^{(1)}{\triangleright}v \otimes \chi^{(2)}{\triangleright}w$ by the action of any invertible element $\chi \in H \otimes H$. Then the requirement that twisting by χ^{-1} in Theorem 2.4.2 gives an ordinary Hopf algebra, which is $\phi((\Delta \otimes \mathrm{id})\chi)\chi_{12} = ((\mathrm{id} \otimes \Delta)\chi)\chi_{23}$, precisely corresponds to the condition in Fig. 9.2 for the forgetful functor to be monoidal. We can push these arguments backwards as well. Thus, if the forgetful functor is monoidal with some natural transformation $\{c_{V,W}\}$, then the latter is necessarily of the form given by the action of an element of $H \otimes H$ (see Example 9.1.1). The condition in Fig. 9.2, considered in the left regular representation and evaluated on $1 \otimes 1 \otimes 1$, tells us that ϕ is a coboundary in the sense of the requirement above. In terms of the non-Abelian coboundary operation (2.13), this requirement just $\phi = \partial \chi$ for some $\chi \in H \otimes H$, provided we realise that ∂ is to be computed using the coproduct of which Δ is the twisting, i.e. using the coproduct $\chi^{-1}(\Delta)\chi$. ∎

This example demonstrates how structures imposed on an algebra correspond directly to properties of its category of representations. We will see this more formally in Section 9.4.1. By a similar calculation, one finds easily that two bialgebras have equivalent module categories, in a way compatible with their forgetful functors, *iff* the two are related by twisting as in Theorem 2.3.4. Similarly for quasibialgebras, via Theorem 2.4.2.

We know also that our constructions have dual counterparts. So, if A is a bialgebra or a Hopf algebra, then the category $\mathcal{M}^A$ of right A-comodules is a monoidal category. The tensor product comodule $V \otimes W$ is defined by the coaction

$$\beta_{V \otimes W}(v \otimes w) = \sum v^{(\bar{1})} \otimes w^{(\bar{1})} \otimes v^{(\bar{2})}w^{(\bar{2})} \tag{9.2}$$

in terms of the coactions on V, W and the product of A. The associativity Φ is the usual vector space one and $\underline{1} = k$ with $\beta_{\underline{1}}(\lambda) = \lambda \otimes 1$. Similarly for right modules and left comodules of Hopf algebras.

Likewise, the right comodules (say) of a dual quasibialgebra or quasi-Hopf algebra form a monoidal category with the same tensor product of comodules as above and

$$\Phi((v \otimes w) \otimes z) = \sum v^{(\bar{1})} \otimes (w^{(\bar{1})} \otimes z^{(\bar{1})})\phi(v^{(\bar{2})} \otimes w^{(\bar{2})} \otimes z^{(\bar{2})}). \tag{9.3}$$

The axioms of a dual quasibialgebra were given at the end of Chapter 2.4. For right modules of a quasibialgebra or left comodules of a dual quasibialgebra, we use similar formulae with ϕ^{-1} in place of ϕ above when defining Φ.

Although they are not the topic of this book, there are, of course, plenty of monoidal categories that have nothing directly to do with Hopf algebras. For example, take any algebra A and consider the category $_A\mathcal{M}_A$ of *bimodules*. These are vector spaces on which A acts from both sides, with the two actions mutually commuting. There is a tensor product $\otimes_A$ over A defined as in Chapter 1.1 for tensor products over a field, but this time quotienting by the relations $v \triangleleft a \otimes w = v \otimes a \triangleright w$ for all $a \in A$. This makes the bimodules into a monoidal category, generalising the category Vec in such a way that the role of the field k is played now by the (possibly noncommutative) algebra A. The maps Φ, etc. are the obvious vector space ones projected down to $\otimes_A$. There are plenty of other monoidal categories besides.

Also, one can make general constructions for monoidal categories as objects in their own right (in the category of monoidal categories). We describe one of these now, namely the notion (due to the author) of the *dual monoidal category* C° of 'representations' of a monoidal category $(C, \otimes)$. The construction grew out of the representation-theoretic self-duality principle described in Chapter 6.4, and says that the axioms of a monoidal category are, in some sense, self-dual (just as the axioms of a Hopf algebra are self-dual.) More precisely, we fix a monoidal category V over which to work (i.e. in which to build our representations) and we show that the category MON/V, consisting of monoidal categories equipped with functors to V, is self-dual in this representation-theoretic sense. Morphisms in MON/V are monoidal functors compatible with the given functors to V.

Theorem 9.1.5 *Let $F : C \to V$ be a monoidal functor between monoidal categories. We define a representation of C in V to be a pair (V, λ_V), where $V \in V$ and $\lambda_V \in \mathrm{Nat}(V \otimes F, F \otimes V)$ is a natural equivalence, i.e. a collection of functorial isomorphisms $\{\lambda_{V,X} : V \otimes F(X) \to F(X) \otimes V\}$, obeying*

$$\lambda_{V,\underline{1}} = \mathrm{id}, \quad \lambda_{V,Y} \circ \lambda_{V,X} = c_{X,Y}^{-1} \circ \lambda_{V,X \otimes Y} \circ c_{X,Y}, \qquad \forall X, Y \in C.$$

The collection of such representations forms a monoidal category C°, the dual of C over V. Explicitly, the morphisms $(V, \lambda_V) \to (W, \lambda_W)$ between representations are morphisms $\phi : V \to W$ such that

$$(\mathrm{id} \otimes \phi) \circ \lambda_{V,X} = \lambda_{W,X} \circ (\phi \otimes \mathrm{id}), \qquad \forall X \text{ in } C.$$

The monoidal product of representations is

$$(V, \lambda_V) \otimes (W, \lambda_W) = (V \otimes W, \lambda_{V \otimes W}), \quad \lambda_{V \otimes W, Z} = \lambda_{V,Z} \circ \lambda_{W,Z}.$$

The unit object is the trivial representation $(\underline{1}, \lambda_{\underline{1}})$, where $\lambda_{\underline{1},X} = \mathrm{id}$. The forgetful functor $C^\circ \to V$ is monoidal.

Proof: In these formulae we omit writing the Φ, l, r morphisms of V, in view of the coherence theorem mentioned above. We also omit unnecessary identity maps. From these formulae it is clear that (V, λ_V) is some kind of 'representation' of the monoidal product in C, and that the morphisms are like 'intertwiners'. This then motivates the formula for their 'tensor product' as the 'pointwise' composition of representations. Once the formulae are known, it is not hard to check directly that they indeed fulfil the axioms of a monoidal category. We have a category C° with a well-defined composition of morphisms because, if ϕ, ψ are morphisms,

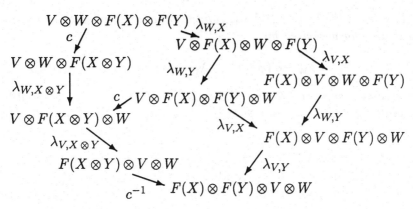

Fig. 9.3. The monoidal product of representations of a monoidal category.

then so is $\phi \circ \psi$ by passing λ first past ψ and then past ϕ. For the monoidal structure, it is clear that $\{\lambda_{V \otimes W,X}\}$ are functorial isomorphisms, since each of the factors $\{\lambda_{V,X}\}, \{\lambda_{W,X}\}$ are. We have to check the representation condition. This is shown in Fig. 9.3; the upper left cell commutes, because this is just the statement that λ_W is a representation. The isomorphisms $c_{X,Y} : F(X) \otimes F(Y) \cong F(X \otimes Y)$ are those that come with the monoidal functor F. (We use the existence and coherence of the $\{c_{X,Y}\}$, rather than the condition in Fig. 9.2 itself). The lower left cell commutes because this is just the statement that λ_V is a representation. The right hand cell commutes because $\lambda_{W,Y}$, $\lambda_{V,X}$ act on different objects. Hence, the two ways to go around the outside of the diagram coincide, i.e. $\lambda_{V \otimes W}$ is a representation. This defines the monoidal product structure for our category $\mathcal{C}^\circ$. It is associative with the same Φ as that of our underlying category $\mathcal{V}$: This is an intertwiner or morphism between the tensor products of any three representations, just from associativity of composition of natural transformations. The monoidal product of morphisms between representations is their underlying monoidal product in $\mathcal{V}$. Finally, it is clear that the functor $\mathcal{C}^\circ \to \mathcal{V}$ which forgets λ, i.e. $F(V, \lambda_V) = V$, is monoidal. The remaining details are easily verified in the same way. Note that we can also make the same definitions without requiring the λ_V to be invertible. ∎

Corollary 9.1.6 *Every monoidal category $\mathcal{C}$ has a dual $\mathcal{C}^\circ$, the dual of $\mathcal{C}$ over itself. Its objects are pairs (V, λ_V), where $V \in \mathcal{C}$ and $\lambda_V = \{\lambda_{V,W} : V \otimes W \to W \otimes V\}$ is a collection of isomorphisms obeying*

$$(\phi \otimes \mathrm{id}) \circ \lambda_{V,W} = \lambda_{V,Z} \circ (\mathrm{id} \otimes \phi), \qquad \forall \phi : W \to Z,$$

$$\lambda_{V,\underline{1}} = \mathrm{id}, \qquad \lambda_{V,Z} \circ \lambda_{V,W} = \lambda_{V,W \otimes Z}, \qquad \forall W, Z \in \mathcal{C}.$$

Morphisms $\phi : (V, \lambda_V) \to (W, \lambda_W)$ and the monoidal product of representations are characterised by

$$(\mathrm{id} \otimes \phi) \circ \lambda_{V,Z} = \lambda_{W,Z} \circ (\phi \otimes \mathrm{id}), \quad \lambda_{V \otimes W, Z} = \lambda_{V,Z} \circ \lambda_{W,Z}, \quad \forall Z \in \mathcal{C}.$$

This special case of Theorem 9.1.5 is also called the center *or* double *of the monoidal category, denoted by $Z(\mathcal{C})$ or $D(\mathcal{C})$.*

Proof: We just take $\mathcal{V} = \mathcal{C}$ and $F : \mathcal{C} \to \mathcal{C}$, the identity functor. We have written the functoriality condition explicitly; in this case, there is a symmetry between the conditions defining λ and the monoidal category structure.

A direct proof of the corollary is also straightforward. For example, the requirement that the monoidal product of λ_V and λ_W obeys the representation condition is

$$\lambda_{V \otimes W, Z} \circ \lambda_{V \otimes W, U} = \lambda_{V,Z} \circ \lambda_{W,Z} \circ \lambda_{V,U} \circ \lambda_{W,U}$$
$$= \lambda_{V,Z} \circ \lambda_{V,U} \circ \lambda_{W,Z} \circ \lambda_{W,U} = \lambda_{V,Z \otimes U} \circ \lambda_{W,Z \otimes U} = \lambda_{V \otimes W, Z \otimes U},$$

for all $U, Z \in \mathcal{C}$, as required. ∎

One can go on to prove a Pontryagin double-duality theorem, construct coadjoint representations, develop Fourier theory, etc. at this categorical level, thinking of $(\mathcal{C}, \otimes)$ as like a group or Hopf algebra. Indeed, we have already mentioned that a functor $F : \mathcal{C} \to \mathcal{V}$ between categories gives $\mathrm{Nat}(F, F)$ an algebra structure, at least in nice cases. If both categories are monoidal and F is monoidal, we also get a coproduct on $\mathrm{Nat}(F, F)$ by regarding it as something like 'functions' on $(\mathcal{C}, \otimes)$. We will study this more formally in Section 9.4 where, at least when $\mathcal{V} = \mathrm{Vec}$, one obtains a Hopf algebra in this way, such that $\mathcal{C}$ is essentially its category of modules. In this case, $\mathcal{C}^\circ$ is its category of comodules. A similar phenomenon holds more generally, using the theory of braided groups.

Example 9.1.7 *Let H be a bialgebra and let $\mathcal{C} = {}_H\mathcal{M} \to \mathrm{Vec}$ be its category of modules as in Example 9.1.3, where the functor is the forgetful functor. Then $\mathcal{C}^\circ$ over Vec (taken without the invertibility condition) is the category ${}^H\mathcal{M}$ of comodules.*

Proof: We work in the familiar category of vector spaces and use the same techniques as in Example 9.1.1, this time establishing a bijection $\mathrm{Lin}(V, H \otimes V) \cong \mathrm{Nat}(V \otimes F, F \otimes V)$, under which the natural transformation λ_V corresponds to a map $\beta : V \to H \otimes V$ by

$$\beta(v) \equiv v^{(\bar{1})} \otimes v^{(\bar{2})} = \lambda_{V,H}(v \otimes 1).$$

Here F is the forgetful functor and H is viewed as an H-module by the left regular representation. That λ_V represents $\otimes$ then corresponds to the comodule property of β, since in this case we have

$$(\mathrm{id} \otimes \beta) \circ \beta(v) = (\mathrm{id} \otimes \lambda_{V,H}) \circ (\lambda_{V,H} \otimes \mathrm{id})(v \otimes 1 \otimes 1)$$
$$= \lambda_{V,H \otimes H}(v \otimes (1 \otimes 1)) = \lambda_{V,H \otimes H}(v \otimes \Delta(1))$$
$$= (\Delta \otimes \mathrm{id}) \circ \lambda_{V,H}(v \otimes 1) = (\Delta \otimes \mathrm{id}) \circ \beta(v).$$

The fourth equality is that λ_V is functorial under the morphism $\Delta : H \to H \otimes H$. Conversely, given a coaction $V \to H \otimes V$, we define

$$\lambda_{V,W}(v \otimes w) = v^{(\bar{1})} \triangleright w \otimes v^{(\bar{2})}$$

and check that it is a natural transformation, etc. It is easy to see that the morphisms in $\mathcal{C}^\circ$ are the linear maps that intertwine the corresponding coactions, and that the tensor product corresponds to the usual tensor product of comodules.

Note that we did not consider here the invertibility condition on the λ (we computed a slightly bigger category). If we insist on this then we have a subcategory of $^H\mathcal{M}$ in which the comodules are invertible in a certain sense. This invertibility is automatic if H has a skew-antipode. ∎

This example demonstrates the sense in which the duality of monoidal categories generalises the familiar duality of Hopf algebras and Abelian groups. On the other hand, we are not at all limited to $\mathcal{V} = \mathrm{Vec}$:

Example 9.1.8 *Let H be a bialgebra and let $\mathcal{C} = {}_H\mathcal{M}$. Then $\mathcal{C}^\circ$ over $\mathcal{C}$ in Corollary 9.1.6 (taken without the invertibility condition on λ) can be identified with the category $^H_H\dot{\mathcal{M}}$ of crossed modules and comodules in Proposition 7.1.6.*

Proof: The condition in Proposition 7.1.6 works for any bialgebra and defines a category $^H_H\dot{\mathcal{M}}$ consisting of vector spaces that are both left H-modules and left H-comodules which are compatible according to this condition. The morphisms are linear maps that intertwine both the action and coaction of H. Our computation to obtain this category is just the same as in the preceding example, except that now everything is required to be H-covariant since we work in the category $_H\mathcal{M}$, rather than in Vec.

Thus, the condition of functoriality tells us, as before, that λ corresponds to a linear map $\beta : V \to H \otimes V$, and the condition that λ represents $\otimes$ tells us that β is a comodule. This time, however, V is also an H-module (not only a vector space) and the $\lambda_{V,W}$ are required to morphisms, i.e. to commute with the action of H. In terms of the corresponding coaction β, this is exactly the compatibility condition in Proposition 7.1.6. Indeed,

$$
\begin{aligned}
h_{(1)}v^{(\bar{1})} \otimes h_{(2)} \triangleright v^{(\bar{2})} &= h \triangleright \lambda_{V,H}(v \otimes 1) \\
&= \lambda_{V,H}(h \triangleright (v \otimes 1)) = \lambda_{V,H}(h_{(1)} \triangleright v \otimes R_{h_{(2)}}(1)) \\
&= (\lambda_{V,H}(h_{(1)} \triangleright v \otimes 1))\,(h_{(2)} \otimes 1) = (h_{(1)} \triangleright v)^{(\bar{1})} h_{(2)} \otimes (h_{(1)} \triangleright v)^{(\bar{2})},
\end{aligned}
$$

where the first equality is the definition of β and the action of H on $H \otimes V$. The second equality is that $\lambda_{V,H}$ is a morphism in $\mathcal{C}$. The final equality uses functoriality under the morphism $R_{h_{(2)}} : H \to H$ given by right-multiplication to obtain the right hand side of the compatibility condition. The converse directions are easier, using the same correspondence as in the preceding example. The morphisms in $\mathcal{C}^\circ$ just correspond to linear maps that intertwine both the action and the coaction of H. Likewise, the monoidal structure of $\mathcal{C}^\circ$ corresponds in ${}_H^H\dot{\mathcal{M}}$ to the tensor product action and tensor product coaction separately. We did not consider here the invertibility condition on the λ (we computed a slightly bigger category); if we insist on this then we have a subcategory of ${}_H^H\mathcal{M}$ in which the comodules are invertible in a certain sense. As before, this invertibility is automatic if H has a skew-antipode. ∎

If H is a finite-dimensional Hopf algebra, then we know from Proposition 7.1.6 that ${}_H^H\mathcal{M}$ can be identified with the category ${}_{D(H)}\mathcal{M}$ of modules of the quantum double $D(H)$. This is how one might come to the quantum double of Chapter 7 from simple categorical ideas of duality: we are taking the dual of H (passing from modules to comodules) but at the same time we are doing it in an H-covariant way, which is why the resulting category is the modules of something containing both H^* and H. Moreover, because the construction in the theorem is quite general, we can also apply it to bialgebras, quasibialgebras etc., and hence define the quantum double (by its category of representations) for these as well. This shows the power of these categorical methods.

Finally, we have preferred the term 'monoidal product' for $\otimes$ rather than 'tensor product' because, at the level of generality in this section, there need not be any field k in the picture. There need not be any direct sums $\oplus$ either. In our algebraic examples connected with Hopf algebras, we do have k-linearity as well as direct sums and other nice properties relating to them (such as good behaviour for exact sequences under $\otimes$). Monoidal categories with a well-behaved direct sum are called *Abelian*. Most of what we do in this chapter does not need direct sums, though these are needed when we want to speak of irreducible representations and Clebsch–Gordan coefficients.

9.2 Quasitensor or braided monoidal categories

So far, we have formulated the associativity and unit properties of $\otimes$ such as we expect them for vector spaces and group representations. It works the same way for Hopf algebra representations using the coproduct Δ to define the tensor product representation. Now we consider a further key property of the tensor product of vector spaces or group representations, namely commutativity of $\otimes$. This is implemented for vector spaces and group representations in a trivial way by the transposition map sending $v \otimes w \mapsto w \otimes v$, the abstract properties of which we take for granted because the map is so simple. For quantum group representations, we will still have isomorphisms $\Psi : V \otimes W \cong W \otimes V$ but they will tend to be nontrivial and their abstract properties need to be formulated. These isomorphisms are a phenomenon for strict quantum groups in the sense of Chapter 2 (with quasitriangular structure). For a general bialgebra or Hopf algebra, the representations $V \otimes W$ and $W \otimes V$ may be quite unrelated.

We begin with the formal definition of the appropriate axioms for Ψ and then show how the reader can easily and intuitively work with them in terms of braid diagrams. The appearance of braids as coming out of our algebraic considerations is quite fundamental. We denote $\otimes^{\mathrm{op}}(V, W) = W \otimes V$ and discuss the sense in which $\otimes$ and $\otimes^{\mathrm{op}}$ should coincide.

Definition 9.2.1 *A* braided monoidal *or* quasitensor *category* $(\mathcal{C}, \otimes, \Psi)$ *is a monoidal category* $(\mathcal{C}, \otimes)$ *as in Section 9.1 which is commutative in the sense that there is a natural equivalence between the two functors* $\otimes, \otimes^{\mathrm{op}} : \mathcal{C} \times \mathcal{C} \to \mathcal{C}$, *i.e. there are given functorial isomorphisms*

$$\Psi_{V,W} : V \otimes W \to W \otimes V, \quad \forall V, W \in \mathcal{C},$$

obeying the hexagon conditions *in Fig. 9.4.*

This just formulates what we might expect for our generalised transposition maps Ψ. If we suppress Φ, then the hexagon conditions are

$$\Psi_{V \otimes W, Z} = \Psi_{V,Z} \circ \Psi_{W,Z}, \quad \Psi_{V, W \otimes Z} = \Psi_{V,Z} \circ \Psi_{V,W} \qquad (9.4)$$

for all objects V, W, Z. These conditions just say that transposing $V \otimes W$ past Z is the same as transposing W past Z and then V past Z, and transposing V past $W \otimes Z$ is the same as first transposing V past W and then V past Z. These are natural properties that we might expect for any reasonable 'transposition' map. From the hexagons one then deduces that Ψ is trivial on the unit object

$$\Psi_{V,\underline{1}} = \mathrm{id}, \quad \Psi_{\underline{1},V} = \mathrm{id} \qquad (9.5)$$

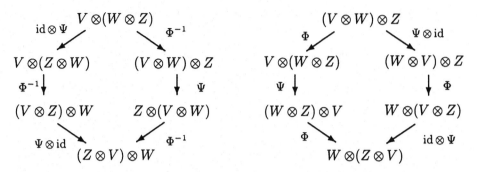

Fig. 9.4. Hexagon conditions for $\otimes$ express compatibility of commutativity and associativity.

as well as a host of other natural identities that one might expect for some kind of transposition.

On the other hand, there is one important property of the usual transposition for vector spaces that we do *not* assume for general Ψ. Namely, we do not assume that $\Psi \circ \Psi = \mathrm{id}$. If this holds then we have a *symmetric* or *tensor* category. These have properties that are very close indeed to the familiar properties of vector spaces. They are also less interesting for the same reason and, moreover, too restrictive for our purposes. In the general quasitensor or braided case that we work with, we must therefore distinguish between $\Psi_{V,W}$ and $(\Psi_{W,V})^{-1}$. They are both morphisms $V \otimes W \to W \otimes V$ so both involve moving V past W to the right, but are distinct. It follows that we should think of the Ψ not so much as transpositions (generating the symmetric group) but as *braids*. For this reason, one says that Ψ is a 'quasisymmetry' or *braiding*.

This leads to the following convenient notation for working with such braidings. We write morphisms pointing generally downwards (say) and denote tensor product by horizontal juxtaposition. Instead of a usual arrow for Ψ, Ψ^{-1} we use the shorthand

$$\Psi_{V,W} = \begin{array}{c} V \quad W \\ \diagdown \\ \diagup \\ W \quad V \end{array}, \qquad (\Psi_{W,V})^{-1} = \begin{array}{c} V \quad W \\ \diagdown \\ \diagup \\ W \quad V \end{array} \qquad (9.6)$$

to distinguish them. We denote any other morphisms as nodes on a string with the appropriate number of input and output legs. In this notation, the hexagons (9.4) and the functoriality of Ψ appear as shown in Fig. 9.5, where the doubled lines in part (a) refer to the composite objects $V \otimes W$ and $W \otimes Z$ in a convenient extension of the notation. The functoriality of Ψ is expressed in part (b) as the assertion that a morphism $\phi : V \to Z$ can be pulled through a braid crossing. Similarly for Ψ^{-1} with inverse braid crossings.

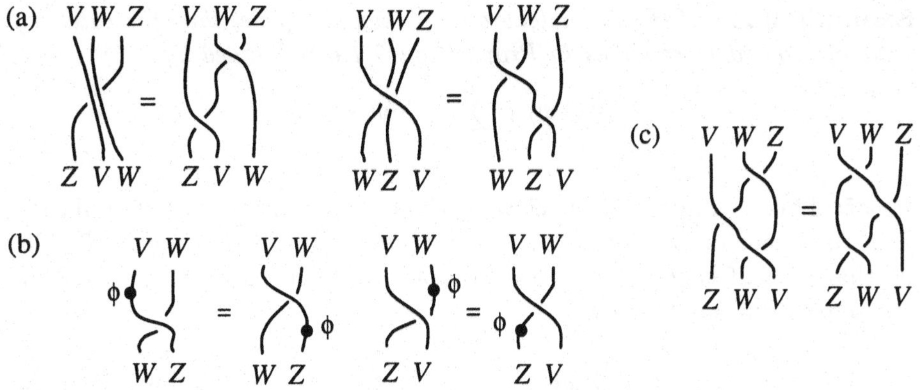

(a)

(b)

(c)

Fig. 9.5. Hexagons (a) and functoriality (b) in the diagrammatic notation. They imply the Yang–Baxter identity or braid relations (c).

The coherence theorem for braided categories can be stated very simply in this notation: if two composite morphisms built from Ψ, Φ correspond to the same braid, then they coincide as morphisms. The reader can easily verify this at lower order. For example, the two braids in Fig. 9.5(c) coincide because $\Psi_{V,W}$ on the left can be pushed up over the Z line. It corresponds to the identity

$$\Psi_{V,W} \circ \Psi_{V,Z} \circ \Psi_{W,Z} = \Psi_{W,Z} \circ \Psi_{V,Z} \circ \Psi_{V,W}, \tag{9.7}$$

which indeed holds by the hexagon identities and functoriality. The general proof in the symmetric or tensor case is due to S. Mac Lane and is based on a presentation of the symmetric group in terms of Ψ. The proof in the braided case is exactly the same with the role of the symmetric group now played by the Artin braid group. The Artin braid group on a given number of strands is the group generated by under and over braid crossings of adjacent strands, regarded as mutually inverse, and the braid relations in Fig. 9.5(c) for three strands. One can also extend the coherence theorem to include branches to ensure that our notation is consistent when we include other morphisms with various numbers of inputs and outputs.

Proposition 9.2.2 *Let C be any monoidal category. The dual monoidal category C° in Corollary 9.1.6 is braided by*

$$\Psi_{(V,\lambda_V),(W,\lambda_W)} = \lambda_{V,W}.$$

Proof: This is clear from the properties of λ stated in Corollary 9.1.6, which imply at once that Ψ as stated obeys the hexagon conditions. ∎

Example 9.2.3 *If H is a bialgebra with skew-antipode, then the category $\overset{H}{{}_H}\mathcal{M}$ of crossed H-modules in Proposition 7.1.6 is braided by*

$$\Psi(v \otimes w) = \sum v^{(\bar{1})} {\triangleright} w \otimes v^{(\bar{2})}.$$

Proof: This is immediate in view of Example 9.1.8 and the formula for λ in the proof of Example 9.1.7. More generally, Proposition 9.2.2 tells us that even if there is no skew-antipode, the subcategory of invertible crossed modules is braided.   ∎

For example, every Hopf algebra acts on itself by the adjoint action Ad and coacts on itself by left regular coaction Δ provided by the coproduct. We have seen in the proof of Example 7.1.8 that they are compatible in the way required for $\overset{H}{{}_H}\dot{\mathcal{M}}$. Hence we have from the corollary a braiding

$$\Psi : H \otimes H \to H \otimes H, \quad \Psi(h \otimes g) = \sum h_{(1)} g S h_{(2)} \otimes h_{(3)}, \qquad (9.8)$$

which necessarily obeys the braid relations or Yang–Baxter equations from Fig. 9.5(c). This works for any Hopf algebra. For $H = U(g)$ it also restricts to the subspace $V = k \oplus g$ as a braiding

$$\Psi : V \otimes V \to V \otimes V, \quad \Psi(\xi \otimes \eta) = [\xi, \eta] \otimes 1 + \eta \otimes \xi, \quad \forall \xi, \eta \in g, \quad (9.9)$$

and $\Psi(1 \otimes \xi) = \xi \otimes 1$ etc., as usual for the identity element. So we have a (nontrivial) braiding or Yang–Baxter operator associated to any nontrivial Lie algebra. We also know that if H is finite dimensional then our category $\overset{H}{{}_H}\dot{\mathcal{M}}$ can be identified with the category of representations of the quantum double. The braiding in this case can also be viewed as an example of the following general type.

Theorem 9.2.4 *If H is a quasitriangular bialgebra or Hopf algebra, then the category $_H\mathcal{M}$ of H-modules is braided with Ψ given by the action of $\mathcal{R}$ followed by the usual transposition. Explicitly,*

$$\Psi_{V,W}(v \otimes w) = \sum \mathcal{R}^{(2)} {\triangleright} w \otimes \mathcal{R}^{(1)} {\triangleright} v.$$

Proof: We verify first that Ψ as stated is indeed a morphism, i.e. that it is an intertwiner for the action of H. We have

$$\Psi(h {\triangleright} (v \otimes w)) = \Psi((\Delta h) {\triangleright} (v \otimes w)) = \tau(\mathcal{R}(\Delta h) {\triangleright} (v \otimes w))$$
$$= \tau((\Delta^{\mathrm{op}} h) \mathcal{R} {\triangleright} (v \otimes w)) = h {\triangleright} \Psi(v \otimes w)$$

as required. We used $\triangleright$ to also denote the action of $H \otimes H$ on $V \otimes W$ in the obvious way. The usual transposition map alone will not in general be an intertwiner: we need to apply $\mathcal{R}$ first.

Next, we verify the hexagon conditions (9.4). Using the explicit notation for $\mathcal{R}$, we have

$$\Psi_{V \otimes W, Z}(v \otimes w \otimes z) = \mathcal{R}^{(2)} \triangleright z \otimes \mathcal{R}^{(1)} \triangleright (v \otimes w)$$
$$= \mathcal{R}^{(2)} \triangleright z \otimes \mathcal{R}^{(1)}{}_{(1)} \triangleright v \otimes \mathcal{R}^{(1)}{}_{(2)} \triangleright w = \mathcal{R}^{(2)} \mathcal{R}'^{(2)} \triangleright z \otimes \mathcal{R}^{(1)} \triangleright v \otimes \mathcal{R}'^{(1)} \triangleright w$$
$$= \Psi_{V,Z}(v \otimes \mathcal{R}'^{(2)} \triangleright z) \otimes \mathcal{R}'^{(1)} \triangleright w = \Psi_{V,Z} \circ \Psi_{W,Z}(v \otimes w \otimes z),$$

$$\Psi_{V, W \otimes Z}(v \otimes w \otimes z) = \mathcal{R}^{(2)} \triangleright (w \otimes z) \otimes \mathcal{R}^{(1)} \triangleright v$$
$$= \mathcal{R}^{(2)}{}_{(1)} \triangleright w \otimes \mathcal{R}^{(2)}{}_{(2)} \triangleright z \otimes \mathcal{R}^{(1)} \triangleright v = \mathcal{R}^{(2)} \triangleright w \otimes \mathcal{R}'^{(2)} \triangleright z \otimes \mathcal{R}'^{(1)} \mathcal{R}^{(1)} \triangleright v$$
$$= \mathcal{R}^{(2)} \triangleright w \otimes \Psi_{V,Z}(\mathcal{R}^{(1)} \triangleright v \otimes z) = \Psi_{V,Z} \circ \Psi_{V,W}(v \otimes w \otimes z),$$

where $\mathcal{R}'$ denotes a second copy of $\mathcal{R}$. We used the form of Ψ and the axioms (2.1) of a quasitriangular structure. One can also do these proofs in the compact notation where numerical suffixes denote the position in a multiple tensor product. For example, the second one is

$$\Psi_{V, W \otimes Z}(v \otimes w \otimes z) = \tau_{23} \circ \tau_{12}(((\mathrm{id} \otimes \Delta)\mathcal{R}) \triangleright (v \otimes w \otimes z))$$
$$= \tau_{23} \circ \tau_{12}(\mathcal{R}_{13} \mathcal{R}_{12} \triangleright (v \otimes w \otimes z)) = \tau_{23} \circ \mathcal{R}_{23} \triangleright ((\tau \circ \mathcal{R} \triangleright (v \otimes w)) \otimes z)$$
$$= \Psi_{V,Z} \circ \Psi_{V,W}(v \otimes w \otimes z).$$

We see that the axioms (2.1) for a quasitriangular structure lead directly to the two hexagons, while the axiom (2.2) leads directly to the intertwiner property. The form of Ψ whereby it is given by an element of $H \otimes H$ acting (and the usual transposition map for vector spaces) ensures functoriality. Finally, the assumption that $\mathcal{R}$ is invertible ensures that the Ψ are invertible. ∎

This tells us that the axioms of a quasitriangular bialgebra are exactly what it takes to make the category of modules have a braiding of this form. Indeed, one can push the preceding proof backwards to derive the axioms for $\mathcal{R}$. The same applies in the quasitriangular quasibialgebra case. The formula for Ψ is just the same and makes the category of modules in Example 9.1.4 braided *iff* $\mathcal{R}$ obeys (2.2) and (2.30). In other words, a quasitriangular structure is just what it takes to make the category of modules of a Hopf algebra braided. We will see this again more formally in the reconstruction theorems in Section 9.4. We also see that the triangular case where $\mathcal{R}_{21} = \mathcal{R}^{-1}$ corresponds precisely to the case where Ψ gives a symmetric or tensor category rather than a truly braided one. We turn now to some elementary examples of braided categories constructed by our theorem.

Example 9.2.5 *Let $\mathbb{Z}'_n$ be the quasitriangular Hopf algebra in Example 2.1.6. Its category $\mathcal{C}_n$ of representations is that of* anyonic vector spaces, *i.e. the objects are vector spaces which are $\mathbb{Z}_{/n}$-graded and the* *morphisms are linear maps that preserve the grading. The category is a braided monoidal one with tensor product defined by adding the grading modulo n and with braiding*

$$\Psi(v \otimes w) = e^{\frac{2\pi i |v||w|}{n}} w \otimes v$$

for homogeneous elements of degree $|v|, |w|$. The category is truly braided for $n > 2$.

Proof: The Hopf algebra is generated by g with $g^n = 1$. Hence its representations decompose as $V = \oplus_{a=0}^{n-1} V_a$, where g acts as $g \triangleright v = e^{\frac{2\pi i a}{n}} v$ for all $v \in V_a$. We say that $v \in V_a$ has degree a. This is the grading. The coproduct is $\Delta g = g \otimes g$, so the decomposition of a tensor product representation is by adding the degrees. The formula for $\mathcal{R}$ in Example 2.1.6 put into Theorem 9.2.4 immediately gives the braiding shown.   ∎

This is why we called this the anyon-generating Hopf algebra in Chapter 2.1: it generates as its representations the category of anyonic vector spaces along with its correct anyonic transposition Ψ. The term is derived from anyonic physics, where such an exchange law is encountered. However, the $n = 2$ case should be familiar. It is exactly the category of $\mathbb{Z}_{/2}$-graded or supervector spaces and even morphisms. Thus, this category is generated by the Hopf algebra $\mathbb{Z}'_{/2}$. Note that this statement could not be made before the introduction of all our theory above: it works because $\mathbb{Z}'_{/2}$ has a nontrivial quasitriangular structure or universal R-matrix.

Example 9.2.6 *Let $\mathbb{Z}_{/2,\alpha}$ be the triangular Hopf algebra generated by g, x in Example 2.1.7. Its category of representations is the category of massless Fredholm modules. Objects are pairs (V, D_V), where V is a $\mathbb{Z}_{/2}$-graded or supervector space and $D_V : V \to V$ is an odd operator such that $D_V^2 = 0$. We learn that this category has a monoidal product and symmetry*

$$(V, D_V) \otimes (W, D_W) = (V \otimes W, D_{V \otimes W}),$$

$$D_{V \otimes W}(v \otimes w) = D_V(v) \otimes w + (-1)^{|v|} v \otimes D_W(w),$$

$$\Psi_{V,W}(v \otimes w) = (-1)^{|w||v|} w \otimes v + \alpha(-1)^{|w|(1-|v|)} D_W(w) \otimes D_V(v),$$

on homogeneous elements.

Proof: Here any representation splits as $V = V_0 \oplus V_1$ according to the eigenvalue of the projection operator $\frac{1}{2}(1 + g)$. We denote by D_V the representation of x on V. It is odd because $gx = -xg$. This means that it takes an off-diagonal form in the above decomposition. A morphism in the category is an even operator that intertwines the corresponding D's. The form of the coproduct and triangular structure in Example 2.1.7 immediately gives the tensor product and braiding as stated. ∎

Note that objects of a similar type with $D^2 = 1$ rather than $D^2 = 0$ play an important role in noncommutative geometry under the heading of Fredholm modules. These examples demonstrate how quite elementary discrete Hopf algebras generate large categories of objects with which we might want to work. This is quite a different role from their role as symmetries. The Hopf algebra serves as a kind of 'template' that encodes the defining features of the category. One may then work with the category without realising that it is the category of representations of something. Many categories used in mathematics are indeed of this form.

We can also give more conventional examples where the role of the Hopf algebra is more like that of a symmetry. The following is a hybrid of these two concepts.

Example 9.2.7 *Let G be a finite group and let $D(G)$ be its quantum double as in Example 7.1.9. Its category of representations as described in Example 7.1.10 consists of vector spaces on which G acts and which are G-graded in a compatible way. The tensor product is given by the tensor product action and the product of the degrees. Morphisms are degree-preserving intertwiners. This category is braided with*

$$\Psi(v \otimes w) = |v| \triangleright w \otimes v$$

on elements $v \in V$ of homogeneous G-degree and $w \in W$.

Proof: The braided category structure comes either from Example 9.2.3 or from Theorem 9.2.4. From the latter point of view, the tensor product of two representations is determined from the coproduct of $D(G)$. Hence it comes out as the obvious one in which G acts on both factors of the tensor product and the G-degrees multiply as $|v \otimes w| = |v||w|$. One can also check directly that the compatibility condition still holds as $|u \triangleright (v \otimes w)| = |u \triangleright v \otimes u \triangleright w| = |u \triangleright v||u \triangleright w| = u|v|u^{-1}u|w|u^{-1} = u|v \otimes w|u^{-1}$ for all $u \in G$, as it must by our general theory. The braiding comes from the quasitriangular structure for the double in Example 7.1.9 as $\Psi(v \otimes w) = \sum_{u \in G}(1 \otimes u) \triangleright w \otimes (\delta_u \otimes e) \triangleright v = \sum_{u \in G} u \triangleright w \otimes \delta_u(|v|)v = |v| \triangleright w \otimes v$. ∎

Finally, we can give examples where the category is conceptually much more like that of representations of a group or Lie algebra. Clearly our quantum groups of Chapter 3 should serve here. These are not really quasitriangular (for generic q), but this does not stop us from using the formula in Theorem 9.2.4 all the same. The universal R-matrix or quasi-triangular structure is given as a formal power series, hence the braiding Ψ is a power series of matrices and can be perfectly convergent when evaluated over $\mathbb{C}$ if we limit ourselves to a suitable subcategory of representations. This works fine for the standard $U_q(g)$ if we limit ourselves to finite-dimensional ones.

Example 9.2.8 *Let $U_q(sl_2)$ be the usual deformation as in Chapter 3.2 and consider the category of representations generated by irreducibles $\{V_j\}$ as described in Proposition 3.2.6. This is a braided monoidal category with*

$$\Psi_{V_{j_1},V_{j_2}}(e^{j_1}_{m_1} \otimes e^{j_2}_{m_2}) = (R^{j_1 j_2})^a{}_{m_1}{}^b{}_{m_2} e^{j_2}_b \otimes e^{j_1}_a,$$

$$(R^{j_1 j_2})^{n_1}{}_{m_1}{}^{n_2}{}_{m_2} = \delta^{n_1+n_2}_{m_1+m_2} q^{m_1 n_2 + m_2 n_1} \frac{(1-q^{-2})^{n_1-m_1}}{[n_1-m_1]!}$$

$$\times \left(\frac{[j_1+n_1]![j_1-m_1]![j_2-n_2]![j_2+m_2]!}{[j_1-n_1]![j_1+m_1]![j_2+n_2]![j_2-m_2]!} \right)^{\frac{1}{2}}$$

for $n_1 \geq m_1$, and zero otherwise.

Proof: This comes from the quasitriangular structure in Example 3.2.1 evaluated in the representations V_{j_1}, V_{j_2}. The indices n_i, m_i run over the appropriate range. ∎

To describe the tensor product of two representations in this last example, we use, of course, the coproduct. This is no longer trivial, so that tensor products $V_{j_1} \otimes V_{j_2}$, etc. are not so easy to describe. One can follow the classical strategy for doing this, but now by means of q-deformed Clebsch–Gordan coefficients. Briefly, we can suppose that the tensor product indeed decomposes into irreducibles (this is true for the generic deformations such as above). So $V_{j_1} \otimes V_{j_2} \cong \oplus_j N_{j_1 j_2}{}^j V_j$ for some multiplicities $N_{j_1 j_1}{}^j$. These are integers and, hence are the same as the usual multiplicities in the case of the standard generic deformations. We want to keep track of the isomorphism too, so we write more explicitly

$$V_{j_1} \otimes V_{j_2} = W_{j_1 j_2}{}^j \otimes V_j,$$

where $W_{j_1 j_2}{}^j$ is a vector space of dimension $N_{j_1 j_2}{}^j$ on which the quantum group acts trivially. Then the *Clebsch–Gordan coefficients* (CGCs) are elements $\begin{bmatrix} j_1 & j_2 & j \\ m_1 & m_2 & m \end{bmatrix} \in W_{j_1 j_2}{}^j$ defined by

$$e^{j_1}_{m_1} \otimes e^{j_2}_{m_2} = \sum_{j,m} \begin{bmatrix} j_1 & j_2 & j \\ m_1 & m_2 & m \end{bmatrix} \otimes e^j_m. \qquad (9.10)$$

The above remarks are quite general (with j a suitable label). For the standard deformations, the CGCs can be described quite explicitly in terms of q-hypergeometric functions.

Using this description of tensor products, one can obviously write the hexagon identities for Ψ in terms of the CGCs and the matrices R. These, being the image of the universal $\mathcal{R}$, obviously obey the generalised QYBE in the form

$$R_{12}^{j_1 j_2} R_{13}^{j_1 j_3} R_{23}^{j_2 j_3} = R_{23}^{j_2 j_3} R_{13}^{j_1 j_3} R_{12}^{j_1 j_2}. \tag{9.11}$$

On the other hand, one can usually work abstractly with the quantum group structure and its coproduct for its action on tensor products, rather than working so explicitly with irreducible representations.

Next, we return to the general theory and give the dual version for comodules. We know, of course, that it is a routine matter to dualise the theory to go from modules to comodules, but this case is sufficiently important that it is worth checking directly.

Exercise 9.2.9 *Let A be a dual quasitriangular bialgebra or Hopf algebra. Show that the braiding in the category $\mathcal{M}^A$ of right A-comodules is*

$$\Psi_{V,W}(v \otimes w) = \sum w^{(\bar{1})} \otimes v^{(\bar{1})} \mathcal{R}(v^{(\bar{2})} \otimes w^{(\bar{2})}),$$

where we denote the coactions explicitly as in Chapter 1.6.

Solution: This is just the dual proof to Theorem 9.2.4. For example, Ψ is an intertwiner because

$$\beta_{W \otimes V} \circ \Psi_{V,W}(v \otimes w) = w^{(\bar{1})(\bar{1})} \otimes v^{(\bar{1})(\bar{1})} \otimes w^{(\bar{1})(\bar{2})} v^{(\bar{1})(\bar{2})} \mathcal{R}(v^{(\bar{2})} \otimes w^{(\bar{2})})$$

$$= w^{(\bar{1})} \otimes v^{(\bar{1})} \otimes w^{(\bar{2})}{}_{(1)} v^{(\bar{2})}{}_{(1)} \mathcal{R}(v^{(\bar{2})}{}_{(2)} \otimes w^{(\bar{2})}{}_{(2)})$$

$$= w^{(\bar{1})} \otimes v^{(\bar{1})} \otimes \mathcal{R}(v^{(\bar{2})}{}_{(1)} \otimes w^{(\bar{2})}{}_{(1)}) v^{(\bar{2})}{}_{(2)} w^{(\bar{2})}{}_{(2)}$$

$$= w^{(\bar{1})(\bar{1})} \otimes v^{(\bar{1})(\bar{1})} \otimes \mathcal{R}(v^{(\bar{1})(\bar{2})} \otimes w^{(\bar{1})(\bar{2})}) v^{(\bar{2})} w^{(\bar{2})}$$

$$= (\Psi_{V,W} \otimes \mathrm{id}) \circ \beta_{V \otimes W}(v \otimes w)$$

as required. We used (2.7). The hexagons (9.4) correspond similarly to (2.6). ∎

The same formula works for the right comodules of a dual quasitriangular dual quasibialgebra.

Example 9.2.10 *Let R be an invertible matrix solution of the QYBE and let $A(R)$ be the associated dual quasitriangular bialgebra constructed in Theorem 4.1.5. There is a braided category $\mathcal{M}^{A(R)}$ of right comodules. The braiding on the fundamental corepresentation $\beta(e_j) = e_a \otimes t^a{}_j$ is*

$$\Psi(e_i \otimes e_j) = e_b \otimes e_a R^a{}_i{}^b{}_j.$$

Proof: From Exercise 9.2.9, we have $\Psi(e_i \otimes e_j) = e_b \otimes e_a \mathcal{R}(t^a{}_i \otimes t^b{}_j)$, which gives the formula stated. Note that the generators x_i of the quantum covector spaces in Chapter 4.5 transform in this way and therefore have this braiding. ∎

In the case of $M_q(2)$ or $SU_q(2)$, the fundamental comodule here is the two-dimensional or spin-$\frac{1}{2}$ comodule. It has the same braiding as obtained from the module point of view in Example 9.2.8 with $j_1 = j_2 = \frac{1}{2}$. As for modules, the spin-$\frac{1}{2}$ comodule generates the whole braided category of finite-dimensional comodules. In fact, starting from any invertible R-matrix, one can generate a braided category $\mathcal{C}(R)$ by using the formula in Example 9.2.10 for the braiding of the $\{e_i\}$ and extending Ψ to tensor products, direct sums, direct summands, etc., by (9.4) and linearity. The morphisms in this category are linear maps such that Ψ is functorial with respect to them. On the other hand, all of this is encoded more explicitly by taking the $A(R)$-comodule point of view, as we have done here.

We conclude this section with an important application of our machinery of braided or quasitensor categories which will be crucial for us in Chapter 10. This is to provide a natural way to define the tensor product of quantum-group covariant algebras. This is the *braided tensor product algebra* and is due to the author.

Definition 9.2.11 *Let $\mathcal{C}$ be a monoidal category. An algebra in $\mathcal{C}$ is an object B of $\mathcal{C}$ equipped with a product morphism $B \otimes B \to B$ and unit morphism $\underline{1} \to B$ obeying the usual associativity and unity axioms (as in Chapter 1.1) but now in the category $\mathcal{C}$.*

We are now in a position to make precise the remark preceding Exercise 1.6.8: from the explanation there, we see that if H is a bialgebra or Hopf algebra then a left H-module algebra is nothing other than an algebra in the monoidal category ${}_H\mathcal{M}$ of left H-modules. The dual statement is that if A is a bialgebra or Hopf algebra then a right A-comodule algebra is nothing other than an algebra in the monoidal category $\mathcal{M}^A$ of right A-comodules. The latter is clear from the tensor product coaction (9.2). These statements work for any Hopf algebra (one needs only a monoidal category to define the notion of an algebra in it), but they become especially useful if H is quasitriangular or if A is dual quasitriangular. The reason is the following general construction, which needs the braiding.

Lemma 9.2.12 *Let B, C be two algebras in a braided monoidal category. Then the object $B \otimes C$ also has the structure of an algebra in the category, denoted by $B\underline{\otimes}C$, the braided tensor product algebra, and defined by*

$$\cdot_{B\underline{\otimes}C} = (\cdot_B \otimes \cdot_C) \circ (\mathrm{id} \otimes \Psi_{C,B} \otimes \mathrm{id})$$

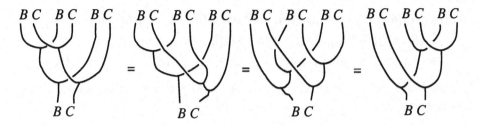

Fig. 9.6. Associativity of the tensor product of two braided algebras.

and the tensor product unit morphism. Moreover, for any three algebras B, C, D in the category, one has $(B\underline{\otimes}C)\underline{\otimes}D \cong B\underline{\otimes}(C\underline{\otimes}D)$ via the underlying associativity Φ.

Proof: As usual, we suppress writing the Φ, l, r explicitly. We have to show that the product is associative. This is done in Fig. 9.6 using the diagrammatic notation explained above for working in braided categories. We write the product morphisms $B \otimes B \to B$ and $C \otimes C \to C$ downwards as $\curlyvee$. The left hand side of the figure is then the product used twice in one order, and the right hand side is the product used twice in the other order. The first equality is functoriality under the second of these morphisms, using Fig. 9.5(b) to push it down over the right copy of B. Then we use associativity in B and C to reorganise the branches. After this, we use functoriality again to push the product morphism of B up and under the left copy of C. The proof of the unit is more trivial. That the construction itself is associative is immediate on writing out the products diagrammatically and using the hexagon identities from Fig. 9.5. This consideration in fact forces the hexagon identities between B, C, D under natural assumptions on their unit morphisms. ∎

This is easy when we use our diagrammatic methods and yet has powerful algebraic consequences.

Corollary 9.2.13 *Let H be a quasitriangular Hopf algebra and let B, C be left H-comodule algebras. Then there is a braided tensor product H-module algebra $B\underline{\otimes}C$ built on $B \otimes C$ with product*

$$(a \otimes c)(b \otimes d) = \sum a(\mathcal{R}^{(2)} \triangleright b) \otimes (\mathcal{R}^{(1)} \triangleright c)d$$

for all $a, b \in B$ and $c, d \in C$, and the tensor product action of H.

Proof: This is immediate from the above lemma and Theorem 9.2.4. It is also easy enough to verify directly if desired. For example, that the new algebra is covariant is

$$h \triangleright ((a \otimes c)(b \otimes d)) = (h_{(1)} \triangleright a)(h_{(2)} \mathcal{R}^{(2)} \triangleright b) \otimes (h_{(3)} \mathcal{R}^{(1)} \triangleright c)(h_{(4)} \triangleright d)$$
$$= (h_{(1)} \triangleright a)(\mathcal{R}^{(2)} h_{(3)} \triangleright b) \otimes (\mathcal{R}^{(1)} h_{(2)} \triangleright c)(h_{(4)} \triangleright d)$$
$$= (h_{(1)} \triangleright a \otimes h_{(2)} \triangleright c)(h_{(3)} \triangleright b \otimes h_{(4)} \triangleright d) = (h_{(1)} \triangleright (a \otimes c))(h_{(2)} \triangleright (b \otimes d)).$$

We used the axiom (2.2) of a quasitriangular structure. Associativity is likewise easy to verify from the axioms (2.1). ∎

Corollary 9.2.14 *Let A be a dual quasitriangular Hopf algebra and let B, C be right A-comodule algebras. Then there is a braided tensor product A-comodule algebra $B \underline{\otimes} C$ built on $B \otimes C$ with product*

$$(a \otimes c)(b \otimes d) = \sum ab^{(\bar{1})} \otimes c^{(\bar{1})} d \mathcal{R}(c^{(\bar{2})} \otimes b^{(\bar{2})})$$

for all $a, b \in B$ and $c, d \in C$, and the tensor product coaction (9.2).

Proof: This is immediate from the lemma and Exercise 9.2.9. One can also verify it directly, if desired. For example, the covariance is

$$\beta((a \otimes c)(b \otimes d)) = (ab^{(\bar{1})})^{(\bar{1})} \otimes (c^{(\bar{1})} d)^{(\bar{1})} \otimes (ab^{(\bar{1})})^{(\bar{2})} (c^{(\bar{1})} d)^{(\bar{2})} \mathcal{R}(c^{(\bar{2})} \otimes b^{(\bar{2})})$$
$$= a^{(\bar{1})} b^{(\bar{1})} \otimes c^{(\bar{1})} d^{(\bar{1})} \otimes a^{(\bar{2})} b^{(\bar{2})}{}_{(1)} c^{(\bar{2})}{}_{(1)} d^{(\bar{2})} \mathcal{R}(c^{(\bar{2})}{}_{(2)} \otimes b^{(\bar{2})}{}_{(2)})$$
$$= a^{(\bar{1})} b^{(\bar{1})} \otimes c^{(\bar{1})} d^{(\bar{1})} \otimes a^{(\bar{2})} c^{(\bar{2})}{}_{(2)} b^{(\bar{2})}{}_{(2)} d^{(\bar{2})} \mathcal{R}(c^{(\bar{2})}{}_{(1)} \otimes b^{(\bar{2})}{}_{(1)})$$
$$= a^{(\bar{1})} b^{(\bar{1})(\bar{1})} \otimes c^{(\bar{1})(\bar{1})} d^{(\bar{1})} \otimes a^{(\bar{2})} c^{(\bar{2})} b^{(\bar{2})} d^{(\bar{2})} \mathcal{R}(c^{(\bar{1})(\bar{2})} \otimes b^{(\bar{1})(\bar{2})})$$
$$= (a^{(\bar{1})} \otimes c^{(\bar{1})})(b^{(\bar{1})} \otimes d^{(\bar{1})}) \otimes a^{(\bar{2})} c^{(\bar{2})} b^{(\bar{2})} d^{(\bar{2})} = \beta(a \otimes c)\beta(b \otimes d)$$

using the axioms of a coaction and the axiom (2.7) of a dual quasitriangular structure. The direct proof of associativity uses the same techniques and axioms (2.6). ∎

The direct proofs here are quite tedious and should convince the reader of the power of our braided category methods, where the general diagrammatic proof was rather easy.

Example 9.2.15 *Let B, C be two anyonic algebras, i.e. $\mathbb{Z}/n$-graded algebras for which the product map is additive in the degree and the unit element has degree 0. There is an anyonic tensor product algebra $B \underline{\otimes} C$ with product*

$$(a \otimes c)(b \otimes d) = e^{\frac{2\pi i |c||b|}{n}} ab \otimes cd$$

and tensor product unit element.

Proof: This is immediate from the braiding in Example 9.2.5. ∎

This example demonstrates how the braided tensor product generalises the notion of $\mathbb{Z}_{/2}$-graded or supertensor product of superalgebras. We will see many more examples in the next section. The general features are the same: one should think of the elements of the algebra as having nontrivial *braid statistics* with respect to another independent algebra in the category. At least in the concrete setting, where objects are vector spaces with additional structure (such as in Corollaries 9.2.13 or 9.2.14), it is clear that $B \equiv B \otimes 1$ and $C \equiv 1 \otimes C$ are subalgebras, since the braiding is trivial on the unit element. If we write $b \equiv b \otimes 1$ and $c' \equiv 1 \otimes c$, then

$$c'b = (1 \otimes c)(b \otimes 1) = \Psi(c \otimes b) = \sum_k (b_k \otimes 1)(1 \otimes c_k) = \sum_k b_k c'_k \quad (9.12)$$

if $\Psi(c \otimes b) \equiv \sum_k b_k \otimes c_k$, say. This means that the two subalgebras B, C fail to commute inside the braided tensor product. The braid statistics are described by Ψ and can be a linear combination rather than simply a phase factor as in Example 9.2.15.

Recall also that we studied the general theory of factorisation of an algebra into subalgebras in the first half of the proof of Theorem 7.2.3. We went on there to consider a coalgebra structure too, but we see also that, if B, C are algebras and $\Psi : C \otimes B \to B \otimes C$ is a linear map obeying conditions (7.10), then there is an algebra, $B \bowtie_\Psi C$, say, built on $B \otimes C$ with

$$(a \otimes c)(b \otimes d) = a\Psi(b \otimes c)d.$$

Moreover, if $X = BC$ is a factorisation of algebras in the sense that

$$B \overset{i}{\hookrightarrow} X \overset{j}{\hookleftarrow} C$$

are subalgebras of an algebra X, and the map $\cdot \circ (i \otimes j) : B \otimes C \to X$ is an isomorphism, then $X \cong B \bowtie_\Psi C$ for a certain Ψ. One can also say this more formally as a universal property.

It should be clear that our braided tensor product is exactly an example of such a factorisation as far as the algebra structure is concerned, with the role of Ψ played now by the braiding $\Psi_{C,B}$. Functoriality with respect to the product combines with the relevant hexagon identity, and ensures precisely that (7.10) hold in our diagrammatic setting. These combinations are all that are needed so far as making an associative algebra $B \underline{\otimes} C$ is concerned, rather than the full structure of a braided category. On the other hand, a braided category *is* the natural setting for our construction if one wants not one specific braided tensor product algebra but the ability to make tensor products quite freely among a collection of algebras in a consistent way. This is the content of the second part of Lemma 9.2.12.

Put another way, one can and should think of $B \underline{\otimes} C$ as the natural generalisation of the *trivial* tensor product of algebras $(a \otimes c)(b \otimes d) =$

$(ab \otimes cd)$, which we take for granted when working with usual algebras. In physical terms, the trivial tensor product of two systems corresponds to making a joint system in which the two subsystems are independent. The supertensor product works the same way and is the way to tensor product independent fermionic systems. Thus, in physical terms, the braided tensor product is a generalisation of the notion of combining independent systems. This is why we think of its noncommutativity as due to braid statistics rather than due to quantisation or some other origin. This point of view will be the theme of Chapter 10, where it is developed further.

At the same time, we see from Theorem 9.2.4 that, whenever a quasitriangular Hopf algebra acts on an algebra, it induces such braid statistics. When ordinary groups do this, the braiding is trivial so we do not see this phenomenon. But when quantum deformations of ordinary groups act they do induce nontrivial braid statistics as a corollary of the deformation. Indeed, two quite different concepts, that of statistics and that of covariance under a symmetry, are unified in the concept of quantum group covariance.

Finally, let us note that while we have suppressed Φ in most of this section, it should be understood in all the general constructions in a braided category, including the diagrammatic ones such as Fig. 9.6 and in later sections. By the coherence theorems we merely have to insert the associator Φ as needed for expressions to make sense; different ways to do this will give the same result.

Exercise 9.2.16 *Let G be an Abelian group, $\phi \in Z^3(G)$, $\mathcal{R}$ a quasi-bicharacter and $(kG, \phi, \mathcal{R})$ the dual quasitriangular dual quasi-Hopf algebra in Example 2.4.3. Its category of comodules is the braided monoidal category of G-graded vector spaces with*

$$\Phi_{V,W,Z}((v \otimes w) \otimes z) = v \otimes (w \otimes z)\phi(|v|, |w|, |z|),$$

$$\Psi_{V,W}(v \otimes w) = w \otimes v\mathcal{R}(|v|, |w|)$$

on elements of homogeneous degree $|v|$, etc.

Solution: The monoidal structure is provided by (9.3); we just drop the $_{(1)}, _{(2)}$, etc. suffixes. The braiding is similarly provided by the dual quasi-Hopf version of Exercise 9.2.9.   ∎

The algebras in this category are associative up to the required Φ, which here means up to ϕ. And using $\mathcal{R}$ they have a braided tensor product $\underline{\otimes}$, and so on. These remarks apply also to the further constructions in the next two sections.

9.3   Duals, quantum dimensions and traces

Another property of vector spaces and group representations that we would also like to generalise is duality. For vector spaces one has the dual vector space V^*. For group representations one has the dual or conjugate representation using the group inverse to turn the natural right action of the group on the dual vector space back into a left action. Now we consider the same construction of the representations of a Hopf algebra. Until now, we have needed only a bialgebra or quasitriangular bialgebra structure: in the present section, we will also need an antipode, as a generalisation of a group inverse.

The question of the existence of dual objects in the category of representations is independent from that of the braiding. It makes sense for any monoidal category, or, in our examples, it exists in the representations of any Hopf algebra independently of whether or not there is a quasitriangular structure. We begin with a formal definition of the appropriate axioms for V^* in a monoidal category. After that we will proceed to the braided case and show how to work with duals in terms of knots and tangle diagrams. The appearance of knots and knot-invariants is a fundamental application of our algebraic constructions.

There are two ways of arriving at the correct notion of a dual object in a monoidal category. The first is to introduce a notion like that of 'linear maps' $\underline{\mathrm{Hom}}(V, W)$ between objects. Then, specialising to $V^* = \underline{\mathrm{Hom}}(V, \underline{1})$ should supply a suitable dual. If we are lucky, we may also have $V^{**} \cong V$. This is how we usually introduce the dual of a vector space, as the vector space of linear maps $V \to k$. Of course, this V^* comes equipped with an evaluation map $\mathrm{ev} : V^* \otimes V \to k$, which gives the canonical pairing between V^* and V. This was the line we took in Chapter 1. Less well-known perhaps is that there is also a *coevaluation map* $\mathrm{coev} : k \to V \otimes V^*$, defined as the map whose dualisation would be the evaluation $V^{**} \otimes V^* \to k$. For finite-dimensional vector spaces, there is a symmetry between V and V^* since we identify the double dual with the original. In this case, we have, explicitly, the maps

$$\mathrm{ev}(f \otimes v) = f(v), \quad \mathrm{coev}(\lambda) = \lambda \sum_a e_a \otimes f^a, \qquad (9.13)$$

for all $v \in V$, $f \in V^*$ and $\lambda \in k$, where $\{e_a\}$ is a basis of V and $\{f^a\}$, a dual basis. The second, and more symmetrical, way of formulating the notion of duals is to specify V^*, ev and coev abstractly. We will develop this approach first since it is easier, and will come to $\underline{\mathrm{Hom}}(V, W)$ later on in the section.

Definition 9.3.1 *Let C be a monoidal category. An object V has a left* dual *or is* rigid *if there is an object V^* and morphisms* $\mathrm{ev}_V : V^* \otimes V \to \underline{1}$, $\mathrm{coev}_V : \underline{1} \to V \otimes V^*$ *such that*

$$
\begin{aligned}
V &\overset{\mathrm{coev}}{\to} (V \otimes V^*) \otimes V \overset{\Phi}{\to} V \otimes (V^* \otimes V) \overset{\mathrm{ev}}{\to} V, \\
V^* &\overset{\mathrm{coev}}{\to} V^* \otimes (V \otimes V^*) \overset{\Phi^{-1}}{\to} (V^* \otimes V) \otimes V^* \overset{\mathrm{ev}}{\to} V^*
\end{aligned}
\tag{9.14}
$$

compose to id_V *and* id_{V^*}, *respectively. If* V, W *have duals and* $\phi : V \to W$ *is a morphism, then*

$$
\phi^* = (\mathrm{ev}_V \otimes \mathrm{id}) \circ (\mathrm{id} \otimes \phi \otimes \mathrm{id}) \circ (\mathrm{id} \otimes \mathrm{coev}_W) : W^* \to V^* \tag{9.15}
$$

is called the dual *or* adjoint *morphism.*

We omit the associativity Φ in (9.15) and in what follows. It is easy to see that if $(V^*, \mathrm{ev}, \mathrm{coev})$ does exist then it is unique up to an isomorphism. Thus, if $(V^{*'}, \mathrm{ev}', \mathrm{coev}')$ is also a dual for V, then one can define a morphism $\theta : V^{*'} \to V^*$ and its inverse by

$$
\theta = (\mathrm{ev}' \otimes \mathrm{id}) \circ (\mathrm{id} \otimes \mathrm{coev}), \quad \theta^{-1} = (\mathrm{ev} \otimes \mathrm{id}) \circ (\mathrm{id} \otimes \mathrm{coev}'),
$$

and we easily see that

$$
\mathrm{ev}' = \mathrm{ev} \circ (\theta \otimes \mathrm{id}), \quad \mathrm{coev}' = (\mathrm{id} \otimes \theta^{-1}) \circ \mathrm{coev}.
$$

If every object in the category has a dual, then we say that $\mathcal{C}$ is a *rigid monoidal category*.

The conditions in Definition 9.3.1 can also be represented in the diagrammatic notation which we found indispensable for working with braided categories. We do not need any braiding yet, but, as before, we suppress writing the unit object and Φ, and we write all morphisms pointing generally downwards. In such a notation, the evaluation and coevaluation appear simply as $\mathrm{ev} = \cup$ and $\mathrm{coev} = \cap$. We write them without marking the node. Then (9.14) in the definition appear as the 'bend-straightening axioms' shown in part (a) of Fig. 9.7. The adjoint morphism ϕ^* is shown in part (b). The remaining part (c) shows that if V, W have duals then $V \otimes W$ does also, with

$$
(V \otimes W)^* = W^* \otimes V^*, \quad \mathrm{ev}_{V \otimes W} = \mathrm{ev}_W \circ \mathrm{ev}_V, \quad \mathrm{coev}_{V \otimes W} = \mathrm{coev}_W \circ \mathrm{coev}_V,
$$

where unnecessary identity maps are suppressed. We take this as the chosen dual of $V \otimes W$ in what follows. In a similar way, if V^* has a dual it is natural to choose it so that $\mathrm{ev}_{V^*} = (\mathrm{coev}_V)^*$ and $\mathrm{coev}_{V^*} = (\mathrm{ev}_V)^*$. One can also see that, if $\mathcal{C}$ is rigid, then $* : \mathcal{C} \to \mathcal{C}$ is a contravariant functor in the sense explained in Section 9.1.

Now we study the more interesting situation where the category is a braided monoidal or quasitensor category, as in Section 9.2. Our notation for ev and coev combines with our previous coherence theorem as

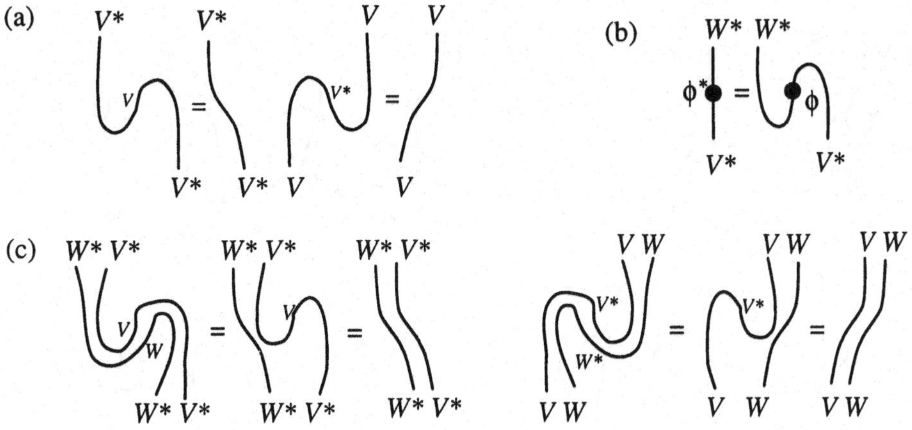

(a)

(b)

(c)

Fig. 9.7. Definition (a) of dual object V^* and (b) of adjoint morphism ϕ^*. Part (c) shows that $(V \otimes W)^* = W^* \otimes V^*$.

described there. So we can slide morphisms through braid crossings, and in the case of $\cap$ and $\cup$ we can straighten the bends of the type shown in the figure. The composite morphism is the same if the diagrammatic picture is the same up to such moves.

Proposition 9.3.2 *In a rigid braided monoidal category there are natural equivalences* $u, v^{-1} \in \mathrm{Nat}(\mathrm{id}, *^2)$ *defined by*

$$u_V = (\mathrm{ev}_V \otimes \mathrm{id}) \circ (\Psi_{V,V^*} \otimes \mathrm{id}) \circ (\mathrm{id} \otimes \mathrm{coev}_{V^*}),$$

$$u_V^{-1} = (\mathrm{id} \otimes \mathrm{ev}_{V^*}) \circ (\Psi_{V^{**},V} \otimes \mathrm{id}) \circ (\mathrm{id} \otimes \mathrm{coev}_V),$$

$$v_V = (\mathrm{ev}_{V^*} \otimes \mathrm{id}) \circ (\mathrm{id} \otimes \Psi_{V,V^*}) \circ (\mathrm{id} \otimes \mathrm{coev}_V),$$

$$v_V^{-1} = (\mathrm{ev}_V \otimes \mathrm{id}) \circ (\mathrm{id} \otimes \Psi_{V^{**},V}) \circ (\mathrm{coev}_{V^*} \otimes \mathrm{id}),$$

obeying

$$u_{V \otimes W} = \Psi_{V,W}^{-1} \circ \Psi_{W,V}^{-1} \circ (u_V \otimes u_W),$$

$$v_{V \otimes W} = \Psi_{V,W}^{-1} \circ \Psi_{W,V}^{-1} \circ (v_V \otimes v_W),$$

and such that

$$(\phi^*)^* = u_W \circ \phi \circ u_V^{-1} = v_W^{-1} \circ \phi \circ v_V, \quad \forall \phi : V \to W.$$

Proof: This is done diagrammatically in Fig. 9.8. We assume that both V and V^* have duals and we define the morphisms u_V, v_V in part (a). If our category is rigid, then we have a whole collection of these, for each

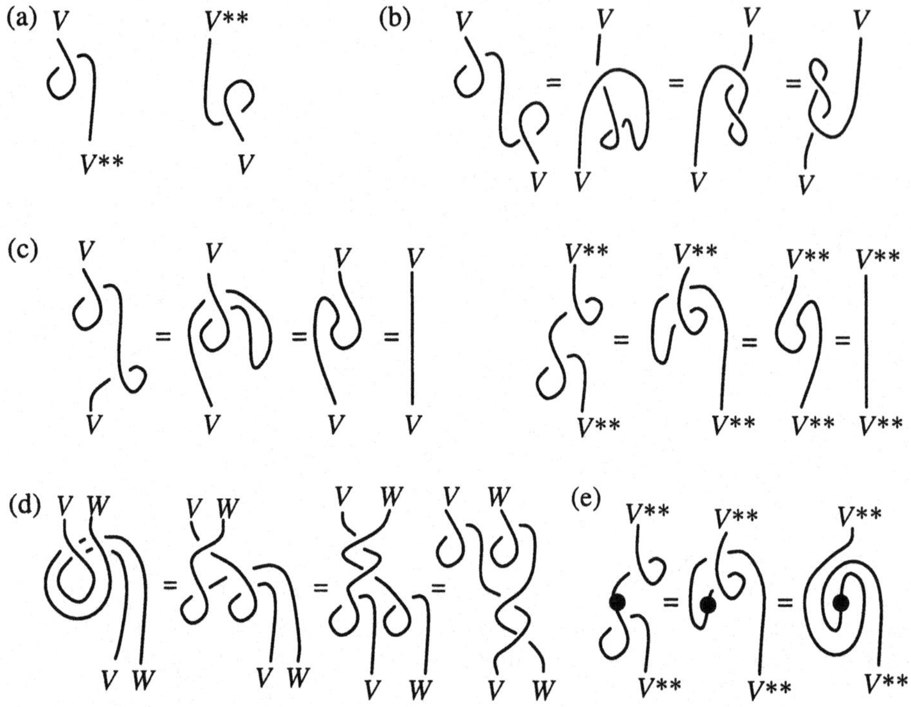

Fig. 9.8. (a) Morphisms u and v and (b) their product $v \circ u$. Parts (c)–(e) are needed in the proof of Proposition 9.3.2.

object V. They are functorial because any other morphism or node ϕ on the string could be pulled through by functoriality of Ψ and elementary properties of ev, coev. Hence they are natural transformations in the sense explained in Section 9.1. Part (c) checks that u is indeed invertible, the lower twist on the left being u^{-1}. The proof for v^{-1} is analogous. Part (d) examines how u behaves on a tensor product. Part (e) computes $u \circ \phi \circ u^{-1}$ and finds ϕ^{**} according to the definition of adjoint morphisms in Fig. 9.7(b). ∎

This completes the definition and basic properties of a dual object in a monoidal or braided monoidal category. One can continue and make various constructions at this categorical level by analogy with familiar constructions for vector spaces. Fig. 9.9 shows two of these. They are both examples of composite morphisms $\underline{1} \to \underline{1}$, which is why they begin in nothing and end in nothing, i.e. they look like knots. The first is called the *categorical dimension* of the object V in a braided category, and the second is called the *categorical trace* of an endomorphism $\phi : V \to V$.

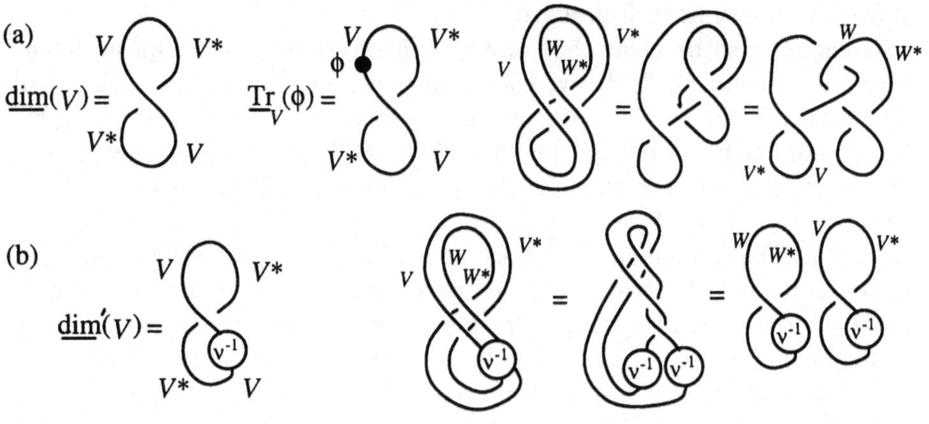

Fig. 9.9. Categorical dimension and trace (a) of an object and endomorphism in a braided category and the breakdown of multiplicativity. Part (b) shows its restoration using a ribbon transformation ν.

Explicitly, the morphisms are

$$\underline{\dim}(V) = \mathrm{ev}_V \circ \Psi_{V,V^*} \circ \mathrm{coev}_V, \tag{9.16}$$

$$\underline{\mathrm{Tr}}_V(\phi) = \mathrm{ev}_V \circ \Psi_{V,V^*} \circ (\phi \otimes \mathrm{id}) \circ \mathrm{coev}_V, \tag{9.17}$$

and they transcribe diagrammatically as shown. The dimension is the trace of the identity morphism. Note that one can only write these down in a braided category. The last part of Fig. 9.9 shows that

$$\underline{\dim}(V \otimes W) \neq \underline{\dim}(V)\underline{\dim}(W),$$

in the general case where $\Psi^2 \neq \mathrm{id}$, with a similar problem for $\underline{\mathrm{Tr}}$. One does have good behaviour with respect to adjoints in the form $\underline{\dim}(V^*) = (\underline{\dim}(V))^*$, as follows easily from our definitions. This is as near as one can get in a general braided category to the usual notions of dimension and trace for finite-dimensional vector spaces and linear operators on them. To go further, one has to assume that the natural transformation $v \circ u$ has a square root $\nu \in \mathrm{Nat}(\mathrm{id}, \mathrm{id})$. This is characterised by

$$\nu_V^2 = v_V \circ u_V, \quad \nu_{V \otimes W} = \Psi_{V,W}^{-1} \circ \Psi_{W,V}^{-1} \circ (\nu_V \otimes \nu_W),$$
$$\nu_{\underline{1}} = \mathrm{id}, \quad \nu_{V^*} = (\nu_V)^*. \tag{9.18}$$

These conditions are not independent (for example, one can conclude the first from the latter three). A *ribbon* or *tortile* category is a rigid braided category equipped with such a *ribbon transformation* ν. In this case, one can restore multiplicativity by using a modified notion of dimension, as shown on the right in Fig. 9.9(b).

It should also be clear that any morphisms which, like <u>dim</u>, are composed only from Ψ, Ψ^{-1}, ev, coev and which start and end with $\underline{1}$ will likewise look like knots. Our coherence theorem then tells us that this morphism depends only on the knot up to the bend-straightening axioms for duals, or up to the cancellation of braids with inverse braids and the braid relations as before. The latter two identifications are called the *second and third Reidemeister moves* for knots. However, we have not said in our coherence theorem that we are allowed to pull tight the twists λ or λ, which would also be possible for knots in three dimensions (the *first Reidemeister move*). Indeed, we know from Proposition 9.3.2 that such morphisms u, v are generally nontrivial. Hence, the net morphism $\underline{1} \to \underline{1}$ is an invariant not exactly of knots in three dimensions (with their usual sense of when two knots are equal), but rather of knots up to *regular isotopy*. Two knots are equal up to regular isotopy if they are related by the second and third Reidemeister moves, which just means that we exclude the straightening of the twists. A second, and related, problem is that not every distinct knot we can write down up to regular isotopy corresponds to some morphism $\underline{1} \to \underline{1}$, i.e. our invariant is only partially defined.

Both problems are resolved in the case of a ribbon category. One then obtains a genuine topological invariant, not exactly of knots but of *framed knots*. By definition, a framed knot is a knot (an embedding of S^1 into S^3 or $\mathbb{R}^3$) and a choice of section of the unit normal bundle to the embedded knot. When a knot is drawn on a piece of paper with under- and over-crossings, there is a canonical framing, the so-called *blackboard framing*, in which the normal vector at each point comes orthogonally out of the page. It is also possible to visualise a framed knot as a ribbon (of sufficiently thin width): one edge of the ribbon is the knot and the other edge is the knot displaced by a finite but sufficiently small amount along the normal vector. The natural topological equivalence or isotopy between framed knots does not involve the first Reidemeister move but instead something weaker, namely $\lambda = \alpha'$ for pieces of a framed knot with the blackboard framing. We will write blackboard-framed knots with thick lines, to remind us of the framing. The weakened form of the first Reidemeister move can easily be checked out with a strip of paper.

Now, in a ribbon category we have the additional transformation ν, and we can use it to define additional evaluation and coevaluation maps $\overline{\mathrm{ev}}_V : V \otimes V^* \to \underline{1}$ and $\overline{\mathrm{coev}}_V : \underline{1} \to V^* \otimes V$ by

$$\overline{\mathrm{ev}}_V = \mathrm{ev}_V \circ (\mathrm{id} \otimes \nu_V^{-1}) \circ \Psi_{V,V^*}, \quad \overline{\mathrm{coev}}_V = \Psi_{V,V^*} \circ (\nu_V^{-1} \otimes \mathrm{id}) \circ \mathrm{coev}_V,$$

such that $(V^*, \overline{\mathrm{ev}}, \overline{\mathrm{coev}})$ is a *right dual* for V. This means that it obeys

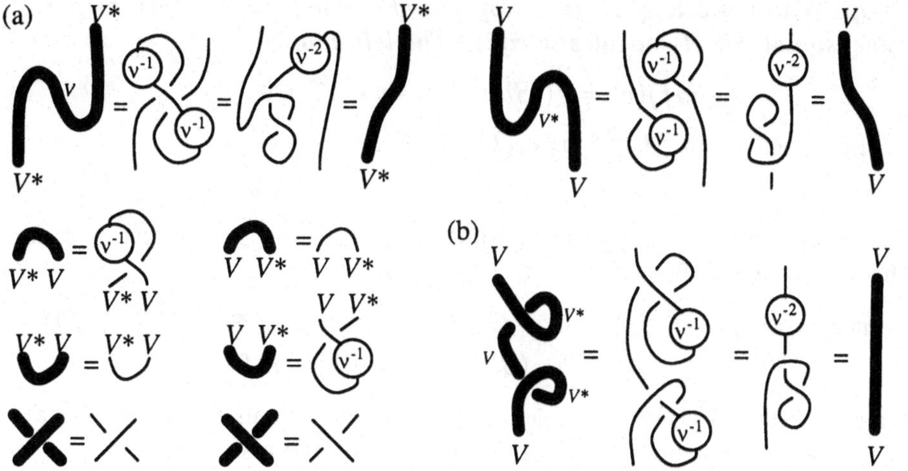

Fig. 9.10. Framed knots or ribbons are drawn with bold lines. We represent them in a ribbon tensor category, as shown to the lower left, and we show that we have (a) right duals and (b) an invariant of framed knots.

axioms as in Definition 9.3.1 with the roles of V, V^* interchanged.

We can add these $\overline{\text{ev}}, \overline{\text{coev}}$ to our diagrammatic notation too, as $\cup, \cap$. They cannot be confused with our existing left handed ev, coev since the labelling by V and V^* is in reversed order. The coherence theorem in this setting is naturally handled now by thinking of the associated diagrams as blackboard-framed knots in three dimensions. First, the axioms of a right dual tell us that we have the right handed bend-straightening axioms as proven using our old notation in Fig. 9.10(a). But we also see in part (b) that our construction is invariant under the weakened form of the first Reidemeister move appropriate to framed knots. In both parts we used $\nu^2 = v \circ u$, as computed in Fig. 9.8(b). Hence two composite morphisms $\underline{1} \to \underline{1}$ are equal if their corresponding framed knots are isotopic.

Moreover, we can read any blackboard-framed knot as a morphism $\underline{1} \to \underline{1}$ in our ribbon category by adding labels V, V^* in such a way that we have a composite of ev, coev, Ψ, Ψ^{-1} or $\overline{\text{ev}}, \overline{\text{coev}}$ when we start at the top and work downwards. One can think of an arc labelled with V as a current of type V flowing (downwards) along the knot and an arc labelled V^* as the same current of type V flowing in the reverse direction. The composite morphism $\underline{1} \to \underline{1}$ associated in this way to a framed knot is an invariant. For example, the modified categorical dimension $\underline{\dim}'$ is just the morphism associated to a blackboard-framed circle. This construction extends to links (disjoint unions of knots) and to tangles in a straightforward way.

Proposition 9.3.3 *If H is a Hopf algebra, then the category of finite-dimensional left H-modules is rigid. The left dual is*

$$(h{\triangleright}f)(v) = f((Sh){\triangleright}v), \quad \forall v \in V,\ f \in V^*,$$

with ev, coev *as in (9.13) for vector spaces.*

Proof: We show that ev, coev are indeed morphisms, i.e. intertwiners for the action of H. Thus,

$$h{\triangleright}\mathrm{coev} = h{\triangleright}(e_a \otimes f^a) = h_{(1)}{\triangleright}e_a \otimes h_{(2)}{\triangleright}f^a = h_{(1)}{\triangleright}e_a \otimes f^a((Sh_{(2)}){\triangleright}(\ ))$$
$$= h_{(1)}(Sh_{(2)}){\triangleright}(\ ) = \epsilon(h)e_a \otimes f^a = \epsilon(h)\mathrm{coev},$$

where we write $V \otimes V^*$ as a linear map. The reader can explicitly evaluate the terms against an element $v \in V$ if desired. We also have

$$\mathrm{ev}(h{\triangleright}(f \otimes v)) = \mathrm{ev}(h_{(1)}{\triangleright}f \otimes h_{(2)}{\triangleright}v) = (h_{(1)}{\triangleright}f)(h_{(2)}{\triangleright}v)$$
$$= f((Sh_{(1)})h_{(2)}{\triangleright}v) = \epsilon(h)\mathrm{ev}(f \otimes v),$$

as required. In both cases we used exactly the antipode axioms (from the two sides). We have (9.14), since these hold for the evaluation and coevaluation in the category of finite-dimensional vector spaces. ∎

We see that the axioms of an antipode in Chapter 1.3 are just what it takes for ev, coev to intertwine with the action on V^* defined by an operator S, i.e. to define the dual or conjugate representation using the dual vector space. If one wants right duals $({}^*V, \overline{\mathrm{ev}}, \overline{\mathrm{coev}})$, one can take similar vector space formulae with ${}^*V$ the predual with basis $\{f^a\}$ and

$$\overline{\mathrm{ev}}(v \otimes f) = v(f), \quad \overline{\mathrm{coev}} = \sum f^a \otimes e_a, \quad v(h{\triangleright}f) = (S^{-1}{\triangleright}v)(f)$$

for all $v \in V$ and $f \in {}^*V$, if S^{-1} exists. In fact, one does not need S itself, but just that this operator denoted S^{-1} is a skew-antipode in the sense of Exercise 1.3.3.

Combining Proposition 9.3.3 with Theorem 9.2.4, we see that the finite-dimensional representations of a quasitriangular Hopf algebra form a rigid braided monoidal category. One also has right duals (since the antipode is necessarily invertible in this case), though they will not necessarily be related to the left duals unless we are in the ribbon case.

Corollary 9.3.4 *If H is a quasitriangular Hopf algebra, then the natural transformations* u, v *in Proposition 9.3.2 in the rigid braided category of finite-dimensional H-modules are given by the action of* u, v *in Proposition 2.1.8. If H is a ribbon Hopf algebra, then the category of modules is ribbon and the natural transformation ν is given by the action of the ribbon element ν in Definition 2.1.10.*

Proof: We work from the definition in Proposition 9.3.2 or Fig. 9.8(a). Thus,

$$u_V(v) = (\mathrm{ev} \otimes \mathrm{id}) \circ \Psi(v \otimes f^a) \otimes E_a = (\mathcal{R}^{(2)} \triangleright f)(\mathcal{R}^{(1)} \triangleright v) \otimes E_a$$
$$= f^a((S\mathcal{R}^{(2)})\mathcal{R}^{(1)} \triangleright v) \otimes E_a = u \triangleright v$$

for all $v \in V$. Here $\{f^a\}$ is a basis of V and $\{E_a\}$ is a dual basis of V^{**}. The result lies in V^{**}. The computation for v is strictly analogous. Hence, if vu has a square root (the ribbon case), we can apply it and define $\nu_V(v) = \nu \triangleright v$ in the same way. That it obeys the condition for $\nu_{V \otimes W}$ follows from the property $\delta\nu$ in Definition 2.1.10. That it obeys the condition for ν_{V^*} corresponds likewise to $S\nu = \nu$ in view of Proposition 9.3.3. ∎

So when we combine these observations with Theorem 9.2.4 or its dual, we see that every finite-dimensional representation of a ribbon Hopf algebra gives an invariant of framed knots. The trivial representation is k, so a morphism $\underline{1} \to \underline{1}$ is a map $k \to k$, i.e. it is described also by an element of k. It is this number which is the invariant. If the Hopf algebra depends on parameters, then so does this number. The same remarks apply for a quasitriangular Hopf algebra and suitable knots up to regular isotopy.

We next compute our invariants for the simplest knots, namely those corresponding to the categorical dimension in Fig. 9.9. Note that, so far, we have considered that a representation of a quantum group leads to a framed-knot invariant, whereas the point of view behind the categorical dimension is just to turn things around and think that each knot gives an invariant of representations of the quantum group. Thus, the trivial knot gives one of the categorical dimensions as explained above, but any other knot also determines a generalised dimension. For example, one has a trefoil dimension of a representation, where we compute the invariant of the trefoil knot in representation V. In short, there is a pairing between quantum group representations and knots, which one can view as a well-defined 'function' on either one or the other collection by fixing either the knot or the representation.

Proposition 9.3.5 *Let H be a quasitriangular Hopf algebra and let V be a finite-dimensional representation. The categorical dimension of V (the corresponding invariant evaluated on the figure of eight) is called the quantum dimension or 'rank'; this and the categorical or quantum trace are given by*

$$\underline{\dim}(V) = \mathrm{Tr}\, u = \mathrm{Tr}\, v = \underline{\dim}(V^*), \quad \underline{\mathrm{Tr}}_V(\phi) = \mathrm{Tr}\, u \circ \phi,$$

where $u, v \in H$ are from Proposition 2.1.8 and are evaluated in the representation V. The trace works for an endomorphism $\phi : V \to V$. In *particular, the multiplicative dimension in the ribbon case is*

$$\underline{\dim}'(V) = \underline{\mathrm{Tr}}_V(\nu_V^{-1}) = \mathrm{Tr}\,\nu^{-1}u = \mathrm{Tr}\,\nu^{-1}v = \underline{\dim}'(V^*).$$

Proof: We proceed from the definition Fig. 9.9(a). Thus,

$$\underline{\dim}(V) = \mathrm{ev} \circ \Psi(e_a \otimes f^a) = (\mathcal{R}^{(2)} {\triangleright} f^a)(\mathcal{R}^{(1)} {\triangleright} e_a) = f^a(u {\triangleright} e_a),$$

where $\{e_a\}$ is a basis of V and $\{f^a\}$, a dual basis. Since $Su = v$, we also check that $\underline{\dim}(V^*) = (u {\triangleright} f^a)(e_a) = f^a(Su {\triangleright} e_a) = \mathrm{Tr}\,v = \mathrm{Tr}\,u$ by cyclicity of the trace. For the quantum trace, we likewise have

$$\underline{\mathrm{Tr}}_V(\phi) = \mathrm{ev} \circ \Psi(\phi(e_a) \otimes f^a) = (\mathcal{R}^{(2)} {\triangleright} f^a)(\mathcal{R}^{(1)} {\triangleright} \phi(e_a)) = f^a(u {\triangleright} \phi(e_a)).$$

For the modified dimension, we take the quantum trace of the morphism $\nu_V^{-1} : V \to V$ according to the definition in Fig. 9.9(b). ∎

One could also try to use these formulae more generally for trace-class operators or as formal power series if the representation is an infinite sum of finite-dimensional ones.

Example 9.3.6 *In the category C_n of anyonic vector spaces, the categorical dimension and trace are*

$$\underline{\dim}(V) = \sum_{a=0}^{n-1} e^{\frac{-2\pi a^2 \imath}{n}}\,\dim(V_a), \quad \underline{\mathrm{Tr}}_V(\phi) = e^{\frac{-2\pi a^2 \imath}{n}} \mathrm{Tr}\,(\phi|_{V_a})$$

if $V = \oplus_a V_a$.

Proof: We gave the formulae for $u = v = \nu$ in this Hopf algebra in Example 2.1.11. The dimension and trace follow at once from the formulae there. It also follows that $\underline{\dim}'$ recovers the usual dimension. ∎

The example precisely generalises the usual superdimension $\dim(V_0) - \dim(V_1)$ and the supertrace, which we recover for $n = 2$. We see that the role of ± 1 in that context is generalised as a Gaussian factor. On the other hand, in the case when our quantum group is a deformation of a usual enveloping algebra, we must obtain for $\underline{\dim}$, etc., a deformation of the usual dimension.

Example 9.3.7 *The categorical or quantum dimensions of the spin-j representations V_j of $U_q(sl_2)$ from Proposition 3.2.6 are*

$$\underline{\dim}(V_j) = q^{-2j(j+1)}[2j + 1], \quad \underline{\dim}'(V_j) = [2j + 1],$$

where $[n] = \frac{q^n - q^{-n}}{q - q^{-1}}$.

Proof: The formula for u is in Proposition 3.2.7, from which we see that

$$\underline{\dim}(V_j) = \sum_{m=-j}^{j} \langle f^{jm}, u \triangleright e_m^j \rangle = \sum_{m=-j}^{j} q^{-2j(j+1)} q^{2m},$$

which sums as stated. Here $\{e_m^j\}$ is a basis of V_j and $\{f^{jm}\}$ is a dual basis. We used the action of q^H from Proposition 3.2.6. The formula for ν in Proposition 3.2.7 immediately tells us that the multiplicative quantum dimension is just $[2j+1]$ as stated. ∎

This recovers the usual dimension of the spin-j representation. Let us note that such q-integers were certainly encountered in abundance when we worked with quantum enveloping algebras in Chapter 3. We see now their conceptual origin as the natural categorical dimensions of representations. Essentially, the concept of dimension of a representation involves a transposition or braiding which is trivial for usual group representations but which shows up for a quantum group.

As an application of our categorical dimension, we note that, if the representation of the quantum group is one of the canonical ones that depend only on the quantum group structure, then our quantum dimension invariant depends only on the quantum group. Thus, one obtains, in this way, invariants of Hopf algebras. The invariant is the same if two Hopf algebras are isomorphic. The problem is just as interesting as deciding if two knots are isomorphic, being a kind of dual problem to that.

Example 9.3.8 *Let H be a finite-dimensional Hopf algebra and let $D(H)$ be its quantum double. The categorical or quantum dimension of the canonical Schrödinger representation of $D(H)$ in Example 7.1.8 is*

$$\underline{\dim}(H) = \operatorname{Tr} S^2,$$

where S is the antipode.

Proof: We know that this is a canonical object in the braided category of $D(H)$-modules. From $\mathcal{R}$ in Theorem 7.1.1, it is obvious that the elements u, v for the quantum double are

$$u = (1 \otimes Se_a)(f^a \otimes 1), \quad v = f^a \otimes Se_a,$$

where $\{e_a\}$ is a basis of H and $\{f^a\}$, a dual basis. The product is in $D(H)$. So we compute its action in the Schrödinger representation as

$$\underline{\dim}(H) = \langle f^b, (1 \otimes Se_a) \triangleright (f^a \otimes 1) \triangleright e_b \rangle = \langle e_{b(1)}, f^a \rangle \langle f^b, (1 \otimes Se_a) \triangleright e_{b(2)} \rangle$$
$$= \langle f^b, (1 \otimes Se_{b(1)}) \triangleright e_{b(2)} \rangle = \langle f^b, (Se_{b(2)}) e_{b(3)} S^2 e_{b(1)} \rangle = \langle f^b, S^2 e_b \rangle.$$

We just used the action of H^*, H in $D(H)$, from Example 7.1.8, and the antipode axioms. ∎

This number, $\mathrm{Tr}\,S^2$, is indeed an important invariant of any finite-dimensional Hopf algebra. We see how it arises here very naturally as the categorical or quantum dimension of a canonical representation of the quantum double.

Another, even more obvious, idea for a canonical representation is the left regular representation. Thus, any quantum group H acts on itself and is therefore an object in its own category of representations. If H is quasitriangular (a strict quantum group), then we define its *quantum order* as

$$|H| = \underline{\dim}(H) = \mathrm{Tr}\,(u) \tag{9.19}$$

in the left regular representation. Obviously, $|kG| = \dim(kG) = |G|$, the usual order (the number of elements) of a finite group G. This is the motivation behind our definition: the usual dimension of the left regular representation exactly counts the number of points in the group, so its categorical or quantum dimension generalises that. It can be quite hard to compute in practice, but it is sure to be an invariant of the Hopf algebra by its categorical definition.

Exercise 9.3.9 *Let H be a finite-dimensional Hopf algebra and let $D(H)$ be its double. Then*

$$|D(H)| = \mathrm{Tr}\,S^2.$$

Proof: This is a harder calculation than that for the dimension of the Schrödinger representation, though the resulting invariant comes out the same. We prefer to compute it as $\mathrm{Tr}\,(v) = \mathrm{Tr}\,(u)$. Then

$$
\begin{aligned}
|D(H)| &= \langle e_b \otimes f^c, (f^a \otimes Se_a)(f^b \otimes e_c)\rangle \\
&= \langle e_b \otimes f^c, f^b{}_{(2)}f^a \otimes (Se_{a(2)})e_c\rangle\langle f^b{}_{(1)}, S^2 e_{a(3)}\rangle\langle f^b{}_{(3)}, Se_{a(1)}\rangle \\
&= \langle e_b, f^b{}_{(2)}(Sf^b{}_{(3)})(Sf^c{}_{(1)})S^2 f^b{}_{(1)}\rangle\langle f^c{}_{(2)}, e_c\rangle \\
&= \langle e_b, (Sf^c{}_{(1)})S^2 f^b\rangle\langle f^c{}_{(2)}, e_c\rangle \\
&= \langle e_b, (Sf^c){}_{(2)}S^2 f^b\rangle\langle (Sf^c){}_{(1)}, S^{-1}e_c\rangle \\
&= \langle e_b, f'^c{}_{(2)}S^2 f^b\rangle\langle f'^c{}_{(1)}, e'_c\rangle,
\end{aligned}
$$

where we used the product in $D(H)$ from Theorem 7.1.1 for the second equality, the antipode axioms for the fourth, that S is an anticoalgebra homomorphism for the fifth and a change of basis to absorb it for the last equality. Comparing the result with the expression for a left integral $\int$

on H^* in Proposition 1.7.3, we see that we have

$$|D(H)| = \langle e'_c, f'^c{}_{(1)} \rangle \int f'^c{}_{(2)} = \int L^*_{e'_c}(f'^c) = \epsilon(e'_c) \int f'^c = \int 1 = \operatorname{Tr} S^2,$$

using invariance under L^*. ∎

For a q-deformation example, we consider the quantum order of the quantum enveloping algebras $U_q(g)$. We know that, for generic q, their representations decompose with the same multiplicities as in the usual case (since these are integers), so we can consider the role of the left regular representation as played by an infinite direct sum of representations with the same multiplicities as classically. For example, we consider

$$V_{\text{left}} = \bigoplus_j (2j + 1)V_j$$

as a model for $U_q(su_2)$ acting on itself. Our motivation is the Peter–Weyl theorem for the decomposition of the left regular representation of the compact group SU_2. One could more conventionally formulate this as a decomposition of the quantum function Hopf *-algebra $SU_q(2)$, on which $U_q(su_2)$ acts via the coproduct. Thinking of V_{left} in this way gives the same answer for the quantum dimension since it is just the conjugate representation to the left multiplication point of view.

Example 9.3.10 *The quantum order of $U_q(su_2)$, defined by the quantum dimension of the left regular representation V_{left}, is*

$$|U_q(su_2)| \equiv \sum_j \dim(V_j)\underline{\dim}(V_j) = \frac{\sum_{n \in \mathbb{z}} q^{-\frac{1}{2}n^2}}{1 - q^{-2}}$$

as a formal power series in q. This can be evaluated for real $q < 1$, where it is finite.

Proof: The first equality is our definition of the categorical dimension as extended by linearity to our infinite sum. The second is an easy calculation given $\underline{\dim}(V_j)$ from Example 9.3.7 above. We consider q here as a formal parameter. However, the result has as numerator a standard Jacobi theta-function, which converges if we now evaluate with $q < 1$. ∎

This is an example of *q-regularisation*, in that we have here a natural definition of the order or 'number of elements' in the quantum group $U_q(su_2)$, which is finite for a suitable range of q. It diverges as $q \to 1$ because, in this limit, we recover the usual order of SU_2, which is infinite. But we see that this infinity becomes, by q-deformation, a pole $(1 - q^{-2})^{-1}$. It is a general feature of q-deformation that certain natural infinities become poles of this type.

The quantum order of a quantum group also has a nice physical interpretation, which helps to explain in part the appearance of number theory in connection with the partition functions of the exactly solvable statistical vertex models in Chapter 4.4. In general, the partition function of a statistical system is a sum over states $\sum_\sigma e^{-\frac{E}{\kappa T}}$, where E is the energy of the state, T is temperature and κ is Boltzmann's constant. The numerator in Example 9.3.10 is of just this form if we consider a quantum particle of mass m confined in a square well potential of width L. Its states are described by natural numbers n, with corresponding energy levels $\frac{\hbar^2\pi^2}{2mL^2}n^2$. We set $q = e^{\frac{\hbar^2\pi^2}{2mL^2\kappa T}}$ to make this interpretation. There is an overcounting factor of 2 since the quantum order counts both n and $-n$.

The quantum order of the general $|U_q(g)|$ was computed by the author and Ya. S. Soibelman using the theory of W-harmonic functions, where W is the Weyl group of g. It is beyond our scope to derive this in detail here. However, the general picture is the same, and one has

$$|U_q(g)| = \frac{\sum_{\Lambda\in P} q^{-(\Lambda,\Lambda)}}{\prod_{\alpha>0}(1 - q^{-2(\rho,\alpha)})} = \frac{Z\binom{\text{quantum particle}}{\text{in an alcove}}}{\prod_{\alpha>0}(1 - q^{-2(\rho,\alpha)})},$$

where P is the weight lattice of g and ρ is half the sum of the positive roots, as in Chapter 3.3. This is a kind of generalised character formula. The product in the denominator is over positive roots $\alpha > 0$. We see that the denominator is again a pole at $q = 1$, with the degree of divergence given by the number $\frac{1}{2}(\dim g - \mathrm{rank}\, g)$ of positive roots. The number of points is again infinite at $q = 1$, but in the approach as $q \to 1$ we see the size of the Lie algebra in the degree of divergence. The numerator meanwhile is again a partition function for a quantum system, namely for a quantum particle confined to an *alcove* of g. This is a bounded domain in the root space t^* in the language of Chapter 3.3 (a fundamental domain for the group generated by the Weyl group and translations on the coroot lattice). Recall that the fundamental weights ω_i form a basis of t^*. The fundamental Weyl chamber is the sector where the coordinates in this basis are all positive. The fundamental alcove is a part of this defined by one more inequality (an upper bound) determined by the maximal root in our partially ordered root system. For su_3, for example, it is an equilateral triangle, and for su_2 it is a tetrahedron. There is also a natural metric on t^*, given, as usual, by the Killing form. The Laplace operator for this metric gives a Hamiltonian, and one can solve the resulting quantum system with Dirichlet boundary conditions. The result is an energy spectrum $2\pi^2\hbar^2(\Lambda, \Lambda)$ for a particle of unit mass, so the numerator of the quantum order can be interpreted as a partition function of such a system. Some care is needed to keep track of overcounting of the states and boundary effects. Another interpretation of the numerator is as the partition function for a particle on t^* with wave-functions which are invariant under W and translations on the coroot lattice, i.e. as a crystal wave-function. Finally, although we have not discussed it here, the quantum order is closely connected with the vacuum expectation value of Wilson loops defined by a quantum non-Abelian gauge theory with Chern–Simons action.

Without going into the details of these physical theories, it is clear that the quantum order indeed 'counts' the points in the group in a quantum sense. The denominator expresses the infinite number of points in G. After factoring out this infinity, we see that the numerator or 'residue' expresses much deeper information about the group not visible in the order at $q = 1$ and corresponding physically to counting the quantum states of a system on the Lie algebra.

Note also that, being a theta-function on the weight lattice, the numerator is a modular function with respect to a suitable subgroup of $SL(2, \mathbb{Z})$. For example, $|U_q(su_2)|$ has as numerator a theta-function $\theta(q^{-\frac{1}{2}})$, which is modular covariant under $\tau \to -\frac{1}{\tau}$, where $q = e^{2\pi i \tau}$. Likewise, $|U_q(e_8)|$ is a modular form of weight 4. Another related feature is that the numerator of $|U_q(g)|$ tends to have multiplicativity properties. For example,

$$|U_{q^{-\frac{3}{2}}}(su_3)| = \frac{1 + 6Z(q)}{(1 - q^3)^2(1 - q^6)},$$

$$Z(q) = q + q^3 + q^4 + 2q^7 + q^9 + q^{12} + 2q^{13} + q^{16} + 2q^{19} + 2q^{21} + q^{25} + q^{27}$$
$$+ 2q^{28} + 2q^{31} + q^{36} + 2q^{37} + 2q^{39} + 2q^{43} + q^{48} + 3q^{49} + 2q^{52} + \cdots.$$

The general formula for the coefficients z_n of q^n has the following remarkable number-theoretic property for the coefficients: if n and m have no common divisors, then

$$z_n z_m = z_{nm}.$$

For example, 4 and 7 have no common factor, the coefficient of q^4 is 1, the coefficient of q^7 is 2 and the coefficient of q^{28} is 1 times 2. The property extends to all orders of q. Functions $Z(q)$ with this property are called *multiplicative* and have Mellin transforms with expressions as infinite products over prime numbers. The arithmetic and modular properties demonstrated in these examples confirm that the quantum order is indeed 'counting' something.

We return now to the general theory of duals in monoidal categories. We know from Section 9.1 that quasibialgebras also generate monoidal categories. We have

Proposition 9.3.11 *If H is a quasi-Hopf algebra, then the category of finite-dimensional left modules is rigid with the same action on V^* as in*

Proposition 9.3.3 but

$$\mathrm{ev}(f \otimes v) = f(\alpha \triangleright v), \quad \mathrm{coev} = \sum_a \beta \triangleright e_a \otimes f^a,$$

where α, β are the elements in the definition (2.31) and (2.32) of a quasi-Hopf algebra.

Proof: We check first that coev, ev are intertwiners. Thus,

$$h \triangleright \mathrm{coev} = h_{(1)} \beta \triangleright e_a \otimes h_{(2)} \triangleright f^a = h_{(1)} \beta \triangleright e_a \otimes f^a((Sh_{(2)}) \triangleright (\ ))$$
$$= h_{(1)} \beta S h_{(2)} \triangleright (\ ) = \epsilon(h) \beta \triangleright e_a \otimes f^a = \epsilon(h) \mathrm{coev},$$
$$\mathrm{ev}(h \triangleright (f \otimes a)) = \mathrm{ev}(h_{(1)} \triangleright f \otimes h_{(2)} \triangleright v) = (h_{(1)} \triangleright f)(\alpha h_{(2)} \triangleright v)$$
$$= f((Sh_{(1)}) \alpha h_{(2)} \triangleright v) = \epsilon(h) \mathrm{ev}(f \otimes v),$$

using the axioms (2.31). The bend-straightening axioms (9.14) come out as

$$(\mathrm{id} \otimes \mathrm{ev}) \circ \Phi(\beta \triangleright e_a \otimes f^a \otimes v) = \phi^{(1)} \beta \triangleright e_a \otimes \mathrm{ev}(\phi^{(2)} \triangleright f^a \otimes \phi^{(3)} \triangleright v)$$
$$= \phi^{(1)} \beta \triangleright e_a (\phi^{(2)} \triangleright f^a)(\alpha \phi^{(3)} \triangleright v) = \phi^{(1)} \beta(S\phi^{(2)}) \alpha \phi^{(3)} \triangleright v = v,$$
$$(\mathrm{ev} \otimes \mathrm{id}) \circ \Phi^{-1}(f \otimes \beta \triangleright e_a \otimes f^a) = \mathrm{ev}(\phi^{-(1)} \triangleright f \otimes \phi^{-(2)} \beta \triangleright e_a) \otimes \phi^{-(3)} \triangleright f^a$$
$$= (\phi^{-(1)} \triangleright f)(\alpha \phi^{-(2)} \beta \triangleright e_a) \otimes \phi^{-(3)} \triangleright f^a = f((S\phi^{-(1)}) \alpha \phi^{-(2)} \beta S \phi^{-(3)} \triangleright (\ )) = f,$$

using the axioms (2.31). Indeed, these axioms are just what it takes for ev, coev the action on V^* of this form to work.   ■

Note that for usual Hopf algebras the freedom in α, β is used to set them equal to unity. But this is not possible when Φ is nontrivial, as one can see from (2.32). In the quasitriangular quasi-Hopf algebra case, we have a braiding and hence a quantum dimension. The same line of calculation as above gives at once

$$\underline{\dim}(V) = \mathrm{Tr}\,(u), \quad u = \sum (S\mathcal{R}^{(2)}) \alpha \mathcal{R}^{(1)} \beta,$$

acting in the representation V.

Of course, we have the dual constructions to all these as well. So, if A is a Hopf algebra, then the category of finite-dimensional right A-comodules is rigid. The dual right comodule is

$$\beta_{V^*}(f) = (f \otimes S) \circ \beta_V, \quad \forall f \in V^*, \tag{9.20}$$

with ev, coev as for vector spaces. Similarly in the dual quasi-Hopf case for suitably modified ev, coev.

The dual formulation is indispensable when one wants to develop the theory of the matrix quantum groups $A(R)$ and their Hopf algebra quotients, as we did in Chapter 4. Recall that for nice and correctly normalised solutions R of the QYBE, one can indeed quotient the dual quasitriangular bialgebra $A(R)$ to a dual quasitriangular Hopf algebra A. Necessary conditions were the biinvertibility of R. In this case we see that the category of right A-comodules is a rigid braided category.

Example 9.3.12 *Let R be a biinvertible solution of the QYBE, and suppose that $A(R)$ has a dual quasitriangular Hopf algebra quotient A as in Proposition 4.2.2. If V is the fundamental covector corepresentation in Example 9.2.10, then the right handed vector corepresentation V^* with dual basis $\{f^i\}$ has coaction $\beta(f^i) = f^a \otimes St^i{}_a$. The braiding of vectors and covectors takes the form*

$$\Psi_{V,V}(e_i \otimes e_j) = e_b \otimes e_a R^a{}_i{}^b{}_j,$$

$$\Psi_{V^*,V^*}(f^i \otimes f^j) = R^i{}_a{}^j{}_b f^b \otimes f^a,$$

$$\Psi_{V,V^*}(e_i \otimes f^j) = \tilde{R}^a{}_i{}^j{}_b f^b \otimes e_a,$$

$$\Psi_{V^*,V}(f^i \otimes e_j) = e_a \otimes f^b R^{-1i}{}_b{}^a{}_j,$$

where $\tilde{R}$ is from (4.12). The quantum dimension of V and quantum trace of matrix $\phi : V \to V$ are

$$\underline{\dim}(V) = \operatorname{Tr} u, \quad \underline{\operatorname{Tr}}(\phi) = \operatorname{Tr} \phi u, \quad u^i{}_j = \tilde{R}^a{}_j{}^i{}_a.$$

Proof: This is similar to the proof of Example 9.2.10, which it subsumes, except that we assume now that the dual quasitriangular structure is well-defined on the generators $\mathbf{t}$ of A. The fundamental covector representation there automatically becomes a coaction of A by pushing out along the projection. If $\{f^i\}$ is a dual basis, then the dual corepresentation from (9.20) is as stated. We used a similar coaction for quantum vectors in Chapter 4.5.2. In this vector corepresentation V^*, we have the braiding from Exercise 9.2.9 as

$$\Psi(f^i \otimes f^j) = \mathcal{R}(St^i{}_a \otimes St^j{}_b) f^b \otimes f^a,$$

$$\Psi(e_i \otimes f^j) = \mathcal{R}(t^a{}_i \otimes St^j{}_b) f^b \otimes e_a,$$

$$\Psi(f^i \otimes e_j) = \mathcal{R}(St^i{}_b \otimes t^a{}_j) e_a \otimes f^b.$$

This gives the results stated, using $\mathcal{R} \circ (S \otimes S) = \mathcal{R}$ and $\mathcal{R}(\mathbf{t}_1 \otimes \mathbf{t}_2) = R$, etc., as in Proposition 4.2.2. The quantum dimension and quantum trace in this setting are obtained, as usual, from our diagrammatic or categorical definitions. We have

$$\underline{\dim}(V) = \operatorname{ev} \circ \Psi(e_a \otimes f^a) = \tilde{R}^b{}_a{}^a{}_b,$$

$$\underline{\operatorname{Tr}}(\phi) = \operatorname{ev} \circ \Psi(\phi(e_a) \otimes f^a) = \phi^b{}_a \tilde{R}^c{}_b{}^a{}_c,$$

where $\phi(e_a) = e_b\phi^b{}_a$. This is as stated, with the contraction of $\tilde{R}$ denoted by u as in Proposition 4.2.2.   ∎

One can also view this construction in entirely matrix terms as a rigid braided category $\mathcal{C}(R, *)$ generated by V, V^*.

We conclude this section with our second approach to duality, based on the concept of internal homomorphisms $\underline{\mathrm{Hom}}$ in a monoidal category. This is actually much more general than the strict notion of duals developed so far. The kind of duality discussed so far is modelled on the properties of finite-dimensional vector spaces, while the notion of internal hom is modelled on the properties of the set of linear maps between vector spaces, finite-dimensional or not. The idea is to be able to define $\underline{\mathrm{Hom}}(V, W)$ for any two objects in such a way that this is itself an object in the category. This is familiar in physics under the name *tensor operators* between group representations. The set of such operators $\underline{\mathrm{Hom}}(V, W)$ transforms under the group action, covariantly in W and contravariantly in V. We generalise some basic elements of this theory to the quantum group case, using categorical methods.

The idea behind $\underline{\mathrm{Hom}}$ is an application of a fundamental idea that works for any category. In any category $\mathcal{C}$, we have, for each object $V \in \mathcal{C}$, a contravariant functor $\mathrm{Mor}(\ , V) : \mathcal{C} \to \mathrm{Set}$ sending $W \in \mathcal{C}$ over to $\mathrm{Mor}(W, V)$. It also maps over morphisms, as $\mathrm{Mor}(\phi, V) = \circ\phi$ for any morphism $\phi : W \to Z$. Precomposition with ϕ maps $\mathrm{Mor}(Z, V)$ over to $\mathrm{Mor}(W, V)$, as it should for a contravariant functor. There is also a similar (covariant) functor $\mathrm{Mor}(V, \ ) : \mathcal{C} \to \mathrm{Set}$ with $\mathrm{Mor}(V, \phi) = \phi\circ$, but we will not need it so much here. Now, a fundamental fact in category theory is that many contravariant functors $\mathcal{C} \to \mathrm{Set}$ are equivalent to functors of the form $\mathrm{Mor}(\ , V)$ for some V. Equivalent means that there are natural isomorphisms $\theta_W : F(W) \cong \mathrm{Mor}(W, V)$, where F is the contravariant functor. One says that the latter is *representable* and that V is its *representing object*. It is determined uniquely up to unique isomorphism.

Now let $\mathcal{C}$ be a monoidal category as in Section 9.1. For any $V, W \in \mathcal{C}$, we define $\underline{\mathrm{Hom}}(V, W)$ as the representing object (when it exists) for the contravariant functor $\mathcal{C} \to \mathrm{Set}$ sending $Z \in \mathcal{C}$ over to $\mathrm{Mor}(Z \otimes V, W)$ and a morphism ϕ to $\circ(\phi \otimes \mathrm{id})$. In other words, $\underline{\mathrm{Hom}}(V, W)$ is defined by the requirement that there are functorial isomorphisms

$$\theta_Z^{V,W} : \mathrm{Mor}(Z \otimes V, W) \cong \mathrm{Mor}(Z, \underline{\mathrm{Hom}}(V, W)) \qquad (9.21)$$

for all objects $Z \in \mathcal{C}$. This is all that the formal definition boils down to. From this definition there nevertheless follow many nice properties of $\underline{\mathrm{Hom}}$ which make it indeed like the space of linear maps between vector spaces.

Proposition 9.3.13 *Let C be a monoidal category with internal hom as defined by (9.21). Then there are morphisms*

$$\underline{\operatorname{Hom}}(Z \otimes V, W) \cong \underline{\operatorname{Hom}}(Z, \underline{\operatorname{Hom}}(V, W)), \tag{9.22}$$

$$\operatorname{ev}_{V,W} : \underline{\operatorname{Hom}}(V, W) \otimes V \to W, \tag{9.23}$$

$$\circ_{V,W,Z} : \underline{\operatorname{Hom}}(V, W) \otimes \underline{\operatorname{Hom}}(Z, V) \to \underline{\operatorname{Hom}}(Z, W), \tag{9.24}$$

for all objects V, W, Z. If C is braided, as in Section 9.2, then we also have the morphisms

$$i_{V,W} : V \to \underline{\operatorname{Hom}}(\underline{\operatorname{Hom}}(V, W), W), \tag{9.25}$$

$$j_{V,W,X,Y} : \underline{\operatorname{Hom}}(V, W) \otimes \underline{\operatorname{Hom}}(X, Y) \to \underline{\operatorname{Hom}}(V \otimes X, W \otimes Y), \tag{9.26}$$

for all objects V, W, X, Y.

Proof: First, we have a sequence of isomorphisms

$$\operatorname{Mor}(U, \underline{\operatorname{Hom}}(Z \otimes V, W)) \cong \operatorname{Mor}(U \otimes (Z \otimes V), W) \cong \operatorname{Mor}((U \otimes Z) \otimes V, W)$$
$$\cong \operatorname{Mor}(U \otimes Z, \underline{\operatorname{Hom}}(V, W)) \cong \operatorname{Mor}(U, \underline{\operatorname{Hom}}(Z, \underline{\operatorname{Hom}}(V, W)))$$

for all objects U. The second isomorphism is defined by precomposition with $\Phi_{U,Z,V}$, while the rest are instances of (9.21). For example, we can take $U = \underline{\operatorname{Hom}}(Z \otimes V, W)$ and the identity morphism, which therefore maps over to a morphism (9.22) as stated. Considering $U = \underline{\operatorname{Hom}}(Z, \underline{\operatorname{Hom}}(V, W))$ and the identity morphism provides the inverse.

Also, and even more basically, we can consider in (9.21) the choice $Z = \underline{\operatorname{Hom}}(V, W)$ and the identity morphism on it. This corresponds on the left to a morphism (9.23) as stated. It is analogous to the evaluation of a linear map on the vector space on which it acts. Of course, at our abstract level, the objects need not be spaces and $\underline{\operatorname{Hom}}$ need not be maps. Given such 'evaluation' morphisms, we next consider $\operatorname{ev}_{V,W} \circ (\operatorname{id} \otimes \operatorname{ev}_{Z,V})$ as an element of the left hand side of

$$\operatorname{Mor}(\underline{\operatorname{Hom}}(V, W) \otimes \underline{\operatorname{Hom}}(Z, V) \otimes Z, W)$$
$$\cong \operatorname{Mor}(\underline{\operatorname{Hom}}(V, W) \otimes \underline{\operatorname{Hom}}(Z, V), \underline{\operatorname{Hom}}(Z, W)),$$

and thereby obtain a 'composition map' (9.24) as stated. This is analogous to the composition of operators. Note that we suppress the associativity morphism Φ here, but it is implicit in the construction.

Finally, we suppose that our category C is not only monoidal but is a braided one as in Section 9.2. Then $\operatorname{ev}_{V,W} \circ \Psi_{V,\underline{\operatorname{Hom}}(V,W)}$ is an element of the left hand side of

$$\operatorname{Mor}(V \otimes \underline{\operatorname{Hom}}(V, W), W) \cong \operatorname{Mor}(V, \underline{\operatorname{Hom}}(\underline{\operatorname{Hom}}(V, W), W))$$

and becomes on the right a morphism (9.25) as stated. Also, we have

$$\mathrm{Mor}(Z, \underline{\mathrm{Hom}}(V, W) \otimes \underline{\mathrm{Hom}}(X, Y)) \cong \mathrm{Mor}(Z, \underline{\mathrm{Hom}}(X, Y) \otimes \underline{\mathrm{Hom}}(V, W))$$
$$\to \mathrm{Mor}(Z \otimes V, \underline{\mathrm{Hom}}(X, Y) \otimes W) \cong \mathrm{Mor}(Z \otimes V, W \otimes \underline{\mathrm{Hom}}(X, Y))$$
$$\to \mathrm{Mor}(Z \otimes V \otimes X, W \otimes Y) \cong \mathrm{Mor}(Z, \underline{\mathrm{Hom}}(V \otimes X, W \otimes Y)),$$

where the first isomorphism is composition with $\Psi_{\underline{\mathrm{Hom}}(V,W),\underline{\mathrm{Hom}}(X,Y)}$ and the third is composition with $\Psi^{-1}_{W,\underline{\mathrm{Hom}}(X,Y)}$. The second and fourth mappings are instances of the general construction

$$\mathrm{Mor}(Z, W \otimes \underline{\mathrm{Hom}}(X, Y)) \to \mathrm{Mor}(Z \otimes X, W \otimes Y), \qquad (9.27)$$

which sends $\phi : Z \to W \otimes \underline{\mathrm{Hom}}(X, Y)$ to $(\mathrm{id} \otimes \mathrm{ev}_{X,Y}) \circ (\phi \otimes \mathrm{id})$. The rest are instances of (9.21). We can then take $Z = \underline{\mathrm{Hom}}(V, W) \otimes \underline{\mathrm{Hom}}(X, Y)$ and the identity morphism, which therefore maps over to a morphism (9.26) as stated. ∎

Finally, we connect our general theory with the theory of left duals or rigidity developed previously, by means of the following lemma.

Lemma 9.3.14 *Let C be a monoidal category and suppose that $V \in C$ is rigid in the sense of Definition 9.3.1. Then*

$$\underline{\mathrm{Hom}}(V, W) = W \otimes V^*$$

is an internal hom for any W in the category. Moreover, in the braided case, the morphism (9.26) becomes an isomorphism.

Proof: If $\phi : Z \otimes V \to W$, we define $\theta_Z^{V,W}(\phi) = (\phi \otimes \mathrm{id}) \circ (\mathrm{id} \otimes \mathrm{coev}_V) : Z \to W \otimes V^*$. In the other direction, if $\phi : Z \to W \otimes V^*$, we define $\theta^{-1}(\phi) = (\mathrm{id} \otimes \mathrm{ev}_V) \circ (\phi \otimes \mathrm{id}) : Z \otimes V \to W$. It is easy to see that these constructions are mutually inverse. As usual, there are implicit Φ morphisms in these formulae. Note also that θ^{-1} is a special case of the more general construction (9.27) already used in the proof of (9.26). If V is rigid, then this map too has an inverse (defined by coev_V in a similar way to θ here). Hence in the braided case we conclude that (9.26) is also an isomorphism when V is rigid. Indeed, in the rigid setting all the morphisms (9.22)–(9.26) themselves become quite straightforward using our previous diagrammatic techniques and this definition for $\underline{\mathrm{Hom}}$. ∎

This shows that the notion of internal hom includes the notion of duals in the sense developed previously. But it is certainly a more general concept, i.e. we might have internal hom without rigidity. Internal hom implies, of course, that we can define $\underline{\mathrm{Hom}}(V, \underline{1})$ and consider it as some kind of dual of V but without a coevaluation coev and the associated isomorphisms in Definition 9.3.1. In the braided case we have a morphism $V \to \underline{\mathrm{Hom}}(\underline{\mathrm{Hom}}(V, \underline{1}), \underline{1})$ as a special case of (9.25), but it need not be an isomorphism as it was in Proposition 9.3.2. It is an isomorphism *iff* V is rigid. In general terms, we keep the properties that we are familiar with for linear maps $V \to W$ but not those that are special to V being finite dimensional.

Example 9.3.15 *Let H be a Hopf algebra. Then the monoidal category* ${}_H\mathcal{M}$ *of H-modules has internal hom*

$$\underline{\mathrm{Hom}}(V, W) = \mathrm{Lin}(V, W), \quad (h \triangleright f)(v) = \sum h_{(1)} \triangleright (f(Sh_{(2)} \triangleright v))$$

for all $h \in H$, $v \in V$ and $f \in \underline{\mathrm{Hom}}(V, W)$. The morphisms (9.21)–(9.24) are the obvious ones as for vector spaces.

Proof: It is easy to see that $\underline{\mathrm{Hom}}(V, W)$ is indeed an object in ${}_H\mathcal{M}$, i.e. that $\triangleright$ as stated is an action. The proof is similar to the proof of the quantum adjoint action in Example 1.6.3. Next we define

$$(\theta_Z^{V,W}(\phi)(z))(v) = \phi(z \otimes v), \quad \forall \phi : Z \otimes V \to W$$

in the obvious way, and check that it indeed maps any morphism $\phi : Z \otimes V \to W$ to a morphism $\theta(\phi) : Z \to \underline{\mathrm{Hom}}(V, W)$ as required. Thus,

$$(h \triangleright (\theta(\phi)(z)))(v) = h_{(1)} \triangleright (\theta(\phi)(z)(Sh_{(2)} \triangleright v)) = h_{(1)} \triangleright (\phi(z \otimes Sh_{(2)} \triangleright v))$$
$$= \phi(h_{(1)} \triangleright z \otimes h_{(2)} Sh_{(3)} \triangleright v) = \phi(h \triangleright z \otimes v) = \theta(\phi)(h \triangleright z)(v)$$

tells us that $\theta(\phi)$ is indeed an intertwiner if ϕ is. On the other hand, θ is invertible, and the collection $\{\theta_Z\}$ is functorial just because these facts are true for the category of vector spaces. Explicitly, the inverse is

$$\left(\theta_Z^{V,W}\right)^{-1}(\phi)(z \otimes v) = \phi(z)(v), \quad \forall \phi : Z \to \underline{\mathrm{Hom}}(V, W).$$

We deduce the morphisms (9.22)–(9.24) from Proposition 9.3.13. Since θ and $\underline{\mathrm{Hom}}$ have the same form as for vector spaces, we know that these maps will also come out in the obvious way as for vector spaces. The new fact is that they intertwine the action of H. For example, the operations $\mathrm{ev}, \circ$ defined by

$$\mathrm{ev}_{V,W}(f \otimes v) = f(v), \quad \circ_{V,W,Z}(f \otimes g)(z) = f(g(z))$$

for $f \in \underline{\mathrm{Hom}}(V, W)$ and $g \in \underline{\mathrm{Hom}}(Z, V)$ commute with our stated actions of H. The ordering of tensor factors is critical here, as one may easily see if one tries to verify these facts directly. ∎

This generalises Proposition 9.3.3 to include infinite-dimensional modules. There is also a right handed version of internal hom corresponding to a skew-antipode and generalising the notion of right duals. The analogue of Corollary 9.3.4 in the quasitriangular case is contained in

Corollary 9.3.16 *Let H be a quasitriangular Hopf algebra. Then the morphisms (9.25) and (9.26) for internal hom in the braided category of left H-modules are given by*

$$i(v)(f) = \sum \mathcal{R}^{(2)} \triangleright \left(f(u\mathcal{R}^{(1)} \triangleright v) \right),$$

$$j(f \otimes g)(v \otimes x) = \sum f(\mathcal{R}^{-(1)} \mathcal{R}^{(1)} \triangleright v) \otimes \mathcal{R}^{-(2)} \triangleright (g(\mathcal{R}^{(2)} \triangleright x)),$$

for all $f \in \underline{\mathrm{Hom}}(V, W)$, $g \in \underline{\mathrm{Hom}}(X, Y)$ and $v \in V$, $x \in X$. Here u is the canonical element of H from Proposition 2.1.8. The second map is an isomorphism in the finite-dimensional case.

Proof: This time, we trace through the latter half of the proof of Proposition 9.3.13 in our concrete setting. We know how H acts on $f \in \underline{\mathrm{Hom}}(V, W)$ from Example 9.3.15, and we put this into Theorem 9.2.4 for the braiding $\Psi(v \otimes f)$ to give

$$\mathrm{ev} \circ \Psi(v \otimes f) = (\mathcal{R}^{(2)} \triangleright f)(\mathcal{R}^{(1)} \triangleright v) = \mathcal{R}^{(2)}{}_{(1)} \triangleright (f((S\mathcal{R}^{(2)}{}_{(2)})\mathcal{R}^{(1)} \triangleright v)),$$

which equals the result stated on using (2.1).

Likewise, the sequence of maps at the end of the proof of Proposition 9.3.13 trace through as follows. We consider a morphism $Z \to \underline{\mathrm{Hom}}(V, W) \otimes \underline{\mathrm{Hom}}(X, Y)$ sending an element $z \mapsto f \otimes g$, say. The first isomorphism in the sequence gives $z \mapsto \Psi(f \otimes g) = \mathcal{R}^{(2)} \triangleright g \otimes \mathcal{R}^{(1)} \triangleright f$. The second isomorphism views this via θ as $z \otimes v \mapsto \mathcal{R}^{(2)} \triangleright g \otimes (\mathcal{R}^{(1)} \triangleright f)(v)$. The third isomorphism is to apply Ψ^{-1} to the output of this, so yielding $z \otimes v \mapsto \mathcal{R}^{-(1)} \triangleright ((\mathcal{R}^{(1)} \triangleright f)(v)) \otimes \mathcal{R}^{-(2)} \mathcal{R}^{(2)} \triangleright g$. The fourth is to view this as $z \otimes v \otimes x \mapsto \mathcal{R}^{-(1)} \triangleright ((\mathcal{R}^{(1)} \triangleright f)(v)) \otimes (\mathcal{R}^{-(2)} \mathcal{R}^{(2)} \triangleright g)(x)$. Finally, we view this as $z \mapsto j(f \otimes g)(v \otimes x)$. Computing the action on f, g from Example 9.3.15 we have

$$j(f \otimes g)(v \otimes x) = \mathcal{R}^{-(1)} \triangleright ((\mathcal{R}^{(1)} \triangleright f)(v)) \otimes (\mathcal{R}^{-(2)} \mathcal{R}^{(2)} \triangleright g)(x)$$

$$= \mathcal{R}^{-(1)} \mathcal{R}^{(1)}{}_{(1)} \triangleright (f(S\mathcal{R}^{(1)}{}_{(2)} \triangleright v)) \otimes (\mathcal{R}^{-(2)} \mathcal{R}^{(2)} \triangleright g)(x)$$

$$= f(S\mathcal{R}^{(1)} \triangleright v) \otimes (\mathcal{R}^{(2)} \triangleright g)(x) = f(S\mathcal{R}^{(1)} \triangleright v) \otimes \mathcal{R}^{(2)}{}_{(1)} \triangleright (g(S\mathcal{R}^{(2)}{}_{(2)} \triangleright x)).$$

The third equality uses the axioms (2.1) of a quasitriangular structure and cancels $\mathcal{R}^{-1} \mathcal{R}$. The result equals the expression stated on using these axioms again. Note that the object Z is irrelevant in this proof since all operations are on a given output $f \otimes g$ of our original map. Alternatively, one can consider that our original map is the identity morphism and that $z = f \otimes g$.

We know from Lemma 9.3.14 that, if V is finite dimensional (and hence rigid from Proposition 9.3.3), then this j is an isomorphism. Making a similar computation to the above for this, one has the inverse as follows. First, we write an element of $\underline{\mathrm{Hom}}(V \otimes X, W \otimes Y) = \mathrm{Lin}(V \otimes X, W \otimes Y)$ as a sum of elements of the form $f \otimes g \in \mathrm{Lin}(V, W) \otimes \mathrm{Lin}(X, Y)$ in the usual trivial way, as in the category of vector spaces. This is where one needs V to be finite dimensional. Then,

$$j^{-1}(f \otimes g)(v)(x) = f(\mathcal{R}^{-(1)} S^2 \mathcal{R}^{(1)} \triangleright v) \otimes \mathcal{R}^{(2)} \triangleright (g(\mathcal{R}^{-(2)} \triangleright x)).$$

It is nontrivial to check directly that this is the inverse and that the maps i, j are indeed intertwiners for the action of H, all of which must be true from our general theory above. ∎

Attempting to verify this corollary directly from the axioms of a quasi-triangular Hopf algebra should convince the reader of the usefulness of these categorical methods. Moreover, the constructions are quite general and apply just as well in the quasi-Hopf setting, for example. We just have to be careful to insert the associativity morphisms Φ which were suppressed above. We take $\underline{\mathrm{Hom}}(V, W) = \mathrm{Lin}(V, W)$ and the action of H just as in Example 9.3.15 but now with

$$\theta_Z^{V,W}(\psi)(z)(v) = \sum \psi(\phi^{-(1)} \triangleright z \otimes \phi^{-(2)} \beta S \phi^{-(3)} \triangleright v),$$
$$\left(\theta_Z^{V,W}\right)^{-1}(\psi)(z \otimes v) = \sum \phi^{(1)} \triangleright \left(\psi(z)((S\phi^{(2)}) \alpha \phi^{(3)} \triangleright v)\right),$$

(9.28)

for all $\psi : Z \otimes V \to W$ and $\psi : Z \to \underline{\mathrm{Hom}}(V, W)$, respectively. One can check from the axioms of a quasi-Hopf algebra in Chapter 2.4 that these maps are mutually inverse and are intertwiners. The calculations are a generalisation of those in the proof of Example 9.3.15. The result generalises Proposition 9.3.11 to the setting of internal hom. The resulting maps (9.22)–(9.24), and (9.25) and (9.26) in the quasitriangular case are likewise modified by ϕ. They can be computed by tracing through the proof of Proposition 9.3.13 in just the same way as above.

As usual, we have the dual theory too, for internal hom in the category of comodules of a Hopf algebra. We have $\underline{\mathrm{Hom}}(V, W) = \mathrm{Lin}(V, W)$ as an object in the category of comodules with coaction β defined by

$$\beta(f)(v) = \sum f(v^{(\bar{1})})^{(\bar{1})} \otimes f(v^{(\bar{1})})^{(\bar{2})} S v^{(\bar{2})},$$

(9.29)

for all $f \in \underline{\mathrm{Hom}}(V, W)$, $v \in V$, where the coactions on V, W are denoted in our usual summation notation. The maps (9.21)–(9.24) are as usual for vector spaces, while, in the dual quasitriangular case, a similar proof to that of Corollary 9.3.16 gives (9.25) and (9.26) as

$$i(v)(f) = \text{ev} \circ \Psi(v \otimes f) = \sum f(v^{(\bar{1})})^{(\bar{1})} u(v^{(\bar{2})}{}_{(1)}) \mathcal{R}(v^{(\bar{2})}{}_{(2)} \otimes f(v^{(\bar{1})})^{(\bar{2})}),$$

$$j(f \otimes g)(v \otimes x) = \sum f(v^{(\bar{1})}) \otimes g(x^{(\bar{1})})^{(\bar{1})} \mathcal{R}(Sv^{(\bar{2})} \otimes g(x^{(\bar{1})})^{(\bar{2})} Sx^{(\bar{2})}).$$

In the dual quasi-Hopf case we have to insert a nontrivial ϕ as well.

We conclude with one application of this notion of internal hom related to the following elementary lemma.

Lemma 9.3.17 *In the category of left modules of a Hopf algebra H, we have*

$$\text{Mor}(V, W) = \underline{\text{Hom}}(V, W)^H,$$

the invariant subspace under the action of H. Moreover, $\underline{\text{Hom}}(V, V)$ is an H-module algebra and $\text{Mor}(V, V)$ is its fixed point subalgebra.

Proof: In $_H\mathcal{M}$ we have seen that $\text{Mor}(V, W)$ consists of those linear maps $V \to W$ that commute with the action of H. If $\phi \in \underline{\text{Hom}}(V, W)$ is a fixed point under H, then $h \triangleright (\phi(v)) = h_{(1)} \triangleright (\phi((Sh_{(2)})h_{(3)} \triangleright v)) = (h_{(1)} \triangleright \phi)(h_{(2)} \triangleright v) = \epsilon(h_{(1)})\phi(h_{(2)} \triangleright v) = \phi(h \triangleright v)$ so ϕ is an intertwiner. We used the Hopf algebra axioms. Conversely, if ϕ is an intertwiner, then $(h \triangleright \phi)(v) = h_{(1)} \triangleright (\phi(Sh_{(2)} \triangleright v)) = h_{(1)} Sh_{(2)} \triangleright (\phi(v)) = \epsilon(h)\phi(v)$, i.e. ϕ is a fixed point under H. For the second part, it is obvious that $\text{Mor}(V, V)$ and $\underline{\text{Hom}}(V, V)$ are algebras by composition. The latter is covariant under the action of H just because $\circ$ is an intertwiner, as we know from our theory above. ∎

For example, let V be a given representation of H, and consider maps $V^{\otimes N} \to V^{\otimes N}$ which commute with the action of H, i.e. self-intertwiners or endomorphisms from the tensor product representation. We identify this as the fixed point subalgebra

$$\text{Mor}(V^{\otimes N}, V^{\otimes N}) = \text{Lin}_H(V^{\otimes N}, V^{\otimes N}) = \underline{\text{Hom}}(V^{\otimes N}, V^{\otimes N})^H.$$

We can also turn things around and consider equally well that this endomorphism algebra is given abstractly as an algebra acting on $V^{\otimes N}$ and that H commutes with it, i.e.

$$H \to \text{Lin}_{\text{Mor}(V^{\otimes N}, V^{\otimes N})}(V^{\otimes N}, V^{\otimes N}).$$

This is a generalisation of the phenomenon known in group theory as *Schur–Weyl duality*. For nice cases (where the category of representations is generated by V), one can expect to be able to reconstruct H entirely as the set of operators that commute with all these endomorphism algebras for all N. This will be clear from the general reconstruction theorems in the next section. It is also familiar in an infinite-dimensional context: for any von Neumann algebra acting on a Hilbert space, the double commutant (the operators that commute with the operators that commute with the von Neumann algebra) can be identified with the von Neumann algebra itself. Our present considerations link up with this latter point of view on taking a suitable limit $N \to \infty$.

On the other hand (when H is quasitriangular), we know very well many examples of elements of these endomorphism algebras, namely the morphisms

$$\psi_i = \mathrm{id} \otimes \cdots \otimes \Psi_{V,V} \otimes \cdots \otimes \mathrm{id},$$

which braid V in the ith position of the tensor product with V in the $i+1$ position. Again, in nice cases these generate the entire endomorphism algebra in question and allow us to compute it rather easily.

For example, for $U_q(sl_2)$ and its spin-$\frac{1}{2}$ representation we have that $\mathrm{Mor}(V^{\otimes N}, V^{\otimes N})$ is the algebra generated by 1 and $N-1$ indeterminates ψ_i modulo the relations

$$\psi_i \psi_{i+1} \psi_i = \psi_{i+1} \psi_i \psi_{i+1}, \quad \psi_i \psi_j = \psi_j \psi_i, \quad \forall |i-j| > 1,$$

$$\psi_i^2 = (q - q^{-1})\psi_i + 1.$$

The first two are the assertion that Ψ generates an action of the Artin braid group, as we have explained in Section 9.2, while the additional relation holds when we look at the spin-$\frac{1}{2}$ representation and its tensor powers. The same endomorphism algebra works for $U_q(sl_n)$ and its fundamental representation. This particular endomorphism algebra is the standard *q-Hecke algebra* known in number theory in the context of q-special functions. The limit $q \to 1$ is the usual permutation group. On the other hand, the quantum group context gives a point of view on such algebras that works quite generally. The endomorphism or 'generalised Hecke' algebra for $U_q(so(n))$ is the Birman–Murakami–Wenzl algebra.

9.4 Reconstruction theorems

So far, we have been formulating the abstract properties of the representations of a Hopf algebra or quasi-Hopf algebra. We have seen that the representations have a tensor product $\otimes$, duals and, in the quasitriangular case, a braiding. In this section, we prove the converse results: every collection of objects which can be strictly identified with vector spaces in a certain sense is equivalent to the representations of some Hopf algebra, which we reconstruct. If the identification is somewhat weaker regarding associativity of $\otimes$, then we get a quasi-Hopf algebra instead. Once this principle is understood, we can just as easily invent other weaker concepts of Hopf algebra tailored to have particular properties for their category of representations. We then prove in the following subsection an even more general diagrammatic reconstruction theorem, where the objects in our collection need not be identified with vector spaces at all and where the reconstructed object is a diagrammatic or braided-Hopf algebra. These will then be studied in detail in the next chapter. This and the quasi-Hopf algebra reconstruction theorem are due to the author, while the theorem for ordinary bialgebras has a classical origin in algebraic geometry.

9.4.1   *Reconstruction in vector spaces*

The idea behind the reconstruction theorems is to build some kind of Hopf algebra of functions on our collection or category of objects. Indeed, let $\mathcal{C}$ be a category and let $F : \mathcal{C} \to \mathrm{Vec}$ be a functor to the category of vector spaces. We have already seen in the discussion preceding Example 9.1.1 how to regard $\mathrm{Nat}(F, F)$ as an 'algebra of flat sections' or 'algebra of covariantly constant functions' on the category. Thus, $h \in \mathrm{Nat}(F, F)$ means a family of maps $\{h_X \in \mathrm{Lin}(F(X), F(X)); \ X \in \mathcal{C}\}$ which are functorial under any morphisms $\phi : X \to Y$ in the sense $h_Y \circ F(\phi) = F(\phi) \circ h_X$. Given two such 'functions' h, g, we define

$$(hg)_X = h_X \circ g_X. \tag{9.30}$$

It is clear that this family of maps $\{(hg)_X\}$ is also functorial, since h, g are. So we have an associative algebra. We also have an identity element η, given by $\eta_X = \mathrm{id}$. Our algebra acts on each vector space $F(X) \ni v$ by $h \triangleright v = h_X(v)$.

The new ingredient is to suppose now that $\mathcal{C}$ is a monoidal category (so that we have a tensor product) and that F is monoidal in the sense that it maps the associativity of $\otimes$ in $\mathcal{C}$ over to the usual vector space associativity. The precise definition of a monoidal functor in Section 9.1 is that there are functorial isomorphisms $c_{X,Y} : F(X) \otimes F(Y) \cong F(X \otimes Y)$ obeying the condition in Fig. 9.2. In this case, we show now that $\mathrm{Nat}(F, F)$ has a coproduct and counit

$$(\Delta h)_{X,Y} = c_{X,Y}^{-1} \circ h_{X \otimes Y} \circ c_{X,Y}, \quad \epsilon(h) = h_{\underline{1}}, \tag{9.31}$$

making it into a 'bialgebra of covariantly constant functions'. We have a bialgebra over a field k if everything is k-linear. Otherwise we have a bimonoid, i.e. a monoid equipped with a compatible comonoid structure.

To explain this coproduct formula, we have to say what we mean by $\mathrm{Nat}(F, F) \otimes \mathrm{Nat}(F, F)$. A little thought shows that, at least if everything is k-linear and the number of objects in our category is finite, we can identify this with $\mathrm{Nat}(F^2, F^2)$, where $F^2 : \mathcal{C} \times \mathcal{C} \to \mathrm{Vec}$ is defined by $F^2(X, Y) = F(X) \otimes F(Y)$. In other words, it consists of 'covariantly constant functions in two variables' on the category with values in vector space endomorphisms. Similarly for functions in three variables, etc. We proceed informally on this basis before giving a more formal proof.

Informally then, Δh means for us a function in two variables constructed as shown. It is functorial or 'covariantly constant' because c, c^{-1} and h are. We have coassociativity of Δ since

$$((\Delta \otimes \mathrm{id}) \circ \Delta h)_{X,Y,Z} = c_{X,Y}^{-1} \circ (\Delta h)_{X \otimes Y, Z} \circ c_{X,Y}$$

$$= c_{X,Y}^{-1} \circ c_{X \otimes Y, Z}^{-1} \circ h_{(X \otimes Y) \otimes Z} \circ c_{X \otimes Y, Z} \circ c_{X,Y},$$

$$((\mathrm{id} \otimes \Delta) \circ \Delta h)_{X,Y,Z} = c_{Y,Z}^{-1} \circ (\Delta h)_{X, Y \otimes Z} \circ c_{Y,Z}$$

$$= c_{Y,Z}^{-1} \circ c_{X, Y \otimes Z}^{-1} \circ h_{X \otimes (Y \otimes Z)} \circ c_{X, Y \otimes Z} \circ c_{Y,Z}.$$

Now, $h_{X \otimes (Y \otimes Z)} \circ F(\Phi_{X,Y,Z}) = F(\Phi_{X,Y,Z}) \circ h_{(X \otimes Y) \otimes Z}$ by functoriality of h under the morphism Φ. Hence, the above expressions are equal (up to the usual vector space associativity) just when the condition in Fig. 9.2 holds. It is also clear that ϵ is a counit for Δ and that

$$\Delta(hg)_{X,Y} = c_{X,Y}^{-1} \circ h_{X \otimes Y} \circ g_{X \otimes Y} \circ c_{X,Y} = (\Delta h)_{X,Y} \circ (\Delta g)_{X,Y},$$

as required for a bialgebra.

Similarly, if $\mathcal{C}$ is braided with braiding Ψ as in Section 9.2, we define $\mathcal{R}$ as a function in two variables by

$$\mathcal{R}_{X,Y} = \tau_{F(X),F(Y)}^{-1} \circ c_{Y,X}^{-1} \circ F(\Psi_{X,Y}) \circ c_{X,Y}, \qquad (9.32)$$

where τ is the usual permutation or transposition map for vector spaces. It is covariantly constant by functoriality of c and the image under F of the functoriality of Ψ. We use these definitions to check half of (2.1) for a quasitriangular structure as

$$((\Delta \otimes \mathrm{id})\mathcal{R})_{X,Y,Z} = c_{X,Y}^{-1} \circ \tau_{F(X \otimes Y),F(Z)}^{-1} \circ c_{Z,X \otimes Y}^{-1}$$

$$\circ F(\Psi_{X \otimes Y, Z}) \circ c_{X \otimes Y, Z} \circ c_{X,Y}$$

$$(\mathcal{R}_{13} \mathcal{R}_{23})_{X,Y,Z} = \tau_{F(X),F(Z)}^{-1} \circ c_{Z,X}^{-1} \circ F(\Psi_{X,Z}) \circ c_{X,Z}$$

$$\circ \tau_{F(Y),F(Z)}^{-1} \circ c_{Z,Y}^{-1} \circ F(\Psi_{Y,Z}) \circ c_{Y,Z},$$

which are equal using that F is monoidal and the equation which is the image under F of half of the hexagon identities (9.4) in Section 9.2. Similarly for the other half. The axiom (2.2) comes out more immediately from the definition of $\mathcal{R}_{X,Y}$ and functoriality of h in the form $F(\Psi_{X,Y}) \circ h_{X \otimes Y} = h_{Y \otimes X} \circ F(\Psi_{X,Y})$.

Finally, if $\mathcal{C}$ is rigid in the sense of Section 9.3 and F is a monoidal functor, then it is immediate that

$$F(X)^{*\prime} = F(X^*), \quad \mathrm{ev}'_{F(X)} = F(\mathrm{ev}_X) \circ c_{X^*,X}, \quad \mathrm{coev}'_{F(X)} = c^{-1}_{X,X^*} \circ F(\mathrm{coev}_X)$$

are a left dual for $F(X)$. Hence, according to the uniqueness of duals up to isomorphism (as explained below Definition 9.3.1), we have induced isomorphisms

$$d_X : F(X^*) \to F(X)^*, \quad d_X = (F(\mathrm{ev}_X) \circ c_{X^*,X} \otimes \mathrm{id}) \circ (\mathrm{id} \otimes \mathrm{coev}_{F(X)})$$

between this and the usual dual. Here, the functor F maps to Vec but the same applies if the target is another monoidal category. In this setting, we have an antipode defined by $((Sh)_X)^* = d_X \circ h_{X^*} \circ d_X^{-1}$, i.e.

$$\begin{aligned}
(Sh)_X &= (\mathrm{id} \otimes \mathrm{ev}_{F(X)}) \circ d_X \circ h_{X^*} \circ d_X^{-1} \circ (\mathrm{coev}_{F(X)} \otimes \mathrm{id}) \\
&= (\mathrm{id} \otimes F(\mathrm{ev}_X) \circ c_{X^*,X}) \circ h_{X^*} \circ (c^{-1}_{X,X^*} \circ F(\mathrm{coev}_X) \otimes \mathrm{id}), \quad (9.33)
\end{aligned}$$

as a new element of $\mathrm{Nat}(F, F)$. The proof that this obeys the antipode axioms is hard to write down (it is best done diagrammatically), but is otherwise straightforward. We first compute

$$\begin{aligned}
((S \otimes \mathrm{id}) \circ \Delta h)_{X,Y} &= (\mathrm{id} \otimes F(\mathrm{ev}_X) \circ_{X^*,X}) \circ c^{-1}_{X^*,Y} \circ h_{X^* \otimes Y} \\
&\quad \circ c_{X^*,Y} \circ (c^{-1}_{X,X^*} \circ F(\mathrm{coev}_X) \otimes \mathrm{id})
\end{aligned}$$

as a map $F(X) \otimes F(Y) \to F(X) \otimes F(Y)$. We then make the product by feeding the output of the $F(Y)$ endomorphism into the input of the $F(X)$ part, setting $Y = X$ and obtaining a single endomorphism $F(X) \to F(X)$. This is the usual composition $\mathrm{End}(F(X) \otimes F(X)) = \mathrm{End}(F(X)) \otimes \mathrm{End}(F(X)) \to \mathrm{End}(F(X))$. We are then in a position to use functoriality of h in the form $F(\mathrm{ev}_X) \circ h_{X^* \otimes X} = h_1 \circ F(\mathrm{ev}_X)$ to collapse the result to $(\cdot (S \otimes \mathrm{id}) \circ \Delta h)_X = h_1 \mathrm{id}$ as required. Similarly for the other half of the antipode axioms, using this time functoriality under coev_X.

These calculations are the content of this section, except that we want to say them more formally and generalise them. Note also that the construction here is the same in spirit as the construction of the Hopf algebra $k(\hat{G})$ of functions on a character group $\hat{G}$, with the role of the latter played by $(\mathcal{C}, \otimes)$. One can think of functions on characters as corresponding to covariantly constant functions defined on the collection of general representations. Now recall that we saw in Corollary 1.5.6 on Fourier transforms that $k(\hat{G}) \cong kG$. This is the Fourier theorem. So we should think of our 'function Hopf algebra' $\mathrm{Nat}(F, F)$ as isomorphic to H if $\mathcal{C}$ was given originally as the category of representations of H. Note that, in practice, one still has to elucidate such an isomorphism since our original Hopf algebra will typically have its own description (such as $U_q(sl_2)$ with its usual generators), and we will want to recognise it in terms of these.

We now proceed with the more formal treatment. As with internal hom in the last section, the key idea is that functors to the category Set are generally representable. Thus, we consider a category $\mathcal{C}$ and functors $F, V \otimes F : \mathcal{C} \to \text{Vec}$, where $(V \otimes F)(X) = V \otimes F(X)$, and we assume that the functor $\text{Vec} \to \text{Set}$ sending a vector space V to $\text{Nat}(V \otimes F, F)$ is representable. This just means that there is some vector space H such that we have functorial isomorphisms

$$\theta_V : \text{Lin}(V, H) \cong \text{Nat}(V \otimes F, F). \tag{9.34}$$

We have, in particular, a collection of linear maps $\{\alpha_X : H \otimes F(X) \to F(X)\}$ as the natural transformation $\theta_H(\text{id})$ in $\text{Nat}(H \otimes F, F)$. On the other hand, $\alpha_X \circ (\text{id} \otimes \alpha_X) : H \otimes H \otimes F(X) \to F(X)$ is then a natural transformation in $\text{Nat}(H \otimes H \otimes F, F)$ and hence corresponds to a map $\cdot : H \otimes H \to H$ under $\theta^{-1}_{H \otimes H}$. It is easy to see that this is associative due to associativity of composition of natural transformations, and that it is cooked up in such a way that each α is an action of H with this algebra structure. The unit element $k \to H$ corresponds under θ_k to the identity natural transformation in $\text{Nat}(F, F)$. This is the formal version of the algebra (9.30). If we begin with the category $_A\mathcal{M}$ of representations of an algebra and take F as the forgetful functor, then we can take as representing object $H = A$, recovering its product correctly. This is the content of Example 9.1.1.

We also assume that the functors sending V to $\text{Nat}(V \otimes F^n, F^n)$ are similarly representable, namely by $H^{\otimes n}$. Here

$$\theta^n_V : \text{Lin}(V, H^{\otimes n}) \cong \text{Nat}(V \otimes F^n, F^n) \tag{9.35}$$

are given by

$$\theta^n_V(\phi)_{X_1, X_2, \ldots, X_n}(v \otimes v_1 \otimes \cdots \otimes v_n) = \sum \phi(v)^{(1)} \triangleright v_1 \otimes \cdots \otimes \phi(v)^{(n)} \triangleright v_n,$$

where $\phi(v) \in H^{\otimes n}$ is given in an explicit summation notation, $v_i \in F(X_i)$, and $\triangleright$ denotes the relevant action α. This is automatic if all our vector spaces here are finite dimensional.

Theorem 9.4.1 *Let $\mathcal{C}$ be a monoidal category and let F be a monoidal functor $F : \mathcal{C} \to \text{Vec}$ obeying our representability conditions. Then $\Delta : H \to H \otimes H$, defined as the inverse image under θ^2_H of the natural transformation $c^{-1}_{X,Y} \circ \alpha_{X \otimes Y} \circ c_{X,Y} : H \otimes F(X) \otimes F(Y) \to F(X) \otimes F(Y)$, makes H into a bialgebra.*

Proof: The strategy is to compute the natural transformation corresponding to both sides by (9.35) of any axiom we want to check. If they coincide, then the maps on the two sides coincide. Thus,

$$\theta_H^3((\Delta \otimes \mathrm{id}) \circ \Delta)_{X,Y,Z} = ((\alpha_X \otimes \alpha_Y) \otimes \alpha_Z) \circ (\Delta \otimes \mathrm{id}) \circ \Delta$$
$$= (c_{X,Y}^{-1} \circ \alpha_{X \otimes Y} \circ c_{X,Y} \otimes \alpha_Z) \circ \Delta = c^{-1} \circ \alpha_{(X \otimes Y) \otimes Z} \circ c^2$$

from the definition of Δ. Similarly for $\theta_H^3((\mathrm{id} \otimes \Delta) \circ \Delta)_{X,Y,Z} = c^{-2} \circ \alpha_{X \otimes (Y \otimes Z)} \circ c^2$. These natural transformations coincide by F monoidal and functoriality of α under $\Phi_{X,Y,Z}$ in just the same way as the direct proof for (9.31).

Likewise, for the homomorphism property of Δ, we compute

$$\theta_{H \otimes H}^2(\Delta \circ \cdot)_{X,Y} = c^{-1} \circ \theta_H(\cdot)_{X \otimes Y} \circ c$$
$$= c^{-1} \circ \alpha_{X \otimes Y} \circ c \circ (\mathrm{id} \otimes c^{-1} \circ \alpha_{X \otimes Y} \circ c)$$
$$= (\alpha_X \otimes \alpha_Y \otimes \alpha_X \otimes \alpha_Y) \circ (\Delta \otimes \Delta),$$

where the first equality recognises $(\alpha_X \otimes \alpha_Y) \circ \Delta = \theta_H^2(\Delta)_{X,Y} = c^{-1} \circ \alpha_{X \otimes Y} \circ c$ leaving $\theta_H(\cdot)$ evaluated there. We use this definition of Δ again for the third equality, where the tensor products refer to position in $H \otimes H \otimes H \otimes H$ being acted upon (the actions on $F(X)$, etc. are still being composed). The necessary transposition maps τ are suppressed. We can then recognise the result as $\theta_{H \otimes H}^2((\cdot \otimes \cdot) \circ \Delta_{H \otimes H})_{X,Y}$. ∎

The definition of Δ in this theorem is such that $c_{X,Y} : F(X) \otimes F(Y) \to F(X \otimes Y)$ becomes an isomorphism of H-modules. So the map $\mathcal{C} \to {}_H\mathcal{M}$ is a monoidal functor, and, in fact, our reconstructed H is the universal bialgebra with this property. Thus, if H' is some other bialgebra which acts on all the vector spaces $F(X)$ in this way, then there is a unique bialgebra homomorphism $H' \to H$ such that these actions are the pullback of our actions α_X of H. Just regard any given collection of actions $\{H' \otimes F(X) \to F(X)\}$ as a natural transformation in $\mathrm{Nat}(H' \otimes F, F)$ and use (9.35) to construct this map. In categorical language, one says that H is universal with the property that $F : \mathcal{C} \to \mathrm{Vec}$ factors through ${}_H\mathcal{M}$ via the forgetful functor.

Proposition 9.4.2 *If $\mathcal{C}$ in the preceding theorem is braided, then $\mathcal{R} \in H \otimes H$, defined as the inverse image under θ_k^2 of the natural transformation $\mathcal{R}_{X,Y}$ in (9.32), makes H a quasitriangular bialgebra in the sense of Chapter 2.1.*

Proof: We use the same technique as in the proof of the preceding theorem. Thus, $\theta_k^3((\Delta \otimes \mathrm{id})\mathcal{R})_{X,Y,Z} = c^{-1} \circ \theta_k^2(\mathcal{R})_{X \otimes Y,Z} \circ c$ and

$$\theta_k^3(\mathcal{R}_{13}\mathcal{R}_{23})_{X,Y,Z} = (\alpha_X \otimes \alpha_Z \otimes \alpha_Y \otimes \alpha_Z)(\mathcal{R} \otimes \mathcal{R}) = \theta_k^2(\mathcal{R})_{X,Z} \circ \theta_k^2(\mathcal{R})_{Y,Z}$$

from the definitions. These are equal for $\theta_k^2(\mathcal{R})$ as stated, in view of the image under F of one of the hexagon identities for Ψ. Similarly for the other half of (2.1). For (2.2) we have

$$\theta_H^2(\mathcal{R}\cdot\Delta)_{X,Y} = (\alpha_X \otimes \alpha_Y \otimes \alpha_X \otimes \alpha_Y)(\mathcal{R} \otimes \Delta) = \theta_k^2(\mathcal{R})_{X,Y} \circ c^{-1} \circ \alpha_{X \otimes Y} \circ c.$$

The calculation for $\theta^2(\Delta^{op}(\ ) \cdot \mathcal{R})_{X,Y} = c^{-1} \circ \alpha_{Y \otimes X} \circ c \circ \theta^2(\mathcal{R})_{X,Y}$ is similar. Our definition of $\theta_k^2(\mathcal{R})$ solves this after using the image under F of functoriality of Ψ. We have suppressed the usual transposition of vector spaces in this proof; they were written more explicitly in (9.32). ∎

Proposition 9.4.3 *If C in Theorem 9.4.1 is rigid, then the map $S : H \to H$, defined as the inverse image under θ_H of the natural transformation $(\mathrm{id} \otimes \mathrm{ev}_{F(X)}) \circ d_X \circ \alpha_{X^*} \circ d_X^{-1} \circ \tau_{H,F(X)}^{-1} \circ (\mathrm{coev}_{F(X)} \otimes \mathrm{id}) : H \otimes F(X) \to F(X)$, makes H into a Hopf algebra.*

Proof: This is best done by conventional diagram-filling, and is shown in Fig. 9.11. The proofs of Theorem 9.4.1 and Proposition 9.4.2 can also be written out more fully in this way. The anticlockwise path from the top is the definition of $\theta_H(\cdot \circ (S \otimes \mathrm{id}) \circ \Delta)_X$. We recognise $\theta_H(\cdot)$ and write it as $\alpha_X \circ (\mathrm{id} \otimes \alpha_X)$ as usual in the cell to the lower left. The cell above this commutes one of these α_X past S. The central lower square is the definition of $\theta_H(S)_X$. The upper central cell commutes coev_X past $\alpha_X \circ \Delta$ (they act on different spaces). Finally, the right hand cell is the definition of $\theta_H^2(\Delta)_{X^*,X}$. The clockwise path, on the other hand, simplifies further using functoriality of α under ev_X in the form $\mathrm{ev}_X \circ \alpha_{X^* \otimes X} = \alpha_{\underline{1}} \circ \mathrm{ev}_X$. Then ev_X combines with coev_X to the identity using one of the rigidity axioms (9.14). But $\alpha_{\underline{1}}\mathrm{id} = \theta_H(\eta \circ \epsilon)$ as required. The proof for the other antipode axiom is strictly analogous. ∎

Exercise 9.4.4 *Let H be a quasitriangular bialgebra, let $C = {}_H\mathcal{M}$ be its category of representations and let F be the forgetful functor as in Example 9.1.1. Then the reconstructed quasitriangular bialgebra $\mathrm{Nat}(F, F)$ is isomorphic to H itself.*

Proof: We have done most of the work already in Example 9.1.1, where we saw that $\mathrm{Nat}(F, F)$ can be identified with $H = \mathrm{Lin}(k, H)$. In the same way, one has now that our representing object in (9.35) can again be taken to be our original H as a vector space. Given a natural transformation $\theta \in \mathrm{Nat}(V \otimes F, F)$, we just evaluate it on the unit element in the left regular representation as before, giving a linear map $\theta_H((\ )\otimes 1) : V \otimes H$. In the other direction, we send $\phi : V \to H$ to the natural transformation

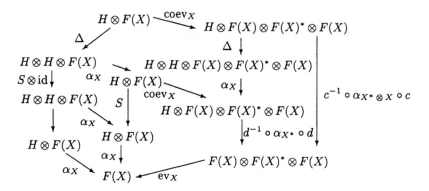

Fig. 9.11. Proof of reconstruction of antipode in Proposition 9.4.3.

$\theta(\phi)_X(v \otimes v_1) = \phi(v) \triangleright v_1$ for all $v_1 \in F(X)$. Here X is an object in $\mathcal{C}$, i.e. it is given as an H-module, and $F(X)$ is its underlying vector space. In our case, F and c are identify maps. We see that $\alpha_X = \triangleright$ as expected, and hence that the reconstructed product is our original one. This was already done (more directly) in Example 9.1.1. The reconstructed coproduct is such that $(\alpha_X \otimes \alpha_Y)(\Delta h)(v \otimes w) = \alpha_{X \otimes Y}(h)(v \otimes w) = h \triangleright (v \otimes w)$ for $v \in X$ and $w \in Y$. But the tensor product representation is defined in Example 9.1.3 using the original coproduct of H, so we see that our coproducts coincide. Finally, the reconstructed quasitriangular structure is defined such that $(\alpha_X \otimes \alpha_Y)(\mathcal{R})(v \otimes w) = \tau^{-1} \circ \Psi_{X,Y}(v \otimes w)$, which coincides with the action of our original $\mathcal{R}$ when Ψ is computed from Theorem 9.2.4. This exercise therefore provides a check on our work. ∎

Likewise, if H is a Hopf algebra and we take the category of finite-dimensional vector spaces, then the reconstruction recovers the antipode. Also, if we take instead a nontrivial $c_{X,Y}$ in this exercise, then we obtain by reconstruction not H but H_χ twisted by a 2-cocycle $\chi \in H \otimes H$ according to Theorem 2.3.4. Here, χ^{-1} just corresponds to the natural transformation c^{-1} viewed as an element of $\mathrm{Nat}(F^2, F^2)$ in the trivial way.

We can also include in this programme more general objects such as quasi-Hopf algebras. These are obtained if we assume that F is *multiplicative* in the sense that there are functorial isomorphisms $c_{X,Y}$: $F(X) \otimes F(Y)$ as before but dropping now the condition in Fig. 9.2. We can see this at our informal 'function algebra' level since now the proof of coassociativity, etc., breaks down. The deficit is conjugation by a natural transformation or 'covariantly constant function of three variables'

$$\phi_{X,Y,Z} = c_{Y,Z}^{-1} \circ c_{X,Y \otimes Z}^{-1} \circ F(\Phi_{X,Y,Z}) \circ c_{X \otimes Y,Z} \circ c_{X,Y}. \tag{9.36}$$

It allows us to recover coassociativity in the weak sense of a quasi-Hopf

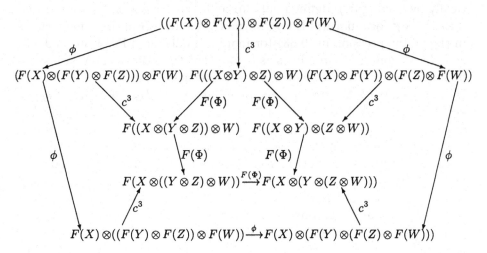

Fig. 9.12. Proof of 3-cocycle condition in the reconstruction of a quasi-Hopf algebra.

algebra. From the more formal point of view, this natural transformation has, as the inverse image under θ_k^3, an element $\phi \in H \otimes H \otimes H$. We still have to show from either point of view that this ϕ obeys the 3-cocycle condition from Chapter 2.3. This is shown in Fig. 9.12. Commutativity of the inner pentagon is Mac Lane's pentagon condition for Φ, as we have explained in Section 9.1. Each of the five cells commutes by the definition of ϕ. As a result, going round the outer pentagon clockwise or anticlockwise gives the same result, which is our desired 3-cocycle condition after using the definition of Δ.

In this weaker setting, we obtain a quasitriangular quasibialgebra when $\mathcal{C}$ is braided, using the same formula for $\mathcal{R}_{X,Y}$. The proof can be done in the same way as our verification of the 3-cocycle condition for ϕ. Finally, if the category $\mathcal{C}$ is rigid, then we can reconstruct S, α, β, forming a quasi-Hopf algebra in the sense of Chapter 2.4. Because the functor is not monoidal, we no longer have that $F(\mathrm{ev}_X) \circ c_{X^*,X}$, etc., make $F(X^*)$ a left dual of $F(X)$. Instead we choose invertible $\alpha, \beta \in \mathrm{Nat}(F, F)$ such that

$$d_X \equiv (F(\mathrm{ev}_X) \circ c_{X^*,X} \otimes \mathrm{id}) \circ \alpha_X^{-1} \circ (\mathrm{id} \otimes \mathrm{coev}_{F(X)}),$$
$$d_X^{-1} \equiv (\mathrm{ev}_{F(X)} \otimes \mathrm{id}) \circ \beta_X^{-1} \circ (\mathrm{id} \otimes c_{X,X^*}^{-1} \circ F(\mathrm{coev}_X))$$

$$(9.37)$$

are mutually inverse. For example, we can take

$$\alpha_X = \mathrm{id}, \quad \beta_X = (\mathrm{id} \otimes F(\mathrm{ev}_X) \circ c_{X^*,X}) \circ (c_{X,X^*}^{-1} \circ F(\mathrm{coev}_X) \otimes \mathrm{id}),$$

or we can make β trivial and α typically nontrivial: we proceed symmetrically by leaving α, β defined only up to a transformation $\alpha \to \alpha \circ U$ and $\beta \to U^{-1} \circ \beta$ for a natural isomorphism U. We take S in the same form as in the first expression in (9.33) (or more formally as in Proposition 9.4.3) and check along the same lines as before that the axioms (2.31) are satisfied. We have extra factors of α, β due to the new definition of d_X. To verify the axioms (2.32), one has to write out $(\sum \phi^{(1)} \beta \otimes S\phi^{(2)} \otimes \alpha\phi^{(3)})_{X,Y,Z}$ etc. from the definitions above and sew up the three copies of $\mathrm{End}(F(X))$ when $Z = Y = X$ by feeding the Z output into the Y input and the Y output into the X input. The result is then trivial on using the image under F of the first half of the rigidity axioms (9.14) for X. Similarly for the second of (2.32) using the second half of (9.14).

As usual, all our constructions have dual ones as well, based on comodules. This actually has some technical advantages over the above module version. Briefly, the representability assumption for this dual setting is that the functor $\mathrm{Vec} \to \mathrm{Set}$ sending V to $\mathrm{Nat}(F, F \otimes V)$ be representable, i.e. there should be functorial isomorphisms

$$\theta_V : \mathrm{Lin}(A, V) = \mathrm{Nat}(F, F \otimes V)$$

for a vector space A. In this case, we let the collection $\{\beta_X : F(X) \to F(X) \otimes A\}$ denote the natural transformation $\theta_A(\mathrm{id})$ and define $\Delta : A \to A \otimes A$ as the inverse image under $\theta_{A \otimes A}$ of the natural transformation $(\beta_X \otimes \mathrm{id}) \circ \beta_X : F(X) \to F(X) \otimes A \otimes A$. We also have a counit as the inverse image under θ_k of the identity natural transformation. This makes A into a coalgebra which coacts by β on each of the $F(X)$. We also need the higher representability

$$\theta_V^n : \mathrm{Lin}(A^{\otimes n}, V) \cong \mathrm{Nat}(F^n, F^n \otimes V),$$

$$\theta_V^n(\phi) = \sum v_1^{(\bar{1})} \otimes v_n^{(\bar{1})} \phi(v_1^{(\bar{2})} \otimes \cdots \otimes v_n^{(\bar{2})}),$$

and use it to define a product $\cdot : A \otimes A \to A$ as the inverse image under θ_A^2 of $c_{X,Y} \circ \beta_{X \otimes Y} \circ c_{X,Y}^{-1} : F(X) \otimes F(Y) \to F(X) \otimes F(Y) \otimes A$. This gives a bialgebra if F is monoidal and a dual quasibialgebra if it is merely multiplicative. If $\mathcal{C}$ is braided we have a dual quasitriangular structure $\mathcal{R}$ as the inverse image under θ_k^2 of $\mathcal{R}_{X,Y}$ from (9.32). Also, if $\mathcal{C}$ is rigid we have an antipode S as the inverse image under θ_A of

$$(\mathrm{id} \otimes \mathrm{ev}_{F(X)}) \circ d_X \circ \beta_{X^*} \circ d_X^{-1} \circ (\mathrm{coev}_{F(X)} \otimes \mathrm{id}). \tag{9.38}$$

As an example of dual reconstruction, one can take the braided category $\mathcal{C}(R)$ generated by a single R-matrix as mentioned below Example 9.2.10. The result is the dual quasitriangular bialgebra $A(R)$ from Chapter 4.1. One can also consider the rigid braided monoidal category generated in the biinvertible case as mentioned below Example 9.3.12, giving a canonical quantum group with antipode associated to R.

9.4.2 Braided reconstruction

In this subsection, we proceed to a much more drastic generalisation of the above reconstruction theorems. This generalisation (also due to the author) is powerful enough to do away with the functor F altogether. More precisely, we still keep a functor $F : \mathcal{C} \to \mathcal{V}$, but we allow $\mathcal{V}$ to be a general braided monoidal category. So we could take $\mathcal{V} = \mathcal{C}$ and $F = \mathrm{id}$ if there was no functor in the picture. This time, the objects which we reconstruct will not be usual Hopf algebras but ones living in the braided category $\mathcal{V}$. These may seem like exotic and unnecessary objects, but, as we will see in Chapter 10, that they are just what is needed in many applications in physics in the context of q-deformation.

We have already covered the theory of algebras B in monoidal categories at the end of Section 9.2 from the point of view of covariant systems. B is an object in the category and its product and unit maps are morphisms. We saw in Lemma 9.2.12 that in the braided case $B \otimes B$ also has a natural structure as an algebra in $\mathcal{V}$, namely the braided tensor product algebra. This lemma is the key to the definition of a braided-Hopf algebra or braided group and was in fact introduced (by the author) for just this purpose.

Definition 9.4.5 *A bialgebra in a braided category, or* braided bialgebra, *is an algebra B in the category (as in Definition 9.2.11) and algebra homomorphisms $\Delta : B \to B \otimes B$ and $\epsilon : B \to \underline{1}$, which form a coalgebra in the category. It is a Hopf algebra in the braided category, or, loosely speaking, a* braided group, *if there is also a morphism $S : B \to B$ obeying the usual axioms but now as morphisms in the category.*

These axioms are summarised in Fig. 9.13(a,b) in the diagrammatic notation of Section 9.2 and Section 9.3. Further axioms of ϵ, η say that these can be grafted with the coproduct or product node without changing the morphism. It turns out that all the elementary Hopf algebra theory of Chapter 1 can be developed in this diagrammatic setting, including dual braided-Hopf algebras, modules, comodules, adjoint actions, cross products, etc. For example, Fig. 9.14 shows the proof of the property

$$S \circ \cdot = \cdot \circ \Psi_{B,B} \circ (S \otimes S), \quad \Delta \circ S = (S \otimes S) \circ \Psi_{B,B} \circ \Delta, \qquad (9.39)$$

which we will need in Chapter 10 as the braided group analogue of Proposition 1.3.1 for ordinary Hopf algebras. The proof grafts on two loops involving S, knowing that they are trivial from Fig. 9.13(b). After some reorganisation, we use Fig. 9.13(a) and then part (b) again for the final result. For the second of (9.39), just turn the book upside down and read the diagrammatic proof again. It is not our goal to digress with this general theory here, limiting ourselves as we do to showing how such

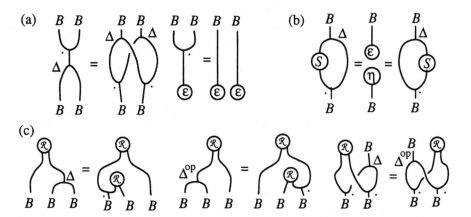

Fig. 9.13. Main axioms of a braided group showing (a) homomorphism property
for a braided coproduct, (b) antipode and (c) quasitriangular structure.

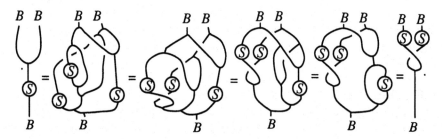

Fig. 9.14. Diagrammatic proof of braided antihomomorphism property of S.

braided-Hopf algebras or braided groups naturally arise by reconstruction.
It is historically how they were first introduced. One can also use the term
'braided group' more strictly to apply to braided-Hopf algebras which are
'braided-commutative' or 'braided-cocommutative' in some sense. We will
see examples of these too.

To make this reconstruction, we follow the same proofs as the formal ones in the preceding subsection, but we work now in a braided
category $\mathcal{V}$ in place of Vec and use the diagrammatic notation which
was explained in Section 9.2. Our assumption is that there is an object
$B \in \mathcal{V}$ such that $\mathrm{Nat}(V \otimes F, F) \cong \mathrm{Mor}(V, B)$ by functorial bijections. Let
$\{\alpha_X : B \otimes F(X) \to F(X); \ X \in \mathcal{C}\}$ be the natural transformation corresponding to the identity morphism $B \to B$. Then using α and the
braiding we get induced maps

$$\theta_V^n : \mathrm{Mor}(V, B^{\otimes n}) \to \mathrm{Nat}(V \otimes F^n, F^n), \qquad (9.40)$$

and we assume that these, likewise, are bijections. This is the *rep-*

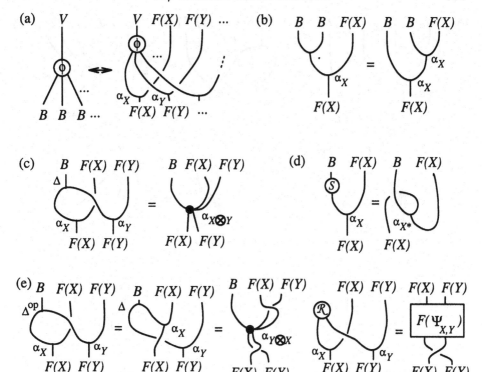

Fig. 9.15. Braided reconstruction theorem showing (a) the identification of morphisms, hence defining (b) product, (c) coproduct, (d) antipode and (e) opposite coproduct and quasitriangular structure.

resentability assumption for modules. The bijections (9.40) are shown in Fig. 9.15(a). We then show in part (b) of the figure the product $\cdot : B \otimes B \to B$ defined in just the same way as we did above, namely as uniquely determined by the requirement that the morphisms α_X are actions. Likewise for the unit morphism $\underline{1} \to B$ as the inverse image under $\theta_{\underline{1}}$ of the identity natural transformation. The proofs are just as easy as before, and do not yet involve the monoidal structure of $\mathcal{C}$ or the braiding in $\mathcal{V}$. We always obtain from a functor $F : \mathcal{C} \to \mathcal{V}$ an algebra B living in $\mathcal{V}$.

Theorem 9.4.6 *If $\mathcal{C}$ is monoidal and $F : \mathcal{C} \to \mathcal{V}$ is a monoidal functor, then B as above becomes a bialgebra in the braided category $\mathcal{V}$ with $\Delta : B \to B \otimes B$ as shown in Fig. 9.15(c), and $\epsilon : B \to \underline{1}$, defined by $\epsilon = \alpha_{\underline{1}}$, makes B into a bialgebra in the braided category $\mathcal{V}$.*

Proof: We define Δ as the inverse image under θ_B^2 of a natural transformation built as before from $\alpha_{X \otimes Y}$. This is shown in Fig. 9.15(c), where

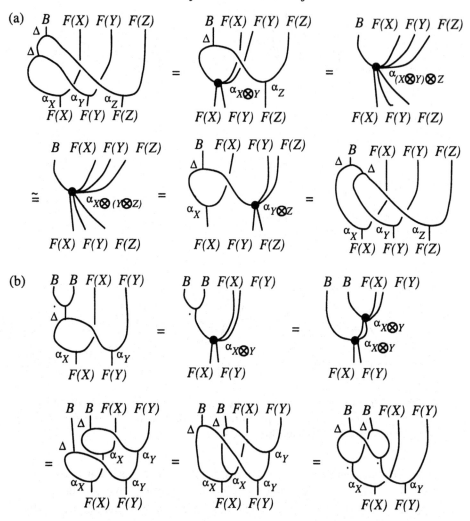

Fig. 9.16. Proof of (a) coassociativity and (b) the homomorphism property for the reconstructed braided coproduct.

we adopt the conventions that the isomorphisms $c_{X,Y}$, d_X, coming from the assumption that F is monoidal, are suppressed. Thus, the solid node marked $\alpha_{X \otimes Y}$ is α on the composite object $X \otimes Y$ but viewed via c as a morphism $B \otimes F(X) \otimes F(Y) \to F(X) \otimes F(Y)$. Similarly for α_{X^*} when we come to the antipode in the next proposition. In this way, all diagrams refer to morphisms in $\mathcal{V}$.

The proof that Δ is coassociative is shown in this notation in Fig. 9.16(a). We use the definition of Δ twice, and then again in reverse, using in the middle that F is monoidal and hence compatible with the (suppressed) associativity in the two categories. The key step is the third equality,

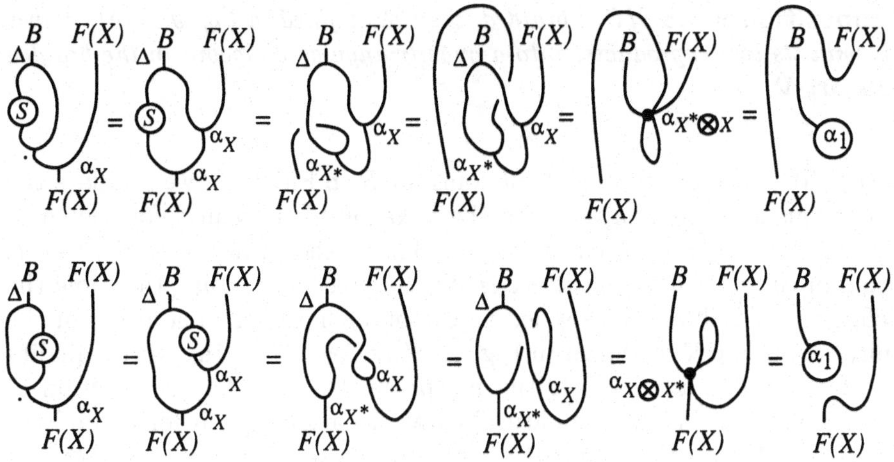

Fig. 9.17. Proof of antipode axiom for reconstructed braided antipode.

which follows from functoriality of α under the associativity morphism $(X \otimes Y) \otimes Z \to X \otimes (Y \otimes Z)$ and from the fact that F is monoidal. If F is not monoidal but merely multiplicative, one has here a quasicoassociative coproduct. Part (b) gives the proof that this Δ is a homomorphism, i.e. that it commutes with the product when $B \otimes B$ has the braided tensor product algebra structure from Lemma 9.2.12. ∎

Proposition 9.4.7 *If C is rigid then S, defined in Fig. 9.15(d), is an antipode for the coproduct Δ.*

Proof: This is shown in Fig. 9.17. The first, second and fourth equalities are the definitions of $\cdot, S, \Delta$. The fifth uses functoriality of α under the evaluation $X^* \otimes X \to \underline{1}$. The result is the natural transformation corresponding to $\eta \circ \epsilon$. Similarly for the second line using functoriality under the coevaluation morphism $\underline{1} \to X \otimes X^*$. ∎

Less useful, but also worth reconstructing, is a quasitriangular structure on B. We expect this when C is braided. The appropriate axioms have been included along with the axioms of a Hopf algebra in Fig. 9.15(d). We require a second coproduct Δ^{op}, and $\mathcal{R} : \underline{1} \to B \otimes B$, relating it by conjugation to Δ. Note that this is a slightly more general concept than the one we are familiar with because we do not insist that the second coproduct is otherwise related to Δ. In our setting, however, we do have a natural choice of 'opposite coproduct', as defined in Fig. 9.15(e), using the opposite monoidal product in C. The proof that this also makes B a bialgebra when C is monoidal is similar to that of Theorem 9.4.6.

Proposition 9.4.8 *If C is braided, then $\mathcal{R}$, defined in Fig. 9.15(e), makes B, with its two coproducts, into a quasitriangular bialgebra in the braided category $\mathcal{V}$.*

Proof: To prove the first of the axioms for $\mathcal{R}$ in Fig. 9.13, we evaluate the definitions and use F applied to the hexagon identity in C for the third equality, and then proceed in reverse. This is shown in Fig. 9.18(a). The proof of the second axiom for $\mathcal{R}$ is strictly analogous, using this time the definition of Δ^{op} and F applied to the other hexagon. The proof of the third axiom for $\mathcal{R}$ is shown in Fig. 9.18(b). We use in the third equality the functoriality of α under the morphism $\Psi_{X,Y}$. The construction of $\mathcal{R}^{-1}$ is based in the inverse natural transformation to that for $\mathcal{R}$, and the proof that this is then inverse in the convolution algebra $\underline{1} \to B \otimes B$ is straightforward using the same techniques.   ∎

   This completes our diagrammatic reconstruction theorem. It gives a kind of *automorphism braided-Hopf algebra* associated to $F : C \to \mathcal{V}$, which acts canonically on each of the $F(X)$. Moreover, the collection of all braided B-modules is itself a monoidal category, and this passage from X to $F(X)$ as a braided module is a monoidal functor. One can define $B = \mathrm{Aut}\,(C, F, \mathcal{V})$ abstractly as the universal braided-Hopf algebra in $\mathcal{V}$ with this property.
   There are many applications of this braided reconstruction theory. One of them is the process of 'superisation' or 'anyonisation', which we will encounter in the next chapter. For the present, we concentrate on the simplest case, where $F = \mathrm{id}$. We see that, to every braided monoidal category C, we have an intrinsic braided group, which we call the *automorphism braided group $U(C)$*. It acts canonically on each of the objects X in C, and is 'braided cocommutative' in the sense that $\Delta^{\mathrm{op}} = \Delta$ and $\mathcal{R} = \eta \otimes \eta$. This makes it more like a classical group algebra or enveloping algebra than a quantum group. Strictly speaking, we may need to take for $\mathcal{V}$ a suitable completion, rather than C itself, in order that our representability conditions above are satisfied.

Example 9.4.9 *Every quasitriangular Hopf algebra H has a braided group analogue $\underline{H} = U({}_H\mathcal{M})$ given by the same algebra, unit and counit as H, and*

$$\underline{\Delta} b = \sum b_{(1)} S\mathcal{R}^{(2)} \otimes \mathcal{R}^{(1)} \triangleright b_{(2)}, \quad \underline{S} b = \sum \mathcal{R}^{(2)} S(\mathcal{R}^{(1)} \triangleright b).$$

The braided coproduct $\underline{\Delta}$ is the covariantised one in Theorem 7.4.2 and is braided cocommutative in the sense

$$\sum \Psi(b_{\underline{(1)}} \otimes Q^{(1)} \triangleright b_{\underline{(2)}}) Q^{(2)} = \sum b_{\underline{(1)}} \otimes b_{\underline{(2)}},$$

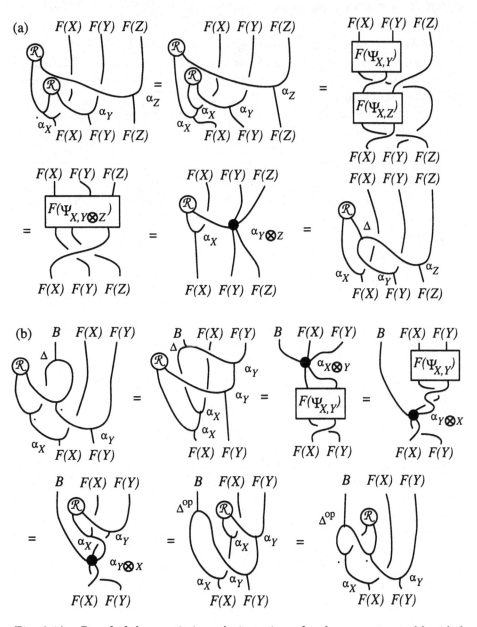

Fig. 9.18.   Proof of the quasitriangularity axioms for the reconstructed braided quasitriangular structure or universal R-matrix.

where $Q^{(2)}$ right-multiplies the output of Ψ. We call $\underline{H}$ the braided group of enveloping algebra type associated to H. It lives in the braided category of left H-modules by the quantum adjoint action in Example 1.6.3.

Proof: We use underlines to distinguish the braided-Hopf algebra structures from the structure maps of our original H. We already know from Theorem 9.2.4 that $\mathcal{C} = {}_H\mathcal{M}$ is braided, and we just apply our reconstruction theory to it with $\mathcal{C} = \mathcal{V}$ and $F = \mathrm{id}$. The first step is to show that we can take $B = H$ as an object in our category by the quantum adjoint action. We denote by B_L the vector space H with the left regular action, which we will also need. Now, given $\phi \in \mathrm{Mor}(V, B)$ (an intertwiner between an H-module V and B), we let $\theta(\phi) \in \mathrm{Nat}(V \otimes \mathrm{id}, \mathrm{id})$ be defined by $\theta(\phi)_X(v \otimes x) = \phi(v) \triangleright x$, where $\triangleright$ is the action of H on X as an object in our category of H-modules. This is just as in Exercise 9.4.4, except that, this time, we have to check that each $\theta(\phi)_X : V \otimes X \to X$ is a morphism in our category if ϕ is. This is true since

$$\theta(\phi)(h \triangleright (v \otimes x)) = \theta(\phi)(h_{(1)} \triangleright v \otimes h_{(2)} \triangleright x) = \phi(h_{(1)} \triangleright v) h_{(2)} \triangleright x$$
$$= (h_{(1)} \triangleright \phi(v)) h_{(2)} \triangleright x = h_{(1)} \phi(v) (Sh_{(2)}) h_{(3)} \triangleright x$$
$$= h\phi(v) \triangleright x = h \triangleright (\theta(\phi)(v \otimes x)),$$

where the appropriate actions are used, notably the quantum adjoint action on $\phi(v) \in B$. In the converse direction, we let $\theta \in \mathrm{Nat}(V \otimes \mathrm{id}, \mathrm{id})$ and define $\phi(v) = \theta_{B_L}(v \otimes 1)$. This is a morphism since

$$h \triangleright \phi(v) = h_{(1)} \theta_{B_L}(v \otimes 1) Sh_{(2)} = \theta_{B_L}(h_{(1)} \triangleright v \otimes h_{(2)}) Sh_{(3)}$$
$$= \theta_{B_L}(h_{(1)} \triangleright v \otimes 1) h_{(2)} Sh_{(3)} = \phi(h \triangleright v),$$

where we used the appropriate actions. The second equality uses that $\theta_{B_L} : V \otimes B_L \to B_L$ is an intertwiner. The third equality uses that θ is functorial under the morphism $B_L \to B_L$ defined by right-multiplication. One can check that these two processes are mutually inverse. Again, the proof is the same as we have seen previously, except that we must check that any maps used are morphisms in our category.

Next we begin our reconstruction. The reconstruction of the algebra does not involve the braiding, and gives our original algebra structure on H as in Exercise 9.4.4. The coproduct is characterised according to Fig. 9.15(c) by

$$\underline{b_{(1)}} \mathcal{R}^{(2)} \triangleright x \otimes (\mathcal{R}^{(1)} \triangleright \underline{b_{(2)}}) \triangleright y = \underline{b_{(1)}} \triangleright x \otimes \underline{b_{(2)}} \triangleright y, \quad \forall\, x \in X,\, y \in Y,$$

where we used the braiding in Theorem 9.2.4 and computed $h \triangleright (x \otimes y)$ in the usual way for the right hand side. Since this is true for all x, y, we conclude that

$$\underline{b_{(1)}} \mathcal{R}^{(2)} \otimes \mathcal{R}^{(1)} \triangleright \underline{b_{(2)}} = \underline{b_{(1)}} \otimes \underline{b_{(2)}}$$

for the braided coproduct $\underline{\Delta}$ in terms of the usual coproduct of H. We can also do this for the braided opposite coproduct $\underline{\Delta}^{\mathrm{op}}$ in Fig. 9.15(e). The first form gives it in terms of $\underline{\Delta}$ as

$$b_{(1)_{\mathrm{op}}} \mathcal{R}^{(2)} \otimes \mathcal{R}^{(1)} \triangleright b_{(1)_{\mathrm{op}}} = \mathcal{R}^{-(1)} b_{(2)} \otimes \mathcal{R}^{-(2)} \triangleright b_{(1)}. \tag{9.41}$$

On the other hand, we know from the general theory above that $\underline{\Delta}^{\mathrm{op}} = \underline{\Delta}$, which then comes out as the braided cocommutativity condition stated. Note that $\underline{\Delta}^{\mathrm{op}}$ in our computation does not come out as the more obvious $\Psi^{-1} \circ \underline{\Delta}$ or $\Psi \circ \underline{\Delta}$ unless we are in the triangular or unbraided case, where the quantum inverse Killing form Q is trivial. Indeed, neither of these are second bialgebra structures for B in general. Finally, the reconstructed $\underline{S}$ according to Fig. 9.15(d) is characterised by

$$\underline{S}b \triangleright x = (\mathrm{id} \otimes \mathrm{ev}_X) \, (\mathcal{R}^{(2)} \triangleright e_a \otimes (\mathcal{R}^{(1)} \triangleright b) \triangleright f^a \otimes x)$$
$$= \mathcal{R}^{(2)} \triangleright e_a \otimes f^a(S(\mathcal{R}^{(1)} \triangleright b) \triangleright x) = (\mathcal{R}^{(2)} S(\mathcal{R}^{(1)} \triangleright b)) \triangleright x,$$

where $\{e_a\}$ is a basis of X and $\{f^a\}$, a dual basis. Since this is true for all $x \in X$ and all X, we conclude the formula as stated. One can also write it, using Proposition 2.1.8, as

$$\underline{S}b = u^{-1}(S\mathcal{R}^{(2)})(Sb)\mathcal{R}^{(1)}. \tag{9.42}$$

It is easy to see that the reconstructed counit is the same, $\underline{\epsilon} = \epsilon$.

This completes our categorical proof. Given these formulae for the braided group $\underline{H}$, one can also check directly that it obeys the axioms in Fig. 9.13 for a braided-Hopf algebra. First, by our construction, all our structure maps are all morphisms, i.e. intertwiners for the action of H. This makes them very useful for making quantum group-covariant constructions. Indeed, we gave a direct proof of the covariance and coassociativity of $\underline{\Delta}$ in Theorem 7.4.2. The new ingredient now is that the usual product of H and this new coproduct fit together to form a braided-Hopf algebra; i.e. $\underline{\Delta} : \underline{H} \to \underline{H \otimes H}$ is an algebra homomorphism, where $\underline{\otimes}$ is the braided tensor product in Corollary 9.2.13. For completeness, we now give a direct proof of this crucial property. Thus,

$$(\underline{\Delta}b) \cdot (\underline{\Delta}c) = b_{(1)}(\mathcal{R}^{(2)} \triangleright c_{(1)}) \otimes (\mathcal{R}^{(1)} \triangleright b_{(2)}) c_{(2)}$$
$$= b_{(1)} \mathcal{R}^{(2)}{}_{(1)} c S \mathcal{R}^{(2)}{}_{(2)} \otimes (\mathcal{R}^{(1)} \triangleright b_{(2)}) c_{(2)}$$
$$= b_{(1)} \mathcal{R}'^{(2)} c_{(1)} S \mathcal{R}^{(2)} \otimes (\mathcal{R}^{(1)} \triangleright (\mathcal{R}'^{(1)} \triangleright b_{(2)})) c_{(2)}$$
$$= b_{(1)} c_{(1)} S \mathcal{R}^{(2)} \otimes (\mathcal{R}^{(1)} \triangleright b_{(2)}) c_{(2)}$$
$$= b_{(1)} c_{(1)} (S \mathcal{R}'^{(2)})(S \mathcal{R}^{(2)}) \otimes (\mathcal{R}^{(1)} \triangleright b_{(2)})(\mathcal{R}'^{(1)} \triangleright c_{(2)})$$
$$= b_{(1)} c_{(1)} S \mathcal{R}^{(2)} \otimes (\mathcal{R}^{(1)}{}_{(1)} \triangleright b_{(2)})(\mathcal{R}^{(1)}{}_{(2)} \triangleright c_{(2)}) = \underline{\Delta}bc,$$

using the definitions and elementary properties of the quasitriangular structure from Chapter 2. We also give a direct proof of the antipode axioms as

$$h \triangleright \underline{S}b = h_{(1)}(\underline{S}b)Sh_{(2)} = h_{(1)}u^{-1}(S\mathcal{R}^{(2)})(Sb)\mathcal{R}^{(1)}Sh_{(2)}$$
$$= u^{-1}(S^2h_{(1)})(S\mathcal{R}^{(2)})(Sb)\mathcal{R}^{(1)}Sh_{(2)}$$
$$= u^{-1}(S(\mathcal{R}^{(2)}(Sh)_{(2)}))(Sb)\mathcal{R}^{(1)}(Sh)_{(1)}$$
$$= u^{-1}(S((Sh)_{(1)}\mathcal{R}^{(2)}))(Sb)(Sh)_{(2)}\mathcal{R}^{(1)}$$
$$= u^{-1}(S\mathcal{R}^{(2)})(S^2h_{(2)})(Sb)(Sh_{(1)})\mathcal{R}^{(1)}$$
$$= u^{-1}(S\mathcal{R}^{(2)})S[h_{(1)}(Sb)Sh_{(2)}]\mathcal{R}^{(1)} = \underline{S}(h \triangleright b),$$

$$b_{(1)}\underline{S}b_{(2)} = b_{(1)}(S\mathcal{R}^{(2)})\underline{S}(\mathcal{R}^{(1)} \triangleright b_{(2)})$$
$$= b_{(1)}(S\mathcal{R}^{(2)})\mathcal{R}'^{(2)}S(\mathcal{R}'^{(1)}\mathcal{R}^{(1)} \triangleright b_{(2)})$$
$$= b_{(1)}(S\mathcal{R}^{(2)}{}_{(1)})\mathcal{R}^{(2)}{}_{(2)}S(\mathcal{R}^{(1)} \triangleright b_{(2)}) = b_{(1)}Sb_{(2)} = \epsilon(b),$$

$$(\underline{S}b_{(1)})b_{(2)} = u^{-1}(S\mathcal{R}^{(2)})(S[b_{(1)}\mathcal{R}'^{(2)}S\mathcal{R}''^{(2)}])\mathcal{R}^{(1)}\mathcal{R}''^{(1)}b_{(2)}\mathcal{R}'^{(1)}$$
$$= u^{-1}(S\mathcal{R}^{(2)})(S^2\mathcal{R}''^{(2)})(S\mathcal{R}'^{(2)})(Sb_{(1)})\mathcal{R}^{(1)}\mathcal{R}''^{(1)}b_{(2)}\mathcal{R}'^{(1)}$$
$$= u^{-1}(S^2(\mathcal{R}^{(2)}\mathcal{R}''^{(2)}))(S\mathcal{R}'^{(2)})(Sb_{(1)})(S\mathcal{R}^{(1)})\mathcal{R}''^{(1)}b_{(2)}\mathcal{R}'^{(1)}$$
$$= u^{-1}(S\mathcal{R}'^{(2)})(Sb_{(1)})b_{(2)}\mathcal{R}'^{(2)} = u^{-1}u\epsilon(b) = \epsilon(b),$$

again using the definitions above and elementary properties of $\mathcal{R}$. ∎

This example shows that the theory of quasitriangular Hopf algebras is contained in the theory of braided groups. It is also interesting because it shifts from the point of view of 'quantum' or noncocommutative objects in the usual category of vector spaces to the point of view of classical or 'cocommutative' objects, albeit in a braided category. This procedure is called *transmutation* because it turns one kind of object into another. It is also a bit like using a metric to determine local coordinates in which the metric then looks linear: we used the quasitriangular Hopf algebra to determine a braided category, and, viewing it there, it looks cocommutative.

The process of transmutation is quite general, and is one of the main reasons for being interested in reconstruction. The point is that, if we can characterise an object entirely by its category $\mathcal{C}$ of representations, then a functor from this to some other category $\mathcal{V}$ allows us to reconstruct our object there. We have in the same way that, if

$$i : H_1 \to H$$

is a Hopf algebra map from a quasitriangular Hopf algebra H_1 to H, then the latter acquires a new (transmuted) braided-Hopf algebra structure $B(H_1, H)$ in the braided category of H_1-modules. We just apply the above reconstruction theory to the functor ${}_H\mathcal{M} \to {}_{H_1}\mathcal{M}$ given by pullback of a representation along i. We take $B = H$ as a vector space acted upon by H_1 via i and the quantum adjoint action. It coincides as an algebra and has

$$\underline{\Delta} b = \sum b_{(1)} i(S\mathcal{R}_1{}^{(2)}) \otimes \mathcal{R}_1{}^{(1)} \triangleright b_{(2)}, \quad \underline{S} b = \sum i(\mathcal{R}_1{}^{(2)}) S(\mathcal{R}_1{}^{(1)} \triangleright b),$$

$$\underline{\mathcal{R}} = \sum \rho^{(1)} i(S\mathcal{R}_1{}^{(2)}) \otimes \mathcal{R}_1{}^{(1)} \triangleright \rho^{(2)}, \qquad \rho = i(\mathcal{R}_1^{-1})\mathcal{R}, \qquad (9.43)$$

$$\sum \Psi(b_{(1)_{op}} \otimes Q_1{}^{(1)} \triangleright b_{(2)_{op}}) i(Q_1{}^{(2)}) = \sum b_{(1)} \otimes b_{(2)}.$$

The last equation can also be inverted to give an explicit formula for $\underline{\Delta}^{op}$. If H is only a bialgebra, we obtain only a braided bialgebra. If H_1 is only a quasitriangular bialgebra, we can still use these transmutation formulae, with the convolution-inverse map i^{-1} in place of $i \circ S$. It is clear that $B(H, H) = \underline{H}$. We will see some other examples in Chapter 10.1.

Finally, we have, as usual, a dual reconstruction theory to the one above, based on comodules rather than modules. Given a monoidal functor $F : \mathcal{C} \to \mathcal{V}$, we require representability of the functor $V \mapsto \mathrm{Nat}(F, F \otimes V)$ and its higher order products, i.e. bijections

$$\mathrm{Mor}(B^{\otimes n}, V) \to \mathrm{Nat}(F^n, F^n \otimes V). \qquad (9.44)$$

This is the *representability assumption for comodules*. This time, we do not have to do all the proofs again because the diagrammatic language is sufficiently powerful that we can literally just turn the above proofs upside down.

By taking the identity functor, we obtain a canonical *automorphism braided group* $B = \mathrm{Aut}(\mathcal{C})$, which coacts on the objects X of $\mathcal{C}$ and is 'braided commutative' in a certain sense, $\cdot^{op} = \cdot$. This makes it more like a function Hopf algebra on a group. More precisely, we may need to take for $\mathcal{V}$ a suitable cocompletion of $\mathcal{C}$. Once again, we have a universal property. It should also be mentioned that the representability condition (9.40), or (9.44), can be written in a more fancy way as the assertion that B is an 'end' or 'coend' respectively in $\mathcal{V}$. In this categorical notation, one writes

$$U(\mathcal{C}) = \int_X X \otimes X^*, \quad \mathrm{Aut}(\mathcal{C}) = \int^X X^* \otimes X, \qquad (9.45)$$

where the first case means a certain subobject or equaliser of an infinite direct limit of $\underline{\mathrm{Hom}}(X, X)$ over all the objects of $\mathcal{C}$ as characterised by (9.40) (so it is like the covariantly constant functions that we began with). It means, in the second case, a certain quotient or coequaliser of an infinite colimit over $\mathcal{C}$ as characterised again by the requirement of naturality. Similarly, in the general case, when $F \neq \mathrm{id}$.

Example 9.4.10 *Every dual quasitriangular Hopf algebra A has a braided group analogue $\underline{A} = \mathrm{Aut}(\mathcal{M}^A)$ with the same coalgebra, counit and unit,* *and*

$$a \underline{\cdot} b = \sum a_{(2)} b_{(2)} \mathcal{R}((Sa_{(1)})a_{(3)} \otimes Sb_{(1)}),$$

$$\underline{S}a = \sum Sa_{(2)} \mathcal{R}((S^2 a_{(3)}) Sa_{(1)} \otimes a_{(4)}).$$

The braided product here is the covariantised one in Theorem 7.4.1, and is braided commutative in the sense

$$\sum \mathcal{R}(a_{(1)} \otimes b_{(1)}) b_{(2)} \underline{\cdot} \mathcal{R}(b_{(3)} \otimes a_{(2)}) a_{(3)}$$

$$= \sum a_{(1)} \underline{\cdot} \mathcal{R}(a_{(2)} \otimes b_{(1)}) b_{(2)} \mathcal{R}(b_{(3)} \otimes a_{(3)}).$$

We call $\underline{A}$ the braided group of function algebra type associated to A. It lives in the braided category of right A-comodules by the right adjoint coaction in Example 1.6.14.

Proof: We use underlines to distinguish the braided-Hopf algebra structures from the structures of our original A. This time, we apply the dual form of Theorem 9.2.4 to $\mathcal{C} = \mathcal{M}^A = \mathcal{V}$ and $F = \mathrm{id}$. So we reconstruct a braided-Hopf algebra B in this category. The proof that we can take $B = A$ as an object in the category by the right coadjoint action and with the same coalgebra is strictly dual to the proof of Example 9.4.9. We content ourselves here with deriving the resulting formulae for $B = \underline{A}$ by these categorical methods, which is how they were first obtained. For the reconstruction, we must use Fig. 9.15 turned upside down and with Δ relabelled as the braided product, α_X as right coactions β_X, etc. These coincide, in our case, with the coactions by which X is an object in our braided category. The braiding in our category is from Exercise 9.2.9. Then, from Fig. 9.15(c), we see that the modified product is characterised by

$$y^{(\bar{1})} \otimes x^{(\bar{1})(\bar{1})} \otimes y^{(\bar{2})}{}_{(2)} \underline{\cdot} x^{(\bar{2})} \mathcal{R}((Sy^{(\bar{2})}{}_{(1)}) y^{(\bar{2})}{}_{(3)} \otimes x^{(\bar{1})(\bar{2})})$$

$$= y^{(\bar{1})} \otimes x^{(\bar{1})} \otimes y^{(\bar{2})}{}_{(2)} \underline{\cdot} x^{(\bar{2})}{}_{(2)} \mathcal{R}((Sy^{(\bar{2})}{}_{(1)}) y^{(\bar{2})}{}_{(3)} \otimes x^{(\bar{2})}{}_{(1)})$$

$$= y^{(\bar{1})} \otimes x^{(\bar{1})} \otimes y^{(\bar{2})} x^{(\bar{2})},$$

where the first equality is that we have a coaction (in our usual shorthand notation) on $X \ni x$, and the second is the definition from Fig. 9.15 as the tensor product coaction on $X \otimes Y$. Since this is for all X, Y, we conclude the general formula stated for $\underline{\cdot}$ in terms of the structure of A. This is the same as the one in Theorem 7.4.1 using the bicharacter property of $\mathcal{R}$.

From Fig. 9.15(e), we derive the opposite product in just the same way as characterised by

$$y^{(\bar{1})} \otimes x^{(\bar{1})} \otimes y^{(\bar{2})}{}_{(2)} \underline{\cdot}^{\mathrm{op}} x^{(\bar{2})}{}_{(2)} \mathcal{R}((Sy^{(\bar{2})}{}_{(1)}) y^{(\bar{2})}{}_{(3)} \otimes x^{(\bar{2})}{}_{(1)})$$

$$= y^{(\bar{1})} \otimes x^{(\bar{1})} \otimes x^{(\bar{2})}{}_{(1)} \underline{\cdot} y^{(\bar{2})}{}_{(2)} \mathcal{R}^{-1}(x^{(\bar{2})}{}_{(2)} \otimes (Sy^{(\bar{2})}{}_{(1)}) y^{(\bar{2})}{}_{(3)})$$

for all X, Y. Hence,

$$a_{(2)}\underline{\cdot}^{\mathrm{op}} b_{(2)} \mathcal{R}((Sa_{(1)})a_{(3)} \otimes b_{(1)}) = b_{(1)}\underline{\cdot} a_{(2)} \mathcal{R}^{-1}(b_{(2)} \otimes (Sa_{(1)})a_{(3)})$$

for the opposite braided product in terms of $\underline{\cdot}$. Expanding the $\mathcal{R}^{-1}$ and rearranging using the elementary properties of dual quasitriangular structures from Chapter 2.2 immediately gives the form stated when $\underline{\cdot}^{\mathrm{op}} = \underline{\cdot}$, as it is here from general grounds. Another form of this braided commutativity is

$$a_{(3)}\underline{\cdot} b_{(3)} \mathcal{R}(Sa_{(2)} \otimes b_{(1)}) \mathcal{R}(a_{(4)} \otimes b_{(2)}) \mathcal{R}(b_{(5)} \otimes Sa_{(1)}) \mathcal{R}(b_{(4)} \otimes a_{(5)}) = b\underline{\cdot}a$$

if one puts all the $\mathcal{R}$ to one side, which looks more like the dual of the braided cocommutativity in Example 9.4.9.

Finally, we derive the braided antipode from Fig. 9.15(d) as

$$x^{(\bar{1})} \otimes \underline{S} x^{(\bar{2})} = e_a \langle f^{a(\bar{1})}, x^{(\bar{1})} \rangle f^{a(\bar{2})}{}_{(2)} \mathcal{R}((Sf^{a(\bar{2})}{}_{(1)}) f^{a(\bar{2})}{}_{(3)} \otimes x^{(\bar{2})})$$

$$= e_a \langle f^a, x^{(\bar{1})(\bar{1})} \rangle (Sx^{(\bar{1})(\bar{2})})_{(2)} \mathcal{R}((S(Sx^{(\bar{1})(\bar{2})})_{(1)})(Sx^{(\bar{1})(\bar{2})})_{(3)} \otimes x^{(\bar{2})})$$

$$= x^{(\bar{1})} \otimes Sx^{(\bar{2})}{}_{(2)} \mathcal{R}((S^2 x^{(\bar{2})}{}_{(3)}) Sx^{(\bar{2})}{}_{(1)} \otimes x^{(\bar{2})}{}_{(4)}),$$

where $\{e_a\}$ is a basis of X and $\{f^a\}$, a dual basis. We used the braiding from Exercise 9.2.9 for the first equality, the coaction on X^* from (9.20) for the second and that we have a coaction on X for the third. Since this is for all X, we conclude the formula for $\underline{S}$ as stated. One can also verify directly that the stated formulae give a braided-Hopf algebra. ∎

We have already developed one concrete example in Chapter 7.4, where we showed that this transmutation or covariantising operation turned the quantum matrices $A(R)$ of Chapter 4.1 to the braided matrices $B(R)$ encountered in Chapters 4.3 and 4.5. More generally, if $p : A \to A_1$ is a map to a dual quasitriangular Hopf algebra A_1, then A acquires a new transmuted product, antipode and dual quasitriangular structure, making it a dual quasitriangular Hopf algebra $B(A, A_1)$ in the braided category of right A_1-comodules. The formulae for its transmuted product and antipode are just as in Example 9.4.10, but with $\mathcal{R}$ replaced by $\mathcal{R}_1 \circ (p \otimes p)$. The braided commutativity now becomes quasicommutativity up to a braided dual quasitriangular structure. It can be written as

$$\sum b_{(1)} \mathcal{R}_1(p(b_{(2)}) \otimes p(a_{(1)})) \underline{\cdot} a_{(2)} \mathcal{R}(a_{(3)} \otimes b_{(3)})$$

$$= \sum \mathcal{R}_1(p(b_{(1)}) \otimes p(a_{(1)})) \mathcal{R}(a_{(2)} \otimes b_{(2)}) \mathcal{R}_1(p(Sa_{(3)}) \otimes p(b_{(3)})) \qquad (9.46)$$

$$a_{(4)} \mathcal{R}_1(p(a_{(5)}) \otimes p(b_{(4)})) \underline{\cdot} b_{(5)},$$

from the expression for the transmuted product $\cdot$ in terms of the original product of A, and the dual quasitriangularity axioms.

Also, one can make precise the sense in which this example is exactly dual to the preceding one. One needs for this the correct categorical formulation of duality in the braided setting. Thus, if a braided-Hopf algebra B is rigid in the sense of Section 9.3, the B^* is also a braided-Hopf algebra with product Δ^*, coproduct $\cdot^*$ and antipode S^*, where $*$ is the adjoint operation in Fig. 9.7(b). This is best proven diagrammatically. In concrete terms, where our objects are vector spaces, it means

$$\mathrm{ev}(b \otimes ad) = \mathrm{ev} \circ \mathrm{ev}(\Delta b \otimes a \otimes d), \quad \mathrm{ev}(bc \otimes a) = \mathrm{ev} \circ \mathrm{ev}(b \otimes c \otimes \Delta a) \quad (9.47)$$

and $\mathrm{ev}(b \otimes Sa) = \mathrm{ev}(Sb \otimes a)$ for all $b, c \in B^*$ and $a, d \in B$. We apply $\mathrm{ev} : B^* \otimes B \to \underline{1}$ in the middle two factors, then in the remaining two. We denote B^* with this dual bialgebra or Hopf algebra structure by $B^\star$. Note that it does not involve any transposition in its definition and so does not in fact reduce in the unbraided case to our usual Hopf algebra duality. Rather, it reduces to the opposite algebra and opposite coalgebra to the usual dual. We have found it natural even for ordinary Hopf algebras at various points in Chapters 6 and 7.

Proposition 9.4.11 *If H is dual to A, then the braided groups $\underline{H}$ and $\underline{A}$ are dual in the categorical sense, $\underline{H} = (\underline{A})^\star = (\underline{A})^{*\mathrm{op}/\mathrm{op}}$.*

Proof: Explicitly, the duality is given by $b \in \underline{H}$ mapping to a linear functional $\langle Sb, (\ ) \rangle$ on $\underline{A}$, where S is the usual antipode of H. Then,

$$\begin{aligned}
\langle Sb, a \underline{\cdot} d \rangle &= \langle Sb_{(2)}, a_{(2)} \rangle \langle Sb_{(1)}, d_{(2)} \rangle \mathcal{R}((Sa_{(1)})a_{(3)} \otimes Sd_{(1)}) \\
&= \langle (S\mathcal{R}^{(1)}{}_{(1)})(Sb_{(2)})\mathcal{R}^{(1)}{}_{(2)}, a \rangle \langle (S\mathcal{R}^{(2)})Sb_{(1)}, d \rangle \\
&= \langle S(b_{(1)}\mathcal{R}^{(2)}), d \rangle \langle S(S^{-1}\mathcal{R}^{(1)}{}_{(2)}b_{(2)}\mathcal{R}^{(1)}{}_{(1)}), a \rangle \\
&= \langle S(b_{(1)}S\mathcal{R}^{(2)}), d \rangle \langle S(\mathcal{R}^{(1)}{}_{(1)}b_{(2)}S\mathcal{R}^{(1)}{}_{(2)}), a \rangle \\
&= \langle (S \otimes S)\underline{\Delta} b, d \otimes a \rangle
\end{aligned}$$

for all $a, d \in \underline{A}$ and $b \in \underline{H}$ as required. The pairing $\langle Sb \otimes Sc, \underline{\Delta} a \rangle = \langle S(cb), a \rangle$ is easier since the product of $\underline{H}$ and the coproduct of $\underline{A}$ coincide with the usual ones. Likewise for the pairing of the units and counits. We also check that the map $\langle S(\ ), \ \rangle$ is indeed a morphism in the category of H-modules,

$$\begin{aligned}
(\langle S(h \triangleright b), \ \rangle)(a) &= \langle S(h \triangleright b), a \rangle = \langle h_{(1)} b Sh_{(2)}, Sa \rangle \\
&= \langle Sh_{(1)}, a_{(3)} \rangle \langle b, Sa_{(2)} \rangle \langle S^2 h_{(2)}, a_{(1)} \rangle \\
&= \langle Sb, a^{(\bar{1})} \rangle \langle Sh, a^{(\bar{2})} \rangle = \langle Sb, (Sh) \triangleright a \rangle \\
&= (h \triangleright \langle Sb, \ \rangle)(a),
\end{aligned}$$

where the natural action on $f \in (\underline{A})^{\star}$ is from Proposition 9.3.3. ∎

This is to be expected. On the other hand, we have already seen directly in Proposition 7.4.3 that the quantum inverse Killing form $Q : A \rightarrow H$ becomes a homomorphism of the covariantised algebras and coalgebras. So we can now give the true meaning of this homomorphism: it is a homomorphism

$$Q : \underline{A} \rightarrow \underline{H}, \qquad Q(a) = (a \otimes \mathrm{id})(\mathcal{R}_{21}\mathcal{R}) \qquad (9.48)$$

of braided groups. When H is factorisable in the strict sense that Q is invertible, we see that $\underline{H} \cong \underline{A}$. In other words, the true meaning of this factorisability condition (which we have encountered many times already) is that, in this case, the associated enveloping algebra braided group and function algebra braided group are isomorphic, i.e. they are self-dual in view of the above proposition.

For example, when R is one of the standard R-matrices, we obtain the quantum function algebra G_q as a quotient of $A(R)$, as we know from Chapter 4.2. Example 9.4.10 applied in this case just gives $BG_q = B(G_q, G_q)$ as a quotient of $B(R)$. On the other hand, these standard cases are essentially factorisable, so, from the self-duality (9.48) we see that this is isomorphic to $BU_q(g) = B(U_q(g), U_q(g))$ from Example 9.4.9. This has the same algebra as $U_q(g)$ in the form generated by $\mathtt{l}$, so we deduce Example 4.3.3 by these conceptual means. This is an important application of the concept of braided groups. It is also not a phenomenon which would be visible in the classical Lie theory. Indeed, when $q \neq 1$, there is only one object, BG_q; taking the limit $q \rightarrow 1$ gives the commutative Hopf algebra $\mathbb{C}(G)$, while using (9.48) and then taking the limit of suitably rescaled generators gives the noncommutative Hopf algebra $U(g)$.

There is also kind of 'converse' or adjoint to Theorem 9.4.6, which tells us that some braided-Hopf algebras (including those in Examples 9.4.9 and 9.4.10) can always be *bosonised* back into equivalent ordinary Hopf algebras.

Theorem 9.4.12 *Any braided-Hopf algebra B in the braided category $_H\mathcal{M}$, for H a quasitriangular Hopf algebra, gives an ordinary Hopf algebra $B \rtimes H$ by making a cross product by the action of H on B as an object in the category and the cross coproduct by the induced coaction from Lemma 7.4.4. We call $B \rtimes H$ the* bosonisation *of B. The modules of B in the braided category correspond to the ordinary modules of $B \rtimes H$.*

Proof: One can obtain this as another application of the categorical constructions above. Namely, consider the monoidal category of modules of B in the braided category of H-modules. Objects are vector spaces on which both H and B act. Taking the forgetful functor all the way to Vec and reconstructing by Theorem 9.4.1, we obtain an ordinary Hopf algebra, which is the abstract definition of bosonisation (i.e. such that its representations are the modules of B in the braided category). To compute it, it is rather easier to construct its braided version first: we can forget only the action of B, giving a forgetful functor to the category of H-modules, and apply the braided reconstruction Theorem 9.4.6. Since B lives in the category of H-modules, it is acted upon covariantly by H, and hence by $\underline{H}$. Because the latter is braided cocommutative in a certain sense, one can make (diagrammatically) a braided cross product $B{\times}\!\!\!\cdot\,\underline{H}$ by this action, knowing that it forms a braided-Hopf algebra with the braided tensor product coalgebra structure, by a braided version of Proposition 6.2.1. This is the braided group reconstructed from Theorem 9.4.6. It is therefore also the braided group reconstructed from the same functor when viewed as going from the category modules of the bosonisation to H-modules; one recognises it as the transmutation (9.43) of an ordinary Hopf algebra inclusion $H \hookrightarrow B{\cdot}\!\!\!\times\, H$, obtaining the required structure of $B{\cdot}\!\!\!\times\, H$ in this way. This outlines how the theorem was first obtained.

On the other hand, once one has the result, it is not hard to give a direct proof. We content ourselves with this here. Since B lives as an algebra in the category of H-modules, we know that H acts covariantly on it as an H-module algebra. Hence we can automatically form the cross product as in Proposition 1.6.6. This is the bosonised algebra. Next, since B is also a coalgebra in the category, we know that it is an H-module coalgebra by this action; hence, by Lemma 7.4.4, it is a left H-comodule coalgebra by $\beta = \mathcal{R}_{21}\triangleright$. Hence we have a cross coproduct coalgebra by Proposition 1.6.18. So,

$$\Delta(b \otimes h) = b_{(1)} \otimes \mathcal{R}^{(2)} h_{(1)} \otimes \mathcal{R}^{(1)}\triangleright b_{(2)} \otimes h_{(2)}, \quad \epsilon(b \otimes h) = \epsilon(b)\epsilon(h) \quad (9.49)$$

is the coalgebra of the bosonised Hopf algebra. It is also a braided tensor coproduct (c.f. Corollary 9.2.13) if we view H in the left regular representation. We check that these constructions fit together to form an ordinary bialgebra when B is a braided bialgebra. Thus,

$$\begin{aligned}
(\Delta(b &\otimes h))(\Delta(c \otimes g)) \\
&= (b_{(1)} \otimes \mathcal{R}^{(2)} h_{(1)})(c_{(1)} \otimes \mathcal{R}'^{(2)} g_{(1)}) \otimes (\mathcal{R}^{(1)}\triangleright b_{(2)} \otimes h_{(2)})(\mathcal{R}'^{(1)}\triangleright c_{(2)} \otimes g_{(2)}) \\
&= b_{(1)}(\mathcal{R}^{(2)}{}_{(1)} h_{(1)}\triangleright c_{(1)}) \otimes \mathcal{R}^{(2)}{}_{(2)} h_{(2)} \mathcal{R}'^{(2)} g_{(1)} \\
&\qquad \otimes (\mathcal{R}^{(1)}\triangleright b_{(2)})(h_{(3)} \mathcal{R}'^{(1)}\triangleright c_{(2)}) \otimes h_{(4)} g_{(2)} \\
&= b_{(1)}(\mathcal{R}''^{(2)} h_{(1)}\triangleright c_{(1)}) \otimes \mathcal{R}^{(2)}\mathcal{R}'^{(2)} h_{(3)} g_{(1)} \\
&\qquad \otimes (\mathcal{R}^{(1)}\mathcal{R}''^{(1)}\triangleright b_{(2)})(\mathcal{R}'^{(1)} h_{(2)}\triangleright c_{(2)}) \otimes h_{(4)} g_{(2)}
\end{aligned}$$

$$= b_{(1)}(\mathcal{R}''^{(2)}h_{(1)}\triangleright c_{(1)}) \otimes \mathcal{R}^{(2)}h_{(3)}g_{(1)}$$
$$\otimes (\mathcal{R}^{(1)}{}_{(1)}\mathcal{R}''^{(1)}\triangleright b_{(2)})(\mathcal{R}^{(1)}{}_{(2)}h_{(2)}\triangleright c_{(2)}) \otimes h_{(4)}g_{(2)}$$
$$= b_{(1)}(\mathcal{R}''^{(2)}h_{(1)}\triangleright c_{(1)}) \otimes \mathcal{R}^{(2)}h_{(3)}g_{(1)}$$
$$\otimes \mathcal{R}^{(1)}\triangleright \left((\mathcal{R}''^{(1)}\triangleright b_{(2)})(h_{(2)}\triangleright c_{(2)})\right) \otimes h_{(4)}g_{(2)}$$
$$= (b(h_{(1)}\triangleright c_{(1)}))_{(1)} \otimes \mathcal{R}^{(2)}h_{(2)}g_{(1)} \otimes \mathcal{R}^{(1)}\triangleright(b(h_{(1)}\triangleright c_{(1)}))_{(2)} \otimes h_{(3)}g_{(2)}$$
$$= \Delta((b\otimes h)(c\otimes g)),$$

where the first equality is the definition of the bosonised Δ as above, the second is the definition of the bosonised product from Proposition 1.6.6 and that the coalgebra is H-covariant, and the third and fourth use the quasitriangularity axioms (2.1) and (2.2). The fifth equality uses that the algebra is H-covariant, while the sixth uses that the coproduct of B is a homomorphism to the braided tensor product algebra from Corollary 9.2.13. Finally, we recognise the result as the bosonised coproduct applied to the bosonised product. There is also an antipode on $B\rtimes H$ given by

$$S(b\otimes h) = (Sh_{(2)})u\mathcal{R}^{(1)}\triangleright\underline{S}b \otimes S(\mathcal{R}^{(2)}h_{(1)}) \tag{9.50}$$

if B has a braided antipode $\underline{S}$. The formula comes from the categorical proof outlined above, but can be verified directly in the same manner as above for the bialgebra structure. The element u is from Proposition 2.1.8.

Finally, a braided module V has, by definition, both an action of H as an object in the category and an action of B. The latter action is a morphism in our category of H-modules, i.e. it is H-covariant. But we know from Corollary 6.1.3 that this is just what it takes for these actions to extend to an action of the semidirect product $B\rtimes H$. The converse is also clear. Finally, this identification respects the tensor product of braided modules (using the braided coproduct of B), which corresponds to the usual tensor product of representations of $B\rtimes H$ using its bosonised coproduct. ∎

As usual, there is a dual theorem that any braided-Hopf algebra B in the braided category of right A-comodules for a dual quasitriangular Hopf algebra A can be bosonised to an ordinary Hopf algebra $A\ltimes B$, where we use the right coaction on B as an object in the category and the induced action from Lemma 7.4.8. This time, the bosonised coalgebra is from Proposition 1.6.16, while the bosonised algebra from Proposition 1.6.10, and the required antipode, are

$$(a\otimes b)(d\otimes c) = \sum ad_{(1)}\otimes b^{(\bar{1})}c\mathcal{R}(b^{(\bar{2})}\otimes d_{(2)}), \tag{9.51}$$

$$S(a\otimes b) = \sum (1\otimes \underline{S}b^{(\bar{1})})(Sb^{(\bar{2})}Sa\otimes 1). \tag{9.52}$$

The algebra has the form of a braided tensor product from Corollary 9.2.14 if we view the coproduct of A as a right comodule algebra structure. Both the cross product point of view and the braided tensor product point of view are useful. One has also

$$A{\bowtie}B^* \cong (B{\bowtie}H)^* \tag{9.53}$$

if A is dual to H. This explains the abstract picture behind some of the results in Chapter 7.4, where we studied Drinfeld's quantum double as the bosonisation $D(H) = A{\bowtie}H$ and its dual as the bosonisation $A{\bowtie}\underline{H}$. We will see further concrete applications in the next chapter, notably to the construction of a q-deformed Poincaré group.

In addition, there are left–right reversed versions of both the module and comodule bosonisation constructions. Thus, a bosonisation $H{\bowtie}B$ for $B \in \mathcal{M}_H$, etc. The left and right versions can even be done simultaneously as follows.

Theorem 9.4.13 *Let B, C be dually paired braided groups in $_H\mathcal{M}$, where H is quasitriangular. Their double bosonisation is the ordinary Hopf algebra $B{\bowtie}H{\bowtie}C^{\mathrm{op}}$ built on the tensor product vector space, containing the bosonisations $B{\bowtie}H$ and $H{\bowtie}C^{\mathrm{op}}$ as sub-Hopf algebras and with the cross relations*

$$b_{\underline{(1)}}\mathcal{R}^{(2)}c_{\underline{(1)}}\mathrm{ev}(\mathcal{R}^{(1)}{\triangleright}b_{\underline{(2)}}, c_{\underline{(2)}}) = \mathrm{ev}(b_{\underline{(1)}}, \mathcal{R}^{(2)}{\triangleright}c_{\underline{(1)}})c_{\underline{(2)}}\mathcal{R}^{(1)}b_{\underline{(2)}}$$

for all $b \in B$ and $c \in C$. When $B = C^*$ the pairing is nondegenerate and the double bosonisation is quasitriangular with

$$\mathcal{R} = \mathcal{R}_H \sum_a \underline{S}e_a \otimes f^a,$$

where $\{e_a\}$ is a basis of C and $\{f^a\}$ a dual basis.

Proof: For the Hopf algebra structure we assume a (possibly degenerate) duality pairing ev : $B \otimes C \to k$ in the sense of (9.47). If B is finite-dimensional we can of course take C as its predual so that $B = C^*$. First of all, we can regard C with its opposite product and a suitable right action as a braided group in $\mathcal{M}_{\bar{H}}$, where $\bar{H}$ denotes H with its conjugate quasitriangular structure $\bar{\mathcal{R}} = \mathcal{R}_{21}^{-1}$. The right action is connected to the given left action by $c{\triangleleft}h = Sh{\triangleright}c$, so that $\mathrm{ev}(h{\triangleright}b, c) = \mathrm{ev}(b, c{\triangleleft}h)$ for all $h \in H$. It is easy to see that C^{op} forms a braided group in this category and we can therefore apply a right handed version of Theorem 9.4.12 to obtain a Hopf algebra $H{\bowtie}C^{\mathrm{op}}$. Viewed inside their respective bosonisations, the cross relations, coproducts and antipodes of our two sub-Hopf algebras are

$$hb = (h_{(1)} \triangleright b)h_{(2)}, \quad hc = (h_{(2)} \triangleright c)h_{(1)},$$

$$\Delta b = b_{(1)} \mathcal{R}^{(2)} \otimes \mathcal{R}^{(1)} \triangleright b_{(2)}, \quad \Delta c = \mathcal{R}^{(2)} \triangleright c_{(1)} \otimes c_{(2)} \mathcal{R}^{(1)}$$

$$Sb = (u\mathcal{R}^{(1)} \triangleright \underline{S}b)S\mathcal{R}^{(2)}, \quad Sc = \mathcal{R}^{-(1)}\underline{S}^{-1}(v^{-1}\mathcal{R}^{-(2)} \triangleright c),$$

where $\triangleright$ denotes the action of H whereby B, C live in the braided category of H-modules. Next, we impose the relations between the B, C subalgebras as stated. The product of the double bosonisation can in fact be given quite explicitly on the vector space $B \otimes H \otimes C$ since the 'cross relations' between B, H, C allow us to reorder expressions in the required way. It remains to verify that these cross relations between B and C are preserved under the above coproducts, which is a long but straightforward direct computation using similar methods to the proof of Theorem 9.4.12, i.e., the definitions and the axioms for $\mathcal{R}$. ∎

There is a categorical construction underlying the theorem, namely given B, C one may define a certain monoidal category $_B\dot{\mathcal{C}}_C$ of compatible left-B and right-C modules in the braided category $\mathcal{C} = {}_H\mathcal{M}$, generalising the representations of the quantum double in Proposition 7.1.6 to the setting of braided groups. If $B = C^*$ then $_B\dot{\mathcal{C}}_C$ is itself a braided category. On the other hand, we can apply the Tannaka–Krein reconstruction in Section 9.4.1 to obtain a (quasitriangular) Hopf algebra $B \bowtie H \ltimes C^{op}$ such that its category of modules can be identified with $_B\dot{\mathcal{C}}_C$. The result is not exactly a braided version of Drinfeld's quantum double but rather of its bosonisation as an ordinary Hopf algebra (unless the category is symmetric, it is not possible to 'untangle' an actual braided group double $D(B)$ from the double bosonisation). This is the conceptual basis for the construction. In Chapter 10 we will see many concrete examples of braided groups of the required kind. In particular, one has an inductive construction for strict quantum groups, including all the quasitriangular Hopf algebras $U_q(g)$, by successively adjoining mutually dual braided groups of additional positive and negative roots.

Finally, we look at a special case of the bosonisation construction in Theorem 9.4.12, namely to braided groups $B \in {}_{D(H)}\mathcal{M}$, the category of modules of the quantum double of a Hopf algebra. Its bosonisation is clearly a Hopf algebra $B \bowtie D(H)$. Since $H \subseteq D(H)$ as a sub-Hopf algebra we might expect a corresponding sub-Hopf algebra, denoted similarly by

$$B \bowtie H \subseteq B \bowtie D(H)$$

This is easily seen to be the case. As an algebra it is simply $B \bowtie H$ by the action of $H \subseteq D(H)$. Meanwhile, the induced coaction of $D(H)$ in

Theorem 9.4.12 is $\beta(b) = \mathcal{R}_{21} \triangleright b = e_a \otimes f^a \triangleright b$, where we use the form of the quasitriangular structure for $D(H)$ in Theorem 7.1.1. This is simply the action of $H^{*\mathrm{op}}$ viewed as a coaction of H and the coalgebra of $B \bowtie D(H)$ is the cross product by this. Then $B \bowtie H$ by this coaction is obviously a subcoalgebra as well. Moreover, we know that $D(H)$ modules are essentially the same thing as crossed modules ${}_H^H\dot{\mathcal{M}}$ (by reformulating an action of $H^{*\mathrm{op}}$ as a coaction of H) so if we assume $B \in {}_H^H\dot{\mathcal{M}}$ from the start then we can build the Hopf algebra $B \bowtie H$ directly without mentioning $D(H)$. This has some advantages in the infinite-dimensional case. There is an interesting converse.

Proposition 9.4.14 *Let H be a Hopf algebra (with invertible antipode, say) and $B \in {}_H^H\dot{\mathcal{M}}$ a braided group. Then*

$$B \bowtie H \underset{i}{\overset{\pi}{\rightleftarrows}} H$$

given by $\pi = \epsilon \otimes \mathrm{id}$ and $i = 1 \otimes$ are Hopf algebra maps with $\pi \circ i = \mathrm{id}$ (a Hopf algebra projection). Conversely, if $H_1 \overset{\pi}{\underset{i}{\rightleftarrows}} H$ is a Hopf algebra projection by maps π, i then there is a braided group $B \in {}_H^H\dot{\mathcal{M}}$ and $H_1 \cong B \bowtie H$. Here $B = H_1^H$, the fixed subalgebra under the right coaction $(\mathrm{id} \otimes \pi)\Delta$, and has crossed module structure, braided coproduct and braided antipode

$$h \triangleright b = \sum i(h_{(1)}) b S \circ i(h_{(2)}), \quad \beta(b) = \pi(b_{(1)}) \otimes b_{(2)},$$

$$\underline{\Delta} b = \sum b_{(1)} S \circ i \circ \pi(b_{(2)}) \otimes b_{(3)}, \quad \underline{S} b = \sum i \circ \pi(b_{(1)}) S b_{(2)}.$$

Proof: We have already given a conceptual proof that $B \bowtie H$ is a Hopf algebra, from bosonisation theory. However, it is easy enough to verify this directly as well; as algebras and coalgebras we make the left cross (co)products from Chapter 1.6 and verify that they fit to form a Hopf algebra. Thus,

$$\begin{aligned}
(\Delta(1 \otimes h)(b \otimes 1)) &= \Delta(h_{(1)} \triangleright b \otimes h_{(2)}) \\
&= h_{(1)} \triangleright b_{(1)} \otimes (h_{(2)} \triangleright b_{(2)})^{(\bar{1})} h_{(3)} \otimes (h_{(2)} \triangleright b_{(2)})^{(\bar{2})} \otimes h_{(4)} \\
&= h_{(1)} \triangleright b_{(1)} \otimes h_{(2)} b_{(2)}{}^{(\bar{1})} \otimes h_{(3)} \triangleright b_{(2)}{}^{(\bar{2})} \otimes h_{(4)} \\
&= (1 \otimes h_{(1)})(b_{(1)} \otimes b_{(2)}{}^{(\bar{1})}) \otimes (1 \otimes h_{(3)})(b_{(2)}{}^{(\bar{2})} \otimes 1) \\
&= (\Delta(1 \otimes h))(\Delta(b \otimes 1))
\end{aligned}$$

using that B is an H-module coalgebra, followed by the crossed module compatibility condition from Proposition 7.1.6. Meanwhile,

$$\Delta(bc \otimes 1) = (bc)_{(1)} \otimes (bc)_{(2)}{}^{(\bar{1})} \otimes (bc)_{(2)}{}^{(\bar{2})} \otimes 1$$

$$= b_{\underline{(1)}}(b_{\underline{(2)}}{}^{(\bar{1})}{\triangleright}c_{\underline{(1)}}) \otimes (b_{\underline{(2)}}{}^{(\bar{2})}c_{\underline{(2)}})^{(\bar{1})} \otimes (b_{\underline{(2)}}{}^{(\bar{2})}c_{\underline{(2)}})^{(\bar{2})} \otimes 1$$

$$= b_{\underline{(1)}}(b_{\underline{(2)}}{}^{(\bar{1})}{\triangleright}c_{\underline{(1)}}) \otimes b_{\underline{(2)}}{}^{(\bar{2})(\bar{1})}c_{\underline{(2)}}{}^{(\bar{1})} \otimes b_{\underline{(2)}}{}^{(\bar{2})(\bar{2})}c_{\underline{(2)}}{}^{(\bar{2})} \otimes 1$$

$$= b_{\underline{(1)}}(b_{\underline{(2)}}{}^{(\bar{1})}{}_{(1)}{\triangleright}c_{\underline{(1)}}) \otimes b_{\underline{(2)}}{}^{(\bar{1})}{}_{(2)}c_{\underline{(2)}}{}^{(\bar{1})} \otimes b_{\underline{(2)}}{}^{(\bar{2})}c_{\underline{(2)}}{}^{(\bar{2})} \otimes 1$$

$$= (b_{\underline{(1)}} \otimes b_{\underline{(2)}}{}^{(\bar{1})})(c_{\underline{(1)}} \otimes c_{\underline{(2)}}{}^{(\bar{1})}) \otimes b_{\underline{(2)}}{}^{(\bar{2})}c_{\underline{(2)}}{}^{(\bar{2})} \otimes 1$$

$$= (\Delta(b \otimes 1))(\Delta(c \otimes 1))$$

using the braided homomorphism property of B for the second equality. For the braiding from Example 9.2.3, this is

$$\underline{\Delta}(bc) = \sum b_{\underline{(1)}}\Psi(b_{\underline{(2)}} \otimes c_{\underline{(1)}})c_{\underline{(2)}} = \sum b_{\underline{(1)}}\,(b_{\underline{(2)}}{}^{(\bar{1})}{\triangleright}c_{\underline{(1)}}) \otimes b_{\underline{(2)}}{}^{(\bar{2})}c_{\underline{(2)}}. \quad (9.54)$$

We then use that B is an H-comodule algebra to recognise the result. The other facts are equally straightforward. In the converse direction we let $B = \{b \in H_1 | b_{(1)} \otimes \pi(b_{(2)}) = b \otimes 1\}$ be the fixed subalgebra under the right regular coaction of H_1 pushed out to a coaction of H by π. From this it is easy to see that the adjoint action of H_1 on itself pulled back along i to an action of H indeed restricts to B. These are the action and coaction of B as stated. On the other hand we know from Example 7.1.8 that H_1 is a crossed module by its regular coaction and adjoint action. Exactly the same computation as in the proof, but with π, i now appearing, proves that the action and coaction on B form a crossed module (we use that π, i are Hopf algebra maps). Next, we can project $\bar{\pi} : H_1 \to B$ by $\bar{\pi}(b) = b_{(1)}S \circ i \circ \pi(b_{(2)})$ for all $b \in H_1$ and we define $\underline{\Delta}b = (\bar{\pi} \otimes \bar{\pi})\Delta b$. This immediately gives the formula stated. It is clearly coassociative since the coproduct of H_1 is, and we have

$$b_{(1)}\,(b_{(2)}{}^{(\bar{1})}{\triangleright}c_{(1)}) \otimes b_{(2)}{}^{(\bar{2})}c_{(2)}$$

$$= b_{(1)}S \circ i \circ \pi(b_{(2)})(\pi(b_{(3)}){\triangleright}(c_{(1)}S \circ i \circ \pi(c_{(2)}))) \otimes b_{(4)}c_{(3)}$$

$$= b_{(1)}c_{(1)}S \circ i \circ \pi(c_{(2)})S \circ i \circ \pi(b_{(2)}) \otimes b_{(3)}c_{(3)} = \underline{\Delta}(bc)$$

as required, using the definitions of the action and coaction. The braided antipode is then determined as stated, with $\underline{\epsilon} = \epsilon$ clearly the appropriate counit. Finally, it is straightforward to verify that $\underline{\Delta}, \underline{S}, \underline{\epsilon}$ are indeed covariant under the action and coaction, i.e., morphisms in ${}^H_H\mathcal{M}$. Hence we have a braided group. The isomorphism $\theta : B{\rtimes}H \to H_1$ is $\theta(b \otimes h) = bi(h)$, with inverse $\theta^{-1}(b) = \sum b_{(1)}S \circ i \circ \pi(b_{(2)}) \otimes \pi(b_{(3)})$ for all $b \in H_1$, which is again a straightforward verification. Note we do not actually need the antipode of H to be invertible in the proof – it is merely the natural setting so that ${}^H_H\mathcal{M}$ is an honest braided category with invertible Ψ. Otherwise the correspondence is with an algebra-coalgebra B obeying (9.54) but without the tools (such as the diagrammatic notation) of the full braided group theory. ∎

There are a number of situations where we have already made such a simultaneous cross product and cross coproduct or *biproduct* (with the action and coaction from the same side). Among them, the bosonisation construction in Theorem 9.4.12 could itself be viewed as resulting in a particular class of biproducts under the monoidal functor $_H\mathcal{M} \hookrightarrow {}^H_H\mathcal{M}$ induced by Lemma 7.4.4. On the other hand, the many nice features of bosonisations (such as the equivalence of their category of representations with those of the original B) do not hold for general biproducts, i.e., this is not necessarily a useful point of view. In fact the biproducts of interest tend to be those essentially obtained by bosonisation and viewed in that way as a means to avoid discussing the quasitriangular structure $\mathcal{R}$ explicitly. There is similarly a double biproduct construction.

Notes for Chapter 9

The material in Section 9.1 is standard and we refer to the text of S. Mac Lane [173] for more of this background. Mac Lane was also the first to prove the coherence theorem for monoidal and symmetric monoidal categories. The pentagon condition for Φ also plays an important role in algebraic topology, from the work of J. Stasheff on spaces which are groups or monoids up to homotopy [174]. The construction of a monoidal category $\mathcal{C}^\circ \to \mathcal{V}$ dual to $\mathcal{C} \to \mathcal{V}$ in Theorem 9.1.5, including the special case in Corollary 9.1.6, is due to the author in 1989 [37]. The motivation was the representation-theoretic self-duality principle described in Chapter 6.4, which led to the search for a general Pontryagin duality theory for monoidal categories, generalising the duality of Abelian groups and Hopf algebras (the point of view in Example 9.1.7). Also to be found in [37] are a functor $\mathcal{C} \to {}^\circ(\mathcal{C}^\circ)$, a categorical coadjoint action of $\mathcal{C}$ on $\mathcal{C}^\circ$ and an associated double cross product category constructed along the lines of Chapter 7.2. Further results are in [140], where $\mathcal{C}^\circ$ is identified, in suitable cases, as the category of representations of the automorphism braided group $\mathrm{Aut}\,(\mathcal{C}, F, \mathcal{V})$ from [38]. The special case $\mathcal{C}^\circ \to \mathcal{C}$ in Corollary 9.1.6, was also known, independently, to V.G. Drinfeld [175] from the quantum double point of view in Example 9.1.8; it can be called the 'centre', $Z(\mathcal{C})$, or 'double', $D(\mathcal{C})$ for this reason. Its braiding in Proposition 9.2.2 is due to Drinfeld. The proof of Example 9.1.8 is taken from [3]. It was recently extended to quasi-Hopf algebras in [54], giving a notion of crossed modules and quantum double $D(H)$ for these.

Braided monoidal categories were formally introduced into category theory by A. Joyal and R. Street [176]. They also arose in the study of the representation theory of quantum groups shortly after Drinfeld's seminal work [22], where the triangular or symmetric case was treated.

One of the first systematic treatments of the quasitriangular or braided case in Theorem 9.2.4 is in the author's lecture notes [1], where they were called *quasitensor categories*, and formulated for this purpose. (The additional attribution at the end of [1] turned out to be incorrect.) Other treatments which appeared at about the same time are in [28], which also covered applications to knot theory, and [40], which also covered the quasi-Hopf algebra case.

Example 9.2.5, of anyonic or $\mathbb{Z}_{/n}$-graded vector spaces, is due to the author [30]. Example 9.2.6 is from [94]. Developed in these works was the role of Hopf algebras as 'generating' categories in which one can pursue other algebraic constructions; for example, the triangular Hopf algebra $\mathbb{Z}'_{/2}$ was identified as the one which generates the category of supervector spaces [94]. Example 9.2.7 was studied [177] in connection with J.H.C. Whitehead's crossed G-sets [144]. The braiding in ${}^H_H\mathcal{M}$ was emphasised in [145], while also following as an application of Drinfeld's quantum double $D(H)$ at least in the finite-dimensional Hopf algebra case [143]. The derivation in Example 9.2.3 (from Proposition 9.2.2) is due to Drinfeld. The braiding of the category of comodules of a dual quasitriangular Hopf algebra in Exercise 9.2.9 formally appeared in [36] and [37], where the notion of dual quasitriangular Hopf algebras was introduced for just this purpose as a preliminary to proving the converse reconstruction theorem; see also [38, 55, 156] and the Notes to Chapters 2 and 4.

The study of algebras in braided categories (especially in the categories of representations of quantum groups) was developed by the author; their braided tensor product in Lemma 9.2.12–Corollary 9.2.14 appeared in [94], [178] and [179] as a key lemma for the formulation of the concept of a braided group (see below). Of course, the supertensor product of superalgebras is well-known, as is the generalisation to symmetric tensor categories [101]. The braided theory provided new techniques, such as the diagrammatic proof of associativity in Fig. 9.6 (or, more nontrivially, the proof of the braided antipode property in Fig. 9.14). The generalised braided tensor product or *algebra factorisation* $B\bowtie_\Psi C$ discussed below (9.12) and in the proof of Theorem 7.2.3 is due to the author in [12] and more explicitly in [148]; it is the algebra part of the $\bowtie$ construction for bialgebra factorisations in [12]. A recent application is in [135]. Moreover, the realisation that the concept of covariance under a quantum group (e.g. the q-deformed Lorentz group in Chapter 7.3) and the concept of generalising vector spaces to supervectors are, mathematically, one and the same is significant for physics [4, 95, 180]. In the present case, it led to Corollaries 9.2.13 and 9.2.14 that (co)module algebras under a (dual) quasitriangular Hopf algebra have a tensor product.

The modern formulation of rigidity and left duals V^* in monoidal categories in Section 9.3 is taken from P. Deligne [181]: the more traditional route is to go through internal hom, as we do later in the section. The situation with both left and right duals (the ribbon or tortile case) was studied in [28], where the construction of knot-invariants from ribbon Hopf algebras was introduced; see also [182].

At about the same time, the notion of categorical dimension and trace in braided categories in Fig. 9.9 and Proposition 9.3.5 was developed by the author [1] as a generalisation of the *rank* in symmetric monoidal categories [183, 184]. The picture in Example 9.3.7 of the q-deformed integers $[n]$ as the categorical dimensions of representations of $U_q(su_2)$ is from [1]. The concept of quantum order and the example of $|U_q(su_2)|$ were also introduced in this work. Exercise 9.3.9 for $|D(H)|$ is due to the author [142]. The general $|U_q(g)|$ and its physical interpretation are due to the author and Ya.S. Soibelman [69, 70]. The assignment of further categorical dimensions for other knots leads a kind of 'duality pairing' between knots up to regular isotopy and quantum groups. The full picture here is due to M.A. Hennings [185, 186], who showed how to construct actual knot invariants, and invariants of three-manifolds obtained by surgery on them, from general finite-dimensional ribbon Hopf algebras. The key ingredient in Hennings' construction is to replace the trace in the left regular representation by an integral on the Hopf algebra, which is unique up to scale. (This is in contrast to the invariants in [77], where rather specific properties of the representations of $U_q^{(r)}(su_2)$ are used.) The theory of <u>Hom</u> in a braided monoidal category is taken from [1], where the direct proof of the isomorphisms in Corollary 9.3.16 can be found. The calculation of the generalised Hecke algebras for $U_q(sl_n)$ and other quantum groups $U_q(g)$ is from [187]. A formulation of Schur–Weyl duality for triangular Hopf algebras is in [188], using the bosonisation theorem (see below).

The 'Tannaka–Krein' reconstruction of a bialgebra in Section 9.4.1 is a classical result, probably due to A. Grothendieck; a full treatment appeared in [183]. The group case is the standard non-Abelian Fourier transform; see, for example, the text of A.A. Kirillov [189]. A more categorical view for algebraic groups is given in [184]. Our treatment in terms of 'covariantly constant functions' is more like the Fourier transform setting. The reconstruction of a (dual) quasitriangular structure in Proposition 9.4.2 in the case when C is braided is due to the author in 1989 [1, 37, 179]. The reconstruction of the antipode S described in Proposition 9.4.3 is due to K. Ulbrich [190], also from 1989.

The generalisation of the reconstruction theorem to the dual quasitriangular dual quasibialgebra case at the end of Section 9.4.1 is due to the author [55]. We include now the details for the quasi-Hopf algebra antipode. The more informal module version is in [53]. Note that, in

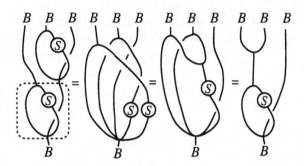

Fig. 9.19. Proof of the braided adjoint action (in box) of a braided group B on itself.

conformal field theories, one usually has something slightly weaker than a multiplicative functor, and hence something weaker than a quasi-Hopf algebra. An application of the quasi-Hopf algebra reconstruction theorem in mathematics is given in [50] by S. Shnider and S. Sternberg; see also their text [52].

The theory of braided groups (or Hopf algebras in braided categories) and the generalisation of the reconstruction theorem to this case are due to the author [38, 156, 178, 179]. Section 9.4.2 is based on these works and their three sequels [34, 146, 154]; see also the two reviews [3] and [4]. The paper [179] gave the generalised reconstruction in Theorem 9.4.6 in module form, and the 1990 works [38, 156], in comodule form. The automorphism braided groups $\mathrm{Aut}(\mathcal{C}, F, \mathcal{V})$ were introduced in these works, precisely in order to make the reconstruction possible when there is no functor to Vec, e.g. $\mathrm{Aut}(\mathcal{C})$, using the identity functor and $\mathcal{V} = \bar{\mathcal{C}}$, a cocompletion of $\mathcal{C}$. See also [191, 192], where the specific realisation (9.45) for this case was considered by V.V. Lyubashenko in connection with a categorical Fourier transform.

The application of these categorical ideas to the theory of quantum groups is also due to the author. The transmutation constructions which turn quantum groups into braided groups $\underline{H}$ and $\underline{A}$ in Examples 9.4.9 and 9.4.10 were introduced in [178, 146] and [38, 156], respectively. We have already made use of them in Chapter 7.4 to obtain results on the quantum double. The self-duality isomorphism $\underline{H}{\cong}\underline{A}$ in the factorisable case is due to the author and Lyubashenko [78]; see also [34].

Perhaps the most remarkable thing about braided groups is that they lie on the interface between algebra and knot theory. One can develop all the usual facts and constructions in group theory and Hopf algebra theory in this setting, with all proofs being done as knot or tangle diagrams in

our braided category. The proof in Fig. 9.14 of the braided antimultiplica-
tivity property (9.39) of the braided antipode is taken from [146]. Also introduced by the author are the braided adjoint action [5, 95], the braided
left coregular action [5, 193], braided group cross products [154], Weyl
algebras [193], the tensor product and dualisation of representations of a
braided group [146], etc., all developed diagrammatically in the language
of Figs. 9.13 and 9.14. For example, Fig. 9.19 shows the proof [95] that
the braided adjoint action is indeed an action. The first equality is functoriality, the second is (9.39) from Fig. 9.14 and the third is the coproduct
axiom in Fig. 9.13(a). The theory of quasitriangular Hopf algebras also
generalises. For example, it is known [146], via diagrammatic methods,
that the subcategory $\mathcal{O}(B, \underline{\Delta}^{\mathrm{op}})$ of certain 'braided cocommutative' representations of a quasitriangular braided group is braided. Also, the ab-
stract properties of the braided adjoint action lead to an axiomatic theory
of *braided-Lie algebras* [5] with the Lie-algebra-like subspaces $\mathcal{L} = \{u^i{}_j\}$ in
Chapter 4.3 appearing now as examples. Braided groups and braided-Lie
algebras also lead to new solutions of the Yang-Baxter equations [194],
and so on. We have not had room here to present this diagrammatic
theory of braided groups, beyond the first steps; it will be taken up in a
sequel to the present volume. Meanwhile, see the above works and the
two reviews [3] and [4]. It should be noted that, while it is a standard
idea to write operations as nodes in a 'flow chart' or 'wiring diagram' (or
in physics, vertices in a Feynman diagram or Penrose spin network), the
novel ingredient in *braided mathematics* is, for the first time, to distinguish between under and over crossings in such 'flow charts'. One has to
choose between them in such a way that constructions go through without
becoming 'tangled up'. This is not a feature of super-Hopf algebras or
Hopf algebras in symmetric categories, for which the generalisation from
the case of usual Hopf algebras is immediate.

The bosonisation theorem, Theorem 9.4.12, is due to the author [154]
as an application of the theory of cross products by braided groups. It
reduces all questions about a braided group B in ${}_H\mathcal{M}$ (or $\mathcal{M}^A$ in the dual
framework) to questions about the ordinary Hopf algebra $B{\rtimes}H$ which
is its bosonisation. That the category of B-modules (in the braided category) can be identified with the category of ordinary $B{\rtimes}H$-modules,
with the forgetful functor becoming the functor to ${}_H\mathcal{M}$ induced by the
inclusion $B{\rtimes}H \supseteq H$, is the abstract characterisation of bosonisation
in [154]; it is defined in such a way that the automorphism braided group
reconstructed from this functor, using Theorem 9.4.6, is the braided group
cross product $B{\rtimes}\underline{H}$. Similarly, a braided cross coproduct $\underline{A}{\blacktriangleright}B$ gives
the comodule version of the bosonisation theory. The double bosonisation
in Theorem 9.4.13 is due to the author in [195]. Its Lie bialgebra version
in the setting of Chapter 8 is also known, using an infinitesimal notion

of braided groups [196]. Applications to construct quantum groups inductively, including non-standard ones, are in [197, 198]. The category $_B\tilde{\mathcal{C}}_C$ in the construction in [195] is related by dualisation to braided group crossed modules studied in [199].

The more explicit way of thinking about bosonisation as a particular class of ordinary cross products and cross coproducts was obtained by the author in the appendix of [34]. Key to this was the functor $_H\mathcal{M} \to {}^H_H\dot{\mathcal{M}}$ of braided categories introduced in [143]. Also in [34] is the required braided groups setting Proposition 9.4.14 for such biproducts. The proposition itself is basically due to D. Radford in [155], who had studied simultaneous cross products and coproducts of thie type, but at a time when braided categories and braided groups had yet to be invented. In particular, B in [155] was an exotic algebra-coalgebra obeying the condition equating the outer two expressions in (9.54) without a theory of braided tensor products in the middle. There is a further double biproduct related similarly to double bosonisation in [195]. Among other recent works, the interaction between bosonisation and the twisting of Chapter 2.3 is in [47]. The response of ${}^H_H\dot{\mathcal{M}}$ to twisting is in [44]. Meanwhile, an introduction to biproducts and bosonisations collecting both left handed and right handed versions of all formulae is [200].

Among related topics that we have not had room to cover in detail, the category ${}^H_H\dot{\mathcal{M}}$ can easily be identified with the category ${}^H_H\mathcal{M}^H_H$ of vector spaces which are both H-bimodules and H-bicomodules with the coactions bimodule maps (i.e., bicovariant bimodules). In one direction, given $V \in {}^H_H\dot{\mathcal{M}}$ the bicovariant bimodule is $V \otimes H$, where the (co)action of H from the left is the tensor product one, from the right the (co)multiplication of H. In the converse direction V is the subspace of the bimodule which is invariant under the right coaction. These observations are part of standard Hopf algebra theory but they were used by S.L. Woronowicz in [91] to identify a bicovariant differential calculus $\Omega^1(H)$ as of this form (here V is the space of right-invariant differential forms on the Hopf algebra H). We have met the universal calculus (on any algebra) at the end of Chapter 6.3; it corresponds to $\ker \epsilon \in {}^H_H\dot{\mathcal{M}}$ under left multiplication and the adjoint coaction (cf. Example 7.1.8 in a dual form). Other calculi correspond to quotients of $\ker \epsilon$. For example, the bicovariant differential calculi on $H = k[x]$ correspond to field extensions $k \subseteq k_\lambda$, where λ extends the field, and have the form [201]

$$\Omega^1(k[x]) = k_\lambda[x], \quad \mathrm{d}f(x) = \frac{f(x+\lambda) - f(x)}{\lambda}.$$

Also, bicovariant calculi on the coordinate algebras of all $U_q(g)$ for generic q have been classified and basically correspond to irreducible representations of g. For example, the (irreducible) calculi on $SU_q(2)$ are labelled by $j \in \frac{1}{2}\mathbb{N}$ and have dimension $(2j + 1)^2$. These classification results are due to the author in mid 1996 [202, 203] and make extensive use of the theory of $D(H)$ in Chapter 7 and the notion of factorisability. (Strictly speaking, there are also certain pathological variants of the basic calculi due to the fact that the relevant quantum groups are not quite factoris- able at the algebaraic level, as known already for some standard examples of calculi in [204].) Returning to the general theory of differential calculi, [91] also used the quantum double braiding in $^H_H\mathcal{M}$ to define a natural exterior algebra $\Omega^{\cdot}(H)$ where the braiding is used in defining the required 'antisymmetry'. Many of these constructions also work for bicovariant calculi on braided groups but with quotients of $\ker \epsilon$ as a representation of the double bosonisation of B rather than of $D(H)$ as in the Hopf alge- bra case. We will also see a different and more explicit approach to the braided group case in the next chapter, based on partial derivatives.

By now, there is an extensive list of very concrete applications of braided groups, some of which we will see in the next chapter. The braided matrices $B(R)$ obtained by transmuting $A(R)$ were already used in Chapters 4.3, 4.5 and 7.4, following the ideas in [94], [34] and [35] re- spectively; see the Notes to Chapters 4 and 7 for details. See also the Notes at the end of the next chapter.

10

Braided groups and q-deformation

This chapter is a kind of epilogue, in which we show how the machinery developed in the preceding chapters, some of it quite mathematical, can be used to provide the beginning of a kind of q-deformed geometry. It turns out that the underlying structure here is not so much a quantum group as one of the exotic braided groups which we have encountered in Chapter 9.4.2. Quantum groups still play a role as the *quantum symmetry* of such q-deformed spaces, but the spaces themselves tend to be braided ones. Thus we will need all the machinery developed so far in this book. Nevertheless, the problem of systematically q-deforming all the geometrical (and other) structures needed in physics is a deep and important one for physics, so we shall try to give here as self-contained and elementary a treatment as possible. It should be possible to come to this chapter directly after Chapter 4, using the intermediate chapters as reference for the mathematical underpinning when required. Also, we cover here only q-deformed or braided versions of $\mathbb{R}^n$, where the theory is fairly complete. Only when this is thoroughly understood could one reasonably expect to move on to define q-manifolds, etc. The further theory of braided geometry is deferred to a sequel to this book.

We have already covered one standard point of view on q-deforming geometrical structures in Chapter 4, namely as some kind of 'quantisation' of an algebra of functions. We consider a manifold in terms of its algebra of functions and allow this to become noncommutative. This point of view works quite well for the quantum group itself as a quantisation of a Poisson bracket or Lie bialgebra structure, as we have seen in Chapter 8. So the point of view of quantisation or noncommutative geometry is quite useful, at least as a mathematical analogy in which a deformation parameter t is treated like Planck's constant. In this usual formulation of noncommutative geometry, the tensor product of algebras (corresponding to direct product of the manifolds) is the usual one in which the factors commute. It is the algebras themselves which become noncommutative. This is an *inner noncommutativity* or *noncommutativity of the first kind* because it is a property within a quantum system or algebra.

The theme in the present chapter is to associate q not with such quantisation but rather with a different *outer noncommutativity* that can exist between independent systems. This is *noncommutativity of the second kind*, which is encountered in quantum physics when we consider fermions: independent fermionic systems anticommute rather than commute. So, the idea is to consider q as a generalisation of the -1 factor for fermions. In mathematical terms, it is the notion of the $\otimes$ product between algebras which we will q-deform and not directly the algebras themselves. These, as far as we are concerned, can remain classical or 'commutative', albeit in a deformed sense appropriate to the noncommutative tensor product.

This is conceptually quite a different role for q than its usual role as quantisation parameter. It turns out to be the key if one wants to q-deform not one algebra but an entire universe of structures: lines, planes, matrices, differentials, etc., in a systematic and consistent way. The reason is that we can use the systematic machinery of the preceding chapter to deform the entire category of vector spaces with its usual $\otimes$ to a braided category with tensor product $\otimes_q$. Most constructions in physics, and many in mathematics, take place in the category of vector spaces, so, by deforming the category itself, we also carry over all our favourite constructions without any further effort.

This braided approach to q-deformation is due to the author, as are most of the results in this chapter. The foundation has already been laid in Lemma 9.2.12, where we saw that two algebras in a braided category have a natural *braided tensor product algebra structure*

$$(a \otimes b)(d \otimes c) = a\Psi(b \otimes d)c,$$

where Ψ is the braiding. Here a, d are in one algebra and b, c are in another. These appear as subalgebras of the braided tensor product, but these two subalgebras no longer commute when Ψ is nontrivial. This noncommutativity is called the *braid statistics* between the two subalgebras. Specifying these cross commutation relations or braid statistics consistently between any two algebras in our universe is the role of the braiding Ψ and the reason that we need a braided category. On the other hand, we will take this concrete line wherever possible by specifying such braid statistics directly, so one does not need to master Chapter 9 to work effectively with this braided approach.

The idea of developing geometry with fermionic statistics (by working with Grassmann variables) is a standard one that comes under the heading of supergeometry. So we will be generalising that. We begin in Section 10.1 with the super-case and its mildest generalisation, where the statistics are given by a phase factor q. Then we cover, in Section 10.2, the more general quantum or braided spaces where the braiding can be a linear combination described by an R-matrix, rather than a simple phase factor. We describe braided vectors and covectors in this setting and formulate a *quantum metric* as an isomorphism between them. The algebras here are just the covariant vector and covector algebras $V(R,\lambda), V\tilde{}(\lambda, R)$ introduced in Chapter 4.5.2, but we have the machinery now to formulate on them an addition law by means of a braided coproduct

$$\underline{\Delta} x_i = x_i \otimes 1 + 1 \otimes x_i,$$

where the $\otimes$ is a braided one. In other words, these natural algebras (which include the so-called quantum plane) form braided groups in the sense of Definition 9.4.5. In Section 10.3, we cover matrices as well in this setting. Here, the braided analogue of the quantum matrices $A(R)$ of Chapter 4.1 are the braided matrices $B(R)$ which we have met in Chapters 4.3, 4.5 and 7.4 as natural covariant algebras. We give direct proofs of their multiplicative braided coproduct $\underline{\Delta} \mathbf{u} = \mathbf{u} \otimes \mathbf{u}$ and also see how they fit into a systematic braided linear algebra. In Section 10.4, we show how to differentiate on braided spaces and outline an integration and Fourier theory as well. With this machinery, one can go right up to the point of classical field theory and Green functions on our braided vector spaces.

Finally, in Section 10.5, we show how this general theory can be put to good use by giving detailed models of q-Euclidean spacetime and q-Minkowski spacetime in our approach. For q-Euclidean space, we use, in fact, a close variant $\bar{M}_q(2)$ of the 2×2 quantum matrices $M_q(2)$ from Chapter 4.1, and a related $*$-structure. For q-Minkowski space, we prefer to use instead 2×2 braided matrices $BM_q(2)$, since these have a matrix coproduct which is compatible with a natural Hermitian $*$-structure, corresponding to the usual description of Minkowski space as 2×2 Hermitian matrices. The two systems are 'gauge equivalent' as algebras by the twisting machinery from Chapter 2.3, but not as $*$-algebras.

Let us note that our point of view on q in this chapter does not preclude the possibility that other physical effects may induce these braid statistics. For example, it may be that $q \neq 1$ is a model of quantum corrections to spacetime geometry, in which case q may indeed involve $\hbar$. Alternatively, q may be considered as a regularisation parameter where the interesting case is $q = 1$ but where some infinities at $q = 1$ can be described as poles $(q-1)^{-1}$ by working in this more general q-geometry. Moreover, there is nothing in this approach which ties us to a single parameter q. Everything we do in this chapter works quite generally for a well-behaved solution R of the QYBE. Thus, what we describe here is more precisely a kind of braided or R-geometry.

10.1   Super and anyonic quantum groups

The simplest example of nontrivial statistics is provided by working with
superspaces and supergeometry. Here the vector spaces are $\mathbb{Z}_{/2}$-graded
into even (bosonic) and odd (fermionic) parts. A *superalgebra* means a
$\mathbb{Z}_{/2}$-graded algebra, i.e. such that the product map adds the degrees
(so the product of two fermionic elements is bosonic) and the identity is
bosonic. We have explained the categorical view on this in Chapter 9.2.
A *supercoalgebra* means, likewise, a linear map $\underline{\Delta}$ for which the sums of
the degrees of the tensor factors in each term of $\underline{\Delta}(b)$ is $|b|$ and which is
coassociative and has a counit $\underline{\epsilon}$. The latter must also preserve the degree
and therefore vanishes on all odd elements since the field is even. Next,
given two superalgebras, there is a natural supertensor product algebra
$\underline{\otimes}$, where the two tensor factors graded-commute according to the degree.
It is the $n = 2$ case of Example 9.2.15.

Definition 10.1.1 *A* superbialgebra *means a superalgebra and superco-*
algebra B with $\underline{\Delta} : B \to B\underline{\otimes}B$ an algebra homomorphism to the superten-
sor product algebra in the sense

$$\underline{\Delta}(bc) = \sum(-1)^{|b_{(2)}||c_{(1)}|}b_{(1)}c_{(1)} \otimes b_{(2)}c_{(2)}.$$

It is a super-Hopf algebra if there is also an antipode $\underline{S}$ which is a degree
preserving linear map obeying the usual axioms. Equivalently, a super-
Hopf algebra is a Hopf algebra in the symmetric category of supervector
spaces, which is the category of $\mathbb{Z}'_{/2}$-modules, where $\mathbb{Z}'_{/2}$ is the quasitri-
angular Hopf algebra in Example 2.1.6.

These axioms are as in Fig. 9.13, with $\Psi = \mathsf{X} = \pm 1$ according to the
degree of the elements being transposed. This is the supertransposition.
From the general theory (9.39), we know that $\underline{S}$ in our case will necessarily
be a graded-antihomomorphism

$$\underline{S}(bc) = (-1)^{|b||c|}(\underline{S}c)(\underline{S}b). \tag{10.1}$$

We can also read off the correct axioms for a superquasitriangular struc-
ture $\underline{\mathcal{R}} \in B\underline{\otimes}B$. This should be of total degree 0 and obey Fig. 9.13(c)
with

$$\underline{\Delta}^{\mathrm{op}}b = \sum(-1)^{|b_{(1)}||b_{(2)}|}b_{(2)} \otimes b_{(1)}. \tag{10.2}$$

Probably the simplest example is the Grassmann plane $\mathbb{C}^{0|1} = \mathbb{C}[\theta]$ with
odd coordinate θ and relations $\theta^2 = 0$. Now, just as the usual polynomials
$\mathbb{C}[X]$ form a Hopf algebra which is like functions on the real line, so
this algebra forms a Hopf algebra, which is, by definition, considered as

'functions on the superline', even there is no underlying usual space here. The coproduct and antipode are

$$\underline{\Delta}\theta = \theta \otimes 1 + 1 \otimes \theta, \quad \underline{S}\theta = -\theta.$$

This is consistent with the relations *provided* the homomorphism $\underline{\Delta}$: $\mathbb{C}[\theta] \to \mathbb{C}[\theta]\underline{\otimes}\mathbb{C}[\theta]$ is defined with $\underline{\otimes}$ the supertensor product algebra. Thus,

$$\begin{aligned}
\underline{\Delta}\theta^2 &= (\theta \otimes 1 + 1 \otimes \theta)(\theta \otimes 1 + 1 \otimes \theta) \\
&= \theta^2 \otimes 1 + \theta \otimes \theta - \theta \otimes \theta + 1 \otimes \theta^2
\end{aligned}$$

is consistent with the relations because of the minus sign when computing $(1 \otimes \theta)(\theta \otimes 1) = -\theta \otimes \theta$. This has a simple extension or deformation as a superquasitriangular Hopf algebra (superquantum group) analogous to Example 2.1.19 in the bosonic case:

Example 10.1.2 $U_\alpha(0|1)$ *is the superquasitriangular super-Hopf algebra with one odd variable* θ, *with* $\theta^2 = 0$ *as above and*

$$\mathcal{R} = 1 - \alpha\theta \otimes \theta$$

for all α.

Proof: We verify directly from the axioms in Fig. 9.13. Thus, $\mathcal{R}\underline{\Delta}\theta = (1 - \alpha\theta \otimes \theta)(\theta \otimes 1 + 1 \otimes \theta) = \underline{\Delta}\theta = \underline{\Delta}^{\mathrm{op}}\theta = (\underline{\Delta}^{\mathrm{op}}\theta)\mathcal{R}$, rather trivially as $\theta^2 = 0$. Similarly for the other axioms. ∎

Another standard construction in physics is the notion of a $\mathbb{Z}_{/2}$-graded or super-Lie algebra g as defined by a degree-preserving linear map $\{\ ,\ ]$: $g \otimes g \to g$ obeying

$$\{\xi, \eta] = (-1)^{|\xi||\eta|+1}\{\eta, \xi],$$

$$\{\xi, \{\eta, \zeta]] = (-1)^{|\xi||\eta|}\{\eta, \{\xi, \zeta]] + \{\{\xi, \eta], \zeta], \quad \forall \xi, \eta, \zeta \in g. \tag{10.3}$$

These are the usual Lie algebra axioms with ± 1 inserted whenever a supertransposition is used.

Example 10.1.3 *Let* g *be a super-Lie algebra. Then there is a super-Hopf algebra* $U(g)$ *defined as generated by* g *and relations* $\xi\eta - (-1)^{|\xi||\eta|}\eta\xi = \{\xi, \eta]$ *and supercoproduct, counit and antipode*

$$\underline{\Delta}\xi = \xi \otimes 1 + 1 \otimes \xi, \quad \underline{\epsilon}\xi = 0, \quad \underline{S}\xi = -\xi,$$

extended as an algebra homomorphism to the supertensor product in the case of $\underline{\Delta}$ *and (10.1) in the case of* $\underline{S}$.

Proof: This generalises Example 1.5.7. We extend $\underline{\Delta}$ to products as an algebra homomorphism to the supertensor product algebra, and have to check that this is consistent. Thus, $\underline{\Delta}(\xi\eta) = (\xi\otimes 1+1\otimes\xi)(\eta\otimes 1+1\otimes\eta) = \xi\eta\otimes 1+\xi\otimes\eta+(-1)^{|\xi||\eta|}\eta\otimes\xi+1\otimes\xi\eta$. Subtracting from this the same calculation for $(-1)^{|\xi||\eta|}\underline{\Delta}(\eta\xi)$ gives $\{\xi,\eta\}\otimes 1+1\otimes\{\xi,\eta\} = \underline{\Delta}\{\xi,\eta\}$ as it should. Similarly for higher orders. ∎

For example, the super-Lie algebra $gl(1|1)$ has even generators C, N and odd generators θ_+, θ_- and bracket

$$\{N, \theta_\pm\} = \pm\theta_\pm, \quad \{\theta_+, \theta_-\} = C,$$
$$\{\theta_\pm, \theta_\pm\} = 0, \quad \{C, N\} = \{C, \theta_\pm\} = 0, \tag{10.4}$$

so the enveloping super-Hopf algebra is defined with commutators $[N, \theta_\pm] = \pm\theta_\pm$ and anticommutator $\{\theta_+, \theta_-\} = C$, $\theta_\pm^2 = 0$ and C central in view of the degrees of the generators.

Another example is the super-Lie algebra $osp(1|2)$ with generators even $H, X_\pm$, odd generators $\theta\pm$ and bracket

$$\{X_+, X_-\} = H, \quad \{H, X_\pm\} = \pm 2X_\pm, \quad \{H, \theta_\pm\} = \pm\theta_\pm,$$
$$\{X_\pm, \theta_\mp\} = \pm\theta_\pm, \quad \{X_\pm, \theta_\pm\} = 0, \tag{10.5}$$
$$\{\theta_\pm, \theta_\pm\} = \pm\frac{1}{2}X_\pm, \quad \{\theta_+, \theta_-\} = -\frac{1}{4}H.$$

The enveloping super-Hopf algebra therefore has relations like $[H, \theta_\pm] = \pm\theta_\pm$ and $\{\theta_+, \theta_-\} = -\frac{1}{4}H$, etc., according to the degrees involved. It is common to regard this enveloping algebra as generated by $\theta_\pm, H$ with these relations and $X_\pm$ defined by $X_\pm = \pm 4\theta_\pm^2$. A typical application is to map this algebra onto the quantum harmonic oscillator $[a, a^\dagger] = 1$ by the realisation $\theta_+ = 8^{-\frac{1}{2}}a^\dagger$, $\theta_- = -8^{-\frac{1}{2}}a$, and $H = a^\dagger a$. The natural Hermiticity properties implied by this mean that the $H, X_\pm$ should be viewed as forming an $su(1, 1)$ subalgebra. One says that $osp(1|2)$ is the *spectrum generating algebra* for the harmonic oscillator.

In addition, many super-Lie algebras also have q-deformations as super-Hopf algebras. Thus there is a q-deformed $U_q(gl(1|1))$ with superquasitriangular structure given by a finite power series in $\theta_\pm$ like that in Example 10.1.2. There is also a q-deformation $U_q(osp(1|2))$ which maps onto the q-harmonic oscillator of Example 3.1.8 and which has an infinite power series for $\mathcal{R}$ like that for $U_q(su_2)$. There is also a theory of superquantum groups of function algebra type and their duality with quantum superenveloping algebras, much as in the bosonic theory in Chapter 4. It is not our goal to present this parallel theory in detail here: we concentrate rather on more conceptual issues related to our braided theory at the end of Chapter 9.4.2.

Proposition 10.1.4 *Let H be a quasitriangular Hopf algebra containing a group-like element g of order 2. Then H has a corresponding superquasitriangular super-Hopf algebra $B = B(\mathbb{Z}'_{/2}, H)$ given by the same algebra and*

$$\underline{\Delta} b = \sum b_{(1)} g^{-|b_{(2)}|} \otimes b_{(2)}, \quad \underline{\epsilon} b = \epsilon b, \quad \underline{S} b = g^{|b|} S b,$$

$$\underline{\mathcal{R}} = \mathcal{R}_g^{-1} \sum \mathcal{R}^{(1)} g^{-|\mathcal{R}^{(2)}|} \otimes \mathcal{R}^{(2)},$$

with the degree defined by $gbg^{-1} = (-1)^{|b|} b$.

Proof: This is an application of the transmutation theory explained below Example 9.4.9. We consider the Hopf algebra $\mathbb{Z}'_{/2}$ which generates the category of superspaces as explained in Example 9.2.5. It is defined by $g^2 = 1$ and $\Delta g = g \otimes g$, so, under our assumptions, we have an inclusion $\imath : \mathbb{Z}'_{/2} \to H$. The formulae stated below Example 9.4.9 immediately give the ones above for the example $B(\mathbb{Z}'_{/2}, H)$. Here, $\mathcal{R}_g$ is the universal R-matrix or triangular structure of $\mathbb{Z}'_{/2}$. The action of g is defined by the adjoint action and determines the degree of a homogeneous element as stated. ∎

This proposition is useful because, if a Hopf algebra has strange nilpotent elements which square to 0, then it may well be that it looks more natural after transmutation to a super-Hopf algebra where such elements are common. For example, the strange nonstandard Hopf algebra in Example 4.2.14 is just $U_q(gl(1|1))$ after a suitable variant of this transmutation theorem. We also have the theory of bosonisation in the other direction as explained at the end of Chapter 9.4.2. In the super-setting, it becomes:

Proposition 10.1.5 *Corresponding to every superquasitriangular super-Hopf algebra B is an ordinary quasitriangular Hopf algebra $B \rtimes \mathbb{Z}'_{/2}$, its bosonisation, consisting of B extended by adjoining an element g with relations, coproduct, counit, antipode and quasitriangular structure*

$$g^2 = 1, \quad gb = (-1)^{|b|} bg, \quad \Delta g = g \otimes g, \quad \Delta b = \sum b_{(1)} g^{|b_{(2)}|} \otimes b_{(2)},$$

$$Sg = g, \quad Sb = g^{-|b|} \underline{S} b, \quad \epsilon g = 1, \quad \epsilon b = \underline{\epsilon} b,$$

$$\mathcal{R} = \mathcal{R}_g \sum \underline{\mathcal{R}}^{(1)} g^{|\underline{\mathcal{R}}^{(2)}|} \otimes \underline{\mathcal{R}}^{(2)},$$

where $\mathcal{R}_g$ is from Example 2.1.7. Moreover, the representations of the bosonised Hopf algebra are precisely the super-representations of the original B.

Proof: This is an immediate corollary of Theorem 9.4.12, with the triangular Hopf algebra $\mathbb{Z}'_{/2}$ generating the category as explained in Example 9.2.5. The cross product by this simply means to adjoin it with cross relations given by the action of the generator g of $\mathbb{Z}'_{/2}$. Since the action is by the degree, we have the cross relations as stated. The cross coproduct is given from the triangular structure $\mathcal{R}_g$ of $\mathbb{Z}'_{/2}$. Likewise for the rest of the Hopf algebra structure. Naturally, one can also verify this proposition directly from the axioms of (super)-Hopf algebras if one does not want to go through the general theory. Note that in Theorem 9.4.12 one can add that if the category is symmetric rather than braided and $\underline{\Delta}^{\mathrm{op}}$ is determined in the obvious way from $\underline{\Delta}$, then a quasitriangular structure of B corresponds to one on the bosonisation. This includes the super-case and gives the formula stated for $\mathcal{R}$. For the second part we note that a super-representation of a superalgebra is a vector space that splits into odd and even parts $V_0 \oplus V_1$ such that the even operators in the superalgebra take the block form $\begin{pmatrix} * & 0 \\ 0 & * \end{pmatrix}$ and the odd operators take the form $\begin{pmatrix} 0 & * \\ * & 0 \end{pmatrix}$. Given such a super-representation of B, we just set $g = \begin{pmatrix} 1 & 0 \\ 0 & -1 \end{pmatrix}$ to obtain a representation of the bosonised $B {\rtimes} \mathbb{Z}'_{/2}$. Conversely, given a representation V of the bosonised Hopf algebra, we decompose V into eigenspaces V_0 and V_1 for the projection operator $\frac{1-g}{2}$.   ∎

This proposition tells us that the theory of super-Lie algebras and super-Hopf algebras is strictly contained in the theory of ordinary Hopf algebras, i.e. we do not have to work with them. As an example, the bosonisation of Example 10.1.1 is the triangular Hopf algebra in Example 2.1.7. This bosonisation can be quite useful if we want to realise the Hopf algebra in a way that avoids the ± 1 statistics in the super-theory. Essentially, the statistics are absorbed into the relations of g. Such a process is an algebraic analogue of the *Jordan–Wigner transform* in physics. For example, one can bosonise $osp(1|2)$.

Next we proceed to our first and simplest generalisation of the theory of super-Hopf algebras to the truly braided case. We simply replace the fermionic anticommutativity factor -1 in the super-case by $e^{\frac{2\pi i}{n}}$. We work in the category $\mathcal{C}_n$ of *anyonic* or $\mathbb{Z}/n$-graded vector spaces with anyonic transposition Ψ defined as in Example 9.2.5 for general n. The category is strictly braided when $n > 2$. We use the anyonic tensor product $\underline{\otimes}$ from Example 9.2.15.

Definition 10.1.6 *An anyonic Hopf algebra is an algebra and coalgebra B with maps that respect the degree additively modulo n, and for which $\underline{\Delta} : B \to B \underline{\otimes} B$ is a homomorphism in the sense*

$$\underline{\Delta}(bc) = e^{\frac{2\pi i |b_{(2)}||c_{(1)}|}{n}} b_{(1)} c_{(1)} \otimes b_{(2)} c_{(2)}$$

for all $b, c \in B$. The antipode should be a degree-preserving linear map obeying the usual axioms. Equivalently, an anyonic Hopf algebra is a Hopf algebra in the braided category of anyonic vector spaces, which is the category $\mathbb{Z}'_{/n}$-modules, where $\mathbb{Z}'_{/n}$ is the quasitriangular Hopf algebra in Example 2.1.6.

The axioms are as in Fig. 9.13 with Ψ now the anyonic braiding. The antipode is then necessarily a braided antihomomorphism as in (9.39),

$$\underline{S}(bc) = e^{\frac{2\pi i |b||c|}{n}} (\underline{S}c)(\underline{S}b). \tag{10.6}$$

Probably the simplest example is as follows.

Example 10.1.7 *The one-dimensional anyonic enveloping algebra or anyonic line $U_n(1)$ is defined by one generator x with*

$$|x| = 1, \quad x^n = 0, \quad \underline{\Delta}x = x \otimes 1 + 1 \otimes x, \quad \underline{S}x = -x$$

extended as a braided-Hopf algebra in C_n, i.e. remembering the anyonic statistics of x. Explicitly,

$$\underline{\Delta}x^m = \sum_{r=0}^{m} \begin{bmatrix} m \\ r \end{bmatrix} x^r \otimes x^{m-r}, \quad \begin{bmatrix} m \\ r \end{bmatrix} = \frac{[m]!}{[r]![m-r]!}, \quad [m] = \frac{e^{\frac{2\pi i m}{n}} - 1}{e^{\frac{2\pi i}{n}} - 1},$$

$$\underline{S}x^m = e^{\frac{\pi i m(m-1)}{n}} (-x)^m.$$

Proof: One can easily check by induction that this is well-defined. For example, using Ψ to take one x past another x, we have $\underline{\Delta}x^2 = (x \otimes 1 + 1 \otimes x)^2 = x^2 \otimes 1 + 1 \otimes x^2 + x \otimes x + \Psi(x \otimes x) = x^2 \otimes 1 + 1 \otimes x^2 + (1 + e^{\frac{2\pi i}{n}})x \otimes x$. For the general case, we use a different notation, namely we compute the anyonic tensor product $U_n(1) \underline{\otimes} U_n(1)$ as generated by x, x', say, for the two independent copies. Then the anyonic tensor product algebras have the relations

$$x'x = e^{\frac{2\pi i}{n}} xx', \quad x^n = 0, \quad x'^n = 0.$$

This follows from Example 9.2.5, as explained there. On the other hand, in these terms $\underline{\Delta}x = x + x'$, as an element of this tensor product algebra. Since it is an algebra homomorphism, we can compute $\underline{\Delta}x^m = (\underline{\Delta}x)^m$ working in this algebra and using the q-binomial theorem as in

Lemma 3.2.2. For the braided antipode we use (10.6) or just check that it obeys the antipode axioms for this anyonic coproduct.   ∎

As before, we have a transmutation theorem: if a Hopf algebra H contains a group-like element g of order n, we consider it as an inclusion $\mathbb{Z}'_{/n} \subseteq H$ and obtain an anyonic Hopf algebra $B = B(\mathbb{Z}'_{/n}, H)$. We have stated the formulae in Proposition 10.1.4 in such a way that they work in this case too with

$$gbg^{-1} = e^{\frac{2\pi i|b|}{n}}b, \quad \underline{\Delta}^{\mathrm{op}}b = \sum e^{-\frac{2\pi i|b_{(1)}||b_{(2)}|}{n}}b_{(2)}g^{-2|b_{(1)}|} \otimes b_{(1)}, \quad (10.7)$$

where the latter is no longer of an obvious form $\Psi^{-1} \circ \underline{\Delta}$ as it was in the super-case.

For example, we can take H as the group algebra of a finite non-Abelian group containing an element g of order n. To be concrete, we take, for our example, the group S_3, the permutation group on three elements, regarded as the symmetries of an equilateral triangle with fixed vertices 0,1,2, numbered clockwise. Let g denote a clockwise rotation of the triangle by $\frac{2\pi}{3}$ and let R_a denote reflections about the bisector through the fixed vertex a. Let $\mathbb{C}S_3$ denote the group Hopf algebra of S_3. It has basis $\{1, g, g^2, R_0, R_1, R_2\}$. Of course, there are many ways to work with S_3: we present it in a way that makes the generalisation to higher n quite straightforward.

Example 10.1.8 *The transmutation BS_3 of $\mathbb{C}S_3$ is the following anyonic group in $\mathcal{C}_3$. Some homogeneous elements are*

$$r_a = \frac{1}{3}\sum_{b=0}^{b=2} e^{-\frac{2\pi i ab}{3}}R_b, \qquad |r_a| = a.$$

Together with $1, g, g^2$ of degree zero, they form a basis of BS_3 as an anyonic vector space. Its anyonic dimension is $\underline{\dim}BS_3 = 2e^{-\frac{\pi i}{3}}$. Its algebra and counit are those of $\mathbb{C}S_3$, but now

$$\underline{\Delta}r_a = \underline{\Delta}^{\mathrm{op}}r_a = \sum_{c=0}^{c=2} e^{-\frac{2\pi i c(a-c)}{3}}r_c \otimes r_{a-c}, \quad \underline{S}r_a = e^{-\frac{2\pi i a^2}{3}}r_a, \quad \underline{R} = \mathcal{R}_g^{-1}.$$

Proof: The reflections have the property that $gR_ag^{-1} = R_{a+1}$ (mod 3). Hence, their inverse Fourier transforms r_a as shown are homogeneous of degree as stated. The Hopf algebra structure on g (of degree zero) is unmodified. The usual coproduct in the remainder of $\mathbb{C}S_3$ is $\Delta R_a = R_a \otimes R_a$, hence $\Delta r_a = \sum_c r_c \otimes r_{a-c}$. This then becomes modified as $\underline{\Delta}r_a = \sum_c r_c g^{c-a} \otimes r_{a-c}$. Now note that, in S_3, $R_ag = gR_ag^{-1} = g\triangleright R_a$

for all a. Hence, $r_a g = g \triangleright r_a = e^{\frac{2\pi \imath a}{3}} r_a$, giving the result shown. Likewise, the original antipode on the R_a is $SR_a = R_a^{-1} = R_a$. Hence, $Sr_a = r_a$ also. From this and $g^{-1}R_a = g \triangleright R_a$ for all a (so that $g^{-1}r_a = e^{\frac{2\pi \imath a}{3}} r_a$) we obtain $\underline{S}$ as shown. The computation for $\underline{\Delta}^{\mathrm{op}}$ is similar to that for $\underline{\Delta}$ and comes out the same. The unmodified $\mathcal{R}$ of $\mathbb{C}S_3$ is $\mathcal{R} = 1 \otimes 1$, so that $\underline{\mathcal{R}} = \mathcal{R}_g^{-1}$. ∎

In the other direction, any anyonic Hopf algebra B gives rise to an ordinary Hopf algebra $B \!\rtimes\! \mathbb{Z}'/_n$. Again the formulae are as stated in Proposition 10.1.5, with cross relations now of the new form with $gb = e^{\frac{2\pi \imath |b|}{n}} bg$. We do not make a claim about the quasitriangular structure in this case as we do not *a priori* have a canonical choice for $\underline{\Delta}^{\mathrm{op}}$. For example, the bosonisation of the anyonic line in Example 10.1.7 is a reduced form of the Weyl algebra in Example 1.3.2; it has the additional relation $x^n = 0$ which appropriate to q an nth root of unity. It is also a q-deformed Borel subalgebra from the point of view in Chapter 3.4.

Finally, we explain the $n = \infty$ limit of the above, i.e. we can work with the category $\mathcal{C}_{\infty,q}$ of $\mathbb{Z}$-graded vector spaces with braiding given by q. This works over any field k and any $q \neq 0$. Objects are vector spaces decomposing as $V = \oplus_{a \in \mathbb{Z}} V_a$, where the $|v| = a$ for $v \in V_a$. The braiding is $\Psi(v \otimes w) = q^{|v||w|} w \otimes v$ and depends on q.

This category can be thought of as the category of representations of a quantum group defined like the above with integers modulo n replaced by the group $\mathbb{Z}$ of integers and with a formally defined quasitriangular structure to play the role of that of $\mathbb{Z}'/_n$. We use the enveloping algebra of $U_q(1)$ of $u(1)$ in Example 2.1.19 with the nonstandard universal R-matrix $\mathcal{R} = q^{\xi \otimes \xi}$, which gives the above braiding but requires us to work over formal power series in a deformation parameter. As usual, to work algebraically it is better not to work with enveloping algebra deformations but dually with function algebras and comodules. We have already seen this in Chapter 4. In our case, the dual of $U_q(1)$ is the quantum group of functions S_q^1, which has as an algebraic model the group algebra $k\mathbb{Z} = k[\varsigma, \varsigma^{-1}]$ generated by a single free element ς and its inverse, equipped with a nonstandard dual quasitriangular structure

$$\mathcal{R}(\varsigma^m \otimes \varsigma^n) = q^{mn}. \tag{10.8}$$

We denote this quantum group $\mathbb{Z}_q$, where the q enters only into the dual quasitriangular structure. It is an example of the construction in Example 2.2.5 with bicharacter defined by q. This time we can generate precisely our category $\mathcal{C}_{\infty,q}$ as the braided category of right comodules of $\mathbb{Z}_q$ using Exercise 9.2.9. The grading corresponds to a coaction $v \mapsto v \otimes \varsigma^{|v|}$. If we do not need to have dual objects in our braided category, then we can take for our background quantum group $\mathbb{N}_q$, consisting of $k[\varsigma] = k\mathbb{N}$ as a bialgebra but again with nonstandard dual quasitriangular structure as above. This time, we generate a category of $\mathbb{N}$-graded spaces, where $\mathbb{N}$ denotes natural numbers, including zero.

Definition 10.1.9 *A q-statistical Hopf algebra is an algebra and coalgebra in the braided category of $\mathbb{Z}_q$-comodules (or $\mathbb{Z}$-graded spaces) with $\underline{\Delta}$ an algebra homomorphism to the braided tensor product algebra in this category, and $\underline{S}$ as usual.*

Explicit formulae are just as in Definition 10.1.6 and (10.6) with q in place of $e^{\frac{2\pi i}{n}}$. As a simple example, we have the *braided line*. It is just the usual algebra $B = k[x]$ of polynomials in x, but we regarded it as a braided-Hopf algebra with

$$\underline{\Delta} x = x \otimes 1 + 1 \otimes x, \quad \underline{S} x = -x, \quad \underline{\epsilon} x = 0,$$

$$|x^n| = n, \quad \Psi(x^m \otimes x^n) = q^{mn} x^n \otimes x^m,$$

(10.9)

and we do not insist that q is a root of unity as we did in Example 10.1.7. So it could, for example, be a real parameter near to $q = 1$. The $\mathbb{Z}$-grading or the coaction of ς is by the power of x. This particular example actually lives in the subcategory of $\mathbb{N}$-graded spaces, since all the degrees are positive. If we think physically, with x in units of length, then the grading and the coaction of ς is just given by the scaling dimension; i.e. one can and should think of ς as the generator of scale transformations (the so-called dilaton). It is the quantum group with respect to which our braided line is covariant, i.e. the braided line lives in its braided category of comodules.

Transmutation from the module point of view converts any Hopf algebra H containing a primitive element ξ to a q-statistical Hopf algebra $B(U_q(1), H)$. The action of ξ is given by the commutator viewed in H, and its eigenvalues are the degrees. The formulae are infinitesimal versions of those in Proposition 10.1.4 with -1 replaced by q. In the reverse direction, we can bosonise from the module point of view of Theorem 9.4.12, in which case we obtain for every q-statistical Hopf algebra B an ordinary Hopf algebra $B \rtimes U_q(1)$ over formal power series. As usual, we can then consider $g = q^\xi$ as a single invertible group-like generator. In this case, the resulting bosonisation formulae are just as in Proposition 10.1.5 with -1 replaced by q. The simplest example $k[x] \rtimes k[g, g^{-1}]$ gives us the Weyl algebra of Example 1.3.2, which is one of the first Hopf algebras that we encountered.

For a more formal derivation of substantially the same bosonisation formulae, we work in a dual setting with B regarded as living in the category of $\mathbb{Z}_q$-comodules. We use the general dual bosonisation formulae for this setting, as given in detail at the end of Chapter 9.4. Computing these for our $\mathcal{R}$ above, we arrive at the same result that any q-statistical Hopf algebra converts to an ordinary bialgebra $\mathbb{N}_q{\bowtie}B$ or Hopf algebra $\mathbb{Z}_q{\bowtie}B$ by bosonisation. The resulting formulae in this case are

$$b\varsigma = q^{|b|}\varsigma b, \quad \Delta\varsigma = \varsigma \otimes \varsigma, \quad \Delta b = \sum b_{(1)} \otimes \varsigma^{|b_{(1)}|}b_{(2)},$$

$$S\varsigma = \varsigma, \quad Sb = (\underline{S}b)\varsigma^{-|b|}, \quad \epsilon\varsigma = 1, \quad \epsilon b = \underline{\epsilon}b, \tag{10.10}$$

which is a right handed version of Proposition 10.1.5 in our setting for general q. The dual of Theorem 9.4.12 ensures not only that this is a Hopf algebra, but also that its comodules are in correspondence with the $\mathbb{Z}$-graded comodules of B. A graded comodule is a graded vector space V with coaction which respects the grading. The corresponding coaction is to first apply the B-coaction and then the coaction of $\mathbb{Z}_q$ determined by the grading.

For example, we could take $k[p]$, a copy of the braided line, coacting on $k[x]$ by the coproduct $\underline{\Delta}$. Then the bosonisation $\mathbb{Z}_q{\bowtie}k[p]$ coacts on $k[x]$ by $x \mapsto x \otimes \varsigma + 1 \otimes p$. It could be thought of as a deformation of the group of translations and scale transformations in one dimension according to the picture above. Meanwhile, double bosonisation of the braided line gives $U_q(sl_2)$.

The obvious generalisation to higher dimensions is with a $\mathbb{Z}^n$-grading and q replaced by the exponential of a bilinear form (in fact, we can take grading by any Abelian group equipped with a bicharacter). Braided groups in this setting are also called q-statistical since the braiding is a simple factor. Important braided groups of this type are $U_q(n_\pm)$ associated to a Cartan matrix. For example, $U_q(n_+)$ for sl_3 is $k\langle E_1, E_2\rangle$ modulo the q-Serre relations

$$E_1^2 E_2 + E_2 E_1^2 - (q + q^{-1})E_1 E_2 E_1 = 0,$$

$$E_2^2 E_1 + E_1 E_2^2 - (q + q^{-1})E_2 E_1 E_2 = 0.$$

The coalgebra and antipode on the generators have the same linear form as above, but the braiding is

$$\Psi(E_i \otimes E_j) = q^{a_{ij}} E_j \otimes E_i,$$

where a_{ij} is the standard Cartan matrix of sl_3. The bosonisation of $U_q(n_+)$ is then $U_q(b_+)$ by similar computations to those above in the 1-dimensional case. Moreover, it can be shown that $U_q(n_+)$ and $U_q(n_-)$ are dually paired as braided groups; their double bosonisation recovers $U_q(g)$.

10.2   Braided vectors and covectors

Now we move on to a class of braided groups of a more complicated kind where the braiding is not just a phase factor $e^{\frac{2\pi i}{n}}$ or q, but possibly some more complicated linear combination provided by an R-matrix. The construction (which is due to the author) precisely generalises the braided line from the last section to the n-dimensional case. The role of $\mathbb{Z}_q$ is played now by the bialgebras $A(R)$ from Chapter 4.1, or, in the regular case, a Hopf algebra $\widetilde{A}$ obtained from it. We emphasise first an explicit matrix picture in which we work in a braided category generated by the R-matrix, and then come to this background quantum group at the end of the section when we study covariance.

Theorem 10.2.1 *Let R be an invertible matrix in $M_n \otimes M_n$ obeying the QYBE and suppose that R' is another invertible matrix such that*

$$R'_{12}R_{13}R_{23} = R_{23}R_{13}R'_{12}, \quad R_{12}R_{13}R'_{23} = R'_{23}R_{13}R_{12}, \qquad (10.11)$$

$$(PR+1)(PR'-1) = 0, \qquad (10.12)$$

where P is the permutation matrix. Then the braided covector algebra $V^{\check{}}(R',R)$ defined by generators $1, x_i$ and relations

$$x_i x_j = x_b x_a R'^a{}_i{}^b{}_j, \quad i.e. \quad \mathbf{x}_1\mathbf{x}_2 = \mathbf{x}_2\mathbf{x}_1 R',$$

forms a braided bialgebra with

$$\underline{\Delta} x_i = x_i \otimes 1 + 1 \otimes x_i, \quad \underline{\epsilon} x_i = 0,$$

$$\Psi(x_i \otimes x_j) = x_b \otimes x_a R^a{}_i{}^b{}_j, \quad i.e. \quad \Psi(\mathbf{x}_1 \otimes \mathbf{x}_2) = \mathbf{x}_2 \otimes \mathbf{x}_1 R,$$

extended multiplicatively with braid statistics. If also

$$R_{21}R' = R'_{21}R, \qquad (10.13)$$

then we have a braided-Hopf algebra with braided antipode $\underline{S}x_i = -x_i$ extended antimultiplicatively with braid statistics. This braided bialgebra or Hopf algebra lives in the braided category generated by R.

Proof: $V^{\check{}}(R',R)$ is by definition an associative algebra. We use here and throughout the compact notation of Chapter 4, where numerical suffixes as in $R'_{12}, \mathbf{x}_1, \mathbf{x}_2$, etc., refer to the position in a matrix tensor product. We have to check that $\Psi, \underline{\Delta}, \underline{S}$ are well-defined when extended to products. First, Ψ extends to tensor products according to the rules of a braiding (R generates a braided category), namely

$$\Psi(\mathbf{x}_1 \otimes (\mathbf{x}_2 \otimes \mathbf{x}_3)) = (\mathrm{id} \otimes \Psi)(\Psi(\mathbf{x}_1 \otimes \mathbf{x}_2) \otimes \mathbf{x}_3)$$

$$= \mathbf{x}_2 \otimes \Psi(\mathbf{x}_1 \otimes \mathbf{x}_3)R_{12} = \mathbf{x}_2 \otimes \mathbf{x}_3 \otimes \mathbf{x}_1 R_{13}R_{12}$$

etc. The extension to products is then in such a way that Ψ is functorial with respect to the product, in the sense

$$\Psi(\mathbf{x}_1 \otimes \mathbf{x}_2\mathbf{x}_3) = (\cdot \otimes \mathrm{id})\Psi(\mathbf{x}_1 \otimes (\mathbf{x}_2 \otimes \mathbf{x}_3)) = \mathbf{x}_2\mathbf{x}_3 \otimes \mathbf{x}_1 R_{13}R_{12}.$$

To see that this extension is well-defined, we compute also

$$\Psi(\mathbf{x}_1 \otimes \mathbf{x}_3\mathbf{x}_2 R'_{23}) = (\cdot \otimes \mathrm{id})\Psi(\mathbf{x}_1 \otimes (\mathbf{x}_3 \otimes \mathbf{x}_2))R'_{23} = \mathbf{x}_3\mathbf{x}_2 \otimes \mathbf{x}_1 R_{12}R_{13}R'_{23},$$

which is consistent with the relation $\mathbf{x}_2\mathbf{x}_3 = \mathbf{x}_3\mathbf{x}_2 R'_{23}$ given the first of conditions (10.11). Hence, $\Psi(\mathbf{x}_1 \otimes (\ ))$ is a well-defined map on the algebra $V^{\check{}}(R', R)$. One can then compute in the same way from functoriality that

$$\Psi(\mathbf{x}_1 \otimes \mathbf{x}_2\mathbf{x}_3 \cdots \mathbf{x}_N) = \mathbf{x}_2\mathbf{x}_3 \cdots \mathbf{x}_N \otimes \mathbf{x}_1 R_{1N} \cdots R_{12}. \tag{10.14}$$

Using this, we compute in a similar way

$$\begin{aligned}
\Psi(\mathbf{x}_1\mathbf{x}_2 \otimes \mathbf{x}_3 \cdots \mathbf{x}_N) &= (\mathrm{id} \otimes \cdot)\Psi((\mathbf{x}_1 \otimes \mathbf{x}_2) \otimes \mathbf{x}_3 \cdots \mathbf{x}_N) \\
&= (\mathrm{id} \otimes \cdot)\Psi(\mathbf{x}_1 \otimes \mathbf{x}_3 \cdots \mathbf{x}_N) \otimes \mathbf{x}_2 R_{2N} \cdots R_{23} \\
&= \mathbf{x}_3 \cdots \mathbf{x}_N \otimes \mathbf{x}_1\mathbf{x}_2 R_{1N} \cdots R_{13}R_{2N} \cdots R_{23} \\
&= \mathbf{x}_3 \cdots \mathbf{x}_N \otimes \mathbf{x}_2\mathbf{x}_1 R'_{12}R_{1N} \cdots R_{13}R_{2N} \cdots R_{23},
\end{aligned}$$

$$\begin{aligned}
\Psi(\mathbf{x}_2\mathbf{x}_1 R'_{12} \otimes \mathbf{x}_3 \cdots \mathbf{x}_N) &= (\mathrm{id} \otimes \cdot)\Psi(\mathbf{x}_2 \otimes \mathbf{x}_3 \cdots \mathbf{x}_N) \otimes \mathbf{x}_1 R_{1N} \cdots R_{13}R'_{12} \\
&= \mathbf{x}_3 \cdots \mathbf{x}_N \otimes \mathbf{x}_2\mathbf{x}_1 R_{2N} \cdots R_{23}R_{1N} \cdots R_{13}R'_{12}.
\end{aligned}$$

Here $R'_{12}R_{1N} \cdots R_{13}R_{2N} \cdots R_{23} = R'_{12}R_{1N}R_{2N} \cdots R_{13}R_{23}$ since matrices living in disjoint tensor factors commute. We can then repeatedly use the second of (10.11) to move R'_{12} to the right to give $R_{2N}R_{1N} \cdots R_{23}R_{13}R'_{12}$ $= R_{23}R_{1N} \cdots R_{13}R'_{12}$. Hence, Ψ is well-defined and functorial with respect to the product. The general form is an array,

$$\begin{aligned}
\Psi(\mathbf{x}_1 &\cdots \mathbf{x}_M \otimes \mathbf{x}_{M+N} \cdots \mathbf{x}_{M+1}) \\
&= \mathbf{x}_{M+N} \cdots \mathbf{x}_{M+1} \otimes \mathbf{x}_M \cdots \mathbf{x}_1 \ R_{1\,M+1}R_{1\,M+2} \cdots R_{1\,M+N} \\
&\qquad\qquad\qquad\qquad\qquad\qquad R_{2\,M+1}R_{2\,M+2} \cdots R_{2\,M+N} \\
&\qquad\qquad\qquad\qquad\qquad\qquad \vdots \qquad\qquad \vdots \\
&\qquad\qquad\qquad\qquad\qquad\qquad R_{M\,M+1}R_{M\,M+2} \cdots R_{M\,M+N},
\end{aligned}$$

as used to describe the dual quasitriangular structure of $A(R)$ in the proof of Theorem 4.1.5. In the multiindex used there, we have a partition function

$$\Psi(\mathbf{x}_I \otimes \mathbf{x}_J) = \mathbf{x}_B \otimes \mathbf{x}_A Z_R(B {\overset{A}{\underset{J}{\bigcirc}}} I), \tag{10.15}$$

where $x_I = x_{i_1}x_{i_2}\cdots$ and $I = (i_1, i_2, \ldots)$, etc.

Next we extend $\underline{\Delta}$ to products in such a way that it is a homomorphism to the braided tensor product,

$$\underline{\Delta}\mathbf{x}_1\mathbf{x}_2 = (\mathbf{x}_1 \otimes 1 + 1 \otimes \mathbf{x}_1)(\mathbf{x}_2 \otimes 1 + 1 \otimes \mathbf{x}_2)$$

$$= \mathbf{x}_1 \mathbf{x}_2 \otimes 1 + 1 \otimes \mathbf{x}_1 \mathbf{x}_2 + \mathbf{x}_1 \otimes \mathbf{x}_2 + \Psi(\mathbf{x}_1 \otimes \mathbf{x}_2)$$

$$= \mathbf{x}_1 \mathbf{x}_2 \otimes 1 + 1 \otimes \mathbf{x}_1 \mathbf{x}_2 + \mathbf{x}_1 \otimes \mathbf{x}_2 + \mathbf{x}_2 \otimes \mathbf{x}_1 R,$$

$$\underline{\Delta} \mathbf{x}_2 \mathbf{x}_1 R' = (\mathbf{x}_2 \otimes 1 + 1 \otimes \mathbf{x}_2)(\mathbf{x}_1 \otimes 1 + 1 \otimes \mathbf{x}_1) R'$$

$$= \mathbf{x}_2 \mathbf{x}_1 R' \otimes 1 + 1 \otimes \mathbf{x}_2 \mathbf{x}_1 R' + \mathbf{x}_2 \otimes \mathbf{x}_1 R' + \mathbf{x}_1 \otimes \mathbf{x}_2 R_{21} R'.$$

Hence, for $\underline{\Delta}$ to be well-defined, we need $\mathbf{x}_1 \otimes \mathbf{x}_2 (R_{21} R' - 1) = \mathbf{x}_2 \otimes \mathbf{x}_1 (R - R')$ or $R_{21} R' - 1 = P(R - R')$, which is (10.12). Here P is the usual permutation matrix $\mathbf{x}_1 \otimes \mathbf{x}_2 P = \mathbf{x}_1 \otimes \mathbf{x}_2$. It is trivial to see that the braiding Ψ is then functorial with respect to the coproduct $\underline{\Delta}$.

Finally, for a Hopf algebra in a braided category, the antipode is braided-antimultiplicative in the sense $\underline{S}(ab) = \cdot \Psi(\underline{S}a \otimes \underline{S}b)$. We define $\underline{S}$ on products of the generators in this way. Then,

$$\underline{S}(\mathbf{x}_1 \mathbf{x}_2) = \cdot \Psi(\underline{S}\mathbf{x}_1 \otimes \underline{S}\mathbf{x}_2) = \mathbf{x}_2 \mathbf{x}_1 R = \mathbf{x}_1 \mathbf{x}_2 R'_{21} R,$$

$$\underline{S}(\mathbf{x}_2 \mathbf{x}_1 R') = \cdot \Psi(\underline{S}\mathbf{x}_2 \otimes \underline{S}\mathbf{x}_1) R' = \mathbf{x}_1 \mathbf{x}_2 R_{21} R'.$$

Thus, $\underline{S}$ here is well-defined by (10.13). More generally, one can compute likewise

$$\underline{S}(\mathbf{x}_1 \cdots \mathbf{x}_N) = (-1)^N \mathbf{x}_N \cdots \mathbf{x}_1$$
$$R_{12} \cdots R_{1N} R_{23} \cdots R_{2N} \cdots R_{N-1\ N}. \tag{10.16}$$

We have given here a direct proof that avoids some of the category theory. ∎

It is quite easy to find suitable matrices R' for a given R. For example, we know that PR obeys some minimal polynomial $\prod_i (PR - \lambda_i) = 0$. For any nonzero eigenvalue λ_i, we can normalise R so that $\lambda_i = -1$, say. Then,

$$R' = P + P \prod_{j \neq i} (PR - \lambda_j) \tag{10.17}$$

clearly solves (10.11)–(10.13). It gives us at least one braided covector space for each nonzero eigenvalue of PR. In the simplest case, where R is Hecke, there are just two eigenvalues, and for each we have a solution with $R' \propto R$.

Example 10.2.2 *The standard* $\mathbb{A}_q^{2|0}$ *quantum plane with relations* $yx = qxy$ *is a braided covector algebra with*

$$\underline{\Delta}x = x \otimes 1 + 1 \otimes x, \quad \underline{\Delta}y = y \otimes 1 + 1 \otimes y,$$

$$\underline{\epsilon}x = \underline{\epsilon}y = 0, \quad \underline{S}x = -x, \quad \underline{S}y = -y,$$

$$\Psi(x \otimes x) = q^2 x \otimes x, \quad \Psi(x \otimes y) = qy \otimes x, \quad \Psi(y \otimes y) = q^2 y \otimes y,$$

$$\Psi(y \otimes x) = qx \otimes y + (q^2 - 1)y \otimes x.$$

Proof: We take

$$R = \begin{pmatrix} q^2 & 0 & 0 & 0 \\ 0 & q & q^2 - 1 & 0 \\ 0 & 0 & q & 0 \\ 0 & 0 & 0 & q^2 \end{pmatrix}, \quad R' = q^{-2}R$$

in Theorem 10.2.1. The algebra was already computed in Example 4.5.8. The braiding Ψ is just a display of the matrix entries of R. ∎

The algebra $\mathbb{A}_q^{2|0}$ with its braided coaddition is called the *braided plane*. The coaddition on products of generators is determined by the braided homomorphism property. Explicitly,

$$\underline{\Delta}(x^m y^n) = \sum_{r=0}^{m} \sum_{s=0}^{n} \begin{bmatrix} m \\ r ; q^2 \end{bmatrix} \begin{bmatrix} n \\ s ; q^2 \end{bmatrix} x^r y^s \otimes x^{m-r} y^{n-s} q^{(m-r)s},$$

where the notation is from Lemma 3.2.2.

Next we describe a convenient shorthand for working with the linear coproducts $\underline{\Delta}$ of the type above. This is a homomorphism from one copy of the algebra to two copies. So, if we denote the generators of the first copy by $x_i \equiv x_i \otimes 1$ and the generators of the second copy by $x_i' \equiv 1 \otimes x_i$, then the assertion that the coaddition $\underline{\Delta}$ of the above linear form is a homomorphism is just that

$$x_i'' = x_i + x_i', \quad \text{i.e.} \quad \mathbf{x}'' = \mathbf{x} + \mathbf{x}', \tag{10.18}$$

obey the same relations of $V^{\check{}}(R', R)$. In other words, we can treat our noncommuting generators x_i like row vector coordinates and have an addition law for them, provided we remember that in the braided tensor product they do not commute but rather obey the *braid statistics*

$$x_i' x_j = x_b x_a R^a{}_i{}^b{}_j, \quad \text{i.e.} \quad \mathbf{x}_1' \mathbf{x}_2 = \mathbf{x}_2 \mathbf{x}_1' R. \tag{10.19}$$

This is the compact way of working with our braided-Hopf algebras. We can add them and treat them like vectors provided we have the appropriate braid statistics between independent copies. In this notation, the essential fact that the coproduct extends to products as a well-defined braided-Hopf algebra is checked as

$$\mathbf{x}_1'' \mathbf{x}_2'' = (\mathbf{x}_1 + \mathbf{x}_1')(\mathbf{x}_2 + \mathbf{x}_2')$$
$$= \mathbf{x}_1 \mathbf{x}_2 + \mathbf{x}_1' \mathbf{x}_2' + \mathbf{x}_1 \mathbf{x}_2' + \mathbf{x}_2 \mathbf{x}_1' R,$$
$$\mathbf{x}_2'' \mathbf{x}_1'' R' = (\mathbf{x}_2 + \mathbf{x}_2')(\mathbf{x}_1 + \mathbf{x}_1') R'$$
$$= \mathbf{x}_2 \mathbf{x}_1 R' + \mathbf{x}_2' \mathbf{x}_1' R' + \mathbf{x}_2 \mathbf{x}_1' R' + \mathbf{x}_1 \mathbf{x}_2' R_{21} R',$$

which coincide by (10.12). This is just as in the proof of Theorem 10.2.1 but in our new shorthand notation. Note that there is a lot more to be checked for a braided-Hopf algebra, as we saw in the formal proof, but this is the most characteristic property.

Example 10.2.3 *The standard quantum plane algebra* $A_q^{2|0}$ *with relations* $yx = qxy$ *is a braided covector algebra with*

$$x'x = q^2 xx', \quad x'y = qyx', \quad y'y = q^2 yy', \quad y'x = qxy' + (q^2 - 1)yx',$$

i.e.

$$(x'', y'') = (x, y) + (x', y')$$

obeys the same relations provided we remember these braid statistics.

Proof: This is just Example 10.2.2 in the shorthand notation. ∎

Example 10.2.4 *The mixed quantum plane* $A_q^{1|1}$ *with relations* $\theta^2 = 0$, $\theta x = qx\theta$ *is a braided covector algebra with*

$$x'x = q^2 xx', \quad x'\theta = q\theta x', \quad \theta'\theta = -\theta\theta', \quad \theta'x = qx\theta' + (q^2 - 1)\theta x',$$

i.e.

$$(x'', \theta'') = (x, \theta) + (x', \theta')$$

obeys the same relations provided we remember these braid statistics.

Proof: We use

$$R = \begin{pmatrix} q^2 & 0 & 0 & 0 \\ 0 & q & q^2 - 1 & 0 \\ 0 & 0 & q & 0 \\ 0 & 0 & 0 & -1 \end{pmatrix}, \quad R' = q^{-2} R$$

in Theorem 10.2.1. The algebra was already computed in Example 4.5.9. We use the effective description for the coaddition $\underline{\Delta}$ and braiding Ψ as explained above. ∎

Example 10.2.5 *The fermionic quantum plane* $A_q^{0|2}$ *with relations* $\theta^2 = 0$, $\vartheta^2 = 0$ *and* $\vartheta\theta = -q\theta\vartheta$ *is a braided covector algebra with*

$$\theta'\theta = -\theta\theta', \quad \theta'\vartheta = -q^{-1}\vartheta\theta', \quad \vartheta'\vartheta = -\vartheta\vartheta', \quad \vartheta'\theta = -q^{-1}\theta\vartheta' + (q^{-2}-1)\vartheta\theta',$$

i.e.

$$(\theta'', \vartheta'') = (\theta, \vartheta) + (\theta', \vartheta')$$

obeys the same relations provided we remember these braid statistics.

Proof: We interchange $R \leftrightarrow -R'$ in Example 10.2.2. This still fulfils the conditions in Theorem 10.2.1 and gives the result shown. This is actually quite a general phenomenon, and indeed we will later define $V^\sim(-R, -R')$ in general as the algebra of differential forms $\theta_i = dx_i$. ∎

These ideas work just as well for vector algebras $\{v^i\}$ with indices up. The general form follows the same pattern as the right-covariant vector algebras $V_R(R, \lambda)$ in Example 4.5.15 but in our new setting with R, R'.

Theorem 10.2.6 *Let* R, R' *be as in Theorem 10.2.1. Then there is also a braided vector algebra* $V(R', R)$ *defined with generators* $1, v^i$ *and relations*

$$v^i v^j = R'^i{}_a{}^j{}_b v^b v^a, \quad i.e. \quad \mathbf{v}_1 \mathbf{v}_2 = R' \mathbf{v}_2 \mathbf{v}_1.$$

This has a braided addition law whereby $\mathbf{v}'' = \mathbf{v} + \mathbf{v}'$ *obeys the same relations if* $\mathbf{v}'$ *is a second copy with braid statistics*

$$v'^i v^j = R^i{}_a{}^j{}_b v^b v'^a, \quad i.e. \quad \mathbf{v}'_1 \mathbf{v}_2 = R \mathbf{v}_2 \mathbf{v}'_1.$$

More formally, it forms a braided-Hopf algebra with

$$\underline{\Delta} v^i = v^i \otimes 1 + 1 \otimes v^i, \quad \underline{\epsilon} v^i = 0, \quad \underline{S} v^i = -v^i,$$

$$\Psi(v^i \otimes v^j) = R^i{}_a{}^j{}_b v^b \otimes v^a, \quad i.e., \quad \Psi(\mathbf{v}_1 \otimes \mathbf{v}_2) = R\mathbf{v}_2 \otimes \mathbf{v}_1.$$

It lives in the braided category generated by R.

Proof: This is similar to Theorem 10.2.1 as far as a direct proof is concerned. In the shorthand notation, the key braided homomorphism or additivity property is checked as

$$\mathbf{v}''_1 \mathbf{v}''_2 = (\mathbf{v}_1 + \mathbf{v}'_1)(\mathbf{v}_2 + \mathbf{v}'_2)$$
$$= \mathbf{v}_1\mathbf{v}_2 + \mathbf{v}'_1\mathbf{v}'_2 + \mathbf{v}_1\mathbf{v}'_2 + R\mathbf{v}_2\mathbf{v}'_1,$$
$$R'\mathbf{v}''_2\mathbf{v}''_1 = R'(\mathbf{v}_2 + \mathbf{v}'_2)(\mathbf{v}_1 + \mathbf{v}'_1)$$
$$= R'\mathbf{v}_2\mathbf{v}_1 + R'\mathbf{v}'_2\mathbf{v}'_1 + R'\mathbf{v}_2\mathbf{v}'_1 + R'R_{21}\mathbf{v}_1\mathbf{v}'_2,$$

which coincide by (10.12). ∎

Example 10.2.7 *The quantum plane* $\mathbb{A}_{q^{-1}}^{2|0}$ *with relations* $wv = q^{-1}vw$ *is a braided vector algebra with braid statistics*

$$v'v = q^2vv', \quad v'w = qwv' + (q^2 - 1)vw', \quad w'v = qvw', \quad w'w = q^2ww',$$

i.e.

$$\begin{pmatrix} v'' \\ w'' \end{pmatrix} = \begin{pmatrix} v \\ w \end{pmatrix} + \begin{pmatrix} v' \\ w' \end{pmatrix}$$

obeys the same relations.

Proof: We take the standard R-matrix as in Example 10.2.2 or 10.2.3. We have already computed the opposite algebra to this vector algebra in Example 4.5.6. To this we now add the braiding and coaddition. ∎

Similarly for the other standard examples $\mathbb{A}_q^{0|2}$, $\mathbb{A}_q^{1|1}$, etc. The possibilities are the same as for the covector case. For example, we have both a covector and vector algebra for each nonzero eigenvalue of PR by our construction (10.17) for R'. We will see in Section 10.4 that the covectors and vectors are dual to each other as braided-Hopf algebras.

Next we consider the covariance properties of our vector and covector algebras. As explained below Example 9.2.10, the category generated by R is essentially the category of comodules of the quantum matrix group $A(R)$. So this plays the role of $\mathbb{Z}_q$ or $\mathbb{N}_q$ at the end of Section 10.1. We will prove this not for the most general solution R, R' of (10.11) but those obtained as follows: consider any matrix $R' \in M_n \otimes M_n$ such that

$$R'\mathbf{t}_1\mathbf{t}_2 = \mathbf{t}_1\mathbf{t}_2 R', \tag{10.20}$$

where $\mathbf{t}$ is the generator of $A(R)$. This is a quick way to ensure (10.11), as we can see just by applying the fundamental representation and conjugate fundamental representations $\rho^{\pm}$ from Proposition 4.1.4 to both sides. In the reverse direction, repeatedly using the conditions (10.11) establishes these relations (10.20) in all tensor powers of these representations and hence essentially corresponds to them abstractly. So (10.20) is a general feature, and we limit ourselves to R' of this general class. If it does not hold, we can always add it to the relations of the $A(R)$, but would then obtain a somewhat smaller quantum group as a consequence.

Proposition 10.2.8 *Let $A(R)$ be the matrix bialgebra with generator $\mathbf{t}$ as in Chapter 4.1. If (10.20) holds (e.g. if PR' is a function of PR) then $V^{\check{}}(R', R)$ lives in the braided category of $A(R)$-comodules. If R is regular then $V^{\check{}}(R', R)$ and $V(R', R)$ live in the braided category of A-comodules. The right coactions are*

$$\mathbf{x} \mapsto \mathbf{x}\mathbf{t}, \quad \mathbf{v} \mapsto \mathbf{t}^{-1}\mathbf{v}, \quad \text{i.e.} \quad x_i \mapsto x_a \otimes t^a{}_i, \quad v^i \mapsto v^a \otimes St^i{}_a.$$

Proof: Given (10.20), we have $x_1t_1x_2t_2 = x_1x_2t_1t_2 = x_2x_1R't_1t_2 = x_2x_1t_2t_1R' = x_2t_2x_1t_1R'$. This means that the algebra is covariant in the sense described in Chapter 4.5, in the present case from the right. Covariance of the coalgebra $\underline{\Delta}$ is immediate on the generators and hence holds in general because the braiding (used in extending $\underline{\Delta}$ to products) is covariant. Covariance means that the relevant structure maps are intertwiners for the quantum group coaction. Finally, we see from Example 9.2.10 that the braiding in this category indeed recovers the Ψ that we have used above. If R is regular, as in Proposition 4.2.2, then the dual quasitriangular structure descends to a suitable quotient A, which has an antipode, so the coaction of this also recovers the correct braiding. In this case, we also have the stated coaction on $V(R', R)$ and a similar calculation that it is covariant. This time, the braiding recovers the one in Theorem 10.2.6 when we compute as in Example 9.3.12 on the generators. ∎

A typical problem is that our R of interest may indeed be regular in a suitable normalisation (the quantum group normalisation as explained in Chapter 4.2), but this may be different from our current normalisation. If λR is the correct normalisation, we say that λ is the *quantum group normalisation constant*. Unfortunately, in our present considerations, we are not free to adjust the normalisation of R in (10.12) since this needs PR to have an eigenvalue -1 for a solution. If this is the only problem, it can be resolved by extending A to a new dual quasitriangular Hopf algebra $\tilde{A}$.

Proposition 10.2.9 *If R is regular in the normalisation λR, i.e. there is a dual quasitriangular structure on A with $\mathcal{R}(t_1 \otimes t_2) = \lambda R$, then we define the* dilatonic extension *$\tilde{A} = \mathbb{Z}_{\lambda^{-1}} \otimes A$, where $\mathbb{Z}_{\lambda^{-1}}$ has the dual quasitriangular structure $\mathcal{R}(\varsigma^m \otimes \varsigma^n) = \lambda^{-mn}$ and $\tilde{A}$ has the tensor product dual quasitriangular structure. Then $V^{\check{}}(R', R)$ and $V(R', R)$ live in the category of $\tilde{A}$-comodules with coaction $\mathbf{x} \mapsto \mathbf{x}t\varsigma$ and $\mathbf{v} \mapsto \varsigma^{-1}t^{-1}\mathbf{v}$.*

Proof: Here $\tilde{A}$ is just A with an invertible group-like and central element ς adjoined. Using it, the dual quasitriangular structure is arranged such that $\mathcal{R}(t_1\varsigma \otimes t_2\varsigma) = \mathcal{R}(\varsigma \otimes t_2\varsigma)\mathcal{R}(t_1 \otimes t_2\varsigma) = \mathcal{R}(\varsigma \otimes \varsigma)\mathcal{R}(\varsigma \otimes t_1)\mathcal{R}(t_1 \otimes \varsigma)$ $\mathcal{R}(t_1 \otimes t_2) = \lambda^{-1}\lambda R = R$, which agrees with the value on the matrix generators of $A(R)$. Now $V^{\check{}}(R', R)$ is $\mathbb{Z}$-graded by the degree of x_i. Thus we have a coaction of $\mathbb{Z}_{\lambda^{-1}}$ by $\mathbf{x} \mapsto \mathbf{x}\varsigma$ (extending to products as an algebra homomorphism so $x_1 \cdots x_N \mapsto x_1 \cdots x_N \varsigma^N$). This coaction measures the degree (or, in physical terms, the scaling dimension) of any homogeneous function of the x_i and $V^{\check{}}(R', R)$ is covariant under it. We already know that A coacts from Proposition 10.2.8 since the covariance itself does not need the correct normalisation of R. Moreover, the two coactions are compatible, and together they give the coaction of $\tilde{A}$, as stated. The induced braiding is then $\Psi(\mathbf{x}_1 \otimes \mathbf{x}_2) = \mathbf{x}_2 \otimes \mathbf{x}_1 \mathcal{R}(\mathbf{t}_1\varsigma \otimes \mathbf{t}_2\varsigma) = \mathbf{x}_2 \otimes \mathbf{x}_1 R$ as required in Theorem 10.2.1. Similarly, ς coacts on v^i with the generators having degree or scaling dimension -1, and the combined coaction as stated gives the correct braiding in Theorem 10.2.6. Note that we adjoin the dilaton ς not to ensure that our covectors and vectors are covariant, which is true under the coaction of A as well, but to ensure that the braiding is correctly induced from the coaction according to Exercise 9.2.9. ∎

There are many advantages of taking the trouble to work in the braided category of comodules of a dual quasitriangular Hopf algebra in this way. For example, we can apply the bosonisation theory of Chapter 9.4 to turn $V^{\check{}}(R', R)$ and $V(R', R)$ into ordinary Hopf algebras. We do one of these here: the other is strictly analogous.

Corollary 10.2.10 *Let $V^{\check{}}(R', R)$ be a braided covector algebra as in Theorem 10.2.1 in the covariant setting of Proposition 10.2.9. Its bosonisation according to the dual form of Theorem 9.4.12 is the ordinary Hopf algebra $\tilde{A}{\bowtie}V^{\check{}}(R', R)$. Explicitly, it has subalgebras $V^{\check{}}(R', R) = \{\mathbf{p}\}$ and $\tilde{A} = \{\mathbf{t}, \varsigma\}$, say, with cross relations and Hopf algebra structure*

$$\mathbf{p}\varsigma = \lambda^{-1}\varsigma\mathbf{p}, \quad \mathbf{p}_1\mathbf{t}_2 = \lambda\mathbf{t}_2\mathbf{p}_1 R, \quad \Delta\mathbf{t} = \mathbf{t} \otimes \mathbf{t}, \quad \Delta\varsigma = \varsigma \otimes \varsigma,$$

$$\Delta\mathbf{p} = \mathbf{p} \otimes \mathbf{t}\varsigma + 1 \otimes \mathbf{p}, \quad \epsilon\mathbf{t} = \mathrm{id}, \quad \epsilon\varsigma = 1, \quad \epsilon\mathbf{p} = 0,$$

$$S\mathbf{t} = \mathbf{t}^{-1}, \quad S\varsigma = \varsigma^{-1}, \quad S\mathbf{p} = -\mathbf{p}\varsigma^{-1}\mathbf{t}^{-1}.$$

Proof: We now denote the generators of $V^{\check{}}(R', R)$ by p_i as they will later play the role of momentum coordinates in our physical application. As we know from Lemma 7.4.8, the right coaction of $\tilde{A}$ is turned by $\mathcal{R}$ into a right action, which comes out as

$$\mathbf{p}_1{\triangleleft}\mathbf{t}_2 = \mathbf{p}_1\mathcal{R}(\mathbf{t}_1\varsigma \otimes \mathbf{t}_2) = \lambda\mathbf{p}_1 R, \quad \mathbf{p}{\triangleleft}\varsigma = \mathbf{p}\mathcal{R}(\mathbf{t}\varsigma \otimes \varsigma) = \mathbf{p}\lambda^{-1}.$$

We also studied such actions directly in Chapter 4: on the generators they are our fundamental representation ρ^+, extended now to the whole covector algebra. Bosonisation in the dual form consists in making a semi-direct product according to Proposition 1.6.10 by this action. This is constructed as in (9.51), on $\tilde{A} \otimes V^{\check{}}(R', R)$ with cross relations $(1 \otimes \mathbf{p}_1)(\mathbf{t}_2 \otimes 1) = (\mathbf{t}_2 \otimes 1)(1 \otimes \mathbf{p}_1{\triangleleft}\mathbf{t}_2)$ due to the matrix form of the coproduct of $\mathbf{t}$, and

$(1 \otimes \mathbf{p})(\varsigma \otimes 1) = (\varsigma \otimes 1)(1 \otimes \mathbf{p})\lambda^{-1}$ as in the one-dimensional case in Section 10.1. The dual bosonisation theorem also tells us to make a cross coproduct of the coalgebra of $V\check{}(R', R)$ by the right coaction of $\tilde{A}$ using Proposition 1.6.16. This is $\Delta(1 \otimes \mathbf{p}) = (1 \otimes \mathbf{p}_{(1)}{}^{(\bar{1})}) \otimes (\mathbf{p}_{(1)}{}^{(\bar{2})} \otimes 1)(1 \otimes \mathbf{p}_{(2)})$, which computes as stated. Here, $\mathbf{p}_{(1)} \otimes \mathbf{p}_{(2)} = \overline{\mathbf{p} \otimes 1 + 1 \otimes \mathbf{p}}$ is the braided coproduct in our case, and $\mathbf{p}^{(\bar{1})} \otimes \mathbf{p}^{(\bar{2})} = \mathbf{p} \otimes t\varsigma$ is the coaction from Proposition 10.2.9. There is also necessarily an antipode given at the end of Chapter 9.4, which comes out as stated in our case. ∎

We also know from the general bosonisation theory in Chapter 9.4 that the covariant (or categorical) corepresentations of the braided-Hopf algebra are just the usual corepresentations of the bosonised ordinary Hopf algebra. In particular, the coaction of the covectors on themselves by $\underline{\Delta}$ is covariant since $\underline{\Delta}$ is a morphism. Hence, we have immediately,

Corollary 10.2.11 *The bosonisation $\tilde{A} {\rtimes} V\check{}(R', R)$ coacts covariantly on $V\check{}(R', R) = \{\mathbf{x}\}$ by $\mathbf{x} \mapsto \mathbf{x}t\varsigma + \mathbf{p}$.*

Proof: This is ensured by a general theory of bosonisation as explained. We have seen the phenomenon already in the one-dimensional case at the end of Section 10.1. In our case, the coaction now comes out as $\beta(\mathbf{x}) = (\beta_{\tilde{A}} \otimes \mathrm{id})\underline{\beta}(\mathbf{x}) = (\beta_{\tilde{A}} \otimes \mathrm{id})(\mathbf{x} \otimes 1 + 1 \otimes \mathbf{p}) = \mathbf{x} \otimes t\varsigma \otimes 1 + 1 \otimes 1 \otimes \mathbf{p}$, where $\beta_{\tilde{A}}$ is the coaction of $\tilde{A}$ on $V\check{}(R', R)$ and $\underline{\beta} = \underline{\Delta}$ is its coaction on itself. The result is written compactly as stated. The general theory of bosonisation ensures not only that the result is a right coaction but that it is an algebra homomorphism. This can also be verified directly in our example as

$$(\mathbf{x}_1 t_1\varsigma + \mathbf{p}_1)(\mathbf{x}_2 t_2\varsigma + \mathbf{p}_2) = \mathbf{x}_1\mathbf{x}_2 t_1 t_2\varsigma^2 + \mathbf{p}_1\mathbf{p}_2 + \mathbf{p}_1\mathbf{x}_2 t_2\varsigma + \mathbf{x}_1 t_1\varsigma\mathbf{p}_2$$
$$(\mathbf{x}_2 t_2\varsigma + \mathbf{p}_2)(\mathbf{x}_1 t_1\varsigma + \mathbf{p}_1)R' = \mathbf{x}_2\mathbf{x}_1 R' t_1 t_2\varsigma^2 + \mathbf{p}_2\mathbf{p}_1 R'$$
$$+ \mathbf{x}_2 t_2\varsigma\mathbf{p}_1 R' + \mathbf{p}_2\mathbf{x}_1 t_1\varsigma R',$$

where $\mathbf{p}_2\mathbf{x}_1 t_1\varsigma R' = \mathbf{x}_1\mathbf{p}_2 t_1\varsigma R' = \mathbf{x}_1 t_1\varsigma\mathbf{p}_2 R_{21} R' = \mathbf{x}_1 t_1\varsigma\mathbf{p}_2 + \mathbf{x}_1 t_1\varsigma\mathbf{p}_2 P(R - R') = \mathbf{x}_1 t_1\varsigma\mathbf{p}_2 + \mathbf{x}_2 t_2\varsigma\mathbf{p}_1 R - \mathbf{x}_2 t_2\varsigma\mathbf{p}_1 R'$, using, in turn, the cross relations in the preceding corollary, the condition (10.12) and the action of the usual permutation P. Using again the cross relations in the corollary, we see that the two expressions coincide, i.e. the transformed $\mathbf{x}t\varsigma + \mathbf{p}$ obeys the same covector algebra relations. It is trivial to see that the linear braided coproduct $\underline{\Delta}$ on $V\check{}(R', R)$ is likewise covariant. ∎

If we think of x_i as space or spacetime generators, then $\tilde{A}$ is the extended rotation or Lorentz group, extended by the scale generator ς. The role of $\mathbb{Z}_{/2}$-grading in the theory of supersymmetry is played now by covariance under this quantum group, i.e. by rotation or Lorentz invariance! This is made possible by the unification of covariance and grading in quantum group theory as explained at the end of Chapter 9.2. Working in our (braided) category means that we are doing everything covariantly, which ensures, for example, that our Hopf algebra $\tilde{A} {\scriptstyle\bowtie} V\check{}(R', R)$ or 'extended Poincaré group' acts covariantly on our spacetime. This unification of the ideas of supersymmetry or grading and Einstein's principle of special relativity is one of the important conceptual advances made possible by quantum groups and is quite interesting even when $q = 1$.

Example 10.2.12 *The bosonisation of* $\mathbb{A}_q^{2|0} = \{x, y\}$ *is the ordinary Hopf algebra* $\widetilde{SL_q(2)} {\scriptstyle\bowtie} \mathbb{A}_q^{2|0}$ *with relations*

$$\varsigma^2 = ad - q^{-1}bc, \quad x\varsigma = q^{\frac{3}{2}}\varsigma x, \quad y\varsigma = q^{\frac{3}{2}}\varsigma y,$$

$$x\begin{pmatrix} a & b \\ c & d \end{pmatrix} = \begin{pmatrix} q^2 a & qb \\ q^2 c & qd \end{pmatrix} x, \quad y\begin{pmatrix} a & b \\ c & d \end{pmatrix} = \begin{pmatrix} qay + (q^2 - 1)bx & q^2 by \\ qcy + (q^2 - 1)dx & q^2 dy \end{pmatrix},$$

where $\begin{pmatrix} a & b \\ c & d \end{pmatrix}$ *have the relations of* $M_q(2)$ *with* $(ad - q^{-1}bc)^{\pm\frac{1}{2}}$ *adjoined. The coproduct is*

$$\Delta x = x \otimes a + y \otimes c + 1 \otimes x, \quad \Delta y = x \otimes b + y \otimes d + 1 \otimes y$$

and the matrix comultiplication for the matrix generators.

Proof: We consider $\mathbb{A}_q^{2|0}$ to be a braided covector algebra with R as in Example 10.2.2. The R-matrix in the quantum group normalisation in Proposition 4.2.4 is λR, where $\lambda = q^{-\frac{3}{2}}$. So this is the value of the quantum group normalisation constant in this example. The quantum group $\widetilde{SL_q(2)}$ consists of $SL_q(2)$ generators $\mathbf{t}$ and a central invertible group-like element ς adjoined as in Proposition 10.2.9. It is convenient to choose generators $\begin{pmatrix} a & b \\ c & d \end{pmatrix} = \mathbf{t}\varsigma$. They have the same commutation relations as the quantum matrices $M_q(2)$ in Example 4.2.5, but, instead of $\det_q(\mathbf{t}) = 1$ as in Proposition 4.2.6, we have the relation $ad - q^{-1}bc = \varsigma^2$. So we can identify $\widetilde{SL_q(2)}$ as a version of $GL_q(2)$ consisting of $M_q(2)$ and $(ad - q^{-1}bc)^{\pm\frac{1}{2}}$ adjoined. The commutation relations are then obtained from Corollary 10.2.10 using the explicit form of R in Example 10.2.2. The coaction and corresponding coproduct are given by $\begin{pmatrix} a & b \\ c & d \end{pmatrix}$ acting on (x, y) from the right as a matrix. $\blacksquare$

This is the quantum group of translations and linear transformations of the quantum plane $\mathbb{A}_q^{2|0}$. We will see more physical four-dimensional examples in Section 10.5, where we obtain by the same constructions natural q-Poincaré groups for q-Euclidean and q-Minkowski spaces.

We also see in this example that it can be convenient to work with $\tilde{\mathbf{t}} = \mathbf{t}\varsigma$ rather than with $\mathbf{t}$ itself. In terms of $\tilde{\mathbf{t}}$, the relations are

$$\mathbf{p}\varsigma = \lambda^{-1}\varsigma\mathbf{p}, \quad \mathbf{p}_1\tilde{\mathbf{t}}_2 = \tilde{\mathbf{t}}_2\mathbf{p}_1 R, \quad \Delta\mathbf{p} = \mathbf{p}\otimes\tilde{\mathbf{t}} + 1\otimes\mathbf{p}.$$

Let us also note that we can just as well use different conventions whereby Corollary 10.2.10 comes out as an ordinary Hopf algebra with R_{21}^{-1} in place of R in the cross relation, which then becomes

$$\mathbf{p}_1\tilde{\mathbf{t}}_2 = \tilde{\mathbf{t}}_2\mathbf{p}_1 R_{21}^{-1}$$

instead of the above. It corresponds to using the conjugate fundamental representation ρ^- for the cross relations. Finally, we know of course that our right comodule constructions are the dual of corresponding left module constructions which we could equally well make if we wanted a q-Poincaré enveloping algebra. For example, the following exercise gives the dual form of Corollaries 10.2.10 and 10.2.11 in the case where A is dual to $U_q(g)$ with matrix generators $\mathbf{l}^{\pm}$ from Chapter 4.

Exercise 10.2.13 *Construct the bosonisation* $V(R',R){>\!\!\!\triangleleft}\widetilde{U_q(g)}$ *and its action on the covector algebra. Show that the bosonisation is the Hopf algebra generated by* $\widetilde{U_q(g)}$ *as a sub-Hopf algebra and* $V(R',R) = \{\mathbf{p}\}$ *with structure*

$$\lambda^\xi \mathbf{p}\lambda^{-\xi} = \lambda^{-1}\mathbf{p}, \quad \mathbf{l}_1^+\mathbf{p}_2 = \lambda^{-1}R_{21}^{-1}\mathbf{p}_2\mathbf{l}_1^+, \quad \mathbf{l}_1^-\mathbf{p}_2 = \lambda R\mathbf{p}_2\mathbf{l}_1^-,$$

$$\Delta\mathbf{p} = \mathbf{p}\otimes 1 + \lambda^\xi\mathbf{l}^-\otimes\mathbf{p}, \quad \epsilon\mathbf{p} = 0, \quad S\mathbf{p} = -\lambda^{-\xi}(S\mathbf{l}^-)\mathbf{p}.$$

Moreover, its covariant action on covectors $V\check{\,}(R',R) = \{\mathbf{x}\}$ *is*

$$\mathbf{l}_2^+{\triangleright}\mathbf{x}_1 = \mathbf{x}_1\lambda R, \quad \mathbf{l}_2^-{\triangleright}\mathbf{x}_1 = \mathbf{x}_1\lambda^{-1}R_{21}^{-1}, \quad \lambda^\xi{\triangleright}\mathbf{x} = \lambda\mathbf{x}, \quad p^i{\triangleright}x_j = \delta^i{}_j.$$

Solution: We use here the original bosonisation Theorem 9.4.12 for braided groups living in the category of left modules of a quasitriangular Hopf algebra. First, if A is dual to $U_q(g)$, then $\tilde{A}$ is dual to $\widetilde{U_q(g)} = U_{\lambda^{-1}}(1)\otimes U_q(g)$, where $U_{\lambda^{-1}}(1)$ is the quantum line in Example 2.1.19. It has a primitive generator ξ, quasitriangular structure and duality pairing

$$\mathcal{R}_\xi = \lambda^{-\xi\otimes\xi}, \quad \langle\varsigma,\xi\rangle = 1, \quad \text{or} \quad \langle\varsigma,\lambda^\xi\rangle = \lambda. \tag{10.21}$$

We work over formal power series, but can also (as usual) regard $\mathbf{l}^{\pm}$, $\lambda^{\pm\xi}$ as abstract generators and work over a field. So the dilatonic extension is just to add this ξ to our quantum enveloping algebra.

Next, the module version of the covariance in Proposition 10.2.9 tells us that the braided vectors and covectors are $\widetilde{U_q(g)}$-module algebras (covariant) under the left action corresponding to the right coaction there by evaluation. For example, on $V(R', R)$ this is

$$l_1^+ \triangleright \mathbf{p}_2 = \langle \varsigma^{-1} St_2, l_1^+ \rangle \mathbf{p}_2 = \langle St_2, l_1^+ \rangle \mathbf{p}_2 = \lambda^{-1} R_{21}^{-1} \mathbf{p}_2,$$
$$l_1^- \triangleright \mathbf{p}_2 = \langle \varsigma St_2, l_1^- \rangle \mathbf{p}_2 = \lambda R \mathbf{p}_2,$$
$$\lambda^\xi \triangleright \mathbf{p} = \langle \varsigma^{-1} St, \lambda^\xi \rangle \mathbf{p} = \langle \varsigma^{-1}, \lambda^\xi \rangle \mathbf{p} = \lambda^{-1} \mathbf{p}.$$

The bosonisation Theorem 9.4.12 tells us to make a cross product from Proposition 1.6.6 by this action, as $(1 \otimes l_1^\pm)(\mathbf{p}_2 \otimes 1) = (l_1^\pm \triangleright \mathbf{p}_2 \otimes 1)(1 \otimes l_1^\pm)$. Similarly for the ξ commutation relations. For the coalgebra we know from Theorem 9.4.12 to make the cross coproduct from Proposition 1.6.18 by the coaction

$$\mathbf{p}^{(\bar{1})} \otimes \mathbf{p}^{(\bar{2})} = \mathcal{R}_{21} \triangleright \mathbf{p} = \mathcal{R}^{(2)} \langle \mathcal{R}^{(1)}, \varsigma^{-1} St \rangle \mathbf{p}$$
$$= 1^- \langle \lambda^{-\xi \otimes \xi}, \mathrm{id} \otimes \varsigma^{-1} \rangle \otimes \mathbf{p} = 1^- \lambda^\xi \otimes \mathbf{p},$$

where we used Corollary 4.1.9 for the part in $U_q(g)$ and (10.21) for the part in $U_q(1)$. Then $\Delta(\mathbf{p} \otimes 1) = (\mathbf{p}_{(1)} \otimes 1)(1 \otimes \mathbf{p}_{(2)}^{(\bar{1})}) \otimes (\mathbf{p}_{(2)}^{(\bar{2})} \otimes 1)$ computes as stated. This gives the structure of our q-Poincaré enveloping algebra as obtained by bosonisation.

Finally, we compute the action of $\widetilde{U_q(g)}$ on $\{\mathbf{x}\}$ in the same way by evaluating against the known coaction from Proposition 10.2.9. This is similar to the above. Thus, $l_2^+ \triangleright \mathbf{x}_1 = \mathbf{x}_1 \langle t_1 \varsigma, l_2^+ \rangle = \mathbf{x}_1 \langle t_1 \varsigma \otimes t_2, \mathcal{R} \rangle = \lambda \mathbf{x}_1 R$, etc. It is also possible to see by direct computation that $V(R', R)$ acts on $V^{\check{}}(R', R)$ as a braided-module algebra by the action $p^i \triangleright x_i = \delta^i{}_j$. It obeys the braided version of our usual covariance condition for actions on algebras (and is best described diagrammatically in the notation of Chapter 9.4). The corresponding bosonised Hopf algebra acts as usual for a module algebra, i.e. its action extends to products using the coproduct as

$$p^i \triangleright (ab) = (p^i \triangleright a)b + (1^{-i}{}_j \lambda^\xi \triangleright a)(p^j \triangleright b) = (p^i \triangleright a)b + \cdot \Psi(p^i \otimes a) \triangleright b,$$

where we have

$$1^{-i}{}_j \lambda^\xi \triangleright a \otimes p^j = \mathcal{R}^{(2)} \lambda^\xi \triangleright a \langle t^i{}_j, S\mathcal{R}^{(1)} \rangle$$
$$= \mathcal{R}^{(2)} \mathcal{R}_\xi^{(2)} \triangleright a \otimes \langle \varsigma^{-1} St^i{}_j, \mathcal{R}^{(1)} \mathcal{R}_\xi^{(1)} \rangle p^j$$
$$= \mathcal{R}^{(2)} \mathcal{R}_\xi^{(2)} \triangleright a \otimes \mathcal{R}^{(1)} \mathcal{R}_\xi^{(1)} \triangleright p^j = \Psi(p^j \otimes a),$$

which is the braiding in the category of $\widetilde{U_q(g)}$-modules according to Theorem 9.2.4. The two points of view, namely p^i acting to give a module algebra with the coproduct Δp as stated, and p^i obeying the 'braided Leibniz rule', are equivalent in this way. We will develop the appropriate theory of 'braided derivatives' in Section 10.4 (in some other conventions suitable for our comodule setting). For the moment, we just verify directly that $p^i \triangleright$ is a well-defined operator when extended in this way. For example, we have

$$p^i \triangleright (x_j x_k) = \delta^i{}_j x_k \otimes x_a R^{-1i}{}_k{}^a{}_j,$$
$$p^i \triangleright (x_b x_a R'^a{}_j{}^b{}_k) = \delta^i{}_b x_a R'^a{}_j{}^b{}_k + x_c R^{-1i}{}_a{}^c{}_b R'^a{}_j{}^b{}_k,$$

which coincide using the relations (10.12). These relations also ensure that the operators p^i obey the relations of the vector algebra in Theorem 10.2.6. The QYBE for R can then be used to check directly that this action of p^i fits together with the action above of $1^\pm, \lambda^\xi$ to form a representation of our q-Poincaré enveloping algebra. Note that we could equally well consider the 'conjugate' bosonisation $V^\smile(R', R_{21}^{-1}) {\rtimes} \widetilde{U_q(g)}$, which has the same algebra and counit as the above but the coproduct and antipode

$$\Delta \mathbf{p} = \mathbf{p} \otimes 1 + \lambda^{-\xi} 1^+ \otimes \mathbf{p}, \quad S\mathbf{p} = -\lambda^\xi (S 1^+) \mathbf{p} \qquad (10.22)$$

instead. This time p^i acts on $\mathbf{x}$ precisely as ∂^i in Definition 10.4.1, extending to products by the braided Leibniz rule in Lemma 10.4.3. ∎

Similarly, one may apply the double bosonisation Theorem 9.2.13 to the braided vectors to obtain a Hopf algebra $V(R', R) {\rtimes} \widetilde{U_q(g)} {\ltimes} V^\smile(R', R)$. This has an additional braided covector of generators. Denoting these by $\mathbf{c} = (c_i)$, the additional cross relations and coproduct are

$$\lambda^\xi \mathbf{c} \lambda^{-\xi} = \lambda \mathbf{c}, \quad 1_1^+ \mathbf{c}_2 = \lambda \mathbf{c}_2 1_1^+ R_{21}, \quad 1_1^- \mathbf{c}_2 = \lambda^{-1} \mathbf{c}_2 1_1^- R^{-1},$$
$$[\mathbf{p}, \mathbf{c}] = \frac{1^+ \lambda^{-\xi} - 1^- \lambda^\xi}{q - q^{-1}}, \quad \Delta \mathbf{c} = \mathbf{c} \otimes 1^+ \lambda^{-\xi} + 1 \otimes \mathbf{c},$$

with the corresponding counit and antipode. Here it is assumed that R depends on a parameter q, say, such that $R_{21} R = \mathrm{id} + O(q - q^{-1})$. This is not needed for the Hopf algebra itself (it is merely a choice of normalisation for the $\mathbf{c}$), but rather for a familiar classical limit as $q \to 1$. We also have, formally, a quasitriangular structure

$$\mathcal{R} = \mathcal{R}_{U_q(g)} \lambda^{-\xi \otimes \xi} \sum_a S e_a \otimes f^a,$$

where $\{e_a\}$ is a basis of the braided vector algebra and $\{f^a\}$ a dual basis. Later on, in Section 10.4, we will interpret the canonical element defined by the summation as a braided exponential or 'plane wave'. Finally, since double bosonisation is a kind of analogue of the quantum double there is also a canonical action of it analogous to Example 7.1.8. It extends the action in Exercise 10.2.13 by the braided adjoint action of the braided covectors on themselves. Explicitly in our case,

$$\mathbf{c}_2 \triangleright \mathbf{x}_1 = \frac{\mathbf{x}_2 \mathbf{x}_1 R - \mathbf{x}_1 \mathbf{x}_2}{q - q^{-1}}.$$

For the q-Euclidean and q-Minkowski spaces in Section 10.5 the double bosonisation becomes the q-conformal group and $\mathbf{c}$ the generators of q-deformed special conformal transformations. Special conformal transformations are therefore a remnant in a $q \to 1$ scaling limit of the braided adjoint action of q-deformed spacetime on itself as an additive braided group.

This describes the vector and covector braided-Hopf algebras and their covariance properties. It can also happen, of course, that our vector and covector algebras are isomorphic. Recall that this is the true meaning of the metric in differential geometry. So, to complete our picture, we define:

Definition 10.2.14 *Let R, R' be as above. A quantum metric for the corresponding vector and covector algebras is a matrix η_{ij} such that $x_i = \eta_{ia} v^a$ is an isomorphism of braided-Hopf algebras. We define its inverse by $\eta_{ja} \eta^{ia} = \delta^i_j = \eta_{aj} \eta^{ai}$ so that $v^i = x_a \eta^{ai}$.*

There are two aspects to this, one for the algebra isomorphism and one for an equivalence of the braiding such that the braided tensor product algebras are also isomorphic. These immediately come out, respectively, as

$$\eta_{ia} \eta_{jb} R'^a{}_k{}^b{}_l = R'^a{}_i{}^b{}_j \eta_{ak} \eta_{bl}, \quad \eta_{ia} \eta_{jb} R^a{}_k{}^b{}_l = R^a{}_i{}^b{}_j \eta_{ak} \eta_{bl}. \tag{10.23}$$

The second of these (the matching of the braiding) follows automatically as long as our identification η is a morphism in the braided category. In the covariant setting of Proposition 10.2.9, it follows from covariance under the action of A. Note that it is not really natural in our geometrical setting to transform vectors and covectors in the same way under the dilaton ς (it is more natural to regard the metric as breaking scaling invariance). So Definition 10.2.14 is not exactly an isomorphism with $V(R', R)$ transforming as usual vectors but rather (the same algebra and braiding) as 'negative-density' vectors transforming like braided covectors under ς. There is then no constraint on η coming from this aspect of the covariance. Covariance under the background quantum group A does impose constraints, however, namely

$$St^i{}_j = \eta_{aj} t^a{}_b \eta^{bi}, \quad \text{i.e.,} \quad \eta_{ab} t^a{}_i t^b{}_j = \eta_{ij}, \tag{10.24}$$

where $\mathbf{t}$ is its matrix generator. This implies the second of (10.23) by evaluating in the fundamental representation from Proposition 4.1.4. It implies other relations as well between η and R which we could take, along with (10.23), as axioms for a strictly matrix characterisation of the quantum metric (much as we took (10.11) in Theorem 10.2.1). These include

$$\eta_{ka}R^i{}_j{}^a{}_l = \lambda^{-2}R^{-1i}{}_j{}^a{}_k\eta_{al}, \quad \eta_{ka}\widetilde{R}^i{}_j{}^a{}_l = \lambda^2 R^i{}_j{}^a{}_k\eta_{al}, \tag{10.25}$$

$$R^a{}_j{}^k{}_l\eta_{ai} = \lambda^{-2}\eta_{ja}R^{-1a}{}_i{}^k{}_l, \quad \widetilde{R}^a{}_j{}^k{}_l\eta_{ai} = \lambda^2\eta_{ja}R^a{}_i{}^k{}_l, \tag{10.26}$$

obtained by evaluating the dual quasitriangular structure of A on various forms of (10.24). Finally, we can also require our metric to be symmetric in some sense. The natural condition is to use the same notion as the sense in which the braided covector and vector algebras are commutative, i.e.

$$\eta_{ba}R'^a{}_i{}^b{}_j = \eta_{ij}. \tag{10.27}$$

The corresponding equations in terms of the inverse metric η^{ij} are

$$\eta^{ia}\eta^{jb}R'^k{}_a{}^l{}_b = R'^i{}_a{}^j{}_b\eta^{ak}\eta^{bl}, \quad \eta^{ia}\eta^{jb}R^k{}_a{}^l{}_b = R^i{}_a{}^j{}_b\eta^{ak}\eta^{bl}, \tag{10.28}$$

$$t^i{}_a t^j{}_b\eta^{ba} = \eta^{ji}, \tag{10.29}$$

$$\eta^{la}R^i{}_j{}^k{}_a = \lambda^{-2}R^{-1i}{}_j{}^l{}_a\eta^{ak}, \quad \eta^{la}\widetilde{R}^i{}_j{}^k{}_a = \lambda^2 R^i{}_j{}^l{}_a\eta^{ak}, \tag{10.30}$$

$$R^i{}_a{}^k{}_l\eta^{aj} = \lambda^{-2}\eta^{ia}R^{-1j}{}_a{}^k{}_l, \quad \widetilde{R}^i{}_a{}^k{}_l\eta^{aj} = \lambda^2\eta^{ia}R^j{}_a{}^k{}_l, \tag{10.31}$$

$$R'^i{}_a{}^j{}_b\eta^{ba} = \eta^{ij}. \tag{10.32}$$

We will see in Section 10.4 how to obtain such quantum metrics as an application of braided differentiation.

10.3   Braided matrices and braided linear algebra

Another corollary of the covariance properties of braided covectors and vectors in Proposition 10.2.9 is that now they manifestly both live in the same category or 'universe', namely that of comodules of a quantum group $\widetilde{A}$: one can ask about other structures living in this same universe. We will see in this section that, as well as vectors and covectors, there are natural 'braided matrices' giving us a fairly complete theory of braided linear algebra in this category. These are provided by the algebras $B(R)$, which we have encountered already in Chapters 4.3, 4.5.2 and 7.4. But their role as braided matrices is the original setting in which these quadratic algebras were introduced and is the reason for their name. We are in a position to explain this now. At first, one might guess that the role of 'matrices' in our braided universe should be the quantum matrices $A(R)$ from Chapter 4.1, but these do not live in our category of $A(R)$-comodules or the comodules of the associated quantum group $\widetilde{A}$. We have already seen this lack of self-covariance already in Chapter 7.4, where we saw how the 'covariantisation' or 'transmutation' procedure (according to Theorem 7.4.1 and Example 9.4.10) solved this problem by replacing quantum matrices by the braided ones $B(R)$. These play the role of the matrix objects in our braided setting. Meanwhile, the quantum groups $A(R)$ or $\tilde{A}$ remain as the background symmetry of our constructions. As before, we give first the direct R-matrix treatment of $B(R)$ and then come to these categorical or covariance properties later. We recall from Chapter 4 that the fundamental property of matrices is that they can be multiplied, which corresponds in terms of coordinates $\{u^i{}_j\}$ to matrix comultiplication.

Theorem 10.3.1 *Let R be a biinvertible solution of the QYBE and let $B(R)$ be the braided matrix algebra with generators $1, \mathbf{u} = \{u^i{}_j\}$ and relations $R_{21}\mathbf{u}_1 R\mathbf{u}_2 = \mathbf{u}_2 R_{21}\mathbf{u}_1 R$ as in Definition 4.3.1. These form a braided bialgebra with*

$$\underline{\Delta} u^i{}_j = \sum_a u^i{}_a \otimes u^a{}_j, \quad \epsilon u^i{}_j = \delta^i{}_j, \quad i.e. \ \underline{\Delta}\mathbf{u} = \mathbf{u} \otimes \mathbf{u}, \quad \epsilon \mathbf{u} = \mathrm{id},$$

$$\Psi(u^i{}_j \otimes u^k{}_l) = u^p{}_q \otimes u^m{}_n R^i{}_a{}^d{}_p R^{-1a}{}_m{}^q{}_b R^n{}_c{}^b{}_l \tilde{R}^c{}_j{}^k{}_d,$$

$$i.e. \quad \Psi(R^{-1}\mathbf{u}_1 \otimes R\mathbf{u}_2) = \mathbf{u}_2 R^{-1} \otimes \mathbf{u}_1 R.$$

In another notation: if $\mathbf{u}'$ is another copy of $B(R)$, then the matrix product

$$\mathbf{u}'' = \mathbf{u}\mathbf{u}'$$

obeys the relations of $B(R)$ also provided $\mathbf{u}'$ has the braid statistics

$$R^{-1}\mathbf{u}'_1 R\mathbf{u}_2 = \mathbf{u}_2 R^{-1}\mathbf{u}'_1 R.$$

Proof: The formulae are obtained from Example 9.4.10 and then verified directly. The relations of $B(R)$ come the braided commutativity property (7.34), as we have seen in Chapter 7.4; we just use Theorem 4.1.5 to evaluate the relations in matrix form. The transmutation theory also gives the braiding, which comes out as a product of four R-matrices as shown, but is also conveniently written in the compact form stated. The extension of Ψ to products is, by definition, in such a way that the product is a morphism in the category generated by this braiding. One can verify directly that this extension is consistent with the relations. To verify directly that the result is indeed a braided bialgebra, we use the second, more compact, description in which we work in the braided tensor product $B(R)\underline{\otimes}B(R)$ with relations as stated. Then,

$$R_{21}\mathbf{u}_1 \mathbf{u}'_1 R\mathbf{u}_2 \mathbf{u}'_2 = R_{21}\mathbf{u}_1 R(R^{-1}\mathbf{u}'_1 R\mathbf{u}_2)\mathbf{u}'_2$$

$$= (R_{21}\mathbf{u}_1 R\mathbf{u}_2)R^{-1}R_{21}^{-1}(R_{21}\mathbf{u}_1' R\mathbf{u}_2')$$
$$= \mathbf{u}_2 R_{21}(\mathbf{u}_1 R_{21}^{-1}\mathbf{u}_2' R_{21})\mathbf{u}_1' R$$
$$= \mathbf{u}_2 R_{21} R_{21}^{-1}\mathbf{u}_2' R_{21}\mathbf{u}_1 \mathbf{u}_1' R = \mathbf{u}_2 \mathbf{u}_2' R_{21}\mathbf{u}_1\mathbf{u}_1' R$$

as required for $\underline{\Delta}$ to extend to $B(R)$ as a bialgebra in a braided category. In each expression, the brackets indicate how to apply the relevant relation to obtain the next expression. This is the most important part of the assertion that $B(R)$ is a braided bialgebra. The other details, such as the counit $\underline{\epsilon}$ and functoriality of Ψ with respect to it, etc., can also be checked in the same explicit way as well as being implied by the general transmutation theory. ∎

The question of when this braided bialgebra has a braided antipode, and hence becomes a braided-Hopf algebra, is strictly analogous to the same question for $A(R)$ as we have studied it in Chapter 4.2. The analogy is strict because of the general transmutation theory (the bialgebra version of Example 9.4.10) that connects the two. The matrix form of the coproduct means that $\underline{S}\mathbf{u} \equiv \mathbf{u}^{-1}$ is required to obey $\mathbf{u}^{-1}\mathbf{u} = \mathrm{id} = \mathbf{u}\mathbf{u}^{-1}$ as well as to be a morphism and hence commute appropriately with Ψ. One generally needs to add to $B(R)$ the determinant and other relations. The braided determinant is characterised by being central and bosonic (in the sense that it has trivial braid statistics with everything else) and group-like in the sense

$$\underline{\Delta}(\underline{\det}(\mathbf{u})) = \underline{\det}(\mathbf{u}) \otimes \underline{\det}(\mathbf{u}), \quad \underline{\epsilon}(\underline{\det}(\mathbf{u})) = 1 \tag{10.33}$$

$$\text{i.e.} \quad \underline{\det}(\mathbf{u}\mathbf{u}') = \underline{\det}(\mathbf{u})\underline{\det}(\mathbf{u}'), \quad \underline{\det}(\mathrm{id}) = 1$$

in the two notations. We will give a general construction for central bosonic elements (from which $\underline{\det}(\mathbf{u})$ can generally be built) later in the section.

We can also make some general comments on the ∗-structure in the case when we work over $\mathbb{C}$. We know already from Proposition 4.3.7 that when R is of real type, then $B(R)$ becomes a ∗-algebra with $\mathbf{u}$ Hermitian.

Proposition 10.3.2 *When we work over $\mathbb{C}$ and R is of real type then the ∗-algebra structure $u^{i*}_j = u^j{}_i$ in Proposition 4.3.7 makes $B(R)$ a braided ∗-bialgebra in the sense*

$$(* \otimes *) \circ \underline{\Delta} = \tau \circ \underline{\Delta} \circ *, \quad \underline{\epsilon} \circ * = * \circ \underline{\epsilon},$$

where τ denotes the usual transposition.

Proof: We know from Proposition 4.3.7 that $B(R)$ in this case is a ∗-algebra. We check now that this is also compatible with the braid statistics and hence with the extension of the matrix comultiplication. Indeed, $B(R)\underline{\otimes}B(R)$ has a natural $*$-structure defined not in the obvious way but by

$$(b \otimes c)^* = c^* \otimes b^*, \quad \forall b, c \in B(R), \quad \text{i.e.} \quad u^i{}_j{}^* = u'^j{}_i, \quad u'^i{}_j{}^* = u^j{}_i$$

in our two notations. In the latter notation we check that this is compatible with the relations of the braided tensor product. Thus,

$$(R^{-1i}{}_a{}^k{}_b u'^a{}_c R^c{}_j{}^b{}_d u^d{}_l)^* = u'^l{}_d R^d{}_b{}^j{}_c u^c{}_a R^{-1b}{}_k{}^a{}_i$$
$$= R^l{}_a{}^j{}_b u^b{}_c R^{-1a}{}_d{}^c{}_i u'^d{}_k = (u^k{}_d R^{-1i}{}_c{}^d{}_a u'^c{}_b R^b{}_j{}^a{}_l)^*,$$

as required. We used the assumption that R is of real type in the sense of Definition 4.2.15 and the braid statistics in Theorem 10.3.1. After this, we just have to note that $\underline{\Delta}$ is a $*$-algebra map on the generators and hence extends to products as a $*$-algebra map. That the counit is a $*$-algebra map $B(R) \to \mathbb{C}$ is more immediate. We can deduce this proposition also from the general transmutation theory cf. Theorem 7.4.2, to which our construction is related at the quotient Hopf algebra level. We have given here a direct proof. ∎

This is not like the axioms for a Hopf $*$-algebra given in Chapter 1.7, but it is what we might expect for a Hermitian matrix. The product of two Hermitian matrices M, N is not Hermitian so we would not expect $\underline{\Delta}$ to be a $*$-algebra homomorphism. Instead, $(MN)^* = NM$, so we might expect it to be a skew-homomorphism as in the above proposition. This turns out to be quite a common feature of braided groups. A natural assumption for the braided antipode, when it exists, is that it should commute with $*$.

Example 10.3.3 *Let R be the standard solution of the QYBE as in Proposition 4.2.4. Then $B(R)$ is the braided matrices $BM_q(2)$ as already computed in Example 4.3.4. It has generators $\mathbf{u} = \begin{pmatrix} a & b \\ c & d \end{pmatrix}$ and relations as given there. Then,*

$$\begin{pmatrix} a'' & b'' \\ c'' & d'' \end{pmatrix} = \begin{pmatrix} a & b \\ c & d \end{pmatrix} \begin{pmatrix} a' & b' \\ c' & d' \end{pmatrix}$$

obey these same relations if the primed copy does and if we use the braid statistics

$$a'a = aa' + (1 - q^2)bc', \quad a'b = ba', \quad a'c = ca' + (1 - q^2)(d - a)c',$$

$$a'd = da' + (1 - q^{-2})bc', \quad b'a = ab' + (1 - q^2)b(d' - a'), \quad b'b = q^2 bb',$$

$$b'c = q^{-2}cb' + (1 + q^2)(1 - q^{-2})^2 bc' - (1 - q^{-2})(d - a)(d' - a'),$$

$$b'd = db' + (1 - q^{-2})b(d' - a'), \quad c'a = ac', \quad c'b = q^{-2}bc',$$

$$c'c = q^2 cc', \quad c'd = dc', \quad d'a = ad' + (1 - q^{-2})bc',$$

$$d'b = bd', \quad d'c = cd' + (1 - q^{-2})(d - a)c', \quad d'd = dd' - q^{-2}(1 - q^{-2})bc'.$$

The elements

$$t = q^{-1}a + qd, \quad \underline{\det}(\mathbf{u}) = ad - q^2 cb$$

are bosonic and central, and $\underline{\det}$ *is group-like in the sense of (10.33). Setting* $\underline{\det}(\mathbf{u}) = 1$ *gives the braided-Hopf algebra* $BSL_q(2)$ *with braided antipode*

$$\underline{S}\begin{pmatrix} a & b \\ c & d \end{pmatrix} \equiv \begin{pmatrix} a & b \\ c & d \end{pmatrix}^{-1} = \begin{pmatrix} q^2 d + (1 - q^2)a & -q^2 b \\ -q^2 c & a \end{pmatrix}.$$

Over $\mathbb{C}$, *the natural* $*$-*structure*

$$\begin{pmatrix} a^* & b^* \\ c^* & d^* \end{pmatrix} = \begin{pmatrix} a & c \\ b & d \end{pmatrix}$$

skew-commutes with $\underline{\Delta}$, *as in Proposition 10.3.2, and commutes with* $\underline{S}$ *in the case with braided antipode.*

Proof: We compute the braiding statistics Ψ from Theorem 10.3.1 using the standard R-matrix as in Example 4.2.5 or Example 10.2.2. We already computed the algebra relations for this case in Example 4.3.4. Note that the normalisation does not affect either of these. From these, we easily verify

$$tx = xt, \quad t'x = xt', \quad x't = tx',$$

$$\underline{\det}(\mathbf{u})x = x\underline{\det}(\mathbf{u}), \quad \underline{\det}(\mathbf{u}')x = x\underline{\det}(\mathbf{u}'), \quad x'\underline{\det}(\mathbf{u}) = \underline{\det}(\mathbf{u})x',$$

for all x, i.e. centrality and the bosonic nature of the elements t and $\underline{\det}$. See Theorem 10.3.8 below for the general picture regarding these. In the $*$-algebra case, these elements are self-adjoint in the sense

$$t^* = t, \quad \underline{\det}(\mathbf{u})^* = \underline{\det}(\mathbf{u}).$$

We already encountered some of these facts in Chapter 7.4 when we considered mass-shells in q-Minkowski space. There, $\frac{1}{2}t$ is the time direction and $\det(\mathbf{u})$ determines the quantum metric. We will develop this interpretation further in Section 10.5.3. The braided antipode is obtained by transmuting the antipode of $SL_q(2)$ in Proposition 4.2.6. It is easily verified directly as the matrix inverse of $\mathbf{u}$.   ∎

As $q \to 1$ in this example, the algebra becomes commutative and the statistics also become commutative so that we return to the case of an algebraic model of the coordinate functions on the set of matrices M_n. On the other hand, our braided technology includes the concept of supersymmetry and supermatrices too.

Example 10.3.4 *Let R be the nonstandard Alexander–Conway solution of the QYBE as in Example 4.2.13. Then $B(R)$ is the braided matrices $BM_q(1|1)$. It has generators $\mathbf{u} = \begin{pmatrix} a & b \\ c & d \end{pmatrix}$ and relations*

$$b^2 = 0, \qquad c^2 = 0, \qquad d - a \text{ central},$$

$$ab = q^{-2}ba, \qquad ac = q^2ca, \qquad bc = -q^2cb + (1 - q^2)(d - a)a,$$

and

$$\begin{pmatrix} a'' & b'' \\ c'' & d'' \end{pmatrix} = \begin{pmatrix} a & b \\ c & d \end{pmatrix} \begin{pmatrix} a' & b' \\ c' & d' \end{pmatrix}$$

obey these same relations if the primed copy does and if we use the braid statistics consisting of $d - a$ bosonic and

$$a'a = aa' + (1 - q^2)bc', \quad b'b = -bb', \quad c'c = -cc', \quad a'b = ba',$$

$$b'c = -cb' - (1 - q^2)(d - a)(d' - a'), \quad b'a = ab' + (1 - q^2)b(d' - a'),$$

$$c'b = -bc', \quad a'c = ca' + (1 - q^2)(d - a)c', \quad c'a = ac'.$$

Over $\mathbb{C}$, we have a braided bialgebra with $\mathbf{u}$ Hermitian as in Proposition 10.3.2.

Proof: Again, a direct computation from Theorem 10.3.1 and Proposition 10.3.2.   ∎

This example is supercommutative in the limit $q \to 1$ with b, c odd and a, d even. The braid statistics also become ± 1 according to the degree. Thus, we recover exactly the superbialgebra $M_{1|1}$ consisting of these generators and their appropriate supercommutation relations. Thus, the notion of braided matrices really generalises both ordinary and super-matrices. Our construction works for any biinvertible R and not just deformations of standard and supermatrices.

Proposition 10.3.5 *In the regular case, $B(R)$ lives in the category of A-comodules by the right $\mathbf{u} \to \mathbf{t}^{-1}\mathbf{ut}$ from Example 4.5.17.*

Proof: We assume that A is a dual quasitriangular Hopf algebra obtained from $A(R)$. We already know from Example 4.5.17 that $B(R)$ is then a comodule algebra of matrix type. Indeed, we already made use of this covariance in Chapter 7.4. Now we check that the braiding Ψ in Theorem 10.3.1 is indeed the one induced by this coaction, i.e. $B(R)$ lives in the braided category of A-comodules. This is easy from the properties of a dual quasitriangular structure and Exercise 9.2.9. Thus,

$$
\begin{aligned}
\Psi(u^i{}_j \otimes u^k{}_l) &= u^c{}_d \otimes u^a{}_b \mathcal{R}((St^i{}_a)t^b{}_j \otimes (St^k{}_c)t^d{}_l) \\
&= u^c{}_d \otimes u^a{}_b \mathcal{R}(St^i{}_a \otimes (St^e{}_c)t^d{}_f)\mathcal{R}(t^b{}_j \otimes (St^k{}_e)t^f{}_l) \\
&= u^c{}_d \otimes u^a{}_b \mathcal{R}(St^i{}_m \otimes St^e{}_c)\mathcal{R}(St^m{}_a \otimes t^d{}_f)\mathcal{R}(t^n{}_j \otimes St^k{}_e)\mathcal{R}(t^b{}_n \otimes t^f{}_l),
\end{aligned}
$$

which gives the formula in Theorem 10.3.1 on using Proposition 4.2.2. Note that the quantum group normalisation constant λ does not enter into the formulae because there are an equal number of R, R^{-1} in the relations and braiding. ∎

This is how the braiding Ψ in Theorem 10.3.1 was first obtained. It fits our construction into the categorical setting of covariant algebras. We can trivially regard $B(R)$ as living also on the braided category of $\tilde{A}$-comodules with trivial coaction under the scale transformation $\mathbb{Z}_{\lambda^{-1}}$. An A-comodule algebra can trivially be regarded as an $\tilde{A}$-comodule algebra just by the inclusion $A \subset \tilde{A}$. In this case, our braided matrices $B(R)$, and our vector and covector algebras from Section 10.2, all live in the same category or 'universe'. Exercise 9.2.9 tells us how to get the braid statistics between any objects in our universe. This is a big advantage over working with these algebras separately and perhaps trying to guess their required commutation relations for particular applications, i.e. it is systematic. We collect the most useful braid statistics as follows.

Proposition 10.3.6 *Let R be regular and let $\mathbf{x}, \mathbf{v}, \mathbf{u}$ be any $\tilde{A}$-comodule algebras of braided covector, vector and matrix type. Their mutual braid-*

ing is given by

$$\Psi(x_i \otimes x_j) = x_n \otimes x_m R^m{}_i{}^n{}_j, \quad \Psi(v^i \otimes v^j) = R^i{}_m{}^j{}_n v^n \otimes v^m,$$

$$\Psi(x_i \otimes v^j) = \tilde{R}^m{}_i{}^j{}_n v^n \otimes x_m, \quad \Psi(v^i \otimes x_j) = x_n \otimes v^m R^{-1i}{}_m{}^n{}_j,$$

$$\Psi(u^i{}_j \otimes x_k) = x_m \otimes u^a{}_b R^{-1i}{}_a{}^m{}_n R^b{}_j{}^n{}_k,$$

$$\Psi(x_k \otimes u^i{}_j) = u^a{}_b \otimes x_m \tilde{R}^n{}_k{}^i{}_a R^m{}_n{}^b{}_j,$$

$$\Psi(u^i{}_j \otimes v^k) = v^m \otimes u^a{}_b R^i{}_a{}^n{}_m \tilde{R}^b{}_j{}^k{}_n,$$

$$\Psi(v^k \otimes u^i{}_j) = u^a{}_b \otimes v^m R^k{}_n{}^i{}_a R^{-1n}{}_m{}^b{}_j,$$

$$\Psi(u^i{}_j \otimes u^k{}_l) = u^p{}_q \otimes u^m{}_n R^i{}_a{}^d{}_p R^{-1a}{}_m{}^q{}_b R^n{}_c{}^b{}_l \tilde{R}^c{}_j{}^k{}_d,$$

where $\tilde{R} = ((R^{t_2})^{-1})^{t_2}$ is the second-inverse as in Proposition 4.2.2.

Proof: The braidings in the proposition are special cases of the braiding in the category of A-comodules. Some of the cases were already computed in Example 9.3.12, which we extend now to include **u** in the adjoint representation. Thus, $\Psi(u^i{}_j \otimes x_k) = x_b \otimes u^m{}_n \mathcal{R}((St^i{}_m)t^n{}_j \otimes t^b{}_k g) = x_b \otimes u^m{}_n \mathcal{R}^{-1}(t^i{}_m \otimes t^b{}_c g)\mathcal{R}(t^n{}_j \otimes t^c{}_k g)$ using the properties of a dual quasi-triangular structure. This computes from Theorem 4.1.5 as stated. Similarly for the other cases. Since **u** transforms as a scalar under the dilaton generator g, it does not enter into any braid statistics involving **u**. The case $\Psi(u^i{}_j \otimes u^k{}_l)$ was already quoted above in Theorem 10.3.1. Since we use only the coactions, the same results apply for any algebras with generators transforming in the same way as our **x, v, u**, as in the general setting of Lemmas 4.5.2, 4.5.14 and 4.5.16, respectively. We need the dilatonic extension only if the normalisation of R is fixed and is not the quantum group normalisation. ∎

As usual, we can also write these braidings in terms of the relations of the corresponding braided tensor product algebras. These braid statistics relations come out as

$$\mathbf{x}_1' \mathbf{x}_2 = \mathbf{x}_2 \mathbf{x}_1' R, \quad \mathbf{v}_1' \mathbf{v}_2 = R \mathbf{v}_2 \mathbf{v}_1',$$

$$\mathbf{x}_1' R \mathbf{v}_2 = \mathbf{v}_2 \mathbf{x}_1', \quad \mathbf{v}_1' \mathbf{x}_2 = \mathbf{x}_2 R^{-1} \mathbf{v}_1',$$

$$\mathbf{u}_1' \mathbf{x}_2 = \mathbf{x}_2 R^{-1} \mathbf{u}_1' R, \quad \mathbf{x}_1' R \mathbf{u}_2 R^{-1} = \mathbf{u}_2 \mathbf{x}_1',$$

$$R^{-1} \mathbf{u}_1' R \mathbf{v}_2 = \mathbf{v}_2 \mathbf{u}_1', \quad \mathbf{v}_1' \mathbf{u}_2 = R \mathbf{u}_2 R^{-1} \mathbf{v}_1',$$

$$R^{-1} \mathbf{u}_1' R \mathbf{u}_2 = \mathbf{u}_2 R^{-1} \mathbf{u}_1' R,$$

where the primes denote the second algebra in the relevant braided tensor product. In the same way, we can figure out from Exercise 9.2.9 the braid statistics with any other objects in the category of A-comodules, i.e. one does not have to guess these every time. We are now in a position to do some general 'braided linear algebra' with our collection of objects.

Proposition 10.3.7 *If $PR = f(PR')$ for some function f, then the map*

$$\mathbf{u} \mapsto \mathbf{vx}' = \begin{pmatrix} v^1 x_1' & \cdots & v^1 x_n' \\ \vdots & & \vdots \\ v^n x_1' & \cdots & v^n x_n' \end{pmatrix}, \quad B(R) \to V(R', R) \underline{\otimes} V^{\sim}(R', R)$$

is a (covariant) algebra homomorphism, where the image is in the braided tensor product algebra.

Proof: We give the proposition here under the mild assumption that PR used for the braided covectors and vectors in Section 10.2 is given as some function of PR'. Since nothing depends on the precise form of the function, it is clear that this is not a strong restriction. As usual, we use the convenient primed notation for the braid statistics. Then we have $R_{21} \mathbf{v}_1 \mathbf{x}_1' R \mathbf{v}_2 \mathbf{x}_2' = R_{21} \mathbf{v}_1 \mathbf{v}_2 \mathbf{x}_1' \mathbf{x}_2' = f(PR') P \mathbf{v}_1 \mathbf{v}_2 \mathbf{x}_1' \mathbf{x}_2' = \mathbf{v}_2 \mathbf{v}_1 f(1) \mathbf{x}_1' \mathbf{x}_2' = \mathbf{v}_2 \mathbf{v}_1 \mathbf{x}_1' \mathbf{x}_2' f(PR') = \mathbf{v}_2 \mathbf{v}_1 \mathbf{x}_2' \mathbf{x}_1' R = \mathbf{v}_2 \mathbf{x}_2' R_{21} \mathbf{v}_1 \mathbf{x}_1' R$. The first and last equalities use the above statistics relations. The middle equalities use the defining relations in the algebras $V(R', R), V^{\sim}(R', R)$. Hence, $\mathbf{vx}'$ is a realisation of the braided matrices $B(R)$. Moreover, this realisation is manifestly covariant, so that (by functoriality) it must be fully consistent with the braiding of $\mathbf{u}$ with other objects in comparison to the braiding of $\mathbf{vx}'$. Note that our mild assumption on the form of R reduces us, in fact, to the slightly easier setting of the covariant algebras $V^{\sim}{}_R(\lambda, R), V_R(R, \lambda)$ in Chapter 4.5.2. If PR is not given explicitly as some function of PR', then the proposition typically still holds but has to be verified directly according to the form in which PR', PR are given. ∎

This says that the tensor product of a braided covector with a braided vector, when treated with the correct braid statistics, is a braided matrix. It is the covariant version of Exercise 4.5.10. This tensor product is a 'rank-one' braided matrix, and, not surprisingly, one generally has

$$\underline{\det}(\mathbf{vx}') = 0, \tag{10.34}$$

as one might expect in this interpretation. This is easy to see for the examples above. It is also easy to see that the 'inner product' element $v'^i x_j$ in the braided tensor product algebra $V(R', R) \underline{\otimes} V^{\sim}(R', R)$ is central and bosonic. More generally, one has

Theorem 10.3.8 *The elements* $c_k = \mathrm{Tr}\,\mathbf{u}^k v$ *are bosonic and central in* $B(R)$, *where* $v^i{}_j$ *is as in Proposition 4.2.2.*

Proof: We use the identities in Proposition 4.2.2 to compute that $\mathrm{Tr}\,\mathbf{u}^k v$ transforms under the coaction of our background dual quasitriangular Hopf algebra to $\mathrm{Tr}\,(St)\mathbf{u}^k t v = \mathrm{Tr}\,\mathbf{u}^k v(S^2 t) \cdot_{\mathrm{op}} St = \mathrm{Tr}\,\mathbf{u}^k vS(tSt) = \mathrm{Tr}\,\mathbf{u}^k vS(1) = \mathrm{Tr}\,\mathbf{u}^k v$. Here, $\cdot_{\mathrm{op}}$ denotes the reverse multiplication in A, and we use that S is an antialgebra map. Clearly, $\mathbf{u}^k$ can be any matrix of generators transforming as a matrix by conjugation as in Lemma 4.5.16. Hence its braiding with everything is trivial, i.e. it is bosonic when we use Exercise 9.2.9 to compute the braiding.

To prove centrality, we use the identities (4.14) for v in Proposition 4.2.2. The relations of $B(R)$ imply by iteration that

$$R_{21}\mathbf{u}_1 R\mathbf{u}_2^k = \mathbf{u}_2 R_{21}\mathbf{u}_1 R\mathbf{u}_2^{k-1} = \cdots = \mathbf{u}_2^k R_{21}\mathbf{u}_1 R. \qquad (10.35)$$

Applying $\mathrm{Tr}\,(\ )v$ to this, we have $\mathrm{Tr}\,_2\mathbf{u}_1 R\mathbf{u}_2^k R^{-1} v_2 = \mathrm{Tr}\,_2 R_{21}^{-1}\mathbf{u}_2^k R_{21}\mathbf{u}_1 v_2$. Computing the left hand side with the aid of the first of (4.14), we have $\mathbf{u}_1 \mathrm{Tr}\,_2 R\mathbf{u}_2^k v_2 \tilde{R} = \mathbf{u}_1 \mathrm{Tr}\,_2\mathbf{u}_2^k v_2 \tilde{R} \cdot_{\mathrm{op}1} R = \mathbf{u}_1 \mathrm{Tr}\,_2\mathbf{u}_2^k v_2$, where $(\tilde{R} \cdot_{\mathrm{op}1} R)^i{}_j{}^k{}_l = \tilde{R}^a{}_j{}^k{}_b R^i{}_a{}^b{}_l = \delta^i_j \delta^k_l$. Similarly, on the right hand side we move v_2 and apply the second of (4.14) (with permuted indices) to obtain $\mathrm{Tr}\,_2\tilde{R}_{21} v_2\mathbf{u}_2^k R_{21}\mathbf{u}_1 = (\mathrm{Tr}\,_2 v_2\mathbf{u}_2^k R_{21} \cdot_{\mathrm{op}1} \tilde{R}_{21})\mathbf{u}_1 = \mathrm{Tr}\,_2\mathbf{u}_2^k v_2\mathbf{u}_1$. ∎

Let us note that this 'quantum trace' $\mathrm{Tr}\,(\ )v$ is nothing other than a version of the abstract category theoretic trace given in Proposition 9.3.5. Related to the invariant trace elements should be a braided determinant. The braided determinant $\underline{\det}(\mathbf{u})$ should be bosonic and central too: it can generally be constructed from elements of this form. Indeed, the central bosonic elements in the braided bialgebra generate a subalgebra (are closed under addition and multiplication).

Corollary 10.3.9 *If R is regular, there is a sequence of mutually commuting elements $\alpha_k \in A(R)$ which are invariant under the right adjoint coaction. They correspond under transmutation to the elements c_k above.*

Proof: This is rather hard to verify directly, but follows at once from the general transmutation theory, whereby $B(R)$ coincides with $A(R)$ as a coalgebra but has a new transmuted or 'covariantised' product. Explicit formulae were already given in (7.37) in Chapter 7.4. Under this mapping, the bosonic and central elements $c_k = \mathrm{Tr}\,\vartheta\mathbf{u}^k$ in Theorem 10.3.8 are the image of some elements α_k in $A(R)$. From (7.37), one finds at once

$$
\begin{aligned}
\alpha_1 &= \vartheta^i{}_j t^j{}_i, \\
\alpha_2 &= \vartheta^i{}_j R^j{}_k{}^m{}_n t^k{}_l \vartheta^l{}_m t^n{}_i, \\
\alpha_3 &= \vartheta^i{}_j R^j{}_k{}^p{}_q R^k{}_l{}^w{}_y t^l{}_m \tilde{R}^m{}_n{}^v{}_w \vartheta^n{}_p R^q{}_s{}^y{}_z t^s{}_u \vartheta^u{}_v t^z{}_i,
\end{aligned}
\tag{10.36}
$$

etc. One can compute all the α_k in a similar way either from (7.37) or from the general formulae in Example 9.4.10. Since the coaction of A is the same on the linear spaces $A(R) = B(R)$, we know that $\alpha_k = c_k$ are invariant under our background quantum group. Also, invariance of the c_k implies that $c_{k \cdot} a = \alpha_k a$ for all a, i.e. that the transmuted product coincides with the original product in $A(R)$ on such elements. This follows from the abstract formula for the transmuted product in Theorem 7.4.1, which can be written in terms of the right adjoint coaction. Hence, $\alpha_k \alpha_l = c_{k \cdot} c_l = c_{l \cdot} c_k = \alpha_l \alpha_k$ is inherited from the centrality of the c_k. Here α_1 is the quantum trace (which is well-known to be ad-invariant), but the higher α_k are also useful, and can be used, for example, to obtain quantum determinants and other expressions. ∎

Also, now that we have introduced our braided matrices $B(R)$, we can use them to act on covectors and vectors, as well as on themselves, in a fully covariant way with respect to the hidden quantum group symmetry $\tilde{A}$. Thus we have braided (covariant) analogues of the results in Chapter 4.5.1 as follows.

Proposition 10.3.10 *If $PR' = f(PR)$ for some function f, then the map $\mathbf{x} \mapsto \mathbf{x}\mathbf{u}$ makes $V^\smile(R', R)$ into a right braided $B(R)$-comodule algebra, i.e. gives a (covariant) algebra map $V^\smile(R', R) \to V^\smile(R', R)\underline{\otimes}B(R)$. Thus, we can act on a braided covector $\mathbf{x}$ by a braided matrix $\mathbf{u}$ in the sense that $\mathbf{x}\mathbf{u}'$ obey the same relations as $\mathbf{x}$ provided we use the braid statistics above.*

Proof: We assume, for convenience, that PR' is given as some function of PR, so that we are sure to be in the covariant setting of Proposition 10.2.9 (the result also holds more generally). Assuming this case, we have $x_1 u'_1 x_2 u'_2 = x_1 x_2 R^{-1} u'_1 R u'_2 = x_2 x_1 R' R^{-1} u'_1 R u'_2 = x_2 x_1 R_{21}^{-1} u'_2 R_{21} u'_1 R'$ $= x_2 u'_2 x_1 u'_1 R'$ as required. Here we used the braid statistics from Proposition 10.3.6 between $\mathbf{x}, \mathbf{u}$ for the first and last equalities, and the relations for $V^\smile(R', R)$ and $B(R)$ in the form $(PR)R^{-1} u_1 R u_2 = R^{-1} u_1 R u_2 (PR)$ for the middle equalities. The $B(R)$ relations imply that

$$
R' R^{-1} u_1 R u_2 = R_{21}^{-1} u_2 R_{21} u_1 R',
\tag{10.37}
$$

which is the transmuted version of our covariance assumption (10.20). It may also be verifiable directly if PR' is not given as a function of PR. The construction is manifestly covariant under the background quantum group $\tilde{A}$ because we used the braided tensor product, which is a covariant construction. ∎

This is easy enough to check out on our standard examples of the quantum (braided) plane and its super-versions.

Example 10.3.11 *The braided matrices $BM_q(2)$ coact from the right on the braided plane $A_q^{2|0}$ in the sense that*

$$(x'' \quad y'') = (x \quad y) \begin{pmatrix} a' & b' \\ c' & d' \end{pmatrix}$$

obeys the quantum plane relations if x, y do and if a', b', c', d' are a copy of the braided matrices, and provided we remember the braid statistics

$$\begin{pmatrix} a' & b' \\ c' & d' \end{pmatrix} x = \begin{pmatrix} xa' + (1-q^2)yc' & q^{-1}xb' + (q-q^{-1})y(a'-d') \\ qxc' & xd' + (1-q^{-2})yc' \end{pmatrix},$$

$$\begin{pmatrix} a' & b' \\ c' & d' \end{pmatrix} y = y \begin{pmatrix} a' & qb' \\ q^{-1}c' & d' \end{pmatrix}.$$

Proof: We use the results above, computing the relevant braid statistics from Proposition 10.3.6 and deducing the coaction of $BM_q(2)$ from Proposition 10.3.10 on the usual quantum plane in Examples 10.2.2 and 10.2.3. One can easily confirm that the transformed x'', y'' obey the same relations $y''x'' = qx''y''$. ∎

Example 10.3.12 *The braided matrices $BM_q(1|1)$ coact from the right on the braided plane $A_q^{1|1}$ in the sense that*

$$(x'' \quad \theta'') = (x \quad \theta) \begin{pmatrix} a' & b' \\ c' & d' \end{pmatrix}$$

obeys the $A_q^{1|1}$ relations if x, θ do and if a', b', c', d' are a copy of $BM_q(1|1)$, and provided we remember the braid statistics

$$\begin{pmatrix} a' & b' \\ c' & d' \end{pmatrix} x = \begin{pmatrix} xa' + (1-q^2)yc' & q^{-1}xb' + (q-q^{-1})y(a'-d') \\ qxc' & xd' + (1-q^2)yc' \end{pmatrix},$$

$$\begin{pmatrix} a' & b' \\ c' & d' \end{pmatrix} \theta = \theta \begin{pmatrix} a' & -q^{-1}b' \\ -qc' & d' \end{pmatrix}.$$

Proof: We use the results above, computing the relevant braid statistics from Proposition 10.3.6 and deducing the coaction of $BM_q(2)$ from Proposition 10.3.10 on the quantum superplane in Example 10.2.4. One can easily confirm that the transformed x'', θ'' obey the same relations $\theta''x'' = qx''\theta''$ and $\theta''^2 = 0$. ∎

Similarly for $BM_q(2)$ acting on the quantum superplane $\mathbb{A}_q^{0|2}$ in Example 10.2.5. It has the same R-matrix in a different normalisation to that in Example 10.3.11, but this does not affect the braid statistics. We can also multiply our braided vectors by braided matrices. This time we assume that we are in the regular case where $A(R)$ has a quotient A. Then the corresponding $B(R)$ has a corresponding braided-Hopf algebra quotient B.

Proposition 10.3.13 *Let R be regular and let B be the braided group obtained by quotienting $B(R)$. Suppose also that $PR' = f(PR)$ for some function f. Then the map $\mathbf{v} \mapsto \mathbf{u}'^{-1}\mathbf{v}$ makes $V(R', R)$ into a right braided B-comodule algebra, i.e. gives a (covariant) algebra homomorphism $B \to V(R', R) \underline{\otimes} B$. Thus, provided $\mathbf{v}, \mathbf{u}$ are treated with the correct braid statistics, $\mathbf{u}'^{-1}\mathbf{v}$ is also a realisation of $V(R', R)$.*

Proof: We assume that we are in the regular setting so that there is a quantum group A obtained from $A(R)$ and, hence, a corresponding B obtained from $B(R)$ (either by transmutation of A or from $B(R)$ with the corresponding additional relations). We then have a braided antipode which, like all the braided group maps, is covariant (a morphism) and hence commutes with Ψ. This means that the braid statistics of $\mathbf{u}^{-1}$ with $\mathbf{v}$ are read off from Proposition 10.3.6, with $\mathbf{u}^{-1}$ in place of $\mathbf{u}$ there. These statistics are essential because the meaning of $\mathbf{u}'^{-1}\mathbf{v}$ is precisely the element $\Psi(\mathbf{u}^{-1} \otimes \mathbf{v})$ in the braided tensor product algebra $V(R', R) \underline{\otimes} B$. We write $\mathbf{u}'^{-1}$ on the left for convenience with regard to its matrix structure, but it officially belongs on the right of the $\mathbf{v}$ after using the cross relations or braid statistics $R^{-1}\mathbf{u}'_1^{-1}R\mathbf{v}_2 = \mathbf{v}_2\mathbf{u}'_1^{-1}$. Then $\mathbf{u}'_1^{-1}\mathbf{v}_1\mathbf{u}'_2^{-1}\mathbf{v}_2 =$
$\mathbf{u}'_1^{-1}R_{21}^{-1}\mathbf{u}'_2^{-1}R_{21}\mathbf{v}_1\mathbf{v}_2 = \mathbf{u}'_1^{-1}R_{21}^{-1}\mathbf{u}'_2^{-1}R_{21}R'\mathbf{v}_2\mathbf{v}_1 = R'\mathbf{u}'_2^{-1}R^{-1}\mathbf{u}'_1^{-1}R\mathbf{v}_2\mathbf{v}_1$
$= R'\mathbf{u}'_2^{-1}\mathbf{v}_2\mathbf{u}'_1^{-1}\mathbf{v}_1$. We used the braid statistics for the first and last equalities, and the relations of $V(R', R)$ and $B(R)$ (in a form like (10.37) for $\mathbf{u}^{-1}$) for the middle equalities. Thus, the $\mathbf{u}'^{-1}\mathbf{v}$ also realise $V(R', R)$. Moreover, our construction is manifestly covariant under our background quantum group $\widetilde{A}$. ∎

Finally, we see that our braided matrices are altogether better behaved than the quantum matrices $A(R)$ and do coact covariantly on themselves, provided we remember the braid statistics.

Proposition 10.3.14 *In the regular case, let B be obtained from $B(R)$ as above. The map $\mathbf{u} \mapsto \mathbf{u}'^{-1}\mathbf{u}\mathbf{u}'$ makes $B(R)$ into a right braided B-comodule algebra, where $\mathbf{u}'$ denotes the generator of B. Thus, provided $\mathbf{u}, \mathbf{u}'$ are treated with the correct braid statistics, $\mathbf{u}'^{-1}\mathbf{u}\mathbf{u}'$ is also a realisation of $B(R)$ and so provides a (covariant) algebra homomorphism $B(R) \to B(R) \underline{\otimes} B$. It is the braided right adjoint coaction of B.*

Proof: As in the previous proposition, the expression $\mathbf{u'}^{-1}\mathbf{u}\mathbf{u'}$ means $\Psi(\mathbf{u'}^{-1} \otimes \mathbf{u})\mathbf{u'}$, where the statistics relations between $\mathbf{u'}^{-1}$ and $\mathbf{u}$ must be used if we want to exhibit this as an element of $B(R)\underline{\otimes}B$ with the $\mathbf{u'}$ parts living in the second factor B. We get the braid statistics for $B(R)$ from either Proposition 10.3.6 or the shorthand form stated below it. Since the braiding is functorial, the same cross relations hold for $\mathbf{u'}^{-1}$ in place of $\mathbf{u'}$, which is the form that we will use. Then

$$R_{21}\mathbf{u'}_1^{-1}\mathbf{u}_1(\mathbf{u'}_1 R\mathbf{u'}_2^{-1})\mathbf{u}_2\mathbf{u'}_2 = R_{21}\mathbf{u'}_1^{-1}(\mathbf{u}_1 R_{21}^{-1}\mathbf{u'}_2^{-1}R_{21})(\mathbf{u'}_1 R\mathbf{u}_2)\mathbf{u'}_2$$
$$= (R_{21}\mathbf{u'}_1^{-1}R_{21}^{-1}\mathbf{u'}_2^{-1})(R_{21}\mathbf{u}_1 R\mathbf{u}_2 R^{-1})\mathbf{u'}_1 R\mathbf{u'}_2$$
$$= \mathbf{u'}_2^{-1}(R^{-1}\mathbf{u'}_1^{-1}R\mathbf{u}_2)R_{21}\mathbf{u}_1(\mathbf{u'}_1 R\mathbf{u'}_2)$$
$$= \mathbf{u'}_2^{-1}\mathbf{u}_2 R_{12}^{-1}\mathbf{u'}_1^{-1}R(R_{21}\mathbf{u}_1 R_{21}^{-1}\mathbf{u'}_2 R_{21})\mathbf{u'}_1 R$$
$$= \mathbf{u'}_2^{-1}\mathbf{u}_2(R_{12}^{-1}\mathbf{u'}_1^{-1}R\mathbf{u'}_2 R_{21})\mathbf{u}_1\mathbf{u'}_1 R = \mathbf{u'}_2^{-1}\mathbf{u}_2\mathbf{u'}_2 R_{21}\mathbf{u'}_1^{-1}\mathbf{u}_1\mathbf{u'}_1 R.$$

We used only the relations for the braided tensor product $B(R)\underline{\otimes}B$ in the form described, applied in each expression to the parts in parentheses to obtain the next expression. Thus, $\mathbf{u'}^{-1}\mathbf{u}\mathbf{u'}$ is a realisation of $B(R)$. As before, all our constructions are covariant under the background quantum group A (or $\tilde{A}$ with g coacting trivially). ∎

Naturally, these braided matrix coactions are compatible with our other constructions above (they are covariant under these braided group coactions). This is a better situation than in the quantum case. Also, we have given direct proofs wherever possible, but the above elementary properties also follow easily from the transmutation theory at the end of Chapter 9.4. For example, this theory derives the braided coactions in Propositions 10.3.10, 10.3.13 and 10.3.14 at once as corresponding under transmutation to the $\tilde{A}$-coactions already in Proposition 10.2.9 and Example 4.5.17. Comodule algebras under $A(R)$ automatically induce braided coactions of $B(R)$. Likewise for coactions of the quantum and braided group quotients.

10.4 Braided differentiation

So far, we have described how to q-deform $\mathbb{R}^n$ as a braided vector space and we have developed the associated concepts of linear algebra. We now proceed to the first steps of 'braided analysis' on such spaces. We concentrate on understanding braided differentiation. This then determines braided exponentials, braided Gaussians, braided integration and braided differential forms, etc., i.e. some of the remaining basic concepts for 'analysis' on $\mathbb{R}^n$.

We work in the R-matrix setting of the braided vector and covector algebras of Section 10.2 and develop algebraic analogues of the these various concepts. As with the preceding section, there is an abstract theory of braided groups from which many of the results here can be deduced in general, but we suppress this here, preferring direct R-matrix proofs where possible. The key idea that we need from Section 10.2 is that our covector algebra $V^{\check{}}(R', R)$ has a coaddition or braided addition law as described in (10.18) and (10.19). We can think of this coaddition equally well as a braided coaction of one copy of the covectors on the other. This is a global translation. If we denote the generators of the coacting copy by $\mathbf{a}$ rather than by $\mathbf{x}'$ as before, then the content of Theorem 10.2.1 is that $\mathbf{a} + \mathbf{x}$ also obeys the relations of a braided covector, provided we remember the braid statistics $\mathbf{x}_1 \mathbf{a}_2 = \mathbf{a}_2 \mathbf{x}_1 R$. We are now ready to follow the ideas of I. Newton and to define differentiation as an infinitesimal translation.

Definition 10.4.1 *In the setting of Theorem 10.2.1, we define the braided differential operators* $\partial^i : V^{\check{}}(R', R) \to V^{\check{}}(R', R)$ *as*

$$\partial^i f(\mathbf{x}) = \left(a_i^{-1}(f(\mathbf{a} + \mathbf{x}) - f(\mathbf{x})) \right)_{\mathbf{a}=0} \equiv \text{coeff of } a_i \text{ in } f(\mathbf{a} + \mathbf{x}).$$

Note that the definition here, taking the linear part in a_i (which is some function of $\mathbf{x}$), is purely algebraic and does not actually depend on inverting a_i or taking a limit. The definition is also easily computable. On monomials we have

$$\partial^i(\mathbf{x}_1 \cdots \mathbf{x}_m) = \text{coeff}_{a_i} \left((\mathbf{a}_1 + \mathbf{x}_1)(\mathbf{a}_2 + \mathbf{x}_2) \cdots (\mathbf{a}_m + \mathbf{x}_m) \right)$$
$$= \text{coeff}_{a_i} \left(\mathbf{a}_1 \mathbf{x}_2 \cdots \mathbf{x}_m + \mathbf{x}_1 \mathbf{a}_2 \mathbf{x}_3 \cdots \mathbf{x}_m + \cdots + \mathbf{x}_1 \cdots \mathbf{x}_{m-1} \mathbf{a}_m \right)$$
$$= \text{coeff}_{a_i} \left(\mathbf{a}_1 \mathbf{x}_2 \cdots \mathbf{x}_m (1 + (PR)_{12} + (PR)_{12}(PR)_{23} + \cdots \right.$$
$$\left. \cdots + (PR)_{12} \cdots (PR)_{m-1,m}) \right)$$
$$= \mathbf{e}^i{}_1 \mathbf{x}_2 \cdots \mathbf{x}_m [m; R]_{1 \cdots m}, \tag{10.38}$$

where $\mathbf{e}^i$ is a basis covector $(\mathbf{e}^i)_j = \delta^i{}_j$ and

$$[m; R] = 1 + (PR)_{12} + (PR)_{12}(PR)_{23} + \cdots$$
$$+ (PR)_{12} \cdots (PR)_{m-1,m} \tag{10.39}$$

is a certain matrix living in the m-fold matrix tensor product. We call such matrices *braided integers* for reasons that will become apparent in the next section. In explicit terms, we have

$$\partial^i x_{i_1} \cdots x_{i_m} = \delta^i{}_{j_1} x_{j_2} \cdots x_{j_m} [m; R]^{j_1 \cdots j_m}_{i_1 \cdots i_m}. \tag{10.40}$$

Proposition 10.4.2 *The operators ∂^i obey the relations of the braided vectors $V(R', R)$, providing an action $V(R', R) \otimes V\check{\,}(R', R) \to V\check{\,}(R', R)$ given by $v^i \otimes f(\mathbf{x}) \mapsto \partial^i f(\mathbf{x})$.*

Proof: We show the identity

$$[m - 1; R]_{2 \cdots m} [m; R]_{1 \cdots m} = (PR')_{12} [m - 1; R]_{2 \cdots m} [m; R]_{1 \cdots m} . \quad (10.41)$$

To do this we use the definition of the braided integers from (10.39), and (10.11) to compute

$$
\begin{aligned}
((PR')_{12} &- 1) [m - 1; R]_{2 \cdots m} [m; R]_{1 \cdots m} \\
&= ((PR')_{12} - 1) [m - 1; R]_{2 \cdots m} (1 + (PR)_{12} [m - 1; R]_{2 \cdots m}) \\
&= ((PR')_{12} - 1)((1 + (PR)_{12}) [m - 1; R]_{2 \cdots m} \\
&\quad + ([m - 1; R]_{2 \cdots m} - 1)(PR)_{12} [m - 1; R]_{2 \cdots m}) \\
&= ((PR')_{12} - 1)(PR)_{23} [m - 2; R]_{3 \cdots m} (PR)_{12} [m - 1; R]_{2 \cdots m} \\
&= ((PR')_{12} - 1)(PR)_{23}(PR)_{12} [m - 2; R]_{3 \cdots m} [m - 1; R]_{2 \cdots m} \\
&= (PR)_{23}(PR)_{12}((PR')_{23} - 1) [m - 2; R]_{3 \cdots m} [m - 1; R]_{2 \cdots m} .
\end{aligned}
$$

Hence, if we assume (10.41) for $m - 1$ as an induction hypothesis, the last expression vanishes, and the result also holds for m. The start of the induction is provided by (10.11) itself.

From (10.38), we have

$$
\begin{aligned}
\partial^i \partial^k \mathbf{x}_1 \cdots \mathbf{x}_m &= \mathbf{e}^k{}_1 \mathbf{e}^i{}_2 \mathbf{x}_3 \cdots \mathbf{x}_m [m - 1; R]_{2 \cdots m} [m; R]_{1 \cdots m} , \\
R'^i{}_a{}^k{}_b \partial^b \partial^a \mathbf{x}_1 \cdots \mathbf{x}_m &= R'^i{}_a{}^k{}_b \partial^b \mathbf{e}^a{}_1 \mathbf{x}_2 \cdots \mathbf{x}_m [m; R]_{1 \cdots m} \\
&= R'^i{}_a{}^k{}_b \mathbf{e}^a{}_1 \mathbf{e}^b{}_2 \mathbf{x}_3 \cdots \mathbf{x}_m [m - 1; R]_{2 \cdots m} [m; R]_{1 \cdots m} \\
&= \mathbf{e}^k{}_1 \mathbf{e}^i{}_2 \mathbf{x}_3 \cdots \mathbf{x}_m (PR')_{12} [m - 1; R]_{2 \cdots m} [m; R]_{1 \cdots m} .
\end{aligned}
$$

These are equal due to our identity (10.41). Hence, ∂^i obey the relations for v^i in Theorem 10.2.6 and are therefore an operator realisation of $V(R', R)$. ∎

This action of braided vectors on braided covectors makes the latter a braided skew module algebra. This is best handled diagrammatically, and we do not want to develop it explicitly here. But this means that our differentials ∂^i obey a braided version of the usual Leibniz rule. This justifies considering these operators as (braided) partial differentiation.

Lemma 10.4.3 *The operators ∂^i obey the braided Leibniz rule*

$$\partial^i(ab) = (\partial^i a)b + \cdot \Psi^{-1}(\partial^i \otimes a)b.$$

Proof: We need here the inverse braiding between braided vectors and braided covectors. This is computed from Proposition 10.3.6 and comes out as $\Psi^{-1}(\partial_2 \otimes \mathbf{x}_1) = \mathbf{x}_1 \otimes R\partial_2$; it extends to products by functoriality as

$$\Psi^{-1}(\partial^i \otimes \mathbf{x}_1 \cdots \mathbf{x}_r) = \mathbf{e}^i{}_1 \mathbf{x}_2 \cdots \mathbf{x}_r \mathbf{x}_{r+1}(PR)_{12} \cdots (PR)_{r,r+1} \otimes \partial_{r+1}.$$
(10.42)

Taking this in the proposition, we compute the right hand side on monomials as

$$(\partial^i \mathbf{x}_1 \cdots \mathbf{x}_r)\mathbf{x}_{r+1} \cdots \mathbf{x}_m + \Psi^{-1}(\partial^i \otimes \mathbf{x}_1 \cdots \mathbf{x}_r)\mathbf{x}_{r+1} \cdots \mathbf{x}_m$$
$$= \mathbf{e}^i{}_1 \mathbf{x}_2 \cdots \mathbf{x}_r \, [r; R]_{1\cdots r} \, \mathbf{x}_{r+1} \cdots \mathbf{x}_m$$
$$+ \mathbf{e}^i{}_1 \mathbf{x}_2 \cdots \mathbf{x}_r \mathbf{x}_{r'+1}(PR)_{12} \cdots (PR)_{r,r'+1} \partial_{r'+1} \mathbf{x}_{r+1} \cdots \mathbf{x}_m$$
$$= \mathbf{e}^i{}_1 \mathbf{x}_2 \cdots \mathbf{x}_r \mathbf{x}_{r+1} \cdots \mathbf{x}_m \, [r; R]_{1\cdots r}$$
$$+ \mathbf{e}^i{}_1 \mathbf{x}_2 \cdots \mathbf{x}_r \mathbf{x}_{r+1}(PR)_{12} \cdots (PR)_{r,r+1} \mathbf{x}_{r+2} \cdots \mathbf{x}_m \, [m-r; R]_{r+1\cdots m} \,,$$

where we use (10.38) to evaluate the differentials. The primed $r'+1$ labels a distinct matrix space from the existing $r+1$ index. These are then identified by the $\mathbf{e}_{r+1}$ brought down by the action of $\partial_{r'+1}$. The resulting expression coincides with $\mathbf{e}^i{}_1 \mathbf{x}_2 \cdots \mathbf{x}_r \mathbf{x}_{r+1} \cdots \mathbf{x}_m \, [m; R]_{1\cdots m}$ from the left hand side of the proposition, since

$$[r; R]_{1\cdots r} + (PR)_{12} \cdots (PR)_{r,r+1} \, [m-r; R]_{r+1\cdots m} = [m; R]_{1\cdots m}. \quad (10.43)$$

This is evident from the definition in (10.39). ∎

Another way to express this braided Leibniz rule is in terms of commutation relations between differentiation and position operators.

Exercise 10.4.4 *Consider the operators* $\hat{x}_i : V^{\check{}}(R', R) \to V^{\check{}}(R', R)$ *consisting of multiplication from the left by* x_i. *Show that*

$$\partial^i \hat{x}_j - \hat{x}_a R^a{}_j{}^i{}_b \partial^b = \delta^i{}_j, \quad i.e. \quad \partial_1 \hat{x}_2 - \hat{x}_2 R_{21} \partial_1 = \mathrm{id}$$

as operators on $V^{\check{}}(R', R)$.

Solution: We just apply the braided Leibniz rule to $\partial^i(x_j f(\mathbf{x}))$. ∎

This is obviously the point of view that could be called 'braided quantum mechanics'. Indeed, the braided action in Proposition 10.4.2 means that we can make a braided cross product $V^{\check{}}(R', R) \rtimes V(R', R)$ as a braided version of our usual cross products in Chapter 1.6. It is generated by $V(R', R) = \{\mathbf{p}\}$ and $V^{\check{}}(R', R) = \{\mathbf{x}\}$ with the cross relations

$$\mathbf{p}_1 \mathbf{x}_2 - \mathbf{x}_2 R_{21} \mathbf{p}_1 = \mathrm{id} \quad (10.44)$$

when we use the action in Proposition 10.4.2. The above Exercise 10.4.4 gives a concrete operator realisation of this 'braided-Weyl algebra'.

Exercise 10.4.5 *If there is a covariant quantum metric η in the sense of Definition 10.2.14 then we can write $\partial_i = \eta_{ia}\partial^a$, which then obey the relations of $V\check{}(R', R)$. Show that in these terms the above braided Leibniz rule becomes*

$$\partial_i\hat{x}_j - \lambda^{-2}\hat{x}_b\partial_a R^{-1b}{}_j{}^a{}_i = \eta_{ij}, \quad i.e. \quad \partial_1\hat{x}_2 - \lambda^{-2}\hat{x}_2\partial_1 R_{21}^{-1} = \eta. \quad (10.45)$$

Solution: We just use covariance identities (10.25) and (10.26). ∎

The braided Leibniz rule is helpful when computing examples because it means that we only have to take ∂^i on generators and extend to products using this rule.

Example 10.4.6 *Let $R = (q)$ be the one-dimensional solution of the QYBE and let $R' = (1)$. We have $V\check{}(R', R) = k[x]$, the braided line as at the end of Section 10.1. In this case, we have*

$$\partial x^m = x^{m-1}(1+q+\cdots+q^{m-1}) = [m; q]\, x^{m-1}, \text{ i.e. } \partial f(x) = \frac{f(qx) - f(x)}{(q-1)x}.$$

The braided Leibniz rule is

$$\partial(x^n x^m) = (\partial x^n)x^m + q^n x^n (\partial x^m), \quad i.e. \quad \partial(fg) = (\partial f)g + (L_q f)\partial g,$$

which is just as in the case of a superderivation, but with q in the role of -1 and a $\mathbb{Z}$-grading in the role of $\mathbb{Z}_{/2}$-grading. The degree of ∂ here is -1, and $L_q(f)(x) = f(qx)$.

Proof: The braiding is $\Psi(\partial \otimes x) = q^{-1}x \otimes \partial$. We use the inverse braiding when computing ∂x^m, giving the q-integer as shown. Or we just use (10.40) with $R = (q)$ and hence $[m; R] = [m; q]$. It is easy enough to verify the braided Leibniz rule here explicitly. We have already encountered it in another context at the end of Section 3.1. ∎

This is where the familiar q-integers in Chapter 3.2 come from in braided geometry, and is the reason that we called (10.39) braided integers. There is a q each time ∂^i passes an x due to the braid statistics between them. It is also the correct point of view on q-differentiation $\partial = \partial_q$ and reproduces easily well-known formulae such as we have already encountered directly in earlier chapters. On the other hand, our formalism works on any higher-dimensional braided space just as well.

Example 10.4.7 *Let R, R' be the standard R-matrices for the quantum plane as a braided covector space in Example 10.2.2. Then ∂^i are*

$$\frac{\partial}{\partial x}x^n y^m = \left[n; q^2\right] x^{n-1}y^m, \quad \frac{\partial}{\partial y}x^n y^m = q^n x^n \left[m; q^2\right] y^{m-1},$$

where $[m; q^2] = \frac{q^{2m}-1}{q^2-1}$. *The derivatives obey the relations*

$$\frac{\partial}{\partial y}\frac{\partial}{\partial x} = q^{-1}\frac{\partial}{\partial x}\frac{\partial}{\partial y}.$$

Proof: From Proposition 10.4.2 we know that relations of the ∂^i are necessarily the braided vector ones for $A_{q^{-1}}^{2|0}$ in Example 10.2.7. The action of the generators is $\partial^i x_j = \delta^i{}_j$, which we then extend using the braided Leibniz rule for the same R as in Example 10.2.2. This gives the above results by an easy induction. Note that the q-derivatives act in the same way as in the one-dimensional case in each variable, except that $\frac{\partial}{\partial y}$ picks up a factor q when it passes x due to some braid statistics. ∎

At another extreme, which works in any dimension, we can let R be any invertible solution of the QYBE and let $R' = P$ be the permutation matrix. Then $V^{\check{}}(P, R)$ is the free algebra $k\langle x_i\rangle$ with no relations. This is the *free braided plane* associated to an R-matrix. It is in a certain sense universal, with the others as quotients. The vector algebra is the free algebra $k\langle v^i\rangle$ with no relations, and is realised by ∂^i acting as in (10.38). The R-matrix is still used, in the braiding.

Our next goal is to prove a braided Taylor's theorem for our braided differential operators. The first step is to understand much better the braided coaddition on our braided covectors in Theorem 10.2.1, namely how it looks on products of the generators. We need q-binomial coefficients as in Example 10.1.7. We are familiar with these in one dimension, but we now have to generalise them to our higher-dimensional R-matrix setting.

Definition 10.4.8 *Let $\left[{m \atop r}; R\right]$ be the matrix defined recursively by*

$$\left[{m \atop r}; R\right]_{1\cdots m} = (PR)_{r,r+1}\cdots(PR)_{m-1,m}\left[{m-1 \atop r-1}; R\right]_{1\cdots m-1}$$
$$+ \left[{m-1 \atop r}; R\right]_{1\cdots m-1},$$
$$\left[{m \atop 0}; R\right] = 1, \qquad \left[{m \atop r}; R\right] = 0 \quad \text{if } r > m.$$

The suffixes here refer to the matrix position in tensor powers of M_n.

This defines in particular

$$\left[{m \atop m}; R\right] = \left[{m-1 \atop m-1}; R\right] = \cdots = \left[{1 \atop 1}; R\right] = 1,$$
$$\left[{m \atop 1}; R\right] = (PR)_{12}\cdots(PR)_{m-1,m} + \left[{m-1 \atop 1}; R\right] = \cdots = [m; R]. \tag{10.46}$$

A similar recursion defines $\left[{}^m_2; R \right]_{1 \cdots m}$ in terms of $\left[{}^m_1; R \right]$ (which is known) and $\left[{}^{m-1}_2; R \right]$, and similarly (in succession) up to $r = m$.

Proposition 10.4.9 *Suppose that* $\mathbf{x}_1 \mathbf{a}_2 = \mathbf{a}_2 \mathbf{x}_1 R$ *as above. Then*

$$(\mathbf{a}_1 + \mathbf{x}_1) \cdots (\mathbf{a}_m + \mathbf{x}_m) = \sum_{r=0}^{r=m} \mathbf{a}_1 \cdots \mathbf{a}_r \mathbf{x}_{r+1} \cdots \mathbf{x}_m \left[{}^m_r; R \right]_{1 \cdots m}$$

Proof: We use induction. Suppose the proposition is true for $m - 1$, then

$$(\mathbf{a}_1 + \mathbf{x}_1) \cdots (\mathbf{a}_m + \mathbf{x}_m)$$
$$= (\mathbf{a}_1 + \mathbf{x}_1) \cdots (\mathbf{a}_{m-1} + \mathbf{x}_{m-1}) \mathbf{a}_m + (\mathbf{a}_1 + \mathbf{x}_1) \cdots (\mathbf{a}_{m-1} + \mathbf{x}_{m-1}) \mathbf{x}_m$$
$$= \sum_{r=0}^{m-1} \mathbf{a}_1 \cdots \mathbf{a}_r \mathbf{x}_{r+1} \cdots \mathbf{x}_{m-1} \mathbf{a}_m \left[{}^{m-1}_r; R \right]_{1 \cdots m-1}$$
$$+ (\mathbf{a}_1 + \mathbf{x}_1) \cdots (\mathbf{a}_{m-1} + \mathbf{x}_{m-1}) \mathbf{x}_m$$
$$= \sum_{r=1}^{m} \mathbf{a}_1 \cdots \mathbf{a}_{r-1} \mathbf{x}_r \cdots \mathbf{x}_{m-1} \mathbf{a}_m \left[{}^{m-1}_{r-1}; R \right]_{1 \cdots m-1}$$
$$+ \sum_{r=0}^{m-1} \mathbf{a}_1 \cdots \mathbf{a}_r \mathbf{x}_{r+1} \cdots \mathbf{x}_m \left[{}^{m-1}_r; R \right]_{1 \cdots m-1}$$
$$= \sum_{r=1}^{m} \mathbf{a}_1 \cdots \mathbf{a}_r \mathbf{x}_{r+1} \cdots \mathbf{x}_m (PR)_{r,r+1} \cdots (PR)_{m-1,m} \left[{}^{m-1}_{r-1}; R \right]_{1 \cdots m-1}$$
$$+ \sum_{r=0}^{m-1} \mathbf{a}_1 \cdots \mathbf{a}_r \mathbf{x}_{r+1} \cdots \mathbf{x}_m \left[{}^{m-1}_r; R \right]_{1 \cdots m-1}$$
$$= \sum_{r=1}^{m} \mathbf{a}_1 \cdots \mathbf{a}_r \mathbf{x}_{r+1} \cdots \mathbf{x}_m \left[{}^m_r; R \right]_{1 \cdots m} + \mathbf{x}_1 \cdots \mathbf{x}_m \left[{}^{m-1}_0; R \right]_{1 \cdots m-1},$$

using the induction hypothesis and Definition 10.4.8. The last term is also the $r = 0$ term in the desired sum, proving the result for m. ∎

In explicit notation, the last proposition reads

$$(a_{i_1} + x_{i_1}) \cdots (a_{i_m} + x_{i_m})$$
$$= \sum_{r=0}^{r=m} a_{j_1} \cdots a_{j_r} x_{j_{r+1}} \cdots x_{j_m} \left[{}^m_r; R \right]_{i_1 \cdots i_m}^{j_1 \cdots j_m} . \qquad (10.47)$$

Also, in terms of the coproduct structure of $V\check{\ }(R', R)$ in Theorem 10.2.1, we see that

$$\Delta(\mathbf{x}_1 \mathbf{x}_2 \cdots \mathbf{x}_m) = \sum_{r=0}^{r=m} \mathbf{x}_1 \cdots \mathbf{x}_r \otimes \mathbf{x}_{r+1} \cdots \mathbf{x}_m \left[{}^m_r; R \right]_{1 \cdots m} . \qquad (10.48)$$

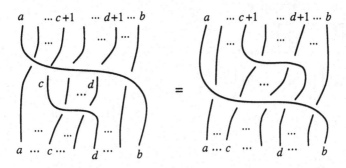

Fig. 10.1. Diagram used in proof of Lemma 10.4.10.

The main theorem is actually to compute these R-binomial coefficient matrices. We need two lemmas.

Lemma 10.4.10 *Denote the monodromy or 'parallel transport' operator by*

$$[a, b; R] = (PR)_{a,a+1} \cdots (PR)_{b-1,b}$$

for $a \le b$ with the convention $[a, a; R] = 1$. It obeys

$$[a, b; R][b, c; R] = [a, c; R] \qquad \forall \, a \le b \le c,$$

$$[a, b; R][c, d; R] = [c + 1, d + 1; R][a, b; R] \qquad \forall \, a \le c \le d < b.$$

Proof: The proof of the second identity follows from repeated use of the QYBE or braid relations in a standard way. This is easily done by writing the monodromies as braided crossings. Thus $[a, b; R]$ is the braid of strand a past the strands up to and including strand b, with each crossing represented by an R-matrix. If the braid $[c, d; R]$ lies entirely inside $[a, b; R]$, then it can be pulled through it in its entirety, giving the commutation relation shown. This is depicted in Fig. 10.1. ∎

Lemma 10.4.11 *The monodromy commutes with the braided binomial coefficients in the sense*

$$[1, m; R] \begin{bmatrix} m - 1 \\ r \end{bmatrix}; R \end{bmatrix}_{1 \cdots m-1} = \begin{bmatrix} m - 1 \\ r \end{bmatrix}; R \end{bmatrix}_{2 \cdots m} [1, m; R].$$

Proof: We proceed by induction. Thus, using Definition 10.4.8 and the result for $m - 2$, we have

$$[1, m; R] \begin{bmatrix} m-1 \\ r \end{bmatrix}_{1\cdots m-1}$$

$$= [1, m; R] \left([r, m-1; R] \begin{bmatrix} m-2 \\ r-1 \end{bmatrix}; R \Big]_{1\cdots m-2} + \begin{bmatrix} m-2 \\ r \end{bmatrix}; R \Big]_{1\cdots m-2} \right)$$

$$= [r+1, m; R][1, m; R] \begin{bmatrix} m-2 \\ r-1 \end{bmatrix}; R \Big]_{1\cdots m-2} + [1, m; R] \begin{bmatrix} m-2 \\ r \end{bmatrix}; R \Big]_{1\cdots m-2}$$

$$= [r+1, m; R] \begin{bmatrix} m-2 \\ r-1 \end{bmatrix}; R \Big]_{2\cdots m-1} [1, m; R] + \begin{bmatrix} m-2 \\ r \end{bmatrix}; R \Big]_{2\cdots m-1} [1, m; R],$$

which equals the right hand side using Definition 10.4.8 again in reverse. For the second equality we used the preceding lemma. ∎

Theorem 10.4.12

$$[r; R]_{1\cdots r} \begin{bmatrix} m \\ r \end{bmatrix}; R \Big]_{1\cdots m} = \begin{bmatrix} m-1 \\ r-1 \end{bmatrix}; R \Big]_{2\cdots m} [m; R]_{1\cdots m}$$

and hence (formally supposing that all the braided integers are invertible)

$$\begin{bmatrix} m \\ r \end{bmatrix}; R \Big]_{1\cdots m} = [r; R]_{1\cdots r}^{-1} \cdots [2; R]_{r-1, r}^{-1} [m-r+1; R]_{r\cdots m} \cdots [m; R]_{1\cdots m}.$$

Proof: We proceed by induction. Suppose the result is true up to $m - 1$. Then, from Definition 10.4.8 and the above lemmas, we have

$$[r; R]_{1\cdots r} \begin{bmatrix} m \\ r \end{bmatrix}; R \Big]_{1\cdots m}$$

$$= [r; R]_{1\cdots r} [r, m; R] \begin{bmatrix} m-1 \\ r-1 \end{bmatrix}; R \Big]_{1\cdots m-1} + [r; R]_{1\cdots r} \begin{bmatrix} m-1 \\ r \end{bmatrix}; R \Big]_{1\cdots m-1}$$

$$= [r, m; R] [r-1; R]_{1\cdots r-1} \begin{bmatrix} m-1 \\ r-1 \end{bmatrix}; R \Big]_{1\cdots m-1}$$

$$\quad + [1, m; R] \begin{bmatrix} m-1 \\ r-1 \end{bmatrix}; R \Big]_{1\cdots m-1} + [r; R]_{1\cdots r} \begin{bmatrix} m-1 \\ r \end{bmatrix}; R \Big]_{1\cdots m-1}$$

$$= [r, m; R] \begin{bmatrix} m-2 \\ r-2 \end{bmatrix}; R \Big]_{2\cdots m-1} [m-1; R]_{1\cdots m-1}$$

$$\quad + \begin{bmatrix} m-1 \\ r-1 \end{bmatrix}; R \Big]_{2\cdots m} [1, m; R] + \begin{bmatrix} m-2 \\ r-1 \end{bmatrix}; R \Big]_{2\cdots m-1} [m-1; R]_{1\cdots m-1}$$

$$= \begin{bmatrix} m-1 \\ r-1 \end{bmatrix}; R \Big]_{2\cdots m} [1, m; R] + \begin{bmatrix} m-1 \\ r-1 \end{bmatrix}; R \Big]_{2\cdots m} [m-1; R]_{1\cdots m-1}$$

$$= \begin{bmatrix} m-1 \\ r-1 \end{bmatrix}; R \Big]_{2\cdots m} [m; R]_{1\cdots m}$$

as required. Here the second equality splits $[r; R]_{1 \cdots r} = [r-1; R]_{1 \cdots r-1} + [1, r; R]$. The $[r, m; R]$ commutes past the first term of this, while, with the second term, it combines to give $[1, m; R]$. The third equality is our induction hypothesis for the outer terms and Lemma 10.4.10 for the middle term. We then used Definition 10.4.8 in reverse to recognise two of the terms to obtain the fourth equality. We then recognise $[1, m; R] + [m-1; R] = [m; R]$ to obtain the result. ∎

Note that one does not really need the braided integers to be invertible here (just as for the usual binomial coefficients). For example, the recursion relation in the theorem implies that

$$\begin{bmatrix} m \\ m-1 \end{bmatrix}_{1 \cdots m} = 1 + (PR)_{m-1,m} + (PR)_{m-1,m}(PR)_{m-2,m-1} + \cdots$$
$$\cdots + (PR)_{m-1,m} \cdots (PR)_{12}, \qquad (10.49)$$

which is a right handed variant of $[m; R]$. This can also be proven directly by using the QYBE repeatedly. One can prove numerous other identities of this type in analogy with usual combinatoric identities. This theorem demonstrates the beginning of some kind of braided-number theory or braided combinatorics. Because it holds for any invertible solution of the QYBE, it corresponds to a novel identity in the group algebra of the braid group. Physically, it corresponds to 'counting' the 'partitions' of a box of braid-statistical particles.

An application of these ideas is to the duality between vectors and covectors, which we promised the reader in Section 10.2. We have

Proposition 10.4.13 *There is a duality pairing of braided vector and covector braided-Hopf algebras by*

$$\langle v^{i_m} \cdots v^{i_2} v^{i_1}, x_{j_1} x_{j_2} \cdots x_{j_r} \rangle = \delta_{m,r}([m; R]!)^{i_1 i_2 \cdots i_m}_{j_1 j_2 \cdots j_m}$$

i.e. $\langle \mathbf{v}_m \cdots \mathbf{v}_2 \mathbf{v}_1, \mathbf{x}_1 \mathbf{x}_2 \cdots \mathbf{x}_r \rangle = \delta_{m,r}[m; R]!$, $\langle f(\mathbf{v}), g(\mathbf{x}) \rangle = \underline{\epsilon} \circ f(\partial) g(\mathbf{x})$,

where

$$[m; R]! = [2; R]_{m-1\,m} [3; R]_{m-2\,m} \cdots [m; R]_{1 \cdots m}.$$

Proof: Clearly, the formula in terms of differentiation is the same in view of (10.38) as the explicit formula in terms of braided integers. On the other hand, it is manifestly well-defined; i.e. the operators ∂^i are well-defined on the $\{x_i\}$ by their very construction in Definition 10.4.1 so the relations are respected on this side of the pairing. The relations of the vector algebra on the other side of the pairing are also respected, by Proposition 10.4.2, which tells us that we had an action of the vector algebra. Finally, our linear map here is a morphism in the category, i.e. an intertwiner for the coaction of the background quantum group $\tilde{A}$ since the ∂^i are constructed covariantly. This is also clear, at least on the generators, in view of Example 9.3.12. We have only to check that our well-defined morphism $V(R', R) \otimes V\check{}(R', R) \to k$ is a (categorical) pairing of braided-Hopf algebras. This means that the product on one side is the braided coproduct of the other. Thus,

$$\langle \mathbf{v}_m \cdots \mathbf{v}_{r+1} \cdot \mathbf{v}_r \cdots \mathbf{v}_1, \mathbf{x}_1 \cdots \mathbf{x}_m \rangle = [m; R]!,$$

$$\langle \mathbf{v}_m \cdots \mathbf{v}_{r+1} \otimes \mathbf{v}_r \cdots \mathbf{v}_1, \underline{\Delta}\mathbf{x}_1 \cdots \mathbf{x}_m \rangle = [r, R]!_{1\cdots r}[m - r, R]!_{r+1\cdots m}\begin{bmatrix} m \\ r \end{bmatrix}; R\Big],$$

where we evaluate the inner $V, V\check{}$ first and then the remaining outer two (the categorical pairing which avoids an unnecessary braiding as explained in (9.47) in Chapter 9.4). The coproduct is from Proposition 10.4.9 in the form (10.48). Our two expressions coincide just by the braided binomial Theorem 10.4.12. Similarly for the coproduct of the $\{v^i\}$ by an analogous computation. The duality pairing with respect to the unit/counit and antipode is clear from the generators. Note that one could also turn this around and recover

$$\partial^i = (\langle v^i, \ \rangle \otimes \mathrm{id}) \circ \underline{\Delta} \tag{10.50}$$

if we start from the knowledge that our vectors and covectors are dual. ∎

The second half of this proof is clear enough by the evident symmetry between the braided vector and braided covector constructions. To give this in detail, one has to redevelop the various steps above under this symmetry. Thus, we have differentiation operators $\overleftarrow{\partial}_i = \frac{\partial}{\partial v^i}$ acting from the right and defined by

$$f(\mathbf{v})\overleftarrow{\partial}_i = f(\mathbf{v} + \mathbf{w})|_{\mathrm{coeff}\ w^i}; \quad \mathbf{w}_1\mathbf{v}_2 = R\mathbf{v}_2\mathbf{w}_1,$$

$$\mathbf{v}_m \cdots \mathbf{v}_2\mathbf{v}_1\overleftarrow{\partial}_i = [m; R]^{\mathrm{op}}\mathbf{v}_m \cdots \mathbf{v}_3\mathbf{v}_2(\mathbf{f_i})_1, \tag{10.51}$$

$$[m; R]^{\mathrm{op}} = 1 + (PR)_{12} + (PR)_{23}(PR)_{12} + \cdots + (PR)_{m-1\,m} \cdots (PR)_{12},$$

where $\mathbf{f_i}$ is a basis vector $(f_i)^j = \delta_i{}^j$. The matrices here are like (10.39) but with the opposite product of matrices. We can equally well write the pairing in Proposition 10.4.13 as

$$\langle \mathbf{v}_m \cdots \mathbf{v}_2\mathbf{v}_1, \mathbf{x}_1\mathbf{x}_2 \cdots \mathbf{x}_r \rangle = \delta_{m,r}[m; R]^{\mathrm{op}}!, \quad \langle f(\mathbf{v}), g(\mathbf{x}) \rangle = \underline{\epsilon} \circ f(\mathbf{v})g(\overleftarrow{\partial}),$$

where

$$[m; R]^{\mathrm{op}}! \equiv [m; R]^{\mathrm{op}}_{1\cdots m} \cdots [3; R]^{\mathrm{op}}_{m-2\,m}[2; R]^{\mathrm{op}}_{m-1\,m} = [m; R]! \tag{10.52}$$

by matrix computations similar to those above. This symmetry between the ∂ and the $\overleftarrow{\partial}$ points of view expresses the symmetry (with left–right reversal) in the axioms of a pairing. It is as important in the braided case as it is in the unbraided case that we have stressed already in earlier chapters.

Also, our pairing between braided vectors and covectors is typically nondegenerate. This is true for standard deformations and other generic R-matrices because it is how the pairing goes in the $q = 1$ case between the functions on $\mathbb{R}^n$ and the enveloping algebra of $\mathbb{R}^n$, namely by differentiation. We know, at least in this case, and hence also near $q = 1$, that the only functions which have zero differentials are constant. We effectively made use of this (for the bosonised one-dimensional braided line) in Example 7.2.9 when computing the quantum double of a q-Borel subalgebra.

In categorical terms, we have an evaluation morphism

$$V(R', R) = (V^{\smile}(R', R))^*, \quad \mathrm{ev} = \langle \, , \, \rangle \qquad (10.53)$$

so that $V^{\smile}$ is the categorical predual of V. Since our algebras are not finite-dimensional objects in the category of $\tilde{A}$-comodules, we do not really expect them to be rigid in the strict sense of Definition 9.3.1 with a corresponding $\mathrm{coev} \in V^{\smile}(R', R) \otimes V(R', R)$. We next give a direct definition of the braided exponential, which, from an abstract point of view, is nothing other than $\exp = \mathrm{coev}$ but developed as a formal power series rather than as an element of the algebraic tensor product. This is the abstract reason that $\exp$ exists on a general braided linear space as a formal power series.

The direct approach to the braided exponential is to seek eigenfunctions of the operators ∂^i. For the moment, we seek these among formal power series in the x_i coordinates, but in our later application to braided Taylor's theorem only finitely many terms will be nonvanishing. The only difference is that we consider both ∂ and $\overleftarrow{\partial}$.

Definition 10.4.14 *We define the* braided exponential $\exp(\mathbf{x}|\mathbf{v})$ *as a formal power series such that*

$$\partial^i \exp(\mathbf{x}|\mathbf{v}) = \exp(\mathbf{x}|\mathbf{v})v^i, \quad (\underline{\epsilon} \otimes \mathrm{id}) \exp(\mathbf{x}|\mathbf{v}) = 1,$$

$$\exp(\mathbf{x}|\mathbf{v}) \overleftarrow{\partial}_i = x_i \exp(\mathbf{x}|\mathbf{v}), \quad (\mathrm{id} \otimes \underline{\epsilon}) \exp(\mathbf{x}|\mathbf{v}) = 1.$$

We also require invariance under the coactions in Proposition 10.2.9.

This is a reasonable definition because it has the right classical limit. It is also possible to consider the 'integrability condition' for the solution of this braided differential equation. Classically it means that we ask for the constraint imposed by commutativity of partial derivatives. Here we have

$$(\partial_1\partial_2 - R'\partial_2\partial_1)\exp(\mathbf{x}|\mathbf{v}) = \partial_1\exp(\mathbf{x}|\mathbf{v})\mathbf{v}_2 - R'\partial_2\exp(\mathbf{x}|\mathbf{v})\mathbf{v}_1$$
$$= \exp(\mathbf{x}|\mathbf{v})(\mathbf{v}_1\mathbf{v}_2 - R'\mathbf{v}_2\mathbf{v}_1) = 0,$$

since the v^i obey the vector algebra. Similarly from the other side in terms of $\overleftarrow{\partial}$. While not a proof, this tells us that, at least generically, an exp should exist which is an eigenfunction with respect to each input.

Example 10.4.15 *If the braided integers $[m; R]$ are all invertible, then exp is given by*

$$\exp(\mathbf{x}|\mathbf{v}) = \sum_{m=0}^{\infty} \mathbf{x}_1 \cdots \mathbf{x}_m \, [m; R]_{1\cdots m}^{-1} \, [m-1; R]_{2\cdots m}^{-1} \cdots [2; R]_{m-1,m}^{-1} \, \mathbf{v}_m \cdots \mathbf{v}_1.$$

Proof: We see at once from (10.38) that

$$\partial^i \exp(\mathbf{x}|\mathbf{v})$$

$$= \sum_{m=0}^{\infty} \partial^i \mathbf{x}_1 \cdots \mathbf{x}_m \, [m; R]_{1\cdots m}^{-1} \left([m-1; R]_{2\cdots m}^{-1} \cdots [2; R]_{m-1,m}^{-1} \right) \mathbf{v}_m \cdots \mathbf{v}_1$$

$$= \sum_{m=0}^{\infty} \mathbf{e}^i{}_1 \left(\mathbf{x}_2 \cdots \mathbf{x}_m \, [m-1; R]_{2\cdots m}^{-1} \cdots [2; R]_{m-1,m}^{-1} \, \mathbf{v}_m \cdots \mathbf{v}_2 \right) \mathbf{v}_1$$

$$= \exp(\mathbf{x}|\mathbf{v})v^i, \tag{10.54}$$

where v^i is the noncommutative eigenvalue. Similarly for $\exp(\mathbf{x}|\mathbf{v})\overleftarrow{\partial}$. ∎

We can take, for example, the free case where $R' = P$ and R is arbitrary. We can always scale R in this case so that the $[m; R]$ are invertible. This corresponds to the condition of 'generic q' in the one-dimensional case in Lemma 3.2.2, and is the correct generalisation of it to higher dimensions. We have to work harder in the more common case where the $[m; R]$ are not all invertible, but the strategy is just the same. We write as ansatz

$$\exp(\mathbf{x}|\mathbf{v}) = \sum_{m=0}^{\infty} \mathbf{x}_1 \cdots \mathbf{x}_m F(m; R)\mathbf{v}_m \cdots \mathbf{v}_1,$$

$$\mathbf{x}_1 \cdots \mathbf{x}_m F(m; R)[m; R]! = \mathbf{x}_1 \cdots \mathbf{x}_m,$$

$$[m; R]!F(m; R)\mathbf{v}_m \cdots \mathbf{v}_1 = \mathbf{v}_m \cdots \mathbf{v}_1, \tag{10.55}$$

$$\mathbf{t}_1\mathbf{t}_2 \cdots \mathbf{t}_m F(m; R) = F(m; R)\mathbf{t}_1\mathbf{t}_2 \cdots \mathbf{t}_m,$$

and solve for F. This ansatz solves Definition 10.4.14 in view of

$$\partial^i \exp(\mathbf{x}|\mathbf{v}) = \sum_{m=0}^{\infty} \mathbf{e}^i{}_1\mathbf{x}_2 \cdots \mathbf{x}_m [m; R]F(m; R)\mathbf{v}_m \cdots \mathbf{v}_1$$

$$= \sum_{m=0}^{\infty} \mathbf{e}^i{}_1 \mathbf{x}_2 \cdots \mathbf{x}_m F(m-1;R)_{2\cdots m}$$

$$[m-1;R]!_{2\cdots m}[m;R]F(m;R)\mathbf{v}_m \cdots \mathbf{v}_1$$

$$= \sum_{m=0}^{\infty} \mathbf{e}^i{}_1 \mathbf{x}_2 \cdots \mathbf{x}_m F(m-1;R)_{2\cdots m}[m;R]!F(m;R)\mathbf{v}_m \cdots \mathbf{v}_1$$

$$= \sum_{m=0}^{\infty} \mathbf{e}^i{}_1 \mathbf{x}_2 \cdots \mathbf{x}_m F(m-1;R)_{2\cdots m}\mathbf{v}_m \cdots \mathbf{v}_1 = \exp(\mathbf{x}|\mathbf{v})v^i$$

and a similar computation for $\overleftarrow{\partial}$ using (10.52). The equations (10.55) correspond to Definition 9.3.1 in a more categorical definition of exp as co-evaluation with respect to the pairing ev $= \langle \, , \, \rangle$ from Proposition 10.4.13. The second and third line are (9.14), and the fourth ensures that exp is a morphism from the unit object, i.e. invariant under $\tilde{A}$.

Example 10.4.16 *If $R' = q^{-2}R$ (so that R is Hecke), then*

$$\exp(\mathbf{x}|\mathbf{v}) = \sum_{m=0}^{\infty} \frac{\mathbf{x}_1 \cdots \mathbf{x}_m \mathbf{v}_m \cdots \mathbf{v}_1}{[m;q^2]!} = e_{q^{-2}}^{\mathbf{x}\cdot\mathbf{v}}.$$

Proof: Since $\mathbf{x}_1\mathbf{x}_2 = \mathbf{x}_1\mathbf{x}_2 PR'$ and $PR'\mathbf{v}_2\mathbf{v}_1 = \mathbf{v}_2\mathbf{v}_1$, we know, without any calculation, that $F(m;R) = F(m;q^2) = [m;q^2]!^{-1}$ in this case. Hence, the braided exponential above collapses in the Hecke case to an ordered ordinary one-dimensional q-exponential. Here the second form follows immediately from $\mathbf{x}_1(\mathbf{x}\cdot\mathbf{v}) = q^2(\mathbf{x}\cdot\mathbf{v})\mathbf{x}_1$, which is valid in the Hecke case as an easy consequence of the braid statistics relations in Proposition 10.3.6 between vectors and covectors. We work with formal power series in the braided tensor product $V^{\check{}}\underline{\otimes}V$ but omit the prime on $\mathbf{v}$ for clarity. Then we have $\mathbf{x}_1(\mathbf{x}_2 \cdot \mathbf{v}_2) = q^{-2}\mathbf{x}_2\mathbf{x}_1 R\mathbf{v}_2 = q^{-2}\mathbf{x}_2\mathbf{x}_1(q^2 R_{21}^{-1} + (q^2 - 1)P)\mathbf{v}_2 = (\mathbf{x}_2 \cdot \mathbf{v}_2)\mathbf{v}_1 + (1 - q^{-2})\mathbf{x}_1(\mathbf{x}_2 \cdot \mathbf{v}_2)$. ∎

We assume then that we have these eigenfunctions $\exp(\mathbf{x}|\mathbf{v})$. Since the ∂^i themselves are a realisation of the vector algebra, we are now able to formulate our braided Taylor's theorem:

Corollary 10.4.17 *Let $\mathbf{x}_1\mathbf{a}_2 = \mathbf{a}_2\mathbf{x}_1 R$, where $\mathbf{x}, \mathbf{a}$ are braided covectors. Then*

$$\exp(\mathbf{a}|\partial)f(\mathbf{x}) = f(\mathbf{a}+\mathbf{x}) = \underline{\Delta}f(\mathbf{x}).$$

We see that ∂ is the infinitesimal generator of the translation corresponding to this braided coproduct.

Proof: This follows at once from the braided binomial theorem developed above. On any polynomial, the power series $\exp(\mathbf{a}|\partial)$ truncates to a finite sum. Computing from (10.38), we have

$$\exp(\mathbf{a}|\partial)\mathbf{x}_1\cdots\mathbf{x}_m$$

$$= \sum_{r=0}^{r=m} \mathbf{a}_{1'}\cdots\mathbf{a}_{r'} F(r;R)_{1'\cdots r'}\partial_{r'}\cdots\partial_{1'}\mathbf{x}_1\cdots\mathbf{x}_m$$

$$= \sum_{r=0}^{r=m} \mathbf{a}_1\mathbf{a}_{2'}\cdots\mathbf{a}_{r'} F(r;R)_{1\cdots r'}\partial_{r'}\cdots\partial_{2'}\mathbf{x}_2\cdots\mathbf{x}_m$$

$$= \sum_{r=0}^{r=m} \mathbf{a}_1\cdots\mathbf{a}_r F(r;R)_{1\cdots r}\mathbf{x}_{r+1}\cdots\mathbf{x}_m\,[m-r+1;R]_{r\cdots m}\cdots[m;R]_{1\cdots m}$$

$$= \sum_{r=0}^{r=m} \mathbf{a}_1\cdots\mathbf{a}_r\mathbf{x}_{r+1}\cdots\mathbf{x}_m F(r;R)[r;R]! \begin{bmatrix} m \\ r \end{bmatrix};R \Big]$$

$$= (\mathbf{a}_1+\mathbf{x}_1)\cdots(\mathbf{a}_m+\mathbf{x}_m) = \underline{\Delta}(\mathbf{x}_1\cdots\mathbf{x}_m).$$

Here the $1',2'$, etc., refer to copies of M_n distinct from the copies labelled by $1\cdots m$, but they are successively identified by the $\mathbf{e}^i$ (which are given by Kronecker δ-functions) resulting from the application of ∂^i. The $\mathbf{x}_{r+1}\cdots\mathbf{x}_m$ commute to the left, and Theorem 10.4.12 and Proposition 10.4.9 allow us to identify the result from (10.48). ∎

There is clearly a corresponding form of Taylor's theorem for $\overleftarrow{\partial}$ recovering the coaddition of braided vectors. Finally, we can apply these theorems to exp itself and deduce its usual bicharacter properties in our braided setting.

Corollary 10.4.18 *Let* $\mathbf{x}_1\mathbf{a}_2 = \mathbf{a}_2\mathbf{x}_1 R$ *as above and let* $\mathbf{w}_1\mathbf{v}_2 = R\mathbf{v}_2\mathbf{w}_1$ *for the braid statistics in the vector case. Then,*

$$\exp(\mathbf{a}+\mathbf{x}|\mathbf{v}) = \exp(\mathbf{x}|\mathbf{v})\exp(\mathbf{a}(\ |\ )\mathbf{v}),$$

$$\exp(\mathbf{x}|\mathbf{v}+\mathbf{w}) = \exp(\mathbf{x}(\ |\ )\mathbf{w})\exp(\mathbf{x}|\mathbf{v}),$$

where $(\ |\ )$ *denotes a space for* $\exp(\mathbf{x}|\mathbf{v})$ *to be inserted in each term of the exponentials.*

Proof: We apply the braided Taylor's theorem just proven,

$$\exp(\mathbf{a}+\mathbf{x}|\mathbf{v}) = \exp(\mathbf{a}|\partial)\exp(\mathbf{x}|\mathbf{v})$$

$$= \sum_{m=0}^{\infty} \mathbf{a}_1\cdots\mathbf{a}_m F(m;R)\exp(\mathbf{x}|\mathbf{v})\mathbf{v}_m\cdots\mathbf{v}_1,$$

using (10.54). The second identity is strictly analogous, using this time the braided Taylor's theorem for the **v** coordinates. Note that, since the braided exponential is invariant under the transformation by $\tilde{A}$, it is bosonic in the sense that its braid statistics with anything else is trivial. So we can also write the above more simply, without the (|), if the appropriate braided tensor product algebra is understood. ∎

These formulae are rather important in physics, where they are key properties of plane waves. On the other hand, in categorical terms they just say that the product in the vector algebra corresponds to the additive coproduct in the covector algebra, and vice versa via the exponential; i.e. they are just the statement that our braided-Hopf algebras are *copaired*, cf. (7.26) in Chapter 7.2 for ordinary Hopf algebras. Corollary 10.4.18 expresses the duality in Proposition 10.4.13 in terms of the coevaluation element rather than the evaluation map. If one starts with $\exp = \mathrm{coev} = \cap$ or (10.55) as the definition, then it is easy to prove the corollary immediately from Proposition 10.4.13 by the diagrammatic methods for braided-Hopf algebras explained in Chapter 9.4, and afterwards deduce the differential properties of exp.

Next we turn to the Gaussian. We proceed in the same direct way by writing down a differential equation that characterises it as a formal power series. The simplest (but not the only) case for which this strategy works is when there is given a covariant metric in the sense explained at the end of Section 10.2. We concentrate on this case for simplicity.

Definition 10.4.19 *Let η be a quantum metric in the sense of Definition 10.2.14. We define the corresponding Gaussian g_η to be a formal power series in $\{x_i\}$ such that*

$$\partial^i g_\eta = -x_a \eta^{ai} g_\eta, \quad \underline{\epsilon}(g_\eta) = 1.$$

As before, we check that this is a good definition by checking that this equation is integrable. We have

$$(\partial^i \partial^j - R'^i{}_a{}^j{}_b \partial^b \partial^a) g_\eta = -\partial^i (x_a \eta^{aj} g_\eta) + R'^i{}_a{}^j{}_b \partial^b (x_c \eta^{ca} g_\eta)$$

$$= (R'^i{}_a{}^j{}_b \eta^{ba} - \eta^{ij}) g_\eta + R'^i{}_a{}^j{}_b x_d R^d{}_c{}^b{}_e \eta^{ca} \partial^e g_\eta - x_c R^c{}_a{}^i{}_d \partial^d \eta^{aj} g_\eta$$

$$= (x_c R^c{}_a{}^i{}_d x_e \eta^{ed} \eta^{aj} - R'^i{}_a{}^j{}_b x_d R^d{}_c{}^b{}_e x_f \eta^{fe} \eta^{ca}) g_\eta$$

$$= (x_c R^c{}_a{}^i{}_d x_e \eta^{ed} \eta^{aj} - R'^i{}_a{}^j{}_b x_d \lambda^{-2} R^{-1d}{}_c{}^f{}_e \eta^{eb} \eta^{ca} x_f) g_\eta$$

$$= (x_c R^c{}_a{}^i{}_d x_e \eta^{ed} \eta^{aj} - \eta^{ai} \eta^{bj} R'^c{}_a{}^e{}_b x_d x_f \lambda^{-2} R^{-1d}{}_c{}^f{}_e) g_\eta$$

$$= (x_c R^c{}_a{}^i{}_d x_e \eta^{ed} \eta^{aj} - \eta^{ai} \eta^{bj} \lambda^{-2} R^{-1e}{}_b{}^c{}_a x_d x_f R'^f{}_e{}^d{}_c) g_\eta$$

$$= (x_c R^c{}_a{}^i{}_d x_e \eta^{ed} \eta^{aj} - \eta^{ai} \eta^{bj} \lambda^{-2} R^{-1e}{}_b{}^c{}_a x_e x_c) g_\eta = 0,$$

where we use the Gaussian equation and the braided Leibniz rule from Lemma 10.4.3 or Exercise 10.4.4. The third equality uses (10.32) and the Gaussian equation again. The fourth uses (10.30), the fifth (10.28) and the sixth (10.13). After that, we use the relations among the $\{x_i\}$ and (10.30) again to obtain zero. While not a proof, this computation suggests that g_η exists at least as a formal power series.

Example 10.4.20 *Suppose further that η obeys the additional conditions*

$$R^i{}_a{}^j{}_b \eta^{ba} = q^{-2}\eta^{ij}, \quad \eta^{ka}R'^l{}_j{}^i{}_a = R'^{-1l}{}_j{}^k{}_a\eta^{ai}. \tag{10.56}$$

Then

$$g_\eta = e_{\lambda-2}^{-[2;q^{-2}]^{-1}x\cdot x}; \quad x\cdot x = x_a x_b \eta^{ba}$$

Proof: We use the braided Leibniz rule from Exercise 10.4.4 to compute

$$
\begin{aligned}
\partial^i\hat{x}\cdot\hat{x} &= \partial^i\hat{x}_a\hat{x}_b\eta^{ba} = \hat{x}_a\eta^{ai} + \hat{x}_c R^c{}_a{}^i{}_d\partial^d\hat{x}_b\eta^{ba}\\
&= \hat{x}_a\eta^{ai} + \hat{x}_c R^c{}_a{}^i{}_b\eta^{ba} + \hat{x}_c R^c{}_a{}^i{}_d x_e R^e{}_b{}^d{}_f\eta^{ba}\partial^f\\
&= (1+q^{-2})\hat{x}_a\eta^{ai} + \lambda^{-2}\hat{x}\cdot\hat{x}\partial^i,
\end{aligned}
$$

where the last equality uses the first of (10.56) and the second of (10.31). Any other factor in place of q^{-2} in (10.56) works too, with the corresponding coefficient in place of $[2;q^{-2}] = 1 + q^{-2}$.

On the other hand, whenever the second of (10.56) holds we have that $\mathbf{x}\cdot\mathbf{x}$ is central in the braided covector algebra since

$$x_i x_a x_b \eta^{ba} = x_d x_c x_b R'^c{}_i{}^d{}_a\eta^{ba} = x_d x_c x_b\eta^{ad}R'^{-1c}{}_i{}^b{}_a = x_d x_a\eta^{ad}x_i$$

using our assumption and the relations in the braided covector algebra. Note also that if the second of (10.56) holds with a factor, then clearly x_i commutes with $\mathbf{x}\cdot\mathbf{x}$ up to that factor, which can be useful on other contexts.

Putting together our two independent facts, it is trivial to see that g_η as stated solves our defining differential equation. There is a similar conclusion in the Hecke case with a factor in the second of (10.56), but this is less useful than the above when it comes to concrete examples. ∎

For an application, we can now use this braided Gaussian to define a translation-invariant integration. This should be a map $\int$ from suitable functions of $\mathbf{x}$ to our field. One might think that integration over $\mathbb{R}^n$ is one thing that cannot be done algebraically since polynomials are not integrable. However, we can reduce it to an algebra if we only want integrals of the form $\int f(\mathbf{x})g_\eta$, where $f \in V^{\check{}}(R',R)$ is a polynomial in the coordinates. Indeed, such integrals are done by parts under the boundary assumption that

$$\int \partial^i(f(\mathbf{x})g_\eta) = 0, \quad \forall f. \tag{10.57}$$

This expresses translation-invariance on our class of functions assuming that the Gaussian 'vanishes at infinity'. One might think that this decay at infinity also requires us to leave the algebraic setting, but it too can be circumvented by looking not at $\int$ but at its ratio with $\int g_\eta$. As is well-known in physics, such ratios are well-defined and algebraically computable objects.

Proposition 10.4.21 *If $\int, g_\eta$ exist as above, then*

$$\mathcal{Z}[\mathbf{x}_1 \cdots \mathbf{x}_m] = \left(\int \mathbf{x}_1 \cdots \mathbf{x}_m g_\eta \right) \left(\int g_\eta \right)^{-1}$$

is a well-defined linear functional on the braided covector algebra. It can be computed inductively by

$$\mathcal{Z}[1] = 1, \quad \mathcal{Z}[x_i] = 0, \quad \mathcal{Z}[x_i x_j] = \eta_{ba} R^a{}_i{}^b{}_j \lambda^2,$$

$$\mathcal{Z}[\mathbf{x}_1 \cdots \mathbf{x}_m] = \sum_{i=0}^{m-2} \mathcal{Z}[\mathbf{x}_1 \cdots \mathbf{x}_i \mathbf{x}_{i+3} \cdots \mathbf{x}_m] \mathcal{Z}[\mathbf{x}_{i+1}\mathbf{x}_{i+2}][i+2, m; R] \lambda^{2(m-2-i)},$$

where $[i + 2, m; R]$ is from Lemma 10.4.10. We call $\mathcal{Z}$ the Gaussian-weighted integral functional on the braided space.

Proof: Assuming $\int, g_\eta$ exist, we compute

$$\int x_{i_1} \cdots x_{i_m} g_\eta = - \int x_{i_1} \cdots x_{i_{m-1}} \eta_{i_m a} \partial^a g_\eta$$

$$= - \int x_{i_1} \cdots x_{i_{m-2}} \eta_{i_m a} \left(-\tilde{R}^b{}_{i_{m-1}}{}^a{}_b g_\eta + \tilde{R}^c{}_{i_{m-1}}{}^a{}_b \partial^b (x_c g_\eta) \right)$$

$$= - \int x_{i_1} \cdots x_{i_{m-2}} \left(-\lambda^2 R^b{}_{i_{m-1}}{}^a{}_{i_m} \eta_{ab} g_\eta + \lambda^2 R^c{}_{i_{m-1}}{}^a{}_{i_m} \eta_{ab} \partial^b (x_c g_\eta) \right)$$

$$= \left(\int x_{i_1} \cdots x_{i_{m-2}} g_\eta \right) \mathcal{Z}[x_{i_{m-1}} x_{i_m}]$$

$$- \left(\int x_{i_1} \cdots x_{i_{m-2}} \eta_{ab} \partial^b (x_c g_\eta) \right) \lambda^2 R^c{}_{i_{m-1}}{}^a{}_{i_m},$$

where we used the braided Leibniz rule (in reverse) for the second equality and (10.25) for the third. When $m = 2$, there is no second term here as it is a total derivative, and we obtain $\mathcal{Z}[x_i x_j]$ as stated. When $m > 2$ we recognise the form shown in terms of $\mathcal{Z}[x_{i_{m-1}} x_{i_m}]$ and an expression similar to our second expression but with a power of $\mathbf{x}$ lower by two to the left of ∂. We now repeat the above process, each time lowering the degree in the residual term by two, until we reach $\int \partial = 0$. All our constructions are covariant under the background quantum group A, so $\mathcal{Z}$ is a well-defined and invariant linear functional on our algebra of the $\{x_i\}$. ∎

The resulting $\mathcal{Z}$ does not require us to construct g_η or any other formal power series. In principle, one can just take Proposition 10.4.21 as an inductive definition of $\mathcal{Z}$, verifying directly from the initial R-matrix data that it is well-defined, even when the Gaussian and $\int$ themselves are not known.

We conclude this introduction to braided geometry with two applications of braided differentiation, namely to the construction of metrics η, as needed above, and the ϵ tensor, as also needed in physics. For the metric we need to solve (10.32) and the invariance condition (10.29) (which implies the others). Such a tensor is immediately obtained if we have an element r^2 of degree two in the $\mathbf{x}$ which is quantum-group invariant; we then just define

$$\eta^{ij} = \partial^i \partial^j r^2. \tag{10.58}$$

Its symmetry in the sense of (10.32) is assured by virtue of Proposition 10.4.2 that the ∂^i obey the braided vector algebra relations. For example, $r^2 = (1 + q^{-2})^{-1}\mathbf{x} \cdot \mathbf{x}$ in Example 10.4.20 is a radius function from which the metric is recovered by (10.58). More generally, we can differentiate any monomial $f(\mathbf{x})$ of degree m, and we will know that

$$v^{i_1 i_2 \cdots i_m} = \frac{\partial}{\partial x_{i_1}} \cdots \frac{\partial}{\partial x_{i_m}} f$$

is an R'-symmetric tensor of rank m in the sense

$$R'^{i_p}{}_a{}^{i_{p+1}}{}_b v^{i_1 \cdots i_{p-1} b a i_{p+2} \cdots i_m} = v^{i_1 \cdots i_m}, \quad \forall p = 1, \ldots, m-1. \tag{10.59}$$

If we want to have tensors with lower indices, we can obtain them also by differentiation. Thus, an R'-symmetric tensor of rank m with lower indices means

$$x_{i_1 \cdots i_{p-1} b a i_{p+2} \cdots i_m} R'^a{}_{i_p}{}^b{}_{i_{p+1}} = x_{i_1 \cdots i_m}, \quad \forall p = 1, \ldots, m-1, \tag{10.60}$$

and such a tensor can be obtained by applying any mth order differential operator built from $\frac{\partial}{\partial x_i}$ to monomials $x_{i_1} \cdots x_{i_m}$.

One of the nice things about braided geometry is that it handles both q-deformed bosonic constructions, such as those above, and q-deformed fermionic ones just as well. Thus, one can think of differential forms dx_i as a realisation of Grassmann or superspace coordinates $\theta_i = dx_i$, as in Section 10.1, and we can apply our braided geometry to them and their q-deformations just as well as to the bosonic x_i above. In order to do this, we impose the further conditions

$$R_{12}R'_{13}R'_{23} = R'_{23}R'_{13}R_{12}, \quad R_{23}R'_{13}R'_{12} = R'_{12}R'_{13}R_{23},$$

$$R'_{12}R'_{13}R'_{23} = R'_{23}R'_{13}R'_{12}, \tag{10.61}$$

in addition to those in Theorem 10.2.1. This ensures that there is a certain symmetry

$$R \leftrightarrow -R', \tag{10.62}$$

and we can then define the braided-Hopf algebras

$$\Lambda(R', R) \equiv V^{\check{}}(-R, -R') = \{\theta_i\}, \quad \Lambda^*(R', R) \equiv V(-R, -R') = \{\phi^i\}.$$

We call these, respectively, the braided *forms* Λ and *coforms* Λ^*. They are covariant under the transformations $\theta \to \theta t\varsigma$ and $\phi \to \varsigma^{-1}t^{-1}\phi$ of a quantum group obtained from $A(-R') = A(R')$. We assume for convenience that this is the same as the quantum group obtained from $A(R)$.

We are now free to apply the same constructions above to the space of forms $\{\theta_i\}$. For example, differentiating $\frac{\partial}{\partial\theta_i}$ in this form-space gives tensors which must be manifestly $-R$-symmetric, i.e. R-antisymmetric in the sense

$$R^{i_p}{}_a{}^{i_{p+1}}{}_b \phi^{i_1 \cdots i_{p-1} b a i_{p+2} \cdots i_m} = -\phi^{i_1 \cdots i_m}, \quad \forall p = 1, \ldots, m-1. \tag{10.63}$$

For example, in nice cases the algebra of forms will have an element of top degree given by $\theta_1 \cdots \theta_n$. We then define

$$\epsilon^{i_1 i_2 \cdots i_n} = \frac{\partial}{\partial\theta_{i_1}} \cdots \frac{\partial}{\partial\theta_{i_n}} \theta_1 \cdots \theta_n = ([n; -R']!)^{i_n \cdots i_1}_{12 \cdots n}, \tag{10.64}$$

and, by the reasoning above, it will be R-antisymmetric. Likewise, an R-antisymmetric tensor with lower indices means

$$\theta_{i_1 \cdots i_{p-1} b a i_{p+2} \cdots i_m} R^a{}_{i_p}{}^b{}_{i_{p+1}} = -\theta_{i_1 \cdots i_m}, \quad \forall p = 1, \ldots, m-1, \tag{10.65}$$

which can be obtained by applying any mth order operator built from $\frac{\partial}{\partial\theta_i}$ to $\theta_{i_1} \cdots \theta_{i_m}$. For example, we define

$$\epsilon_{i_1 \cdots i_n} = \frac{\partial}{\partial\theta_n} \cdots \frac{\partial}{\partial\theta_1} \theta_{i_1} \cdots \theta_{i_n} = ([n; -R']!)^{12 \cdots n}_{i_1 \cdots i_n}. \tag{10.66}$$

Its total R-antisymmetry is inherited this time from the antisymmetry of the θ_i coordinates in form-space.

Such epsilon tensors form the next layer of braided geometry. One can use them in conjunction with a metric to define the Hodge $*$-operator $\Lambda \to \Lambda$ along the usual lines. There are also concrete applications to quantum group theory, such as a general R-matrix formula

$$\det(\mathbf{t}) \propto \epsilon_{i_1 \cdots i_n} t^{i_1}{}_{j_1} \cdots t^{i_n}{}_{j_n} \epsilon^{j_n \cdots j_1} \tag{10.67}$$

for a q-determinant in our background quantum group $A(R)$. We recall that formulae for such determinants were introduced by hand in Chapter 4.2 on a case by case basis; the general formula here needs the above braided geometry.

Also, for a geometrical picture, we can consider both the coordinates $\{x_i\}$ and the forms $\{\theta_i\}$ together with $\theta_i = dx_i$. Thus we consider the *exterior algebra*

$$\Omega = \Lambda \underline{\otimes} V^{\check{}}, \quad x_1\theta_2 = \theta_2 x_1 R, \quad \theta_1\theta_2 = -R\theta_2\theta_1, \quad x_1x_2 = R'x_2x_1, \quad (10.68)$$

where the product is the braided tensor product with the cross relations from Proposition 10.3.6. We then define the components Ω^p of form-degree p and the exterior derivative

$$d : \Omega^p \to \Omega^{p+1}, \quad d(\theta_{i_1} \cdots \theta_{i_p} f(\mathbf{x})) = \theta_{i_1} \cdots \theta_{i_p} \theta_a \frac{\partial}{\partial x_a} f(\mathbf{x}). \quad (10.69)$$

In our conventions it obeys a right handed $\mathbb{Z}/2$-graded Leibniz rule

$$d(fg) = (-1)^p (df)g + f dg, \quad \forall f \in \Omega, \ g \in \Omega^p, \quad (10.70)$$

and gives us a differential graded algebra when

$$d^2 = 0, \quad \text{i.e.} \quad \theta_1\theta_2\partial_2\partial_1 = 0. \quad (10.71)$$

The latter holds quite generally because ∂^i are symmetric and θ_i are antisymmetric. It is immediate when $PR' = f(PR)$ for some function f, with $f(-1) \neq 1$ and can also be verified in other cases according to the form of R and R'.

This is the construction of the exterior differential calculus on a quantum or braided vector space coming out of braided geometry. The resulting R-matrix formulae are just

$$dx_1 dx_2 = -dx_2 dx_1 R, \quad x_1 dx_2 = dx_2 x_1 R. \quad (10.72)$$

These relations can also be deduced by consistency arguments and in an axiomatic approach to exterior algebras that we have not taken here. Combined with the Hodge $*$-operation coming from ϵ, we clearly have the basis for the theory of electromagnetism in our setting of general braided vector and covector spaces. At least in principle, we can write down such things as the Maxwell action $\int F^*F$, where A is a 1-form, $F = dA$, and the corresponding Maxwell equations. The braided integration and braided exponential allow one to also write down Green functions and Fourier transforms.

10.5   Examples of braided addition

We have given above a fairly complete outline of the first steps in braided geometry, along with some elementary examples, such as the braided line and braided (or quantum) plane. We conclude now with some more detailed examples that might serve in a realistic approach to q-deformed spacetime and (ultimately) q-deformed physics in four dimensions. There are many reasons why one might want to do this from a physical point of view, as discussed in the introduction. Also, this is not the only interesting way to q-deform Euclidean and Minkowski spaces, but it is the one that arises naturally in the framework of braided geometry. It is based on the idea (as in twistor theory) that Minkowski space should be thought of as 2×2 Hermitian matrices with metric given by the determinant. In fact, there are a great many algebras encountered in the context of quantum groups and q-deformations which have a natural braided addition law of the type to which the theory in the last sections applies. We begin with one of these, which also serves as a warm-up to our braided approach to q-deforming spacetime.

10.5.1   Coaddition on quantum matrices

The reader knows very well by now that the quantum matrices $A(R)$ in Chapter 4 have a comultiplication $\Delta \mathbf{t} = \mathbf{t} \otimes \mathbf{t}$ corresponding in the language of coordinate functions $t^i{}_j$ to the multiplication in the set M_n of matrices. But matrices can also be added, so what about an additive coproduct (coaddition) for $A(R)$ corresponding to this? We address this here.

Proposition 10.5.1 *Let R be a solution of the QYBE of Hecke type in the sense $(PR - q)(PR + q^{-1}) = 0$. Then, the usual quantum matrices $A(R)$ in Example 4.1.3, form a braided covector algebra with addition law*

$$t''^i{}_j = t^i{}_j + t'^i{}_j, \quad i.e. \quad \mathbf{t}'' = \mathbf{t} + \mathbf{t}'$$

obeying the same relations of $A(R)$ provided $\mathbf{t}'$ is a second copy with braid statistics

$$\mathbf{t}'_1 \mathbf{t}_2 = R_{21} \mathbf{t}_2 \mathbf{t}'_1 R.$$

More formally, $A(R)$ is a braided-Hopf algebra with

$$\Psi(\mathbf{t}_1 \otimes \mathbf{t}_2) = R_{21} \mathbf{t}_2 \otimes \mathbf{t}_1 R, \quad \underline{\Delta} \mathbf{t} = \mathbf{t} \otimes 1 + 1 \otimes \mathbf{t}, \quad \underline{\epsilon} \mathbf{t} = 0, \quad \underline{S} \mathbf{t} = -\mathbf{t}.$$

Moreover, the coaddition $\underline{\Delta}$ is compatible with the usual matrix comultiplication Δ in the sense

$$(\mathrm{id} \otimes \cdot) \circ (\mathrm{id} \otimes \tau \otimes \mathrm{id})(\Delta \otimes \Delta) \circ \underline{\Delta} = (\underline{\Delta} \otimes \mathrm{id}) \circ \Delta,$$

$$(\cdot \otimes \mathrm{id}) \circ (\mathrm{id} \otimes \tau \otimes \mathrm{id})(\Delta \otimes \Delta) \circ \underline{\Delta} = (\mathrm{id} \otimes \underline{\Delta}) \circ \Delta,$$

where τ denotes the usual transposition map.

Proof: We give a direct proof. By definition, $\underline{\Delta}$ extends to products as an algebra homomorphism to the braided tensor product algebra. We have to show that this is consistent, i.e. that $\mathbf{t}''$ in our shorthand notation obeys the same algebra relations. This is

$$R(\mathbf{t}_1 + \mathbf{t}'_1)(\mathbf{t}_2 + \mathbf{t}'_2) = R\mathbf{t}_1\mathbf{t}_2 + R\mathbf{t}'_1\mathbf{t}'_2 + RR_{21}\mathbf{t}_2\mathbf{t}'_1 R + R\mathbf{t}_1\mathbf{t}'_2,$$
$$(\mathbf{t}_2 + \mathbf{t}'_2)(\mathbf{t}_1 + \mathbf{t}'_1)R = \mathbf{t}_2\mathbf{t}_1 R + \mathbf{t}'_2\mathbf{t}'_1 R + R\mathbf{t}_1\mathbf{t}'_2 R_{21}R + \mathbf{t}_2\mathbf{t}'_1 R,$$

which are equal because $R_{21}R = 1 + (q - q^{-1})PR$ and $RR_{21} = 1 + (q - q^{-1})RP$ from the q-Hecke assumption. For a full picture here, we also have to check that Ψ shown here on the generators extends consistently to products in such a way as to be functorial. The strategy is like that in Theorem 10.2.1 and reduces to the QYBE for R. Explicitly, one has

$$\Psi(\mathbf{t}_1\mathbf{t}_2 \cdots \mathbf{t}_M \otimes \mathbf{t}_{M+1} \cdots \mathbf{t}_{M+N})$$

$$= R_{M+1\,M} \cdots R_{M+1\,1} \qquad\qquad R_{1\,M+N} \cdots R_{1\,M+1}$$

$$\vdots \qquad\qquad \vdots \qquad \mathbf{t}_{M+1} \cdots \mathbf{t}_{M+N} \otimes \mathbf{t}_1 \cdots \mathbf{t}_M \qquad \vdots \qquad\qquad \vdots$$

$$R_{M+N\,M} \cdots R_{M+N\,1} \qquad\qquad R_{M\,M+N} \cdots R_{M\,M+1}$$

where the blocks are to be multiplied in the order shown.

Finally, we prove the codistributivity conditions, which hold trivially on the generators. On products $\mathbf{t}_1\mathbf{t}_2$ of generators, we have, for the first condition,

$$(\mathrm{id} \otimes \cdot)(\mathrm{id} \otimes \tau \otimes \mathrm{id})(\Delta \otimes \Delta)(\mathbf{t}_1\mathbf{t}_2 \otimes 1 + 1 \otimes \mathbf{t}_1\mathbf{t}_2 + \mathbf{t}_1 \otimes \mathbf{t}_2 + R_{21}\mathbf{t}_2 \otimes \mathbf{t}_1 R)$$
$$= (\mathrm{id} \otimes \cdot)\tau_{23}(\mathbf{t}_1\mathbf{t}_2 \otimes \mathbf{t}_1\mathbf{t}_2 \otimes 1 \otimes 1 + 1 \otimes 1 \otimes \mathbf{t}_1\mathbf{t}_2 \otimes \mathbf{t}_1\mathbf{t}_2 + \mathbf{t}_1 \otimes \mathbf{t}_1 \otimes \mathbf{t}_2 \otimes \mathbf{t}_2$$
$$\qquad + R_{21}\mathbf{t}_2 \otimes \mathbf{t}_2 \otimes \mathbf{t}_1 \otimes \mathbf{t}_1 R)$$
$$= \mathbf{t}_1\mathbf{t}_2 \otimes 1 \otimes \mathbf{t}_1\mathbf{t}_2 + 1 \otimes \mathbf{t}_1\mathbf{t}_2 \otimes \mathbf{t}_1\mathbf{t}_2 + \mathbf{t}_1 \otimes \mathbf{t}_2 \otimes \mathbf{t}_1\mathbf{t}_2 + R_{21}\mathbf{t}_2 \otimes \mathbf{t}_1 \otimes \mathbf{t}_2\mathbf{t}_1 R$$
$$= (\mathbf{t}_1\mathbf{t}_2 \otimes 1 + 1 \otimes \mathbf{t}_1\mathbf{t}_2 + \mathbf{t}_1 \otimes \mathbf{t}_2 + R_{21}\mathbf{t}_2 \otimes \mathbf{t}_1 R) \otimes \mathbf{t}_1\mathbf{t}_2$$
$$= (\Delta \otimes \mathrm{id})(\mathbf{t}_1\mathbf{t}_2 \otimes \mathbf{t}_1\mathbf{t}_2)$$

because of the relations in $A(R)$. τ_{23} denotes transposition in the middle two factors. One proves the general case in a similar way by induction. Similarly for codistributivity from the other side. ∎

    We have given here a direct proof of the coaddition structure. Alternatively, we can put it more explicitly in the braided covector algebra form in Theorem 10.2.1 and thereby apply the various other results from Sections 10.2–10.4 quite directly. In this case, we work with the covector notation $t_I = t^{i_0}{}_{i_1}$ for the generators, where $I = (i_0, i_1)$, $J = (j_0, j_1)$, etc., are multiindices. Then

$$R\mathbf{t}_1\mathbf{t}_2 = \mathbf{t}_2\mathbf{t}_1 R \Leftrightarrow t_I t_J = t_B t_A \mathbf{R}'^A{}_I{}^B{}_J, \quad \mathbf{R}'^I{}_J{}^K{}_L = R^{-1}{}^{j_0}{}_{i_0}{}^{l_0}{}_{k_0} R^{i_1}{}_{j_1}{}^{k_1}{}_{l_1},$$

$$\mathbf{t}'_1\mathbf{t}_2 = R_{21}\mathbf{t}_2\mathbf{t}'_1 R \Leftrightarrow t'_I t_J = t_B t'_A \mathbf{R}^A{}_I{}^B{}_J, \quad \mathbf{R}^I{}_J{}^K{}_L = R^{l_0}{}_{k_0}{}^{j_0}{}_{i_0} R^{i_1}{}_{j_1}{}^{k_1}{}_{l_1},$$

put $A(R)$ explicitly into the form of a braided covector algebra with n^2 generators. We use the bold multiindex $\mathbf{R}, \mathbf{R}'$ matrices built from our original R. They obey the conditions (10.11)–(10.13) and also the additional (10.61), in virtue of the QYBE and q-Hecke condition on R.

So, all the machinery of Sections 10.2–10.4 applies. The associated braided vector algebra from Theorem 10.2.6 comes out as $\mathbf{v}_1 \mathbf{v}_2 R = R \mathbf{v}_2 \mathbf{v}_1$, where $v^I = v^{i_1}{}_{i_0}$ is a matrix of generators. The associated partial derivatives from Proposition 10.4.2 are

$$\partial^I = \partial^{i_1}{}_{i_0} = \frac{\partial}{\partial t^{i_0}{}_{i_1}}, \quad \partial_1 \partial_2 R = R \partial_2 \partial_1, \quad \partial^i{}_j \hat{t}^k{}_l - R^a{}_j{}^k{}_b \hat{t}^b{}_c R^c{}_l{}^i{}_d \partial^d{}_a = \delta^i{}_l \delta^k{}_j$$

obeying the vector relations and the braided Leibniz rule in the form from Exercise 10.4.4.

The background symmetry of our construction is a right coaction by $A(R)^{\mathrm{cop}} \otimes A(R)$. The covariance of the algebra was already explained in Chapter 4.5.1 as bicovariance under two copies of $A(R)$, coacting from the left and right by matrix multiplication. We view the left coacting copy as $A(R)^{\mathrm{cop}}$ coacting from the right. It is easy to see that the braiding Ψ in Proposition 10.5.1 is just that obtained from Exercise 9.2.9 when $A(R)^{\mathrm{cop}} \otimes A(R)$ is equipped with its natural dual quasitriangular structure obtained from that on $A(R)$. If one wants the background covariance to be under a Hopf algebra, one can assume the regular case so that $A(R)$ has a dual quasitriangular quotient A. Then $A(R)$ becomes an $A^{\mathrm{cop}} \otimes A$-comodule algebra. It is convenient to view this equally well (via the antipode) as an $A^{\mathrm{op}} \otimes A$-comodule algebra. Then the coaction is

$$\mathbf{t} \mapsto \sigma^{-1} \mathbf{t} \tau, \quad R \tau_1 \tau_2 = \tau_2 \tau_1 R, \quad \sigma_1 \sigma_2 R = R \sigma_2 \sigma_1, \quad [\tau_1, \sigma_2] = 0 \qquad (10.73)$$

in terms of matrix generators σ, τ of $A^{\mathrm{op}} \otimes A$. The normalisation in Proposition 10.5.1 is fixed, and if it is not the quantum group normalisation then one needs to extend to $\widetilde{A^{\mathrm{op}}} \otimes A$ if one wants to induce the correctly normalised braiding in this way. It is also possible to formulate our background quantum group as a quotient of $A(\mathbf{R})$ with generators $\Lambda^I{}_J$, say, and various relations. It connects with our $A^{\mathrm{op}} \otimes A$ point of view via $\Lambda^I{}_J \mapsto S\sigma^{j_0}{}_{i_0} \otimes \tau^{i_1}{}_{j_1}$.

Next, if there is a suitable $A^{\mathrm{op}} \otimes A$-invariant quadratic element, we can differentiate it to obtain a metric η^{IJ}. Then we can lower indices and use Exercise 10.4.5 to write the braided Leibniz rule more compactly in terms of ∂_I as a matrix $\partial^{i_0}{}_{i_1}$. We can also then define translation-invariant integration $\int$ on $A(R)$. In another direction, we can follow the constructions outlined at the end of Section 10.4 and write down differential forms $\theta^i{}_j = dt^i{}_j$ with relations

$$\mathrm{dt}_1 \mathrm{dt}_2 = -R_{21} \mathrm{dt}_2 \mathrm{dt}_1 R, \quad \mathbf{t}_1 \mathrm{dt}_2 = R_{21} \mathrm{dt}_2 \mathbf{t}_1 R. \qquad (10.74)$$

These follow automatically from Proposition 10.5.1 since we know from (10.68) that they have just the same form as the braiding. We can also differentiate in form-space and obtain a q-epsilon tensor on $A(R)$ to go with this differential calculus. Clearly, writing $A(R)$ as a braided covector space allows us to perform many crucial constructions needed for its picture as functions on the linear space of matrices. Moreover, codistributivity as in the proposition ensures that our constructions are also compatible with the matrix multiplication, allowing us, for example, to write left-invariant vector fields along the usual lines in our coordinates $t^i{}_j$. Some, but not all, of these constructions also pass to the level of A to define its quantum geometry.

Example 10.5.2 *The standard quantum matrices $M_q(2)$ in Example 4.2.5 have a braided addition law whereby*

$$\begin{pmatrix} a'' & b'' \\ c'' & d'' \end{pmatrix} = \begin{pmatrix} a & b \\ c & d \end{pmatrix} + \begin{pmatrix} a' & b' \\ c' & d' \end{pmatrix}$$

also obeys the relations of $M_q(2)$ provided the second primed copy has the braid statistics

$$a'a = q^2 aa', \quad b'b = q^2 bb', \quad c'c = q^2 cc', \quad d'd = q^2 dd',$$

$$a'b = qba', \quad a'c = qca', \quad a'd = da', \quad b'd = qdb', \quad c'd = qdc',$$

$$b'a = qab' + (q^2 - 1)ba', \quad b'c = cb' + (q - q^{-1})da',$$

$$c'a = qac' + (q^2 - 1)ca', \quad c'b = bc' + (q - q^{-1})da',$$

$$d'b = qbd' + (q^2 - 1)db', \quad d'c = qcd' + (q^2 - 1)dc',$$

$$d'a = ad' + (q - q^{-1})(cb' + bc') + (q - q^{-1})^2 da'.$$

Moreover, this addition law distributes in the expected way over multiplication of quantum matrices. The background covariance is under $SU_q(2)^{\mathrm{op}} \otimes SU_q(2)$, with a dilatonic extension for the correct braid statistics.

Proof: We take the standard R as in Proposition 4.2.4 without the normalisation factor there; i.e. we take R in the normalisation as required for the q-Hecke condition in Proposition 10.5.1. This is not relevant to the algebra, but is needed for the correct braiding. We compute from the formulae above.   ∎

This example has a natural metric η obtained by differentiating the q-determinant $ad - q^{-1}bc$. Also, like all braided-Hopf algebras in a comodule category, it yields an ordinary Hopf algebra by bosonisation. Likewise for the rest of the general theory above.

There are several further directions in which we can go from examples of this type. One direction is the analogous result for rectangular quantum matrices $A(R_1, R_2)$ in Chapter 4.5.1. These include quantum planes. In general, the background symmetry now is $A(R_1)^{\mathrm{cop}} \otimes A(R_2)$ and its quotients. We have induced braid statistics

$$\Psi(\mathbf{t}_1 \otimes \mathbf{t}_2) = (R_1)_{21}\mathbf{t}_2 \otimes \mathbf{t}_1 R_2$$

with respect to which we can add the generators. Another direction is to make a twisting of the covariant system in Proposition 10.5.1, using a quantum 2-cocycle as explained at the end of Chapter 2.3. This gives us a family of examples, of which the standard 2×2 case is a natural q-Euclidean space. Or we can proceed to make a transmutation of the $A(R)$ system and obtain this time (as we know) the braided matrices $B(R)$, equipped now with a braided addition law. The standard 2×2 case makes a good q-Minkowski space. We describe these now, as well as their corresponding q-Poincaré groups.

10.5.2 q-Euclidean space

Next we describe a general R-matrix construction which is a variant of Proposition 10.5.1 and which, in the four-dimensional case, intersects the more obvious approach to q-Euclidean spaces based on $SO_q(N)$-quantum planes. There is nothing stopping us applying our constructions in Sections 10.2–10.4 to the latter, which gives a different point of view in higher dimensions from our matrix one here.

Proposition 10.5.3 *We define $\bar{A}(R)$ to be the algebra generated by a matrix of generators $\{x^i{}_j\}$ and relations $R_{21}\mathbf{x}_1\mathbf{x}_2 = \mathbf{x}_2\mathbf{x}_1 R$. It forms a braided covector algebra if R is a Hecke solution of the QYBE, with addition law and braid statistics*

$$\mathbf{x}'' = \mathbf{x} + \mathbf{x}', \quad \mathbf{x}'_1\mathbf{x}_2 = R\mathbf{x}_2\mathbf{x}'_1 R.$$

More formally, it forms a braided-Hopf algebra with

$$\Psi(\mathbf{x}_1 \otimes \mathbf{x}_2) = R\mathbf{x}_2 \otimes \mathbf{x}_1 R, \quad \underline{\Delta}\mathbf{x} = \mathbf{x} \otimes 1 + 1 \otimes \mathbf{x}, \quad \underline{\epsilon}\mathbf{x} = 0, \quad \underline{S}\mathbf{x} = -\mathbf{x}.$$

Proof: The proof is very similar to the first half of the proof of Proposition 10.5.1. We check that $\underline{\Delta}$ extends to products as an algebra homomorphism to the braided tensor product algebra, i.e. that $\mathbf{x}''$ obeys the same relations. This is

$$R_{21}(\mathbf{x}_1 + \mathbf{x}'_1)(\mathbf{x}_2 + \mathbf{x}'_2) = R_{21}\mathbf{x}_1\mathbf{x}_2 + R_{21}\mathbf{x}'_1\mathbf{x}'_2 + R_{21}R\mathbf{x}_2\mathbf{x}'_1 R + R_{21}\mathbf{x}_1\mathbf{x}'_2,$$
$$(\mathbf{x}_2 + \mathbf{x}'_2)(\mathbf{x}_1 + \mathbf{x}'_1)R = \mathbf{x}_2\mathbf{x}_1 R + \mathbf{x}'_2\mathbf{x}'_1 R + R_{21}\mathbf{x}_1\mathbf{x}'_2 R_{21} R + \mathbf{x}_2\mathbf{x}'_1 R,$$

which are equal by the q-Hecke assumption, as in the proof of Proposition 10.5.1. As before, we also have to check that Ψ extends consistently to products of the generators in such a way as to be functorial. This reduces to the QYBE for R along the lines given in Theorem 10.2.1. ∎

The usual matrix comultiplication of $\mathbf{x}$ forms neither a quantum group nor a braided one but something in between. On the other hand, as before, we can put the coaddition explicitly into our usual braided covector form by working with the multiindex notation $x_I = x^{i_0}{}_{i_1}$. We have

$$R_{21}\mathbf{x}_1\mathbf{x}_2 = \mathbf{x}_2\mathbf{x}_1 R \quad \Leftrightarrow \quad x_I x_J = x_B x_A \mathbf{R'}^A{}_I{}^B{}_J,$$

$$\mathbf{x}_1'\mathbf{x}_2 = R\mathbf{x}_2\mathbf{x}_1' R \quad \Leftrightarrow \quad x_I' x_J = x_B x_A' \mathbf{R}^A{}_I{}^B{}_J;$$

$$\mathbf{R'}^I{}_J{}^K{}_L = R^{-1}{}^{l_0}{}_{k_0}{}^{j_0}{}_{i_0} R^{i_1}{}_{j_1}{}^{k_1}{}_{l_1}, \quad \mathbf{R}^I{}_J{}^K{}_L = R^{j_0}{}_{i_0}{}^{l_0}{}_{k_0} R^{i_1}{}_{j_1}{}^{k_1}{}_{l_1},$$

and use these composite $\mathbf{R}, \mathbf{R'}$ for our general braided vector and covector constructions. All the general theory of Sections 10.2–10.4 once again applies. For example, we have the vector algebra as $R\mathbf{v}_2\mathbf{v}_1 = \mathbf{v}_1\mathbf{v}_2 R_{21}$ in terms of $v^I = v^{i_1}{}_{i_0}$. The associated partial derivatives from Proposition 10.4.2 obey the relations of $\bar{A}(R)$ again:

$$\partial^I = \partial^{i_1}{}_{i_0} = \frac{\partial}{\partial x^{i_0}{}_{i_1}}, \quad R_{21}\partial_1\partial_2 = \partial_2\partial_1 R, \quad \partial^i{}_j \hat{x}^k{}_l - R^k{}_b{}^a{}_j \hat{x}^b{}_c R^c{}_l{}^i{}_d \partial^d{}_a = \delta^i{}_l \delta^k{}_j$$

from the vector relations and the braided Leibniz rule in the form given in Exercise 10.4.4.

The background symmetry of our construction is different now, namely $A(R) \otimes A(R)$ or $A \otimes A$ in the regular case. If $\mathbf{s}, \mathbf{t}$ are the matrix generators of the latter, then the coaction is

$$\mathbf{x} \mapsto \mathbf{s}^{-1}\mathbf{x}\mathbf{t}, \quad R\mathbf{s}_1\mathbf{s}_2 = \mathbf{s}_2\mathbf{s}_1 R, \quad R\mathbf{t}_1\mathbf{t}_2 = \mathbf{t}_2\mathbf{t}_1 R, \quad [\mathbf{s}_1, \mathbf{t}_2] = 0. \quad (10.75)$$

Again, we generally need to pass to the dilatonic extension $A \widetilde{\otimes} A$ if we want the braiding to be induced by a coaction of this form with the correct normalisation in Proposition 10.5.3. We use the tensor product dual quasitriangular structure. It is also possible to formulate our background quantum group as a quotient of $A(\mathbf{R})$ with generators $\Lambda^I{}_J$, say, and connecting with the $A \otimes A$ point of view via

$$\Lambda^I{}_J \mapsto S s^{j_0}{}_{i_0} \otimes t^{i_1}{}_{j_1}. \quad (10.76)$$

As an immediate corollary of this covariance, we have at once a q-Poincaré group by bosonisation. This is generated according to Corollary 10.2.10 by $\mathbf{p}$ (a copy of $\mathbf{x}$ but regarded as momentum) and Λ with cross relations from $\mathbf{R}$. We also have a spinorial version following the same lines with the spacetime rotations generated now by $\mathbf{s}, \mathbf{t}$ and cross relations from R. To compute it we just follow the bosonisation procedure from the end of Chapter 9.4. We put the dual quasitriangular structure of $A \widetilde{\otimes} A$ into Lemma 7.4.8 to convert the coaction (10.75) into a right action of $A \widetilde{\otimes} A$ given by

$$\mathbf{p}_1 \triangleleft \mathbf{t}_2 = \mathbf{p}_1 \lambda R, \quad \mathbf{p}_1 \triangleleft \mathbf{s}_2 = \lambda^{-1} R^{-1} \mathbf{p}_1, \quad \mathbf{p} \triangleleft g = \lambda^{-2} \mathbf{p}, \tag{10.77}$$

where λR is the quantum group normalisation. Then $(A \widetilde{\otimes} A) \bowtie \bar{A}(R)$, from Proposition 1.6.10, has the corresponding cross relations

$$\mathbf{p}_1 \mathbf{t}_2 = \lambda \mathbf{t}_2 \mathbf{p}_1 R, \quad \mathbf{p}_1 \mathbf{s}_2 = \mathbf{s}_2 \lambda^{-1} R^{-1} \mathbf{p}_1, \quad \mathbf{p} \varsigma = \lambda^{-2} \varsigma \mathbf{p}. \tag{10.78}$$

The coproduct is the usual cross coproduct from Proposition 1.6.16 by the coaction (10.75). It comes out as

$$\Delta \mathbf{p} = \mathbf{p} \otimes \mathbf{s}^{-1}(\ ) \mathbf{t} \varsigma + 1 \otimes \mathbf{p}. \tag{10.79}$$

where the indices of $\mathbf{p}$ have to be inserted into the space. There is also a counit $\epsilon \mathbf{p} = 0$ and antipode.

Also, if there is a suitable $A \otimes A$-invariant quadratic element, we can then differentiate it to obtain a metric η^{IJ}. Exercise 10.4.5 then gives a more compact form $R \partial_2 \hat{\mathbf{x}}_1 R - \lambda^{-2} \hat{\mathbf{x}}_1 \partial_2 = R \eta_{21} R$ of the braided Leibniz rule in terms of ∂_I written as matrices. We can also define translation-invariant integration $\int$ on $\bar{A}(R)$. Even without a metric, we can define differential forms $\theta^i{}_j = \mathrm{d}x^i{}_j$ with relations

$$\mathrm{d}\mathbf{x}_1 \mathrm{d}\mathbf{x}_2 = -R \mathrm{d}\mathbf{x}_2 \mathrm{d}\mathbf{x}_1 R, \quad \mathbf{x}_1 \mathrm{d}\mathbf{x}_2 = R \mathrm{d}\mathbf{x}_2 \mathbf{x}_1 R, \tag{10.80}$$

as follows immediately from the braiding in Proposition 10.5.3 or the form of $\mathbf{R}$. Differentiating the top form gives, in general, a q-epsilon tensor on $\bar{A}(R)$ to go with this differential calculus.

Finally, to complete our general theory, let us note that the construction of $\bar{A}(R)$ with its addition law is clearly very similar to the one for $A(R)$ explained above. The following exercise makes this precise using the theory of twisting from Chapter 2.3.

Exercise 10.5.4 *Show that the covariant system* $(A \otimes A, \bar{A}(R))$ *is the twisting according to the dual of Proposition 2.3.8 of the covariant system* $(A^{\mathrm{op}} \otimes A, A(R))$ *discussed previously by the 2-cocycle*

$$\chi((a \otimes b) \otimes (c \otimes d)) = \mathcal{R}^{-1}(a \otimes c)\epsilon(b)\epsilon(d).$$

Solution: For the sake of discussion, we suppose here that R is regular. We know from the dual of Example 2.3.6 that the dual quasitriangular structure on A^{op} (which is $\mathcal{R}^{-1}$ in terms of that of A) is a 2-cocycle, and that twisting by it in the sense of the dual of Theorem 2.3.4 turns A^{op} into A. We do this now to the first factor of $A^{\mathrm{op}} \otimes A$; i.e. this has a cocycle as stated, and twisting by it turns our background quantum group symmetry of $A(R)$ into $A \otimes A$. On the other hand, our $A(R)$ is no longer covariant under this twisted quantum group. We know from the dual (2.27) of Proposition 2.3.8 that we must also twist any algebra on which the quantum group coacts if we want it to remain covariant. Using (2.27), we have the new product

$$t^i{}_j \cdot_\chi t^k{}_l = t^a{}_b t^c{}_d \chi^{-1}((S\sigma^i{}_a \otimes \tau^b{}_j) \otimes (S\sigma^k{}_c \otimes \tau^d{}_l)) = \lambda R^i{}_a{}^k{}_b t^a{}_j t^b{}_l,$$

after evaluating on the matrix generators. Similarly for higher products of generators. If we denote $\mathbf{t}$ with the new product by $\mathbf{x}$, then the relation between the two is just

$$\mathbf{x} = \mathbf{t}, \quad \mathbf{x}_1 \mathbf{x}_2 = \lambda R \mathbf{t}_1 \mathbf{t}_2,$$

etc. This gives the relations of $\bar{A}(R)$ as in Proposition 10.5.3, as well as its covariance properties in (10.75). In principle, we also twist the coproduct, according to (2.28), so that $\bar{A}(R)$ is the twisting of $A(R)$ as an additive braided group. This does not change the additive from on the generators, however.   ∎

This theory is quite general. We describe now the results for our standard R-matrix as in Proposition 4.2.4 as a reasonable notion of q-deformed Euclidean spacetime, but one can put in multiparameter or nonstandard R-matrices from Chapter 4 just as easily.

Example 10.5.5 *The standard sl_2 R-matrix as in Proposition 4.2.4 with the Hecke normalisation gives* $\mathbf{x} = \begin{pmatrix} a & b \\ c & d \end{pmatrix}$ *as*

$$ba = qab, \quad ca = q^{-1}ac, \quad da = ad, \quad db = q^{-1}bd \quad dc = qcd,$$

$$bc = cb + (q - q^{-1})ad,$$

which we call the algebra of q-Euclidean space $\mathbb{R}^4_q = \bar{M}_q(2)$. *It has a braided addition law whereby*

$$\begin{pmatrix} a'' & b'' \\ c'' & d'' \end{pmatrix} = \begin{pmatrix} a & b \\ c & d \end{pmatrix} + \begin{pmatrix} a' & b' \\ c' & d' \end{pmatrix}$$

also obeys the relations of $\bar{M}_q(2)$ provided the second primed copy has the braid statistics

$$a'a = q^2 aa', \quad b'b = q^2 bb', \quad c'c = q^2 cc', \quad d'd = q^2 dd',$$

$$a'b = qba', \quad c'a = qac', \quad c'b = bc', \quad c'd = qdc', \quad d'b = qbd',$$

$$a'c = qca' + (q^2 - 1)ac', \quad a'd = da' + (q - q^{-1})bc',$$

$$b'c = cb' + (q - q^{-1})(ad' + da') + (q - q^{-1})^2 bc',$$

$$b'a = qab' + (q^2 - 1)ba', \quad b'd = qdb' + (q^2 - 1)bd',$$

$$d'a = ad' + (q - q^{-1})bc', \quad d'c = qcd' + (q^2 - 1)dc'.$$

The background covariance is under $SU_q(2) \otimes SU_q(2)$, with a dilatonic extension for the correctly normalised braid statistics.

Proof: We take the same R as we did in Example 10.5.2 and compute from the R-matrix formulae above. ∎

We now see how the rest of the general theory above looks in this important example. First, the relationship with $A(R)$ is so close in this example that we have, in fact, an algebra isomorphism $\mathbb{R}_q^4 \cong M_q(2)$ by the permutation of generators

$$\begin{pmatrix} a & b \\ c & d \end{pmatrix} \leftrightarrow \begin{pmatrix} c & d \\ a & b \end{pmatrix}. \tag{10.81}$$

This is the effect of twisting in this case. It also relates the braidings in the two examples.

Next, we have by bosonisation the associated four-dimensional Euclidean q-Poincaré group $SU_q(2) \widetilde{\otimes} SU_q(2) \ltimes \mathbb{R}_q^4$ from (10.78). Its explicit structure is similar to two copies of Example 10.2.12. The generators $\Lambda^I{}_J$ provide a vectorial picture of the Euclidean rotation group $SO_q(4)$ in our framework with corresponding q-Poincaré group $\widetilde{SO_q(4)} \ltimes \mathbb{R}_q^4$. Both versions coact on $\mathbb{R}_q^4$, which becomes a comodule algebra. The scaling parameter is $\lambda = q^{-\frac{1}{2}} \neq 1$, so a dilaton is needed.

The element $ad - qcb$ is central and invariant under $SU_q(2) \otimes SU_q(2)$. Hence we take it as a natural radius function. The corresponding metric by differentiation is

$$\eta^{IJ} = \begin{pmatrix} 0 & 0 & 0 & 1 \\ 0 & 0 & -q & 0 \\ 0 & -q^{-1} & 0 & 0 \\ 1 & 0 & 0 & 0 \end{pmatrix} = \frac{\partial}{\partial x_I} \frac{\partial}{\partial x_J} (ad - qcb). \tag{10.82}$$

Here, $(1 + q^{-2})^{-1} x_A x_B \eta^{BA} = ad - qcb$ recovers the radius function as in Example 10.4.20. Both the conditions (10.56) hold so the Gaussian has a simple form as well.

When we work over $\mathbb{C}$ with q real, we also have a natural $*$-structure on $\mathbb{R}_q^4$, inherited from the one on $M_q(2)$ in Example 4.2.16. It is given by

$$\begin{pmatrix} a^* & b^* \\ c^* & d^* \end{pmatrix} = \begin{pmatrix} d & -q^{-1}c \\ -qb & a \end{pmatrix}. \tag{10.83}$$

The same applies more generally when R is of real type. It is then natural to make a change of variables

$$t = \frac{a - d}{2i}, \quad x = \frac{c - qb}{2}, \quad y = \frac{c + qb}{2i}, \quad z = \frac{a + d}{2} \tag{10.84}$$

to generators which are self-adjoint in the sense $t^* = t, x^* = x, y^* = y, z^* = z$. They are the physical spacetime coordinates, and

$$ad - qcb = \left(\frac{1 + q^2}{2}\right) t^2 + x^2 + y^2 + \left(\frac{1 + q^2}{2}\right) z^2, \tag{10.85}$$

so that the signature of the metric is the Euclidean one.

The rest of the structure is routinely computed from the R-matrix formulae above. The vector algebra of derivatives is

$$\frac{\partial}{\partial b}\frac{\partial}{\partial a} = q\frac{\partial}{\partial a}\frac{\partial}{\partial b}, \quad \frac{\partial}{\partial c}\frac{\partial}{\partial a} = q^{-1}\frac{\partial}{\partial a}\frac{\partial}{\partial c}, \quad \frac{\partial}{\partial d}\frac{\partial}{\partial a} = \frac{\partial}{\partial a}\frac{\partial}{\partial d},$$

$$\frac{\partial}{\partial d}\frac{\partial}{\partial b} = q^{-1}\frac{\partial}{\partial b}\frac{\partial}{\partial d}, \quad \frac{\partial}{\partial d}\frac{\partial}{\partial c} = q\frac{\partial}{\partial c}\frac{\partial}{\partial d}, \quad \frac{\partial}{\partial b}\frac{\partial}{\partial c} = \frac{\partial}{\partial c}\frac{\partial}{\partial b} + (q - q^{-1})\frac{\partial}{\partial a}\frac{\partial}{\partial d}.$$

Their braided Leibniz rule is read off from the braiding listed in Example 10.5.5.

The algebra of forms is

$$dada = 0, \quad dbdb = 0, \quad dcdc = 0, \quad dddd = 0,$$

$$dbda = -q^{-1}dadb, \quad dcda = -dadcq, \quad dddb = -dbddq,$$

$$dcdb = -dbdc, \quad dddc = -q^{-1}dcdd, \quad ddda = -(q - q^{-1})dbdc - dadd.$$

These relations generate a finite-dimensional algebra, with the dimension at each degree being the same as in the classical case. The commutation relations of forms with position coordinates again follows the braiding in Example 10.5.5.

The q-epsilon tensor is

$$\epsilon_{abcd} = -\epsilon_{acbd} = \epsilon_{adbc} = -\epsilon_{adcb} = \epsilon_{bcad} = -\epsilon_{bcda} = 1,$$

$$-\epsilon_{cbad} = \epsilon_{cbda} = -\epsilon_{dabc} = \epsilon_{dacb} = -\epsilon_{dbca} = \epsilon_{dcba} = 1,$$

$$\epsilon_{acdb} = -\epsilon_{cdba} = -\epsilon_{dcab} = q, \quad -\epsilon_{abdc} = -\epsilon_{bacd} = \epsilon_{bdca} = \epsilon_{dbac} = q^{-1},$$

$$-\epsilon_{cadb} = \epsilon_{cdab} = q^2, \quad \epsilon_{badc} = -\epsilon_{bdac} = q^{-2}, \quad -\epsilon_{adad} = \epsilon_{dada} = q - q^{-1}.$$

We see that there are a few unexpected entries, which are zero at $q = 1$.

Finally, the Gaussian-weighted integral on degree two is $\mathcal{Z}[x_I x_J] = q^{-4}\eta_{IJ}$. Its nonzero values on degree four are

$$\mathcal{Z}[abcd] = -q^{-11}, \quad \mathcal{Z}[acbd] = -q^{-7},$$

$$\mathcal{Z}[a^2 d^2] = q^2 \mathcal{Z}[b^2 c^2] = q^{-2}\mathcal{Z}[c^2 b^2] = \mathcal{Z}[bc^2 b] = \mathcal{Z}[cb^2 c] = q^{-10}(q^2 + 1),$$

along with the other cases implied by the relations in $\mathbb{R}_q^4$. In terms of the spacetime coordinates, we have

$$\mathcal{Z}[t^2] = \mathcal{Z}[z^2] = \frac{1}{2}q^{-4}, \quad \mathcal{Z}[x^2] = \mathcal{Z}[y^2] = \frac{[2;q^2]}{4}q^{-4},$$

$$\mathcal{Z}[t^4] = \mathcal{Z}[z^4] = \frac{3[2;q^2]}{8}q^{-10}, \quad \mathcal{Z}[x^4] = \mathcal{Z}[y^4] = \frac{[3;q^2]!}{8}q^{-10},$$

which show a degree of spherical symmetry even in the noncommutative case.

10.5.3 q-Minkowski space

Finally, we follow the same constructions as in the preceding subsections, not for quantum matrices but for the braided matrices $B(R)$, which we have encountered many times previously. In this case, the Hermitian $*$-structure is natural because it is compatible (when R is real-type) with the matrix coproduct, which means that the 2×2 case provides an obvious choice for q-Minkowski space within braided geometry.

We already know from the abstract transmutation theory in Chapter 9.4 and the more explicit Theorem 10.3.1 that $B(R)$ has a (braided) coproduct $\underline{\Delta}\mathbf{u} = \mathbf{u} \otimes \mathbf{u}$ corresponding to multiplication of matrices. But what about the addition of braided matrices? The following proposition due to U. Meyer answers this question.

Proposition 10.5.6 *Let R be a biinvertible Hecke solution of the QYBE. Then the usual braided matrices $B(R)$ in Definition 4.3.1, form a braided covector algebra with addition law and additive braid statistics*

$$\mathbf{u}'' = \mathbf{u} + \mathbf{u}', \quad R^{-1}\mathbf{u}_1' R\mathbf{u}_2 = \mathbf{u}_2 R_{21}\mathbf{u}_1' R.$$

More formally, $B(R)$ is a braided-Hopf algebra with

$$\Psi(R^{-1}\mathbf{u}_1 \otimes R\mathbf{u}_2) = \mathbf{u}_2 R_{21} \otimes \mathbf{u}_1 R,$$

$$\underline{\Delta}\mathbf{u} = \mathbf{u} \otimes 1 + 1 \otimes \mathbf{u}, \quad \underline{\epsilon}\mathbf{u} = 0, \quad \underline{S}\mathbf{u} = -\mathbf{u}.$$

Proof: We show that $\underline{\Delta}$ extends to products as an algebra homomorphism to the braided tensor product algebra, i.e. that $\mathbf{u}''$ obeys the same relations. This is

$$R_{21}(\mathbf{u}_1 + \mathbf{u}_1')R(\mathbf{u}_2 + \mathbf{u}_2') = R_{21}\mathbf{u}_1 R\mathbf{u}_2 + R_{21}\mathbf{u}_1' R\mathbf{u}_2'$$
$$+ R_{21}R\mathbf{u}_2 R_{21}\mathbf{u}_1' R + R_{21}\mathbf{u}_1 R\mathbf{u}_2',$$

$$(\mathbf{u}_2 + \mathbf{u}_2')R_{21}(\mathbf{u}_1 + \mathbf{u}_1')R = \mathbf{u}_2 R_{21}\mathbf{u}_1 R + \mathbf{u}_2' R_{21}\mathbf{u}_1' R$$
$$+ R_{21}\mathbf{u}_1 R\mathbf{u}_2' R_{21}R + \mathbf{u}_2 R_{21}\mathbf{u}_1' R,$$

which are equal by the q-Hecke assumption. Functoriality of Ψ under the product map can also be checked explicitly by these techniques, as well as the antipode and other properties needed for a braided-Hopf algebra. ∎

This gives a direct proof of the coaddition structure. We can put it explicitly into the braided covector form of Theorem 10.2.1 by working with the multiindex notation $u_I = u^{i_0}{}_{i_1}$ and

$$\mathbf{R}'^{I}{}_{J}{}^{K}{}_{L} = R^{-1d}{}_{k_0}{}^{j_0}{}_a R^{k_1}{}_b{}^a{}_{i_0} R^{i_1}{}_c{}^b{}_{l_1} \widetilde{R}^c{}_{j_1}{}^{l_0}{}_d,$$

$$\mathbf{R.}^{I}{}_{J}{}^{K}{}_{L} = R^{j_0}{}_a{}^d{}_{k_0} R^{-1a}{}_{i_0}{}^{k_1}{}_b R^{i_1}{}_c{}^b{}_{l_1} \widetilde{R}^c{}_{j_1}{}^{l_0}{}_d,$$

$$\mathbf{R}^{I}{}_{J}{}^{K}{}_{L} = R^{j_0}{}_a{}^d{}_{k_0} R^{k_1}{}_b{}^a{}_{i_0} R^{i_1}{}_c{}^b{}_{l_1} \widetilde{R}^c{}_{j_1}{}^{l_0}{}_d.$$

Then we have

$$R_{21}\mathbf{u}_1 R\mathbf{u}_2 = \mathbf{u}_2 R_{21}\mathbf{u}_1 R, \quad \Leftrightarrow \quad u_J u_L = u_B u_A \mathbf{R}'^{A}{}_{J}{}^{B}{}_{L},$$

$$\mathbf{u}'' = \mathbf{u}\mathbf{u}', \quad R^{-1}\mathbf{u}_1' R\mathbf{u}_2 = \mathbf{u}_2 R^{-1}\mathbf{u}_1' R, \quad \Leftrightarrow \quad u_J' u_L = u_B u_A' \mathbf{R.}^{A}{}_{J}{}^{B}{}_{L},$$

$$\mathbf{u}'' = \mathbf{u} + \mathbf{u}', \quad R^{-1}\mathbf{u}_1' R\mathbf{u}_2 = \mathbf{u}_2 R_{21}\mathbf{u}_1' R, \quad \Leftrightarrow \quad u_J' u_L = u_B u_A' \mathbf{R}^{A}{}_{J}{}^{B}{}_{L}.$$

The matrix $\mathbf{R.}$ expresses the *multiplicative braiding* needed for the braided comultiplication in Theorem 10.3.1. We have included it here for comparison with the additive braiding $\mathbf{R}$ needed for the coaddition. It is easy to see that $\mathbf{R}', \mathbf{R}$ obey the conditions (10.11)–(10.13) needed for our braided covector space as well as the supplementary ones (10.61) for the coaddition of forms.

So, the general theory of Sections 10.2–10.4 applies once again. For example, we have the vector algebra as $Rv_2\tilde{R}v_1 = v_1\tilde{R}_{21}v_2R_{21}$ in terms of $v^I = v^{i_1}{}_{i_0}$. The associated partial derivatives from Proposition 10.4.2 are

$$\partial^I = \partial^{i_1}{}_{i_0} = \frac{\partial}{\partial u^{i_0}{}_{i_1}}, \quad R\partial_2\tilde{R}\partial_1 = \partial_1\tilde{R}_{21}\partial_2R_{21},$$

$$\partial^i{}_j\hat{u}^k{}_l - \hat{u}^m{}_nR^k{}_a{}^d{}_iR^j{}_b{}^a{}_mR^n{}_c{}^b{}_e\tilde{R}^c{}_l{}^f{}_d\partial^e{}_f = \delta^i{}_l\delta^k{}_j,$$

obeying the vector relations and the braided Leibniz rule from Exercise 10.4.4.

The background symmetry of our construction is now the double cross product $A(R)\bowtie A(R)$, from Example 7.2.12, or $A\bowtie A$ in the regular case, as in Example 7.3.3. If s, t are the matrix generators of the latter, then the coaction is

$$u \mapsto s^{-1}ut, \quad Rs_1s_2 = s_2s_1R, \quad Rt_1t_2 = t_2t_1R, \quad Rt_1s_2 = s_2t_1R. \quad (10.86)$$

Again, we generally need to pass to the dilatonic extension $\widetilde{A\bowtie A}$ if we want the braiding to be induced by a coaction of this form with the correct normalisation in Proposition 10.5.3. We use the dual quasitriangular structure $\mathcal{R}_L$ as in Example 7.3.3. It is also possible to formulate our background quantum group as a quotient of $A(\mathbf{R})$ with generators $\Lambda^I{}_J$, say, and connecting with the $A\bowtie A$ point of view via (10.76) as before.

As an immediate corollary of this covariance, we have at once a q-Poincaré group by bosonisation. This is generated according to Corollary 10.2.10 by $\mathbf{p}$ (a copy of $\mathbf{u}$ but regarded as momentum) and Λ with cross relations from $\mathbf{R}$. We also have a spinorial version following the same lines, with the Lorentz rotations generated now by s, t. We again just follow the bosonisation procedure from the end of Chapter 9.4. Thus we put the dual quasitriangular structure of $\widehat{A\bowtie A}$ into Lemma 7.4.8 to convert the coaction (10.86) into a right action of $\widehat{A\bowtie A}$ with the result

$$\mathbf{p}_1\triangleleft t_2 = \lambda^2R_{21}\mathbf{p}_1R, \quad \mathbf{p}_1\triangleleft s_2 = R^{-1}\mathbf{p}_1R, \quad \mathbf{p}\triangleleft\varsigma = \lambda^{-2}\mathbf{p}, \quad (10.87)$$

where λR is the quantum group normalisation. Then $(\widetilde{A\bowtie A})\bowtie\bar{A}(R)$, from Proposition 1.6.10 and Proposition 1.6.16, has the structure

$$\mathbf{p}_1t_2 = \lambda^2t_2R_{21}\mathbf{p}_1R, \quad \mathbf{p}_1s_2 = s_2R^{-1}\mathbf{p}_1R, \quad \mathbf{p}\varsigma = \lambda^{-2}\varsigma\mathbf{p},$$

$$\Delta\mathbf{p} = \mathbf{p}\otimes s^{-1}(\ )t\varsigma + 1\otimes\mathbf{p}, \quad (10.88)$$

where the matrix indices of $\mathbf{p}$ have to be inserted into the space. There is also a counit $\epsilon\mathbf{p} = 0$ and antipode.

Also, if there is a suitable $A\bowtie A$-invariant quadratic element, we can then differentiate it to obtain a metric η^{IJ}. Exercise 10.4.5 then gives a more compact appearance $\partial_2 R_{21} \mathbf{u}_1 R - \lambda^{-2} R^{-1} \mathbf{u}_1 R \partial_2 = \eta^{(2)} R_{21} \eta^{(1)} R$ for the braided Leibniz rule in terms ∂_I written as matrices. Here $\eta^{(1)} \otimes \eta^{(2)}$ is $\eta_{IJ} = \eta^{i_0}{}_{i_1}{}^{j_0}{}_{j_1}$ as an element of $M_n \otimes M_n$. We can also define translation-invariant integration $\int$ on $B(R)$. Even without a metric, we can define differential forms $\theta^i{}_j = d u^i{}_j$ with relations

$$R^{-1} d\mathbf{u}_1 R d\mathbf{u}_2 = -d\mathbf{u}_2 R_{21} d\mathbf{u}_1 R, \quad R^{-1} \mathbf{u}_1 R d\mathbf{u}_2 = d\mathbf{u}_2 R_{21} \mathbf{u}_1 R, \quad (10.89)$$

as follows immediately from the braiding in Proposition 10.5.6 or the form of $\mathbf{R}$. Differentiating the top form gives, in general, a q-epsilon tensor on $B(R)$ to go with this differential calculus.

Finally, we already know from Chapters 7.4 and 9.4 that the braided matrices $B(R)$ are closely related to $A(R)$ by transmutation. One can check that this connects the additive structure above to that of $A(R)$ in Section 10.5.1. It is also related to $\bar{A}(R)$ in Section 10.5.2 by twisting as a comodule algebra.

Exercise 10.5.7 *Show that the covariant system* $(A \bowtie A, B(R))$ *is the twisting according to the dual of Proposition 2.3.8 of the covariant system* $(A \otimes A, \bar{A}(R))$ *in Section 10.5.2 by the 2-cocycle*

$$\chi((a \otimes b) \otimes (c \otimes d)) = \epsilon(a) \mathcal{R}^{-1}(b \otimes c)\epsilon(d).$$

Solution: We suppose that R is regular. We know from Proposition 7.3.1 that dual-twisting $A \otimes A$ by this cocycle gives us the double cross product Hopf algebra $A \bowtie A$ which we have used above. We must twist or 'gauge transform' any comodule algebra using (2.27) so that it remains covariant. Applying this to $\bar{A}(R)$, we have

$$x^i{}_j \cdot_\chi x^k{}_l = x^a{}_b x^c{}_d \chi^{-1}((Ss^i{}_a \otimes t^b{}_j) \otimes Ss^k{}_c \otimes t^d{}_l) = x^i{}_a x^b{}_l \lambda^{-1} \tilde{R}^a{}_j{}^k{}_b,$$

using Proposition 4.2.2 to evaluate on the matrix generators. Similarly for higher products of generators. If we denote $\mathbf{x}$ with the new product by $\mathbf{u}$, then the relation between the two is just

$$\mathbf{u} = \mathbf{x}, \quad \mathbf{u}_1 R \mathbf{u}_2 = \lambda^{-1} \mathbf{x}_1 \mathbf{x}_2,$$

etc. The relations among the $\mathbf{u}$ implied by those of $\mathbf{x}$ in Proposition 10.5.3 are just those of the braided matrices $B(R)$. The additive braided coproduct also twists, using (2.28), but retains its additive form. So this is really a twisting of additive braided groups. Note also that combining this exercise with the twisting from $A(R)$ to $\bar{A}(R)$ in Exercise 10.5.4 gives the correct transmutation formula $\mathbf{u}_1 R \mathbf{u}_2 = R \mathbf{t}_1 \mathbf{t}_2$ in (7.37). In other words, transmutation can be expressed as two twistings of comodule algebras: first by the 2-cocycle in Exercise 10.5.4 and then by the present 2-cocycle.

∎

As before, the above constructions are quite general. We can apply them to our standard R-matrix from Proposition 4.2.4, as well as to other multiparameter and nonstandard ones.

Example 10.5.8 *We define q-Minkowski space* $\mathbb{R}_q^{1,3} = BM_q(2)$ *as the* 2×2 *braided Hermitian matrices computed in Example 4.3.4 and 10.3.3 with* $\mathbf{u} = \begin{pmatrix} a & b \\ c & d \end{pmatrix}$ *and relations*

$$ba = q^2 ab, \quad ca = q^{-2} ac, \quad da = ad, \quad bc = cb + (1 - q^{-2})a(d - a),$$

$$db = bd + (1 - q^{-2})ab, \quad cd = dc + (1 - q^{-2})ca.$$

It has a braided addition law whereby

$$\begin{pmatrix} a'' & b'' \\ c'' & d'' \end{pmatrix} = \begin{pmatrix} a & b \\ c & d \end{pmatrix} + \begin{pmatrix} a' & b' \\ c' & d' \end{pmatrix}$$

also obeys the relations of $BM_q(2)$ *provided the second primed copy has the additive braid statistics*

$$a'a = q^2 aa', \quad a'b = ba', \quad b'b = q^2 bb', \quad a'c = ca'q^2 + (q^2 - 1)ac',$$

$$a'd = da' + (q^2 - 1)bc' + (q - q^{-1})^2 aa', \quad b'a = (q^2 - 1)ba' + ab'q^2,$$

$$b'c = cb' + (1 - q^{-2})(da' + ad') + (q - q^{-1})^2 bc' - (2 - 3q^{-2} + q^{-4})aa',$$

$$b'd = db' + (q^2 - 1)bd' + (q^{-2} - 1)ba' + (q - q^{-1})^2 ab', \quad c'a = ac',$$

$$c'b = bc' + (1 - q^{-2})aa', \quad c'c = q^2 cc', \quad c'd = dc'q^2 + (q^2 - 1)ca',$$

$$d'a = ad' + (q^2 - 1)bc' + (q - q^{-1})^2 aa', \quad d'b = bd'q^2 + (q^2 - 1)ab',$$

$$d'c = cd' + (q^2 - 1)dc' + (q - q^{-1})^2 ca' + (q^{-2} - 1)ac',$$

$$d'd = dd'q^2 + (q^2 - 1)cb' + (q^{-2} - 1)bc' - (1 - q^{-2})^2 aa'.$$

The background covariance is the q-Lorentz group $SU_q(2) \bowtie SU_q(2)$ *with dilatonic extension for the correctly normalised braid statistics.*

Proof: We take the same R as in Examples 10.5.2 and 10.5.5, which is the one for $M_q(2)$ in Example 4.2.4 but without the normalisation there. The normalisation is needed for the correct braiding using the formulae above. ∎

We now see how the rest of the general theory above looks in this example. We have already seen the parallels provided by transmutation between $BM_q(2)$ and usual quantum matrices $M_q(2)$, even though they are different algebras. Thus, the usual q-determinant of $M_q(2)$ transmutes into the braided determinant $\underline{\det}(\mathbf{u}) = ad - q^2 cb$, as we have explained in Example 4.3.4. We now also have parallels with q-Euclidean space $\mathbb{R}_q^4 = \bar{M}_q(2)$, made precise as twisting in Exercise 10.5.7 above. The twisting of comodule algebras in this exercise is therefore an algebraic model of *quantum Wick rotation* in our picture.

Next, we have by bosonisation the associated Minkowski q-Poincaré group $SU_q(2)\widetilde{\bowtie}SU_q(2){\gtrdot}\mathbb{R}_q^{1,3}$ from (10.78). The multiindex generators $\Lambda^I{}_J$ provide a vectorial picture of the q-Lorentz group $SO_q(1,3)$ in our framework, with corresponding q-Poincaré group $\widetilde{SO_q(1,3)}{\gtrdot}\mathbb{R}_q^{1,3}$ The scaling parameter is $\lambda = q^{-\frac{1}{2}} \neq 1$, so a dilaton is needed.

The braided determinant $\underline{\det}(\mathbf{u})$ being q-Lorentz invariant and central provides our natural radius function. The corresponding metric, by differentiation, is

$$\eta^{IJ} = \frac{\partial}{\partial u_I}\frac{\partial}{\partial u_J}(ad - q^2 cb) = \begin{pmatrix} q^{-2}-1 & 0 & 0 & 1 \\ 0 & 0 & -q^2 & 0 \\ 0 & -1 & 0 & 0 \\ 1 & 0 & 0 & 0 \end{pmatrix}. \quad (10.90)$$

Here $(1+q^{-2})^{-1}u_A u_B \eta^{BA} = ad - q^2 cb$ recovers the radius function as in Example 10.4.20. The conditions (10.56) hold, so the Gaussian has a simple form.

We also know already from Proposition 4.3.7 and Example 10.3.3 that $\mathbb{R}_q^{1,3}$ has a natural Hermitian $*$-structure when we work over $\mathbb{C}$ with q real. We have the natural self-adjoint generators t, x, y, z in (7.44), and we know that the distance function in terms of them in (7.45) is a deformation of the Minkowski signature. We came across this interpretation $\mathbb{R}_q^{1,3} = BM_q(2)$ in the context of the quantum double, and we reach the same conclusion now from braided geometry.

The rest of the structure is routinely computed from the R-matrix formulae above. The vector algebra of differentiation operators is

$$\frac{\partial}{\partial d}\frac{\partial}{\partial b} = q^{-2}\frac{\partial}{\partial b}\frac{\partial}{\partial d}, \quad \frac{\partial}{\partial d}\frac{\partial}{\partial c} = \frac{\partial}{\partial c}\frac{\partial}{\partial d}q^2, \quad \frac{\partial}{\partial d}\frac{\partial}{\partial a} = \frac{\partial}{\partial a}\frac{\partial}{\partial d},$$

$$\frac{\partial}{\partial c}\frac{\partial}{\partial a} = \frac{\partial}{\partial a}\frac{\partial}{\partial c} + \frac{\partial}{\partial c}\frac{\partial}{\partial d}(q^2-1), \quad \frac{\partial}{\partial b}\frac{\partial}{\partial a} = \frac{\partial}{\partial a}\frac{\partial}{\partial b} + \frac{\partial}{\partial b}\frac{\partial}{\partial d}(q^{-2}-1),$$

$$\frac{\partial}{\partial b}\frac{\partial}{\partial c} = \frac{\partial}{\partial c}\frac{\partial}{\partial b} + \frac{\partial}{\partial d}\frac{\partial}{\partial d}(q^{-2}-1) + \frac{\partial}{\partial a}\frac{\partial}{\partial d}(q^2-1).$$

Their Leibniz rule comes from the braiding listed in Example 10.5.8.

The algebra of forms is

$$dcdc = 0, \quad dada = 0, \quad dbdb = 0, \quad dbda = -dadb,$$

$$dcda = -dadc, \quad dcdb = -dbdc, \quad dddd = dbdc(1 - q^{-2}),$$

$$dddc = -dcddq^{-2} + dadc(1 - q^{-2}), \quad dddb = -dbddq^2 - dadb(q^2 - 1),$$

$$ddda = -dbdc(q^2 - 1) - dadd.$$

These relations generate a finite-dimensional algebra, with the dimension at each degree being the same as in the classical case. The commutation relations of forms with position coordinates again follows the braiding in Example 10.5.8.

The q-epsilon tensor is

$$\epsilon_{addd} = -\epsilon_{bdcd} = -\epsilon_{dadd} = \epsilon_{dbdc} = \epsilon_{ddad} = -\epsilon_{ddda} = 1 - q^{-2},$$

$$-\epsilon_{adad} = -\epsilon_{cdbd} = \epsilon_{dada} = \epsilon_{dcdb} = q^2 - 1,$$

$$\epsilon_{abcd} = -\epsilon_{acbd} = \epsilon_{adbc} = -\epsilon_{adcb} = -\epsilon_{bacd} = \epsilon_{bcad} = -\epsilon_{bcda} = \epsilon_{cabd} = 1,$$

$$-\epsilon_{cbad} = \epsilon_{cbda} = -\epsilon_{dabc} = \epsilon_{dacb} = \epsilon_{dbac} = -\epsilon_{dbca} = -\epsilon_{dcab} = \epsilon_{dcba} = 1,$$

$$\epsilon_{acdb} = -\epsilon_{cadb} = \epsilon_{cdab} = -\epsilon_{cdba} = q^2,$$

$$-\epsilon_{abdc} = \epsilon_{badc} = -\epsilon_{bdac} = \epsilon_{bdca} = q^{-2},$$

which has even more nonzero elements than in the Euclidean case.

Finally, the Gaussian-weighted integral on degree two is $\mathcal{Z}[u_I u_J] = q^{-4}\eta_{IJ}$. Its nonzero values on degree four are

$$\mathcal{Z}[abcd] = -q^{-12}, \quad \mathcal{Z}[acbd] = -q^{-8}, \quad \mathcal{Z}[bcd^2] = -q^{-10}(1 - q^{-4}),$$

$$\frac{1}{2}\mathcal{Z}[d^4] = -\mathcal{Z}[cbd^2] = q^{-8}(1 - q^{-2})^2, \quad \mathcal{Z}[ad^3] = q^{-10}(1 + 2q^2)(1 - q^{-2}),$$

$$\mathcal{Z}[a^2 d^2] = q^4 \mathcal{Z}[b^2 c^2] = \mathcal{Z}[c^2 b^2] = q^2 \mathcal{Z}[bc^2 b] = q^2 \mathcal{Z}[cb^2 c] = q^{-10}(q^2 + 1),$$

along with various other cases implied by the relations in $\mathbb{R}^{1,3}_q$. In terms of spacetime coordinates (7.44), we have

$$\mathcal{Z}[t^2] = \frac{[2; q^2]}{4} q^{-4}, \quad \mathcal{Z}[x^2] = \mathcal{Z}[y^2] = \mathcal{Z}[z^2] = -\frac{[2; q^2]}{4} q^{-6},$$

$$\mathcal{Z}[z^4] = \frac{3[2; q^2]}{8} q^{-12}, \quad \mathcal{Z}[t^4] = q^4 \mathcal{Z}[x^4] = q^4 \mathcal{Z}[y^4] = \frac{[3; q^2]!}{8} q^{-10}.$$

We see that the Gaussian-weighted integral $\mathcal{Z}$ is quite similar to the Euclidean one in its values, except for the sign in the spacelike directions. We also note that the negative sign in the spacelike directions agrees with the sign obtained in physics by Wick rotation to make sense of Gaussian integrals in Minkowski space.

Notes for Chapter 10

The theory of braided groups and the main results in this chapter are due to the author. Some of the applications are with collaborators. There are some 50 or so research papers in which the basics of this approach to q-deformation was developed, of which some of the more explicit are covered in the present chapter. Some existing reviews by the author are [3] and [4], with [205, 206, 207] as introductions.

The papers [38, 146, 154, 156, 178, 179] introduced the notion of braided groups or Hopf algebras in braided categories, proved basic lemmas and the main theorems to date regarding transmutation and bosonisation. We refer to the Notes of Chapter 9 for details. Super-Hopf algebras and Hopf algebras in symmetric (but not braided) categories are much older; see, for example [101, 208].

Anyonic and q-statistical braided groups were introduced in [30] and [31], where Proposition 10.1.4–Example 10.1.8 can be found. The theory in Section 10.1 was applied to supersymmetry [209, 210], where transmutation was related to the Jordan–Wigner transform, and [87, 211], in which the nonstandard quantum group in Example 4.2.14 was understood by superisation as $U_q(gl(1|1))$ and its superquasitriangular structure thereby found. We have not covered these applications in detail in Section 10.1 due to lack of space. For recent works, see [212] and [213]. The paper [31] also covered the bosonisations of the braided line and its higher-dimensional analogues as $U_q(b_+)$ in Example 1.3.2 and higher-dimensional Weyl algebras. More general q-statistical braided groups where $\mathbb{Z}_{/n}$ is replaced by a general Abelian group equipped with a bicharacter were also considered here. The examples $U_q(n_\pm)$ are from [214] (the algebra f) and are clearly braided groups of this q-statistical type. The appendix of [215] contains the theory of group-graded quantum matrices, obtained by transmutation. Recent applications also allow for a nontrivial 3-cocycle, defining a nontrivial associator. For example, the category of $(\mathbb{Z}_2)^3$-graded spaces with cocycle and (quasi) bicharacter [56]

$$\phi(\vec{x}, \vec{y}, \vec{z}) = (-1)^{(\vec{x} \times \vec{y}) \cdot \vec{z}}, \quad \mathcal{R}(\vec{x}, \vec{y}) = \begin{cases} 1 & \text{if } \vec{x} = 0 \text{ or } \vec{y} = 0 \text{ or } \vec{x} = \vec{y} \\ -1 & \text{otherwise} \end{cases}$$

is a (symmetric) monoidal one in which the octonions live as an associative algebra in the category. We write the group elements as $\mathbb{Z}_2$-valued vectors.

The octonion product in a graded basis is [56]

$$e_{\vec{x}} \cdot e_{\vec{y}} = e_{\vec{x}+\vec{y}}(-1)^{\sum_{i \leq j} x_i y_j + y_1 x_2 x_3 + x_1 y_2 x_3 + x_1 x_2 y_3}.$$

The notion of braided coaddition and the construction of braided vectors and covectors in Section 10.2 was introduced in [150]. Theorem 10.2.1 through to Exercise 10.2.13 can be found there. The shorthand notation in (10.18) and (10.19) was emphasised in its sequel [193]. The application of the covariance to construct a q-Poincaré group, as in Corollary 10.2.10, was the main result of [150], which also contained the result that a dilaton ς is needed when the quantum group normalisation constant λ is not 1, i.e. when the braided covector normalisation (with eigenvalue -1 for the braiding) is not the quantum group normalisation. Some special cases (although not the general R-matrix theory) of Corollary 10.2.10 were found by hand [216]. Corollary 10.2.11, in which the Poincaré group coacts covariantly on the covector space, was missing in such previous attempts, but is an immediate application of the braided approach based on bosonisation. A second main result from this approach is the correct formulation of $*$-structures on such q-Poincaré algebras [47, 217], which we have not covered here. One shows that the second conjugate coproduct (10.22) is twisting-equivalent to the opposite of the first one in Exercise 10.2.13, while manifestly connected to it by $*$ [47, 217]. The further extension of these ideas to q-conformal algebras after Exercise 10.2.13 is from [197]. The formulation of the quantum metric in Definition 10.2.14 is due to U. Meyer [151]. Again, some examples of matrices which could reasonably be called quantum metrics were known previously without the abstract picture made possible by the theory of braided vectors and covectors.

The notion of braided matrices $B(R)$ as a quadratic algebra and with braided comultiplication was introduced in 1990 [94, 156]; $BM_q(2)$ and $BM_q(1|1)$ in Examples 10.3.3 and 10.3.4 and the covariance in Proposition 10.3.5 were basic features in these papers; see also the Notes to Chapter 4. That the braided matrix relations $R_{21}\mathbf{u}_1 R\mathbf{u}_2 = \mathbf{u}_2 R_{21}\mathbf{u}_1 R$ map onto those of $1^+ S 1^-$ was known for some time [82, 90] and became the theory of braided-Lie algebras [5, 34, 194]; see again the Notes to Chapter 4. The conjunction of braided matrices $B(R)$ and an earlier version of quantum vectors and covectors to form a general braided linear algebra was introduced in [95]. The construction of the braid statistics from the dual quasitriangular structure in Proposition 10.3.6 and the bosonic central elements $\mathrm{Tr}\,\mathbf{u}^k v$ in Theorem 10.3.8 can be found there. The latter provide Casimirs $\mathrm{Tr}\,(1^+ S 1^-)^k v$, known already for standard $U_q(g)$ [82]. Our own application in Corollary 10.3.9 is [218], where we construct a class of bicovariant differential forms on quantum groups. The $*$-structure on braided groups, as in Proposition 10.3.2, was introduced in [35], where also the interpretation of $BM_q(2)$ as q-Minkowski space was developed. The remaining results of Section 10.3 are also from [95]. Proposition 10.3.7 introduced there, gives the relation between $B(R)$ as q-Minkowski space in the braided approach and an earlier approach to q-Minkowski space in [153] and [219] based on the tensor product of two quantum planes. This point was elaborated in detail [220], and was used to obtain the matrix form of the q-Poincaré group in Section 10.5.3. The *-structures on more general linear braided spaces are analysed in [221], depending on the R-matrix. In some cases there is no natural *-algebra structure but rather one should consider the algebra as the 'holomorphic' part of a larger system. This includes the standard sl_n quantum planes of Example 4.5.7; see [222, 223].

Differentiation on braided spaces was introduced in [193] as an infinitesimal coaddition. Section 10.4 is based heavily on this paper. This includes Definition 10.4.1–Corollary 10.4.18 in some form. Example 10.4.6 recovered the usual one-dimensional q-derivative as studied already in Chapter 3.1, while Example 10.4.7 recovered the q-derivative on the quantum plane introduced by other means by J. Wess and B. Zumino [224]. The form of the Leibniz rule in Example 10.4.6 was also known for these standard sl_n R-matrices [222]. The braided approach [193] made contact with these cases and worked quite generally. The braided Gaussian in Definition 10.4.19, the general form of the braided exponential and the general theory of Gaussian-weighted integration were introduced in [11]. This paper then proceeded to a theory of Fourier transform on braided covector spaces which we have not had space to cover here. It makes contact with Fourier theory on general braided groups [10, 78]. Gaussian integration on $SO_q(N)$-covariant spaces introduced by other means in [225] fits into this scheme as well. The one-dimensional Jackson integral

$$\int_0^x f = (1-q)x \sum_{n=0}^{\infty} f(q^n x)q^n$$

inverse to the q-derivative also has a natural characterisation in terms of coaddition on the braided line, namely [11]

$$\int_0^{x+y} f = \int_0^x f((\ ) + y) + \int_0^y f = \int_0^x f + \int_0^y f(x + (\ ))$$

when $yx = qxy$. This is somewhat different from the more usual picture as a version of a Riemann sum with sample points given by scaling by powers of $q \in (0,1)$. Finally, the approach to the braided epsilon tensor on braided covector spaces at the end of Section 10.4 is due to the author [226, 227]. We have included it because it also makes contact with the formulae (10.72) for differential forms on quantum spaces which are usually derived [224] by consistency arguments within the axiomatic

framework for differential forms in [91]. The braided approach constructs
d starting from coaddition and partial differentiation rather than by introducing the symbols dx_i abstractly.

The coaddition on quantum matrices $A(R)$ and $A(R_1 : R_2)$ in Section 10.5.1 is due to the author [102, 103], and allows us to apply the
theory of braided linear spaces to these cases. The required braiding for
coaddition on $A(R)$ corresponds to the relations (10.74) among its differential forms introduced in [228]. We have already met quantum matrices
in Chapter 4, and we use the rectangular form to define $\bar{A}(R) = A(R_{21} :
R)$, which we consider as q-Euclidean space in Section 10.5.2, following
the work [229] by the author. This was a novel approach in [229], but
makes contact in four dimensions with work on $SO_q(N)$-covariant spaces
in [230], where the q-epsilon tensor was found by hand. The matrix version of the corresponding q-Poincaré group in Section 10.5.2 is due to the
author [229].

The main result in the paper [102] by the author and M. Markl is the
formulation of a gluing operation,

$$
R_1 \oplus_q R_2 = \begin{pmatrix} R_1 & 0 & 0 & 0 \\ 0 & 1 & (q - q^{-1})P & 0 \\ 0 & 0 & 1 & 0 \\ 0 & 0 & 0 & R_2 \end{pmatrix}, \tag{10.91}
$$

for Hecke R-matrices. It obeys $A(R_1 \oplus_q R_2 : R_3) = A(R_1 : R_3) \otimes_{R_3} A(R_2 :
R_3)$, etc. Among other things, this solved, in the Hecke case, the problem
of how to tensor product two braided groups. The answer is that one
must $\oplus_q$ their braidings too. This is needed in physics when one wants to
make infinite tensor products [180] of braided vector spaces in order to define braided wave-functions and braided path integration [103]. Another
application of gluing is to write the exterior algebra associated to any R-
matrix of Hecke type as a $2 \times n$ rectangular quantum matrix $A(R_\Omega : R)$,
where R_Ω is a twisting of the nonstandard R-matrix in Example 4.2.13.
This has a number of immediate consequences, such as a braided addition
law and a left–right bicovariance [87, 215]. The gluing is also interpreted
in [102] as a convolution operation among the corresponding knot invariant. The double bosonisation construction [195] in Theorem 9.4.13 and
after Exercise 10.2.13 also allows one to build up $U_q(sl_n)$ inductively by
successively adjoining quantum planes and presumably generalises the
above ideas to the non-Hecke case.

The coaddition on braided matrices $B(R)$ in Section 10.5.3 is due to
Meyer [151]. This then made it possible to apply the constructions of
Section 10.4 to q-Minkowski space in the form of Example 10.5.8, which
is also due to Meyer. The differential forms for this algebra and a lot of
concrete relations for the q-Poincaré algebra were found in [219] and [231],

and a braided coaddition in [150]. As well as putting those works into a general R-matrix formulation, the braided approach provided, for the first time, the existence of the braided exponential, the Gaussian, integration [11], the epsilon tensor and Hodge *-operator [232] on q-Minkowski space, and more. The latter paper also contains some first implications for q-electrodynamics, while the further geometry of q-Minkowski space in the braided approach is in Meyer's Ph.D. thesis. There are many further points of contact between the extensive q-deformation literature and the braided approach in the present chapter; it is not practical to mention all of them here. The twisting constructions in Exercises 10.5.4 and 10.5.7 (quantum Wick rotation) are due to the author [229]. They are extended in [47] to the theorem that the corresponding q-Poincaré groups are related by twisting of Hopf algebras. These twisting results tell us that the system which we have called Euclidean and the one which we have called Minkowski are 'gauge equivalent' at the algebraic level; their only real difference is their choice of *-structures, which are different even when the Minkowski theory is 'untwisted' so as to coincide algebraically with the Euclidean one. This is already clear from the form of * in the 'Lorentz' sector in Proposition 7.3.4. The q-conformal quantum group for q-Euclidean and q-Minkowski space is in [197].

There is a great deal more braided group theory that we have not had room to include here. Perhaps the most important is the theory of braided-Lie algebras [5], which was mentioned above and in the Notes for Chapters 4 and 9. One application of such finite-dimensional objects is to build further solutions of the QYBE [194, 233]. The principal result, however, is that, for every such braided-Lie algebra $\mathcal{L}$, there is a braided bialgebra $U(\mathcal{L})$. Perhaps most remarkable for physics is the quantum–geometry transformation [5, 207]

$$U(gl_{2,q}) \cong \mathbb{R}_q^{1,3} \tag{10.92}$$

as braided bialgebras. This isomorphism is singular at $q = 1$, i.e. it is a *purely quantum phenomenon*. When $q \to 1$, the right hand side of course tends to the commutative algebra of functions on Minkowski space, while the left hand side tends to the noncommutative algebra $U(su_2 \oplus u(1))$. We have covered this in concrete terms in Chapter 4.3.

References

[1] S. Majid. Quasitriangular Hopf algebras and Yang-Baxter equations. *Int. J. Mod. Phys. A*, 5:1–91, 1990.

[2] T. Brzeziński and S. Majid. Quantum group gauge theory on quantum spaces. *Commun. Math. Phys.*, 157:591–638, 1993 and 167:235, 1995.

[3] S. Majid. Algebras and Hopf algebras in braided categories. In *Advances in Hopf Algebras, Lec. Notes in Pure and Applied Math.*, 158:55–105, 1994. Marcel Dekker.

[4] S. Majid. Beyond supersymmetry and quantum symmetry (an introduction to braided groups and braided matrices). In M-L. Ge and H.J. de Vega, editors, *Quantum Groups, Integrable Statistical Models and Knot Theory*, pp. 231–282. World Sci., 1993.

[5] S. Majid. Quantum and braided Lie algebras. *J. Geom. Phys.*, 13:307–356, 1994.

[6] G.I. Kac and V.G. Paljutkin. Finite ring groups. *Trans. Amer. Math. Soc.*, 15:251–294, 1966.

[7] M.E. Sweedler. *Hopf Algebras*. Benjamin, 1969.

[8] E. Abe. *Hopf Algebras*. CUP, Cambridge, UK, 1977.

[9] R.G. Larson and D.E. Radford. Semisimple cosemisimple Hopf algebras. *Amer. J. Math.*, 110:187–195, 1988.

[10] V.V. Lyubashenko and S. Majid. Fourier transform identities in quantum mechanics and the quantum line. *Phys. Lett. B*, 284:66–70, 1992.

[11] A. Kempf and S. Majid. Algebraic q-integration and Fourier theory on quantum and braided spaces. *J. Math. Phys.*, 35:6802–6837, 1994.

[12] S. Majid. Physics for algebraists: Non-commutative and non-cocommutative Hopf algebras by a bicrossproduct construction. *J. Algebra*, 130:17–64, 1990.

[13] S. Majid. Quantum and braided diffeomorphism groups. *J. Geom. Phys.*, 28:94–128, 1998.

[14] S. Majid. More examples of bicrossproduct and double cross product
      Hopf algebras. *Isr. J. Math*, 72:133–148, 1990.

[15] S.L. Woronowicz. Compact matrix pseudo groups. *Commun. Math.
      Phys.*, 111:613–665, 1987.

[16] M. Enock and J-M. Schwartz. Une dualite dans les algebres de von
      Neumann. *Bull. Soc. Math. Fr. Mem.*, 44:144, 1975.

[17] S. Stratila. *Modular Theory of Operator Algebras*. Abacus-Kent, UK,
      1975.

[18] S. Majid. Hopf-von Neumann algebra bicrossproducts, Kac algebra
      bicrossproducts, and the classical Yang-Baxter equations. *J. Funct.
      Analysis*, 95:291–319, 1991.

[19] S. Baaj and G. Skandalis. Unitaires multiplicatifs et dualite pour les
      produits croisse de C^*-algebres. *Ann. Sci. Ecol. Norm. Sup.*,
      26:425–488, 1993.

[20] S. Majid. Quantum random walks and time reversal. *Int. J. Mod. Phys.
      A*, 8:4521–4545, 1993.

[21] D.E. Evans and Y. Kawahigashi. *Quantum Symmetries on Operator
      Algebras*. OUP, Oxford, UK, 1998.

[22] V.G. Drinfeld. Quantum groups. In A. Gleason, editor, *Proceedings of
      the ICM*, pp. 798–820. AMS, Rhode Island, 1987.

[23] E.K. Sklyanin. Some algebraic structures connected with the
      Yang-Baxter equations. *Funct. Anal. Appl.*, 16:263–270, 1982.

[24] E.K. Sklyanin. Some algebraic structures connected with the
      Yang-Baxter equations. Representations of quantum algebras. *Funct.
      Anal. Appl.*, 17:273–284, 1982.

[25] P.P. Kulish and N.Yu. Reshetikhin. Quantum linear problem for the
      Sine-Gordon equation and highest weight representations. *J. Sov.
      Math.*, 23:2435–2441, 1983.

[26] M. Jimbo. A q-difference analog of $U(g)$ and the Yang-Baxter equation.
      Lett. Math. Phys., 10:63–69, 1985.

[27] L.D. Faddeev and L.A. Takhtajan. The quantum inverse scattering
      method of the inverse problem and the Heisenberg XYZ model. *Russ.
      Math. Surv.*, 34:11, 1979.

[28] N.Yu. Reshetikhin and V.G. Turaev. Ribbon graphs and their invariants
      derived from quantum groups. *Commun. Math. Phys.*, 127:1–26, 1990.

[29] V.G. Drinfeld. On almost-cocommutative Hopf algebras. *Leningrad
      Math. J.*, 1:321–342, 1990.

[30] S. Majid. Anyonic quantum groups. In Z. Oziewicz *et al*, editors,
      *Spinors, Twistors, Clifford Algebras and Quantum Deformations
      (Proceedings of the 2nd Max Born Symposium, Wroclaw, 1992)*,
      pp. 327–336. Kluwer.

[31] S. Majid. C-statistical quantum groups and Weyl algebras. *J. Math. Phys.*, 33:3431–3444, 1992.

[32] D. Radford. On the quasitriangular structures of a semisimple Hopf algebra. *J. Algebra*, 141:354–358, 1991.

[33] N.Yu. Reshetikhin and M.A. Semenov-Tian-Shansky. Quantum *R*-matrices and factorization problems. *J. Geom. Phys.*, 5:533, 1988.

[34] S. Majid. Braided matrix structure of the Sklyanin algebra and of the quantum Lorentz group. *Commun. Math. Phys.*, 156:607–638, 1993.

[35] S. Majid. The quantum double as quantum mechanics. *J. Geom. Phys.*, 13:169–202, 1994.

[36] S. Majid. Quantum groups and quantum probability. In *Quantum Probability and Related Topics VI (Proceedings of the Trento conference, 1989)*, pp. 333–358. World Sci.

[37] S. Majid. Representations, duals and quantum doubles of monoidal categories. *Suppl. Rend. Circ. Mat. Palermo, Ser. II*, 26:197–206, 1991.

[38] S. Majid. Braided groups. *J. Pure and Applied Algebra*, 86:187–221, 1993.

[39] V.G. Drinfeld. Quasi-Hopf algebras and Knizhnik-Zamolodchikov equations. In A.U. Klimyk *et al*, editors, *Problems of Modern Quantum Field Theory*, pp. 1–13. Springer, 1989.

[40] V.G. Drinfeld. Quasi-Hopf algebras. *Leningrad Math. J.*, 1:1419–1457, 1990.

[41] V.G. Drinfeld. On quasitriangular quasi-Hopf algebras and a certain group closely connected with Gal($\overline{\mathbb{Q}}/\mathbb{Q}$). *Leningrad Math. J.*, 2:829–860, 1991.

[42] V.G. Drinfeld. On the structure of the quasitriangular quasi-Hopf algebras. *Funct. Anal. Appl.*, 26:63–65, 1992.

[43] V.G. Drinfeld. On constant quasiclassical solutions of the Yang-Baxter equations. *Sov. Math. Dokl.*, 28:667–671, 1983.

[44] S. Majid and R. Oeckl. Twisting of quantum differentials and the Planck-scale Hopf algebra. *Comm. Math. Phys.*, 205:617–655, 1999.

[45] D.I. Gurevich and S. Majid. Braided groups of Hopf algebras obtained by twisting. *Pac. J. Math.*, 162:27–44, 1994.

[46] S. Majid. Cross product quantization, nonabelian cohomology and twisting of Hopf algebras. In V.K. Dobrev *et al*, editors, *Generalised Symmetries in Physics*, pp. 13–41. World Sci., 1994.

[47] S. Majid. Quasi-* structure on *q*-Poincaré algebras. *J. Geom. Phys.*, 22:14–58, 1997.

[48] M.E. Sweedler. Cohomology of algebras over Hopf algebras. *Ann. Math.*, pages 205–239, 1968.

[49] M. Gerstenhaber and S.D. Schack. Bialgebra cohomology, deformations and quantum groups. *Proc. Natl. Acad. Sci. USA*, 87:478–481, 1990.

[50] S. Shnider and S. Sternberg. The cobar resolution and a restricted deformation theory for Drinfeld algebras. *J. Algebra*, 169:343–366, 1994.

[51] M. Markl and J.D. Stasheff. Deformation theory via deviations. *J. Algebra*, 170:122–155, 1994.

[52] S. Shnider and S. Sternberg. *Quantum Groups*. International Press, Boston, 1993.

[53] S. Majid. Quasi-quantum groups as internal symmetries of topological quantum field theories. *Lett. Math. Phys.*, 22:83–90, 1991.

[54] S. Majid. Quantum double for quasi-Hopf algebras. *Lett. Math. Phys.*, 45:1–9, 1998.

[55] S. Majid. Tannaka-Krein theorem for quasi-Hopf algebras and other results. *Contemp. Maths*, 134:219–232, 1992.

[56] H. Albuquerque and S. Majid. Quasialgebra structure of the octonions. *J. Algebra*, 220:188–224, 1999.

[57] E. Celeghini, R. Giachetti, E. Sorace, and M. Tarlini. The quantum Heisenberg group $H(1)_q$. *J. Math. Phys.*, 32:1155–1158, 1991.

[58] C. Gomez and G. Sierra. Quantum harmonic oscillator algebra and link invariants. *J. Math. Phys.*, 34:2119–2131, 1993.

[59] A. Macfarlane. On q-analogues of the quantum harmonic oscillator and the quantum group $SU(2)_q$. *J. Phys. A*, 22:4581, 1989.

[60] L.C. Biedenharn. The quantum group $SU_q(2)$ and the q-analogue of the boson operators. *J. Phys. A*, 22:L873, 1989.

[61] G.E. Andrews. q-Series: Their development and application in analysis, number theory, combinatorics, physics and computer algebra. Technical Report 66, AMS, 1986.

[62] T. Masuda, K. Mimachi, Y. Nakagami, M. Noumi, Y. Saburi, and K. Ueno. Unitary representations of the quantum group $SU_q(1,1)$, I,II. *Lett. Math. Phys.*, 19:187–204, 1990.

[63] T.H. Koornwinder. Orthogonal polynomials in connection with quantum groups. In P. Nevai, editor, *Orthogonal Polynomials: Theory and Practice*, NATO ASI Series C, 294:257–292. Kluwer, 1990.

[64] N. Burroughs. The universal R-matrix for $U_q(sl_3)$ and beyond! *Commun. Math. Phys.*, 127:128, 1990.

[65] M. Rosso. An analogue of the P.B.W. theorem and the universal R-matrix for $U_h sl(N+1)$. *Commun. Math. Phys.*, 124:307–318, 1989.

[66] S.Z. Levendorskiĭ and Ya. S. Soibelman. Some applications of quantum Weyl group. *J. Geom. and Phys.*, 7:1–14, 1991.

[67] A.N. Kirillov and N.Yu. Reshetikhin. q-Weyl group and a multiplicative formula for the universal R-matrices. *Commun. Math. Phys.*, 134:421–431, 1990.

[68] M. Rosso. Comparison des groupes $SU(2)$ quantique de Drinfeld et de Woronowicz. *CR. Acad. Sci. Paris*, 304:323–326, 1987.

[69] S. Majid and Ya. S. Soibelman. Rank of quantized universal enveloping algebras and modular functions. *Commun. Math. Phys.*, 137:249–262, 1991.

[70] S. Majid and Ya. S. Soibelman. Chern-Simons theory, modular functions and quantum mechanics in an alcove. *Int. J. Mod. Phys. A*, 6:1815–1827, 1991.

[71] S. Majid and Ya. S. Soibelman. Bicrossproduct structure of the quantum Weyl group. *J. Algebra*, 163:68–87, 1994.

[72] V.G. Drinfeld. A new realization of Yangians and quantized affine algebras. *Sov. Math. Dokl.*, 36/296:212–216, 1988.

[73] V. Chari and A. Pressley. Quantum affine algebras. *Commun. Math. Phys.*, 142:261–283, 1991.

[74] V.V. Lyubashenko. Real and imaginary forms of quantum groups. In *Proceedings of the Euler Institute, St. Petersberg, 1990, Lec. Notes in Math.*, 1510:67–78. Springer.

[75] E. Twietmeyer. Real forms of $U_q(g)$. *Lett. Math. Phys.*, 24:49–58, 1992.

[76] G. Lusztig. Quantum groups at roots of 1. *Geometricae Dedicata*, 35:89–114, 1990.

[77] N.Yu. Reshetikhin and V.G. Turaev. Invariants of 3-manifolds via link polynomials and quantum groups. *Invent. Math.*, 103:547–597, 1991.

[78] V.V. Lyubashenko and S. Majid. Braided groups and quantum Fourier transform. *J. Algebra*, 166:506–528, 1994.

[79] M. Kashiwara. On crystal bases of the q-analogue of uninversal enveloping algebras. *Duke Math J.*, 63:465–516, 1991.

[80] G. Lusztig. Canonical bases arising from quantized enveloping algebras. *J. Amer. Math. Soc.*, 3:447–498, 1990.

[81] R.J. Baxter. *Exactly Solvable Models in Statistical Mechanics*. Academic, 1982.

[82] L.D. Faddeev, N.Yu. Reshetikhin, and L.A. Takhtajan. Quantization of Lie groups and Lie algebras. *Leningrad Math. J.*, 1:193–225, 1990.

[83] S. Majid. Quantum group duality in vertex models and other results in the theory of quasitriangular Hopf algebras. In L-L. Chau and W. Nahm, editors, *Proc. XVIIIth DGM, Tahoe, 1989, Nato-ASI Series B*, 245:373–386. Plenum.

[84] R.G. Larson and J. Towber. Two dual classes of bialgebras related to the concepts of 'quantum group' and 'quantum Lie algebra'. *Commun. Algebra*, 19:3295–3345, 1991.

[85] E.E. Demidov, Yu.I. Manin, E.E. Mukhin, and D.V. Zhdanovich. Non-standard quantum deformations of $GL(n)$ and constant solutions of the Yang-Baxter equations. *Prog. Theo. Phys. Suppl.*, 102:203–218, 1990.

[86] M-L. Ge, N. Jing, and Y-S. Wu. A new quantum group associated with "non-standard" braid group representation. *Lett. Math. Phys.*, 21:193–203, 1991.

[87] S. Majid and M.J. Rodriguez-Plaza. Nonstandard quantum groups and superisation. *J. Math. Phys.*, 36:7071–7097, 1995.

[88] J. Hietarinta. All constant solutions of the quantum Yang-Baxter equations on two dimensions. *Phys. Lett. B*, 165:245, 1992.

[89] L. Hlavaty. Unusual solutions of the Yang-Baxter equations. *J. Phys. A*, 20:1661, 1987.

[90] N.Yu. Reshetikhin and M.A. Semenov-Tian-Shansky. Central extensions of quantum current groups. *Lett. Math. Phys.*, 19:133–142, 1990.

[91] S.L. Woronowicz. Differential calculus on compact matrix pseudogroups (quantum groups). *Commun. Math. Phys.*, 122:125–170, 1989.

[92] B. Jurco. Differential calculus on quantized simple Lie groups. *Lett. Math. Phys.*, 22:177–186, 1991.

[93] P. Schupp, P. Watts, and B. Zumino. Bicovariant quantum alegbras and quantum Lie algebras. *Commun. Math. Phys.*, 157:305–329, 1993.

[94] S. Majid. Examples of braided groups and braided matrices. *J. Math. Phys.*, 32:3246–3253, 1991.

[95] S. Majid. Quantum and braided linear algebra. *J. Math. Phys.*, 34:1176–1196, 1993.

[96] I.V. Cherednik. *Theor. Mat. Phys.*, 61:55, 1984.

[97] E.K. Sklyanin. Boundary conditions for the integrable quantum systems. *J.Phys. A*, 21:2375–2389, 1988.

[98] Yu.I. Manin. Quantum groups and non - commutative geometry. Technical Report, Centre de Recherches Math, Montreal, 1988.

[99] A. Sudbery. Matrix-element bialgebras determined by quadratic co-ordinate algebras. *J. Algebra*, 158:375–399, 1994.

[100] M. Takeuchi. Matrix bialgebras and quantum groups. *Isr. J. Math.*, 72:232–251, 1990.

[101] D.I. Gurevich. Algebraic aspects of the quantum Yang-Baxter equation. *Leningrad Math. J.*, 2:801–828, 1991.

[102] S. Majid and M. Markl. Glueing operation for R-matrices, quantum groups and link invariants of Hecke type. *Math. Proc. Camb. Phil. Soc.*, 119:139–166, 1996.

[103] S. Majid. On the addition of quantum matrices. *J. Math. Phys.*, 35:2617–2633, 1994.

[104] S. Wang. Quantum symmetry groups of finite spaces. *Commun. Math. Phys.*, 195:195–211, 1998.

[105] S.A. Joni and G.-C. Rota. Coalgebras and bialgebras in combinatorics. *Studies in Applied Mathematics*, 61:93–139, 1979.

[106] P. Doubilet. *Studies in Partitions and Permutations*. PhD thesis, 1974. See also *J. Algebra*, 28:127–132, 1974.

[107] D.B. Benson. Bialgebras: Some foundations for distributed and concurrent computation. *Fundamenta Informaticae*, 12:427–486, 1989.

[108] A. Dür. Möbius functions, incidence algebras and power series representations. In *Lec. Notes in Math.*, 1202, 1986. Springer.

[109] R. Grossman and R.G. Larson. Hopf-algebraic structure of families of trees. *J. Algebra*, 126:184–210, 1989.

[110] W.R. Schmitt. Hopf algebras of combinatorial structures. *Can. J. Math.*, 45:412–428, 1993.

[111] L. Accardi, M. Schürmann, and M. von Waldenfels. Quantum independent increment processes on superalgebras. *Math. Z.*, 198:451–477, 1988.

[112] M. Schürmann. White noise on involutive bialgebras. In *Quantum Probability and Related Topics VI (Proceedings of the Trento conference, 1989)*, pp. 401–419. World Sci.

[113] J.M. Lindsay and K.R. Parthasarathy. The passage from random walks to diffusion in quantum probability II. *Indian J. Stat. A*, 50:151–170, 1988.

[114] P. Biane. Some properties of quantum Bernoulli random walks. In *Quantum Probability and Related Topics VI (Proceedings of the Trento conference, 1989)*, pp. 193–203. World Sci.

[115] K.R. Parthasarathy. *An Introduction to Quantum Stochastic Calculus*. Birkhauser, 1992.

[116] S. Majid. *Non-commutative-geometric Groups by a Bicrossproduct Construction*. PhD thesis, Harvard, 1988.

[117] S. Majid. Principle of representation-theoretic self-duality. *Phys. Essays*, 4:395–405, 1991.

[118] S. Majid. Hopf algebras for physics at the Planck scale. *J. Classical and Quantum Gravity*, 5:1587–1606, 1988.

[119] O. Bratteli and D. W. Robinson. *Operator Algebras and Quantum Statistical Mechanics, II*. Springer-Verlag, 1979.

[120] S. Majid. Matched pairs of Lie groups associated to solutions of the Yang-Baxter equations. *Pac. J. Math.*, 141:311–332, 1990.

[121] S. Majid and H. Ruegg. Bicrossproduct structure of the κ-Poincaré group and non-commutative geometry. *Phys. Lett. B*, 334:348–354, 1994.

[122] S. Majid. Duality principle and braided geometry. In *Strings and Symmetries, Lec. Notes in Phys.*, 447:125–144, 1995. Springer.

[123] E. Beggs, J. Gould, and S. Majid. Finite group factorisations and braiding. *J. Algebra*, 181:112–151, 1996.

[124] E. Beggs and S. Majid. Quasitriangular and differential structures on bicrossproduct Hopf algebras. *J. Algebra*, 219:682–727, 1999.

[125] W. Singer. Extension theory for connected Hopf algebras. *J. Algebra*, 21:1–16, 1972.

[126] M. Takeuchi. Matched pairs of groups and bismash products of Hopf algebras. *Commun. Alg.*, 9:841, 1981.

[127] K. Mackenzie. Double Lie algebroids and second-order geometry, I. *Adv. Math.*, 94:180–239, 1992.

[128] S. Majid. Bicrossproduct structure of affine quantum groups. *Lett. Math. Phys.*, 39 243-252, 1997.

[129] Y. Doi and M. Takeuchi. Cleft comodule algebras and Hopf modules. *Commun. Alg.*, 14:801–817, 1986.

[130] Y. Doi. Equivalent crossed products for a Hopf algebra. *Commun. Algebra*, 17:3053–3085, 1989.

[131] H-J. Schneider. Principal homogeneous spaces for arbitrary Hopf algebras. *Isr. J. Math.*, 72:167–195, 1990.

[132] T. Brzeziński and S. Majid. Quantum differentials and the q-monopole revisited. *Acta Appl. Math.*, 54:185–232, 1998.

[133] P. Hajac and S. Majid. Projective module description of the q-monopole. *Commun. Math. Phys.*, 1999.

[134] P. Podleś. Quantum spheres. *Lett. Math. Phys.*, 14:193–202, 1987.

[135] T. Brzezinski and S. Majid. Coalgebra bundles. *Commun. Math. Phys.*, 191:467–492, 1998.

[136] S. Majid. Diagrammatics of braided group gauge theory. *J. Knot Theor. Ramif.*, 8:731–771, 1999.

[137] S. Majid. Quantum and braided group Riemannian geometry. *J. Geom. Phys.*, 30:113–146, 1999.

[138] H.D. Doebner and J. Tolar. Quantum mechanics on homogeneous spaces. *J. Math. Phys.*, 16:975–984, 1975.

[139] T. Yamanouchi. The intrinsic group of Majid's bicrossproduct Kac algebra. *Pac. J. Math.*, 159:185–199, 1993.

[140] S. Majid. Braided groups and duals of monoidal categories. *Canad. Math. Soc. Conf. Proc.*, 13:329–343, 1992.

[141] F. W. Lawvere. Intrinsic boundary in certain mathematical toposes exemplify 'logical' operators not passively preserved by substitution. Technical Report, Buffalo, November 1989.

[142] S. Majid. Representation-theoretic rank and double Hopf algebras. *Commun. Algebra*, 18:3705–3712, 1990.

[143] S. Majid. Doubles of quasitriangular Hopf algebras. *Commun. Algebra*, 19:3061–3073, 1991.

[144] J.H.C. Whitehead. Combinatorial homotopy, II. *Bull. Amer. Math. Soc.*, 55:453–496, 1949.

[145] D.N. Yetter. Quantum groups and representations of monoidal categories. *Math. Proc. Camb. Phil. Soc.*, 108:261–290, 1990.

[146] S. Majid. Transmutation theory and rank for quantum braided groups. *Math. Proc. Camb. Phil. Soc.*, 113:45–70, 1993.

[147] Y. Doi and M. Takeuchi. Multiplication alteration by two-cocycles – the quantum version. *Commun. Alg.*, 22:5715–5732, 1994.

[148] S. Majid. Some remarks on the quantum double. *Czech. J. Phys.*, 44:1059–1071, 1994.

[149] D. Radford. Minimal quasitriangular Hopf algebras. *J. Algebra*, 157:285–315, 1993.

[150] S. Majid. Braided momentum in the q-Poincaré group. *J. Math. Phys.*, 34:2045–2058, 1993.

[151] U. Meyer. q-Lorentz group and braided coaddition on q-Minkowski space. *Commun. Math. Phys.*, 168:249–264, 1995.

[152] P. Podleś and S.L. Woronowicz. Quantum deformation of Lorentz group. *Commun. Math. Phys*, 130:381–431, 1990.

[153] U. Carow-Watamura, M. Schlieker, M. Scholl, and S. Watamura. A quantum Lorentz group. *Int. J. Mod. Phys. A*, 6:3081–3108, 1991.

[154] S. Majid. Cross products by braided groups and bosonization. *J. Algebra*, 163:165–190, 1994.

[155] D. Radford. The structure of Hopf algebras with a projection. *J. Algebra*, 92:322–347, 1985.

[156] S. Majid. Rank of quantum groups and braided groups in dual form. In *Proceedings of the Euler Institute, St. Petersberg, 1990, Lec. Notes in Math.*, 1510:79–89. Springer.

[157] V.G. Drinfeld. Hamiltonian structures on Lie groups, Lie bialgebras and the geometric meaning of the classical Yang-Baxter equations. *Sov. Math. Dokl.*, 27:68, 1983.

[158] M. Jimbo. Quantum R-matrix related to generalized toda system: an algebraic approach. *Lec. Notes in Phys.*, 246:335–361, 1986. Springer.

[159] W. Michaelis. The dual Poincaré-Birkhoff-Witt theorem. *Adv. Math.*, 57:93–162, 1985.

[160] A.A. Belavin and V.G. Drinfeld. On the solutions of the classical Yang-Baxter equations for simple Lie algebras. *Funct. Anal. Appl.*, 16:1–29, 1982.

[161] A. Stolin. On rational solutions of the classical Yang-Baxter equations. Maximal orders in the loop algebra. *Commun. Math. Phys.*, 141:533–548, 1991.

[162] A. Stolin. On rational solutions of the classical Yang-Baxter equations for $sl(n)$. *Math. Scand.*, 69:57–80, 1991.

[163] E. Beggs and S. Majid. Matched pairs of topological Lie algebras corresponding to Lie bialgebra structures on $diff(S^1)$. *Ann. Inst. H. Poinc. Theor. Phys.*, 53:15–34, 1990.

[164] M.A. Semenov-Tian-Shansky. What is a classical R-matrix. *Funct. Anal. Appl.*, 17:17, 1983.

[165] Y. Kosmann-Schwarzbach and F. Magri. Poisson-Lie groups and complete integrability, I: Drinfeld bigebras, dual extensions and their canonical representations. *Ann. Inst. H. Poinc. Theor. Phys.*, 49:433–460, 1988.

[166] J-H. Lu and A. Weinstein. Poisson Lie groups, dressing transformations and Bruhat decompositions. *J. Diff. Geom.*, 31:501–526, 1990.

[167] J. Lukierski, A. Nowicki, H. Ruegg, and V.N. Tolstoy. q-Deformation of Poincaré algebra. *Phys. Lett. B*, 264,271:331,321, 1991.

[168] S. Majid. On q-regularisaion. *Int. J. Mod. Phys. A*, 5:4689–4696, 1990.

[169] A.G. Reyman and M.A. Semenov-Tian-Shansky. Reduction of Hamiltonian systems, affine Lie algebras and Lax equations I. *Invent. Math.*, 54:81–100, 1979.

[170] A.G. Reyman and M.A. Semenov-Tian-Shansky. Reduction of Hamiltonian systems, affine Lie algebras and Lax equations II. *Invent. Math.*, 63:423–432, 1981.

[171] M.A. Semenov-Tian-Shansky. Dressing transformations and Poisson group actions. *Publ. RIMS (Kyoto)*, 21:1237–1260, 1985.

[172] L.D. Faddeev and L.A. Takhtajan. *Hamiltonian Methods in the Theory of Solitons*. Springer-Verlag, 1987.

[173] S. Mac Lane. *Categories for the Working Mathematician*. Springer, 1974.

[174] J.D. Stasheff. On the homotopy associativity of H-spaces, I and II. *Trans. Amer. Math. Soc.*, 108:275–292 and 293–312, 1963.

[175] V.G. Drinfeld. Letter to the author, February, 1990.

[176] A. Joyal and R. Street. Braided monoidal categories. Mathematics Reports 86008, Macquarie University, 1986.

[177] P. Freyd and D. Yetter. Braided compact closed categories with applications to low dimensional topology. *Adv. Math.*, 77:156–182, 1989.

[178] S. Majid. Braided groups and algebraic quantum field theories. *Lett. Math. Phys.*, 22:167–176, 1991.

[179] S. Majid. Reconstruction theorems and rational conformal field theories. *Int. J. Mod. Phys. A*, 6:4359–4374, 1991.

[180] S. Majid. Infinite braided tensor products and 2D quantum gravity. *Int. J. Mod. Phys. A (Proc. Suppl)*, 3A:294–297, 1993.

[181] P. Deligne. Catégories Tannakiènnes. In P. Cartier *et al*, editors, *Grothendieck Festschrift, Vol 2*, pp. 111-195. Birkhauser, 1991.

[182] A. Joyal and R. Street. Tortile Yang-Baxter operators in tensor categories. *J. Pure Applied Algebra*, 71:43–51, 1991.

[183] N. Saavedra Rivano. Catégories Tannakiennes. *Lec. Notes in Math.*, 265, 1972. Springer.

[184] P. Deligne and J.S. Milne. Tannakian categories. *Lec. Notes in Math.*, 900, 1982. Springer.

[185] M.A. Hennings. Hopf algebras and regular isotopy invariants for link diagrams. *Math. Proc. Camb. Phil. Soc.*, 109:59–77, 1991.

[186] M.A. Hennings. Invariants of links and 3-manifolds obtained from Hopf algebras. Technical Report, Cambridge University, 1989.

[187] N.Yu. Reshetikhin. Quantized universal enveloping algebras, the Yang-Baxter equations and invariants of links, I and II. Technical Reports, LOMI, 1988.

[188] D. Fischman M. Cohen and S. Westreich. Schur's double centralizer theorem for triangular Hopf algebras. *Proc. Amer. Math. Soc.*, 122:19–29, 1994.

[189] A.A. Kirillov. *Elements of the Theory of Representations.* Springer-Verlag, Heidelberg, 1976.

[190] K.-H. Ulbrich. On Hopf algebras and rigid monoidal categories. *Israel J. Math*, 72:252–256, 1990.

[191] V.V. Lyubashenko. Tangles and Hopf algebras in braided categories. *J. Pure and Applied Algebra*, 98:245–278, 1995.

[192] V.V. Lyubashenko. Modular transformations for tensor categories. *J. Pure and Applied Algebra*, 98:279–327, 1995.

[193] S. Majid. Free braided differential calculus, braided binomial theorem and the braided exponential map. *J. Math. Phys.*, 34:4843–4856, 1993.

[194] S. Majid. Solutions of the Yang-Baxter equations from braided-Lie algebras and braided groups. *J. Knot Th. Ram.*, 4:673–697, 1995.

[195] S. Majid. Double bosonisation and the construction of $U_q(g)$. *Math. Proc. Camb. Phil. Soc.*, 125:151–192, 1999.

[196] S. Majid. Braided-lie bialgebras. *Pacific J. Math.*, 1999.

[197] S. Majid. Braided geometry of the conformal algebra. *J. Math. Phys.*, 37:6495–6509, 1996.

[198] S. Majid. New quantum groups by double-bosonisation. *Czech J. Phys.*, 47:79–90, 1997.

[199] Yu. N. Bespalov. Crossed modules and quantum groups in braided categories. *Appl. Categ. Str.*, 5:155–204, 1997.

[200] S. Majid. Some comments on bosonisation and biproducts. *Czech J. Phys.*, 47:151–171, 1997.

[201] S. Majid. Quantum geometry of field extensions. *J. Math. Phys.*, 40:2311–2323, 1999.

[202] S. Majid. Advances in quantum and braided geometry. In H.-D. Doebner and V.K. Dobrev, editors, *Quantum Group Symposium at Group XXI, Goslar, 1996*, pp. 11–26. Heron Press, Sofia.

[203] S. Majid. Classification of bicovariant differential calculi. *J. Geom. Phys.*, 25:119–140, 1998.

[204] K. Schmüdgen and A. Schüler. Classification of bicovariant differential calculi on quantum groups. *Commun. Math. Phys.*, 170:315–335, 1995.

[205] S. Majid. Braided groups and braid statistics. In L. Accardi et al, editors, *Quantum Probability and Related Topics VIII (Proceedings of the Delhi conference, 1990)*, pp. 281–295. World Sci.

[206] S. Majid. Braided geometry: A new approach to q-deformations. In R. Coquereaux *et al*, editors, *First Caribbean Spring School of Mathematics and Theoretical Physics, Guadeloupe, 1993*, pp. 190–204. World Sci.

[207] S. Majid. Lie algebras and braided geometry. *Adv. Appl. Clifford Algebras (Proc. Suppl.)*, 4:61–77, 1994.

[208] J. Milnor and J.C. Moore. On the structure of Hopf algebras. *Ann. Math.*, 81:211, 1969.

[209] A. Macfarlane and S. Majid. Spectrum generating quantum group of the harmonic oscillator. *Int. J. Mod. Phys. A*, 7:4377–4393, 1992.

[210] A. Macfarlane and S. Majid. Quantum group structure in a fermionic extension of the quantum harmonic oscillator. *Phys. Lett. B*, 268:71–74, 1991.

[211] S. Majid and M.J. Rodriguez-Plaza. Quantum and super-quantum group related to the Alexander-Conway polynomial. *J. Geom. Phys.*, 11:437–443, 1993.

[212] S. Majid and M.J. Rodriguez-Plaza. Random walk and the heat equation on superspace and anyspace. *J. Math. Phys.*, 35:3753–3769, 1994.

[213] S. Majid and M.J. Rodriguez-Plaza. Anyonic FRT construction. *Czech. J. Phys.*, 44:1073–1080, 1994.

[214] G. Lusztig. *Introduction to Quantum groups*. Birkhauser, 1993.

[215] S. Majid. Introduction to braided geometry and q-Minkowski space. In L. Castellani *et al*, editors, *Proceedings of the International School of Physics 'Enrico Fermi' CXXVII, Varenna, 1994*, pp. 267–348. IOS Press, Amsterdam.

[216] M. Schlieker, W. Weich, and R. Weixler. Inhomogeneous quantum groups. *Z. Phys. C*, 53:79–82, 1992.

[217] S. Majid. Some remarks on the q-Poincaré algebra in R-matrix form. In H.-D. Doebner *et al*, editors, *Nonlinear, Deformed and Irreversible Quantum Systems*, pp. 352–384. World Sci., 1995.

[218] T. Brzeziński and S. Majid. A class of bicovariant differential calculi on Hopf algebras. *Lett. Math. Phys.*, 26:67–78, 1992.

[219] U. Carow-Watamura, M. Schlieker, M. Scholl, and S. Watamura. Tensor representation of the quantum group $SL_q(2, \mathbb{C})$ and quantum Minkowski space. *Z. Phys. C*, 48:159, 1990.

[220] S. Majid and U. Meyer. Braided matrix structure of q-Minkowski space and q-Poincaré group. *Z. Phys. C*, 63:357–362, 1994.

[221] S. Majid. *-structures on braided spaces. *J. Math. Phys.*, 36:4436–4449, 1995.

[222] W. Pusz and S.L. Woronowicz. Twisted second quantization. *Rep. Math. Phys*, 27:231, 1989.

[223] A. Kempf. Quantum group-symmetric Fock-spaces and Bargmann-Fock space: Integral kernels, green functions, driving forces. *J. Math. Phys.*, 34:969–987, 1993.

[224] J. Wess and B. Zumino. Covariant differential calculus on the quantum hyperplane. *Proc. Suppl. Nucl. Phys. B*, 18:302, 1990.

[225] G. Fiore. The $SO_q(N, R)$-symmetric harmonic oscillator on the quantum Euclidean space R_q^N and its Hilbert space structure. *Int. J. Mod. Phys. A*, 26:4679–4729, 1993.

[226] S. Majid. q-epsilon tensor for quantum and braided spaces. *J. Math. Phys.*, 36:1991–2007, 1995.

[227] S. Majid. Braided approach to exterior algebra on q-Euclidean and q-Minkowski spaces. In J. Lukierski *et al*, editors, *Quantum Groups: Proceedings of the 30th Winter School of Theoretical Physics, Karpacz, 1994*, pp. 131–148. Polish Sci.

[228] A. Sudbery. The algebra of differential forms on a full matric bialgebra. *Math. Proc. Camb. Phil. Soc.*, 114:111–130, 1993.

[229] S. Majid. q-Euclidean space and quantum Wick rotation by twisting. *J. Math. Phys.*, 35:5025–5034, 1994.

[230] G. Fiore. Quantum groups $SO_q(N), Sp_q(N)$ have q-determinants, too. *J. Phys. A*, 27:3795–3802, 1994.

[231] O. Ogievetsky, W.B. Schmidke, J. Wess, and B. Zumino. q-Deformed
       Poincaré algebra. *Commun. Math. Phys.*, 150:495–518, 1992.

[232] U. Meyer. Wave equations on q-Minkowski space. *Commun. Math.
       Phys.*, 174:457–475, 1996.

[233] M. Wambst. Quantum Koszul complexes for Majid's braided Lie
       algebras. *J. Math. Phys.*, 35:6213–6223, 1994.

Symbols

Symbol	Element	Page	Typical usage
k	λ, μ	2	field
$\mathbb{C}, \mathbb{R}$	λ, μ		complex, real numbers
$\mathbb{Z}, \mathbb{Z}_{/n}, \mathbb{N}$	n, m		integers, modulo n, natural numbers incl. 0
$\imath$			$\sqrt{-1}$
V, W	v, w	2	vector space
V^*	ϕ, ψ	3	linear dual
e_a, f^a			basis, dual basis
$\mathrm{Lin}, \mathrm{End}$		3,16	linear maps, endomorphisms
$\oplus$		2	direct sum
$\otimes$		2	tensor product
A	a, b	3	algebra
C	c, d	5	coalgebra
H	h, g	6,72	quantum group, e.g. enveloping type
A	a, b	6,108	quantum group, e.g. function type
G	u, v	1	group
M	s, t		manifold, set, sometimes group
g	ξ, η	15	Lie algebra
m	ϕ, ψ		Lie algebra
op		9	opposite product or Lie bracket
cop		9	opposite coproduct
Δ	$h_{(1)} \otimes h_{(2)}$	5	coproduct
$\cdot$		1	product
ϵ		5	counit
S		7	antipode
τ		3	transposition
$\langle \, , \rangle$		11	duality pairing
kG		12	group Hopf algebra
$k(G)$		12	group function algebra
$U(g)$		15	enveloping algebra
χ		13,56	character or cocycle

625

Symbol	Element	Page	Typical usage
α	$h\triangleright,\ \triangleleft h$	16	left, right action
β	$v^{(\bar{1})}\otimes v^{(\bar{2})}$	22	coaction or bicharacter
Ad		18,24	adjoint action or coaction
Ad*		18,25	coadjoint action or coaction
L, R		17,24	left, right regular action or coaction
L^*, R^*		18,24	left, right action or coaction on dual
$\rtimes, \ltimes$		19	left, right cross product or sum
$\blacktriangleright\!\!\triangleleft, \blacktriangleright\!\!\ltimes$		25	left, right cross coproduct or cobracket
$\vert\ \vert$		20	grading
$\int$		28	integral functional
Λ		28	integral element
W		29	fundamental operator
$*$		31,89	antilinear antiinvolution
$(\ ,\ )$		31	inner product
$\mathcal{H}$	$\vert\ \rangle$	79	Hilbert space
$B(\mathcal{H})$		31,204	bounded operators on
$\mathcal{R}$	$\mathcal{R}^{(1)}\otimes\mathcal{R}^{(2)}$	39	quasitriangular structure
u, v		43	elements implementing square antipode
ν		45	ribbon element
Q		46	quantum inverse Killing form
$\mathbb{Z}'_{/n}$	g	41	anyon-generating quantum group
$U_q(1)$	ξ	49	quantum line enveloping algebra
$\mathbb{Z}_q$	ς	53	quantum line function algebra
$\mathcal{R}(\ )$		51	dual quasitriangular structure
u( ), v( )		52	functionals implementing square antipode
$Q(\ )$		55	quantum inverse Killing form
∂		56	coboundary
$Z(\ ,\ )$	γ, χ, ϕ	56	cocycle
γ, F		58,67	cochain of degree 1, 2 for gauging
$\mathcal{H}(\ ,\ )$		58	cohomology space
$(\ )_\chi$		58	twisting by 2-cocycle
ϕ		65	3-cocycle of quasi-Hopf algebra
α, β		66	antipode elements of quasi-Hopf algebra
$a, a^\dagger$		74	harmonic oscillator generators
q, t		73	deformation parameter
$U_q(sl_2), U_q(g)$	$H, X_\pm$	86,97	quantum enveloping sl_2, g
$\partial, \partial_q, e_q$		83,88	q-differential, q-exponential
$[\ ], [\ ]_q, [\ ; q]$		86,88,97	symmetric, nonsymmetric q-integer
K, t, Λ		95	Killing form, Cartan subalgebra, weight
$a_{ij}, \alpha_i, \check{\alpha}_i, d_i$		95	Cartan matrix, root, coroot, length
W	w, s_i	96	Weyl group
ρ		97	half sum of positive roots
$U_q(g)$	g_i, E_i, F_i	99	other generators of quantum enveloping g
T_i		100	Lusztig automorphism
$U_q^{(r)}(sl_2), u_q(g)$		102	quantum enveloping algebra at root of 1

Symbol	Element	Page	Typical usage
i, j, a, b		111	indices, repeated indices
I, J, A, B		147	multiindices
R, R_{21}	$R^i{}_j{}^k{}_l$	110	R-matrix and its transpose
$\widetilde{R}$		129	second-inverse of R-matrix
P	$\delta^i{}_l \delta^k{}_j$	126,155	permutation matrix
λ_i		126,155	skew-eigenvalue of R-matrix
u, v	$u^i{}_j, v^i{}_j$	129	quantum trace matrices
$A(R)$	$\mathbf{t}, t^i{}_j$	113	quantum matrix $R\mathbf{t}_1\mathbf{t}_2 = \mathbf{t}_2\mathbf{t}_1 R$
$M_q(2)$		133	standard 2×2 quantum matrix
G_q		135	standard matrix quantum group
$\rho^{\pm}$		114	(conjugate) fundamental representation
$\widetilde{U}(R)$	$\mathbf{L}^{\pm}, L^{\pm i}{}_j$	119	enveloping bialgebra
$\mathbf{l}^{\pm}$	$l^{\pm i}{}_j$		FRT generators in quantum group
ρ		119	canonical representations
$B(R)$	$\mathbf{u}, u^i{}_j$	144	braided matrices $R_{21}\mathbf{u}_1 R\mathbf{u}_2 = \mathbf{u}_2 R_{21}\mathbf{u}_1 R$
$BM_q(2)$		146	standard 2×2 braided matrix
BG_q		513	standard matrix braided group
$\mathcal{L}$	$\mathbf{1}, l^i{}_j$	146	matrix braided-Lie algebra
$V_L(R, \lambda)$	$\mathbf{v}, v^i$	156	left-covariant quantum vector
$V_R(R, \lambda)$	$\mathbf{v}, v^i$	164	right-covariant quantum vector
$V_L^*(\lambda, R)$	$\mathbf{x}, x_i$		left-covariant quantum covector
$V_R^*(\lambda, R)$	$\mathbf{x}, x_i$	156	right-covariant quantum covector
$A(R_1 : R_2)$	$\mathbf{t}, t^\mu{}_\alpha$	161	bicovariant rectangular quantum matrix
$\mathbb{A}_q^{2|0}$	x, y	156	quantum plane
Ω	x, y	178	universal set or probability space
$\vee, \wedge$		180	Boolean or, and
$L^\infty(\Omega), \mathbb{C}(\Omega)$	a, b	196	functions on
$\langle\ \rangle$		193	expectation value
ρ, ϕ		192	probability density, state
T		197	Markov transition
t		195,226	time parameter
$\hbar$		74,223	Planck's constant
$\mathcal{S}(\phi, \psi)$		211	relative entropy
$C^\infty(M), \mathrm{Vect}(M)$	f, g	226	smooth functions, vector fields
G, m_P, x_P		288	gravitational constant, Planck mass, length
p, H		206,428	momentum, Hamiltonian
$\bowtie$		255,315	double cross product or sum
$\blacktriangleright\!\!\triangleleft, \triangleright\!\!\blacktriangleleft$		240,397	left–right, right–left bicrossproduct
$H^*, A^\star$		243	categorical dual
$D(H), D(\mathfrak{g})$		304,383	quantum, classical double
$\sigma(,\ )$		320	skew-pairing
$\blacktriangleright\!\!\blacktriangleleft$		333	double cross coproduct or cobracket
$\rtimes, \ltimes$		513,519	left, right biproduct
$BU_q(\mathfrak{g})$	$\mathbf{1}, l^i{}_j$	352	standard braided group enveloping algebra

Symbol	Element	Page	Typical usage
δ	$\xi_{[1]} \otimes \xi_{[2]}$	366	Lie cobracket
r, r_1, r_2	$r^{[1]} \otimes r^{[2]}$	370	classical quasitriangular structure, as maps
$r_\pm, r_M$		371	antisymmetric, symmetric part and ratio
$[\,,\,]$		368	Schouten bracket defines CYBE
g^*, G^*		367,426	dual Lie bialgebra and associated group
u, U		379	real form and associated group
$\widetilde{\xi}$	$L_{u*}(\xi)$	408	left-invariant vector field
$\widetilde{\xi}$	$R_{u*}(\xi)$	408	right-invariant vector field
d		408	exterior differential
ω^L, ω^R		408	left-, right-covariant Maurer–Cartan form
$\{\,,\,\}, \gamma$		420	Poisson bracket and associated 2-tensor
$\mathcal{C}, \mathcal{V}$	V, W	438	category and objects in
Mor	ϕ, ψ	438	set of morphisms between objects
F, G		439	functor between categories
Nat$(\,,\,)$	θ, λ	440	natural transformation between functors
$\underline{1}, l, r$		442	unit object and natural isomorphisms
Φ		442	associativity natural isomorphisms
Ψ		451	braided transposition or quasisymmetry
ev, coev		466	evaluation, coevaluation
$\underline{\dim}, \underline{\mathrm{Tr}}$		469	categorical dimension, trace
$|H|$		476	quantum order or rank
$\underline{\mathrm{Hom}}(\,,\,)$		482	internal hom
ψ_i		489	Hecke algebra generator
$\underline{\otimes}$		460	braided tensor product algebra
B, C	b, c	460,499	braided group or algebra
$\underline{H}, \underline{A}$		345,504	braided group made from quantum group
$B(\,,\,)$		509,509	transmutation construction
$\{\,,\,]$		531	super-Lie bracket
$\mathbb{Z}'_{/2}$	g	456,533	as generator of superstatistics
$\mathbb{Z}'_{/n}$	g	455,535	as generator of anyonic statistics
$\mathbb{Z}_q$	ς	537	quantum line as dilaton
$V^{\check{}}(R', R)$	$\mathbf{x}, x_i$	540	braided covector $\mathbf{x}_1\mathbf{x}_2 = \mathbf{x}_2\mathbf{x}_1 R'$
$V(R', R)$	$\mathbf{v}, v^i$	545	braided vector $\mathbf{v}_1\mathbf{v}_2 = R'\mathbf{v}_2\mathbf{v}_1$
$\mathbb{C}_q, \mathbb{C}_q^{2|0}$	x, y	538,542	braided line, braided plane
$\mathbf{x}', \mathbf{u}'$	x', a'	543	second factor in braided tensor product
λ		129,547	quantum group normalisation constant
$\widetilde{A}$	t, ς	547	dilaton extended quantum functions
$\widetilde{U_q(g)}$	$\mathbf{l}^\pm, \xi$	551	dilaton extended quantum enveloping
η	η_{ij}, η^{ij}	554	braided metric, transposed-inverse
$[\,; R], [\,; R]!$		569,577	braided integer matrix, factorial
$\exp(\,|\,)$		579	braided exponential or coevaluation
g_η, Z		583,585	braided Gaussian, weighted integration
$\epsilon^{i_1 \cdots i_n}$		587	braided epsilon tensor
$\Lambda, \Omega, \mathrm{d}$	$\theta_i = \mathrm{d}x_i$	588	braided forms, braided exterior algebra

Index

2-Index	quantum group	quasi-Hopf	group	Lie bialgebra	Poisson–Lie group	category
action	16	444	2,19	19,398	428	520
adjoint action	18		19	19		
antipode	7	66	12	15		466
axioms	6	65	1	366	421	438
bicross	240		430	397	430	
braided	144,499		513	174		451
coaction	22		24	399		
coadjoint action	18		236	402		520
cocycle	56,263	68	57	64	422	
cohomology	58,265	68	57	64	422	
complexification	303		434	403	428	
cross (co)product	19,25		19	391		
crossed module	309	520	312			447
double cross	315		255	396	427	520
double	304	520	311	383	428	447
duality	11	69	13	367	426	446
dual quasitr.	51	69	53	374	426	459
example	8,86	68	536	375	425	441
extension	267		276	278		
factorisable	46,513			370		
homomorphism	8		2	367	420	439
matched pair	315		247	395	427	520
module	16	444	2	398		520
opposite	9		407	383		451
quasitriangular	39	65	48	370	423	454
real form, *	31,47		382	379	426	
reconstruction	493	497				504
regular repn	17		34	408		
representations	21,443	444	13			446
self-duality	10,513		13	367		294
triangular	46			370		452
twisting	58,64	67		372		496